THE DETERMINATION OF CHEMICAL ELEMENTS IN FOOD

BICENTENNIAL 1807 WILEY 2007 BICENTENNIAL

The Wiley Bicentennial–Knowledge for Generations

Each generation has its unique needs and aspirations. When Charles Wiley first opened his small printing shop in lower Manhattan in 1807, it was a generation of boundless potential searching for an identity. And we were there, helping to define a new American literary tradition. Over half a century later, in the midst of the Second Industrial Revolution, it was a generation focused on building the future. Once again, we were there, supplying the critical scientific, technical, and engineering knowledge that helped frame the world. Throughout the 20th Century, and into the new millennium, nations began to reach out beyond their own borders and a new international community was born. Wiley was there, expanding its operations around the world to enable a global exchange of ideas, opinions, and know-how.

For 200 years, Wiley has been an integral part of each generation's journey, enabling the flow of information and understanding necessary to meet their needs and fulfill their aspirations. Today, bold new technologies are changing the way we live and learn. Wiley will be there, providing you the must-have knowledge you need to imagine new worlds, new possibilities, and new opportunities.

Generations come and go, but you can always count on Wiley to provide you the knowledge you need, when and where you need it!

William J. Pesce
President and Chief Executive Officer

Peter Booth Wiley
Chairman of the Board

THE DETERMINATION OF CHEMICAL ELEMENTS IN FOOD: APPLICATIONS FOR ATOMIC AND MASS SPECTROMETRY

Sergio Caroli
National Institute of Health (ISS), Rome, Italy

WILEY-INTERSCIENCE
A JOHN WILEY & SONS, INC., PUBLICATION

Published by John Wiley & Sons, Inc., Hoboken, New Jersey
Published simultaneously in Canada

Library of Congress Cataloging-in-Publication Data:

Caroli, Sergio, 1943–
The determination of chemical elements in food: applications for atomic and mass spectrometry / Sergio Caroli.
p. cm.
ISBN-13: 978-0-471-68784-9
ISBN-10: 0-471-68784-7
1. Food–Analysis 2. Mass spectrometry. 3. Atomic spectroscopy. I. Title.
TX547.C37 2007
664′.07- -dc22 2006038881

Printed in the United States of America

10 9 8 7 6 5 4 3 2 1

In memory of
Professor Francesco Pocchiari,
a man of science, wisdom and
goodwill

CONTENTS

PREFACE

Confidence in the quality and safety of food is an undisputed priority worldwide. The presence in food of undesired chemicals (basically residues and contaminants) as well as the lack of essential chemical substances at the required concentrations can pave the way to very serious consequences for human health. Chemical elements (be they major, minor, or trace elements) have a place of their own in this context. Needless to say, assessing the safety of food from both viewpoints demands entirely reliable experimental information, which in turn is based on the availability of fit-for-purpose powerful analytical techniques.

This multiauthored book aims at highlighting the role played by atomic and mass spectrometry (with particular reference to atomic absorption spectrometry, inductively coupled plasma atomic emission spectrometry and inductively coupled plasma mass spectrometry) in supporting and promoting research and control of foodstuffs and food commodities as regards both essential and potentially toxic chemical elements. The progress made so far in this field is overviewed and emphasis is put on the open problems that require further investment and development in the public and private sectors.

The 22 chapters that form this book are written by prominent scientists and cover primary issues such as: (1) fundamental aspects and method development, also as regards sampling, sample storage, and sample pretreatment; (2) challenges posed by the quantification of chemical elements (mostly trace elements) in food matrices; (3) existing legal provisions at the national, community and international level; (4) expected development of instrumental spectrometric techniques that can be exploited to better protect consumers' health, with a full account of the most promising trends of spectrometric instrumentation and ancillary apparatuses in food analysis; (5) quantification of the various chemical species of elements in foodstuffs primarily through hyphenation of spectrometric equipments with chromatographic instrumentation; and (6) adoption of quality schemes along with the proper use of reference materials and proficiency testing schemes. All these topics are subdivided into three main sections, namely: *Fundamentals and method development*, *Selected applications*, and *Speciation analysis*, where the various aspects listed above are approached in a holistic manner so as to be of interest to the widest readership.

This book greatly benefits from the enthusiastic participation and support of all authors who patiently collaborated with the Editor and tolerated his endless requests. A special word of thanks goes to Monica Brocco[1] for revising the style of the various versions of each chapter and to Paola De Castro,[1] Cosimo Curianò,[1] and Marco Cornacchia[1] for their precious assistance in preparing artwork.

Last, but by no means least, the generous support of PerkinElmer Italiana in planning this book is gratefully acknowledged: without their assistance this project would have never been possible.

SERGIO CAROLI

Editor

[1]Istituto Superiore di Sanità, Rome.

CONTRIBUTORS

Yetunde Aregbe, European Commission, Joint Research Centre, Institute for Reference Materials and Measurements, Retieseweg 111, 2440 Geel, Belgium

Daniel A. Batistoni, Comisión Nacional de Energía Atómica, Centro Atómico Constituyentes, Unidad de Actividad Química, Avenida General Paz 1499, B1650KNA San Martín, Provincia de Buenos Aires, Argentina

M. Bettinelli, C.M.B. Sistemi integrati per i laboratori, la qualità e l'ambiente, Via G. Landi 77, 29100 Piacenza, Italy

F. Bolle, Scientific Institute of Public Health, Section of Food, rue Juliette Wytsman 14, 1050, Brussels, Belgium

Daniel L. G. Borges, Departamento de Química, Universidade Federal de Santa Catarina, 88040-900, Florianópolis, SC, Brazil

K. Boutakhrit, Scientific Institute of Public Health, Section of Food, rue Juliette Wytsman 14, 1050, Brussels, Belgium

B. Bouyssiere, UMR 5034 CNRS, Hélioparc, 2 Avenue President Angot, 64053 Pau, France

Roberta Cammarone, Istituto Superiore di Sanità, Rome, Viale Regina Elena 299, 00161, Italy

Joseph A. Caruso, Department of Chemistry, University of Cincinnati, OH 45221-0172, USA

Sergio Caroli, Istituto Superiore di Sanità, Rome, Viale Regina Elena 299, 00161, Italy

Tue Christensen, National Food Institute, Technical University of Denmark, Department of Nutrition, Mørkhøj Bygade 19, DK-2860 Søborg, Denmark

Francesco Cubadda, National Centre for Food Quality and Risk Assessment, Istituto Superiore di Sanità Viale Regina Elena 299, 00161 Rome, Italy

Marilena D'amato, Istituto Superiore di Sanità, Viale Regina Elena 299, 00161 Rome, Italy

J. M. Degroodt, Scientific Institute of Public Health, Section of Food, rue Juliette Wytsman 14, 1050 Brussels, Belgium

Mihály Dernovics, Department of Applied Chemistry, Faculty of Food Science, Corvinus University of Budapest, Hungary

Joakim Engman, Livsmedelsverket, PO Box 622, SE-751 26 Uppsala, Sweden

María Luisa Fernández sánchez, Department of Physical and Analytical Chemistry, Faculty of Chemistry, University of Oviedo, C/Julián Clavería 8, Oviedo, E-33006 Spain

Rafael R. de la Flor St. Rèmy, Department of Physical and Analytical Chemistry, Faculty of Chemistry, University of Oviedo, C/ Julián Clavería 8, Oviedo, E-33006 Spain

Péter Fodor, Department of Applied Chemistry, Faculty of Food Science, Corvinus University of Budapest, Hungary

Chiara Frazzoli, Istituto Superiore di Sanità, Rome, Viale Regina Elena 299, 00161, Italy

A. Gatti, Laboratory for Environmental and Toxicological Measurements, S. Maugeri Foundation, Via Ferrata 8, 27100 Pavia, Italy

L. Goeyens, Scientific Institute of Public Health, Section of Food, rue Juliette Wytsman 14, 1050 Brussels, Belgium

Uwe Heitmann, ISAS – Institute for Analytical Sciences, Department for Interface Spectroscopy, Albert-Einstein-Str. 9, 12489 Berlin, Germany

Simon Hird, Food Safety and Quality Group, Central Science Laboratory, Sand Hutton, York YO41 1LZ, UK

Lars Jorhem, Livsmedelsverket, PO Box 622, SE-751 26 Uppsala, Sweden

Jason Kirby, Ecochemistry Laboratory, Institute of Applied Ecology, University of Canberra, ACT 2601, Australia

Frank Krikowa, Ecochemistry Laboratory, Institute of Applied Ecology, University of Canberra, ACT 2601, Australia

Petra Krystek, Formerly at the National Institute for Public Health and the Environment (RIVM), The Netherlands

Erik H. Larsen, National Food Institute, Technical University of Denmark, Departmente of Food Chemistry, Mørkhøj Bygade 19, DK-2860 Søborg, Denmark

John Lewis, Food Safety and Quality Group, Central Science Laboratory, Sand Hutton, York YO41 1LZ, UK

Joris Van Loco, Scientific Institute of Public Health, J. Wytsmanstraat 14, B-1050 Brussels, Belgium

William A. Maher, Ecochemistry Laboratory, Institute of Applied Ecology, University of Canberra, ACT 2601, Australia

Henryk Matusiewicz, Poznań University of Technology, Department of Analytical Chemistry, 60-965 Poznań, Poland

Bernhard Michalke, GSF National Research Center for Environment and Health, Institute for Ecological Chemistry, Neuherberg/München, Ingolstädter Landstrasse 1 D-85764, Germany

C. Minoia, Laboratory for Environmental and Toxicological Measurements, S. Maugeri Foundation, Via Ferrata 8, 27100 Pavia, Italy

A. Polatajko, UMR 5034 CNRS, Hélioparc, 2 Avenue President Angot, 64053 Pau, France

Thomas Prohaska, University of Natural Resources and Applied Life Sciences, Muthgasse 18, A-1190 Vienna, Austria

Rob Ritsema, National Institute for Public Health and the Environment (RIVM), PO Box 1, 3720 BA Bilthoven, The Netherlands

Piotr Robouch, European Commission, Joint Research Centre, Institute for Reference Materials and Measurements, Retieseweg 111, 2440 Geel, Belgium

C. Roggi, Department of Health Applied Sciences, Section of Human Nutrition, University of Pavia, Pavia, Italy

Inge Rokkjær, Danish Veterinary and Food Administration, Regional Laboratory, Sønderskovvej 5, DK-8520 Lystrup, Denmark

A. Ronchi, Laboratory for Environmental and Toxicological Measurements, S. Maugeri Foundation, Via Ferrata 8, 27100 Pavia, Italy

Alfredo Sanz-Medel, Department of Physical and Analytical Chemistry, Faculty of Chemistry, University of Oviedo, C/ Julián Clavería 8, Oviedo, E-33006 Spain

Patricia Smichowski, Comisión Nacional de Energía Atómica, Centro Atómico Constituyentes, Unidad de Actividad Química, Avenida General Paz 1499, B1650KNA San Martín, Provincia de Buenos Aires, Argentina

Joseph Sneddon, Department of Chemistry, McNeese State University, Lake Charles, LA 70609, USA

S. Spezia, Laboratory for Environmental and Toxicological Measurements, S. Maugeri Foundation, Via Ferrata 8, 27100 Pavia, Italy

J. Szpunar, UMR 5034 CNRS, Hélioparc, 2 Avenue President Angot, 64053 Pau, France

G. Turconi, Department of Health Applied Sciences, Section of Human Nutrition, University of Pavia, Pavia, Italy

Bernhard Welz, Departamento de Química, Universidade Federal de Santa Catarina, 88040-900, Florianópolis, SC, Brazil

Katarzyna Wrobel, Instituto de Investigaciones Científicas, Universidad de Guanajuato, L. De Retana 5, 36000 Guanajuato, Mexico

Kazimierz Wrobel, Instituto de Investigaciones Científicas, Universidad de Guanajuato, L. De Retana 5, 36000 Guanajuato, Mexico

Gyula Záray, Department of Inorganic and Analytical Chemistry, Eötvös University, Budapest, Hungary

SECTION 1

FUNDAMENTALS AND METHOD DEVELOPMENT

1

IMPROVEMENT IN PRETREATMENT AND ANALYSIS WITH SPECTROMETRIC METHODS: A TYPICAL APPLICATION TO ROUTINE ANALYSIS

KHALID BOUTAKHRIT, FABIAN BOLLE, JEAN-MARIE DEGROODT, AND LEO GOEYENS

Scientific Institute of Public Health, Section of Food, rue Juliette Wytsman 14, 1050 Brussels, Belgium, E-mail: Khalid.Boutakhrit@iph.fgov.becc

The Determination of Chemical Elements in Food: Applications for Atomic and Mass Spectrometry. Edited by Sergio Caroli

1.1 PREPARATION OF SAMPLE IN ROUTINE ANALYSIS

In most cases, sample preparation includes several stages, such as drying, homogenization and mineralization. Some of these steps, as performed in the laboratory, may be the source of contamination phenomena, essentially due to the type of vessels and purity of reagents used. They become more and more troublesome as the concentrations to be determined decrease.

1.1.1 Purity of Reagents

The purity of the reagents used in the laboratory, mainly HNO_3 and H_2O_2, is permanently checked when quantifying two procedural blanks. In the ranges used for calibrating, reagent purity has never been a problem.

For instance, HNO_3 is guaranteed at a purity of better than 0.4 $\mu g\,l^{-1}$ in lead and 0.08 $\mu g\,l^{-1}$ in cadmium. These values must be compared with dilution or concentration factors resulting from sample preparation. The water used in the laboratory (>18 MΩ) has always yielded to totally insignificant blanks. Therefore, if one has 2.5 mL HNO_3 for a total volume of 10 mL, the maximum contamination can be 0.1 $\mu g\,l^{-1}$ Pb. As the range used goes from 1 to 40 $\mu g\,l^{-1}$, the potential contamination from acid can be considered negligible. It is worth noting that the use of a lower range could pose such problems. Considering the nature of routine analyses, the range adopted is the best compromise.

The organization of the analytical process is another point. A systematic contamination is unveiled when analyzing the procedural blanks. Because of the systematic nature of the procedure, one can reasonably assume that, given an equal treatment of all the samples, the same error occurs for every sample. In this way, the error can be brought under control and corrected.

1.1.2 Contamination from Contact Materials

In this chapter, procedures are outlined for defining and showing the lowest reliable determination of Pb in food by electrothermal atomization atomic absorption Spectrophotometry (ET-AAS), especially as regards milk and infant formulae, where the maximum acceptable level is 20 $\mu g\,kg^{-1}$ fresh weight as defined by Directive 92/46/EEC and Directive 91/321/EEC, respectively [1].

However, the lowest possible measurement of Pb is limited by the contribution of the element from several sources of contamination [2], for example, reagents, vessels and laboratory environment. The contamination coming from ambient air is hard to control unless it is at the price of substantial investment, which sometimes cannot be afforded. The most important contribution of Pb found in this work was the one due to leaching of Pb traces from the walls of materials, mainly digestion tubes and equivalent. This contribution increases with temperature, pressure and time contact between the digested solution and the wall of the digestion tubes. The last parameter (wall of the tube) may not be reduced and the use of wet digestion depends strongly on it. The only way out is to look for other materials of vessel digestion that release less trace Pb from the walls.

This kind of contamination was carefully evaluated by using various labware: glass and quartz tubes, polytetrafluoroethylene (PTFE) vessels, fluorinated ethylene propylene (FEP) and perfluoroalkoxy (PFA) tubes. Ideally, all trace element determinations should be carried out in "clean" rooms in which the incoming air is filtered and special clothing is worn.

1.1.3 Sample Manipulation

The key steps in the analysis of foodstuffs for their trace metal content are basically the acquisition of a representative sample, destruction of the organic matter and end determination. Each of these steps is vital to the success of the overall analysis. Foodstuffs are generally heterogeneous and sample preparation is obviously very important to ensure the quality of the entire analysis. The preparation of the sample depends on its nature (matrix, fresh or dry foods, solid or liquid, fatty compounds, etc.) and on the destruction of the organic matter (wet digestion or dry ashing). To reduce the contamination risks during these steps, a number of rules must be respected [3], for example, ensure a clean environment in the laboratory, minimize handling, avoid filtration and transfer of solutions unless they are absolutely necessary, and carefully clean all vessels by soaking them in acid. In this chapter, sample contamination during sampling and storage is not addressed, although it plays a crucial role.

In the case at hand, samples are first homogenized by mechanical mixing in a mortar for dry foods or a high-speed blender for wet foods. Then 0.5 g of wet product or 0.2 g of dry product is placed directly in PFA digestion tubes equipped with Teflon-lined screw caps. Liquids are probably the easiest material to sample, provided that they have been thoroughly mixed. All materials used are immersed overnight in 10 percent HNO_3 before use and PFA digestion tubes are decontaminated by introducing 5 mL concentrated HNO_3, then heated at 180°C for 2 h and rinsed three times with Milli Q water.

It is well known that the extreme hydrophobic and anti-adhesive properties of PFA allow quick and simple acidic cleaning procedures to be adopted when preparing labware for trace analysis [4]. On the other hand, preparing conventional quartz vessels for trace analysis requires considerably longer, more powerful and more costly cleaning methods (e.g., boiling or streaming for many hours with HNO_3) [5].

1.2 ORIGINALITY AND APPLICABILITY OF THE PROCEDURE

1.2.1 Routine Analysis of Trace Element

For the routine analysis of food contaminants, atomic spectrometry has an obvious advantage over chromatographic methods in terms of detection time.

However, the mineralization of sample is time-consuming for application in routine analyses. The analysis duration can be reduced by investigation of the sample digestion. Table 1.1 shows that is possible to reduce the analysis duration by resorting to wet digestion with PFA tubes.

In ET-AAS, the electrothermal atomization is an obvious advantage, considering the complexity of some matrices. Thanks to the capabilities of this technique, one can better control the effects of the matrix. As sample preparation is less difficult, the time between mineralization and actual measuring out is reduced, which in turn reduces contamination risks.

The control of matrices in ET-AAS also enables a major application for many types of samples. Previous experiments revealed that the behaviors of vegetal and animal matrices were not significantly different. Therefore, at least in principle, one can use the same procedure for both. In practice, and for many determinations in organic chemistry, one and the same operator for monitoring the entire analytical chain is preferable. However, in routine analysis, the tasks of preparing the sample can be differentiated from those of measuring out. As long as the digestion process is under control, the level of knowledge needed is not the same as for the handling of spectrometers.

1.2.2 Advantage of Zeeman ET-AAS Over Inductively Coupled Plasma Mass Spectrometry

The use of low volumes in ET-AAS is an obvious advantage in routine analysis. The risk of an insufficient quantity of digested sample is almost nil. In the concrete case of control during analysis, the mode of the sample injection has several advantages; it allows the apparatus calculation of the recovery rate on each sample without supplementary sample preparation. The addition being directly done by the automatic sampler, the sample does not need further handling. On the other hand, depending on the injection technique, in the case of vegetal samples with a large amount of insoluble silica, filtering is certainly useful. Manipulation is therefore reduced and this is all the more convenient since filtration is a potential source of contamination.

It should be also pointed out that the robustness of electrothermal atomization enables one to avoid the use of high dilution factors. The sensitivity of ET-AAS to a high percentage of dissolved salts, major elements and/or acids is relatively controllable. Manipulation can be also reduced. Calibration is therefore possible at concentrations where contamination phenomena can be better mastered. Multi-element atomic spectrometers have additional advantages of saving time and resources by quantifying simultaneously Cd and Pb.
On the other hand, multi-element ET-AAS shows some weak aspects, namely:

(1) Linear ranges are element dependent.
(2) Electrothermal programs may differ widely depending on the elements so that it is impossible to use them in the simultaneous mode.

TABLE 1.1. Sample Analysis Duration from Digestion to End of the Analysis Depends Strongly on the Mineralization Procedures and the Labware Used

	Digestion (h)	Comparison of Time and Digestion for Various Procedures Analysis Time (min)	Digestion Rate (%)	Samples/ Run	Comparison of Sample Run Capacities Digestion (h)	Time/Sample (min)
Glass or quartz tubes	7	10	98	40	7	10.5
Microwave oven	2	10	92	10	2	12
Dry ashing	12	10	99	10	12	72
Bomb	3	10	94	20	3	9
PFA tubes	2	10	92	40	2	3

(3) There is an increase in the difficulties of checking parameters with the number of elements for analyses.

1.2.3 Simplicity and Efficacy at the Same Time

The search for efficiency in the analysis process can often coincide with the search for simplification of the analytical process. It is understandable that some requirements must be respected, such as the digestion level of the samples and the control of the matrix effects. This leads to:

- Greater execution speed because fewer operations are required.
- Greater flexibility because the reaction speed is greater.
- Greater analytical pace because analyses are quicker.
- Better control of the parameters because they are less numerous.
- Fewer risks of contamination, because the analysis time and the number of manipulations are reduced.
- Better task assessment.
- Better quality control because processes are shorter and hence better defined.

1.3 DOMAIN OF APPLICATION

1.3.1 Council Directive 96/23

The aim of the Council Directive 96/23 of 29 April 1996 is to keep the percentage of food contamination at an affordable level. An effective reduction in the levels of residues and contaminants can guarantee a better level of health protection, especially when sensitive groups of population are taken into account.

Another major aim of this Directive is to harmonize legal provisions regarding maximum percentages of contaminants in some foodstuffs to avoid possible barriers to trade caused by differences in standards among the Member States.

Lead can be a serious danger to public health as quoted in the Council Directive 96/23 of 29 April 1996. Lead can slow down cognitive development, reduce the intellectual abilities of children and increase cardiovascular disease and high blood pressure in adults. One should note that, over the last few years, the lead percentage in foodstuffs has considerably decreased thanks to public awareness of the problem and efforts to reduce lead levels. A significant role was played in this context by the quality and reliability of chemical analyses.

The Scientific Committee on Food (Directorate General, "Health and Consumer Protection," European Commission) concluded, in its statement of 19 June 1992, that the average percentage of lead in foodstuffs did not seem alarming, but that longer term action was necessary to further reduce it.

As far as cadmium is concerned, this metal can accumulate in the human body and cause kidney, bone, and reproduction pathologies. Carcinogenic effects are also possible. The Scientific Committee for Food, in its statement of 2 June 1995, recommended further reduction of food exposure to cadmium because food is the main exposure mode in humans. Hence, for cadmium, maximum affordable percentages should be set at the lowest possible level.

1.3.2 Applications in Control Laboratories

The Belgian Food Agency is in charge of the control, analysis, and appraisal of raw materials at every step of the food chain, that is, production, transformation, storage, transport, trade, import, and export. Within this control body there are various units composed of all the external departments entrusted with the control of animals, which was previously the responsibility of several administrations or entities (Veterinary Appraisement Institute, General Inspection of Foodstuffs). While effective control is ensured by the Federal Agency for Foodstuffs, the development and implementation of control measures and programs in order to ensure the protection of consumer health is the task of another administration depending on the Ministry of Public Health.

In their control role, laboratories must ensure quick analysis of the samples taken by the control bodies, aiming at a prompt reaction in cases of crisis and possessing flexibility in their analytical capacity in terms of rapidity (acute crisis), throughput (large number of determinations), and reliability (different types of determinations). Measurement accuracy must obviously be guaranteed.

In the authors' laboratory the goal is to submit results on the same day, as long as the sample is brought in the morning. The speed in processing samples has increased substantially in order to be able to cope with crisis situations, and the laboratory can now manage 60 samples a day for simultaneous determination of Cd and Pb with ET-AAS.

1.4 MATERIALS AND METHODS

1.4.1 Mineralization and Analytical Blank

Most samples require a procedure to get the matrix into solution before analysis by atomic absorption spectrometry. The two most commonly used techniques to accomplish this task are dry ashing at a defined temperature and wet digestion with mineral acids. Dry ashing is generally rather time-consuming: It usually takes a day or more, but very little attention from the analyst is necessary. Contamination and loss of analyte during ashing are possible. Wet digestion methods are generally more rapid; analyses may be completed within a few hours and the methods are less sensitive to contamination, volatilization, or analyte loss.

In the authors' laboratory many acidic mixtures are used to digest organic matter in food. Containers of various materials are also adopted for the digestion of the samples so as to minimize the analytical blank. The magnitude of the latter, in fact, and the degree of uncertainty associated with it, usually limits the ability to perform reliable quantitative determinations at low concentration levels [6].

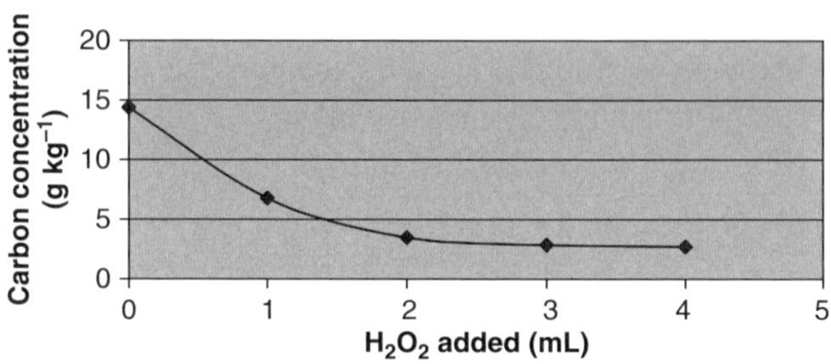

Figure 1.1. Dependence of the digestion process on the amount of H_2O_2 (from 0 to 4 mL) after digestion of the sample with HNO_3 only for 2 h at a temperature of 180°C.

The combination of HNO_3 and $HClO_4$ followed by H_2O_2 is the most commonly used method for food sample digestion. $HClO_4$ is an excellent oxidizing medium, but sometimes it produces violent explosions, mainly with food powder. It should not be used alone and the mixture should not be evaporated to dryness since inorganic perchlorates are explosive [7]. Therefore, HNO_3 is often preferred and increasing amounts of H_2O_2, from 0 to 4 mL, are generally added. After a rapid digestion, the residual carbon can range from 14.4 g kg^{-1} (0 mL H_2O_2) to 2.7 g kg^{-1} (4 mL H_2O_2), which is equivalent to 97 percent of destruction (see Fig. 1.1). In the case of skim milk powder (a BCR Certified Reference Material), for example, use was made of $HNO_3 + H_2O_2$ and the ratio of organic matter digestion was estimated to be 99.1 percent.

Plant tissues were tested as well to understand the effect of HF in the digestion mixture. There is good evidence that dissolving silica with HF does not improve the quality of the Pb analysis (see Table 1.2). The use of HF should be discouraged because of the risks it brings, all the more so in the case of routine analysis.

To achieve more reliable determinations at lower concentration levels by reducing the analytical blank, several types of labware (glass, quartz, PTFE, FEP, and PFA) were tested to determine their leachibility and cleanliness. All materials were kept in 10 percent HNO_3, then different digestion procedures were carried out using the same amounts of reagents (5 mL HNO_3 + 3 mL H_2O_2) and with the use of the heating block reactor, microwave, electric oven, or Bombs. The results obtained for the analytical blanks are summarized in Table 1.3.

TABLE 1.2. Comparison of Lead Values in Certified Reference Materials Obtained with and Without Adding HF in the Wet Digestion Process

Materials	Matrix	Certified Value (μg kg^{-1})	Value Without HF	Value with HF
SRM 1570a	Spinach leaves	200[a]	198.7 ± 4.6	200.6 ± 3.7
BCR 62	Olive leaves	25000 ± 1500	26149 ± 345	26388 ± 410
IMEP 19	Rice flour	415.9 ± 7.1	444.4 ± 9.3	438.3 ± 10.7

[a]No certified value.

TABLE 1.3. Estimation of the Detection Limit by Determining the Blank Level for Each Digestion Labware Piece

Vessels	Blank Absorbance Value Due to Adsorption	[Blank] ($\mu g\ l^{-1}$)	L.D ($\mu g\ kg^{-1}$)
Glass tubes	0.0073 ± 0.0009	2.47 ± 0.30	34
Quartz tubes	0.0026 ± 0.0018	0.88 ± 0.61	27
PTFE (microwave)	0.0056 ± 0.0010	2.27 ± 0.34	33
PTFE (bomb)	0.0038 ± 0.0008	1.19 ± 0.27	20
FEP tubes	0.0012 ± 0.0010	0.40 ± 0.34	14
PFA tubes	0.0012 ± 0.0004	0.40 ± 0.14	8

The results confirmed that the ancillary materials used in an analytical procedures might contribute to blank values more than reagents themselves do, especially glassware and microwave digestion systems. Although the latter have been extensively employed to shorten the time required for sample dissolution, some problems have been ascribed to this procedure, as reported by Lima et al. [8].

Digestion bombs attain satisfactory blank values if properly handled, although experimental data turn out to be less accurate. Recovery rates are in general exceedingly high due to lead migration in the Teflon cup walls. In fact, PTFE shows some degree of inhomogeneity, a very rough surface with deep pores and protrusions [9], which renders decontamination more difficult. On the other hand, PFA and FEP wares have a smooth surface with the lowest roughness (Fig. 1.2).

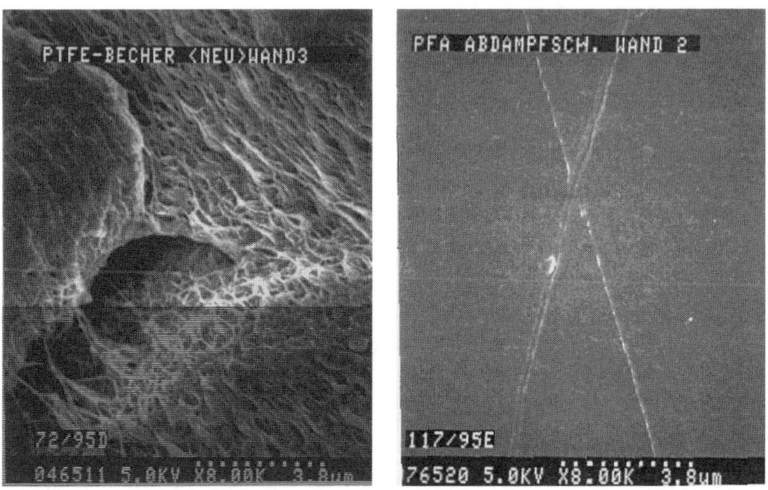

Figure 1.2. Scanning Electron Microscopy (SEM) image of surface of PTFE beaker (left) and a PFA evaporating dish (right). Reproduced from [9] with kind permission of VITLAB GmbH, Grossostheim, Germany.

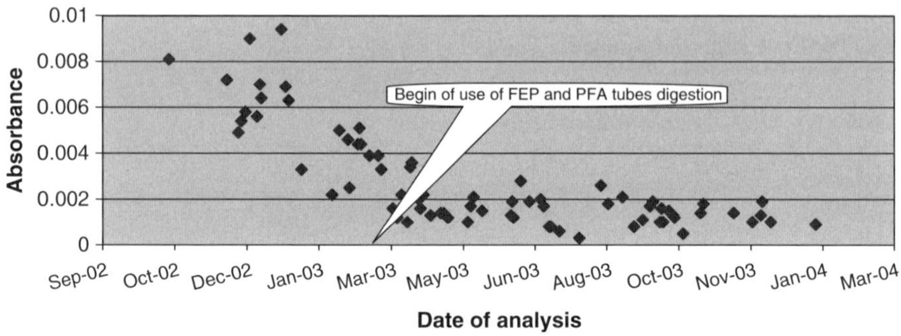

Figure 1.3. Improvement in the procedural blank response due to adsorption by using different vessels. In the order: glass tubes, quartz tubes, FEP tubes, and PFA tubes.

The improvement of the procedural blank signal using FEP and PFA tubes instead of glass and quartz tubes is shown in Fig. 1.3. The last shows a better decrease of the signal (blank absorbance) and therefore a lower concentration level will be reached. Fluoromaterials possess a series of unique properties such as wide operational temperature range (-200 to $+260$°C) and high chemical resistance, thus being particularly useful in trace and ultratrace work. Needless to say, a low analytical blank can substantially improve the limits of detection (LoDs) and the accuracy of the method [10].

1.4.2 Determinations

The development of analytical methods for reliable lead determinations in foodstuffs requires the utmost care in consideration of the extremely low levels of this metal. Each single sample preparation steps deserves attention because contamination must always be strictly controlled [11], and problems related to inadequate LoDs can impair the analysis. In this framework, an analytical method was set up for the determination of lead at the trace level and its application to baby food was investigated where the maximum accepted level for the metal is 20 $\mu g\ kg^{-1}$ fresh weight. Samples (about 0.5 g) were digested using $HNO_3 + H_2O_2$ (5 mL + 3 mL) in FEP and PFA tubes. For the determination of Pb, a Perkin–Elmer atomic absorption spectrometer model SIMAA 6000 was used. It is equipped with longitudinal Zeeman effect background correction and a transversely heated graphite atomizer (THGA). The best matrix modifier was a mixture of $NH_4H_2PO_4$ (0.5 percent w/v) and $Mg(NO_3)_2$ (0.03 percent w/v). The furnace conditions were 500°C ash temperature and 1650°C atomization temperature. Further details about the GFAAS program are given in Table 1.4.

Calibration was performed using aqueous standard solutions containing 40 $\mu g\ l^{-1}$ Pb in 2 percent HNO_3 prepared by diluting a Perkin–Elmer Standard of 1.0 $g\ l^{-1}$. The measuring cycle was as follows: 20 μL of sample solution and 10 μL of matrix modifier solution were pipetted by the auto-injector on the L'vov platform in the graphite tube and the THGA program was started. All

TABLE 1.4. Programme for ET-AAS Analysis

Steps	T (°C)	Ramp Time (s)
Injection	20	—
Drying	110–130	10
Ashing	500	10
Atomization	1650	0
Clean out	2400	3

measurements were run in duplicate. Sample content calculations are based on peak areas of the background corrected absorption signal and the calibration graph was drawn up for each series of determinations, up to 40 $\mu g\ l^{-1}$ Pb.

According to IUPAC, the LoD is defined as three times the standard deviation of the mean of the blank determinations added to the mean of the blank measures [12]. The LoD obtained for the PFA procedure (Table 1.3) is 8.2 $\mu g\ kg^{-1}$ Pb. This value is low enough to quantify lead in the vast majority of foodstuffs. This method was successfully applied for various food samples, vegetable and animal tissues. No interferences from other elements have been observed. Matrix effects were seen with some samples, but were eliminated simply by diluting the samples.

1.5 VALIDATION

The selection of an analytical technique and the development of procedures and operating conditions are only one part of the process for achieving reliable measurements. Among the first steps of the validation figure the assessment of the method precision; this parameter must be assessed between analysts and days. Appropriate precision, although necessary for reliable work, is not sufficient to guarantee accurate results: Evaluation of sample matrix effect, development, and analysis of spiked samples, analysis of Certified Reference Materials (CRM) of similar composition for which the Pb concentration has been established by other laboratories and comparison of results between laboratories (proficiency testing) are equally important [13].

1.5.1 Precision (Repeatability and Reproducibility)

The precision measures include the determination of the repeatability (Sr) and the reproducibility (SR). A number of determinations were made on spiked meat and vegetable matrix samples. As shown in Table 1.5, on the basis of six runs in six consecutive analytical series (different days and calibrations) and for different analytical levels of Pb, average values of $Sr = 1.9$ percent, $SR = 5$ percent and $Sr = 1.9$ percent, $SR = 4$ percent; were obtained for meat and vegetables, respectively.

From these results, it seems that, as the concentration of analyte in a sample approaches the LoD, the relative standard deviation associated with that determination increases, as reported in the literature [14]. It is worth noting that these

TABLE 1.5. Precision of the Procedure for Meat and Vegetable Matrices

	Repeatability (*Sr*)		Reproducibility (*SR*)	
Lead ($\mu g\ kg^{-1}$)	SD ($\mu g\ kg^{-1}$)	CV (%)	SD ($\mu g\ kg^{-1}$)	CV (%)
		Meat		
50	1.98	3.96	4.88	9.76
100	2.86	2.86	8.80	8.80
300	3.98	1.33	9.42	3.14
500	5.28	1.06	12.90	2.58
750	9.04	1.21	22.25	2.97
1000	8.79	0.88	29.40	2.94
		Vegetables		
50	1.98	3.96	2.84	5.69
100	1.64	1.64	3.13	3.13
200	2.92	1.46	6.41	3.20
300	1.61	0.54	14.84	4.95

results were obtained by digestion in FEP tubes. In glass tubes, random contamination at low levels is the major factor of variation, which can be 5 times higher for 50 $\mu g\ kg^{-1}$ Pb.

Further improvement of precision at low levels would require the reduction of blank variation; in practice, an additional source of variation may arise from sample inhomogeneity.

1.5.2 Accuracy

CRMs are often the best tools for assessing accuracy. The results obtained for FEP and PFA tubes digestion are in agreement with the certified contents for various types of materials, while with other digestion vessels, the lead content was higher than the certified value (Table 1.6).

Checking the FAPAS (S7R40), the largest analytical chemistry proficiency testing scheme in the food sector, lead content with the PFA tube procedure provided the result expected, whereas that obtained previously with glass tubes turned out to be higher.

1.5.3 Internal Quality Control

Once the analytical parameters have been determined from the method development and the method has proven suitable for routine measurements, internal quality control (IQC) procedures must be established to maintain the validity of the analytical scheme and to better monitor potential sources of errors. The IQC used includes pre- and post-digestion controls, blank determination, half range of the calibration graph checking, and recovery rate of the samples. The stability of the recovery rate with time (Fig. 1.4) shows that the method is robust after using

TABLE 1.6. Certified and Found Values for Reference Materials Obtained with Different Vessels

Certified Reference Material	Code	Certified Concentration (μg kg^{-1})	Pb Content Found with Different Vessels (μg kg^{-1})					
			Glass Tubes	Quartz Tubes	PTFE[a]	PTFE[b]	FEP Tubes	PFA Tubes
Bovine muscle	BCR 184	239 ± 11	330 ± 8	280 ± 16	351 ± 20	317 ± 52	250 ± 3	245 ± 5
Bovine liver	BCR 185R	172 ± 9	344 ± 33	191 ± 41	260 ± 30	259 ± 42	167 ± 10	191 ± 3
Pig kidney	BCR 186	306 ± 11	413 ± 37	—	464 ± 41	476 ± 165	334 ± 6	330 ± 4
Mussel tissue	BCR 278R	2000 ± 40	2390 ± 160	2240 ± 126	—	2089 ± 121	2109 ± 76	2079 ± 63
Cod muscle	BCR 422	85 ± 15	209 ± 23	143 ± 8	193 ± 14	219 ± 28	124 ± 18	—
Tuna fish	FAPAS 740	9.21	21 ± 2	—	—	—	—	8.8 ± 2

[a]Microwave digestion.
[b]Bomb digestion.

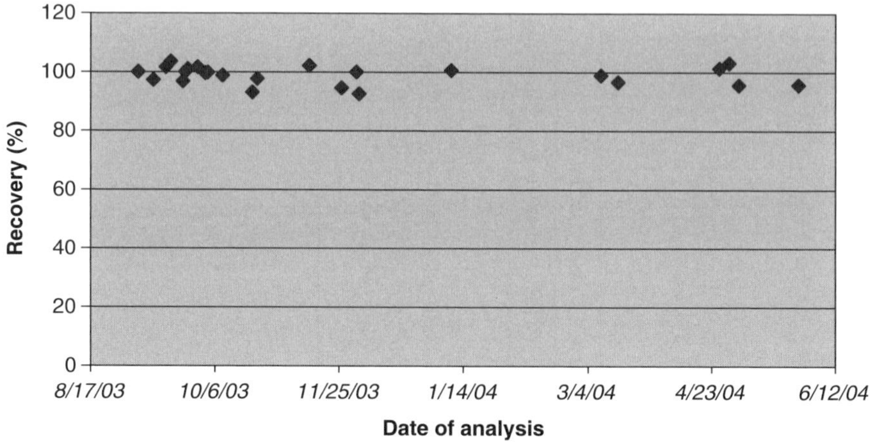

Figure 1.4. Recovery of Pb in foodstuff routine samples.

it for about 6 months for the analysis of various foodstuffs matrices. IQC also includes CRMs and PT analyses, control charts, internal and external audit. Therefore IQC comprises the routine practical procedures which enable the analytical chemist to accept either a result or a group of results as fit for purpose or to reject them and repeat the analysis.

In the method described, the procedural blank determination is essential for the analytical procedure since it allows the estimation of all kinds of contaminations. To assess the matrix effect for each sample, recovery rates were determined. The checking of the half range of the calibration graph for each 10 runs of samples allowed the evaluation of calibration graph deviation.

1.6 CONCLUSIONS

For foodstuff matrices, the use of PFA labware and H_2O_2 reagent improved the conventional analytical parameters (LoDs, level of blanks, stability, recovery rate, and also the speed of analyses. The quantification of lead at low concentration was possible down to 9 μg kg^{-1}, whereas in the recent past the LoD in the same laboratory was at about 35 μg kg^{-1}, thus making participation in the previous FAPAS S7R40 PT impossible. Thanks to the improved procedure, more parameters of quality control could be integrated in the program, that is, recovery rate is calculated for each run of sample and this without lengthening analysis time.

The method described was included among those covered by the accreditation granted by BELAC, that is, the official organization in charge of quality management in Belgium. It can be concluded that reliable results in trace lead analysis may be affected by the purity and surface properties of labware, which can impact on the quality and accuracy of the method in ultra-trace analysis.

REFERENCES

1. T. Berg, D. Licht, International legislation on trace elements as contaminants in food, *Food Addit. Contam.*, **19** (2002), 916–927.
2. G. Capar, A. Subjoc, Defining a lowest level of reliable measurement for lead in foods, *J. Assoc. Off. Anal. Chem.*, **65** (1982), 1025–1029.
3. M. Hoenig, Preparation steps in environmental trace element analysis, facts and traps, *Talanta*, **54** (2001), 1021–1038.
4. K. Englert, G. Giebenhain, H.-J. Mosh, N. Muller, Stability of diluted standard reference solution of mercury (Hg) and precious metal in PFA-vessels, *GIT Fachz. Lab.*, **1997**, 32.
5. P. Tschöpel, L. Kotz, S. Schulz, M. Veber, G. Tölg, Causes and elimination of systematic errors in the determination of elements in aqueous solutions in the ng/mL and pg/mL range, *Fresenius' Z. Anal. Chem.*, **302** (1980), 1–14.
6. R. K. Skogerboe, The analytical blank: sources and effects on lead analyses, *J. Assoc. Off. Anal. Chem.*, **65** (1982), 957–964.
7. N. T. Crosby, Determination of metals in foods, *Analyst*, **102** (1977), 222–268.
8. C. Lima, F. Barbosa, J. Krug, Comparison of ultrasound-assisted extraction, slurry sampling and microwave-assisted digestion for cadmium, copper and lead, *J. Anal. Atom Spectrom.*, **15** (2000), 995–1000.
9. J. Dahmen, K. Englert, G. Giebenhain, Properties and production of labware from fluorinated hydrocarbons and their advantages for ultratrace analysis, *Int. Lab. (Pacific Rim Ed)*, **12** (1997 April/May), 30–32.
10. M. J. Baxter, J. A. Burrell, H. M. Crews, R. C. Massey, D. J. McWeeny, A procedure for determination of lead in green vegetables at concentrations down to 1 μg/kg, *Food Addit Contam.*, **6** (1989), 341–349.
11. Y. Schmitt, Influence of preanalytical factors on the atomic absorption spectrometry determination of trace elements in biological samples, *J. Trace Elem. Electrolytes Health Dis.*, **1** (1987), 107–114.
12. *IUPAC Compendium of Chemical Terminology*, 1997, 2nd Ed; http:/www.chemsoc.org/chembytes/goldbook/.
13. D. A. Bass, D. Hickok, D. Quig, K. Urek, Trace element analysis in hair: Factors determining accuracy, precision, and reliability, *Alternative Med. Rev.*, **6** (2001), 472–482.
14. R. Satzger, E. Bonnin, L. Fricke, Chemical contaminants monitoring, *J. Assoc. Off. Anal. Chem.*, **67** (1984), 1138–1140.

2

SOLUBILIZATION: TRENDS OF DEVELOPMENT IN ANALYTICAL ATOMIC SPECTROMETRY FOR ELEMENTAL FOOD ANALYSIS

HENRYK MATUSIEWICZ

Poznań University of Technology, Department of Analytical Chemistry, 60-965 Poznań, Poland, E-mail: Henryk.Matusiewicz@put.poznan.pl

2.1 INTRODUCTION

Since food is the primary source of essential elements for humans, the accurate and precise analysis of food stuff and commodities is critical for consumers and health care professionals to take decisions and make recommendations based on the

The Determination of Chemical Elements in Food: Applications for Atomic and Mass Spectrometry. Edited by Sergio Caroli

composition of foods. On the other hand, a wide range of elements occur in foods, either naturally or intentionally added, as well as elements arising by adventitious contamination. These mineral elements can generally be classified as nutritionally essential major elements (Ca, Cl, K, Mg, N, Na, and P), nutritionally essential minor and trace elements (As, B, Br, Co, Cr, Cu, F, Fe, I, Mn, Mo, Se, Si, Sn, V, and Zn), and those regarded as toxic or with both essential and toxic aspects (As, Au, Cd, Co, Cr, Fe, Hg, Li, Mn, Mo, Ni, Pb, Pd, Se, Sn, Tl, V, and Zn).

The most suitable techniques for the rapid, accurate determination of the elemental content of foods are based on analytical atomic spectrometry, for example, atomic absorption spectrometry (AAS), atomic emission spectrometry (AES), and mass spectrometry, the most popular modes of which are flame (F), electrothermal atomization (ET), and hydride generation (HG) AAS, inductively coupled plasma (ICP), microwave-induced plasma (MIP), direct current plasma (DCP) AES, and ICP-MS. Challenges in the determination of elements in food include a wide range of concentrations, ranging from ng/g to percent levels, in an almost endless combination of analytes with matrix specific matrices.

Sample digestion and extraction remain among the most time-consuming and error-prone steps in the food analysis process. Samples are usually presented to the atomic spectrometers as solutions. Depending on the sample matrix and analytes of interest, there are several procedures to get samples (or analytes) into solution. These include extraction, dry ashing (dissolution), and wet ashing (digestion). In some cases, a simple analyte extraction or leaching procedure using basic (alkaline) solutions or tertiary amines can be effectively employed for the alkaline extraction of elements from food samples. The success of these methods depends on the analytes of interest and the sample matrix.

In this chapter, the description of applications highlights the selection of the instrumentation used as well as the alkaline solubilization sample preparation (often the most critical part of a complete analytical method). The main characteristics of the suspension sample introduction method are also emphasized. A brief discussion on subgroups of these methods, identified by the atomic spectrometric instrumental approach, is finally presented.

2.2 NOMENCLATURE

For some methods of analysis, it may be required that the food sample be in liquid form – the sample solution. Thus, standard procedures are required to convert solid (or solid-containing) samples to solutions prior to detection. However, this conventional terminology is often imprecise or even misleading with respect to the actual mechanism of the process. Several very different terms are sometimes applied to a single technique, which presents a substantial obstacle for anyone (particularly a nonspecialist) interested in acquiring a quick overview of systems applicable to a specific task. The terms solubilization with bases, dissolution/solubilization, alkaline solubilization/extraction/digestion, solubilization/digestion/leaching/dissolution with tetramethylammonium hydroxide (TMAH), base digestion and solubilization all refer to this process. Hereafter the term used for obtaining the resulting sample solution (liquid, suspension) is *alkaline solubilization.*

2.2.1 Analytical Atomic Spectrometric Techniques

The choice of the food sample preparation procedure has an impact on the performance of the quantitation technique used, the behavior of the sought-after element and the amenability of the sample matrix to proper digestion. A convenient way to introduce solid (liquid) material into the measurement cell of an atomic spectrometer is to prepare a suspension or a slurry.

2.2.1.1 Atomic Absorption Spectrometry AAS has probably been the most widely used technique for the determination of elements (major, minor, and trace) in food. F-AAS, in particular, is a well-established technique for the determination of elements in food samples. The relatively low cost and the commercial availability of a variety of instruments make this procedure attractive to many laboratories interested in the analysis of foods. Because of the robustness (i.e., maturity) of the technique, relatively few new F-AAS methods are being published in the analytical literature.

Liver and kidney homogenate samples were prepared for F-AAS quantification by digestion with an appropriate concentration of aqueous tetramethylammonium hydroxide TMAH [1]. The tissue levels of Cu, Cd, Pb, and Zn, the reproducibility of the analyses, and the recovery of added element standards compare favorably with the results obtained by standard wet ashing procedures using concentrated HNO_3 or $HNO_3{-}HClO_4$.

Zhou et al. [2] described a microwave (MW) assisted digestion system with a closed PTFE vessel to improve the leaching of inorganic constituents from the Certified Reference Materials (CRMs) NIST SRM 1577b Bovine Liver and NIES CRM no. 1 Pepperbush with TMAH. The effects of MW parameter settings and the quantities of TMAH used on leaching efficiency were evaluated. The major (Ca, K, Mg, Na) and minor (Fe, Mn, Zn) elements in the digests were determined by F-AAS. The analytical results were in good agreement with the certified values.

F-AAS is a powerful technique [1], but it may not always provide the necessary sensitivity for the determination of trace elements present at extremely low concentrations. AAS procedures utilizing electrothermal atomization in place of flame atomization give potential for much increased sensitivity. ET-AAS is a highly sensitive technique for element determination, typically characterized by 10–100 times lower limits of detection (LoDs) than conventional F-AAS.

Tsalev et al. [3] explored the advantages of using V(V) as a chemical modifier in ETAAS, particularly when determining As. This modifier is compatible with alkaline sample solutions, samples solubilized in TMAH, and slurries.

An MW-digestion system with a closed PTFE vessel was used to improve the leaching of inorganic constituents from biological samples with TMAH [2]. This new digestion method has been applied to the NIST SRM 1577b Bovine Liver

CRM. Minor elements (Al, Cu, Ni, Pb, and Zn) in the digests were quantified by ET-AAS. The analytical results were in good agreement with the certified values.

High-pressure homogenization was evaluated for the preparation of slurries suitable for the determination of Cr, Cu, Fe, Mn, Ni, and Se by ET-AAS in soft organ tissues (liver and kidney), CRMs of biological origin, and animal feeds [4]. The sample was blended with ethanol–water (1 + 9 v/v) containing 0.25 percent m/m TMAH and the resulting mixture was homogenized. The resulting solution was suitable for analysis by ET-AAS. The aqueous standards could be used to calibrate the instrumental response.

The TMAH-based digestion method has been proposed for the determination of the total Se content in wheat and mussel samples [5]. The results achieved with this digestion procedure were compared with those obtained after MW-assisted acid decomposition. ET-AAS was used for the final Se determination. HG-AAS analyses showed that TMAH digestion could also be exploited for speciation purposes.

An alternative method to acid digestion was proposed for the preparation of marine biological tissues and botanical samples for trace metal determination by ET-AAS [6]. Samples were solubilized using methanolic solutions of TMAH, either overnight at room temperature or by the application of heat and/or ultrasonic agitation. Following dilution with water, aliquots were sampled into a graphite furnace; external calibration was used for all analyte (Cd, Cr, Cu, Mn, Ni, and Pb) determinations. Samples prepared with TMAH can be stored for up to three years at room temperature without any change in analyte concentrations.

Lelièvre et al. [7] developed a rapid method for the direct determination of Cu and Fe in butter by ET-AAS. Butter was dissolved in a butylamine (BTA)–water–tetrahydrofuran (THF) mixture. The solution (10 μL) was pipetted directly into the graphite tube. The calibrations were compared with the matrix-matched standard and standard addition calibration.

TMAH was used to solubilize the DORM-1 dogfish muscle CRM as a model substance for the determination of As, Cd, Pb, and Se by ET-AAS [8]. The sample was mixed with a small amount of TMAH and heated to 60°C for 10 min in a water bath. The calibration was performed with aqueous solutions in 0.2 percent v/v HNO_3. Results from the determination of these elements in the DORM-1 CRM were within the 95 percent confidence interval of the certified values.

A method to prepare milk powder, bovine liver, and bovine muscle samples for analysis by ET-AAS was proposed by Martins et al. [9]. Samples were mixed with a small amount of TMAH and a stable and homogeneous slurry was produced in *ca* 2 h with heating at 60–70°C. After such sample preparation and dilution with water, trace elements (Ag, Cd, Cr, Cu, Ni, and Pb) were determined in CRMs. External calibration was used for every analyte.

The feasibility of Ru as a permanent modifier for the determination of Cd in biological samples treated with TMAH by ET-AAS was investigated [10]. The sample dissolution procedure was very simple: a sample aliquot was mixed with a small volume of a 25 percent TMAH solution, the volume was made up to 50 mL and the mixture was kept at 60°C for 1 h. Four biological CRMs (DOLT-2, DORM-1, LUST-1, and TORT-2) were analyzed and the obtained Cd concentrations were within the 95 percent confidence interval of the certified values.

2.2.1.2 Inductively Coupled Plasma Atomic Emission Spectrometry Liquid samples offer the best opportunity for aspiration into the plasma sources with minimal pretreatment. Therefore, introduction of solubilized samples into the ICP torch has received and still is receiving much attention [11–18].

The combined use of TMAH and EDTA was effective for leaching inorganic major and minor constituents from rice flour and liver samples for their determination by ICP-AES [11].

To avoid a positive systematic error in ICP-AES determination of Si, a new MW-assisted dissolution procedure incorporating tertiary amine addition was developed [12]. The determination of low Si concentration levels in food was significantly improved by adding water-soluble tertiary amines to the solution after dissolution with HF. Better accuracy and precision, and an improved LoD (75 ng/mL of Si), were obtained with amine addition than without.

A simple ICP-AES analysis procedure for biological (liver, total diet, egg) materials was developed, based on the extraction with a commercial, water-soluble, tertiary amine solution [13]. CRMs were dispersed in a 10 percent (v/v) amine mixture at pH 8 and the resulting dispersion (suspension) was stirred for 30 min at room temperature. After centrifugation, Al, Ca, Cu, Fe, K, Mg, Mn, Na, and Zn were determined by ICP-AES in the supernatant.

The direct determination of some major elements (Ca, K, Mg, Na, and P) and Zn by ICP-AES was performed in powdered milk [14]. Samples were diluted with a 5 or 10 percent (v/v) water-soluble, mixed tertiary amine reagent at pH 8. This reagent mixture dissociated casein micelles and stabilized liquid phase cations. No decrease in analyte emission intensities was observed. Reference solutions were prepared in 10 percent (v/v) mixed amine solution, and no internal reference element was needed for ICP-AES. The direct technique is as fast as slurry approaches, without particle size effects or sensitivity losses.

Sample cuvette – tungsten boat furnace (TBF) – ICP-AES was investigated for the direct determination of trace amounts of Cd and Pb in biological materials (pepperbush, rice flour, oyster tissue) [15]. Into each small sample cuvette made from W, a mixture of the ground biological sample and powdered diammonium hydrogenphosphate was placed after accurate weighing. After the TMAH solution was injected, the cuvette was kept at a wet-digestion temperature to decompose the solid sample. After the digestion was completed, the temperature was progressively increased to generate the analyte vapor to be introduced into the plasma. If an aqueous standard solution is used the standardization procedure is easier.

A nonoxidative alkaline sample digestion procedure using TMAH and a high-pressure, MW-assisted autoclave-digestion system was developed [16]. The Si concentrations of the solubilized samples were measured by ICP-AES. The method was developed and tested using Si-spiked ascorbic acid and applied to samples of animal tissue and organs.

A fully alkaline procedure for the determination of total Se in biological samples was described [17], that includes alkaline fusion, prereduction of Se(VI) to Se(IV) in alkaline medium, and H_2Se generation in an alkaline solution, and ICP-AES determination. Alkaline fusion coupled with alkaline mode HG effectively avoided volatile losses of Se by changing it into soluble salts such as selenate or selenite and largely eliminated the chemical interferences from the co-existing transition metals Cu, Fe, and Mn. The method was verified with a CRM and was successfully applied to real samples such as egg and animal feeds.

A simple method for the solubilization of powdered milk and soluble coffee using TMAH was proposed [18]. Solubilization at 80°C takes place within 10–30 min. The concentrations of Ca, Cu, Fe, Mg, Mn, Na, P, Se, Sn, and Zn were determined by ICP-AES. When applied to CRMs and to commercial samples the proposed method showed good results and compared well with the other two preparation methods.

Moeller et al. [19] considered the TMAH–HNO_3 wet digestion procedure for the determination of various elements by ICP-AES.

2.2.1.3 Microwave-Induced Plasma–Atomic Emission Spectrometry A simple, alternative method to acid digestion was presented for the preparation of biological and food materials for major and trace elements by MIP-AES [20]. CRMs were solubilized using a methanolic solution of TMAH by ultrasonic agitation. Following dilution with water, suspensions were pumped to a V-groove Babington-type nebulizer to be presented to the Ar plasma. Standard addition procedures were used to minimize possible matrix interferences. Measured concentrations were in satisfactory agreement with certified values for the biological and food CRMs (lobster, milk, liver).

2.2.1.4 Inductively Coupled Plasma–Mass Spectrometry ICP-MS has been accepted as a powerful technique for elemental and isotopic analysis [21–25]. It is characterized by a linear dynamic range of five to six orders of magnitude, rapid multielemental analysis, comparative freedom from spectral interferences, and LoDs in the range of 10–100 pg mL^{-1} for most elements. This makes it ideal for the multielemental analysis of food samples on a routine basis.

A semiquantitative ICP-MS analysis protocol for various food samples was developed to include water-soluble tertiary amines with MW-heated sample preparation [21]. CRMs digested with HNO_3, H_2O_2, and HF required addition of H_3BO_3 and a solution of tertiary amines (CFA-C reagent) to dissolve insoluble fluorides and neutralize free fluorides. Measurement parameters including m/z range, spectrometer sensitivity, and number of elements in the external calibration were evaluated. The accuracy of the semiquantitative analysis improved when the standard solution was matrix matched; 38 elements were determined in six CRMs and food samples with 1–20 percent precision and 1–50 percent accuracy. Aluminium, K, Na, and Si determinations turned out to be unreliable.

Water-soluble tertiary amines enhance signals and decrease polyatomic chloride interferences in the ICP-MS determination of As and Se in food samples [22]. Arsenic and Se ICP-MS determination parameters have been optimized by a simplex procedure with amines in Ar plasma. A simple, direct, quantitative procedure for As and Se determination in food samples was set up, that provides good accuracy and fit-for-purpose LoDs.

A simple ICP-MS analysis procedure for biological materials was developed based on the extraction with a commercial, water-soluble tertiary amine solution [13]. Arsenic, Ba, Ca, Cr, Cu, Mn, Mo, Pb, Rb, Se, and Zn were determined.

The direct determination of trace elements (Al, Ba, Cu, I, Mn, Mo, Pb, Rb, Se, Sr, and Zn) by ICP-MS in powdered milk was reported [14]. Samples were diluted with a 5 or 10 percent (v/v) water-soluble, mixed tertiary amine reagent at pH 8. This reagent mixture dissociated casein micelles and stabilized liquid phase cations. Mass intensity losses were not observed. The quantitative ICP-MS procedure was applied the standard additions method with a Y internal reference. This direct technique is as fast as the slurry approach without particle size effects or sensitivity losses.

In recent years the electrothermal volatilization (ETV) technique has developed into a popular tool for the determination of trace elements. This technique allows solubilized samples to be used and separates matrix constituents using particular temperature and time components before transportation of the analyte into the plasma excitation source via a stream of gas. Several papers on solubilized sampling ETV coupled with ICP-MS have been published [23–25]. In a method concerned with the preparation of biological samples to be analyzed by ETV-ICP-MS [23], a solution or slurry was formed by mixing a sample aliquot (20–100 mg) with a small volume (10–200 μL) of a 25 percent (m/v) TMAH solution. For animal tissues, complete dissolution was obtained, whereas for plant and whole egg, slurries were obtained. External calibration was used preferentially, but in some instances, the standard additions method was employed. Good results were obtained for CRMs. In a further extension of the work an isotopic dilution method was proposed for the determination of Ag, Cd, Hg, Pb, and Tl in biological materials by ETV-ICP-MS [24]. The sample was simply dissolved with TMAH or acid digested in an MW oven, with an on-line matrix separation. The on-line pre-concentration was performed in a flow injection (FI) system. The isotopic dilution calibration was applied to two CRMs (bovine liver and dogfish muscle), dissolved with TMAH or acid digested, and to another two CRMs (corn bran and rice flour), acid digested and subjected to analyte preconcentration. The obtained concentration values agree with the certified ones, showing that this calibration procedure leads to accurate results in the determination of low concentrations of volatile elements. The newly conceived ETV system using a TBF sample cuvette was designed for the direct analysis of solid samples with ICP-MS detection [25]. Into this small sample cuvette, a solid mixture of the biological samples and diammonium hydrogenphosphate powder as a fusion flux was placed into a TBF. The TMAH solution was added to the mixture. After the digestion had been completed, the analyte in the cuvette was vaporized and introduced into the ICP torch of the mass spectrometer. The solid samples were analyzed using a calibration curve prepared from aqueous standard solutions.

2.3 SPECIATION

The chemical speciation of an element is the determination of the individual physical–chemical forms of the element, which together make up its total concentration in sample. The current interest in speciation arises from the fact that the form or *species* of a particular element present in foods may affect its nutritional value or toxicological properties. However, there is a keen need for much more work in this field as the determination of total analyte concentrations is still a common practice.

A tissue solubilizer (TMAH) was used to dissolve fish samples without altering the chemical forms of the alkyl-Pb species [26]. The various alkyl-Pb species and Pb(II) were isolated quantitatively by chelation extraction and then determined by a gas chromatography (GC)–AAS method.

A method was described for the selective determination of selenomethionine in wheat samples [27], where pretreatment of selenomethionine with TMAH did not induce degradation of the selenoamino acid. The procedure constitutes a suitable sample pretreatment for the selective determination of selenomethionine. Of the four detection techniques (FPD, MIP-AES, AAS, MS) compared, only MS provided LoDs sufficiently good for the determination of selenomethionine in real samples.

MW-assisted extraction was used for the extraction of As species from fish tissue [28]. Quantitative extraction of As from the spiny dogfish muscle (DORM-2) CRM was achieved using 5 percent TMAH solution with MW heating at 65°C in a closed-vessel MW system. The extracted As species were separated using both ion-exchange and ion-pair chromatography with ICP-MS detection. The DORM-2 CRM was analyzed for As along with three different varieties of fish purchased from a local market. In all samples, most of the As present was in the form of arsenobetaine, a basically nontoxic chemical species.

2.4 METHOD DEVELOPMENT FOR INDIVIDUAL ELEMENTS

2.4.1 Iodine

To avoid acid digestion is beneficial for the determination of volatile and difficult-to-determine elements like I. The development and evaluation of an FI-ICP-MS method for the determination of Cu and I in milk and milk powder has been described [29]. The sample preparation consists simply in the dilution of the sample by an alkaline solution containing KOH and TMAH. Four different milk powder CRMs were analyzed and the concentrations were found to be in good agreement with the certified values, thus indicating that the method was unbiased.

Extraction of I by TMAH solution at high temperatures in a closed vessel was reported in connection with isotope dilution (ID)-ICP-MS for accurate and precise determinations of I traces in food samples (milk powders, infant food, bovine liver) [30].

A simple alkaline extraction method, at elevated temperatures (90°C) in a sealed vessel was applied to the determination of I, employing TMAH and ICP-MS [31]. All the I present in the samples (milk powder, egg powder, cod muscle, pig kidney) was extracted completely by TMAH. The results were confirmed by decomposition of the samples.

Knapp et al. [32] developed an extraction (leaching) method using TMAH for the final determination of I concentrations in food samples by ICP-MS and a catalytic technique. TMAH has proven to be an excellent medium for use with ICP-MS, and its high alkalinity makes it a very effective reagent for extraction.

In a method for the determination of I in dietary foods [33], this element was quantified by ICP-MS after extraction with the alkaline reagent TMAH.

A rapid, robust ICP-MS method was described for the determination of I in food for human consumption as well as for pets [34]. The sample preparation was made by alkaline hydrolysis with TMAH using either MW heating or the high-pressure asher. Method validation was carried out using seven food CRMs with certified I content and by cross-validation with GC, neutron activation analysis (NAA) and colorimetry. The method has been proven to give accurate and repeatable results for a range of fortified food commodities for human and animal consumption.

2.4.2 Mercury

The interest of many groups continues to focus on optimizing sample pretreatment methods to obtain maximal recovery of both inorganic and organic Hg species from food matrices for quantitative determination by different spectroscopic techniques. Mercury may be present in food in three different forms, that is, as elemental Hg, Hg(II), and alkyl-Hg.

A simple technique was described for the rapid determination of methyl-Hg in fish tissue [35]. Following simple dissolution in methanolic KOH solution, aqueous phase ethylation by derivatization with $NaB(C_2H_5)_4$, cryogenic trapping on a packed chromatographic column, and GC separation, volatile Hg species were detected by AAS. Analysis of CRMs demonstrates the accuracy, precision, and reproducibility of the analytical method.

A novel approach was described for the speciation analysis of Hg, namely methyl-Hg and Hg(II), in fish tissue using GC-MIP-AES [36]. Focused MW-assisted digestion was applied for sample preparation (tissues were dissolved with TMAH, 25 percent aqueous solution), a technique which enables mild, quick, and complete dissolution of the sample. The method was validated by analysis of the BCR 464 freeze-dried tuna fish CRM.

A series of studies have been carried out by Sturgeon et al. [37–40]. A rapid method for the determination of total and inorganic Hg in biological tissues was presented using ETV-ICP-MS. Samples were solubilized using TMAH [37]. A sensitive and interference-free method was presented for the quantification of inorganic Hg and methyl-Hg species in biological tissues using purge-and-trap injection GC-AES. Samples were solubilized with TMAH and the ionic species were purged from aqueous solution after ethylation with sodium tetraethylborate [38]. A simple, rapid, and reliable method was developed for the determination of total Hg in biological samples. Samples were solubilized using TMAH. The organically bound Hg was cleaved and converted to inorganic Hg by on-line addition of $KMnO_4$. The decomposed Hg substances and the inorganic Hg originally present in samples were determined by FI cold vapor (CV) AAS after reduction to elemental Hg vapor using $NaBH_4$ [39].

In another similar method, the inorganic Hg was released by on-line addition of L-cysteine and then reduced to metallic Hg by $SnCl_2$ [40]. The proposed methods were validated by the analysis of a series of biological marine CRMs (DORM-2, DOLT-2, TORT-2) [37–40].

A rapid and simple MW-assisted digestion method with alkaline solution (TMAH or methanolic KOH solution) was developed for speciation analysis of inorganic Hg and methyl-Hg in biological tissues [41]. Extracts with quantitative recoveries of Hg species after the alkaline dissolution of the sample were directly analyzed by an automated on-line hyphenated system incorporating aqueous HG, cryogenic trapping, GC, and detection by AAS. The proposed method was validated by the analysis of biological CRMs (CRM 463, DORM-1, TORT-1) and one BCR sample from an interlaboratory study (Tuna Fish 2).

A multivessel system for the determination of Hg by CV-AAS was developed [42]. The performance of the proposed device was tested by determining total Hg in quality-control samples of fish following alkaline solubilization.

A simple and rapid procedure was developed for the simultaneous determination of methyl-Hg and Hg(II) in fish CRMs [43]. The procedure was based on a rapid MW-assisted solubilization of biomaterial with TMAH, simultaneous quantitative ethylation-extraction of the Hg species into hexane, flash isothermal separation using minicolumns, and final detection by MIP-AES. The method was validated for speciation analysis for HG using the BCR 463 and 464 Tuna Fish CRMs.

The detection of methyl-Hg species in fish tissue was investigated [44]. Samples were digested with KOH–methanol and acidified prior to extraction with methylene chloride. After the extraction, methyl-Hg was back-extracted into water, phenylated, then the methylphenylated Hg species was extracted with toluene and injected into the GC-MIP-AES system.

A simple and rapid MW-assisted alkaline digestion procedure was developed in combination with high-performance liquid chromatography (HPLC)–UV post-column oxidation–CV–AFS detection for methyl-Hg determination in biological tissues [45]. The accuracy of the method was evaluated by the analysis of the BCR 463 Tuna Fish CRM.

A method was described for the accurate and precise determination of monomethyl-Hg by species specific isotope dilution calibration using SPME in combination with GC separation and ICP-MS detection [46]. Samples were digested with methanolic KOH, derivatized in aqueous solution with sodium tetrapropylborate and headspace sampled with a polydimethylsiloxane SPME fused silica fiber. The analyte was then directly transferred from the fiber to the head of the GC column for desorption.

2.5 ANALYTICAL FIGURES OF MERIT

The performance of analytical methods based on atomic spectrometry is characterized by figures of merit, such as LoDs and limits of quantification (LoQs), linear dynamic range, and precision and accuracy of measurements.

It is a common practice to give the LoD pertaining to a particular technique or method, and to draw comparisons between the LoDs obtained using similar techniques. The expected LoDs for atomic spectrometric techniques are summarized in Table 2.1. The range of reported values ensues from differences in instrumental

TABLE 2.1. Analytical Characteristics of Solubilized Forms in Element Determination by Analytical Atomic Spectrometric Techniques

Element	Detection Method	$LOD^a_{3\sigma}$			LOQ^a (ng g^{-1})	Characteristic Mass (pg)	Precision (RSD)[b]	References
		ng mL^{-1}	μg g^{-1}	ng				
Cd, Cu, Zn	F-AAS		0.5(2σ)					[1]
Pb	F-AAS		1.0(2σ)					[1]
Se	ET-AAS	2.42			8.08 ng mL^{-1}		1.5	[5]
Cd, Cr, Cu, Mn, Ni, Pb	ET-AAS		0.02, 0.06, 0.14, 0.02, 0.5, 0.1			0.5, 3.1, 7.3, 0.8, 22, 11	0.3–11	[6]
Cu, Fe	ET-AAS		21.4, 3.3			6.5, 5.5	5.5, 4	[7]
As, Cd, Pb, Se	ET-AAS		0.4, 0.005, 0.04, 0.6			13, 0.9, 7.6, 20	0.1–2.5	[8]
Ag, Cd, Cr, Cu, Ni, Pb	ET-AAS		0.053, 0.008, 0.015, 0.029, 0.051, 0.069				<10	[9]
Cd	ET-AAS	0.05				1.3	<3	[10]
Si	ICP-AES	75					6	[12]
Si	ICP-AES		2				2–24	[16]
Ca, Cu, Fe, K, Mg, Mn, Na, P, Se, Sn, Zn	ICP-AES		2.1, 0.065, 0.11, 103, 0.088, 0.011, 1.5, 2.1, 1.1, 0.66, 0.11				<10	[18]
Ca, Cu, Fe, K, Mg, Na, Sr, Zn	MIP-AES		1.1, 1.5, 3.0, 560, 2.5, 12, 0.8, 1.4				6–12	[20]
As, Se			0.008, 0.12					[22]
15 trace elements	ETV-ICP-MS	0.004–0.99						[23]
Ag, Cd, Hg, Pb, Tl	ETV-ICP-MS		0.009, 0.03, 0.041, 0.009, 0.0003					[24]
Pb	ETV-ICP-MS		0.010				6.5	[25]
I	FI-ICP-MS	0.45						[29]
I	ICP-IDMS		0.008				0.6–2.8	[30]
I	ICP-MS				30(9σ)			[31]

(*Continued*)

TABLE 2.1. *Continued*

Element	Detection Method	$LOD^{a}_{3\sigma}$			LOQ^{a} (ng g^{-1})	Characteristic Mass (pg)	Precision $(RSD)^{b}$	References
		ng mL^{-1}	μg g^{-1}	ng				
I	ICP-MS				15–25(10σ)		5	[34]
Pb	GC-AAS		0.075					[26]
Selenomethionine	GC-MS			0.015	0.05		9	[27]
CH_3Hg Hg(II)	GC-AAS			4 75				[35]
CH_3Hg	GC-AES		3.0 pg g^{-1}				6–13	[36]
Hg(II)			12.5 pg g^{-1}					
Total Hg	ETV-ICP-MS		0.05				7.8	[37]
CH_3Hg	GC-FAPES		0.0014	0.007			<10	[38]
Hg(II)			0.0002	0.001				
Total Hg	FI-CV-AAS	0.1					1.3	[39]
Inorganic Hg	FI-CV-AAS	0.2					<2	[40]
Total Hg		0.1						
CH_3Hg	HG-GC-ETAAS		0.05				7	[41]
Hg(II)			0.05					
Hg	CV-AAS	0.11						[42]
CH_3Hg	GC-MIP-AES		0.08				1–2	[43]
Hg(II)			0.02					
CH_3Hg	GC-MIP-AES		0.1	0.0004	2.5 pg			[44]
CH_3Hg	HPLC-CV-AFS		0.01				<8	[45]
CH_3Hg	GC-ICP-MS		0.002				2.5–7.7	[46]

[a] LoDs and LoQs were calculated according to IUPAC rules, based on a σ_{blank} criterion.
[b] Relative standard deviation.

applications and variability of the atomic spectrometric technique capabilities. LoDs are presented in terms of both mass and concentration or characteristic mass to simplify comparison. The data of Table 2.1 refer mainly to the determination of elements in solutions containing the analyte of interest. Because of this simplification, any application of the data to practical trace analysis must be subjected to some restrictions. The LoD is only one of the several figures of merit characterizing a technique and should not be used alone as a criterion of choice. Nevertheless, this compilation of data can be useful as an initial survey of the effectiveness of solubilized sampling analytical atomic spectrometric techniques with respect to the determination of major and trace levels of the analytes of interest. Moreover, there is much confusion over the definition of the term LoD. Users of such data as shown in Table 2.1 and in the literature should always check the definition applied in the original papers. Some relative standard deviation (RSD) values reported for the analysis of samples are also summarized in Table 2.1.

Acceptable precision in most instances, reported as RSDs, usually range from 0.1 percent to slightly higher than 20 percent, with the most frequent value centered at 5–10 percent. RSD is only an informative value and does not allow the solubilization techniques to be distinguished.

The accuracy of the methods described above for the analysis of biological and food samples has been checked by different approaches. These include: recovery test and standard addition, use of independent analytical methods of proven validity, and verification of the method by means of CRMs, the latter two approaches being the most applied. In the specific case of biological and food samples, a great variety of CRMs, such as those produced by NIST, BCR, NRC, and IAEA, are available. The reliability of the techniques mentioned in this survey has been mainly checked against these CRMs. There is much evidence in the literature that the capabilities of the solubilization sampling technique compare favorably with those of other techniques.

2.6 PRACTICAL APPLICATIONS

Illustrative applications of solubilization sampling in atomic spectrometric techniques have been summarized in Table 2.2. These applications are listed for sample type (matrix) and elements and species determined, the analytical atomic spectrometry instrumental mode, and the methodological approach. The aim of this section is to examine published material not merely to present potential users with established methods, but rather to point to the reasons why solubilization sampling has been used to solve specific problems and to stimulate further interest in its application. The references cited may contain additional determinations or trials for a particular food sample type. A wide range of applications is clearly evident, showing that the solubilization approach is widely applicable throughout biological and foodstuff analysis. The variety and number of samples indicate that future studies involving sample solubilization combined with analytical atomic spectrometric techniques would be readily applicable to the analysis of more complex samples.

TABLE 2.2. Analysis of Foodstuff Materials

Elements Determined/Species	Sample/Matrix	Analytical Atomic Spectrometric Technique	Sample Mass[a] (g)	Alkaline Solubilization Procedure	References
Cd, Cu, Pb, Zn	Liver, kidney	F-AAS	1	Samples were subjected to sonication procedure in water, 10% TMAH was added	[1]
Al, Ca, Cu, Fe, K, Mn, Na, Ni, Pb, Zn	NIES CRM-1,6,7, NIST SRM 1577b	F-AAS, ET-AAS	0.1	Reference materials were diluted with 25% TMAH and 200 mM EDTA ammoniacal solution was added. An MW digestion system with a closed PTFE vessel was used for sample decomposition	[2]
As	NIES CRM-6, IAEA MA-A-2	ET-AAS	0.1	Samples were heated with 5 mL of 10% TMAH in a quartz tube at 70°C for 30 min with occasional vortex mixing	[3]
Cr, Cu, Fe, Mn, Ni, Se	Liver, kidney, dogfish muscle, DORM-1, DORM-2	ET-AAS		0.25% TMAH in 10% C_2H_5OH. Slurries were stable for at least 6 days. $NH_4H_2PO_4$ and NH_4NO_3 were used as modifiers. Contamination from the homogenizer (Cu, Pb) caused high blanks, but accurate results were obtained with CRMs	[4]
Se	Mussels, wheat flour SRM	ET-AAS HG-AAS	0.1	TMAH digestion was compared with acid mineralization. The use of TMAH in speciation studies was also evaluated	[5]

Cd, Cr, Cu, Mn, Ni, Pb	TORT-2, DORM-2, DOLT-2	ET-AAS	0.25	Samples were weighed directly into bottles and 4 mL of 25% (m/v) TMAH in methanol were added. The reaction between the TMAH solution and the biological tissue samples occurred at room temperature (18 h), or alternatively by heating in an ultrasonic bath for 45 min. Following the reaction, the samples were diluted to 25 mL with water	[6]
Cu, Fe	Butter	ET-AAS	1.5	Samples were weighed into a glass tube and 6 mL of BTA–water (80 : 20) were added and vortexed. The mixture was warmed for 30 min at 37°C, then 4 mL of THF was added. The solution contained 15% (w/w) butter	[7]
As, Cd, Pb, Se	DORM-1	ET-AAS	0.3	Certified reference material was weighed and 1.2 mL of TMAH was added. The screw-capped vessel was heated to 60°C for 10 min in a water bath. After cooling, the sample was made up to a volume of 25 mL with water	[8]
Ag, Cd, Cr, Cu, Ni, Pb	NBS CRM 1577a, NIST CRM 1549 and 8414	ET-AAS	0.125–1.0	For the sample aliquot placed in flask, a volume of TMAH was added. Subsequently, the mixture was heated in a water bath (60–70°C) for 60 min, being manually agitated from time to time. Then the volume was made up to 10 mL with water	[9]

(*Continued*)

TABLE 2.2. *Continued*

Elements Determined/Species	Sample/Matrix	Analytical Atomic Spectrometric Technique	Sample Mass[a] (g)	Alkaline Solubilization Procedure	References
Cd	DOLT-2, DORM-2, LUST-1, TORT-2	ET-AAS	0.1	Sample aliquots were weighed and mixed with 2 mL of a 25% (m/v) TMAH solution, except for the DORM-1, for which about 0.250 g were mixed with 4 mL of a 25% (m/v) TMAH solution. Then, the volume was made up to 50 mL with water and the mixture was kept in a water bath at 60°C for 1 h	[10]
Al, Ca, Cu, Fe, K, Mg, Mn, Na, Ni, P, Sr, Zn	NIES CRMs: No. 1, 6, 7, NIST SRM 1577	ICP-AES	0.05	Powdered samples were weighed into a vial and mixed (magnetic stirrer) with 1 mL of 25% TMAH and 200 mM EDTA. The vial with 2 mL water was placed in a PTFE vessel inserted in a stainless-steel jacket. The whole vessel was heated for 4 h in a thermostated oven at 120°C. After cooling the vessel to room temperature, 2 mL water was added to the inner vial and the pH of the sample suspension was adjusted to 8–9. Prior to ICP-AES measurements, the suspension was transferred to a glass tube and centrifuged for 3 min at 3000 rpm (1400*g*). The supernatant thus obtained served as the sample solution	[11]

Si	Canned meat, fruit, milk, NIST SRM 1566	ICP-AES	1	Food samples were freeze-dried, homogenized and weighed into a Teflon PFA vessel followed by the addition of HNO_3, H_2O_2, and HF. The samples were MW heated. After digestion, 2 mL of solution of CFA-C reagent was added (pH 7.5–8.0). Solutions were diluted to 10 mL with water	[12]
Al, Ca, Cu, Fe, K, Mg, Mn, Na, Zn	CRMs: bovine liver, oyster tissue, total diet, whole egg powder	ICP-AES	0.2	Samples were weighed and 9.5 mL of 10% mixed amines were added. The dispersion was stirred for 30 min at room temperature with a magnetic stirrer and the resulting extract was centrifuged at 5000 rpm for 5 min. Elements were determined in the supernatant	[13]
Ca, K, Mg, Na, P, Zn	Milk	ICP-AES	0.1	Milk samples were dissolved in 10 mL of 10% (v/v) mixed amine reagent at pH 8. Elements were determined in the resulting dispersions	[14]
Cd	NIES CRM No. 1,10b,10c, NIST SRM 1566a	ETV-ICP-AES	0.01	A mixture of sample powder and diammonium hydrogenphosphate (2 + 1) was weighed in the sample cuvette. One of the cuvettes was then superimposed on the TBF and the TMAH solution was added into it. Then, the cuvette temperature was kept at a wet-digestion temperature to decompose the solid sample. After the digestion was completed, the temperature was progressively increased to generate the analyte vapor, which was then presented to the plasma	[15]

(*Continued*)

TABLE 2.2. *Continued*

Elements Determined/Species	Sample/Matrix	Analytical Atomic Spectrometric Technique	Sample Mass[a] (g)	Alkaline Solubilization Procedure	References
Si	Pork liver, NIST SRM 1577b	ICP-AES	0.05–0.1	Samples were weighed into PTFE tubes and 2 mL of 25% TMAH and 2.5 mL of water were added. The digestion mixtures were pretreated for 30 min in an ultrasonic bath. Finally, the MW-assisted autoclave digestion system was used	[16]
Se	Pork liver, NIST SRM 1577b, IRMM BCR 184	ICP-AES	0.05–0.1	A 2 mL aliquot of 25% TMAH was added to samples with MW heating to achieve dissolution. The digestion mixtures were pretreated for 30 min in an ultrasonic bath. The digestion temperature was varied between 60 and 200°C	[17]
Ca, Cu, Fe, K, Mg, Mn, Na, P, Se, Sn, Zn	Coffee, milk	ICP-AES	0.35	Samples were weighed into glass digester flasks and 1 mL of 25% TMAH was added. The flasks were heated to 80°C in a heated block digester. Solutions were transferred to a 25 mL volumetric flask and topped up with water	[18]
Ca, Cu, Fe, K, Mg, Na, Sr, Zn	TORT-1, NIST 1577, IAEA-153	MIP-AES	0.25	Samples of CRMs were weighed into 15 mL cups and 2 or 4 mL of 25% (m/v) TMAH were added. Following the reaction of the tissue with the TMAH for	[20]

				ca 2 min, 2 mL of water were added. Suspensions were pretreated by sonication for 2–4 min. After sonication, the samples were diluted to 10 mL with water. Aerosols, produced from suspensions by the V-groove nebulizer, were immediately carried on by the Ar plasma support gas through the spray chamber into the plasma	
31 elements	SRM NIST 1548, 1549, 1577b, 1566, TORT-1	ICP-MS	1	Samples were weighed into a PFA vessel and 5 mL of HNO_3, 0.5 mL of HF and 2 mL of H_2O_2 were added. Vessels were heated in an MW system. After cooling, 5 mL of 4% H_3BO_3 were added and the solution was neutralized to pH 7–8 with *ca* 5 mL of the tertiary amine reagent and diluted to 50 mL with water	[21]
As, Se	NIST SRM 1548, 1549, 1577b, 1566, TORT-1	ICP-MS	1 (0.5)	Samples were weighed into a Teflon-PFA vessel and 5 mL of HNO_3 and 2 mL of H_2O_2 were added. Vessels were heated in an MW system. After cooling, 2 mL of CFA-C solution were added and the mixture was diluted to 10 mL with water	[22]
15 elements	NIST SRM 1566a, NIST RM 8414, 8415	ETV-ICP-MS	0.02–0.1	The samples were prepared in 25 flasks. To the sample aliquot, 10–200 μL of a 25% (w/w) TMAH solution were added. After a few minutes, the volume was made up to 10 mL	[23]

(*Continued*)

TABLE 2.2. *Continued*

Elements Determined/Species	Sample/Matrix	Analytical Atomic Spectrometric Technique	Sample Mass[a] (g)	Alkaline Solubilization Procedure	References
Ag, Cd, Hg, Pb, Tl	NBS SRM 1577a, DORM-1	ETV-ICP-MS	0.1–1	The bovine liver and the dogfish muscle were dissolved with TMAH. To the sample aliquot placed in a flask, a volume of a 25% (m/v) TMAH solution was added. The mixture was heated in a water bath (100°C) for 30 min and let to stand for 24 h, being manually stirred from time to time. The volume was made up to 15 mL with water for the dogfish samples and to 5 mL for the bovine liver samples	[24]
Pb	NIST SRM 1568, 1577, 1566a	ETV-ICP-MS	*ca* 0.001	To the mixture of sample powder and diammonium hydrogenphosphate the TMAH solution was added	[25]
As, Ba, Ca, Cr, Cu, Pb, Mn, Mo, Rb, Se, Zn	NIST SRM 1577b, 1566a, 1548, 8415	ICP-MS	0.2	Samples were weighed and 9.5 mL of 10% (v/v) mixed amines were added. The dispersion was stirred for 30 min at room temperature with a magnetic stirrer and the resulting extract was centrifuged at 5000 rpm for 5 min. Extracts were diluted 10-fold before measurements	[13]
Al, Ba, Cu, I, Mn, Mo, Pb, Rb, Se, Sr, Zn	Milk	ICP-MS	0.2	Milk powder was dissolved in 49.5 mL of 5% (v/v) mixed amine reagent (pH 8). The elements were determined in the resulting dispersions	[14]

Pb(II), alkyllead species	Fish	GC-AAS	2	Fish samples were homogenized in a commercial meat grinder. The homogenized paste was digested in 5 mL of TMAH solution in a water bath at 60°C for 1–2 h. After cooling, the solution was neutralized with 50% HCl to pH 6–8. The mixture was extracted with 3 mL of benzene for 2 h in a mechanical shaker after addition of 2 g of NaCl and 3 mL of NaDDTC. After centrifugation of the mixture, 1 mL of benzene was transferred to a glass-stoppered vial and butylated with 0.2 mL of *n*-BuMgCl. The organic layer was injected to the GC-AAS system for the analysis	[26]
Selenomethionine	Wheat	GC-MS	0.2	Wheat samples were weighed and placed in a glass flask. A 5 mL aliquot of a 25% (m/v) TMAH solution was added and the flask was placed in a water-bath and shaken at 60°C for 4 h. After solubilization, water was added up to 50 mL and the pH of the solution was adjusted to 2.5 by adding concentrated HCl. The solution was centrifuged at 3000 rpm for 15 min to separate the carbohydrates that precipitate when the solution is acidified. The precipitate was discarded and the supernatant kept for the clean-up step, the derivatization, and the measurements	[27]

(*Continued*)

TABLE 2.2. *Continued*

Elements Determined/Species	Sample/Matrix	Analytical Atomic Spectrometric Technique	Sample Mass[a] (g)	Alkaline Solubilization Procedure	References
As, arsenobetaine	Fish, DORM-2	HPLC-ICP-MS	0.1	Extraction of As species from samples were performed with a 5% TMAH solution	[28]
I	BCR CRM 63, 63R, 150, milk	FI-ICP-MS	0.2	In a volumetric flask reference milk powder was dissolved in 50 mL of the KOH/TMAH solution. For raw milk 0.5 mL milk was diluted with 9.5 mL TMAH/KOH solution directly in the tubes	[29]
I	BCR CRM 63, 150, 151, BCR CRM 185	ICP-ID-MS	0.1–0.8	Samples were weighed into a Teflon-PFA vessel and added with 4.5 mL of water and 1 mL of 25% TMAH. Extraction of I was carried out in closed vessels (3 h at 90°C) and dilution with 25 mL water followed	[30]
I	NIST SRM 8415, BCR CRM 63R, 150, 186	ICP-MS	0.2–0.5	Samples were weighed into a flask, 1 mL of 25% TMAH was added, the bottle was closed gas-tight and placed in a dry oven at 90°C. After 3 h the sample was allowed to cool before being diluted with water to a volume of 25 mL	[31]
I	BCR CRM 150, 186, NIST SRM 8415, 1566a, diet soya, milk pap	ICP-MS	0.2–0.5	Samples were weighed into a bottle, 1 mL of 25% TMAH was added, the bottle was sealed and placed in a drying oven at 90°C. After 3 h the sample was cooled and filled with water to a volume of 25 mL	[32]

I	BCR CRM 150, NIST SRM 1566a, dietary foods	ICP-MS	0.2–0.5	Samples were weighed in a bottle, 1 mL of 25% TMAH was added, the bottle was placed in a oven at 90°C. After 3 h the samples were cooled and filled with water to a volume of 25 mL	[33]
I	NIST SRM 8415, 8433, 8435, 1549, BCR CRM 150, 151, culinary products, pet foods	ICP-MS	0.4	Solid or liquid products were weighed into 50 mL flasks. Then, 10 mL of water and 2 mL of TMAH were added. The flasks were closed and heated in the MW oven digestion system. After 30 min in the MW system, the flasks were allowed to cool to room temperature and made up to volume 50 mL with water	[34]
Hg species	TORT-1, DOLT-1, DORM-1, fish	GC-AAS	0.03–0.08	Samples of pulverized freeze-dried fish tissue were weighed and transferred into a 30 mL Teflon bottle. Then 10 mL of 25% KOH–methanolic solution were added, and the samples was shaken in an ultrasonic bath for 2 h. The samples were then diluted with 10 mL of methanol and stored in the dark until analysis. For the Hg species determination, the speciation trap is precooled to −196°C with liquid nitrogen. Then 20 mL of water were placed in the reaction vessel and the pH was adjusted to 4.5 with 2 mL of acetate buffer	[35]

(Continued)

TABLE 2.2. *Continued*

Elements Determined/Species	Sample/Matrix	Analytical Atomic Spectrometric Technique	Sample Mass[a] (g)	Alkaline Solubilization Procedure	References
Hg species	BCR CRM 464	GC-MIP-AES	0.2	Fish was weighed into the extraction container, 5 mL of 25% TMAH were added, and the mixture was treated for 20 min at 20 W MW power. It was then neutralized with 1.3 mL of 99% acetic acid, buffered with 20 mL of a sodium acetate buffer at pH 5.0, and allowed to cool to room temperature in a water bath. Then, 1 mL of $NaBEt_4$ solution and 1 mL of hexane were added and the mixture was shaken for 7 min. To enable phase separation the mixture was centrifuged for 2 min, and the organic phase was transferred into a container and stored in a cool, dark place until analysis	[36]
Hg	DORM-2, DOLT-2, TORT-2	ETV-ICP-MS	0.25	Samples were weighed into a beaker and 4 mL of TMAH (30% in methanol) were added. Following the reaction of the tissue with the TMAH for *ca* 5 min, the solution was transferred to a 25 mL calibrated flask and diluted to volume with water	[37]

Hg species	TORT-2, DORM-2, DOLT-2	GC-FAPES	0.25	Samples were placed in screw-capped bottles and 4 mL of 25% TMAH were added. The slurry was allowed to stand for a few hours and sufficient water added to bring the volume to 25 mL. Ionic species were purged from aqueous solution after ethylation with $NaBEt_4$	[38]
Hg	DORM-2, DOLT-2, TORT-2	FI-CV-AAS	0.25	Samples were weighed into 50 mL screw-capped bottles and 4 mL of TMAH (30% in methanol) were added. Following the reaction of the tissue with the TMAH for *ca* 5 min, water was added to bring the volume to 25 mL. The resulting samples were ready to be analyzed in 30 min	[39]
Hg species	DORM-2, DOLT-2, TORT-2	FI-CV-AAS	0.25	Samples were weighed into 50 mL screw-capped bottles and 4 mL of TMAH (30% in methanol) added. Following the reaction of the tissue with the TMAH for *ca* 5 min, water was added to bring the volume to 25 mL	[40]
Hg species	DORM-1, TORT-1, BCR CRM 463	HG-GC-ETAAS	0.1–0.5	An MW-assisted digestion method with alkaline solution (TMAH or methanolic KOH solution) for speciation analysis was developed	[41]
Hg	Fish	CV-AAS		Alkaline digestion	[42]

(*Continued*)

TABLE 2.2. *Continued*

Elements Determined/Species	Sample/Matrix	Analytical Atomic Spectrometric Technique	Sample Mass[a] (g)	Alkaline Solubilization Procedure	References
Hg species	DORM-1, BCR CRM 463, 464	GC-MIP-AES	0.1–0.2	The samples and 5 mL of 25% TMAH solution were placed in a reaction vessel and exposed to the MW irradiation at 45 W for 2.5 min. After MW digestion, samples were diluted with 15 mL of water. The pH was adjusted to 4 using 1.5 mL of concentrated acetic acid and 5 mL of buffer. Then, 1 mL of the $NaBEt_4$ solution (1%) and 2 mL of hexane were added and the mixture was shaken for 5 min. The sample was centrifuged for 2 min at 2000 rpm. An aliquot of the supernatant was injected into the GC column	[43]
Hg species	DORM-2, BCR CRM 464, fish	GC-MIP-aES	0.05–0.5	After the alkaline digestion (sample, 2 mL KOH–methanol, 3 h, 75°C), extraction (of MeHg into CH_2Cl_2/H^+), separation	[44]

				(using separating funnel), back extraction (of MeHg into 80 mL H_2O by CH_2Cl_2 evaporation at 50°C), aqueous phase derivatization (using $NaBPh_4$, pH 5), extraction (of MePhHg into 2 mL toluene phase) followed	
Hg species	TORT-2, BCR CRM 463	HPLC-CV-AFS	0.15	Samples were weighed into glass tubes, 6 mL of 25% (m/v) methanolic KOH were added and the tubes were capped. The slurry was homogenized by magnetic agitation and submitted to MW irradiation for 30–180 s at 48–132 W. The alkaline extract was mechanically shaken in a glass separating funnel for 10 min with 6 mL of dichloromethane and 4.5 mL of concentrated HCl	[45]
Hg species	DORM-2, DOLT-2, DOLT-3	GC-ID-ICP-MS	0.25	Samples and 20 mL of 25% (m/v) methanolic KOH solution were shaken for 5 h and then stored at 4°C until analysis	[46]

[a]Amount of sample refers to original test portion used in the analytical procedure.

2.7 CONCLUSIONS AND FUTURE PROSPECTS

It is evident that there is considerable research being done into the alkaline solubilization of samples combined with analytical atomic spectrometric methods. The most advantageous designs are probably those in which the determination of elements from solubilized samples produces reliable analytical data and uses less time for analysis since the full sample digestion (decomposition/dissolution) step is skipped. The alkaline solubilization of the food sample is very simple and convenient, as only a small amount of this reagent is needed, special apparatuses such as MW or conventional ovens are not required, sample handling is minimized, and preparation time is reasonable. It is particularly useful for food sample preparation when an ultrasonic probe or bath is available for the stabilization of suspension. Consequently, sample pretreatment is reduced to the solubilization preparation procedure. The technique is inexpensive and holds considerable promise, although it still needs to be firmly established.

Very few observations have been reported concerning the speciation of solubilized samples, which makes it difficult to draw any conclusion on this matter, although encouraging preliminary results were obtained for the speciation of analytes in solubilized samples.

The technique may be subject to a number of positive and/or negative systematic errors, depending on the element to be determined, the instrumental technique used, the matrix composition, and still other factors. However, as shown in Table 2.2, there is a tendency towards the use of the standard additions method and CRMs to minimize some possible matrix effects and to ensure validity of results. Nevertheless, it appears from the survey of the literature that the solubilization sampling introduction technique compares favorably with other atomic spectrometric methods for the determination of trace elements in a variety of matrices.

The solubilization technique should be used more frequently in combination with atomic spectrometric methods. Further research to improve its analytical performance in terms of LoDs, precision, and accuracy is required.

Another area of growing interest is the chemical speciation of elements to provide more useful information than that afforded by the total concentrations. In this context, alkaline solubilization procedures are especially suitable for speciation work, because this approach allows samples to be directly analyzed without prior preparation (destruction of the matrix), thus preserving the chemical forms of the analytes of interest.

Accordingly, it would be desirable for the alkaline solubilization sampling atomic spectrometric procedures to be accepted both as a routine quality control technique and/or as a screening approach in different processes involving foodstuffs. In this respect, the reduced sample manipulation, the low turnaround time and the relative low cost are primary merits. While single-element determinations using methods such as AAS will continue to play an important role in this field, multielement and multiisotope measurements will play an ever more important role in food science in the years to come.

REFERENCES

1. L. Murthy, E. E. Menden, P. M. Eller, H. G. Petering, Atomic absorption determination of zinc, copper, cadmium, and lead in tissues solubilized by aqueous tetramethylammonium hydroxide, *Anal. Biochem.*, **53** (1973), 365–372.
2. C. Y. Zhou, M. K. Wong, L. L. Koh, Y. C. Wee, Microwave digestion of biological samples with tetramethylammonium hydroxide and ethylenediaminetetraacetic acid for element determination, *Talanta*, **43** (1996), 1061–1068.
3. D. L. Tsalev, T. A. Dimitrov, P. B. Mandjukov, Study of vanadium(V) as a chemical modifier in electrothermal atomisation atomic absorption spectrometry, *J. Anal. Atom. Spectrom.*, **5** (1990), 189–194.
4. Y. Tan, J.-S. Blais, W. D. Marshall, Slurry preparation by high-pressure homogenization for the determination of heavy metals in zoological and botanical certified reference materials and animal feeds by electrothermal atomic absorption spectrometry, *Analyst*, **121** (1996), 1419–1424.
5. C. Brunori, M. B. de la Calle, M. Angelone, R. Morabito, Determination of total selenium in mussel and wheat samples after tetramethylammonium hydroxide digestion, *Ann. Chim.*, **89** (1999), 873–880.
6. R. G. L. Silva, S. N. Willie, R. E. Sturgeon, R. E. Santelli, S. M. Sella, Alkaline solubilization of biological materials for trace element analysis by electrothermal atomic absorption spectrometry, *Analyst*, **124** (1999), 1843–1846.
7. C. Lelièvre, D. Hennequin, J. F. Lequerler, D. Barillier, A rapid method for the direct determination of copper and iron in butter by GFAAS, *Atom. Spectrosc.*, **21** (2000), 23–29.
8. M. B. O. Giacomelli, M. C. Lima, V. Stupp, R. M. de Carvalho, Júnior, J. B. B. da Silva, P. Bermejo Barrera, Determination of As, Cd, Pb and Se in DORM-1 dogfish muscle reference material using alkaline solubilization and electrothermal atomic absorption spectrometry with Ir + Rh as permanent modifiers or Pd + Mg in solution, *Spectrochim. Acta*, **57B** (2002), 2151–2157.
9. P. Martins, D. Pozebon, V. L. Dressler, G. A. Kemieciki, Determination of trace elements in biological materials using tetramethylammonium hydroxide for sample preparation, *Anal. Chim. Acta*, **470** (2002), 195–204.
10. J. B. B. da Silva, D. L. G. Borges, M. A. M. S. da Veiga, A. J. Curtius, B. Welz, Determination of cadmium in biological samples solubilized with tetramethylammonium hydroxide by electrothermal atomic absorption spectrometry, using ruthenium as permanent modifier, *Talanta*, **60** (2003), 977–982.
11. T. Uchida, H. Isoyama, K. Yamada, K. Oguchi, G. Nakagawa, H. Sugie, C. Iida, Determination of twelve elements in botanical samples with inductively coupled plasma atomic emission spectrometry after leaching with tetramethylammonium hydroxide and ethylenediaminetetraacetic acid, *Anal. Chim. Acta*, **256** (1992), 277–284.
12. A. P. Krushevska, R. M. Barnes, Determination of low silicon concentrations in food and coral soil by inductively coupled plasma atomic emission spectrometry, *J. Anal. Atom. Spectrom.*, **9** (1994), 981–984.
13. J. A. Nóbrega, Y. Gélinas, A. Krushevska, R. M. Barnes, Determination of elements in biological and botanical materials by inductively coupled plasma atomic emission and mass spectrometry after extraction with a tertiary amine reagent, *J. Anal. Atom. Spectrom.*, **12** (1997), 1239–1242.

14. J. A. Nóbrega, Y. Gélina, A. Krushevska, R. M. Barnes, Direct determination of major and trace elements in milk by inductively coupled plasma atomic emission and mass spectrometry, *J. Anal. Atom. Spectrom.*, **12** (1997), 1243–1246.
15. Y. Okamoto, T. Fujiwara, T. Kumamaru, Direct determination of trace elements in biological materials by inductively coupled plasma atomic emission spectrometry with a tungsten boat furnace, *Anal. Sci.*, **13** (1997), 25–26.
16. S. Hauptkorn, J. Pavel, H. Seltner, Determination of silicon in biological samples by ICP-OES after non-oxidative decomposition under alkaline conditions, *Fresenius J. Anal. Chem.*, **370** (2001), 246–250.
17. F. Xu, Z. Ni, D. Qiu, J. You, W. Hu, P. Yang, Alkaline fusion-alkaline mode HG-ICP spectrometry for selenium determination, *J. Anal. Atom. Spectrom.*, **17** (2002), 534–536.
18. A. S. Ribeiro, A. L. Moretto, M. A. Z. Arruda, S. Cadore, Analysis of powdered coffee and milk by ICP-OES after sample treatment with tetramethylammonium hydroxide, *Mikrochim. Acta*, **141** (2003), 149–155.
19. A. Moeller, R. F. Ambrose, S. S. Q. Hee, A comparison of techniques for preparing fish fillet for ICP-AES multielemental analysis and the microwave digestion of whole fish, *Food Addit. Contam.*, **18** (2001), 19–29.
20. H. Matusiewicz, B. Golik, Determination of major and trace elements in biological materials by microwave induced plasma optical emission spectrometry (MIP-OES) following tetramethylammonium hydroxide (TMAH) solubilization, *Microchem. J.*, **76** (2004), 23–29.
21. A. Krushevska, A. Lásztity, M. Kotrebai, R. M. Barnes, Addition of tertiary amines in the semiquantitative, multi-element inductively coupled plasma mass spectrometric analysis of biological materials, *J. Anal. Atom. Spectrom.*, **11** (1996), 343–352.
22. A. Krushevska, M. Kotrebai, A. Lásztity, R. M. Barnes, D. Amarasiriwardena, Application of tertiary amines for arsenic and selenium signal enhancement and polyatomic interference reduction in ICP-MS analysis of biological samples, *Fresenius J. Anal. Chem.*, **355** (1996), 793–800.
23. D. Pozebon, V. L. Dressler, A. J. Curtius, Determination of trace elements in biological materials by ETV-ICP-MS after dissolution or slurry formation with tetramethylammonium hydroxide, *J. Anal. Atom. Spectrom.*, **13** (1998), 1101–1105.
24. D. Pozebon, V. L. Dressler, A. J. Curtius, Determination of volatile elements in biological materials by isotopic dilution ETV-ICP-MS after dissolution with tetramethylammonium hydroxide or acid digestion, *Talanta*, **51** (2000), 903–911.
25. Y. Okamoto, Direct determination of lead in biological samples by electrothermal vaporization inductively coupled plasma mass spectrometry (ETV-ICP-MS) after furnace-fusion in the sample cuvette-tungsten boat furnace, *Fresenius J. Anal. Chem.*, **367** (2000), 300–305.
26. Y. K. Chau, P. T. S. Wong, G. A. Bengert, J. L. Dunn, Determination of dialkyllead, trialkyllead, tetraalkyllead, and lead(II) compounds in sediment and biological samples, *Anal. Chem.*, **56** (1984), 271–274.
27. M. Beatriz de la Calle-Guntiñas, C. Brunori, R. Scerbo, S. Chiavarini, P. Quevauviller, F. Adams, R. Morabito, Determination of selenomethionine in wheat samples: comparison of gas chromatography–microwave-induced plasma atomic emission spectrometry, gas chromatography–flame photometric detection and gas chromatography- mass spectrometry, *J. Anal. Atom Spectrom.*, **12** (1997), 1041–1046.

28. K. L. Ackley, C. B'Hymer, K. L. Sutton, J. A. Caruso, Speciation of arsenic in fish tissue using microwave-assisted extraction followed by HPLC-ICP-MS, *J. Anal. Atom. Spectrom.*, **14** (1999), 845–850.

29. S. Stürup, A. Büchert, Direct determination of copper and iodine in milk and milk powder in alkaline solution by flow injection inductively coupled plasma mass spectrometry, *Fresenius J. Anal. Chem.*, **354** (1996), 323–326.

30. G. Rådlinger, K. G. Heumann, Iodine determination in food samples using inductively coupled plasma isotope dilution mass spectrometry, *Anal. Chem.*, **70** (1998), 2221–2224.

31. P. A. Fecher, I. Goldmann, A. Nagengast, Determination of iodine in food samples by inductively coupled plasma mass spectrometry after alkaline extraction, *J. Anal. Atom. Spectrom.*, **13** (1998), 977–982.

32. G. Knapp, B. Maichin, P. Fecher, S. Hasse, P. Schramel, Iodine determination in biological materials: options for sample preparation and final determination, *Fresenius J. Anal. Chem.*, **362** (1998), 508–513.

33. P. Fecher, C. Walther, J. Sondermann, Determination of iodine in dietary foods by ICP-MS, *Dtsch. Lebensm.-Rundsch.*, **95** (1999), 133–142.

34. D. Andrey, A routine quality control method for the determination of iodine in human and pet food by ICP-MS, *Atom. Spectrosc.*, **22** (2001), 299–305.

35. R. Fisher, S. Rapsomanikis, M. O. Andreae, Determination of methylmercury in fish samples using GC/AA and sodium tetraethylborate derivatization, *Anal. Chem.*, **65** (1993), 763–766.

36. C. Gerbersmann, M. Heisterkamp, F. C. Adams, J. A. C. Broekaert, Two methods for the speciation analysis of mercury in fish involving microwave-assisted digestion and gas chromatography-atomic emission spectrometry, *Anal. Chim. Acta*, **350** (1997), 273–285.

37. S. N. Willie, D. C. Grégoire, R. E. Sturgeon, Determination of inorganic and total mercury in biological tissues by electrothermal vaporization inductively coupled plasma mass spectrometry, *Analyst*, **122** (1997), 751–754.

38. M. S. Jimenez, R. E. Sturgeon, Speciation of methyl- and inorganic mercury in biological tissues using ethylation and gas chromatography with furnace atomization plasma emission spectrometric detection, *J. Anal. Atom. Spectrom.*, **12** (1997), 597–601.

39. G. Tao, S. N. Willie, R. E. Sturgeon, Determination of total mercury in biological tissues by flow injection cold vapour generation atomic absorption spectrometry following tetramethylammonium hydroxide digestion, *Analyst*, **123** (1998), 1215–1218.

40. G. Tao, S. N. Willie, R. E. Sturgeon, Determination of inorganic mercury in biological tissues by cold vapor atomic absorption spectrometry following tetramethylammonium hydroxide solubilization, *J. Anal. Atom. Spectrom.*, **14** (1999), 1929–1931.

41. C. M. Tseng, A. de Diego, F. M. Martin, D. Amouroux, O. F. X. Donard, Rapid determination of inorganic mercury and methylmercury in biological reference materials by hydride generation, cryofocusing, atomic absorption spectrometry after open focused microwave-assisted alkaline digestion, *J. Anal. Atom. Spectrom.*, **12** (1997), 743–750.

42. G. R. Boaventura, A. C. Barbosa, G. A. East, Multivessel system for cold-vapor mercury generation: determination of mercury in hair and fish, *Biol. Trace Elem. Res.*, **60** (1997), 153–161.

43. I. R. Pereiro, A. Wasik, R. Łobiński, Determination of mercury species in fish reference materials by isothermal multicapillary gas chromatography with atomic emission detection after microwave-assisted solubilization and solvent extraction, *J. Anal. Atom. Spectrom.*, **13** (1998), 743–747.
44. H. E. L. Palmieri, L. V. Leonel, Determination of methylmercury in fish tissue by gas chromatography with microwave-induced plasma atomic emission spectrometry after derivatization with sodium tetraphenylborate, *Fresenius J. Anal. Chem.*, **366** (2000), 466–469.
45. E. Ramalhosa, S. Río Segade, E. Pereira, C. Vale, A. Duarte, Microwave treatment of biological samples for methylmercury determination by high performance liquid chromatography-cold vapour atomic fluorescence spectrometry, *Analyst*, **126** (2001), 1583–1587.
46. L. Yang, Z. Mester, R. E. Sturgeon, Determination of methylmercury in fish tissues by isotope dilution SPME-GC-ICP-MS, *J. Anal. Atom. Spectrom.*, **18** (2003), 431–436.

3

CHEMICAL ELEMENTS IN FOOD AND THE ROLE OF ATOMIC AND MASS SPECTROMETRY. ADVANTAGES AND DRAWBACKS OF THE DETERMINATION OF SELECTED TRACE ELEMENTS IN FOODSTUFFS BY ATOMIC ABSORPTION SPECTROMETRY

LARS JORHEM AND JOAKIM ENGMAN

Livsmedelsverket, PO Box 622, SE-751 26 Uppsala, Sweden,
E-mail: lajo@slv.se

The Determination of Chemical Elements in Food: Applications for Atomic and Mass Spectrometry.
Edited by Sergio Caroli

3.1 INTRODUCTION

The term *trace elements*, when foodstuffs or food supplements are concerned, is perceived as something positive and essential for the well-being regardless of the elements involved. Heavy metals, on the other hand, are considered toxic and dangerous, regardless of the nature of the metal. *Trace* has been defined in earlier analytical nomenclature as the concentration range 10^{-4}–10^{2} $\mu g\ g^{-1}$, followed by microtrace between 10^{-7} and 10^{-4} $\mu g\ g^{-1}$. In up-to-date nomenclature the term *trace* is not used, but popular expressions die hard. In turn, the term *heavy metals* is difficult to define, but it is just as popular. It is, however, much debated and an agreement on its meaning is not in sight [1].

Atomic Absorption Spectrometry (AAS) has been used in analytical laboratories since the early 1960s. The technique has developed considerably over the decades and today it can be considered mature. AAS is therefore quite likely to remain as a standard technique for the determination of elements in foodstuffs and other biological samples for many years to come.

Some analysts, as well as laboratory heads, see AAS as a rather simple analytical technique that does not require special competence. In many laboratories, the analyst shares time between different analytical techniques. This may perhaps be seen as cost-effective, but it usually means that the analyst does not gain a deeper knowledge of the science of AAS. This is unfortunate because analysis by AAS is not that simple, especially electrothermal atomization (ET) AAS, an analytical technique that makes use primarily of the graphite furnace (GF) to vaporize and atomize the sample. Not only the technique as such, but also the laboratory environment in which the analysis is carried out, should be mastered since contamination is a major problem, even if this is often overlooked. The fact that the determination of elements in food is no simple matter is clearly indicated by the results of proficiency tests (PTs) on elements in food matrices. The results frequently cover several orders of magnitude and even factors of 10–100 are not uncommon!

A number of elements will be used to exemplify different aspects of food analysis with AAS as the common denominator: Cd, Cr, and Pb as examples of metals that may be difficult to quantify and usually occurring at very low concentrations, this requiring the use of ET-AAS; As, Se, and Hg as examples of elements requiring specific techniques to determine low concentrations in foodstuffs; Sn and Zn as examples of elements usually occurring at fairly high concentrations in various foods, either naturally (Zn), or as a contaminant from the packaging material (Sn); Ca and Na as examples of metals usually occurring at high concentrations, but which may call for special requirements regarding the analytical technique.

Prior to the determination of elements by AAS, the samples have to be brought into solution. There are several ways in which this can be achieved, all having both positive and negative facets.

3.2 DIGESTION TECHNIQUES USED FOR FOOD SAMPLES

3.2.1 General Aspects

The composition of foodstuffs may vary enormously. It may be 100 percent fat, proteins, or carbohydrates, or any mixture of the major components. In addition, foodstuffs contain varying amounts of salts and other minerals. This means that sample digestion can be quite a challenge. Often a laboratory uses only one digestion method, usually because that method has *always* been used and because it may take a considerable effort to introduce a new technique. If the analyst can resort to several techniques for sample digestion, it may be worth giving it some extra thought, since the digestion mode can have an impact on the results of the analysis in more than one way. In this context, the two most commonly used digestion techniques for the analysis of foodstuffs are dry ashing and wet digestion.

3.2.2 Dry Ashing

This technique is losing popularity, partly because it is, rather unfairly, considered to result in loss of some analytes (Cd and Pb in particular) during the ashing procedure [2, 3]. In addition, it is assumed to be very sensitive to contamination since the ashing vessels are exposed to the laboratory environment for most of the ashing process. The latter aspect is true to a certain extent. Not all laboratories, in fact, are suitable for dry ashing. Regular analysis of sample blanks will demonstrate if the laboratory environment is suitable for trace element analysis. For elements to be determined by flame (F) AAS, for example, Ca, Na, Sn, and Zn, there are normally few, if any, contamination problems. At any rate, their occurrence is usually significant only at the lowest concentrations.

Elements will be lost if the applied temperature is too high, or if the rate of temperature increase is too rapid (this may create hot-spots or ignition of the sample, with no control on temperature as a result). On the other hand, if a collaboratively tested method is used and the temperature programme adhered to, the risk of loss of analyte is virtually none. In most cases when loss of element during ashing has been reported, it can be traced back to poor temperature control [4]. Another pitfall that can be ascribed to dry ashing is cross contamination during the ashing process [2]. This is undeniably possible (and actually happened in the authors' laboratory when samples with an extremely high Cd content were ashed), but it is not considered to be a frequent problem. Foods with very high contents of fat or sugar are unsuitable for dry ashing as they may self-ignite or hot spots may evolve, even if the temperature is increased at a slow rate. Under such conditions, the temperature in the sample can no longer be controlled.

Ashing in Pt crucibles may lead to some contamination of the samples by Pt. This is, however, not entirely negative, since it works as a matrix modifier, in a similar way to Pd. The Pt in crucibles is usually alloyed either with Au or Ir, which give the alloys somewhat different properties. An interference in the determination of Fe and Zn has been noted by the authors when phosphate-rich foodstuffs, such as milk, are ashed in Pt–Au crucibles, but not in Pt–Ir crucibles. This is probably due to the formation of a Pt phosphide.

Sometimes ashing aids, for example, $Mg(NO_3)_2$ or HNO_3, are used to speed up the ashing process, but they should be used with caution, as they may contaminate the samples, as well as causing additional background interferences; the consequences will be particularly noticeable in ET-AAS.

When samples are dry ashed for Sn determination, it is theoretically possible to volatilize the metal during the process, just as for Cd and Pb. This does not seem to be the case in practical work, though. Results from dry ashing have compared very well with results obtained with other techniques, such as HCl extraction or the dithizone method. It is crucial, however, that the ashes are dissolved in HCl and not HNO_3.

Dry ashing may also be applied to samples to be analyzed for the rather volatile elements As and Se. If suitable ashing aids, for example, MgO and/or $Mg(NO_3)_2$, are used, no volatilization of these elements will occur. The increase in background caused by ashing aids is no problem when the detection is done by the hydride generation (HG) method.

Dry ashing thus brings about a number of advantages, such as:

- Better limits of detection (LoDs) as a consequence of the higher concentration of analytes in the samples.
- High sample throughput with little sample manipulation and possibility of treating large sample masses (hence, lesser problems with inhomogeneity).

On the other hand, disadvantages of dry ashing are:

- Occurrence of contamination problems.
- Volatilization of analytes if precautions are not taken.
- Long duration of the ashing process (even up to several days).

3.2.3 Wet Digestion

Acid digestion is the other main technique for getting food samples into solution. To this end, H_2SO_4 and $HClO_4$ have been much used, in combination with HNO_3. $HClO_4$ easily forms explosive perchlorates with organic materials. Thus, it has largely disappeared from use for safety reasons. Since quite large acid volumes may be necessary to complete the digestion, contamination from the acids is sometimes a problem. In the past, the digestion was usually carried out in open vessels, which made the solutions prone to atmospheric contamination as well. The digestion process could also be rather time-consuming.

Ease of use and analytical speed have been the driving force behind development in this field. The most commonly used technique today is closed digestion under pressure using HNO_3, alone or in combination with H_2O_2. Vessels of plastic or glass/quarts contained in stainless steel cylinders (the so-called bombs) are often used and heated in an autoclave or some other heating device. A drawback with steel bombs is that they cool down slowly.

The dominant technique today for heating is undoubtedly sample digestion in closed plastic or glass/quartz vessels by means of microwave (MW) irradiation. There are several producers of MW ovens for laboratory use on the market. It is worth noting in this context that, for safety reasons, household MW ovens are not recommended. A nice feature of digestion in closed vessels is that the same digest can be used for the determination of both volatile and more refractory elements. A general drawback is the small sample masses afforded, which may result in homogeneity problems for certain food products.

The development of the MW technique has been towards higher pressures and temperatures, which is important in applications where the complete destruction of all organic material is needed. Residues of organic material can interfere with the determination of the analytes of interest through altered viscosity of the sample solution or chemical interferences during the determination. Exhaustive information on MW digestion can be found in, for example, the book by Kingston and Jassie [5].

The MW digestion has several advantages:

- Speed of operation (samples can be received and analyzed in one day).
- Low risk of contamination.
- Good reproducibility of the analytical results.

On the other hand, the disadvantages of the MW digestion are:

- Large dilution factor (which implies poor LoDs).

- Small sample masses, typically less than 0.5 g dry matter (with the ensuing problems of inhomogeneity).
- Labor-intensive procedure.

3.2.4 Other Approaches

Sometimes the need for sample digestion can be circumvented. One example is the extraction with HCl. The method is based on hot HCl and is specifically developed for extraction of Sn from canned fruit and vegetables. It is quite rapid and particularly useful for screening purposes. Another example is solid sampling. This is a technique specifically suited for ET-AAS and therefore not generally applicable. It is useful when homogeneity at the milligram level is of interest, for example, in the production of Certified Reference Materials (CRMs).

In this framework, it should be carefully taken into account that there are many causes of potential chemical contamination in the laboratory. Among these, of primary importance are dust (due to unfiltered air and/or lower air pressure in the laboratory), smokers' hands, abrasion of metallic surfaces through glassware and porcelain, flakes of paint, corrosion on metallic surfaces due to the use of strong acids, carry-over effects from uncontrolled dishwashing, and contaminated chemicals.

A wealth of information on the subject of sample destruction can be found in a comprehensive book on sample preparation for trace element analysis [3].

3.3 FEATURES OF ATOMIC ABSORPTION SPECTROMETRY

3.3.1 General Aspects

After the early development of AAS as a possible analytical technique [6], it was only at the beginning of the1960s that the first commercial FAAS instrument was placed on the market. In the 1970s the technique was well established and probably it is nowadays the most widely used technique for the determination of trace elements in foodstuffs. The later introduction of the GF for ET-AAS greatly improved the LoDs. This technique is, however, more affected by analytical interferences, making background correction (BC) extremely important. Deuterium (D_2) continuum lamps have been widely used for BC both in FAAS and ET-AAS. It is crucial that the D_2 lamp is aligned with the source lamp and that their energies match. The Zeeman-effect BC, which is easier to use, has taken over much of the BC in ET-AAS applications and is probably the dominant technique today. The Smith–Hieftje BC, which uses pulsing of the lamp energy to produce self-reversal, has not found widespread acceptance.

A problem with AAS is that the detection power differs greatly among elements. The LoD for Cd, for example, is approximately 20 times better than for

Pb (see Table 3.1), thus making the former easier to determine at very low levels. Some other elements, such as As and Hg, have very poor LoDs by FAAS. Other AAS techniques thus have to be applied.

FAAS is widely used for food analysis and is the standard technique for trace element determination in most laboratories. Since the sensitivity varies among elements, there is no firm threshold for the concentrations above which FAAS is fit for purpose. In general, concentrations above 0.5–1 $\mu g\ g^{-1}$ can be determined by FAAS with reasonable precision, which means that Cd and Pb in food can rarely be quantified by this technique. It should always be taken for granted that BC is necessary at lower concentrations, unless there is evidence for the opposite. As the concentration of the analyte increases, the relative effect of the background interference decreases. Elements like Zn usually give fewer problems in the determination and are quantified using an oxidizing (fuel lean) air–acetylene flame. Zinc is usually present at microgram per gram quantities in most foodstuffs, which means that ET-AAS is seldom needed.

Tin is of interest as it is released from cans into food. The maximum EU limit for Sn in canned food is currently 200 $\mu g\ g^{-1}$, these levels being well suited for FAAS. A reducing (fuel-rich) nitrous oxide–acetylene flame is required to properly atomize Sn. As this type of flame is more unstable than air–acetylene, it may result in somewhat poorer repeatability. The quantification of naturally occurring levels of Sn in foodstuffs is quite different and would require ET-AAS.

Sodium and other alkali metals are easily ionized and need special caution. Early instruments were usually fitted with an air–propane burner that yielded a cooler flame, that is, less energy rich, hence giving rise to lesser ionization. Modern instruments do not usually have this facility; rather, they use an air–acetylene flame that results in a standard upward curve as the ionization decreases with increasing concentration of the analyte. In such cases, ionization has to be counteracted by modifying the sample solution. Another metal with a high ionization potential is added in large quantities, for example, 1000 $\mu g\ g^{-1}$, to the

TABLE 3.1. Instrumental LoDs in $\mu g\ l^{-1}$. From *Guide to Atomic Spectroscopy Techniques and Applications*, Perkin–Elmer, 2000

Element	FAAS	Cold Vapor/Hydride Generation	ET-AAS	ICP-AES	ICP-MS
As	150	0.03	0.05	2	0.0006
Ca	1.5		0.01	0.05	0.0002
Cd	0.8		0.002	0.1	0.00009
Hg	300	0.2	0.6	1	0.016
Na	0.3		0.005	0.5	0.0003
Pb	15		0.05	1	0.00004
Se	100	0.03	0.05	4	0.0007
Sn	150		0.1	2	0.0005
Zn	1.5		0.02	0.2	0.0003

sample and standard solutions (it is important to reach the same concentration in all solutions that will be analyzed by AAS). This results in a large number of free electrons that suppress the ionization of the analyte. As an example, an excess of K can be added to the sample solution when Na is determined, and vice versa. If one uses Cs for this purpose, it is possible to quantify both K and Na in the same solution.

In the determination of Ca in, for example, milk, there is often a phosphate interference that results in too low readings, owing to the formation of Ca phosphate. This interference may be overcome by the addition of La, which reacts with the phosphate and eliminates the interference. An alternative approach could be to use the hotter nitrous oxide-acetylene flame.

The importance of BC cannot be overemphasized. To attempt trace element analysis by FAAS at low concentrations without verifying that BC is not required probably results in a waste of both time and money. In older flame instruments with no BC installed, or where the BC does not work any longer, it is still possible to apply an approximate correction. To this end, samples are first run at the specified wavelength, thus obtaining the total absorbance. Then a nearby nonabsorbing line (emitted by the filling gas, an alloy metal, or an impurity in the cathode) is selected, either from the same lamp or from another lamp and samples are run again at this wavelength. This gives the background absorbance. Then the background absorbance is subtracted from the total absorbance and a corrected absorbance reading is achieved [7]. This is obviously not a technique for high-precision analysis, but may be helpful in avoiding some very gross errors.

In turn, ET-AAS is mostly used for concentrations below those suitable for FAAS, that is, $<0.5-1\ \mu g\ g^{-1}$, which is usually the case when elements such as Cd, Cr, and Pb are quantified in foodstuffs. The ET-AAS technique is considerably more complicated than FAAS, a fact which the analyst is not always aware of. One reason is that the analytical signal is not continuous as in the flame-based mode, this implying that the transient signal yields both a peak height and a peak area, of which the area is the preferred mode of evaluation, although requiring some experience from the analyst. Since the analyte is present at very low concentrations and the background effect is more or less the same as when FAAS is used, the ratio between the analyte and the background becomes very unfavorable and the importance of BC is even greater. Figure 3.1 shows an example of the relationship between the Pb signal and the background signal in fish muscle after dry ashing. The need for BC is obvious!

3.3.2 Background Correction

It is often taken for granted that BC systems always correct the analytical signal in a proper fashion, but this is not necessarily true, particularly when D_2 is used. The function of a BC system is, however, reasonably easy to check with a fine wire-mesh (mosquito net or similar), giving a total absorbance of 0.4–0.5 at the Pb wavelength of 283.3 nm. If the BC is properly set, it should give a signal for the wire-mesh of zero absorbance units. A sample solution (e.g.,

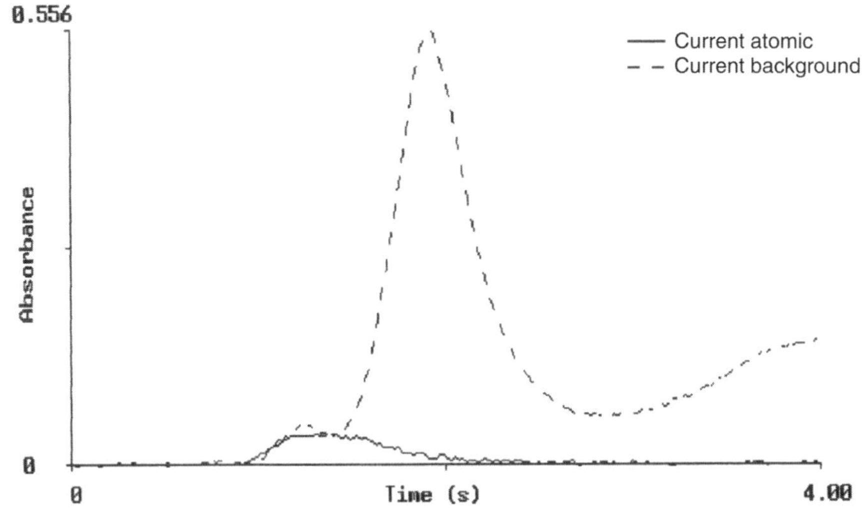

Figure 3.1. Platform atomization of Pb in the CRM DORM-2 *Fish muscle*.

NaCl) giving a total absorbance of approximately 0.5 absorbance units for 20 μL can also be used for BC checking. After the drying step, the salt (sample) is heated (ashed) to approximately 1000°C, this removing all traces of Pb. Finally, during the atomization stage it will be clear whether the background interference is properly corrected for. As can be seen in Figure 3.2*a*, the background, caused by NaCl, is fully compensated for. Figure 3.2*b* shows that only the electronic noise remains. Since the transient atomization signals can be very rapid, the background may, in some cases, be overcorrected, which may result in negative readings.

A refractory metal like Cr behaves quite differently from Cd and Pb. The high atomization temperature makes it possible to eliminate some, or most, of the matrix components during the ashing stage, prior to atomization, as shown in Figure 3.3*a*. To what degree the background can be reduced depends, of course, on the sample matrix. It should also be noted that, at the wavelength of Cr, that is, 357.9 nm, a D_2 background correction system is operating at, or sometimes beyond, its working limit and that background interferences may remain uncorrected.

3.3.3 Temperature Programming

Each metal has specific requirements of ashing and atomization temperatures. Temperature programs are normally provided by the instrument manufacturers and are often very general. As the mineral composition of foods varies widely, different sample matrices may require specifically designed temperature programs to yield optimum results. It is a simple process to optimize the ashing and atomization temperatures. In the first step, the ashing temperature is fixed (at a suitably low temperature) and the atomization temperature is increased stepwise until an

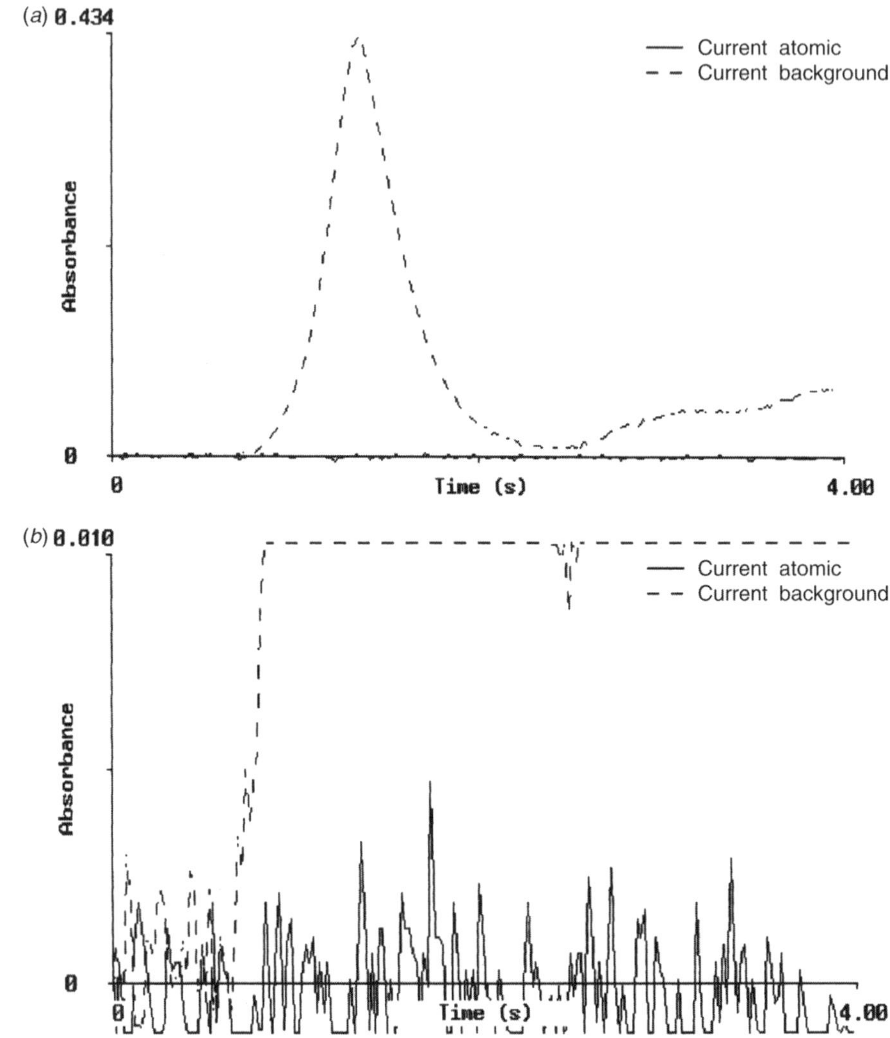

Figure 3.2. (*a*) Control of the background correction at the Pb wavelength using a solution of NaCl ashed at 1000°C and atomized at 1700°C. (*b*) Lead in NaCl. A blow-up of the baseline from (*a*) shows the complete elimination of the interfering background.

optimum signal is identified. Next, the atomization temperature is fixed at the optimum and the ashing temperature is increased stepwise until the analyte begins to vaporize. The optimum ashing temperature is just before the volatilization begins, although the behavior of the matrix should also be considered. Figure 3.4 shows an example of an ash-atomization curve established for Pb in spiked muscle tissue from salmon. The graphite tube will last considerably longer if one avoids setting the atomization temperature higher than necessary. The difference

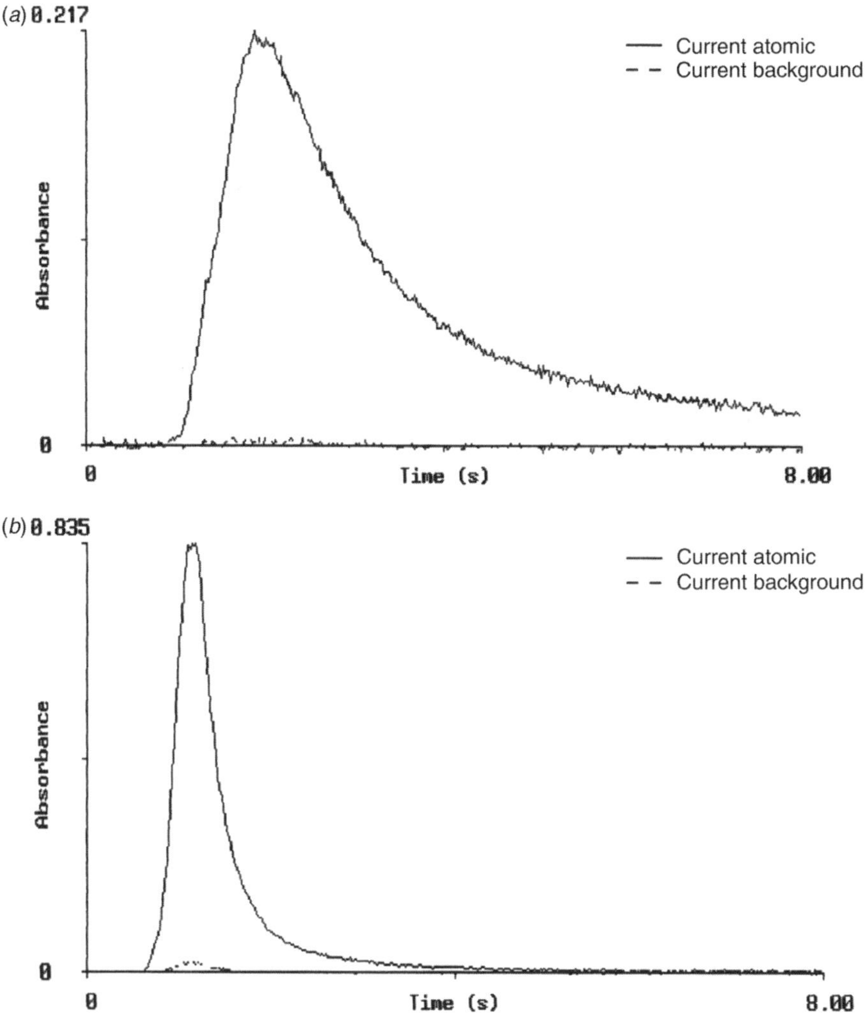

Figure 3.3. (*a*) Chromium in the CRM DORM-2 *Fish muscle*. Atomization from platform. (*b*) Chromium in the CRM DORM-2 *Fish muscle*. Atomization from the graphite tube wall.

in behavior between peak height and peak area in the atomization step is central. As the temperature is increased, peak height gives taller and thinner peaks, whereas the peak area remains virtually unchanged.

3.3.4 Matrix Modification

By using well-selected matrix modifiers, background signals may be reduced, or even eliminated, before the atomization step takes place. Modifiers can be used in two different ways: they either facilitates the vaporization of the interfering

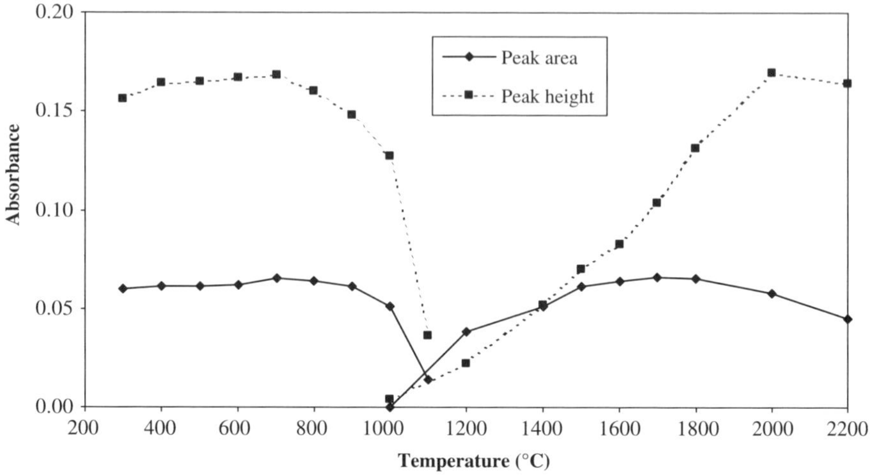

Figure 3.4. Ash and atomization curve for Pb in a sample solution of spiked salmon.

compound so that it disappears during the ashing stage, or they raise the boiling point of the interfering compound so that it does not vaporize during atomization.

Modifiers are, however, used in a rather indiscriminate way in many laboratories. If used carelessly they can contaminate the sample solution with the element that is being determined and they can actually add to the background interference which one intends to reduce. By carefully optimizing the ashing and atomization temperatures for specific food matrices, as described above, the use of matrix modifiers can be reduced to the cases when they are really necessary. An additional benefit of matrix modification is that the sample and standard matrix are made very similar, this often making the standard addition method unnecessary. How this is carried out is described in detail in most instrument manuals and in specific textbooks. Commonly used modifiers are ammonium nitrate, ammonium phosphate, Mg nitrate, Pd nitrate, and ascorbic acid.

3.3.5 Platform Atomization

The traditional atomization of an analyte from the graphite tube wall results in some unwanted effects. The peak evolves very rapidly from the hot tube wall into a cooler environment, which in turn expands rapidly. This results in skewed, tailing peaks. The introduction of the platform (basically, the type devised by L'vov), from which the analyte is atomized into an already hot environment, resulted in peaks that had a more gaussian profile. For metals like Cd and Pb this was a great improvement. Figure 3.5*a*, *b* shows examples of Cd atomized from the tube wall and platform. This curve also demonstrates that a well-timed cut-off can be used to eliminate some of the background. Today the platform is the standard approach. For refractory metals like Cr, requiring much higher atomization temperatures, the platform is not only an asset, since it may result in drawn-out,

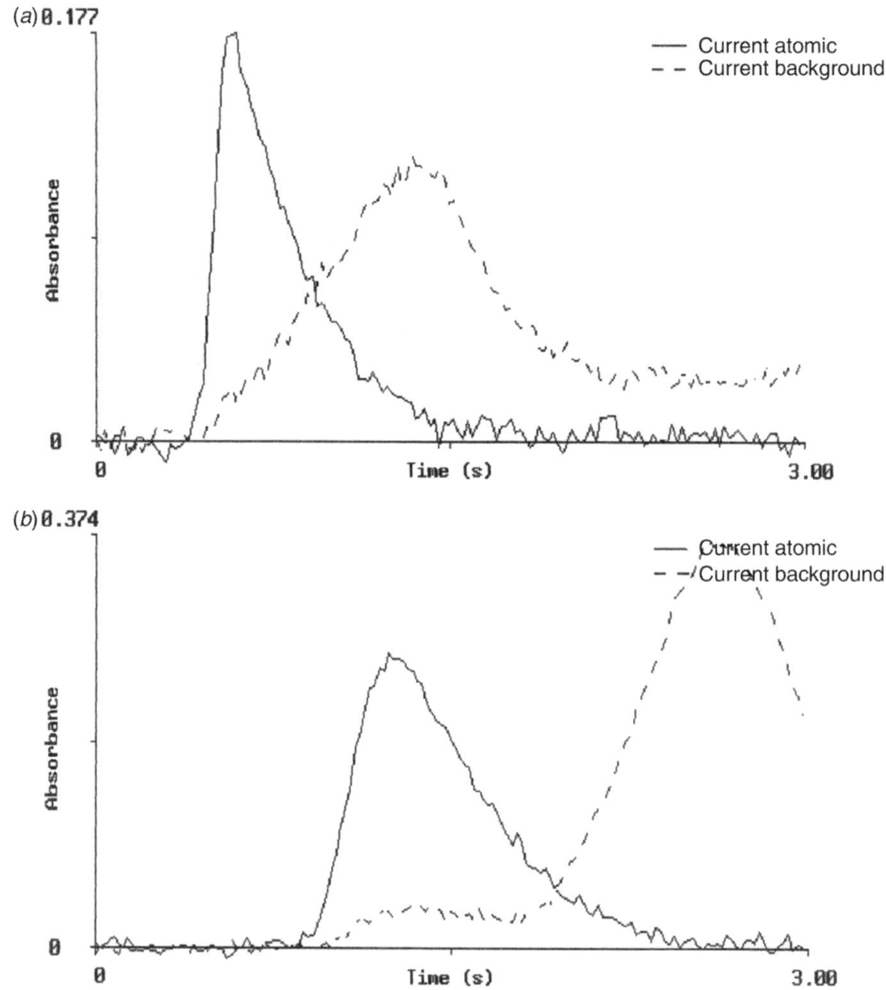

Figure 3.5. (*a*) Cadmium in the CRM DORM-2 *Fish muscle*. Atomization from the tube wall. (*b*) Cadmium in CRM DORM-2 *Fish muscle*. Atomization from platform.

tailing, peaks with considerably reduced peak height. This can be seen in Figure 3.3*a*, *b*. Here the analyst must make a decision on which technique to use, since the choice is not necessarily obvious.

3.3.6 Method of Standard Addition – Matrix-Matched Standards

The ET-AAS technique is especially prone to nonspectral interferences, that is, effects on the formation of atoms. Such interferences cannot be eliminated by BC. Some of these effects can, however, be corrected by the use of the standard addition procedure. To investigate whether standard addition is necessary, the slope of a standard calibration curve is compared to the slope of standards added to the sample. If the slopes are significantly different, the standard addition method has to be used.

A limitation of the technique is that all readings must fall within the linear part of the working range. This means that samples with high concentrations must be diluted. Moreover, the highest addition should be three to five times higher than the concentration in the sample. This is illustrated by Figure 3.6*a*, which also shows the difference in slope between a standard calibration curve and the standard addition curve. The method uses extrapolation, which increases the uncertainty. Figure 3.6*b* illustrates the effect of too small a difference in

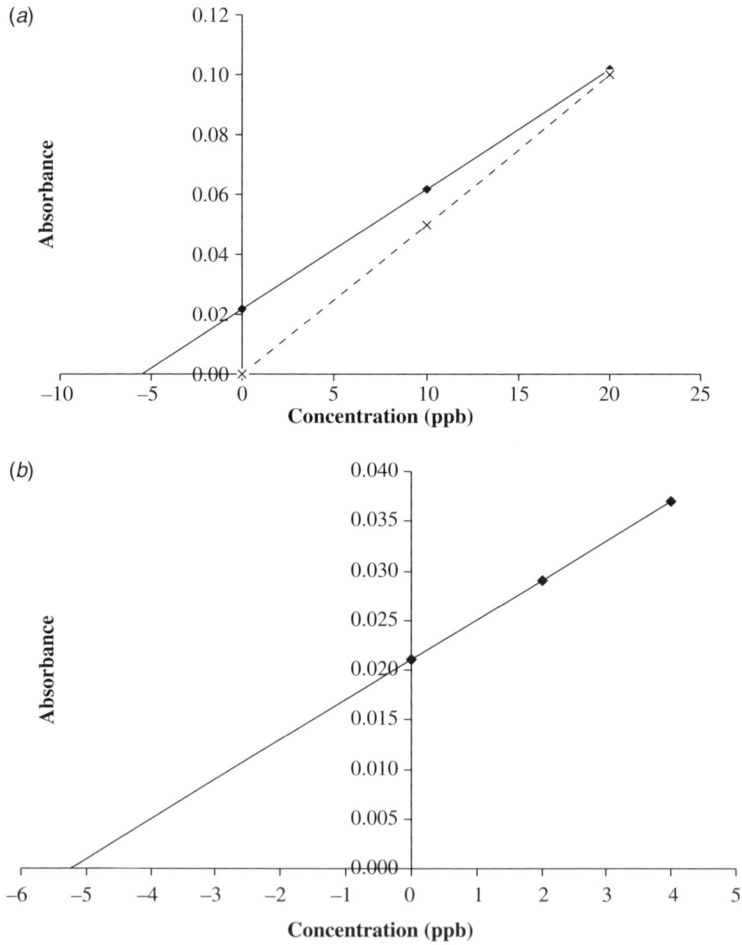

Figure 3.6. (*a*) Example of a well-proportioned standard addition curve, as well as of the possible difference between a traditional standard curve and a standard addition curve. (*b*) Example of a disproportional standard addition curve, which may result in a large uncertainty.

concentration between the sample solution and the standards. A slight shift in slope will cause a large difference in result. The method of standard addition thus leads to better trueness, but may well decrease the precision of the measurements.

A simplified version of the method of standard additions can be used when a longer series of samples with the same matrix, volume, and sample mass is to be analyzed. Then the standard addition curve is shifted to pass through zero and hence becomes a matrix-matched standard calibration curve.

3.3.7 Hydride Generation AAS for As and Se

After dry ashing or pressure digestion, the As and Se ions are converted into their hydride forms, usually by $NaBH_4$ in an acidic solution. The hydride is transferred to the measurement cell by a gas stream and atomized. This is a highly specific technique with very little spectral interference. If dry ashing is used, an ashing aid in the form of $MgO{-}Mg(NO_3)_2$ is needed in order not to volatilize the analytes. The mixing of the (dry) sample powder and the ashing aid is facilitated by the addition of a small amount of, for example, methanol, which lowers the surface tension and thereby improves the contact between the sample powder and the ashing aid. The HG-AAS technique has good LoDs, down to 0.03 $\mu g\ l^{-1}$ (see Table 3.1) depending on the sample mass and dilution. It can thus be applied to most foodstuffs, although the main source of As in the diet is fish and seafood.

3.3.8 Cold Vapor AAS for Hg

This is, like HG-AAS, a very specific technique and probably the most used technique today for the determination of Hg. The detection power is superior to that achieved by FAAS, and even somewhat better than the performance of ET-AAS. Mercury vapor is released from the digested sample solution (dry ashing is not an option here) by a strong reducing agent, such as $NaBH_4$ or $SnCl_2$, and carried by an inert gas to the absorption cell. It is advisable to check for background interference, since water vapor may give rise to molecular absorption. By accumulating Hg on Au (amalgamation) and then releasing the trapped Hg by heating the Au trap, the LoD can be considerably improved.

A wealth of information on the various aspects of AAS is contained in the comprehensive book by Welz and Sperling [8].

3.4 VALIDATION OF EXPERIMENTAL DATA

The literature frequently reports very high concentrations of elements, for example, Cd and Pb, in supposedly uncontaminated foodstuffs that would normally contain extremely low levels of these elements. It is rather unusual to see that those high and unusual results are checked in any way, which, of course, renders such surveys highly unreliable. Validation of results, both normal and abnormal, should always be performed, also in the interests of the analyst.

All experimental data obtained by AAS are incorrect, to some extent. This is mostly hidden by measurement uncertainty, which can be calculated and quantified. In contrast to measurement uncertainty, there are uncertain measurements, or analytical errors, which cannot be estimated since the error is largely unknown. How can one know whether an analytical error has occurred? Normally one does not know, which is the inherent problem with errors! This is why a healthy dose of suspicion is never out of place. If a single result, or a series of results, seem unusual, that is, deviating from the expected values, this may well be very important or just a blunder. In both cases, the experimental data need to be checked. This can be done in different ways and is quite simple (obvious controls, such as calibrants and dilutions, are not discussed here). The suspected sample solution can be diluted (e.g., $1 + 1$ and $1 + 3$) and reanalyzed. If these results (after taking into account the dilution factor) remain at the level of the first result, there is probably no error. Another way of checking results is to reanalyze the sample at other wavelengths that are sufficiently sensitive, for example, for Pb, 283.3 and 217.0 nm. If there is no significant difference between the two results, the measurement is probably correct. If doubts still linger, the sample should be analyzed by another laboratory for cross checking. If a CRM analyzed along with the sample gives satisfactory results, this provides some evidence that the determination is correct, although there is still no guarantee.

3.5 APPLICATION OF AAS TO FOODSTUFFS

3.5.1 General Aspects

When looking into the literature, there is no doubt that AAS, in its various forms, is the dominant technique for the determination of elements in foodstuffs. In a literature survey of Cd, Cr, and Pb in muscle tissue from domestic animals (horse excluded), fish muscle (tuna excluded), and milk, 103 published papers could be selected that met certain (quality) criteria. In those papers AAS was used in 93 percent of the cases, whereas the percentage of methods based on ICP-AES or ICP-MS was 11 percent and that of voltametric methods 3 percent (the total percentage exceeds 100 percent as more than one technique was used in some studies) [9]. The period covered was 1993–2003.

Lead contamination of food has decreased dramatically over the last decades, which is one of the, sadly few, success stories regarding environment and health in recent times. This is primarily due to the phasing out of leaded gasoline. In addition, the replacement of food cans with Pb-soldered side-seams by cans with welded side seams or two-piece cans has resulted in a substantial reduction of the intake of Pb via food. The introduction of analytical quality control (QC) systems has resulted in higher awareness of analytical problems, which has led to fewer over-estimations of Pb levels. Thus, there are lower Pb levels in the samples analyzed today than in earlier days. This has as a consequence that, the lower the levels are, the more demanding the analysis becomes. On the other hand, as long as the toxicologist discovers negative effects at lower and lower intakes, the struggle to obtain reliable results at very low concentration levels goes on.

The contamination of food during sampling or sample pretreatment in the laboratory is a general problem. Eliminating contamination during sampling may be difficult, since the analyst has usually little or nothing to do with that step of an investigation. Extensive use of blanks will show how severe the contamination situation is in the laboratory, but since it can be very erratic, it may be difficult to keep it under full control. Some elements, for example, Cr, are virtually impossible to eliminate as contaminants as they are present everywhere in daily life, for example, in stainless steel appliances and coins, just to mention a few.

3.5.2 Legal Limits for Cd, Hg, and Pb in Foodstuffs

Since Cd, Hg, and Pb are toxic even at very low levels of ingestion, their occurrence in food is regulated by both national and international legislation. In addition, recommendations are issued on how to limit their intake via food. Internationally, the Codex Alimentarius prescribes maximum levels for these elements in certain foodstuffs [10]. These are, however, not legislated levels for the occurrence in foods, but recommendations referred to in international trade. The legislation in the European Union will serve as an example, primarily because it is well known in many parts of the world, but also because it takes into account a large variety of foodstuffs. The maximum limits (MLs) are based on toxicological assessments and analytical data from the Member States. Most currently available data on elements in foodstuffs are based on AAS. If such data are to be the basis for legislation, it is obviously important that they are produced and reported in a way that ensures their reliability. Since quality assurance (QA) procedures are not yet fully operative everywhere, there is a risk that the analytical quality of the data cannot always been guaranteed. This may have led to inadvertent errors in some of the MLs. Since analytical problems in most cases lead to too high results, it is reasonable to assume that analytical errors have resulted in an overestimation of MLs. As analytical quality awareness increases and new quality-assured analytical data appear, MLs are periodically being revised.

The MLs for Cd and Pb in certain foodstuffs in the EU Member States [11] are summarized in Tables 3.2 and 3.3. For the official control to be enforced, it is mandatory for the laboratories to have accreditation for the methods used. It is also necessary that the methods comply with the following criteria [12, 13]:

- LoDs better than one-tenth of the ML (for milk and infant formulae, one fifth).
- Precision of less than 1.5 Horwitz ratio (HORRAT) (in collaborative tests).
- Recovery of 80–120 percent (in collaborative tests). For Sn, recovery should be 80–105 percent.

These regulations are under revision. In the new regulation which covers all four metals and is expected to be published in 2007, it is suggested that, for metals, the HORRAT is changed to 2 and that results from certified reference materials are used instead of recovery.

TABLE 3.2. Summary of the Maximum Levels of Pb in Foodstuffs (EC Regulation No. 1881/2006). This Table is a Simplification of the Original

Product	Maximum Level (mg kg^{-1} Wet Weight)
Cow's milk, infant formulae	0.02
Fruit juices, concentrated fruit juices (for direct consumption), and fruit nectars	0.05
Meat of bovine animals, sheep, pig, and poultry. Vegetables, excluding brassica, leaf vegetables, fresh herbs, and all fungi. For potatoes the maximum level applies to peeled potatoes. Fruit, excluding berries and small fruit. Fats and oils	0.1
Berries and small fruit. Cereals, legumes, and pulses. Wine, cider, aromatized wine	0.2
Brassica, leafy vegetables, and cultivated fungi. Muscle meat of fish	0.3
Offal of most domestic animals. Crustaceans, excluding brown meat of crab and head and thorax of lobster	0.5
Bivalve molluscs	1.0
Cephalopods (without viscera)	1.5

TABLE 3.3. Summary of the Maximum Levels of Cd in Certain Foodstuffs (EC Regulation No. 1881/2006). This Table is a Simplification of the Original

Product	Maximum Level (mg kg^{-1} Wet Weight)
Meat of most domestic animals and most fish. Vegetables and fruits, excluding leafy vegetables, fresh herbs, fungi	0.050
Meat of other fish. Cereals, excluding bran, germ, wheat, and rice. Stem vegetables, pine nuts, root vegetables, and potatoes	0.1
Horse meat. Bran, germ, wheat and rice. Soybeans. Leaf vegetables, fresh herbs, celeriac, and all cultivated fungi	0.2
Meat of swordfish	0.3
Liver of most domestic animals. Crustaceans, excluding brown meat of crab and head and thorax of lobster	0.5
Kidney of most domestic animals. Bivalve molluscs. Cephalopods (without viscera)	1.0

The official control of Pb in milk, which has an ML of 0.020 mg kg^{-1} thus requires a method with a LoD of 0.004 mg kg^{-1}. This creates problems as very few laboratories can reach such an LoD with an AAS method. In addition, the general requirement of a HORRAT better than 1.5 is generally no problem when it comes to Cd because of the low LoD of the technique for this metal. For Pb, however, it causes problems, since in most collaborative tests the precision for this analyte will be at or above 1.5 HORRAT in one or several matrices. Finally, the requirement of 80–120 percent recovery is not relevant for this type of method. Recovery will be discussed in Section 3.6.

For Hg, the MLs apply only to fish and fishery products [11]. For most fish species the ML is 0.5 mg kg^{-1}. In certain other species, mostly predatory such as pike and tuna, the ML is set at 1 mg kg^{-1}. It is generally accepted that over 90 percent of Hg in fish and shellfish is present as methylmercury, that is, a much more toxic species.

For Sn, the EU MLs apply to canned foods of all kinds [11]. For canned food in general the ML is 200 mg kg^{-1} wet weight, for canned beverages 100 mg kg^{-1}, and for canned foods for infants, babies, and young children 50 mg kg^{-1}. In practice, however, whenever the ML is exceeded, this occurs almost exclusively in canned fruit and vegetables. High levels of Sn in foods may cause gastric irritation, but they are not known to cause any chronic effect. For enforcement of the legislation on Sn the methods used in the official control within the EU must comply with the following criteria [14]:

- LoDs no worse than 5 mg kg^{-1}.
- Precision of less than 1.5 HORRAT (in collaborative tests).
- Recovery of 80–105 percent (in collaborative tests).

3.5.3 Recommendations for Intake Limitation of Cd, Hg, and Pb

An expert group from WHO/FAO has issued recommendations for intake limitations based on the weekly intake (provisional tolerable weekly intake, PTWI) of Cd, Hg, and Pb by a general adult population [14]. PTWIs are based on the intake per kilogram of body weight and are set to 0.007, 0.025, and 0.005 mg kg^{-1}, respectively. For a person weighing 70 kg the PTWI is calculated to be 0.49 mg for Cd, 1.75 mg for Pb, and 0.35 mg for Hg. The weekly intake in the EU Member states is much lower than the PTWI, however. In a SCOOP report from 2004 [15] the average weekly intake by adults was estimated to 0.101 mg for Cd, 0.039 mg for Hg, and 0.296 mg for Pb.

3.5.4 Labelling and Health Claims

There are no MLs in force for elements such as As, Ca, Mg, Na, and Se (although As is under discussion) in foodstuffs. Since those elements may have both positive and adverse properties, it is important that concentration levels stated on food packaging labels, as well as claims of health effects at certain concentrations, are checked for reliability by the competent authorities.

3.5.5 Application of AAS to Food of Animal Origin

For the purpose of comparison, data on Cd and Pb in several types of animal-based foodstuffs, assumed to be uncontaminated, are shown in Table 3.4. Sampling of tissues from domestic animals, for the determination of Cd and Pb, can be a problem if the person who performs the sampling is unexperienced in trace element determination. If muscle tissue comes into contact, for example, with liver or kidney, the risk of cross contamination is considerable. Sampling of milk requires clean equipment as well as clean hands. Bone fragments in meat samples can be difficult to avoid, but may have a great impact on the results of Cd and Pb, particularly if the sample portion is very small. For crustaceans meticulous care is necessary to avoid contamination of the muscle tissue from the hepatopancreas. This is crucial, since the hepatopancreas contains far higher levels of many trace elements, Cd in particular.

In general, the levels of Cd and Pb are extremely low in muscle tissue and milk of animals from uncontaminated environments (see Table 3.4), which means, in turn, that sample pretreatment in the laboratory becomes critical. It is not always possible to have clean room facilities available, but it is certainly mandatory to work in a laboratory and with equipment that are as clean as possible. Nevertheless, there is a limit to what can be achieved in a conventional laboratory with standard equipment. Thus, the idea must be accepted that the contamination level establishes the LoDs.

If the procedures described above are adhered to, analytical problems can normally be circumvented. As the levels of most trace elements in muscle tissue and milk are very low, it may well happen that they are often at or below the LoDs of the technique resorted to, which for ET-AAS is usually reported to be in the order of 5–25 $\mu g\ kg^{-1}$, or even lower. Cadmium is usually present at levels below 5 $\mu g\ kg^{-1}$ (with the exception of horse meat), often below 1 $\mu g\ kg^{-1}$. Since the detection power for Cd is much better than for Pb, the LoD is typically in the order of <1–5 $\mu g\ kg^{-1}$. For control purposes this is entirely sufficient. For intake calculations however, results below the LoDs are impossible to quantify. For this purpose, better LoDs are needed. Since this may be very difficult to achieve,

TABLE 3.4. Examples of Cd and Pb Levels in Uncontaminated Foodstuffs of Animal Origin (Based on Results from References [17–27])

Foodstuff	mg kg^{-1} Fresh Weight	
	Cd	Pb
Meat (excluding horse), most fishes. Dairy products	≤0.005	≤0.01
Certain fish (e.g., herring). Eggs	≤0.025	≤0.01
Horse meat	≤0.2	≤0.02
Liver and kidney from domestic animals	≤0.5	≤0.2
Mussels and oysters	≤1	≤0.5
Horse kidney	>10	≤0.2

other ways should be tested, for example, by injecting into the GF larger sample volumes, or multiple volumes, with drying in between. However, results obtained in this way should be watched closely to check that the background does not reach levels that cannot be compensated for.

It should be noted that some animal tissues can contain extremely high Cd levels, for example, crab hepatopancreas (the so-called "brown meat") and horse kidney, and are therefore analyzed by FAAS. Often the same tissue is analyzed for Pb using ET-AAS. After a number of firings when determining Pb, the GF becomes severely contaminated with Cd and is very difficult to clean. This will obviously be a problem when other samples with low Cd levels are to be analyzed, especially so if Pb also is to be quantified. The best solution is to use two separate instruments, which obviously is not possible in many laboratories.

The level of As is very low in most foods. The exception is seafood, for example, plaice, which often has an As concentration of >10 mg kg^{-1}. As mentioned above, FAAS is not the first choice for this kind of determination. Dry ashing in combination with HG-AAS is well suited for this purpose and has sufficient detection power for the analysis of concentrations well below 0.05 mg kg^{-1}, levels that can be found in pork and beef. ET-AAS after MW digestion is also suitable for seafood, but lacks the detection power necessary for the analysis of other foodstuffs. Contaminated chemicals can be a problem, and it is advisable to have new batches checked before use.

The reason why Hg has MLs established for fish only is, of course, that this is the only food in which it is normally found. In other foodstuffs Hg levels are usually so low that the analysis is worth trying only if specific questions need an answer. Systems for CV-AAS are generally available as ancillary to FAAS instruments. MW digestion is probably the most suitable way to get the sample tissue into solution. Contamination by Hg is, in most laboratories, very low and is seldom a problem.

3.5.6 Application of AAS to Food of Vegetable Origin

For the purpose of comparison, Table 3.5 shows the levels of Cd and Pb found in several types of foods of vegetable origin that are assumed to be uncontaminated. Vegetable foods are often contaminated by the environment in which they are produced. It may be by soil particles from the fields, the uptake of metals by the root system, or atmospheric fallout. Before an analysis is started it is important to decide what question should be answered. If it is an environmental question, perhaps the samples should be analyzed without any prior rinsing to remove contamination, or perhaps analyzed both rinsed and unrinsed to ascertain the degree of contamination. If it is a food question, how is it normally eaten? Are, for example, apples or potatoes eaten with or without peel? Contamination during sampling is probably not as big a problem with vegetable foods as with food of animal origin. Conversely, cross contamination is an ever-present danger. It is easy to contaminate vegetables grown in a greenhouse by letting them come into

TABLE 3.5. Examples of Cd and Pb Levels in Uncontaminated Vegetable Foods (Based on Results from References [17–22 and 25–27]

Foodstuff	mg kg^{-1} Fresh Weight	
	Cd	Pb
Most vegetables	≤0.025	≤0.05
Rye flour	≤0.025	≤0.05
Oats	≤0.025	≤0.05
Wheat flour	≤0.05	≤0.05
Potatoes	≤0.05	≤0.01
Leafy vegetables	≤0.05	≤0.2
Most wild mushrooms. Wheat bran	≤0.2	≤0.5
Sunflower and flax seeds	≤0.5	<0.025
Certain mushrooms (e.g., *Agaricus Augustus)*	>10	≤1

contact with products grown outdoors, or take samples of, for example, lettuce in a way that results in contamination through soil.

A particular contamination phenomenon is canned food, which is contaminated by Sn from the tin can in which it is preserved. This occurs primarily in canned fruit and vegetables. The highest Sn levels are found in cans where the Sn layer is left unlacquered, which is mostly the case with canned fruit. The concentration of the metal in unlacquered cans is a function of storage time and temperature, pH, and several other factors. In cans with a lacquered inside, the increase in Sn concentration is mostly due to damage to the lacquer layer, which makes the rate of increase unpredictable. It is thus just a matter of time before the Sn level approaches the legislated ML of 200 mg kg^{-1}. Food in tin cans may be contaminated by Cr (used in the manufacture of cans), but not at a very high level. Foodstuffs in tin cans may also be contaminated by Fe. High Fe levels are usually a sign that the tin layer is more or less dissolved and that the underlying sheet iron is exposed.

As for meat-based foodstuffs, vegetables pose no specific analytical problems as long as some basic recommendations are adhered to. Many vegetable samples leave a silicate residue after digestion. Generally, in the determination of Cd and Pb, such a residue does not affect the results of the analysis. In analyses where silicate residues are suspected to interfere, volatilization by heating with HF (not in glass beakers) is recommended. Levels of Cd, Cr, and Pb are mostly higher in vegetable foods than in animal foods (see Tables 3.4 and 3.5), but usually not high enough to make FAAS an option.

Results from recent proficiency tests on Sn in vegetable test materials have shown an unexpectedly large variation. This is likely to be a general problem with the determination of Sn in foodstuffs. An additional problem is that the available methods which are collaboratively tested and approved may not meet current requirements for accuracy. Thus, there is presently an uncertainty as to which method to choose for the determination of Sn in foods.

3.6 ANALYTICAL QUALITY ASSURANCE AND QUALITY CONTROL PROCEDURES

3.6.1 General Aspects

Analytical quality issues have gained a lot of ground over recent decades. Awareness of the importance of providing some sort of guarantee that a result is credible is becoming more and more popular in food laboratories worldwide. Today, the ISO/IEC Standard 17025 regulates much of what is mandatory or recommended for analytical laboratories in the preparation of the quality manual, which is in turn the first step towards accreditation. The main pillars of an analytical QC system (Fig. 3.7) are CRMs and PTs. Both are essential in the laboratory, since they satisfy different needs. There are several uses for both CRMs and PTs. Some points are described in the following sections.

3.6.2 Certified Reference Materials

These materials are of great importance to check that the analytical procedure is under control in the daily work of the laboratory. A suitable CRM is analyzed regularly together with the samples and the result is plotted in a control chart, in which warning and action levels are marked. These levels should be based on the standard deviation and not the 95 percent confidence interval (CI) given in the accompanying certificate. Then there are general rules as to what deviations and trends are acceptable. This is probably the most important use of CRMs.

The other major use of CRMs is to ascertain the trueness of the results found. A common (but unregulated) practice is to assume that the results for the samples are correct if the result of the CRM, analyzed along with the samples, is within the 95 percent CI. It should be kept in mind that the analyst knows the certified levels of the analytes in the CRM prior to the analysis, which unavoidably makes him/her biased. Therefore, good CRM results may indicate that the results of the samples the CRM was analyzed with are just as good, but are not useful as a guarantee for the trueness of results.

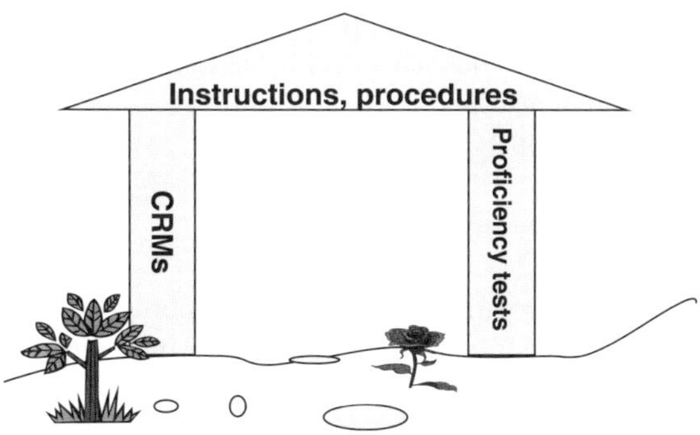

Figure 3.7. The building blocks of a quality system.

Another important use of CRMs is for the validation of in-house reference materials. Since CRMs are fairly expensive and perhaps not as specific as necessary for certain analysis, it would be cost-effective to prepare in-house reference materials when possible. By analyzing a specific sample (that is homogenous and available in a reasonably large quantity) along with a similar CRM, traceability of the found value is established to the certified value. Then this in-house material can be used as a complement to the CRM. A recently published procedure describes how to get the most complete information out of CRMs [16].

3.6.3 Proficiency Testing

As a quality tool, PTs are as important as CRMs. The main difference between PTs and CRMs is that the concentration of the analyte in a PT sample is unknown by the analyst at the time of analysis. Another significant difference is that PT samples are not produced with the same rigorous metrological control as the CRMs. PTs are, however, the only independent means by which an analysts' competence can be estimated. One PT result only is not enough to make such a judgement, as this is just a "snapshot" at a specific time. After a number of PT results (e.g., more than three for a given element), the picture becomes more complete and the performance can be evaluated. How frequently a laboratory should take part in PT programs is not easy to say. Probably more frequent PT results are needed when competence is being built up than when good competence has been established.

The z-score evaluation procedure, used in most PT programs, is simple and informative. It is nevertheless useful for PT participants to take a look at the assigned value and the uncertainty that are used in the calculation of the z-score, to find out what they are based on. The assigned value of an analyte in a PT sample can be established in different ways. One of the more common procedures is based on the results submitted by the PT participants, which may render the assigned value sensitive to skewed distributions of results when many inexperienced analysts participate. Outliers can be eliminated by statistical standard procedures, but only to a certain extent before the assigned mean value can be calculated. Another evaluation procedure that is gaining ground is robust statistics, which may lead to a somewhat different assigned value. The measurement uncertainty (target σ) may also be established in different ways and this may have a great effect on the width of the z-score.

3.6.4 Recovery

This is a procedure that is regularly given more importance than deserved in the efforts to guarantee the trueness of experimental results when determining trace elements in foodstuffs. The information value of recovery tests in element analysis by AAS in food is very limited. Recovery information is important in chemical analysis where there is an extraction step in order to elucidate the efficiency of the extraction. If the recovery, by using such a method, is limited to, for example, 70 percent, it should be decided whether a correction factor should be applied to the result or not. It does not indicate that the analytical method is inadequate. In the analysis of trace elements in foodstuffs the digestion methods result in total destruction of the organic matter, but all the inorganic matter is still there. It is thus assumed to yield 100 percent recovery $\pm$ measurement uncertainty. If a result is outside the acceptable recovery interval (see, e.g., the EU requirements mentioned above), there is probably an analytical error, which should be identified and eliminated, but not corrected for.

Moreover, it is very difficult to interpret the result of recovery tests in trace element analysis. The recovery is the difference between the original level of the analyte found in the sample and the analyte added (spike), usually in the form of a standard solution of a suitable concentration. However, since the spike and the original analyte may be present in different chemical forms, one cannot know whether they behave identically during the digestion step and the determination. If the original analyte (partly) volatilizes during sample digestion and the spike does not, recovery will be appear acceptable! If the spike volatilizes and the original analyte does not, this will be interpreted as a loss of analyte! Furthermore, the added spike will not reveal whether there is contamination of the sample or uncorrected interferences. As a consequence, results may be obtained that are obviously much too high, although the recovery is 100 percent. Therefore, CRMs and PT are crucial quality tools in the laboratory, whereas recovery is not.

3.7 COMPARISON OF THE PERFORMANCE OF AAS, ICP-AES, AND ICP-MS

3.7.1 Limits of Detection

The instrumental LoDs set forth in Table 3.1 shows that inductively coupled plasma–mass spectrometry (ICP-MS) is the technique that yields the best values. It should be considered that these limits are based on pure solutions analyzed under optimal conditions. Dealing with real foodstuffs will dramatically change the picture, owing to the complexity of the matrices and contamination phenomena in the laboratory. This means that in food laboratories without clean-room facilities (which is the vast majority) the practical difference in LoDs for ET-AAS and ICP-MS will be of minor importance. The relatively poor LoDs for inductively coupled plasma–atomic emission spectrometry (ICP-AES) when compared to those of ET-AAS implies that this technique is not fit for low-level determination of many elements, for example, Cd and Pb.

3.7.2 Interferences

A limitation in ICP-MS is that spectral interferences (overlapping masses) can considerably worsen the LoDs. For example, $^{40}Ar^{35}Cl$ in Cl-rich foodstuffs interferes with the analysis of ^{75}As and $^{40}Ar^{12}C$ interferes with ^{52}Cr. C interference is a result of incomplete digestion. When using wet digestion methods it can thus be important to make sure that the digestion is complete, that is, that no C remains. These interferences can be minimized by using instruments with dynamic reaction cells or collision cells which can reduce the Ar-based polyatomic ions.

The ICP-AES and ICP-MS techniques may also suffer from matrix effects, such as spray chamber effects caused by the different viscosity of the samples and the calibration standards. The careful choice of internal standards can reduce this problem. The effects caused by high amounts of easily ionized elements may be solved by internal standardization or by the use of matrix-matched calibration curves. An additional specific problem with ICP-AES is the risk of spectral overlaps.

3.7.3 Other Factors

An obvious advantage of ICP-AES and ICP-MS is the great linear dynamic range as compared to that of ET-AAS. The dynamic range cannot always be fully used within a batch, since low- and high-level samples may be mixed, which results in rather long rinsing times after a high-concentration sample solution has been analyzed. If the analysis of many different elements in a long series of samples is required, the great ability of ICP-AES and ICP-MS for simultaneous analysis is a major advantage. Another factor to be considered is the ability to tolerate dissolved solids. The amount of total dissolved solids that ICP-MS can handle is in the order of 0.5–1 percent, for ICP-AES up to 10 percent, and for FAAS 4–5 percent. In contrast, it is possible to use solid sampling with ET-AAS (although very small sample masses), for certain applications.

3.7.4 Economy

The cost of running an ICP-MS is quite high. Ar is expensive, especially so because it is used in large volumes. Moreover, spare parts like skimmer and sampler cones, torches, and detectors have a limited lifetime. ICP-AES and ICP-MS have similar Ar consumptions. On the other hand, the former technique is somewhat cheaper to run since the replacement rate of consumables is lower. The different forms of AAS require various amounts of different gases. These are, however, far less than the volumes consumed by the competing techniques. The cost for other items such as graphite tubes and cones is fairly moderate. All three techniques require service at regular intervals, which is in itself rather expensive.

3.7.5 Complexity

AAS in its various forms is not so simple as many are led to believe, but it is a fairly rugged technique once the analyst has become sufficiently trained and has learned the "tricks of the trade." The running of an ICP-MS requires highly trained personnel and even then the interpretation of results from such various matrices as foodstuffs is not an easy task. The complexity of the ICP-AES technique is perhaps more like that of AAS rather than that of ICP-MS, or may lie somewhere in between.

3.8 CONCLUSIONS

It can be concluded that, for the average food laboratory where only a few elements are determined in each sample, sample series are limited, and legislated MLs have to be controlled at levels well below 0.1 mg kg^{-1}, AAS will defend its leading position well into the future, although ICP-AES and ICP-MS will narrow the gap existing between the two types of techniques.

As regards AAS, one limiting factor is the dynamic range. For flame applications this can be extended by the use of automatic dilutors, which dilute samples to the useful range, thereby avoiding tedious and time-consuming manual dilutions. This is already in practical use, but even more sophisticated approaches can be foreseen in the near future.

The multielement function of the plasma-based techniques has been a source of challenges in the AAS field. This has resulted in the fast sequential technique, which is a simple way to mimic the multi-element function. However, it only works for FAAS applications. Moreover, simultaneous multielement ET-AAS systems for analysis have also been placed on the market, although there are some spectral limitations. It also has the drawback of using the same time and temperature programme for all elements. In the future, further developments in the multielement technique can be envisaged which will resort to continuum sources as well as CCD and other multiwavelength detectors.

REFERENCES

1. J. H. Duffus, "Heavy metals" – A meaningless term?, *Pure Appl. Chem.*, **74** (2002), 793–807.
2. P. Fecher, G. Ruhnke, Cross contamination of lead and cadmium during dry ashing of food samples, *Anal. Bioanal. Chem.*, **373** (2002), 787–791.
3. Z. Mester, R. Sturgeon (Eds), *Sample Preparation for Trace Element Analysis*, Volume XLI in Wilson & Wilson's Comprehensive Analytical Chemistry, Elsevier, Amsterdam, 2003.
4. L. Jorhem, Dry ashing, sources of error and performance evaluation in AAS, *Microchim. Acta*, **119** (1995), 211–218.
5. H. M. Kingston, L. B. Jassie (Eds), *Introduction to Microwave Sample Preparation, Theory and Practice*, American Chemical Society, Washington, DC, 1988.

6. A. Walsh, The application of atomic absorption spectra to chemical analysis, *Spectrochim. Acta*, **7** (1955), 108–117.
7. D. Skoog, *Principles of Instrumental Analysis*, CBS College Publishing. Saunders College/Holt, Rinehart and Winston, 3rd edn, 1985.
8. B. Welz, M. Sperling, *Atomic Absorption Spectrometry*, Wiley-VCH, Weinheim, 3rd completely revised edn, 1999.
9. L. Jorhem, J. Engman, B. Sundström, A. Nilsson, Evaluation of measurement data for Cd, Cr and Pb in certain uncontaminated foodstuffs published in surveys: analytical quality vs. uncertainty of measurements, *J. Accred. Qual. Assur.*, **10** (2006), 647–658.
10. www.codexalimentarius.net
11. European Commission No. 1881/2006 of 19 December 2006, Setting maximum levels of certain contaminants in foodstuffs, *Off. J. Eur. Commun.*, **L364**, 20 December 2006.
12. European Commission No. 22/2002 of 8 March 2001, Methods for sampling and analysis, *Off. J. Eur. Commun.*, **L77**, 16 March 2001.
13. European Commission No. 16/2004 of 12 February 2004, Methods for sampling and analysis – Tin, *Off. J. Eur. Commun.*, **L42**, 13 February 2004.
14. World Health Organization, Evaluation of Certain Food Additives and Contaminants, *WHO Technical Report Series*, No. 837, WHO, Geneva, 1993.
15. European Commission Task for Scientific Co-operation (SCOOP), Project 3.2.11 for assessment of the dietary exposure to Pb, Cd, Hg and As of the population of EU member states, 2004.
16. NMKL Procedure No. 9, *Evaluation of Results from the Analysis of Certified Reference Materials*, Nordic Committee on Food Analysis, Oslo, 2001.
17. L. Jorhem, B. Sundström, Levels of lead, cadmium, zinc, copper, nickel, chromium, manganese and cobalt in foods on the Swedish market, 1983–1990, *J. Food Comp. Anal.*, **6** (1993), 223–241.
18. L. Jorhem, B. Sundström, Levels of some trace elements in edible fungi, *Z. Lebensm. Unters. Forsch.*, **201** (1995), 311–316.
19. B. Hardy, *The 1996 Australian Market Basket Survey*, Australia New Zealand Food Authority, Canberra, 1998.
20. N. Marro, *The 1994 Australian Market Basket Survey*, Australia New Zealand Food Authority, Canberra, 1996.
21. G. Ysart, P. Miller, H. Crews, P. Robb, M. Baxter, C. de L'Argy, S. Lofthouse, C. Sargent, N. Harrison, Dietary exposure estimates of 30 elements from the UK total diet study, *Food Addit. Contam.*, **16** (1999), 391–403.
22. G. Ysart, P. Miller, M. Croasdale, H. Crews, P. Robb, M. Baxter, C. de L'Argy, N. Harrison, 1997 UK total diet study: aluminium, arsenic, cadmium, chromium, copper, lead, mercury, nickel, selenium, tin and zinc, *Food Addit. Contam.*, **17** (2000), 775–786.
23. J. Engman, L. Jorhem, Toxic and essential elements in fish from Nordic waters, with the results seen from the perspective of analytical quality assurance, *Food Addit. Contam.*, **15** (1998), 884–892.
24. L. Jorhem, Lead and cadmium in tissues from horse, sheep, lamb and reindeer in Sweden, *Z. Lebensm. Unters. Forsch.*, **208** (1999), 106–109.

25. E. H. Larsen, N. L. Andersen, A. Møller, A. Petersen, G. K. Mortensen, J. Petersen, Monitoring the content and intake of trace elements from food in Denmark, *Food Addit. Contam.*, **19** (2002), 33–46.

26. S. Karavoltsos, A. Sakellari, M. Dimopoulos, M. Dasenakis, M. Scoullos, Cadmium content in foodstuffs from the Greek market, *Food Addit. Contam.*, **19** (2002), 954–962.

27. S. P. Dolan, G. C. Capar, Multi-element analysis of food by microwave digestion and inductively coupled plasma-atomic emission spectrometry, *J. Food Comp. Anal.*, **15** (2002), 593–615.

4

HIGH-RESOLUTION CONTINUUM SOURCE AAS AND ITS APPLICATION TO FOOD ANALYSIS

BERNHARD WELZ AND DANIEL L. G. BORGES

Departamento de Química, Universidade Federal de Santa Catarina, 88040-900 Florianópolis, SC, Brazil, E-mail: w.bernardo@terra.com.br

UWE HEITMANN

ISAS – Institute for Analytical Sciences, Department for Interface Spectroscopy, Albert-Einstein-Str. 9, 12489 Berlin, Germany

The Determination of Chemical Elements in Food: Applications for Atomic and Mass Spectrometry.
Edited by Sergio Caroli

4.1 INTRODUCTION

The first atomic absorption spectroscopes that were assembled by Bunsen and Kirchhoff and a few others in the second half of the nineteenth century used a continuum source (CS), as this was the only reliable light source available at that time. This was certainly also one of the main reasons why atomic emission spectrometry (AES) was preferred over atomic absorption spectrometry (AAS) during the first half of the twentieth century, when atomic spectroscopy was increasingly used for quantitative analysis. It is obviously much easier to detect a weak emission signal against a dark background than a slight reduction in emission intensity over a small spectral interval against a bright background. For the same reason, in 1952 Walsh [1, 2] came to the conclusion that line sources (LS) with the sharpest possible emission lines are required for measurements with AAS, as a resolution of approximately 2 pm would be necessary if a continuum source was used. Walsh also concluded that it is necessary to modulate the emission of the primary radiation source, and to use a selective amplifier tuned to the same frequency in order to eliminate any emission from the atomizer, which is obviously not modulated, and hence disregarded by the amplifier. This conclusion resulted in half a century of LS AAS with all its advantages and limitations.

Some of the advantages of LS AAS are: (1) the relatively simple and inexpensive design of the equipment, as no high-resolution monochromators are required; (2) the high selectivity and specificity due to the use of an element-specific LS and the modulation principle; (3) the high freedom from spectral interferences due to line overlap with other elements, compared with AES, because significantly fewer lines appear in absorption than in emission; and (4) the relatively high tolerance of flame and graphite tube atomizers towards matrix constituents, compared with excitation sources for AES, particularly the inductively coupled plasma (ICP).

Among the drawbacks of LS AAS, the most frequently mentioned one is without doubt that it is a one-element-at-a-time technique, which means that it is quite tedious if a large number of elements have to be determined in a set of samples. However, the greatest limitation, although only rarely mentioned, is the fact that absorption is measured only over the very narrow spectral range corresponding to the width of the atomic line emitted by the radiation source, that is, over a few picometers, and no information is available about the spectral environment. However, the absorption measured at the analytical line might well consist not only of atomic absorption due to the element of interest, but also to background absorption caused by matrix constituents [3]. With LS it is only possible to measure total absorption. The background absorption then has to be determined in a second measurement and subtracted from the total absorption to arrive at the net atomic absorption. These two measurements have to be made in rapid sequence, particularly in the case of time-dependent signals, in order to minimize artifacts and errors. This has resulted in the design of a variety of background correction (BC) systems, using deuterium arc lamps as a second source [4], high current pulsing of the primary source [5], or the Zeeman effect to split the absorption profile [6]. All these approaches have made LS AAS equipment complex and expensive, and, although the latter system has in the meantime reached a high level of performance [7], none of the solutions is ideal, and the first two approaches in particular have serious limitations in practical analysis.

The idea of using CS for AAS, however, has never been given up, and the results obtained by this technique have become ever more promising over the years. In 1989 Hieftje [8] wrote in a review article: "for AAS to remain viable in the face of strong competition from alternative techniques will require novel instrumentation approaches. Among the novel concepts that have been introduced are those involving continuum sources and high-resolution spectral-sorting devices ... and entirely new detection approaches." Only a decade later did the progress made in CS AAS cause Harnly [9] to forecast: "The future appears bright for CS AAS. Whereas previously, CS AAS was striving for parity with LS AAS, it is now reasonable to state that it is CS AAS, which is setting the standard." This forecast was at least in part based on the results of the pioneering work done by the research group under Becker-Ross at the ISAS (Institute for Analytical Sciences, formerly Institute of Spectrochemistry and Applied Spectroscopy, Department Berlin, Germany), which is reviewed in Welz et al. [10].

4.2 INSTRUMENTATION FOR HR-CS AAS

This breakthrough of high-resolution (HR)-CS AAS was possible only because this group in Berlin has been working systematically on the solution of all the problems that are associated with this technique. The first equipment, where all essential components were optimized for HR-CS AAS, was described in 1996 by Heitmann et al. [11]. Firstly, there was the notoriously weak emission in the far UV below 250 nm of commercially available continuum radiation sources, such as xenon arc lamps, which are for example used for stadium illumination. The problem was solved by the development of a xenon short-arc lamp with an electrode distance of $\leq$1 mm and a gas pressure of about 15 bar (cold), which is shown in Figure 4.1. Under normal operating conditions the pressure in the lamp increases to about 50 bar, and a hot-spot develops with a diameter of less than 0.2 mm with a temperature of about 10,000 K. This results in significantly higher radiation intensity, particularly in the far UV, where the intensity is some two orders of magnitude higher than that of conventional xenon arc lamps, and about three orders of magnitude higher than that of a deuterium arc lamp, as shown in Figure 4.2.

Secondly, when a CS is used for AAS, it is necessary to utilize a high-resolution monochromator in order to avoid loss of sensitivity and excessive curvature of the calibration function, and also to avoid spectral interferences. Becker-Ross et al. [12] have shown for several elements that the sensitivity continuously increases with increasing resolution until the spectral bandwidth is in the order of the width of the atomic absorption line, and that no further improvement in sensitivity is possible beyond that level. Becker-Ross et al. [13] also determined the half-width of the absorption lines of a large number of elements, and found that a monochromator with a resolution $\lambda/\Delta\lambda$ of about 100,000 is necessary for HR-CS AAS. Then

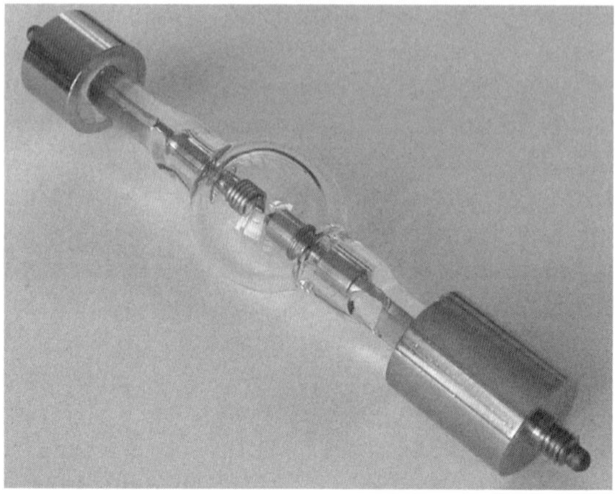

Figure 4.1. Xenon short-arc lamp for HR-CS AAS (XBO 301, 300 W, GLE, Berlin, Germany).

Heitmann et al. [11] designed a very compact double monochromator, consisting of a 300 mm prism pre-monochromator and a 400 mm echelle grating monochromator, both in Littrow mounting, which is shown schematically in Figure 4.3. The prism monochromator selects the part of the spectrum that is of interest, and the echelle monochromator provides the high dispersion of the selected spectral interval, which is better than 2 pm per pixel at 200 nm (see Welz et al. [10]).

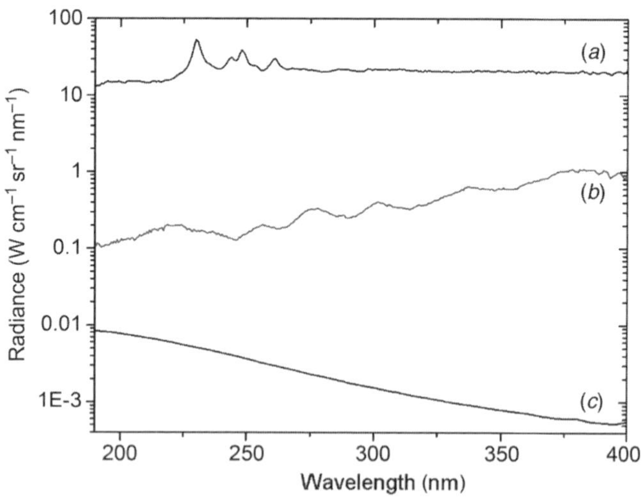

Figure 4.2. Intensity distribution over the spectral range used in AAS for (*a*) the xenon short-arc lamp used for HR-CS AAS (XBO 301, 300 W, GLE, Berlin, Germany), compared to (*b*) a conventional xenon arc lamp (L 2479, 300 W, Hamamatsu, Japan) and (*c*) a deuterium arc lamp (MDO 620, 30 W, Heraeus, Germany).

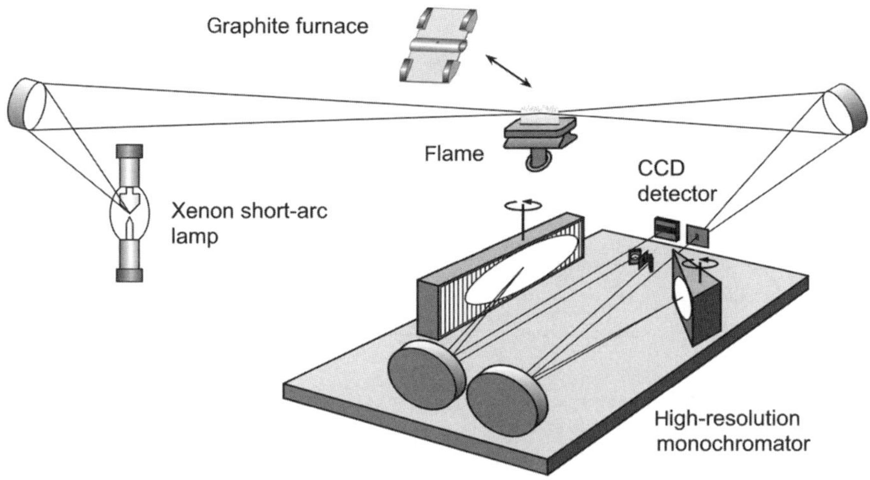

Figure 4.3. Instrumental concept for HR-CS AAS.

Obviously, such a high-resolution monochromator requires active wavelength stabilization in order to avoid drift problems. This has been accomplished through an internal neon lamp, mounted on an adjustable stand in front of the intermediate slit between the pre- and echelle-monochromator, so that it can be moved into the beam automatically if necessary. The neon lamp emits many relatively narrow lines in the 580–720 nm range, and, in the absence of any pre-selection, these are separated by the echelle grating into various superimposed orders. This means that without pre-dispersion at least two neon lines for every grating position surely fall on the detector, and can be used for stabilization. The precision of this stabilization is only limited by the stepper motor for grating adjustment, and is better than one-tenth of a pixel width (see Welz et al. [10]).

Thirdly, a UV-sensitive, thinned back-illuminated Charge-Coupled Device (CCD) array detector array detector with 512×58 pixels, size 24×24 μm, operating in full vertical binning mode, has been chosen as it offers a whole series of new features that will be discussed in detail later. Thus, the instrument is actually operating with 512 fully independent detectors, 200 of which are normally used for analytical purposes. This means that the entire spectral environment about ± 0.2 nm on both sides of the analytical line becomes "visible" around 200 nm, and about ± 0.5 nm in the visible range. This makes it possible to present the measured data in three different ways. One is the conventional recording of absorbance over time, which is typically used in electrothermal (ET) AAS (Fig. 4.4*a*). In comparison with conventional LS AAS, the major difference is that this scan has been obtained with one single pixel, that is, at high resolution, and similar scans can be obtained for all the other pixels of the CCD array, providing detailed information about the spectral environment of the analytical line. Another mode of presenting the data is in wavelength integrated over time (Fig. 4.4*b*). Here any kind of absorption is shown that has been measured over the spectral interval of

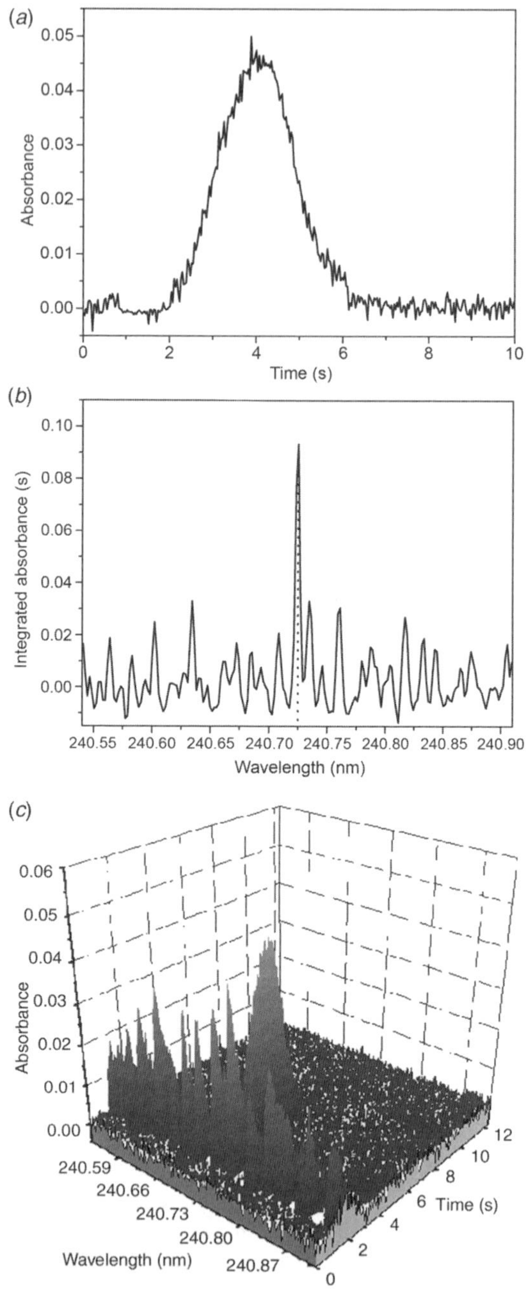

Figure 4.4. The three ways of presenting the measured data in HR-CS AAS. Determination of Co in the NIST 1577a Bovine Liver SRM at 240.725 nm using ET AAS: (*a*) absorbance-over-time measured at the center pixel (line core) only; (*b*) absorbance over wavelength integrated over time (the dotted line represents the center pixel); (*c*) three-dimensional plot of absorbance over wavelength and time.

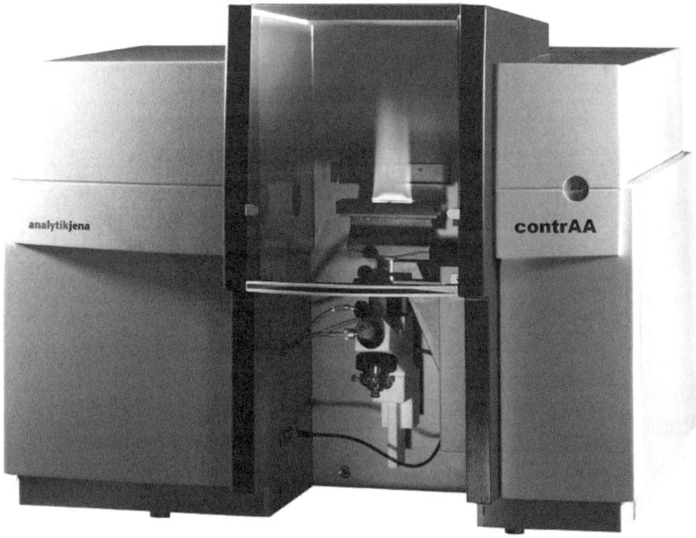

Figure 4.5. ContrAA 300, the first commercial instrument for HR-CS AAS (Analytik Jena AG, Jena, Germany).

interest. A third mode – especially for techniques that produce time-dependent signals, such as ET AAS – is the three-dimensional image, that is, absorbance over time and over wavelength (Fig. 4.4*c*). In other words, a wealth of additional information becomes available that can be used for much more efficient method development and to avoid interferences. In addition, as all pixels are illuminated and read out simultaneously, the instrumental concept offers the first simultaneous double-beam system and the first simultaneous BC system in AAS, as will also be discussed later. The first commercially available instrument based on this technology is shown in Figure 4.5.

4.3 SPECIAL FEATURES OF HR-CS AAS

4.3.1 The Modulation Principle

When Walsh started to think about using AAS for analytical purposes back in 1952, one of his key conclusions was that, in order to carry out absorption measurements on luminous atomic vapors, it would be necessary to employ a modulated light source and a synchronously tuned detection system, so that any radiation emitted by the sample would produce no signal at the output of the detection system [2]. This modulation principle, using either an AC-operated radiation source or a chopper in the radiation beam, and a selective amplifier tuned to the same modulation frequency, has ever since been applied in all commercially available atomic absorption spectrometers. It has been considered one of the major advantages of

AAS over AES, which, along with the element-specific radiation source, guarantees the selectivity and specificity of atomic absorption measurements.

However, such "luminous atomic vapors" or any other emission from the atomizer, such as that from the red-hot graphite tube, can cause a problem in AAS only when it is within the spectral bandwidth that is used for measurement, and when its intensity becomes significant in comparison to that of the primary radiation source. Neither condition is met in HR-CS AAS. Firstly, the intensity of the radiation source used for HR-CS AAS is typically two orders of magnitude higher than that of a conventional line source for LS AAS, which reduces the risk for emission interference. Secondly, as the detectors used for LS AAS integrate the radiation over the spectral bandwidth, that is, typically over 0.2–2 nm, any atomic emission within this spectral range would be recorded and amplified. The situation might be even worse for broadband emission, which is integrated over the entire bandwidth, making it some two orders of magnitude greater than it is on the analytical wavelength. In HR-CS AAS, any atomic emission that does not coincide directly with the analytical line is not measured by the analytical pixel(s), and the intensity of any broadband emission is some two orders of magnitude lower on the analytical pixel(s) compared to its integrated intensity in LS AAS. This means that it becomes negligible, particularly when the higher lamp intensity is considered as well. Last but not least, in the very unlikely case that there is some extremely intense emission from the atomizer that contributes to the measured signal, the BC systems – to be discussed in detail later – will compensate for this potential interference as well.

This means that HR-CS AAS, due to its special features, does not need any modulation of the source or any selective amplifier. This also means that a potential source of noise has been eliminated, as both AC operation of hollow cathode lamps and the mechanical choppers are contributing to noise in LS AAS. In addition, other problems that are associated with strong emission of the atomizer source in LS AAS – such as the "emission noise" caused by the nitrous oxide – acetylene flame in the determination of Ba and Ca due to the CN band emission [3] – are equally absent in HR-CS AAS for the same reasons, that is, the higher intensity of the primary radiation source, and the high resolution.

4.3.2 Simultaneous Double-Beam Concept

In LS AAS it is normal to distinguish between single-beam and double-beam spectrometers. In a single-beam spectrometer the primary radiation is conducted through the absorption volume without geometric beam splitting, while in a double-beam spectrometer the beam is divided in two sections. A portion of the radiation, the *sampling radiation* (or sample beam), passes through the absorption volume (flame, furnace, etc.), while the other portion, the *reference radiation* (or reference beam), bypasses the absorption volume. The geometric splitting and recombining of the beam can be done by means of a rotating chopper made from a partially mirrored rotating quartz disk, by means of semi-transparent mirrors, or by combinations of these. Typical switching (modulation) frequencies are between 50 and 100 Hz [3].

Single-beam spectrometers have the advantage that they contain fewer optical components and thus radiation losses are lower and the optical conductance is higher. The major advantage claimed for *double-beam spectrometers* is better long-term stability, since they compensate for changes in the intensity of the source and the sensitivity of the detector. Nevertheless, this advantage is frequently overvalued, since during the warm-up phase not only the radiant intensity of the source changes, but also the line profile, and thereby the sensitivity. The double-beam system is incapable of recognizing or compensating for changes in the atomizer, such as flame drift during the warm-up phase of the burner. Nowadays it has been accepted that a double-beam spectrometer has clear advantages for routine flame (F) AAS in order to avoid frequent checks for baseline drift and recalibration. For ET AAS, where the baseline is usually re-set automatically before each atomization stage, using a baseline offset correction (BOC) routine, long-term stability is not of interest, so that the better optical conductance of a single-beam spectrometer is of advantage [3].

Spectrometers for HR-CS AAS are optical single-beam instruments, as there is no geometric splitting or recombining of the beam included. However, HR-CS AAS actually offers a far superior stabilization system that can take over all the functions of an optical double-beam instrument with the great advantage of working simultaneously, and not sequentially, as optical systems do. The CCD array detector of an HR-CS AA spectrometer typically consists of several hundred pixels, which can be considered as independent detectors. They are all illuminated by the same radiation source and read out simultaneously. As only a few of these pixels are typically used to measure atomic absorption, any other pixel or set of pixels may be used to correct for fluctuations in lamp intensity, as these fluctuations are obviously the same for all pixels, and as all pixels are illuminated and read out simultaneously, not sequentially, even the fastest changes in emission intensity will be perfectly corrected, as shown in the example of Figure 4.6.

Moreover, as the entire radiation passes through the atomizer, any temporal change in the transmission, for example of the flame gases, is corrected in the same way, a feature that is obviously not available in any optical double-beam system. This correction capability of HR-CS AAS is shown in Figure 4.7 in the case of Bi determination at the 306.772 nm line doublet in an air–acetylene flame. This line doublet directly overlaps with a molecular absorption band of the OH radical. As this molecular absorption is not stable over time, but subject to fluctuations, it contributes significantly to the overall noise level of the measurement, as is shown in Figure 4.7*a*. In HR-CS AAS it is possible to record the molecular spectrum of OH over the spectral range under consideration by measuring the flame absorption without any analyte. This spectrum can be stored in the software and subtracted from the measurement of a real sample using a least-squares algorithm, as will be discussed in detail in the following section about background measurement and correction. The measurement corrected in this way is shown in Figure 4.7*b*, where all the flame noise has disappeared, resulting in a significantly improved signal-to-noise ratio (SNR) for the measurement.

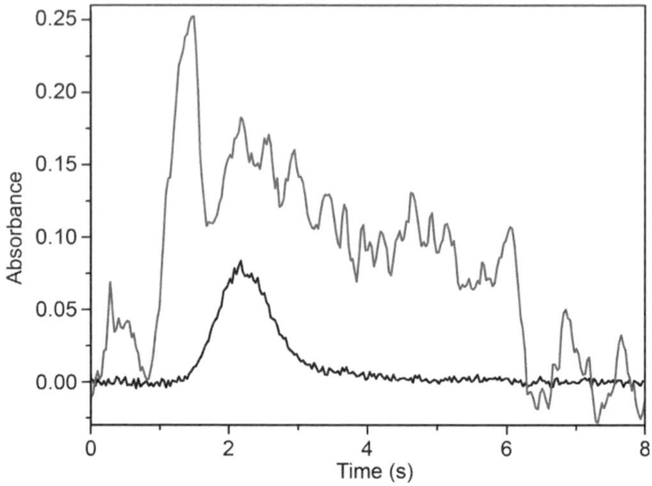

Figure 4.6. Absorbance-over-time for cobalt in the TORT-2 Lobster Hepatopancreas CRM at 240.725 nm using HR-CS ET AAS; gray line: uncorrected signal; black line: corrected signal (from Ribeiro et al. [14]).

4.3.3 Background Measurement and Correction

In practical analytical work radiation is not only absorbed by atoms, but also by gaseous molecules and radicals, and radiation may also be scattered by particles in the atomizer. Molecular absorption spectra in the visible and UV range are due to electron excitation and exhibit a pronounced rotational fine structure, or they are due to molecule dissociation and are then "continuous," that is, without fine structure, and they do not change, or change monotonously within the observed spectral range. Molecular absorption may be caused by concomitants in the sample or by the atomizer itself, for example, the flame gases, as shown in the previous section. Radiation scattering may be observed mainly in graphite tube atomizers, when refractory matrix components cannot be removed during the pyrolysis stage. Obviously radiation scattering is also "continuous," that is, without fine structure. All this attenuation of radiation, which is not specific to the analyte, is usually termed "background" or "non-specific" absorption.

There are at least two distinct differences in the measurement of background in HR-CS AAS compared to LS AAS. Firstly, due to the wavelength-resolved detection of the absorbance, the background, its nature and spectral distribution become visible (refer to Fig. 4.4*c*), which makes it orders of magnitude easier to take the appropriate action for its removal or correction. Secondly, the measurement of atomic and background absorption is strictly simultaneous, so that no artifacts are observed, even for the most rapidly changing background signals, which are characteristic of the sequential measurement of total and background absorption in LS AAS.

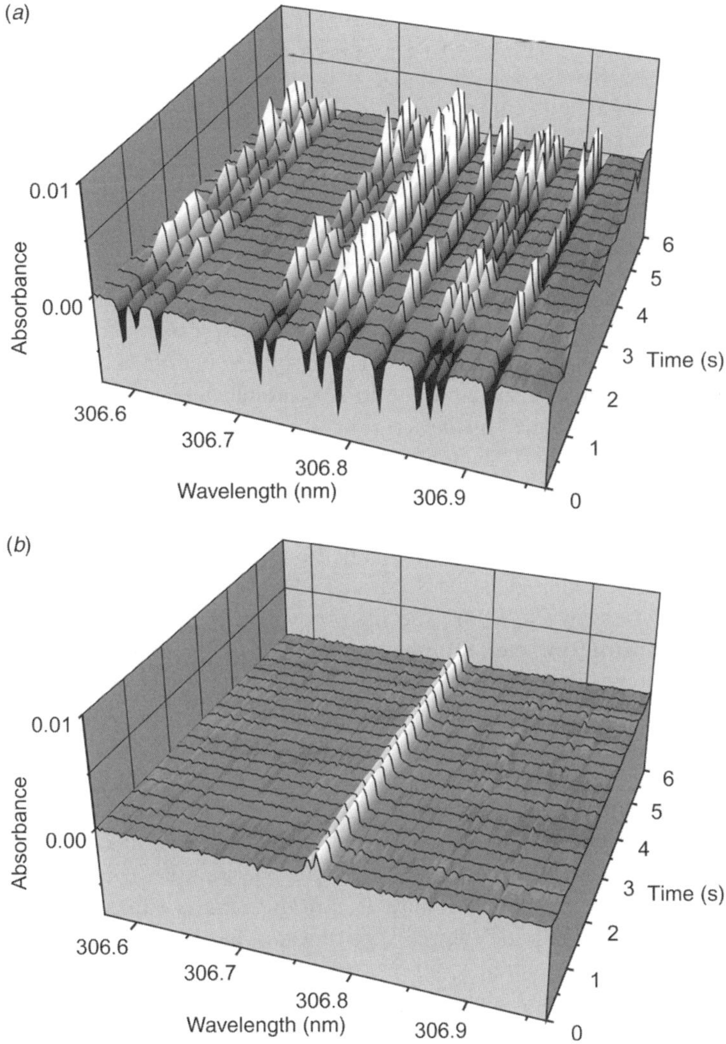

Figure 4.7. Absorbance over wavelength and time plot for the determination of Bi at 306.772 ± 0.3 nm in an air–acetylene flame: (*a*) without correction; (*b*) with least-squares BC (from Welz et al. [10]).

The computer-aided signal processing of an analytical cycle in HR-CS AAS always has the following simplified scheme. Before the atomization stage of a graphite furnace or the aspiration of a sample into a flame, a given number of sequential reference intensity scans (BOC scans) without any analyte absorption is recorded. These scans are averaged and used as the reference intensity signal. Then each analytical intensity scan is normalized to the reference intensity signal, giving an intensity correction factor. In this way, all systematic errors, such as pixel errors and different illumination of the spectrometer, are removed from the analytical scans. The intensity correction factor represents the non-specific broadband background absorption, leading to a decrease in the total signal intensity. Finally, the absorbance is calculated from the corrected analytical intensity scans, resulting in a three-dimensional plot of absorbance versus wavelength and time, as shown in Figure 4.4*c*. In total, three different types of signals could be observed: the analyte absorbance, the fine-structured atomic and molecular absorption from concomitants, and the broadband background absorption. All of them are recorded simultaneously, but the last one will be removed automatically during the calculation step and cannot be seen in the final plots.

Once the broadband absorbance is eliminated as described, the rotational fine structure of molecular absorption due to electron excitation becomes visible, as well as possible atomic absorption due to concomitant elements. In most cases it is not necessary to take further action in order to remove this residual fine-structured background, as long as it is not directly overlapping with the analyte absorption in wavelength and time. Again, the decision can usually be made easily by inspecting the time- and wavelength-resolved absorbance readings. Obviously, interference occurs only when the background absorption is on the analytical pixel(s), and if it is not separated in time from the analyte absorption (in the case of transient signals). The first case is relatively rare due to the high resolution of the spectrometer, and the second case can be influenced by optimization of the graphite furnace program.

For the case that analyte and background absorption coincide both spectrally and temporally, it is possible to correct for the structured background originating from the electron excitation spectra of molecules. The software of the instrument offers the possibility to measure and store reference spectra, as shown for the spectrum of the OH radical in the previous section. These spectra may then be subtracted from the spectrum measured for the actual sample using a least-squares algorithm. The mathematical procedure used in this case is a linear fit of the reference spectrum to every single sample spectrum that has been measured over time, typically from several hundred to a few thousand per measurement. The reference spectrum will increase or decrease by multiplication with a magnification factor. The differences between the reference and the sample spectrum as well as their squares will be calculated pixel by pixel, and the sum of the square values over all pixels will be added up. After that the said magnification factor will be varied in order to minimize the sum of the squares or, in other words, to find the "least squares." Using this procedure, specifically that part of the background will be eliminated that corresponds to the fine structure of the reference spectrum. A linear combination of more than one reference spectrum can be used for the same purpose.

Figure 4.8 shows an example for this mode of correction applied for the determination of Pb in pig kidney reference material, using direct solid sample analysis. The three-dimensional plot in Figure 4.8*a* shows that a strong molecular absorption with pronounced fine structure appears short after the atomic absorption signal. Figure 4.8*b* shows the time-integrated absorbance spectrum of PO,

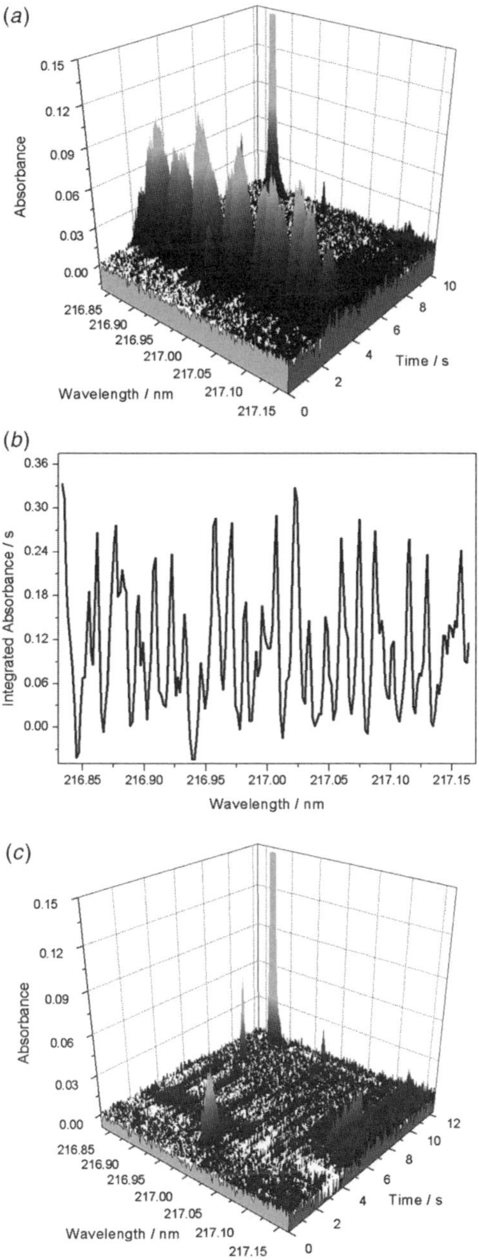

Figure 4.8. Least-squares BC for molecular spectra with rotational fine structure: determination of Pb in the BCR 186 Pig Kidney CRM at 217.001 nm using HR-CS ET AAS and direct solid sample analysis; (*a*) absorbance over time and wavelength after correction for continuous absorption; (*b*) reference spectrum: absorbance over wavelength integrated over time for $NH_4H_2PO_4$ (the dotted line represents the center pixel); (*c*) absorbance over time and wavelength after subtraction of the reference spectrum using least-squares BC.

which has been obtained by atomizing a solution of $NH_4H_2PO_4$ in the same wavelength interval. After subtraction of this reference spectrum, using least-squares BC, the three-dimensional plot shown in Figure 4.8*c* was obtained, which demonstrates that the molecular structures are almost exclusively due to the PO molecule, and are removed with this algorithm.

It is obviously of advantage to know the kind of molecule that is causing this spectral interference, but this knowledge is not necessary. Firstly, there are only a very limited number of diatomic molecules that can exist at temperatures typically used for atomization in flames and graphite tubes at a high enough concentration to produce a detectable molecular absorption spectrum. Secondly, the spectra of most of these molecules are known and published in the literature [10], so that it is not too difficult to find out the source of interference. Thirdly, there will be no harm at all if, by chance, the wrong molecular spectrum is chosen for correction purposes, as the software will simply disregard it. This means that several reference spectra might be applied until the proper one has been found.

4.3.4 Selection of Analytical Lines

Another well-known limitation of LS AAS is that a separate lamp is usually purchased for each element that has to be determined, which becomes very expensive when a large number of elements are involved, and when some of them are to be determined only very infrequently. In HR-CS AAS obviously only one lamp is required, and this lamp emits all lines, even for those elements for which no line source is available.

Usually, in LS AAS, the most sensitive analytical line is used for the determination of an element, because AAS is mostly applied for trace and ultra-trace analysis, which obviously requires the highest sensitivity. Another reason for using the most sensitive line is that it makes it possible to apply higher dilution in case of complex sample matrices, and hence avoid potential interferences. On occasions, however, the most sensitive line is not recommended in LS AAS, as it does not provide the best SNR, as in the case with the 217.001 nm Pb line. Another reason might be a strongly nonlinear working curve due to the presence of other lines in the lamp spectrum that cannot be excluded even with a 0.2 nm bandwidth [3].

All these limitations do not exist in HR-CS AAS. Firstly, the radiation intensity of the source is always high enough to provide a significantly better SNR than the LS in conventional AAS. In a first approximation the radiation intensity, and hence the SNR for all lines, is of the same magnitude, although it degrades somewhat in the far UV. Secondly, the resolution of the monochromator is such that only the center of the line is detected by the analytical pixel. As the source emits a continuum, and the analytical pixel is adjusted in such a way that it is always in the line center, none of the phenomena associated with LS exist, for example, line shift, self absorption, or the presence of other lines emitted by the lamp [3].

Although AAS is predominantly used as a technique for trace analysis, a significant demand for a reduction of sensitivity has been created by the technique of direct solid sample analysis using ET AAS. As solid samples cannot be easily "diluted," and the sample mass can only be varied within about one order of magnitude without deteriorating precision, the use of alternative, less sensitive lines is the most obvious way to reduce sensitivity for the determination of higher concentrations. This demand for alternative lines will increase with HR-CS AAS, as this technique has already been shown to be particularly suited for the direct analysis of solid samples [15]. In LS AAS the use of secondary lines has often been discouraged mostly for two reasons. Firstly, secondary lines were often "weak" lines; that is, the emission intensity was low, compared to the primary line(s), resulting in the deterioration of the SNR, and hence in measurement precision. Secondly, these lines were often poorly investigated, and little was known about potential spectral interferences. Hence, a general warning was often issued to be particularly careful with spectral interferences when using secondary lines.

Again, in HR-CS AAS these problems are essentially nonexistent for the same reasons as given above. Firstly, because of the relatively constant, very intense emission of the primary radiation source, there are no more weak lines; that is, the same high SNR will be obtained on all analytical lines, regardless of their spectral origin. The only factors that will have an influence will be the absorption coefficient and the population of the low excitation level in case nonresonance lines are used. Secondly, because of the high resolution of the monochromator, and the "visibility" of the entire spectral environment of the analytical line in HR-CS AAS, potential spectral interferences can easily be detected, and in addition cannot influence the actual measurement, except in the rare case of direct line overlap. However, even in this case, HR-CS AAS provides an appropriate solution, as discussed in the previous section.

One of the characteristics of HR-CS AAS is that no element-specific radiation source is required, and that in principle absorption can be measured at any wavelength, including molecular absorption bands. Huang et al. [16] investigated the possibility of using PO molecular absorption bands in an air-acetylene flame for analytical purposes. Although molecular absorption and emission of PO and HPO had previously been proposed for the determination of phosphorus, all this work was done with low-resolution equipment, making it highly susceptible to all kind of spectral interference. As shown in Figure 4.9, many of the PO bands consist of a very strong and sharp band head, followed by a weak and complex band shoulder. It is obvious that, with a high-resolution instrument, measurement should be made at the band head – the half width of which is not much different from that of an atomic line – and not at the whole band in order to obtain high sensitivity and avoid spectral interferences. The best SNR was obtained at the 324.619 nm band head, where the authors reached a detection limit of 1.3 mg l^{-1} P in a slightly fuel-rich air – acetylene flame, and a linear range of more than three orders of magnitude. The same authors also investigated the possibility of determine S using the molecular absorption bands of the CS radical in the range between 257 and 260 nm, shown in Figure 4.10. The best limit of detection of about 2.5 mg l^{-1} was obtained at the 258.056 nm line using a fuel-rich

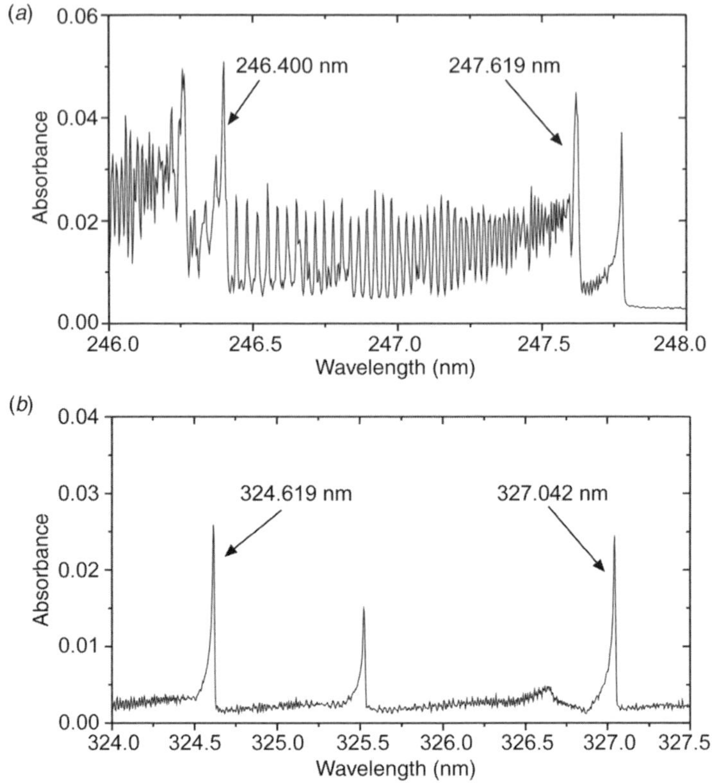

Figure 4.9. PO molecular absorption obtained for 400 mg l^{-1} P in an air–acetylene flame at different wavelengths with the band head at (*a*) 246.400 and 247.619 nm; (*b*) 324.619 and 327.042 nm (from Huang et al. [16]).

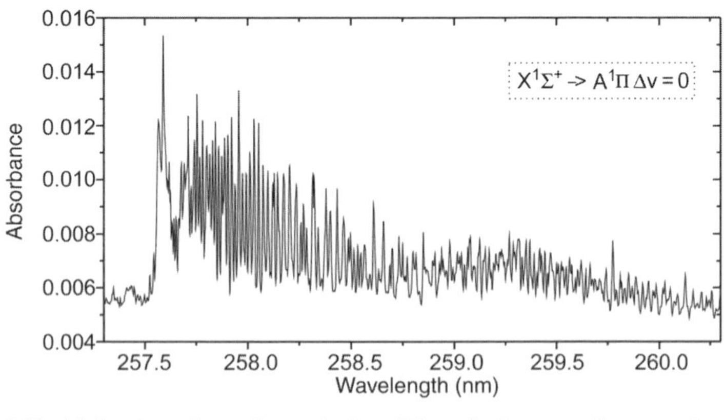

Figure 4.10. Molecular absorption of the CS radical over the wavelength range 257–260 nm (from Welz et al. [10]).

air–acetylene flame, and the working range covered more than three orders of magnitude [17].

4.3.5 Sensitivity and Working Range

As in AAS the absorbance A depends on the *ratio* of the incident radiant power Φ_0 and the transmitted radiant power Φ_{tr} according to the Beer–Lambert law

$$A \equiv \log \Phi_0/\Phi_{tr} = 0.43\, Nl\kappa$$

the sensitivity is independent of the intensity of the radiation source. For this reason the same sensitivity should be expected for LS AAS and HR-CS AAS. This is also found in the first approximation in practical analysis. However, some deviations will be discussed below, which are mainly caused by the difference in overlap of the emitted radiation with the absorption line due to line shift, hyperfine structure, and so on, which results in a different absorption coefficient κ.

A major detraction for LS AAS has always been the relatively short linear region of the calibration curves, typically not more than two orders of magnitude in concentration. The limits of the linear working range arise from stray radiation and the finite width of the emission lines of the radiation source, which is not monochromatic and just three to five times narrower than the absorption profile. With HR-CS AAS, there is no theoretical limit to the calibration range, only the practical limits imposed by the size of the array detector, the increasing possibility of spectral interferences, and the ability to clean the atomizer after extremely high analyte concentrations have been introduced.

Heitmann et al. [18] have demonstrated the performance of a HR-CS AAS instrument for two elements, Ag at 328.068 nm with a very narrow absorption line, which is barely resolved by the spectrometer, and In at 303.935 nm, which extends clearly over more than five pixels, as shown in Figure 4.11*a* and *b*. As mentioned above, using HR-CS AAS, signal registration cannot only be done at the center pixel **M**, corresponding to the line core, as in conventional LS AAS, but also using the volume $\mathbf{M_x}$ of the absorption peak with an increasing number of pixels, or the side pixels $\mathbf{N_x}$ only, as shown in Figure 4.12. In the first case of peak volume detection, that is, when an increasing number of pixels is used for detection, the sensitivity can be improved, depending on the peak width. In the case of a narrow absorption line, as is shown for the example of Ag, the sensitivity obtained for the center pixel (CP) only is comparable with that obtained with a line source, as shown in Figure 4.13*a*. Using three pixels, that is, CP$\pm$1, results in a significant increase in sensitivity and in the linear working range, but a further extension of the peak volume detection does not make sense, as the signal drops rapidly, and only the noise would be integrated. The situation changes in case of a broader absorption line, as shown in Figure 4.13*b* for the case of indium, where the sensitivity obtained in HR-CS AAS is significantly lower than that in LS AAS when the CP only is used for measurement. In this

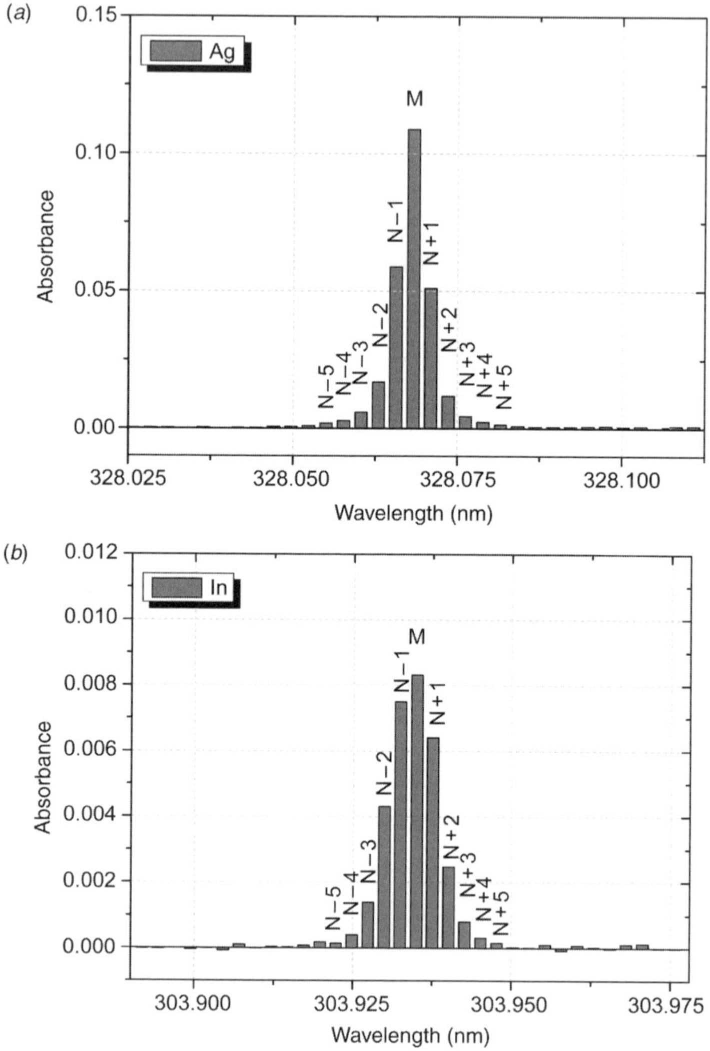

Figure 4.11. Typical line profiles of 1 mg l^{-1} calibration solutions: (*a*) Ag at 328.068 nm; (*b*) In at 303.935 nm (from Heitmann et al. [18]).

case a significant improvement in sensitivity can be obtained for HR-CS AAS not only by using CP $\pm$ 1, but also by further increasing the peak volume to CP$\pm$2 and CP $\pm$ 3. On the other hand, using side pixels only at increasing distance from the line core for evaluation results in reduced sensitivity and increasing linearity, as shown in Figure 4.14*a* and *b*. Although the increase in the linear range, as compared to LS AAS, is clearly more pronounced for narrow absorption lines,

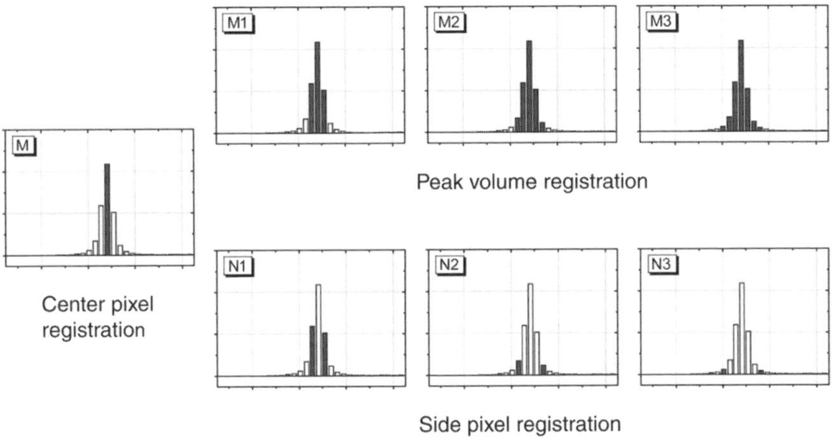

Figure 4.12. Definition of the peak volume detection **M** and the side pixel detection **N** in dependence on pixel number (from Heitmann et al. [18]).

mostly because of the more pronounced effect of line shifts in LS AAS (Fig. 4.14*a*), the overall trend is the same in HR-CS AAS, independently of the line shape, that is, measurement at the line wings might be used without compromising the determination of high analyte concentrations, extending the working range of HR-CS AAS to at least five orders of magnitude in concentration, as claimed for ICP AES and ICP-MS.

4.3.6 Signal-to-Noise Ratio, Precision, and Limit of Detection

As the "signal," that is, the sensitivity, has been fully treated in Section 4.3.5, only the "noise" component of the above terms has to be discussed here. In addition, only the contribution of instrumental components to the noise will be treated, not the "analytical" part, such as imprecision in sample introduction, flame noise, or the inhomogeneity of the sample itself, which obviously are important contributions as well.

For a CCD detector the absorbance noise is independent of the spectral bandwidth, but it depends on the number of measurement pixels n_{sam} and reference pixels n_{ref} in such a way that n_{sam} should be as small as possible and n_{ref} should be larger than n_{sam}. The other component that influences the noise is the intensity I of the radiation source, in that the absorbance noise is inversely proportional to the square root of I [12]. As the intensity of the radiation source in CS AAS is in some cases up to two orders of magnitude higher than that of a typical LS for conventional AAS, an improvement in the SNR and limits of detection (LoD) by factors of 3–10 could be expected, unless other factors, such as flame noise, become dominant. The values given in Table 4.1 show that this expectation has in fact been realized for the majority of the elements.

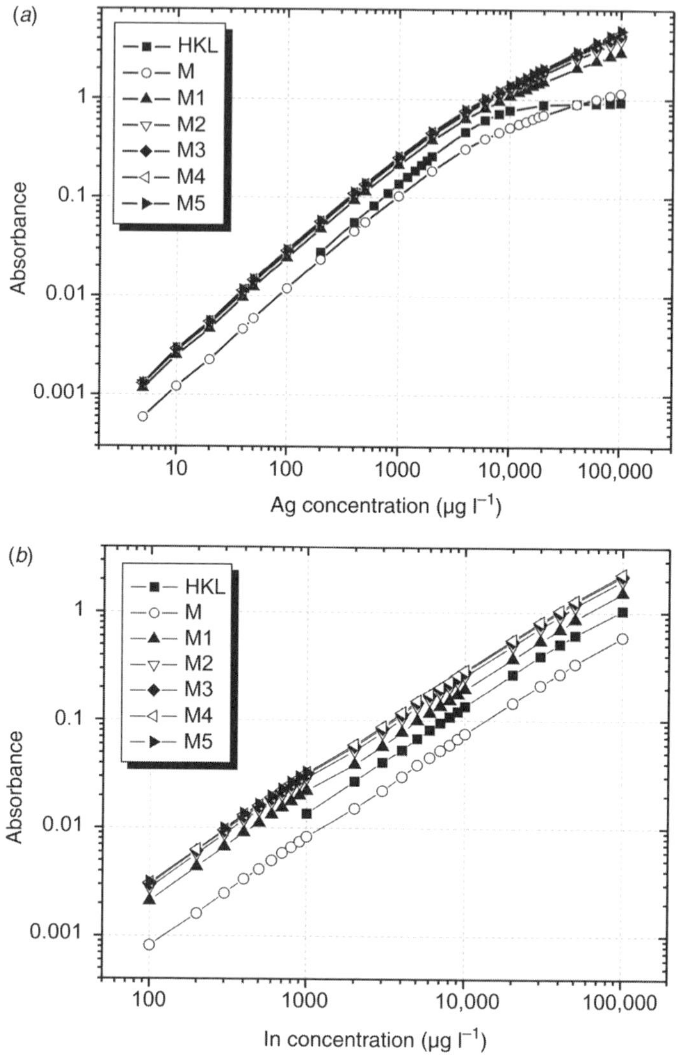

Figure 4.13. Calibration curves for (*a*) Ag and (*b*) In in logarithmic scale obtained for an increasing number of pixels using peak volume detection, compared to that obtained with a Hollow Cathode Lamp (HCL) (from Heitmann et al. [18]).

Besides the intensity of the radiation source, its stability obviously also plays an important role, and arcs are notorious for their instability. However, as this instability results for a small wavelength interval of some tenths of a nanometer in a wavelength-independent, "white" noise, which causes an identical noise pattern on all the pixels of the CCD array, it can easily be cancelled out with the use of reference pixels, as already discussed.

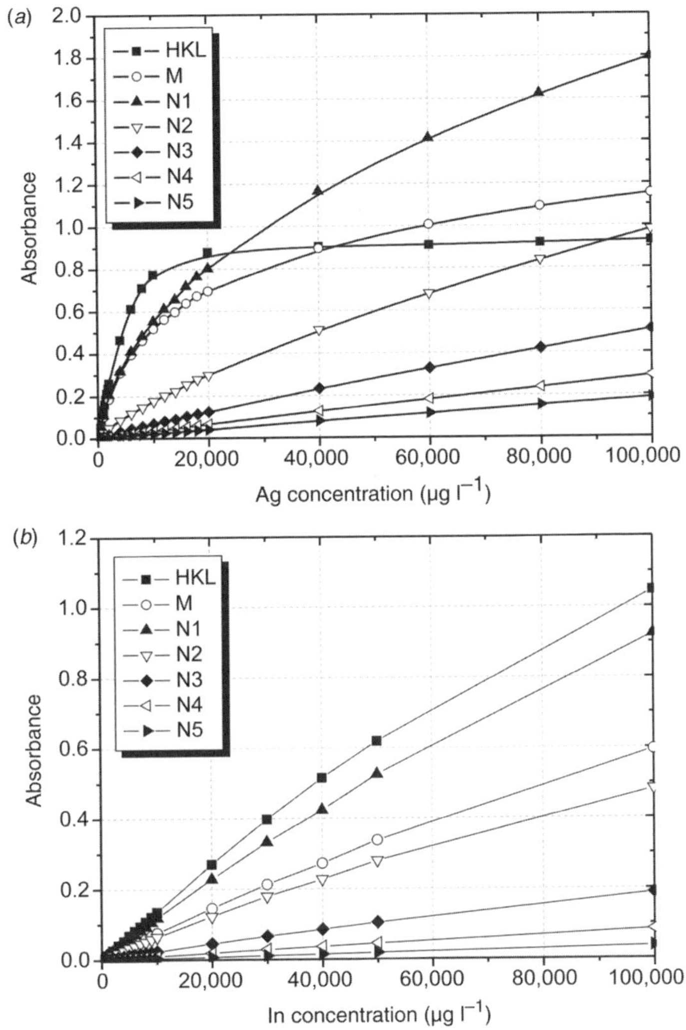

Figure 4.14. Linearity of calibration curves for (*a*) Ag and (*b*) In using different side pixels for calibration, compared to that obtained with LS (HKL) (from Heitmann et al. [18]).

As a result of the almost perfect correlation of the spectral intensity values within the small range of observation, the minimum detectable absorbance signal is determined only by statistical variations of the intensity between the neighboring pixels (shot noise). This means that an increase in radiation intensity or of the measurement time by a factor of 4 will reduce the absorbance noise by a factor of 2 (square root of 4).

The first effect, that is, the influence of the radiation intensity, is shown in Figure 4.15 [19]. Although the intensity distribution of the xenon short-arc lamp

TABLE 4.1. Comparison of Limits of Detection for Selected Elements, Obtained with LS AAS and HR-CS AAS, Using the Same Nebulizer-Burner System (AAS 6 Vario, Analytik Jena, Germany) and the Same Integration Time of 5 s

Element	Wavelength (nm)	LoD ($\mu g\ l^{-1}$)		Improvement Factor CS/LS
		LS AAS	HR-CS AAS	
Ag	328.068	1.5	0.6	3
As	193.696	150	330	0.5
Cd	228.802	1.2	0.4	3
Cr	357.869	5	1	5
Fe	248.327	4	1	4
K	766.491	15	0.07	200
Mn	279.482	3	0.3	10
Na	588.995	5	0.03	170
Ni	232.003	4	1.2	3
Pb	217.001	13	5	3
Rb	780.027	3	0.2	15
Zn	213.857	1.4	0.7	2

is relatively flat over the entire spectral range, there are increasing losses of radiation intensity in the far UV range, particularly below 200 nm in any optical system that cannot be avoided. Nevertheless, as shown in Figure 4.15*a*, even for Se at 196.026 nm, an absorption noise as little as 0.000114 can be obtained using a measurement time of only 5 s. For the wavelength of Cd at 228.802, which is still in the far UV region, the noise is already reduced to a value of 0.000038, that is, a factor of 3 lower, due to the higher radiation intensity that reaches the detector, as shown in Figure 4.15*b*. This tendency continues towards longer wavelengths, although it becomes less and less pronounced.

The second effect, that is, the influence of the measurement time is shown in Figure 4.16 for Pb at 217.001 nm. Simply by increasing the measurement time from 5 to 45 s, that is, by a factor of 9, the noise level is reduced by a factor of 3, which is exactly according to expectation for short-noise limited measurements, and the LoD is improved by the same factor.

4.3.7 Multi-Element Atomic Absorption Spectrometry

Harnly [9, 20] has been speculating repeatedly about simultaneous multi-element AAS, because the idea of combining the analytical capabilities of HR-CS AAS with simultaneous multi-element detection is intriguing. In essence, any instrument that uses a high-resolution echelle spectrometer and a sufficiently intense CS, such as a xenon short-arc lamp, is capable of simultaneous multi-element AAS measurement. A spectrometer of this type has actually been built by Becker-Ross and co-workers [21] as a tool for studies of structured background in FAAS, although it did not cover the entire wavelength range of AAS. The major

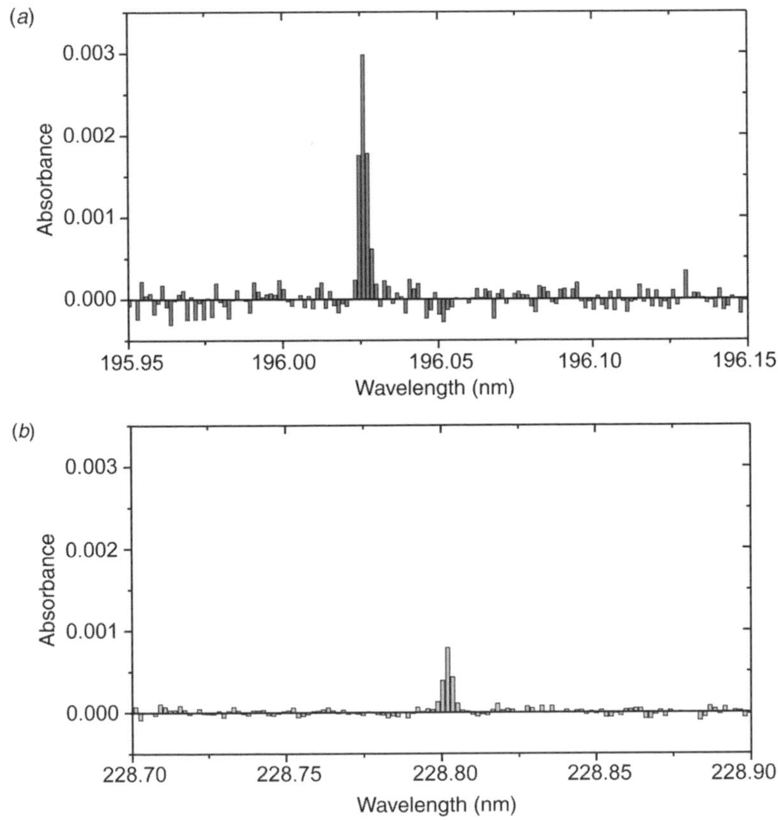

Figure 4.15. Absorbance noise for different wavelength regions, using a measurement time of 5 s each: (*a*) 1 mg l^{-1} Se at 196.026 nm; (*b*) 0.01 mg l^{-1} Cd at 228.802 nm (from Heitmann et al. [19]).

limitation of this instrument is the readout time of the detector of 2 s per spectrum; in order to obtain a good SNR the authors used an alternate measurement of 50 spectra of both sample and blank solution, resulting in a total measurement time of several minutes. This might be acceptable for F AAS, although it would not bring any advantages in speed of analysis compared to conventional LS AAS using fast sequential techniques, and it cannot be used at all for transient signals, as they are produced in ET AAS. In conclusion, the readout speed of currently available detectors is the limiting factor for simultaneous multi-element AAS, unless essential features of HR-CS AAS are given up.

However, any spectrometer that uses CS and a double monochromator with an echelle grating makes it possible to reach any line within an extremely short period of time of much less than 1 s, as both the grating and the prism are stepper-motor controlled. This feature allows a fast sequential multi-element determination to be performed with the great advantage that flame conditions and burner

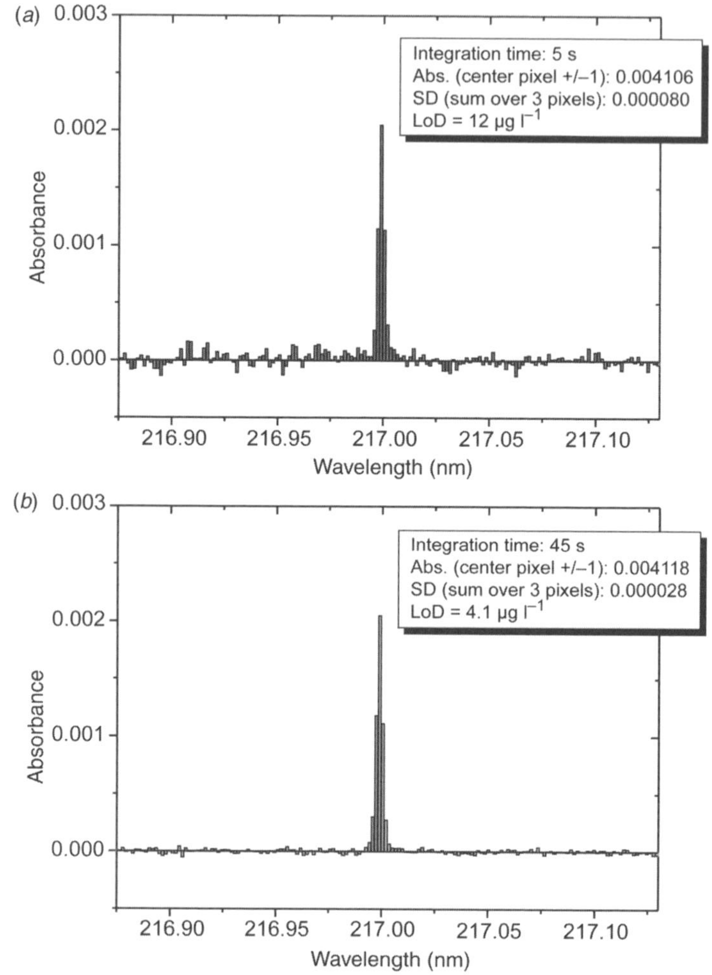

Figure 4.16. Influence of measurement time on the noise level and LoD for Pb at 217.001 nm; sample 0.07 mg l^{-1} Pb: (*a*) measurement time 5 s; (*b*) measurement time 45 s.

height can be optimized as well under computer control for each element. This means that each element can be determined under optimized and not under compromised conditions, as would be the case with simultaneous multi-element determination. The high intensity of the xenon short-arc lamp, which results in high SNR, further supports fast sequential multi-element FAAS measurements, as it makes possible very short measurement times, so that stabilization of the flame becomes the rate-determining factor in cases where a change of flame stoichiometry or of flame gases (air to nitrous oxide) is necessary. Nevertheless, the total time period required for changing the wavelength, adjusting flame stoichiometry and burner height, and for measuring the absorbance signal, should typically be only a few seconds. Obviously, the initial calibration of the equipment, and a possible re-calibration after a set of measurements, will require slightly more time than the following determination.

The typical routine determination of a number of elements in a set of similar sample solutions will therefore no longer be the determination of element A in all the samples, followed by a change of radiation source, wavelength, flame conditions, burner height and so on, and determination of element B in the same samples, and so on, as it is common practice in LS FAAS. In HR-CS FAAS it will rather consist of a calibration of the instrument for all the elements of interest, followed by a determination of all elements in sample 1, all elements in sample 2, all elements in sample 3, and so on. It might be worth mentioning that, at least for a limited number of elements, the total analysis time required for HR-CS FAAS will be even shorter than that for a simultaneous ICP AES measurement, because of the much shorter equilibration time required for a typical AAS burner after changing the sample solution, compared to the spray chambers used in ICP AES.

Another option that is available with HR-CS AAS to a much greater extent than with LS AAS is the use of secondary, less sensitive analytical lines, as already discussed in Section 4.3.4, because all lines have essentially the same high SNR; that is, there are no more "weak lines," although the emission intensity and the detector sensitivity are going down at wavelengths <220 nm. This means that two or three analytical lines of the same element with significantly different sensitivity can be chosen in the fast sequential mode in cases where the analyte concentration in the samples is unknown, and might vary over several orders of magnitude. This way the analyte can always be determined in the optimum working range of an analytical line without the need for any additional dilution of the measurement solution. This is obviously time saving, and avoids the risk of dilution errors.

Yet, another feature that becomes available with the fast sequential mode of operation in HR-CS AAS is the use of the "reference element technique," that is, the use of an "internal standard." This technique has rarely been described in LS AAS [3] for a number of obvious reasons. Firstly, the technique requires a dual- or multi-channel spectrometer, and there have been only very few spectrometers of this type available commercially over the past decades. Secondly, the reference element technique is ideally suited to correct for nonspecific interferences, such as transport interferences, but it is notoriously difficult to find an appropriate reference element for element-specific interferences. Thirdly, the most successful multi-channel LS AAS equipment, the Perkin-Elmer Model SIMAA 6000, which has been available for a number of years, was designed for ET AAS only, a technique that does not typically exhibit nonspecific interferences. The number of publications using this technique is therefore very limited.

In HR-CS AAS, although the simultaneous measurement of two elements or more is only possible within the simultaneous recorded wavelength interval, the reference element technique can be used essentially without compromises for FAAS, where nonspecific interferences are quite common. The fast-sequential mode of operation can handle this type of interference without problems, as it does not change with time. The most widespread interference of this type is transport interferences caused by changes in the viscosity between sample and calibration solutions and from sample to sample due to variations in the matrix composition. In LS FAAS these interferences are usually eliminated by "matrix matching" if the composition of the sample matrix is known and constant for a large number of samples, or by the analyte addition technique in case the sample matrix is unknown or different from sample to sample. The former technique usually requires a high concentration of ultra pure chemicals in order to keep the risk of introducing high blank values as low as possible. The latter technique is very labor-intensive and time-consuming, strictly limited to the linear part of the calibration curve, and characterized by inferior precision, because the analyte content is determined by extrapolation, not interpolation. All these limitations are absent when the reference element technique is used to eliminate nonspecific interferences in FAAS, whereby the time-saving aspect is probably the most significant one. Another advantage of the reference element technique is that dilution errors can be recognized and eliminated when the reference element is added at an early stage of sample preparation before the final dilution. Moreover, by using the reference element technique, quantitative analysis could actually be done without accurate dilution and without using volumetric flasks, and so on, which will save more time and eliminate potential errors in sample preparation.

4.4 APPLICATION OF HR-CS AAS TO FOOD ANALYSIS

It is worth stressing that everything that has ever been published on the application of AAS to food analysis can be done at least as well with HR-CS AAS. Since the same flames and burners, and the same type of electrothermal atomizers, are used in both systems, and only the spectrometer part from the radiation source to the detector has been re-designed, it is much more appropriate to talk about the improvements brought about by this change, and about the simplifications and the additional features that have become available this way.

Talking about flame atomization, there is first of all the improvement in LoD (refer to Section 4.3.6), which will extend the application of flame methods to lower concentration ranges, and/or make possible higher dilution of sample solutions, hence reducing potential matrix effects. Whenever a greater number of elements have to be determined, the fast sequential mode of operation will clearly be of advantage (refer to Section 4.3.7), including the possibility to determine the same analyte routinely at two lines of different sensitivity in case a concentration range of several orders of magnitude has to be covered. Yet, another feature available in this mode is the use of the reference-element technique (internal standardization) instead of the analyte addition technique in order to compensate for nonspecific interferences, with a tremendous saving in time and labor. Last but not least, as any analytical line of interest may be chosen because of the continuum radiation source, new elements might come into reach (refer to Section

4.3.4), as Huang et al. [16] demonstrated for the determination of P using a PO molecular absorption line. The authors successfully determined P in the 1575 Pine Needles SRM (NIST, Gaithersburg, MD, USA), which indicated that the method might be applied for other plant materials as well.

In the case of ET AAS, it is the far superior simultaneous BC and the "visibility" of the spectral environment (refer to Section 4.3.3) that dramatically contributes to an overall simplification of the procedure and to an increase in accuracy of the analysis. All things that are visible are much easier to understand and control than things that remain in the dark, as is the case with LS AAS. This is particularly true for all kinds of spectral events that might result in interference, although this risk is significantly reduced in HR-CS AAS thanks to the high resolution of the spectrometer.

For quite a number of elements there are no spectral interferences to be expected after conventional method optimization, which includes a high enough pyrolysis temperature, typically $\geq$700°C, in order to eliminate the bulk of the organic matrix prior to atomization, if it has not been previously destroyed with acid digestion. Obviously, stabilized temperature platform furnace (STPF) conditions [22] should be applied in any case in order to minimize spectral and nonspectral interferences. These STPF conditions include atomization from a graphite platform in a transversely heated graphite tube, integration over peak area for signal evaluation, and usually also a chemical modifier to stabilize the analyte to a high enough pyrolysis temperature.

Under these conditions spectral interferences become predictable, as the wavelength ranges are well documented [10], in which the few diatomic molecules that have been observed in ET AAS exhibit their electron excitation spectra. An example of an interference-free spectral environment is shown in Figure 4.17 for the determination of Cr in whole egg powder at 357.869 nm. The second line within the spectral window at 358.119 nm is due to Fe, and obviously causes no interference in HR-CS AAS. It also causes no interference in LS AAS unless deuterium BC is used with a spectral band pass $\geq$0.5 nm.

Another possibility to avoid spectral interferences, at least for the most volatile elements, is to select an atomization temperature that is high enough to atomize the analyte in a short time period, but which is low enough to avoid excessive volatilization of the inorganic matrix. This is shown in Figure 4.18: Cd determination in dogfish muscle using an atomization temperature of 1600°C. The phosphates that are the source of the PO molecule become volatile only at temperatures $\geq$1700°C, so that the determination of Cd remains uninfluenced, although its resonance line is in the range of a strong PO absorption band.

The situation is clearly different in the case of Pb determination, in particular at the most sensitive analytical line at 217.001 nm, as shown in Figure 4.19*a* and *b*. Lead requires an atomization temperature of at least 1700°C, and the phosphates are already significantly volatilized under these conditions, resulting in a pronounced molecular absorption. Fortunately, in the case of lead, the atomic absorption pulse is well resolved from the molecular structures, both in wavelength and in time, as shown in Figure 4.19*c*. There are no molecular structures

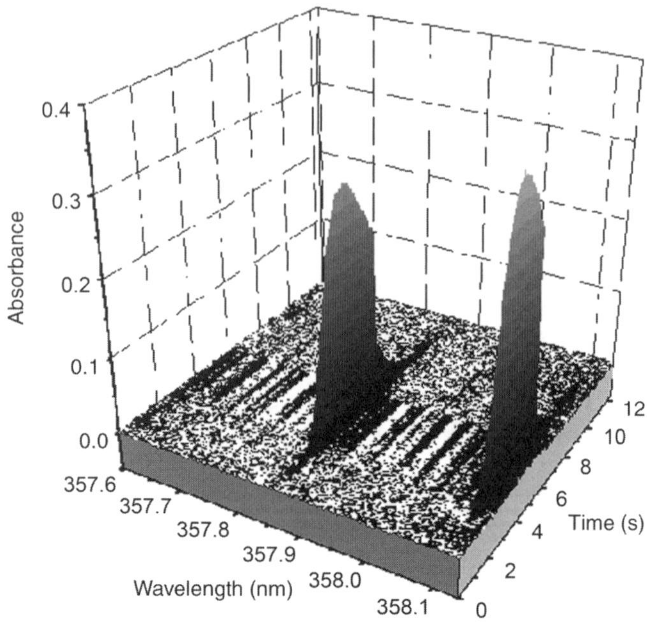

Figure 4.17. Three-dimensional graph for Cr in the NIST 8415 Whole Egg Powder SRM in the vicinity of the 357.87 nm line; pyrolysis temperature 700°C, atomization temperature 2500°C; direct solid sampling analysis.

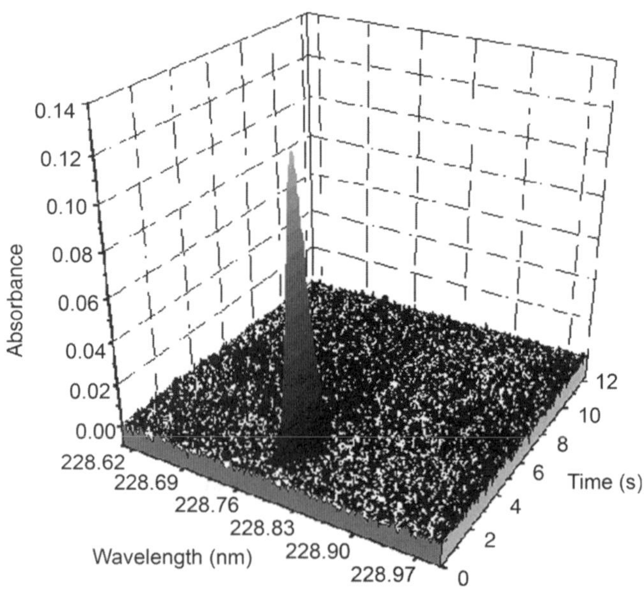

Figure 4.18. Three-dimensional graph for Cd in the DORM-1 Dogfish Muscle CRM in the vicinity of the 228.805 nm line; pyrolysis temperature 700°C, atomization temperature 1600°C; Ir permanent modifier; direct solid sampling analysis.

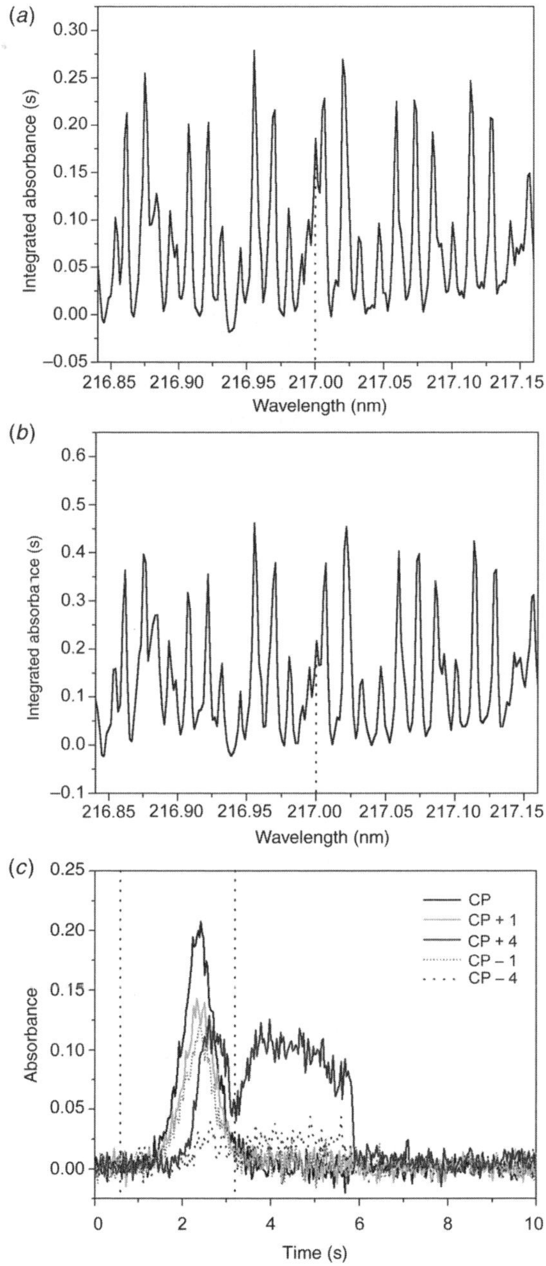

Figure 4.19. Determination of Pb in biological reference materials in the vicinity of the 217.001 nm line; pyrolysis temperature 700°C, atomization temperature 1700°C; Ru permanent modifier; direct solid sampling analysis; (*a*) wavelength integrated over time for the NIST 8414 Bovine Muscle SRM; (*b*) wavelength integrated over time for NIST 8415 Whole Egg Powder; (*c*) absorbance over time for Whole Egg Powder, recorded at selected pixels in the vicinity of the analytical line.

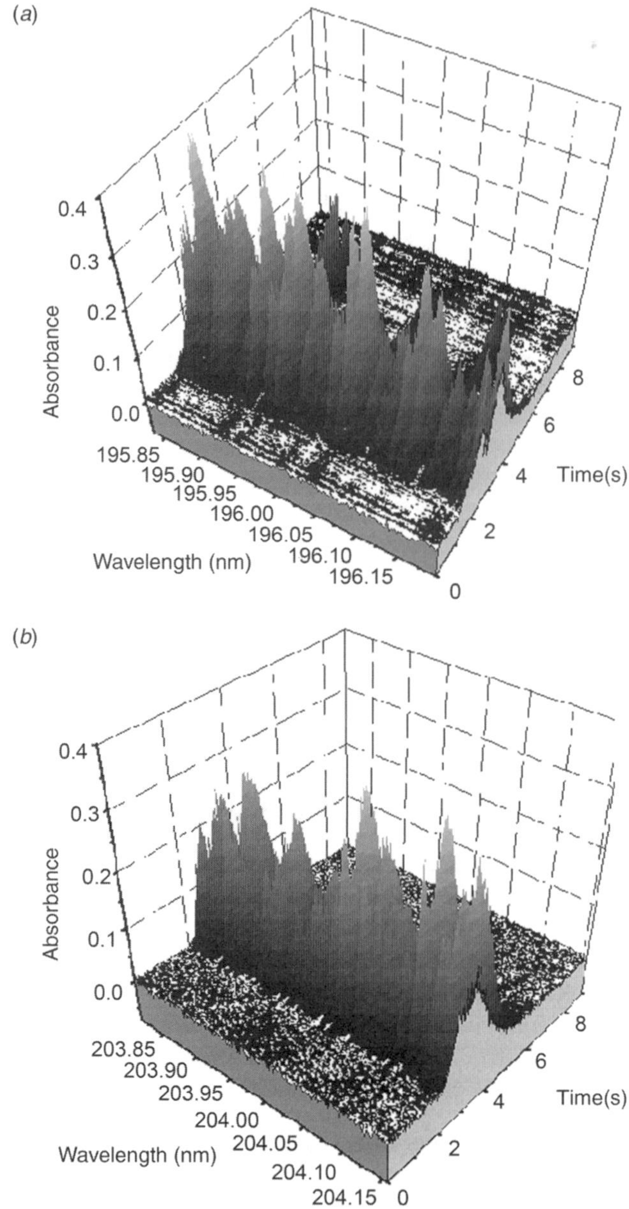

Figure 4.20. Three-dimensional graphs for Se in the BCR 186 Pig Kidney CRM; pyrolysis temperature 800°C, atomization temperature 2000°C; Ir permanent modifier; direct solid sampling analysis; (*a*) in the vicinity of the 196.026 nm line; (*b*) in the vicinity of the 203.985 nm line.

visible at the center pixel and the two neighboring pixels, so that no additional measures are required when the program is optimized. While this is possible in the case of HR-CS AAS, where the environment of the analytical line is visible, it is much more difficult when using LS AAS. In addition, none of the BC systems available for LS AAS is able to correct for this kind of fine-structured background satisfactorily. Even Zeeman-effect BC is unable to handle this situation, as the PO molecule also exhibits Zeeman splitting; that is, the background with magnetic field is different from that without. HR-CS AAS still offers least-squares BC, in case a complete separation of the atomic and the molecular absorption in time and/or wavelength is not possible. By comparing Figure 4.19*a* (bovine muscle) with Figure 4.19*b* (whole egg powder), it becomes obvious that the molecular structures are identical, which is actually true for all animal-based food samples, and this is due to the PO molecule only. This makes treatment of the potential interference straightforward, as the spectrum produced by any phosphate might be used for least-squares BC.

A sort of worst-case situation is shown in Figure 4.20*a* and *b* for the determination of Se in pig kidney at 196.026 and 203.985 nm, respectively. As Se requires an atomization temperature of at least 2000°C, the analyte absorption is completely shifted in time into the molecular absorption structures. The only absorption lines that are separated in time, and appearing after the molecular absorption, are the well-known Fe lines at 195.950, 196.059, and 196.145 nm that cause severe spectral interference in LS AAS when deuterium BC is used, even in the absence of any phosphate [23]. While this Fe interference can be eliminated using Zeeman-effect BC, the interference due to PO molecular absorption cannot, for the reasons discussed in the previous paragraph. The molecular absorption in the vicinity of 196.026 nm (Fig. 4.20*a*) is mostly due to PO, although NO also starts to absorb below 196.0 nm, however, without overlapping with the Se line [24]. This means that the situation at this most sensitive wavelength for selenium can be solved by subtracting the PO spectrum only, using least-squares BC, and leaving some of the structures due to NO at lower wavelengths, as they have no influence on the measurement. In the vicinity of the 203.985 nm line, however, we observe a complete overlap of the molecular absorption spectra of PO and NO, making it necessary that both spectra are considered for least-squares BC. However, this line is seldom used, as the sensitivity is lower by about a factor of 4 compared to the main resonance line.

4.5 CONCLUSIONS

In essence, HR-CS AAS offers only advantages over LS AAS, and eliminates most of its notorious limitations. Firstly, as the instrument uses a CCD array with several hundred pixels, it is equipped with hundreds of independent detectors (200 of which are typically used for analytical purposes), which are all illuminated and read out simultaneously. Since only a few pixels are necessary to measure atomic absorption, the rest of them are available for other purposes. The first duty of these detectors is to correct for all the spectral events that occur simultaneously on all pixels, such as lamp flicker noise and changes in the transmission of the atomizer that are independent of wavelength. This correction creates a simultaneous double-beam system that results in shot-noise limited readout; that is, the noise is inversely proportional to the square root of the integration time. Obviously, the system also corrects for any background absorption that is continuous within the spectral window controlled by the detector. In addition, measurement of atomic and background absorption is strictly simultaneous so that even a very high and rapidly changing background can be corrected without problems.

Secondly, because of the much higher radiation intensity of the CS compared to conventional LS, the LoD obtained with HR-CS AAS are typically a factor of 5 better than those of LS AAS. On the other hand, it is possible to reduce the sensitivity by using pairs of pixels further away from the center pixel to measure the absorbance on the line wings only. This way several working curves of significantly different sensitivity can be established simultaneously, extending the linear working range to 5–6 orders of magnitude.

Thirdly, while in LS AAS only the absorbance (with flame atomizers) or the absorbance over time (with electrothermal atomizers) is measured, the CCD array detector adds the wavelength as a third dimension in HR-CS AAS. This three-dimensional imaging obviously enhances the information about the spectral environment of the analytical line dramatically, which is of particular advantage in the case of dynamic signals, as they are produced in electrothermal atomizers. Potential spectral interferences due to other atomic lines, and particularly due to electron excitation spectra of diatomic molecules, which exhibit a pronounced fine structure, can now be easily recognized, and often avoided by program optimization. In cases where this is not possible, there is the option to store complete molecular spectra in the software and subtract them from the measured spectrum using a least-squares algorithm, so that only the net atomic absorption signal is left.

Fourthly, as a continuous radiation source is used in HR-CS AAS, any line of the spectrum is available, and even molecular absorption "lines" can be used for quantitative determination, as shown for the determination of P at PO bands and sulfur at CS bands using FAAS.

Fifthly, although simultaneous multi-element measurement is not yet possible because of the limited readout rate of currently available CCD detectors, especially with the transient signals of electrothermal atomizers, fast sequential measurements are possible using flame atomizers. The double monochromator with echelle grating supports a rapid wavelength change, and the sequential approach has the clear advantage that flame stoichiometry and observation height can be adjusted automatically for each element, avoiding "compromised conditions," typical of the simultaneous approach.

It is obvious that HR-CS AAS will redefine not only AAS, but the entire field of atomic spectroscopy, as it combines the simplicity, ease of operation, relatively low cost and freedom from interferences of classical AAS with a number of features unavailable until now, or available only with much more sophisticated instrumentation. This is also only the beginning of a new era, so that much more might be expected with future instrumentation in this field.

ACKNOWLEDGMENT

The authors are grateful to Conselho Nacional de Desenvolvimento Cientifico e Tecnològico (CNPq) for financial support: B. W. and D. L. G. B have research scholarship from CNPq. The financial support by the Senatverwaltung für Wissenschaft, Forschung und Kultur des Landes Berlin and the Bundesministerium für Bildung und Forschung is gratefully acknowledged.

REFERENCES

1. A. Walsh, The application of atomic absorption spectra to chemical analysis, *Spectrochim. Acta*, **7** (1955), 108.
2. A. Walsh, Atomic absorption spectroscopy – stagnant or pregnant? *Anal. Chem.*, **46** (1974), 698A.
3. B. Welz, M. Sperling, *Atomic Absorption Spectrometry*, 3rd edn, Wiley-VCH, Weinheim, New York, 1999.
4. S. R. Koirtyohann, E. E. Pickett, Background correction in long path atomic absorption spectrometry, *Anal. Chem.*, **37** (1965), 601.
5. S. B. Smith, G. M. Hieftje, A new background-correction method for atomic-absorption spectrometry, *Appl. Spectrosc.*, **37** (1983), 419.
6. M. T. C. de Loos-Vollebregt, L. de Galan, Theory of Zeeman atomic absorption spectrometry, *Spectrochim. Acta*, **33B** (1978), 495.
7. H. Gleisner, K. Eichardt, B. Welz, Optimization of analytical performance of a graphite furnace atomic absorption spectrometer with Zeeman-effect background correction using a variable magnetic field strength, *Spectrochim. Acta*, **58B** (2003), 1663.
8. G. M. Hieftje, Atomic absorption spectrometry – has it gone or where is it going? *J. Anal. Atom. Spectrom.*, **4** (1989), 117.
9. J. M. Harnly, The future of atomic absorption spectrometry: a continuum source with a charge coupled array detector, *J. Anal. Atom. Spectrom.*, **14** (1999), 137.
10. B. Welz, H. Becker-Ross, S. Florek, U. Heitmann, *High-resolution Continuum Source AAS – The Better Way to do Atomic Absorption Spectrometry*, Wiley-VCH, Weinheim, New York, 2005.
11. U. Heitmann, M. Schütz, H. Becker-Ross, S. Florek, Measurements on the Zeeman splitting of analytical lines by means of a continuum source graphite furnace atomic absorption spectrometer with a linear charge-coupled device array, *Spectrochim. Acta*, **51B** (1996), 1095.
12. H. Becker-Ross, S. Florek, U. Heitmann, R. Weisse, Influence of the spectral bandwidth of the spectrometer on the sensitivity using continuum source atomic absorption spectrometry, *Fresenius J. Anal. Chem.*, **355** (1996), 300.
13. H. Becker-Ross, S. Florek, U. Heitmann, M. Schütz, An echelle spectrometer with external prism order separation designed for continuum source AAS, Poster 575, *FACSS*, 1997, Book of Abstracts, p. 208.

14. A. S. Ribeiro, M. A. Vieira, A. F. Silva, D. L. G. Borges, B. Welz, U. Heitmann, A. J. Curtius, Determination of cobalt in biological samples by line-source and high-resolution continuum source graphite furnace atomic absorption spectrometry using solid sampling or alkaline treatment, *Spectrochim. Acta*, **60B** (2005), 693.
15. A. F. Silva, D. L. G. Borges, B. Welz, M. G. R. Vale, M. M. Silva, A. Klassen, U. Heitmann, Method development for the determination of thallium in coal using solid sampling graphite furnace atomic absorption spectrometry with continuum source, high-resolution monochromator and CCD array detector, *Spectrochim. Acta*, **59B** (2004), 841.
16. M. D. Huang, H. Becker-Ross, S. Florek, U. Heitmann, M. Okruss, Determination of phosphorus by molecular absorption of phosphorus monoxide using a high-resolution continuum source absorption spectrometer and an air-acetylene flame, *J. Anal. At. Spectrom.*, **21** (2006), 338.
17. M. D. Huang, H. Becker-Ross, S. Florek, U. Heitmann, M. Okruss, Determination of sulfur by molecular absorption of carbon monosulfide using a high-resolution continuum source absorption spectrometer and an air-acetylene flame, *Spectrochim. Acta*, **61B** (2006), 181.
18. U. Heitmann, S. Florek, M. D. Huang, Sensitivity, linearity and working range of a modern continuum-source atomic absorption spectrometer, *Seventh Rio Symposium on Atomic Spectrometry, Book of Abstracts, Florianópolis, SC, Brazil*, 7–12 April 2002, p. 200.
19. U. Heitmann, H. Becker-Ross, Atomabsorptions-Spektrometrie mit einem Kontinuumstrahler (CS-AAS), *GIT Labor-Fachzeitschrift*, **7** (2001), 728.
20. J. M. Harnly, Instrumentation for simultaneous multielement atomic absorption spectrometry with graphite furnace atomization, *Fresenius J. Anal. Chem.*, **355** (1996), 501.
21. H. Becker-Ross, M. Okruss, S. Florek, U. Heitmann, M.D. Huang, Echelle-spectrograph as a tool for studies of structured background in flame atomic absorption spectrometry, *Spectrochim. Acta*, **57B** (2002), 1493.
22. W. Slavin, D. C. Manning, G. R. Carnrick, The stabilized temperature platform furnace, *Atom. Spectrosc.*, **2** (1981), 137.
23. F. J. Fernandez, M. M. Beaty, Elimination of spectral interferences in the determination of Se in nickel-base alloys using Zeeman background correction, *Spectrochim. Acta*, **39B** (1984), 519.
24. H. Becker-Ross, S. Florek, U. Heitmann, Observation, identification and correction of structured molecular background by means of continuum source AAS – determination of selenium and arsenic in human urine, *J. Anal. Atom. Spectrom.*, **15** (2000), 137.

5

DETERMINING THE GEOGRAPHICAL ORIGIN OF FOODS: CONSIDERATIONS WHEN DESIGNING EXPERIMENTAL PROTOCOLS AND CHOOSING ANALYTICAL APPROACHES

JOHN LEWIS AND SIMON HIRD

Food Safety and Quality Group, Central Science Laboratory, Sand Hutton, York YO41 1LZ, UK, E-mail: j.lewis@csl.gov.uk; s.hird@csl.gov.uk

The Determination of Chemical Elements in Food: Applications for Atomic and Mass Spectrometry.
Edited by Sergio Caroli

5.1 INTRODUCTION

The aim of this chapter is to examine the factors that should be considered when designing analytical, experimental protocols for food-based, geographical origin studies. It is not intended as a detailed review of specific methods used in this area of analytical chemistry, although an overview of the techniques that are routinely used is given. The chapter also seeks to provide an insight into the historical reasons for the concern over the authenticity of food, focusing specifically on the challenge of verifying claims made regarding a product's regional provenance. Examples of such studies are presented, showing the power of the techniques that have been developed and the robustness of the data that has been generated in order to meet that challenge.

Historically, the adulteration of food has been an issue of importance for both economic and food safety reasons. Whenever a commodity achieves an economic advantage or can demand a premium price, an opportunity simultaneously arises for an unscrupulous manufacturer to commit a fraudulent act. This can either be by offering a falsely labeled, inferior product for sale, or by increasing the volume of the commodity by "extending" it with another material. Examples of food "extension," reported from Victorian Britain, are:

- Powdered glass in sugar
- White lead [a mixture of Pb_2CO_3 and $Pb(OH)_3$] in flour
- Red lead (Pb_3O_4) in tea
- Mahogany shavings in tea.

These sorts of adulteration issues led local and national governments to introduce legislation, designed to protect both the consumer and the scrupulous manufacturer [1]. The earliest example in English law was the "Assize of Bread," brought in by King Henry III (1266), which defined not only the type of cereal to be used, but also the quality and quantity of the finished product.

> Assisa Panis (Assize of Bread): When a quarter of wheat is sold for 12d., then Wastel Bread of a farthing shall weigh £6 and 16s. But Bread Cocket of a farthing of the same grain and bultel, shall weigh more than Wastel by 2s. And Cocket Bread made of grain of lower price, shall weigh more than Wastel by 5s. Bread made into a Simnel shall weigh 2s. less than Wastel. Bread made of the whole wheat shall weigh a Cocket and a half, so that a Cocket shall weigh more than a Wastel by 5s. Bread of Treet shall weigh 2 Wastels. And bread of common wheat shall weigh two great Cockets.

Later amendments were introduced which defined lists of ingredients that were not permitted for use in bread making. However, it is not clear how these regulations could be enforced without some means of identifying when fraudulent activities had occurred. The literature would suggest that it was not really until the nineteenth century, with the development of reasonable quality microscopes, that a more analytical approach, other than just squeezing or sniffing the bread, could be adopted.

Following on from these early days, individual governments have introduced legislation to address a whole host of food-related issues; some based on ensuring that criminal fraud does not occur, and others focused on the safety aspects of food production. As of 1 January 2005, Europe combined many of these pieces of legislation into a single set of regulations, designed to address all aspects of food safety, including food labeling (Regulation EC/178/2002).

However, for such legislation to be successful it also has to be enforceable; that is, with regard to food labeling claims, the authorities have to be able to prove that a fraudulent or inaccurate claim has been made. This is where the food scientists play their part, by providing legislators with techniques and procedures capable of characterizing foods in such a way as to make their adulteration or authenticity clearly measurable.

In this chapter, an attempt is made to show the range of analytical challenges faced by the food scientist when dealing with food authenticity. The questions that can be asked do not just cover potential adulteration with other substances, but include identification of geographical origin, the means of production and varietal provenance of a food. This chapter focuses on the approaches that have been applied to the specific issue of proving the geographical provenance of foods. However, as no single analytical technique has yet been established which can address all of the facets relating to this challenge, the text, by necessity, describes not only those techniques involved in the measurement of inorganic analytes, that is, trace element profiles and "heavy" isotope ratios, but includes also other analytical strategies which have been proven to be of use in such investigations.

5.2 CONSIDERATIONS WHEN DESIGNING A FOOD AUTHENTICITY STUDY

5.2.1 General Considerations

Generally, when investigating food provenance issues, it is essential to consider all of the factors that may have influenced the composition of the commodity. For example, when attempting to establish the geographical origin of a wine, a simplistic approach might be just to measure the levels of trace elements (as an indicator of the underlying geology on which the vine was grown) or a number of light isotope ratios (to establish the latitude/altitude at which the vine was located). However, as was highlighted by Kwan et al. [2], there is a strong possibility that these measurements would also reflect the geographical origin of other ingredients added to the grape juice, as part of the wine making process, and so would not be truly representative of the wine itself.

Therefore, when using the trace element profile of a food to assign its geographical origin, one also has to consider what other factors (apart from the underlying geology) may have influenced the profile.

5.2.2 Varietal Differences

Coni et al. [3] reported differences in the multielement composition of milk originating from several species of animal (some species being quite closely related, i.e., sheep and goats). Similar effects were reported [4, 5] with regard to the uptake of minerals by different varieties of rice. However, Kwan et al. [2] showed their awareness of this potential pitfall when designing an experiment to investigate differences in trace element profiles between wines made in two different regions in America and one region in France. They ensured that the grapes were all of the *Vitis vinifera* cv. Pinot Noir variety, so as to eliminate this variable from their considerations.

5.2.3 Age (the Stage in the Food Organism's Life Cycle at Which it is Harvested or Slaughtered)

Wu et al. [6] reported that rice plants, in the later stages of their growth cycle, accumulate Cd in their leafy parts to a much greater extent than is observed in the earlier stages. It may be argued that this trend would not affect the trace element profile measured in the ripened rice grains, as they would only be harvested at a single time-point in the plants growth cycle. However, if the foliage was taken for fodder, or if this trend was common throughout the plant kingdom, animals consuming such plant material would be subjected to differing intakes of minerals at different time points throughout the year. The result of this would be that the trace element markers might possibly exhibit a seasonality in their applicability to identifying, or characterizing a particular foodstuff, thereby complicating the issue of proving its provenance.

5.2.4 Seasonality

Following on from the above, behavioural differences, for example, reproductive cycle or annual variations in animal husbandry, such as the wintering of livestock, are also likely to be significant factors influencing the chemical characteristics of the end product. An example of this might be where a milk-producing herd of cattle is brought inside for the winter months, and is fed pre-harvested forage that has been bought by the farmer rather than having been produced on his own farm. There is a strong possibility that differences in the mineral profile of the summer and winter forage will be reflected in the composition of the resulting milk. This hypothesis was borne out by researchers in Italy [7], who investigated the factors that affected the trace element content of milk and associated products. They reported higher levels of Fe in summer milk, but lower levels of many other elements, including Ba, Cd, and Co, compared to milk obtained during the winter.

Similarly, with regard to isotopic ratios of the "light" elements, O and C, a German team [8] reported a summer increase in the $^{18}O:^{16}O$ ratio and a decrease in the $^{13}C:^{12}C$ ratio measured in samples of butter from the same geographical region. The researchers concluded that the decrease observed in the C ratio was probably due to the cattle feed being supplemented, during the winter months, with products made from maize, which, as will be discussed later, has evolved a different metabolic pathway for the incorporation of C into its biomass than is used by normal grass. The authors did not comment on the cause of the alteration observed in the isotopic profile of O, but it is highly likely to be due to seasonal variations in precipitation affecting the composition of the animals' drinking water.

5.2.5 Animal Transportation

The UK's most recent outbreak of foot and mouth disease highlighted the extent to which livestock are routinely transported over quite significant distances. The literature does not yet contain data from investigations into the effect of the commercial movement of animals on parameters normally used to identify the geographical origin of a food. However, it is likely that this would represent a significant challenge in terms of developing methods capable of tracking such agricultural practices via purely analytical means.

5.2.6 Sample Database Construction

In addition to the above factors, construction of a statistically valid, representative database of authenticated samples appears to be a major omission from many of the studies published in the scientific literature. Often, researchers focus only on the methodologies used and refer to their sampling strategy as: "A number of samples were purchased from a local supermarket," or "Information relating to the country of origin was obtained from the labels" or even, "one sample of rice from Vietnam." Considering the comments made in the previous sections, if fraudulent labeling is the issue that the researchers are investigating, how could they be sure that the commercially bought samples were truly authentic? Similarly, it is highly unlikely that, for instance, one sample of rice can be said to be truly representative of all the different varieties of rice grown in Vietnam, or be representative of all the geological environments present in a country of 330,000 km^2.

The cost implications associated with the analysis of sufficient samples as to ensure that a database is truly representative of all possible variability, for example, seasonal, year-on-year, varietal, geological, and climatic, are obvious (and substantial). However, if a database is to be used in a court of law, either in support of legislation or in industrial litigation, such considerations must be made or, at least, experimental data should be available to show that these factors do not affect the particular matrix/analyte combinations being presented. In reality, given the limited funds that are usually available for this sort of investigation, a compromise position is often what will be arrived at. This, at worst, would involve the construction of a sufficiently large database which, although not able to answer every question, would contain sufficient information so as to persuade a fraudulent producer, or retailer, that they are likely to be caught if they continue their illegal practises.

5.3 CONCLUSIONS TO THE EXPERIMENTAL DESIGN SECTION

From the previous section, it is hopefully clear that a significant number of factors need to be considered when designing an investigation into the geographical origin of food and that its success will only be assured if the interrelationship between the geosphere and the biosphere is well understood and has been taken into account when constructing the sampling plan/database. Similarly, it is ultimately important that the database is constructed under the close supervision of a statistician, experienced in experimental design, and that full consideration is given to the statistical validity of the resulting data. However, this particular facet of experimental design cannot be included in this chapter because there are so many variables to be considered, that the whole of the text would need to be devoted purely to that one topic.

In the following sections, an overview of some of the techniques commonly used in food authentication studies is given, along with an explanation as to why a particular technique (or combination of techniques) might be chosen.

5.4 TECHNICAL APPROACHES USED IN GENERAL FOOD AUTHENTICATION INVESTIGATIONS

5.4.1 General Aspects

In the preceding 20–30 years, a wide array of analytical techniques have been utilized in the development of methodologies for the characterization of foods, as part of their authentication. Parameters that have been used include bulk properties, such as pH, total N content, and trace element levels. The isotopic ratios of certain elements have also been used, including those of light elements such as C, H, N, O, and S (often also referred to as bioelements due to their involvement in biological processes), and heavy elements such as Pb and Sr (often referred to as geoelements due to their involvement in geochemical processes). In the remainder of the text, the terms light and heavy will be used when referring to the isotopic ratios of the aforementioned elements.

All of the above have proven useful in identifying either fraudulent labeling or the incidence of adulteration. However, as the analytical complexity used by the enforcement agencies increases, so does the awareness and technical sophistication exhibited by the fraudulent perpetrator. Day et al. [9] reflected on the historical struggle between food chemists and those manufacturers of fruit juices who were intent on obtaining financial gain through the selling of adulterated product. They described how early authentication methods were dependent on parameters such as the K and P content, and the formol value (indicative of the total levels of amino acids). These were circumvented by the unscrupulous suppliers adding K and NH_4 phosphates and/or glycine. Following this response, the food chemists developed analytical protocols based on the relative proportions of components such as organic acids, sugars, and free amino acids. However, Day reported that, by the late 1980s, the fraudulent manufacturers had again discovered ways of concealing their activities, which resulted in the researchers having to develop even more technically demanding methods, based on the determination of the isotopic abundances of C in the juice sugars.

This example shows how manufacturers do not just give up their fraudulent ways. Hence, it is important that the ongoing development of methods capable of identifying new fraudulent strategies is maintained The following section will outline some of the main techniques used in food authentication studies, with specific focus on geographical provenance investigations. Through the text, it will hopefully be seen how, over time, analytical techniques have developed, providing the food chemist with increasingly specific information with which to identify illegal practices.

5.4.2 Elemental Analysis (Trace Element Profiling)

The primary source of nutrients required for the healthy and successful propagation of a plant (and the animals that feed on it) is the soil and water on which it has grown. This, in turn, is a reflection of the underlying geology of the area within the organisms immediate vicinity. Therefore, it is not surprising that the mineral composition of a food is widely used as a means of identifying its geographical origin, as for example, in the case of wine [2,10,11], rice [12,13], onions [14], dairy products [8], wheat [15], and orange juice [16].

However, it is not just the presence or absence of an element that is useful (as most elements will be present at some concentration), but it is the relative variation in the trace element profile that is the parameter that provides the major discriminatory power. McHard et al. [16] were possibly some of the first researchers to apply a normalization procedure to multielement data in order to maximize the differences between two sets of samples. Their approach, which is now accepted as being a standard tool for use in chemometric investigations, was to identify an element whose concentration was constant, irrespective of the geographical origin of the samples, and then to normalize all other elemental data against it. In McHard's study on fruit juice, they used Zn. The authors of this chapter used Ca in an egg authenticity study, where eggshells were used as the sample matrix (unpublished data); and Mg was used in a study of Welsh onions [14].

The simple calculation, used by McHard to ratio elemental data from the fruit juice, is as follows:

$$(I_{AY}/I_{AR})/(I_{BY}/I_{BR})$$

where A is the element under consideration, B is the reference element, I is the intensity reading for the respective element, Y is the sample being analyzed, and R is the reference sample (the analyst must identify a suitable sample)

This approach has three main advantages:

- It acts as an internal standardization procedure, negating the need for calculations to be performed to compensate for other sample attributes, such as solids content, specific gravity, matrix interferences, and so on.

- It simplifies data, allowing elements present at low concentrations to be easily visualized alongside elements present at many orders of magnitude higher, for example, K vs Li in orange juice (1575 vs 0.0015 μ g^{-1}).
- It reduces the possibility of introducing systematic statistical bias due to computing packages automatically assigning a greater significance (weighting) to numbers of greater magnitude.

In the early days of elemental analysis, the analyst would have been restricted in the numbers of elements that could be determined. The limitations would have been in terms of both the sensitivity of the instrumentation and the amount of time available for collecting and processing the data, as many of the techniques were only capable of measuring a single element at a time. However, the advent of automated systems and multielement techniques provided the analyst with a realistic capability for measuring a significant range of elements, to appropriate levels of detection, in the sorts of numbers of samples that would be required to construct a statistically meaningful database of "authentic" commodities.

During the late 1970s and early 1980s, atomic emission spectrometry (AES) instrumentation was complemented by the development of atomic absorption spectrometry (AAS) as a means of elemental detection. This advance, coupled with the development of alternative flame-gas mixtures, for example, nitrous oxide–acetylene, provided the analyst with reduced matrix interferences, improved limits of detection (LoDs), and provided access to several elements that were not easily measurable using more conventional flame-gas mixtures, for example, air–acetylene, because of their refractory nature. Some of these elements (including Al, Si, and the lanthanides) were later proven to be important in the authentication of foods, indicating the potential of multielement profiling in this particular area of food chemistry. Other technical advances, such as the electrothermal atomization (ET) AAS, led to further improvements in detection power and sample throughput. However, the most significant instrumental development was that of the inductively coupled plasma (ICP), fitted with either AES or mass spectrometry (MS) detection systems, as it enabled high-throughput, simultaneous, multielement determinations to be made. Of the two ICP-based approaches, ICP-MS has been the most widely used in food authentication studies due to its extensive multielement capabilities, wide dynamic range, sensitivity, and, in more recent work, its ability to measure isotopic ratios.

One of the earliest applications of ICP-based techniques in food authenticity was the determination of the trace element composition of orange juice [16]. This work successfully showed that, using ICP-AES, it was possible to distinguish between juice produced in Florida and Brazil, utilising only five of the 32 elements measured.

5.4.3 Heavy Isotope Ratios

Since the early days of ICP, there have been many investigations, showing the usefulness of trace element profiling. However, in recent times, the use of mass spectrometry-based ICP systems has shown an additional advantage when applied to the determination of the isotopic ratios of elements such as lead and strontium.

5.4.3.1 How Heavy Isotopic Fractionation Occurs As with trace element profiling, certain isotopic ratios ($^{11}B:^{9}B$, $^{87}Sr:^{86}Sr$, and $^{204}Pb:^{206}Pb$) reflect characteristics of the underlying geological and hydrological environment, and have therefore been used in the geographical tracing of foods.

Boron undergoes fractionation of its isotopes due to differences in the exchange reactions between $B(OH)_3$ and $B(OH)_4$, and effects which occur during borate mineral crystallization. Lead has four stable, naturally occurring isotopes: ^{204}Pb (1.4%), ^{206}Pb (24.1%), ^{207}Pb (22.1%), and ^{208}Pb (52.4%). ^{206}Pb, ^{207}Pb, and ^{208}Pb are all radiogenic, and are the end products of complex decay chains that begin at ^{238}U, ^{235}U, and ^{232}Th, respectively.

The heavier isotope of Sr, ^{87}Sr, is increased due to the radiogenic (β) decay of rocks containing ^{87}Rb. Hence, as the ^{86}Sr isotope is not affected by this process, the ratio of ^{87}Sr to ^{86}Sr increases. IUPAC state that the natural abundance variation in geological systems is between 0.69469 and 0.73231.

Unfortunately, the measurement of these heavy isotopes is analytically challenging due to:

- The precision that is required to discriminate between geographical sources.
- The presence of chemical interferences, for example, ^{87}Rb on ^{87}Sr, and ^{204}Hg on ^{204}Pb.

For these reasons, the literature does not contain as many references relating to their use as indicators of geographical origin as have been written about the use of trace element profiling. However, it could be said that, as a single measurement, an isotope ratio can have much greater discriminating power than that of any single elemental concentration measurement. Therefore, Table 5.1 has been constructed so as to present food authenticity-based references where heavy isotopic ratios have been used. This table includes the ratio measured, the food matrix investigated, the instrumentation used, and the date of publication.

5.4.3.2 Heavy Isotope Ratio Measurement In the past, this area of analytical chemistry had been seen as being only available to the users of thermal ionization (TI) MS instrumentation, due to its inherent precision and accuracy. The basis of the technique is the formation of atomic or molecular ions on the thermally elevated surface of a metal filament. The high precision of the technique is partly attributable to it having a very stable ion source (whereas, for example, the plasma of an ICP is an extremely dynamic source, introducing significant temporal variation in ion beam intensity). Through the use of isotope dilution (ID) analysis, TIMS can also achieve a high degree of accuracy (although this analytical approach is also applicable to any MS-based technique). The accuracy also results from the sample introduction constraint of TIMS, which requires an extremely

TABLE 5.1. An Overview of the Use of "Heavy" Isotopic Ratio Analysis in Studies to Determine the Geographical Origin of Foods

Isotope Ratio Measured	Food Commodity	Instrument Used	Author	Date of Publication
$^{87}Sr:^{86}Sr$	Cheese	MC-ICP-MS	Fortunato (21)	2004
$^{206}Pb:^{207}Pb$ and $^{208}Pb:^{206}Pb$	Wine	MC-ICP-MS	Larcher (18)	2003
$^{87}Sr:^{86}Sr$	Rice	MC-ICP-MS	Kawasaki (22)	2002
$^{87}Sr:^{86}Sr$	Wine	MC-ICP-MS	Barbaste (23)	2002
$^{11}B:^{10}B$ and $^{87}Sr:^{86}Sr$	Rice	MC-ICP-MS	Oda (24)	2001
$^{87}Sr:^{86}Sr$	Wine	ICP-Q-MS	Almeida (25)	2001
$^{87}Sr:^{86}Sr$	Salmon	TIMS	Kennedy (26)	2000
$^{87}Sr:^{86}Sr$	Butter	TIMS	Rossmann (8)	2000
$^{87}Sr:^{86}Sr$	Salmon	TIMS	Ingram (27)	1999
$^{87}Sr:^{86}Sr$	Wine	TIMS	Horn (19)	1993

pure form of the analyte, free from chemical interferents. The downside of this is that extensive, and often time-consuming, cleanup procedures are required prior to analysis. This makes the technique somewhat impractical for routine, high-throughput use. Therefore, as can be seen from Table 5.1, there has been a noticeable trend towards the use of ICP-MS for the analysis of heavy isotope ratios.

With ICP-MS, there are significant advantages to be had, in terms of sample throughput, when compared to TIMS. However, because of the dynamic nature of the plasma ion source, instrumental precision will never be as good as that routinely achieved from TIMS, making the choice of instrumentation an area of considerable contention. Horn et al. [19] commented that "in this pilot study [investigating the geographical origin of wine] the precision of ICP-MS would not have been sufficient to recognize differences in the $^{87}Sr:^{86}Sr$ ratio [of the different wines]." However, in a later paper describing a similar investigation [20], the researchers felt it appropriate to revise their comments as follows: "Precision and accuracy of ICP-MS analyses for isotope abundance ratios . . . very recently, have reached the standards set by TIMS."

There are certain considerations that need to be made when using ICP-MS:

- Is the required precision beyond the capabilities of a scanning mass spectrometer? If so, then an ICP-MS fitted with a multicollector array will be the minimum requirement.
- If the samples have not been subjected to strenuous chemical cleanup, or there are plasma-based spectral interferences to be considered, then high-resolution MS detection and/or interference correction equations may be required, rather than using quadrupole MS instrumentation. On the other hand, these two approaches are likely to introduce their own influence on the precision/accuracy of the measured isotopic ratio.

However, if the proposed analysis is not affected by either of the above, then ICP-MS (multicollector or high resolution) may well provide a useable alternative to TIMS. To provide support for this conclusion, Table 5.1 includes the date-of-publication so as to highlight the trend toward the use of ICP-MS systems in this important area of analytical chemistry.

5.4.4 Light Isotope Ratios (H, C, N, S, O)

The term light isotopes relates to the isotopes of elements, at the lower end of the mass range, which, because of their pivotal role in biological systems, are sometimes better referred to as "bioelements." The reason that they are of interest to the food chemist is that their isotopic abundance ratios are not constant in nature due to their tendency to undergo fractionation during their involvement in various physical, chemical, and biological processes.

5.4.4.1 How Light Isotopic Fractionation Comes About Fractionation of the isotopes of H and O is mainly brought about due to differences in the evaporation rates exhibited by the various isotopomers of water, for example, $H_2^{16}O$ evaporates more readily than the heavier isotopomer $H_2^{18}O$ [28]. This difference in evaporation rates is amplified by the effect of latitude, with the lighter form evaporating more rapidly at the equator than at the poles. When this meteoric water falls at higher latitudes and altitudes, and is taken up by plants, its characteristic isotopic profile is also transferred. Figure 5.1 shows the worldwide distribution of $^{18}O : {}^{16}O$ ratios, as measured in rainfall.

In addition to climatic processes, other interactions, such as transpiration through the leaf stomata, will also affect water overall isotopic ratio, enriching with regard to the heavier isotopomer. Therefore, it can be seen how the combination of climate, topography, and plant physiology can all contribute to the isotopic profile of both the H and the O available for inclusion into plant material.

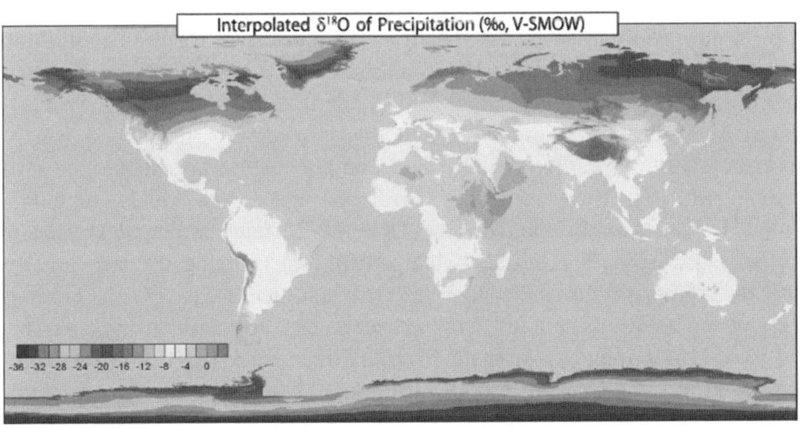

Figure 5.1. $^{18}O : {}^{16}O$ ratio map of the world.

With respect to C, the isotopic abundance is affected by climatic parameters such as temperature and aridity, which, in theory, can be used to identify the geo-climatic region in which the food was produced. However, a factor that has a greater effect than this is the natural alteration of the $^{13}C:^{12}C$ ratio that occurs during the involvement of the element in a plant's C cycle.

Over time, specific groups of plants have evolved in such a way that they now utilize completely different C fixation pathways. The C_3 (Calvin) pathway (the most widely used in the plant kingdom) uses carboxylates ribulose-1,5-diphosphate to produce phosphoglyceric acid (a three-carbon compound). The C_4 (Hatch–Slack) pathway brings about the incorporation of C via carboxylate phosphoenolpyruvate to produce oxalacetate (a 4-C compound), and is mainly used by tropical grasses such as maize and sugar cane. A third photosynthetic class of plants use the crassulacean acid metabolism (CAM) pathway. The isotopic fractionation of C comes about because the enzyme-catalyzed reaction used by the C_3 plants discriminates against ^{13}C, with the result that less $^{13}CO_2$ is incorporated into the biomass compared to that of the nondiscriminatory, C_4 plants. This natural variation in the element's isotopic abundance complicates the elucidation of a food's geographical provenance by diluting down, or superseding, any alteration in the ratio due to the climatic conditions under which the product was grown. However, situations may arise where this measurement could be used to identify the country of origin for animal-based commodities, if there were cultural or traditional differences in the husbandry techniques used to produce the foodstuff, that is, where animals were habitually fed maize products rather than normal herbage. It is unlikely that this is a common-place occurrence, hence it is given as an example of what might happen, rather than a documented case history.

5.4.4.2 Light Isotope Ratio Measurement Measurement of the "bulk" $^{18}O:^{16}O$ and D:H ratio of a food have become routine determinands in geographical origin investigations, and are measured using Isotope Ratio Mass Spectrometry (IRMS). In this method, the element has to be measured in the form of a gas, therefore requiring either combustion (C, N, and S), equilibration (O), or pyrolysis (H) prior to introduction into the mass spectrometer. The value for the isotopic ratio is presented not as an absolute value, but as a difference (δ value, expressed in parts per million) from an internationally recognized standard.

However, as mentioned previously, methods have been developed which can measure these isotopic ratios in individual compounds and at specific sites within those compounds. The measurement of C isotope ratios in individual compounds can be achieved using gas chromatography (GC) IRMS, but compound-specific information for other isotopes is usually obtained by first isolating the individual components, for example, remove the sugar fraction, and measure its $^{18}O:^{16}O$ ratio [29].

Site-specific measurement of isotope ratios can be made using site-specific natural isotope fractionation–nuclear magnetic resonance (SNIF-NMR). This was first proposed in 1981 by Martin and Martin [30], who then applied it to the adulteration of food (wine) in 1988 [31]. This technique allows D:H ratios for specific H atoms within a molecule to be measured and can provide additional

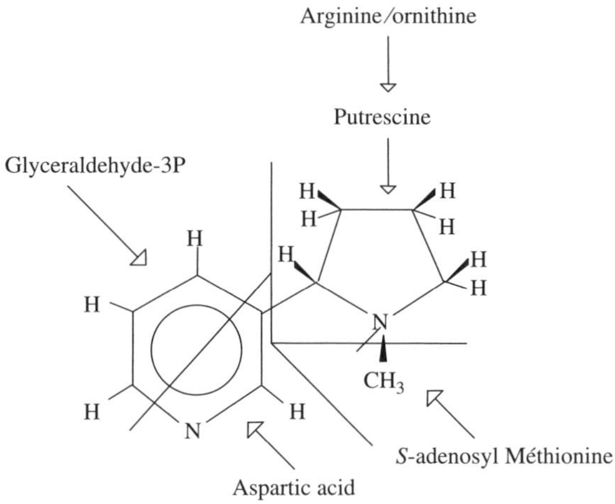

Figure 5.2. A schematic representation of the nicotine molecule showing the biosynthetic origins of the various H-containing fragments.

information on the biosynthesis history. Figure 5.2 shows how this technique was used to characterize the site-specific D : H ratios in nicotine and the biochemical pathways by which the individual H-containing fragments had come to be incorporated into the molecule. The authors, Jamin et al. [32] went on to use the information, combined with ratios of C and N, to successfully classify tobacco from four continents (achieving greater than 97 percent confidence).

However, as can be seen from Figure 5.1, the distribution of a specific O ratio, worldwide, is not confined to a particular geographical location, but can be observed in many regions across several continents. In addition to this, the ratios of O and H exhibit seasonal and longer-term variability [8]. Hence, although they are very powerful discriminands under well-defined conditions, it might be suggested that they are best used when combined with factors that are fixed, for example, trace element profiles or heavy isotope ratios.

Therefore in the concluding section, a brief overview is given of studies where light isotopic ratios and/or heavy isotopic ratios and/or trace element profiling have been combined in order to optimize the technologies available for investigating the geographical provenance of foods.

5.5 MULTITECHNIQUE APPROACH TO GEOGRAPHICAL ORIGIN INVESTIGATION

So far, in this chapter, individual techniques have been taken into account that have been used in the elucidation of the geographical origin of foods. It has been seen how trace element profiling and heavy isotope ratio measurements reflect the underlying geohydrological environment on which the food was produced. However, it has been shown how quite a wide range of chemical and biological processes, related to the foodstuff itself, can affect the robustness of this approach. It has also been shown that the use of light isotope ratios is subject to certain limitations resulting from climatic and physiological variables.

Therefore, in an attempt to maximize the discrimination of geographical provenance, several researchers have combined data from a number of techniques. This is obviously not an option that is open to all institutes, due to the significant financial investment required in terms of analytical equipment. However, as it possibly represents the ultimate approach to determining the geographical origin of a food, those studies where combinations of trace element profiling and/or light and/or heavy isotope ratios have been employed, are highlighted in Table 5.2. The reasons for choosing particular instruments or analytes is reviewed, and used to consider the benefits of this multitechnique approach.

A recent example of the application of multiple techniques is reported by Brescia et al. [33], who utilized four technical approaches: high performance liquid chomatography (HPLC) for Ca, K, Li, Mag, and Na, ICP-AES for Al, Ba, Cr, Cu, Fe, Mn, Ni, Zn, Pb, and Se, NMR (D : H ratio), and IRMS (C and N ratios). When discussing the data, they systematically appraised the factors that appeared to indicate a strong discriminatory capability. For example, the measurement of Na could be used to discriminate between the milk produced in the two geographical regions. On the other hand, once made into mozzarella, the addition of salt (as part of the cheese manufacturing process) masked this effect. Similarly, for the C ratio, the researchers identified that the significant difference observed between cheeses from the two areas was due to differences in the feeding strategies routinely used in the two agricultural regions. Surprisingly, they concluded that this could be deemed to be a legitimate marker of geographical origin. However, one would say that their conclusion could only be justified if the animal's feeding strategy was a fixed component in the Protected Designation of Origin (PDO) statement defining the cheese and its method of production.

TABLE 5.2. Publications Describing the Use of Combinations of Techniques in Geographical Provenance Investigations

Food Commodity	Author	Date of Publication	Analytes
Mozzarella	Brescia (33)	2005	10 elements plus H, N, and C ratios
Cheese	Pillonel (17)	2003	9 elements, C, N, O, H, and Sr ratios plus 3 radioisotopes
Wheat	Branch (15)	2003	2 elements plus Pb, Sr, C, and N ratios
Rice	Kelly (13)	2002	50 elements plus O and C ratios
Wine	Day (27)	1995	4 elements plus the H ratio

Pillonel et al. [17] used the most extensive range of techniques so far noted in the literature, including IRMS, TIMS, ET-AAS, ICP-MS, α-spectrometry, and γ-spectrometry. Their aim was to identify the technique, or techniques, most applicable to the characterization of markers-of-origin for Emmental cheese. They discussed the seasonality of certain of the light isotope ratios, and stated that this facet would be given greater attention in a follow-on study. The isotopic ratio of Sr showed the expected relationship with the underlying geology of the different regions within Germany, but was not useable for regional classification of cheeses from within Finland due to the country's homogenous geological profile. Molybdenum was a strong discriminand for cheeses of Swiss origin, and will be the subject of future investigation to identify whether it was a true indicator of geographical origin or just of the method of production. The workers concluded that, with a database of only 20 samples (covering France, Switzerland, Germany, and Austria), a much more thorough investigation, using a statistically representative sample set, would be their priority.

Branch et al. [15] used a combination of two specifically selected elements (Cd and Se), heavy isotopic ratios (Pb and Sr), and light isotopic ratios (C and N) in an investigation into the geographical origin of wheat. The elements were chosen for the following reasons: Cd is reported to be a good indicator of anthropogenic activity and natural, geological, occurrence, making it a useful determinand in geographical origin investigations; the Se content of plant material grown on the North American continent is known to be elevated. With regard to Sr ratios, the values measured in foods from the united States are reported to be significantly lower than those in other parts of the world (0.703–0.706 vs 0.712–0.730, United States vs Korea) [22]. When reflecting on the choice of C and N ratios, the researchers commented that fractionation of the elements was influenced by environmental and varietal differences. They concluded that, as predicted, the total Se content was an extremely good indicator of the wheat having come from the North American continent (193–636 vs 25–45 ng g^{-1}; United States and Canada vs Europe). However, they also commented on the hazards associated with the over use of statistics on limited data sets and suggested that "over-fitting" of some of the data may have occurred. This shows a good awareness of the potential pitfalls of using too few samples, but, as they stated, the study was only a preliminary investigation, and that the issue of ensuring a representative database would be addressed in any future study.

Kelly et al. [13], when describing their work on the geographical origin of rice, also addressed the issue of over-fitting of data, and proposed a "stepwise" approach, selecting the minimum number of variables (from the 52 measured) in order to maximize the separation whilst ensuring that statistical over-fitting was minimized.

In surprising contrast to the findings of Branch et al. [15] (who measured Se levels in wheat samples from different geographical sources), Kelly reported only marginal differences in the levels of total Se (with significant overlaps of values) from rice samples originating from the United States and two continents (Asia vs Europe vs United States: 5–293 vs 0–178 vs 64–398 ng g^{-1}). For B, the data were much clearer (Asia vs Europe vs United States: 354–874 vs 97–1513 vs 2887–9649 ng g^{-1}). This shows the advantage of using the full capacity of a multielement technique, such as ICP-MS, as it allows a suite of elements (up to 75) to be accessed, rather than trying to predict which elements might be good discriminands (as Branch et al. did) and thereby missing out on possibly useful data.

Kelly and coworkers concluded that a combination of only seven (of the 50) elements measured, and the isotopic ratios of C and O, gave 100 percent discrimination between rice samples from any of the three continental land masses, and that measurement of just B and Mg was sufficient to allow a reasonable indication to be gained as to whether the rice had originated in the United States. This statement underlines the comment made in an earlier section of the chapter, regarding the importance of establishing the question-to-be-asked, or the answer-that-is-required, when designing an authenticity study.

In the final (and earliest) example to be presented in this section, Day et al. [29] utilized SNIF-NMR (individual D : H ratios on the methyl and ethyl sites of the ethanol molecule), IRMS (O and D : H ratio in the wine water; C ratio in the wine distillate); ET-AAS (10 elements) and ICP-AES (Ba only). The concluding section of their paper sums up most of what has been written in this chapter and is strongly recommended for further reading, particularly as it represents possibly one of the earliest publications to describe the use of a multitechnique approach.

They also commented on the concept of terroir, which serves to describe or characterize the environment in which the vineyard resides. They reported that, using a combination of certain trace elements and isotope ratios it was possible for them to discern wine from large production areas around France and also said that it was possible to identify smaller regions, in which typical Appellations could be discerned. The workers commented that the work had been performed in only one season, but were confident that the techniques, if not that specific database, would be applicable to other vintages. They also noted that different grape varieties (even when grown in close proximity), would exhibit subtle differences in their light isotopic ratios, due to variations in the climatic conditions prevalent during the different growing and ripening times of the individual varieties. Finally, they commented on the expectation that trace element profiles of different wines would be dependent on the behavior of vine root-stock on which the grapes were grown.

5.6 CONCLUSIONS

It is hoped that the reader has obtained an insight into the range of factors that should be considered when designing an investigation into the geographical origin of a foodstuff. With careful consideration of the parameters that may affect the results, it should be possible to construct a strategy that is not too expensive and is fit-for-purpose.

Finally, it should be noted that there are several other techniques that can provide data and are of great use when performing food authenticity studies, such as liquid chromatography MS, GC-MS, and DNA fingerprinting. However, they were not covered in this text because they are mainly used to measure components in foods that reflect the quality of a commodity rather than the geographic/climatic conditions under which the food was produced.

REFERENCES

1. P. B. Hutt, P. B. Hutt (II), A history of government regulation of adulteration and misbranding of food, *Food, Drug Cosm. Law J.*, **39** (1984), 2–73.
2. W-O. Kwan, B. R. Kowalski, R. K. Skogerboe, Pattern recognition analysis of elemental data. Wines of *vitis vinifera* cv. Pinot noir from France and the United States, *J. Agric. Food Chem*, **27** (1979), 1321–1326.
3. E. Coni, A. Bocca, D. Ianni, S. Caroli, Preliminary evaluation of the factors influencing the trace element content of milk and dairy products, *Food Chem.*, **52** (1995), 123–130.
4. T. Arao, N. Ae, Genotypic variations in cadmium levels of rice grains, *Soil Sci. Plant Nutr.*, **94** (2003), 473–479.
5. J. Liu, Q. Zhu, Z. Zhang, J. Xu, J. Yang, M. H. Wong, Variations in cadmium accumulation among rice cultivars and types and the selection of cultivars for reducing cadmium in the diet, *J. Sci. Food Agric.*, **85** (2005), 147–153.
6. Q. T. Wu, L. Chen, G. S. Wang, Differences in cadmium uptake and accumulation among rice cultivars and its mechanism, *Acta Ecol. Sinica*, **19** (1999), 104–107.
7. E. Coni, A. Bocca, P. Coppolelli, S. Caroli, C. Cavallucci, M. Trabalza Marinucci, Minor and trace element content in sheep and goat milk and dairy products, *Food Chem.*, **57** (1995), 253–260.
8. A. Rossmann, G. Haberhauer, S. Holzl, P. Horn, F. Pichelmayer, S. Voerkelius, The potential of multielement stable isotope analysis for regional origin assignment of butter, *Eur. Food Res. Technol.*, **211** (2000), 32–40.
9. M. P. Day, P. Correia, D. A. Hammond, ^{13}C-IRIS: an improved method to detect the addition of low levels of C_4-derived sugars to juices, *J. Assoc. Off. Anal. Chem.*, **84** (2001), 957–963.
10. M. J. Baxter, H. M. Crews, M. J. Dennis, I. Goodall, D. Anderson, The determination of the authenticity of wine from its trace element composition, *Food Chem.*, **60** (1997), 443–450.
11. J-P. Perez-Trujillo, M. Barbaste, B. Medina, Contents of trace and ultratrace elements in wines from the Canary Islands (Spain) as determined by ICP-MS, *J. Wine Res.*, **13** (2002), 243–256.
12. A. Yasui, K. Shindoh, Determination of the geographic origin of brown-rice with trace-element composition, *Bunseki Kagaku*, **49** (2000), 405–410.
13. S. Kelly, M. Baxter, S. Chapman, C. Rhodes, J. Dennis, P. Brereton, The application of isotopic and elemental analysis to determine the geographical origin of premium long grain rice, *Eur. Food Res. Technol.*, **214** (2002), 72–78.
14. K. Aryiama, H. Horita, A. Yasui, Application of tinorganic element ratios to chemometrics for determination of the geographical origin of Welsh onions, *J. Agric. Food Chem.*, **52,** (2004), 5803–5809.

15. S. Branch, S. Burke, P. Evans, B. Fairman, C. S. J. Wolff Briche, A preliminary study in determining the geographical origin of wheat using isotope ratio inductively coupled plasma mass spectrometry with ^{13}C, ^{15}N mass spectrometry, *J. Anal. Atom. Spectrom.*, **18** (2003), 17–22.

16. J. A. McHard, J. F. Foulk, J. D. Winefordner, A comparison of the trace element content of Florida and Brazil orange juice, *J. Agric. Food Chem.*, **27** (1979), 1326–1328.

17. L. Pillonel, R. Badertscher, P. Froidevaux, G. Haberhauer, S. Holzl, P. Horn, A. Jakob, E. Pfammatter, U. Piantini, A. Rossmann, R. Tabacchi, J. O. Bosset, Stable isotope ratios, major, trace and radioactive elements in Emmental cheeses of different origins, *Lebensm.-Wiss. U.-Technol.*, **36** (2003), 615–623.

18. R. Larcher, G. Nicolini, P. Pangrazzi, Isotope ratios of lead in Italian wines by inductively coupled plasma mass Spectrometry, *J. Agric. Food Chem.*, **51** (2003), 5956–5961.

19. P. Horn, P. Schaaf, B. Holbach, L. Hölzl, H. Eschnauer, $^{87}Sr/^{86}Sr$ from rock and soil into vine and wine, *Z. Lebensm. Unters. Forsch.*, **196** (1993), 407–409.

20. P. Horn, L. Hölzl, W. Todt, D. Matthies, Isotope abundance ratios of Sr in wine provenance determination, in a tree-root activity study, and of Pb in a pollution study on tree-rings, *Isotopes Environ. Health Studies*, **34** (1998), 31–42.

21. G. Fortunato, K. Mumic, S. Wunderli, L. Pillonel, J. O. Bosset, G. Gremaud, Application of strontium isotope ratios measured by MC-ICP-MS for food authentication, *J. Anal. Atom. Spectrom.*, **19** (2004), 227–234.

22. A. Kawasaki, H. Oda, T. Hirata. Determination of rice provenance using strontium isotope ratio and levels of selected trace elements of rice, *Trans. World Congr. Soil Sci.*, (2002), 2019/1–2019/9.

23. M. Barbaste, K. Robinson, S. Guilfoyle, B. Medina, R. Lobinski, Precise determination of the strontium isotope ratios in wine by inductively coupled plasma sector field multicollector mass spectrometry (ICP-SF-MC-MS), *J. Anal. Atom. Spectrom.*, **17** (2002), 135–137.

24. H. Oda, A. Kawasaki, T. Hirata., Determining of the geographic origin of brown-rice with isotope ratios of $^{11}B/^{10}B$ and $^{87}SR/^{86}Sr$, *Anal. Sci.*, **17**(suppl), (2001), 1627–1630.

25. C. M. Almeida, M. T. S. D. Vasconcelos, ICP-MS determination of strontium isotope ratios in wine in order to be used as a fingerprint of its regional origin, *J. Anal. Atom Spectrom.*, **16** (2001), 607–611.

26. B. P. Kennedy, J. D. Blum, C. L. Folt, K. H. Nislow, Using natural strontium isotopic signatures as fish markers: methodology and application, *Can. J. Fish. Aquat. Sci.*, **57** (2000), 2280–2292.

27. B. L. Ingram, P. K. Weber, Salmon origin in California's Sacramento–San Joaquin river system as determined by otolith strontium isotopic composition *Geology*, **27** (1999), 851–854.

28. G. J. Martin, M. L. Martin, Climatic significance of isotope ratios, *Phytochem. Rev.*, **2** (2003), 179–190.

29. M. P. Day, B. L. Zhang, G. J. Martin, Determination of the geographical origin of wine using joint analysis of elemental and isotopic composition. II—Differentiation of the principal production zones in France for the 1990 vintage, *J. Sci. Food Agric.*, **67** (1995), 113–123.

30. G. J. Martin, M. L. Martin, Deuterium labelling at the natural abundance level as studied by high field quantitative ^{2}H NMR, *Tetrahedron Lett.*, **22,** (1981), 3525–3528.

31. G. J. Martin, C. Guillou, M. L. Martin, M.-T. Canabis, Y. Tep, J. Aeny, Natural factors of isotope fractionation and the characterisation of wines, *J. Agric. Food Chem.*, **36,** (1988), 316–322.

32. E. Jamin, N. Naulet, G. J. Martin, Multi-element and multi-site isotopic analysis of nicotine from tobacco leaves, *Plant, Cell, Environ*, **20** (1997), 589–599.

33. M. A. Brescia, M. Monfrea, A. Buccolieri, C. Carrino, Characterisation of the geographical origin of buffalo milk and mozzarella cheese by means of analytical and spectroscopic determinations, *Food Chem.*, **89** (2005), 139–147.

6

METHOD VALIDATION FOR FOOD ANALYSIS: CONCEPTS AND USE OF STATISTICAL TECHNIQUES

JORIS VAN LOCO

Scientific Institute of Public Health, J. Wytsmanstraat 14, B-1050 Brussels, Belgium, E-mail: joris.vanloco@iph.fgov.be

The Determination of Chemical Elements in Food: Applications for Atomic and Mass Spectrometry.
Edited by Sergio Caroli

6.1 INTRODUCTION

The ISO 17025 requires that every method is validated to demonstrate fitness for purpose [1]. Analytical methods can be validated during a collaborative trial or in a single laboratory validation experiment. Single laboratory method validation has become common practice for every analytical chemist. Several guidance documents have been published concerning the validation of analytical methods [2, 3]. Methods used for the official control of foodstuffs are validated according to the European decision EC 2002/657 [4].

During a single laboratory validation study, the following performance characteristics of an analytical method are assessed: trueness, recovery, precision (repeatability and within-laboratory reproducibility), selectivity, specificity, calibration range, detection capability, stability, and applicability/ruggedness. This chapter gives guidance to the food chemist on the concepts of method validation. Emphasis is put on the evaluation of the data obtained during method validation. It is illustrated how method performance characteristics are obtained and assessed to demonstrate fitness for purpose. In particular, the statistical methods used to determine and assess trueness, precision, detection capability, and calibration range are explained in detail. These statistical techniques are described in a simplified way for nonstatisticians.

6.2 CALIBRATION

6.2.1 Background Information

Before assessing the other validation parameters (trueness, recovery, precision, selectivity, specificity, detection capability, stability, and applicability/ruggedness), the appropriateness of the calibration model should be evaluated. The correctness of the analytical determination of elements in food and food products depends indeed on the choice and the evaluation of the calibration model. The calibration model gives the mathematical relationship between the signal of the measuring system and the concentration in the sample. Several authors have published guidelines concerning calibration in analytical chemistry [5–7].

The most commonly used regression model in analytical chemistry is the simple linear model. This model is given by the following equation:

$$Y_i = Y_i = \alpha + \beta X_i + \varepsilon_i + \beta X_i + \varepsilon_i \tag{6.1}$$

with Y_i the value of the response variable (i.e., the response of the analytical equipment); α, β the true regression parameters; X_i the value of the independent variable (i.e., the concentration of the standard solution); and ε_i a random error term.

This model is called "simple" since there is only one independent variable (X), and "linear" because it is linear in the parameters. This means that no parameters appear as exponents or are multiplied or divided by another parameter (i.e., X^p). The term "linear" can cause some confusion in nonstatistical literature since second or higher order polynomial models are also called linear models (but in the terminology of "linear in the parameters" and not as a straight line). To avoid any confusion, here the term "straight-line model" is used for the above-described simple linear model. For an in-depth discussion of linear models, the reader is referred to an appropriate statistical manual [8].

The parameters α and β are the true regression parameters. These parameters can be estimated by using, for example, the least-squares method. The best fit of the regression line is obtained by minimizing the residuals (e_i):

$$e_i = Y_i - \hat{Y}_i \tag{6.2}$$

with $\hat{Y}_i = a + bX_i$.

The regression parameters intercept (a) and slope (b) are determined via the following equations:

$$a = \overline{Y} - b\overline{X} \tag{6.3}$$

$$b = \frac{\sum (X_i - \overline{X})(Y_i - \overline{Y})}{\sum (X_i - \overline{X})^2} \tag{6.4}$$

with $\overline{Y}$ the mean of all Y_i, and $\overline{X}$ the mean of all X_i.

The residual standard error of a straight-line model is given by: $S_y = \sqrt{[e_i^2/(n-2)]}$.

An example of a calibration curve for Cd with a Zeeman electrothermal atomization atomic absorption spectrometry (ET-AAS) is presented in Figure 6.1. A simple linear regression model is fitted through the data points. The response of the ET-AAS is placed on the ordinate and the concentration of the injected standard solutions on the abscissa. The concentration of the unknown samples can be calculated back as $X_i = (Y_i - a)/b$.

Although calibration and regression are closely related, they do not coincide with each other. With regression, the response Y_i is estimated from the independent variable X_i by means of the mathematical model: $Y_i = f(X_i) + \varepsilon_i$. On the other hand, in analytical chemistry the independent variable (that is, the concentration) is predicted from the response variable (i.e., the instrument response). One should be aware that, from a statistical viewpoint, this is not correct.

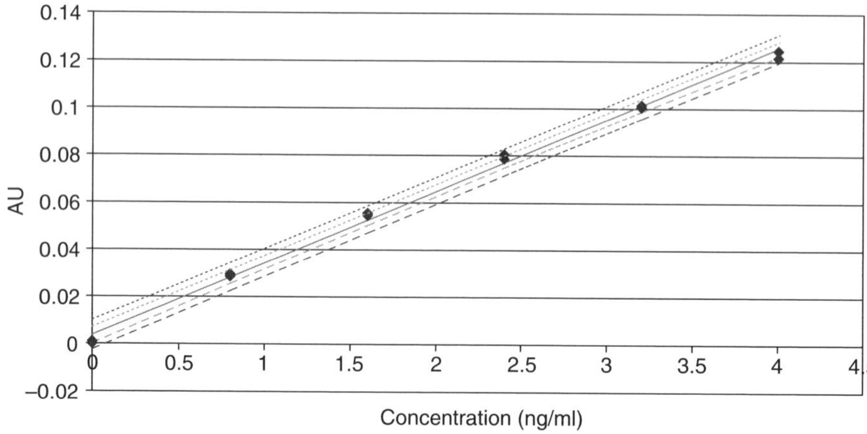

Figure 6.1. The calibration curve of Cd is obtained by fitting a simple linear regression line through the data points. The response of the Zeeman ET-AAS is plotted against the concentration of the standard solutions. The dotted lines represent the confidence (inner lines) and the prediction (outer lines) intervals.

However, the error introduced is in most cases negligible when compared to the analytical errors in residue analysis. To deal with this problem, inverse regression and regression with errors on both axes have been proposed in the literature [9].

A straight-line model is the most used, but also the most misused, model in analytical chemistry. The analytical chemist should check five basic assumptions during method validation before deciding whether to use a straight-line regression model for calibration purposes. These five assumptions are described in detail by MacTaggart and Farwell [6] and basically are linearity, error-free independent variable, random and homogeneous error, uncorrelated errors, and normal distribution of the error. The evaluation of these assumptions and the remedial actions are discussed hereafter.

6.2.2 Linearity

6.2.2.1 General Aspects It is useless to fit a straight line through the experimental data if one knows that the response of the detector is not linear. The response of most detectors is not linear when considering the entire range. However, for most detectors a part of the response range is (nearly) linear.

The linearity of (a part of) the range should be evaluated to check the appropriateness of the straight-line model. This can be achieved by a graphical evaluation of the residual plots or by using statistical tests. It is strongly recommended to use the residual plots in addition to the statistical tests. Mostly, the lack-of-fit test and Mandel's fitting test are used to evaluate the linearity of the regression line [8, 10]. The ISO 8466 describes in detail the statistical evaluation of the linear calibration function [11].

6.2.2.2 Graphical Evaluation The easiest way to evaluate the appropriateness of the chosen regression model is to check how much the regression line fits the

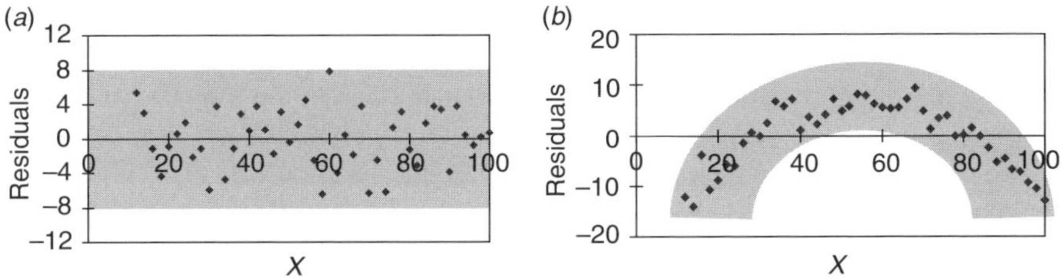

Figure 6.2. (*a*) A symmetric bar-shaped spread of the residuals confirms the appropriateness of the regression model. In this case there is a linear relationship between the response variable and the independent variable. (*b*) A U-shaped spread of the residuals indicates a curvilinear relationship between the response and the independent variable.

data points. When one carefully looks at the calibration graph in Figure 6.1, one can see that the data points in the middle of the range are located above the calibration line, while the data points on the outer limits of the range are beneath the calibration line. One should conclude for this calibration curve that a curvilinear regression model would provide a better fit.

A much better way to evaluate the fitness of the regression model is by evaluating the residual plots. The residuals (e_i) are plotted versus X or versus $\hat{Y}$. Both graphs provide equivalent information for straight-line models.

The residual plot in Figure 6.2*a* represents a linear relationship between the response and the independent variable. The residuals are randomly scattered around the centerline. The U-shaped residual plot of Figure 6.2*b* indicates that a curvilinear regression model should be fitted through the data points. Any departure from the symmetric bar shape of Figure 6.2*a* may indicate that the chosen regression model is inappropriate. The residual plots are also used to detect violations of other basic assumptions. This is discussed further in the text.

6.2.2.3 Statistical Evaluation

Lack-of-Fit Test The best-known statistical test to evaluate the appropriateness of the chosen regression model is the lack-of-fit test [6]. A prerequisite for this test is the availability of replicate measurements. An analysis of variance (ANOVA) approach is used in this test. The total sum of squares (SS_{T}) can be written as follows:

$$SS_{\mathrm{T}} = \sum_{i=1}^{k}\sum_{j=1}^{n_i}(Y_{ij} - \overline{Y})^2$$

$$= \underbrace{\underbrace{\sum_{i}^{k}\sum_{j}^{n_i}(Y_{ij} - \overline{Y}_i)^2}_{SS_{\mathrm{PE}}} + \underbrace{\sum_{i}^{k} n_i\left(\overline{Y}_i - \hat{Y}_i\right)^2}_{SS_{\mathrm{LOF}}}}_{SS_{\mathrm{R}}} + \underbrace{\sum_{i}^{k} n_i\left(\hat{Y}_i - \overline{Y}\right)^2}_{SS_{\mathrm{REG}}} \tag{6.5}$$

with k, the number of concentration levels; n_i, the number of replicates per concentration level; SS_{PE}, the pure error sum of squares; SS_{LOF}, the lack-of-fit sum of squares; SS_R, the residual sum of squares; and SS_{REG}, the regression sum of squares.

The SS can be decomposed in regression *SS* and residual SS. The latter can be decomposed again in the lack-of-fit *SS* and the pure error. The mean square (MS) error is obtained by dividing the *SS* by the corresponding degrees of freedom (d.f.). The *SS*, the MS and the corresponding d.f. are presented in Table 6.1.

The idea of the lack-of-fit test is to compare the pure error of the regression line with the error due to the use of an inappropriate regression model. The MS_{LOF}, which is a measure for the spread of the mean response per concentration from the regression line, is divided by the MS_{PE}, which is a measure for the spread of the instrument response due to experimental variation. The obtained F-value ($F = MS_{LOF}/MS_{PE}$) is compared with the F-distribution with $k-2$ and $n-k$ d.f.

Mandel's Fitting Test The idea of Mandel's fitting test is to compare the standard error (S_y) of a straight-line regression model with the standard error of a second-order polynomial regression model [10]. If the standard error of the second order model is significantly lower than the standard error of the straight-line model, the linearity of the regression line should be rejected. This test is performed by fitting a straight line ($Y = a + bX$) and a second order ($Y = a' + b'X + c'X^2$) regression model through the data set. The standard errors of the two regression curves are calculated as:

$$S_{y1} = \sqrt{\frac{\sum (a + bX_i - i)^2}{n-2}} \tag{6.6}$$

$$S_{y2} = \sqrt{\frac{\sum (a' + b'X_i + c'X_i^2 - Y_i^2)^2}{n-3}} \tag{6.7}$$

The obtained F-value is compared with the F-distribution with 1 and $n-3$ degrees of freedom.

$$F = \frac{(n-2)S_{y_1}^2 - (n-3)S_{y_2}^2}{S_{y_2}^2} \tag{6.8}$$

TABLE 6.1. Regression ANOVA Table

Source of Variation	SS	d.f.	MS	F-test
Regression	SS_{REG}	1	MS_{REG}	MS_{REG}/MS_R
Residual	SS_R	$n-2$	MS_R	
Lack of fit	SS_{LOF}	$k-2$	MS_{LOF}	MS_{LOF}/MS_{PE}
Pure error	SS_{PE}	$n-k$	MS_{PE}	
Total	SS_T	$n-1$		

This test has the advantage that replicate measurements are not mandatory. Instead of using the second-order model, one can use any other curvilinear or nonlinear model for comparison.

Coefficient of Correlation (r) The coefficient of correlation (r) is given by the following equation:

$$r = \frac{\sum (X_i - \overline{X})(Y_i - \overline{Y})}{\sqrt{\sum (X_i - \overline{X})^2 \sum (Y_i - \overline{Y})^2}} \tag{6.9}$$

The coefficient of correlation is not a correct tool for evaluating the linearity of the regression line. This has been extensively described in the literature [6, 12, 13]. Data sets with a slight curvilinearity are easily misjudged as linear when looking only at the coefficient of correlation. The residual plot in Figure 6.2*b* represents a regression line with a high coefficient of correlation ($r = 0.991$), while it is obvious that a curvilinear model should be preferred.

The lack-of-fit test or Mandel's fitting test are more appropriate to evaluate the linearity of the calibration lines. In Figure 6.3*a* and *b*, the lack-of-fit test and Mandel's fitting test, are compared with the coefficient of correlation. The F-value of the lack-of-fit test and of Mandel's fitting test and the coefficient of correlation are calculated for several Pb (0–50 ng ml^{-1}) and Cd (0–4.0 ng ml^{-1}) calibration lines obtained with ET-AAS. The calibration lines are all curvilinear. Even calibration lines with a coefficient of correlation up to 0.9995 are judged to be nonlinear at a significance level of 95 percent. Using an arbitrary value for the coefficient of correlation (i.e., $r > 0.999$) to assess the linearity of calibration lines may lead to significant measurement errors. It was estimated that for the Cd calibration curves a supplementary bias of 4 percent is introduced for measurements in the mid-range of the calibration curve, when curvilinearity was ignored [13].

Other Statistical Tests for Linearity Several other tests are for linearity have been described in the literature: the *test for significance of the quadratic*

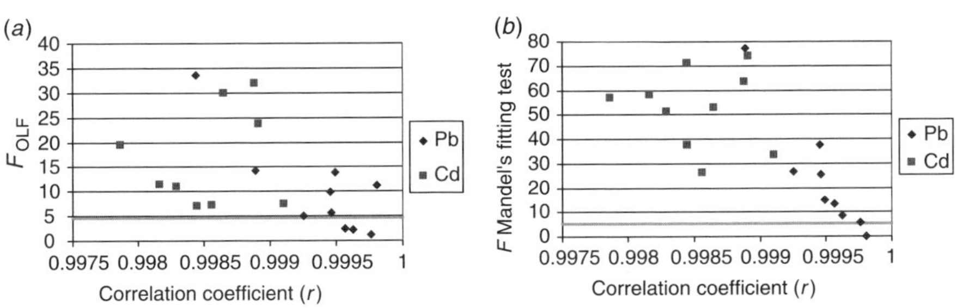

Figure 6.3. (*a*) Plot of r versus the lack-of-fit (LOF) test. The bold line represents the critical F-value at the 95 percent confidence level. (*b*) Plot of r versus Mandel's fitting test. The bold line represents the critical F-value at the 95 percent confidence level.

regression coefficient also compares a straight-line regression model with a second-order regression model. A second-order regression curve is fitted through the data points. Nonlinearity is detected when the second-order regression coefficient is significantly larger than zero [14]. Vankeerbergen et al. have proposed the *slope ranking method* and the *quality coefficient* to evaluate appropriateness of the straight line model [15, 16].

6.2.2.4 Remedial Actions When linearity has been rejected by one or more of the above-mentioned tests, the regression plot can be linearized or one can make use of polynomial or nonlinear regression models.

Transformations Transformation of the variables X and Y are used to correct for nonlinearity, but are also used to correct for heteroskedasticity (i.e., the variance is not constant over the concentration range, see below). Transformations of the X variable (i.e., $X^* = \log_{10}X$ or $X^* = \sqrt{X^*}$) are used to linearize concave regression curves, while transformations of $X^* = X^2$ or $X^* = e^X$ are used for convex curves.

Transformations of the Y variable (i.e., $Y^* = \sqrt{y}$, $Y^* = \log_{10}Y$ or $Y^* = 1/Y$) are merely used to correct for nonnormality or heteroskedasticity [8, 14].

Xiaoning et al. have proposed a methodology to linearize atomic spectroscopy regression curves by using a transformation of the form $Y^* = Y^p$. The value of p is determined by minimizing the quality coefficient [17]. Excellent results have been obtained by minimizing the sum of the squared residuals ($\mathrm{MIN}\lfloor\Sigma(Y_i - \hat{Y}_i)^2\rfloor$). This approach can even be used for standard addition curves.

Polynomial Regression Polynomial regression is a special case of multiple linear regression. The regression function is given by:

$$Y_i = \alpha_0 + \alpha_1 X_i + \alpha_2 X_i^2 + \cdots + \alpha_m X_i^m + \varepsilon_i \tag{6.10}$$

By using matrix notation, the expression of the equations can be kept simple. The following matrices are defined:

$$\mathbf{Y} = \begin{bmatrix} Y_1 \\ Y_2 \\ \cdot \\ \cdot \\ \cdot \\ Y_n \end{bmatrix}, \quad \mathbf{X} = \begin{bmatrix} 1 & X_1 & X_1^2 & \cdots & X_1^m \\ 1 & X_2 & X_2^2 & \cdots & X_2^m \\ \cdot & \cdot & \cdot & \cdots & \cdot \\ \cdot & \cdot & \cdot & \cdots & \cdot \\ \cdot & \cdot & \cdot & \cdots & \cdot \\ 1 & X_n & X_2^m & \cdots & X_n^m \end{bmatrix}, \quad \mathbf{A} = \begin{bmatrix} a_0 \\ a_1 \\ \cdot \\ \cdot \\ \cdot \\ a_m \end{bmatrix} \tag{6.11}$$

with $\mathbf{Y}$ the vector of the responses, $\mathbf{X}$ the matrix of the concentration data set, and A the vector of the regression parameters.

The matrix of the regression parameters is estimated by using the least squares approach $\mathbf{A} = (\mathbf{X}^T\mathbf{X})^{-1}(\mathbf{X}^T\mathbf{Y})$, where the index T means the transposed matrix and -1 is for the inversed matrix. A detailed description of multiple linear regression is given by Neter et al. [8].

A second-order regression curve ($Y_i = a_0 + a_1X_i + a_2X_i^2$) is in most cases sufficient to deal with curvilinear regression curves. The concentration of the analyte in the sample can be back-calculated from the response. Nowadays, most instrument software packages include the use of different types of regression models and the concentration of the samples is automatically back-calculated.

$$X_i = \frac{-a_1 \pm \sqrt{\Delta}}{2(a_0 - Y_i)} \tag{6.12}$$

with

$$\Delta = a_0^2 - 4(a_0 - Y_i)a_2$$

6.2.3 Error-Free Independent Variable

As mentioned earlier, error is only allowed for the response variable, when fitting a linear regression model through the data points. The independent variable (the concentration) should be free of error. This means that the preparation of the standard solutions should be free of any error, which is, of course, impossible. In most cases the error associated with the preparation of the standard solutions is significantly lower than the error associated with sample preparation, digestion, dilution, injection, signal amplification, etc., which all contribute to the error of the response. Hence, the use of linear regression models is in most cases justified in analytical chemistry. When the error of the independent variable is not negligible compared with the error of the response – which might be the case with standard addition curves – other regression models should be preferred. MacTaggart and Farwell have discussed in detail the techniques used for the regression analysis of data with error on both variables [9]. The calculations for these regression models are more complex and seldom standardly available in equipment software or spreadsheet programs.

6.2.4 Homoskedasticity

6.2.4.1 General Aspects The third basic assumption in regression analysis is that the variance should be constant over the calibration range. This is called homoskedasticity.* In analytical chemistry, the variance often increases with increasing concentration. When the variance is not constant over the calibration range, the regression parameters for the slope and the intercept are still unbiased and consistent, but the least squares solution is not efficient anymore. This means that the standard

*The terms heteroskedasticity and homoskedasticity were introduced in 1905 by Karl Pearson. Pearson wrote: "If . . . all arrays are equally scattered about their means, I shall speak of the system as a homoscedastic system, otherwise it is a heteroscedastic system." The words derive from the Greek skedastos (capable of being scattered). J. Huston McCulloch, On Heteroskedasticity, *Econometrica*, **53** (1985), 483.

errors and the confidence intervals are incorrect. This is of a particular importance when determining the detection capability of a method with a regression experiment [18]. It has been demonstrated by Van Loco et al. that heteroskedasticity has a larger impact on the detection capability than nonlinearity [19].

Heteroskedasticity can be evaluated by means of several statistical tests (modified Levene, White, or Breusch–Pagan test) or by looking at the residual plots [20–22].

6.2.4.2 Graphical Evaluation The residuals (e_i) are plotted versus X or versus $\hat{Y}$. If the error is homogeneously distributed the residuals are symmetrically and randomly plotted around the centerline (Fig. 6.2*a*). When heteroskedasticity is present, the variance is not constant. In Figure 6.4, the variance is larger at higher concentrations. A triangular spread of the residuals is visible.

Statistical Evaluation of Heteroskedasticity The modified Levene test [20] is a robust test, which is quite insensitive against serious departures from non-normality. The test is based on the variability of the residuals. The larger the variance is, the larger the variability of the residuals is. If heteroskedasticity is present in analytical chemistry, the variance is mostly concentration-dependent. In addition, the number of data points is typically limited. To perform the Levene test the data set is divided in two groups, that is, the lower and the higher concentration range (if an hourglass-type spread of the residuals is suspected, dividing into three groups is recommended).

The test statistic t_L^* follows approximately a t-distribution with $n - 2$ degrees of freedom:

$$t_L^* = \frac{\bar{d}_1 - \bar{d}_2}{S\sqrt{\frac{1}{n_1} + \frac{1}{n_2}}} \tag{6.13}$$

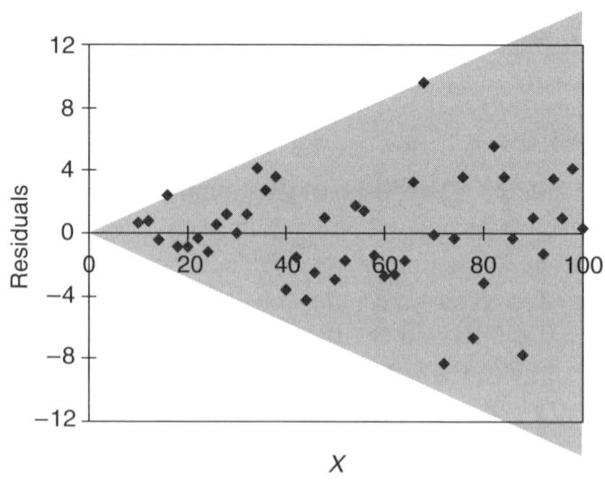

Figure 6.4. In analytical chemistry, heteroskedasticity is mostly present as an increase in variance at higher concentration levels.

with $n = n_1 + n_2$ and n_1, n_2 the data size of the two groups;

$$S^2 = \frac{\sum (d_{i1} - \bar{d}_1) + \sum (d_{i2} - \bar{d}_2)}{n - 2} \tag{6.14}$$

with $\bar{d}_1$ and $\bar{d}_2$ the means of all the d_{i1} and d_{i2}, respectively. d_{i1} and d_{i2} are the absolute deviations of the residual (e_{i1}, e_{i2}) from the median residual in their group ($\tilde{e}_1$ and $\tilde{e}_2$). The absolute deviations are obtained as follows: $d_{i1} = |e_{i1} - \tilde{e}_1|$ and $d_{i2} = |e_{i2} - \tilde{e}_2|$.

6.2.4.3 Remedial Actions

Transformations Transformation of the data set can be used to correct for heteroskedasticity or for linearization of the regression line. A major drawback of transformations is that, when correcting for heteroskedasticity, the assumption of linearity may become violated. Therefore, weighted regression should be preferred for correcting heteroskedasticity.

Weighted Regression Heteroskedasticity can be solved by introducing weighing factors (W_i) in the regression model. The weighing factors are inversely related to the variance: $W_i = 1/S_{Yi}^2$.

The intercept and the slope are calculated by using the following equations:

$$b = \frac{\sum W_i(X_i - \overline{X}_w)(Y_i - \overline{Y}_w)}{\sum W_i(X_i - \overline{X}_w)^2} \tag{6.15}$$

$$a = \overline{Y}_w - b\overline{X}_w \tag{6.16}$$

with

$$\overline{X}_w = \frac{\sum W_i X_i}{\sum W_i} \quad \text{and} \quad \overline{Y}_w = \frac{\sum W_i Y_i}{\sum W_i}$$

The above equations have already been described in the literature [8]. It is not the evaluation of heteroskedasticity, nor the application of the above equations, that causes problems in analytical chemistry, but the choice of the weighing factors. If you have a large number of replicates at each concentration level, the weighing factors are easily estimated by calculating the variance of the replicates at each concentration level. However, the number of data points available during a calibration experiment is limited (i.e., six concentration levels with two or three replicates at each level).

To solve this problem, some assumptions should be made on the relationship between the error of the regression line and the concentration. As a rule, one assumes that the error of the regression line is proportional to the concentration. The variance function Var(X) is obtained by plotting the standard error vs. the concentration. The function is consequently estimated with the least-squares method: $\text{Var}(X) = S_x^2 = (c + d^*\text{conc})^2$. An alternative approach is described in the ISO 11483-2 standard, which uses an iterative procedure to estimate the variance function [18].

Ignoring heteroskedasticity in an ordinary calibration experiment will not lead to major errors when measuring in the middle of the calibration line since the estimated regression parameters are unbiased (it should be recalled that the regression estimators are inefficient, but unbiased). However, the uncertainty of the results is much larger in the lower part of the calibration line when incorrect or no weighing factors have been applied to correct for heteroskedasticity. It should be stressed that extreme errors are introduced by using the calibration line for the estimation of the detection capability, if heteroskedasticity is not corrected.

6.2.5 Uncorrelated Errors

The error terms might be correlated with each other or with an external parameter (i.e., time, injection sequence). It is always a good idea to plot the residuals vs. time and injection sequence to evaluate possible correlation effects. Detector drift, carry-over or other effects can be easily detected with these plots. For a correct evaluation of the independence of the error, however, the calibration solutions should be injected in a randomized manner and correction for heteroskedasticity and/or curvilinearity should be applied on the calibration model.

6.2.6 Normal Distribution of the Error

Small departures of normality do not significantly influence the use of the calibration model in residue analysis. However, major departures of normality are mostly related to analytical or instrumental problems. The use of an inappropriate calibration model can give rise to nonnormality of the residuals. In this case also, one or more of the other four basic assumptions have been violated. Normality can be evaluated by means of several statistical tests (i.e., Kolgomorov–Smirnov, Shapiro–Wilk *W*) or by constructing normal probability plots [8].

6.3 MATRIX INTERFERENCES

6.3.1 General Aspects

Using standard solutions for quantifying concentrations in an unknown sample may give rise to measurement errors due to the influence of food matrix remnants in the injection solution. Ion suppression or enhancement is a typical matrix effect seen in mass spectrometry. Matrix-matched standards are generally used in order to avoid such possible matrix interferences. Standard addition is a valid alternative for dealing with matrix effects. On the other hand, standard addition or matrix based calibration curves require more manipulation, which is time- and money-consuming, whereas a greater risk of manipulation errors ensues.

6.3.2 Detecting Matrix Interferences

Matrix interferences can be easily detected by comparing the slope of the standard-solution-based calibration line (b_s) with the slope of a matrix-based calibration line (b_a). This matrix-based calibration line is established by adding known concentrations of the analyte to the final extract/digestion solution. Both lines are plotted on the same graph. If the curves are parallel, a standard-solution-based calibration line may be used. If this is not the case, matrix interferences are present and the use of standard-solution-based calibration plots should be avoided.

A t-test can also be applied for comparing the slopes of the two calibration lines (under the assumption of linearity and homoskedasticity, normality, and independence of the error):

$$t(b) = \frac{|b_s - b_a|}{Sp\sqrt{\dfrac{1}{\sum(X_{i,s} - \overline{X}_s)^2} + \dfrac{1}{\sum(X_{i,a} - \overline{X}_a)^2}}}$$

$$Sp = \sqrt{\frac{(n_s - 2)S^2_{y,s} + (n_k - 2)S^2_{y,a}}{n_s + n_a - 4}}$$

If the slopes of both curves do not differ significantly [$t(b) < t^{\alpha}_{\mathrm{diff}}$ with d.f. $= n_s + n_a - 4$], matrix effects are not present and a standard-solution-based calibration line may be used. It is noted that, for calibration lines having a very small residual standard deviation (Sy), matrix interferences have often been detected based on the statistical significance while the lines are nearly parallel. The contribution of the error of this small matrix effect is often negligible compared to the total measurement error. Therefore, it is strongly recommended to perform a visual interpretation of the parallelism of the lines in conjunction with this t-test.

6.3.3 Standard Addition

Standard addition is used to quantify the concentration in unknown samples when matrix interferences are present. The use of standard addition has been extensively discussed by Rodriguez et al., Cardone, and Honorato et al. [23–26]. Exact amounts of the analyte in increasing concentrations are added to the sample. The response (Y) is plotted vs. the added concentration (X). A straight line is regressed through the data points. The concentration of the analyte in the sample is given by the intercept on the X-axis ($X_{\mathrm{sample}} = a/b$). Standard addition is probably the best way to correct for matrix effects. Rodriguez et al. have described the statistical techniques that can be used for the validation of analytical methods with standard addition [23].

6.4 ACCURACY

6.4.1 Background Information

Analytical results are subject to two types of errors, namely, systematic and random. Both will influence the accuracy of a test result. Accuracy has been defined in the ISO 3534-1 standard as the closeness of agreement between the test result and the accepted reference value [27]. During method validation the magnitudes of both errors are assessed.

Systematic errors are associated with the fact that the mean of a large number of replicate measurements of the same sample will deviate from the true value. The extent of the systematic error is explained by the bias of the method. The bias is defined as the difference between the expectation of the test result and the accepted result.

Random errors are associated with the fact that, when measurements are repeated, each result will deviate from other results. These deviations cannot be associated to a systematic tendency and are randomly (mostly normally) distributed with a mean equal to zero. The random errors are accounted for by the precision of the method.

6.4.2 Trueness

The bias of an analytical method can be assessed by the use of reference materials (RMs), or certified reference materials (CRMs) by comparing the method with a reference method or through a proficiency study.

Well-designed proficiency studies provide a good estimate of the method bias. Several protocols and statistical methodologies have been developed for assessing this bias, for example, ISO 5725-4 [28], International Harmonized Protocol for Proficiency studies [29], and Youden plot [30].

The EU Commission Decision EC 2002/657 prescribes the use of RMs or CRMs [4]. Eurachem has published a document that provides instructions to analytical chemists on the selection and use of RMs [31]. Different types of RMs exist, namely, pure substances, standard solutions, matrix RMs, physical–chemical RMs, and reference objects or artifacts. The difference between RMs and CRMs lies in that the values or properties of a CRM are certified during a collaborative trial according to a well-defined protocol, whereas the values or properties of an RM are defined, but have not been established through a collaborative trial. The following classes are generally used for RMs: primary, secondary, and in-house or working RMs. The uncertainty associated with primary RMs is the smallest, while in-house RMs have the largest uncertainty.

To assess the method bias of an analytical method, the appropriate RMs should be chosen. A list of available BCR RMs can be found on the website of the Institute for Reference Materials and Measurements (IRMM) of the Joint Research Centre, European Commission, Geel Establishment (www.irmm.jrc.be/rm/intro.html). The validity of assessing the method bias by means of RMs depends on the assumption that the analytical method is otherwise unbiased as well as on the range of appropriate matrix RMs available. It should not be overlooked, however, that there may be a matrix mismatch between the test material and the most appropriate RMs available [32].

By preference, an RM is analyzed under reproducibility conditions. Since the reproducibility standard deviation is not known in a single laboratory validation study, the RM is analyzed n times under within-laboratory reproducibility conditions. The mean and the standard deviation are determined. The bias is estimated by $\delta = y - \mu_{\text{ref}}$, with μ_{ref} the stated reference value and $\bar{y}$ the mean of the results. The significance of the bias can be evaluated with this simplified Student's t-test:

$$t = \frac{\bar{y} - \mu_{\text{ref}}}{\sqrt{S_{\text{ref}}^2 + \frac{S^2}{n}}} \tag{6.18}$$

where S_{ref} is the stated standard error of the RM [33].

The bias is significant if the calculated Student's t-value is larger than the tabulated critical $t_{1-\alpha/2,n-1}$. The bias should be included in the measurement uncertainty if significant.

Example. Evaluation of the Bias of the Analysis of Hg with the Automatic Analyzer AMA254 The CRM-422 Cod Muscle (with a concentration of Hg of 0.559 ± 0.016 μg g^{-1}) has been analyzed 12 times. The results are presented in Table 6.2. Student's t can be calculated from the experimental data and the reference value and uncertainty of the CRM. The standard error associated with the reference value is derived from the uncertainty of the CRM: $S_{\text{ref}} = U_{\text{ref}}/2$. The uncertainty of the CRM is the 95 percent confidence interval.

The Student's t test indicates that the bias of the method is significant.

$$t = \frac{0.52378 - 0.559}{\sqrt{\left(\frac{0.016}{2}\right)^2 + \frac{0.00904^2}{12}}} = -4.18$$

An equivalent approach for assessing the bias of atomic spectroscopy methods was published by Kurfurst [34].

TABLE 6.2. Analytical Results of the Analysis of Hg in the CRM-422 (Results are Expressed in μg g^{-1})

0.535	0.521
0.5364	0.515
0.5259	0.5186
0.5213	0.5223
0.5317	0.5138
0.5349	0.5095
Mean	0.5239
S	0.0090
N	12

The BCR [35] proposes the following equation for assessing the method bias with RMs:

$$-a_1 - 2\sigma_L < \bar{y} - \mu_{ref} < a_2 + 2\sigma_L \tag{6.19}$$

σ_L being the long-term within-laboratory reproducibility and a_1, a_2 adjustment values chosen by the user for technical, economical, or other limitations.

When RMs or proficiency studies are unavailable, the bias of the analytical method is mostly assessed using spiking experiments. As a rule, the bias is estimated by using blank material that is fortified with a known amount of the analyte. The samples are analyzed n times and the mean is calculated:

$$\text{Apparent recovery}(\%) = 100 \cdot \frac{\text{mean observed concentration}}{\text{reference value}} \tag{6.20}$$

Thornburn Burns et al. define a difference between recovery, which is the extraction yield, and the apparent recovery, which represents the observed value/reference value ratio [36]. The apparent recovery for a method using isotopic dilution is nearly 100 percent, while the recovery (extraction yield) can be much lower, 70 percent.

The representativeness of a spiking experiment to asses the method bias is an important issue in food analysis. It has been questioned whether one can evaluate the extraction of substances in a matrix by means of a spiking experiment. The extraction efficiency of incurred and spiked analytes is seldom the same. For some substances, it is hard to find real blank materials. Furthermore, several parameters have influence on the uncertainty of the experimentally determined method bias [32]. It should be stressed that, as it may be concentration-dependent, the method bias should be evaluated for the entire concentration range of the analytical method.

The precision has been defined in the ISO 5725-1 standard as the closeness of agreement between independent test results obtained under stipulated conditions [28]. Many factors may contribute to the variability of test results obtained with the same method on identical samples, including (but not limited to) the operator(s), the equipment, the reagents, the RM(s), the environment, the time between measurements, and the laboratories. The maximal variability of test results is explained by the reproducibility (R) of the method. All factors that have influence on the variability of a test method should be taken into consideration when assessing the reproducibility. The repeatability (r) is assessed by keeping all the above-mentioned factors constant (e.g., same operator, same equipment, same laboratory, short time interval). It is a measure of the minimum variability of a method. The intermediate precision is situated between the two extreme measures of precision: repeatability and reproducibility. The terms within-laboratory reproducibility (w), long-term precision, and so on, are often used to demonstrate the intermediate precision of a method. For a correct interpretation of the intermediate precision, the factors that have been taken into account should be known.

Repeatability, within-laboratory reproducibility, and reproducibility can be evaluated using experimental designs [28]. In the following example three factors

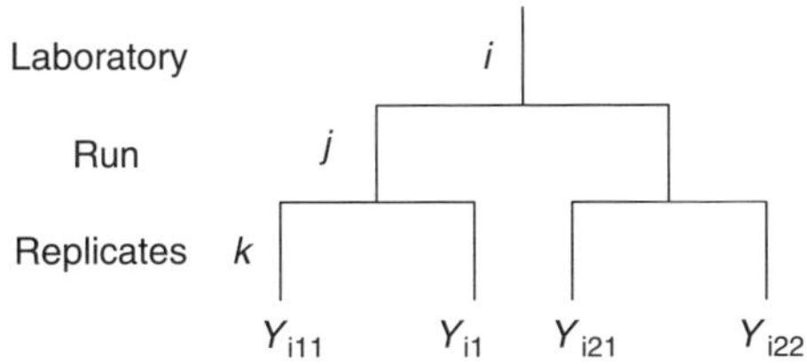

Figure 6.5. Experimental design for the assessment of the repeatability, intermediate precision and the reproducibility of an analytical method.

haven been used: the samples are analyzed by i different laboratories on two different analytical runs and in duplicate. The scheme is illustrated in Figure 6.5.

The total variance of the results, obtained by the different laboratories, during different runs and in duplicate, can be decomposed by means of ANOVA in a variance due to the duplicate measurements, in a run variance component and in a laboratory variance component. These variances, which are explained by the respective SS, are summarized in Table 6.3.

The values S_r, S_{run}, and $S_{laboratory}$ can be estimated from their respective MSs.

$$S_r^2 = MS_r$$

$$S_{run}^2 = \frac{1}{n_r}(MS_{run} - MS_r)$$

$$S_{lab}^2 = \frac{1}{n_{run} \cdot n_r}(MS_{lab} - MS_{run})$$

The repeatability, within-laboratory reproducibility, and reproducibility variance are derived from:

$$S_r^2 = MS_r$$

$$S_w^2 = S_r^2 + S_{run}^2$$

$$S_R^2 = S_r^2 + S_{run}^2 + S_{lab}^2$$

TABLE 6.3. ANOVA Table for a Two-Factor Balanced Design with Replicates

Source	SS	d.f.	MS	Expected Means Square
Laboratory	$SS_{laboratory}$	$n_{laboratory} - 1$	$MS_{laboratory} = SS_{laboratory}/(n_{laboratory} - 1)$	$\sigma_r + n_r \cdot \sigma_{run} + n_r \cdot n_{run} \cdot \sigma_{laboratory}$
Run	SS_{run}	$n_{laboratory} \cdot (n_{run} - 1)$	$MS_{run} = SS_{run}/[n_{laboratory} \cdot (n_{run} - 1)]$	$\sigma_r + n_r \cdot \sigma_{run}$
Replicate	SS_r	$n_{laboratory} \cdot n_{run}(n_r - 1)$	$MS_r = SS_r/[n_{laboratory} \cdot n_{run}(n_r - 1)]$	σ_r
Total	SS_T	$n_T - 1$	$MS_T = SS_T/(n_T - 1)$	

TABLE 6.4. Repeatability and Within-Laboratory Reproducibility Data of the Analysis of Cd with ET-AAS

Day	Result 1 ($ng\ g^{-1}$)	Result 2 ($ng\ g^{-1}$)	Mean ($ng\ g^{-1}$)
1	70	72	71
2	72.2	72.8	72.5
3	76.4	76.4	76.4
4	74.9	75	74.95
5	75.9	76.6	76.25
6	78.9	77.7	78.3
Overall mean ($ng\ g^{-1}$)		74.9	
Recovery		99.87%	
S_r ($ng\ g^{-1}$)/CV_r%		0.72/1.0%	
S_W ($ng\ g^{-1}$)/CV_W%		2.75/3.7%	

Example. The Repeatability and Within-Laboratory Reproducibility of the Analysis of Cd with ET-AAS is Assessed by Means of a Precision Experiment A blank sample has been spiked at a level of 75 $ng\ g^{-1}$ Cd. The samples have been analyzed in duplicate on six different analytical runs according to the prescribed analytical procedure (SOP 22/F/1701) [37]. The results are reported in Table 6.4.

The repeatability and within-laboratory reproducibility (expressed as CV percent) are evaluated against the Horwitz equation according to the EU Commission Decision [4]. The Horwitz equation is an empirical relationship between the concentration of the analyte and the precision of the method. Initially the Horwitz equation was developed from data obtained during collaborative trials [38, 39]. The following equation for the maximum reproducibility CV is valid. The maximum repeatability is between one-half and two-thirds of the CV_R (percent).

$$CV_R(\%) = 2C^{-0.1505} \tag{6.21}$$

with C the concentration level expressed as fraction.

Thompson has suggested that at concentrations below 120 $ng\ g^{-1}$ the Horwitz equation systematically overestimates the reproducibility CV [41].

Grubbs' test is recommended by ISO [28]. This test has been described by Grubbs and Beck [41]. The single Grubbs test evaluates whether the largest or the lowest result in a series of results should be considered an outlier. The Grubbs statistic (G) is calculated by

$$G = \frac{|x_i - \bar{x}|}{S}$$

x_i being the suspected outlying result, $\bar{x}$ the mean, and S the standard deviation of the complete dataset. The double Grubbs test evaluates whether the two largest or lowest results are outliers. The data are arranged in order of increasing magnitude.

The Grubbs statistic is obtained as

$$G = \frac{\sum_{i=3}^{n} (x_i - \bar{x})^2}{\sum_{i=1}^{n} (x_i - \bar{x})^2} \quad \text{or} \quad G = \frac{\sum_{i=1}^{n-2} (x_i - \bar{x})^2}{\sum_{i=1}^{n} (x_i - \bar{x})^2}$$

with $\bar{x}$ the mean of the complete dataset.

Grubbs' statistic is compared with the critical values in Table 6.5. If the Grubbs statistic is significant at a probability of 5 percent, the outlying values are considered stragglers. Values are considered outliers if significant at the 1 percent level.

TABLE 6.5. Two-Sided Critical Values for the Single and Double Grubbs Test. Single Grubbs Test: The Suspected Value is an Outlier if the Calculated *G* is Above the Critical Value. Double Grubbs Test: The Suspected Values are Outliers if *G* is Below the Critical Value

	Single Grubbs' Critical Values		Double Grubbs' Critical Values	
N	1%	5%	1%	5%
3	1.155	1.155		
4	1.496	1.481	0.0000	0.0002
5	1.764	1.715	0.0018	0.0090
6	1.973	1.887	0.0116	0.0349
7	2.139	2.020	0.0308	0.0708
8	2.274	2.126	0.0563	0.1101
9	2.387	2.215	0.0851	0.1492
10	2.482	2.290	0.1150	0.1864
11	2.564	2.355	0.1448	0.2213
12	2.636	2.412	0.1738	0.2537
13	2.699	2.462	0.2016	0.2836
14	2.755	2.507	0.2280	0.3112
15	2.806	2.549	0.2530	0.3367
16	2.852	2.585	0.2767	0.3603
17	2.894	2.620	0.2990	0.3822
18	2.932	2.651	0.3200	0.4025
19	2.968	2.681	0.3398	0.4214
20	3.001	2.709	0.3585	0.4391
21	3.031	2.733	0.3761	0.4556
22	3.060	2.758	0.3927	0.4711
23	3.087	2.781	0.4085	0.4857
24	3.112	2.802	0.4234	0.4994
25	3.135	2.822	0.4376	0.5123
26	3.157	2.841	0.4510	0.5245
27	3.178	2.859	0.4638	0.5360
28	3.199	2.876	0.4759	0.5470
29	3.218	2.893	0.4875	0.5574
30	3.236	2.908	0.4985	0.5672

Dixon's test is a parametric test for detecting outliers. At least three results are needed. The dataset is arranged in order of increasing magnitude: x_1, $x^2 \ldots x_n$. Dixon's Q-statistic is calculated. The equation depends on the number of results.

Q-statistic for three to seven observations:

$$Q = \text{MAX}\left[\frac{x_2 - x_1}{x_n - x_1}, \frac{x_n - x_{n-1}}{x_n - x_1}\right] \tag{6.22a}$$

Q-statistic for eight to 12 observations:

$$Q = \text{MAX}\left[\frac{x_2 - x_1}{x_{n-1} - x_1}, \frac{x_n - x_{n-1}}{x_n - x_2}\right] \tag{6.22b}$$

TABLE 6.6. Critical Values for Dixon's Test. The Suspected Value is an Outlier if the Calculated Q-Statistic is Above the Critical Value

	Dixon Critical Values	
N	1%	5%
3	0.994	0.970
4	0.926	0.829
5	0.821	0.710
6	0.740	0.628
7	0.680	0.569
8	0.717	0.608
9	0.672	0.564
10	0.635	0.530
11	0.605	0.502
12	0.579	0.479
13	0.697	0.611
14	0.670	0.586
15	0.647	0.565
16	0.627	0.546
17	0.610	0.529
18	0.594	0.514
19	0.580	0.501
20	0.567	0.489
21	0.555	0.478
22	0.544	0.468
23	0.535	0.459
24	0.526	0.451
25	0.517	0.443
26	0.510	0.436
27	0.502	0.429
28	0.495	0.423
29	0.489	0.417
30	0.483	0.412

Q-statistic for more than 12 observations:

$$Q = \text{MAX}\left[\frac{x_3 - x_1}{x_{n-2} - x_1}, \frac{x_n - x_{n-2}}{x_n - x_3}\right] \quad (6.22c)$$

The Q-statistic is compared with the critical values in Table 6.6. The lowest or largest value in the dataset is identified as a straggler or as an outlier if the Q-statistic is larger than the critical value with a statistical certainty of 95 or 99 percent, respectively.

An excellent review on outlier treatment is given by Beckman and Cook [42], and by Miller [43]. Some scientists do prefer the use of robust statistics instead of outlier detection and rejection [45]. Whether one prefers the use of statistical tests or chooses to use robust statistics, one should be critical of the dataset. Data points should not be eliminated on the basis of statistical significance only. A cause analysis should be performed before discarding outliers.

6.5 DETECTION CAPABILITIES

6.5.1 Background Information

One of the most important performance characteristics in residue analysis is certainly the detection capability of a method. Unfortunately, many different definitions with regard to the detection and quantifying capability of an analytical method are found in the literature. Attempts have been made to harmonize the definitions with regard to the limit of detection (LoD) and the limit of quantification (LoQ). For the sake of consistency, here reference is made only to the harmonized definitions of IUPAC and ISO for the detection decision (L_C, critical value), LoD (minimum detectable value) and LoQ (minimum quantifiable value) (45–47). ISO terminology is given between brackets.

6.5.2 The Detection Decision

The L_C value is used to distinguish between a detected and a nondetected signal (concentration or amount) of the chemical parameter in a sample. Samples with a signal (concentration or amount) larger than L_C should be considered as contaminated (in the meaning of containing a certain concentration or amount of the chemical parameter). Note that the term "larger" is used. It is a one-sided test. Under the assumption of normality and homoskedasticity, L_C can be calculated with the equation $L_C = t_{1-\alpha \text{d.f.}} S_0$, S_0 being the standard deviation of a number of replicate measurements of a blank sample. For a huge number of replicates, the above equation is simplified to $L_C = 1.645 S_0$. A probability of 95 percent ($\alpha = 0.05$) is mostly chosen in analytical chemistry.

L_C can be used to discriminate between detectable quantities and nondetectable quantities. When the measured signal (concentration or amount) of a sample is larger than L_C, one might say with a statistical certainty of 95 percent that the sample is contaminated. Two types of risk are involved in hypothesis testing. One is the risk of erroneously declaring that the concentration of the analyte in the sample is larger than in the blank sample. This is called type I error or α-error in statistics (also called a "false positive" by the chemists). The type II error or β-error is associated with the risk of not detecting, while the sample is truly contaminated (false negatives). This β-error is not considered in the definition of the detection decision. On the other hand, the definition of L_D takes into account this β -risk.

6.5.3 The Detection Limit

The LoD is used to discriminate between a true blank sample and a real contaminated sample. Both concepts of false positive and false negative samples are considered in the determination of the LoD. The LoD is calculated by $L_D = \delta_{\alpha,\beta,\text{d.f.}}\sigma_0$ with $\delta_{\alpha\beta,\text{d.f.}}$ the critical value of the noncentral t-distribution (note that both the α- and β-errors are considered). The $\delta_{\alpha\beta,\text{d.f.}}$ can be approximated by $2t_{1-\alpha,\text{d.f.}}\ \text{d.f.}/(\text{d.f.}+1)$ for $\alpha = \beta$ and d.f. < 25. The population statistic δ_0 is mostly not known and in practice it is estimated by the standard deviations

$$S_0 = \sqrt{\frac{\sum (X_i - \overline{X})^2}{n-1}}$$

Even for estimates of σ_0 based on a large number of data, the associated uncertainty is quite important, as explained by Currie [47, 48]. As a consequence, this will have a major impact on the estimated LoD.

The above equations can be applied in spectrochemical analysis directly on data related to the detector response (i.e., the signal). However, most people are interested in having LoDs expressed as a concentration. Signals are converted to concentrations by means of the calibration function. This has been extensively described in the literature. An excellent document for the determination of the L_C (critical value) and the LoD (minimum detectable value) is the ISO standard 11843-2 [18].

6.5.4 The Quantification Limit

Unlike L_C and LoD, LoQ is not related to statistical probabilities (α- and β-error). LoQ is the minimum signal (concentration or amount) that can be quantified. The ability to quantify in analytical chemistry is related to a signal (concentration or amount) that can be readily reproduced. Therefore, the residual standard deviation (RSD_Q) is included in the definition $l_Q = (1/\text{RSD}_\text{Q})\sigma_Q$. The IUPAC default value for RSD_Q is 0.1. Using the default value, one obtains $L_Q = 10\sigma_Q$.

6.5.5 ISO 11843: Capability of Detection

The methodology for assessing LoDs (i.e., the minimum detectable quantity in the ISO terminology) in the case of a linear regression model (LRM) has been extensively described in the ISO 11843 [18, 45]. To evaluate LoDs of an analytical method, a set of five or more samples is prepared at and above the expected values for the LoDs. The concentration of the samples should be approximately equidistant in range between the largest and the smallest concentrations. When a blank sample is available it should be included in the experiment. The nature of the prepared or reference samples should be as close as possible to the nature of the routinely analyzed samples. The samples are analyzed at least in duplicate.

Under the assumption of linearity, normality, independence, and homoskedasticity, the LoD (since it is expressed in terms of concentration, x_D and x_C are used instead of LoD and L_C) is given by:

$$x_\text{D} = \delta_{\alpha,\beta,\text{d.f.}} \frac{Sy}{b} \sqrt{\frac{1}{K} + \frac{1}{I \cdot J} + \frac{\bar{x}^2}{J \cdot \sum (x_i - \bar{x})^2}} \tag{6.23}$$

The detection decision (x_C) is calculated with the following equation:

$$x_\text{C} = t_{1-\alpha,\text{d.f.}} \frac{Sy}{b} \sqrt{\frac{1}{K} + \frac{1}{I \cdot J} + \frac{\bar{x}^2}{J \cdot \sum (x_i - \bar{x})^2}} \tag{6.24}$$

The assumption of heteroskedasticity is frequently rejected in analytical chemistry. Weighted regression can be used to correct for heteroskedasticity. The equations to calculate the detection decision and LoD in case of weighted linear regression models (WLRM) are given by:

$$x_\text{D} = \frac{\delta_{\alpha,\beta,\text{d.f.}}}{b} \sqrt{\frac{S^2_{x_\text{D}}}{K} + \left[\left(\frac{1}{J \cdot \sum w_i}\right) + \frac{\bar{x}_w^2}{J \cdot \sum w_i (x_i - \bar{x}_w)^2}\right] S_y^2} \tag{6.25}$$

$$x_\text{C} = \frac{t_{1-\alpha,\text{d.f.}}}{b} \sqrt{\frac{S^2_{x_\text{D}}}{K} + \left[\left(\frac{1}{J \cdot \sum w_i}\right) + \frac{\bar{x}_w^2}{J \cdot \sum w_i (x_i - \bar{x}_w)^2}\right] S_y^2} \tag{6.26}$$

with b the estimate of the slope, t the Student t parameter with $n - 2$ degrees of freedom, δ the noncentrality parameter associated with the α- and β-error and $n - 2$ degrees of freedom, I the number of reference states, that is, number of replicates per concentration for the spiked or reference samples, $i = 1, 2, \ldots, I$, J the number of preparations for the reference states, that is, the number of concentrations for the spiked or reference samples, K the number of preparations of the actual state, that is, the number of replicates for the unknown samples, $n = I \cdot J$, Sy the standard error of the estimate, Sx_D the residual standard deviation at $x = x_\text{D}$, w_i applied weights ($w_i = 1$ in the case of unweighted regression), x_i the spiked concentration, and $\bar{x}$ the mean of the concentrations, $\bar{x}_w = (\sum w_i x_i)/\sum w_i$.

The linear range of most detectors is limited. In case of rejection of linearity by the LOF- or Mandel's fitting test, LoDs can be estimated with linearized calibration curves or by using curvilinear regression models. A better approach to the estimation of LoDs is to reduce the concentration range of the prepared samples. In some cases, this will also eliminate the problem of heteroskedasticity. It has been shown that the impact of nonlinearity on LoDs is small compared to the impact of heteroskedasticity [19]. Testing of, and correction for, heteroskedasticity should be performed in all cases when estimating LoDs. Ignoring the heteroskedasticity of calibration lines will mostly lead to a serious overestimation of LoDs. Torrades and Garcia Raurich have developed a methodology for determining LoDs in case of heteroskedasticity [49].

Example. The Determination of Cd and Pb in Tissues of Animal Origin with ET-AAS LoDs for this method were estimated by using the following experiment. Blank material of meat, liver, and fish were spiked at five concentration levels. The samples were analyzed in duplicate on a Perkin–Elmer SimAA. This experiment was repeated on several occasions. Linearity and homoskedasticity of the calibration curves were not rejected with the Mandels fitting test and visual interpretation of the plot of the residuals. The results described below were obtained.

LoDs shown in Table 6.7 were compared with those of Table 6.8 obtained two years earlier based on three times the signal-to-noise ratio (S/N). It should be noted that this old definition of LoD based on 3 S/N is not equivalent to the IUPAC definition of LoD. One might conclude that the results of LoDs calculated with the ISO 11843 methodology are slightly higher than, but still comparable with, LoDs based on 3 S/N.

6.5.6 The Decision Limit (CCα) and the Detection Capability (CCβ) According the EU Commission Decision EC 2002/657

The determination of LoDs and LoQs is not required by the EU Commission Decision EC 2002/657 (4). Instead, the decision limit (CCα) and the detection capability (CCβ) are the required performance characteristics.

CCα is used to discriminate between compliant and noncompliant samples. It is defined as the limit at, and above which it can be concluded with an error probability of α that a sample is noncompliant. CCβ means the smallest content of the

TABLE 6.7. LoDs for Cd and Pb in Tissues of Animal Origin with ET-AAS (X_D) Calculated Using the ISO 11843 Approach. Results are Expressed in μg kg^{-1} Fresh Weight

Cd			Pb		
Liver	Fish	Meat	Liver	Fish	Meat
4.4	3.6	2.4	46	30	30
3.2	3.4	4.8	26	36	20
3.8	2.8	3.8	22	38	38

TABLE 6.8. LoDs for Cd and Pb in Tissues of Animal Origin with ET-AAS Based on 3 S/N Ratio. Results are Expressed in μg kg^{-1} Fresh Weight

Cd	Pb
1.4	24

substance that can be detected, identified, and/or quantified in a sample with an error probability of β. In the case of substances for which no permitted limit has been established, this quantity is the lowest concentration at which a method is able to detect truly contaminated samples with a statistical certainty of $1 - \beta$. In the case of substances with an established permitted limit, this means that the detection capability is the concentration at which the method is able to detect permitted limit concentrations with a statistical certainty of $1 - \beta$.

The statistical concepts behind CCα and CCβ are identical to the concepts of L_C and L_Ds, respectively. However, due to the different experimental design, the different interpretation and the legal framework, CCα and CCβ should not be considered as alternative definitions of decision limit and LoDs. A graphical representation of CCα and CCβ in the case of an established limit is given in Figure 6.6.

CCα is obtained by analyzing at least 20 blank materials per matrix fortified with the analyte(s) at the permitted limit. The concentration at the permitted limit plus 1.64 times the corresponding standard deviation equals the decision limit ($\alpha = 5$ percent). CCβ is obtained by analyzing at least 20 blank materials per matrix fortified with the analyte(s) at the decision limit. The value of the decision limit plus 1.64 times the standard deviation of the within-laboratory reproducibility of the measured content equals the detection capability ($\beta = 5$ percent).

In accordance to the EU Commission Decision, 40 samples per matrix are needed to asses CCα and CCβ. B. Jülicher et al. have developed an in-house validation protocol to asses the detection capabilities [50]. This approach is especially interesting for validating multi-matrix methods.

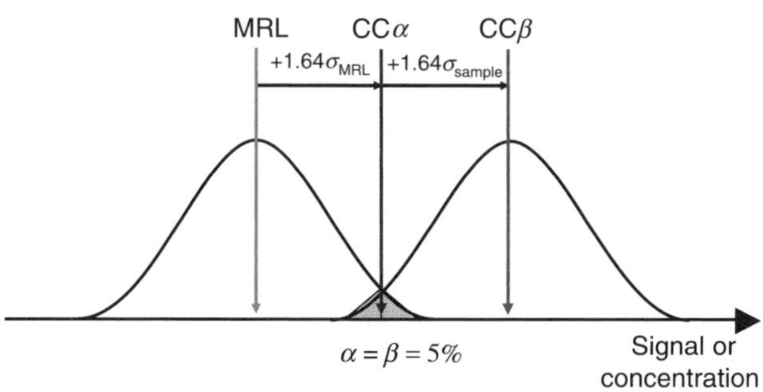

Figure 6.6. Graphical representation of CCα and CCβ.

ACKNOWLEDGMENTS

Dr F. Bolle and Dr K. Bouthakrit (both from the Scientific Institute of Public Health, Brussels, Belgium) are gratefully acknowledged for providing their experimental data.

REFERENCES

1. ISO/IEC 17025 Standard, *General Requirements for the Competence of Calibration and Testing Laboratories*, ISO, Geneva, 1999.
2. Eurachem, *The Fitness for Purpose of Analytical Methods: a Laboratory Guide to Method Validation and Related Topics*, 1998.
3. M. Thompson, S. L. R. Ellison, R. Wood, Harmonized guidelines for single-laboratory validation of methods of analysis, *Pure Appl. Chem.*, **74** (2002), 835–855.
4. European Commission, Commission Decision 2002/657/CE, *Off. J. Eur. Commun.*, **L221** (2002), 8–36.
5. J. N. Miller, Basic statistical methods for analytical chemistry. Part 2. Calibration and regression methods. A review, *Analyst*, **116** (1991), 3–14.
6. D. MacTaggart, S. O. Farwell, Analytical use of linear regression, Part I. Regression procedures for calibration and quantitation, *J. AOAC Int.*, **75** (1992), 594–607.
7. K. Danzer, L. A. Currie, Guidelines for calibration in analytical chemistry. Part I. Fundamentals and single component calibration (IUPAC Recommendations 1998), *Comm. Gen. Asp. Anal. Chem.*, **70** (1998), 993–1014.
8. J. Neter, M. H. Kutner, C. J. Nachtsheim, W. Wasserman, *Applied Linear Statistical Models*, 4th ed, Boston, MA, Irwin, 1996.
9. D. MacTaggart, S. O. Farwell, Analytical use of linear regression. Part II. Statistical error in both variables, *J. AOAC Int.*, **75** (1992), 608–614.
10. J. Mandel, *The Statistical Analysis of Experimental Data*, New York, Interscience, 1964.
11. ISO 8466-1, *Water Quality – Calibration and Evaluation of Analytical Methods and Estimation of Performance Characteristics- Part 1: the Statistical Evaluation of the Linear Calibration Function*, Geneva, ISO, 2001.
12. Analytical Methods Committee, Uses (proper and improper) of correlation coefficients, *Analyst*, **113** (1988), 1469–1471.
13. J. Van Loco, M. Elskens, C. Croux, H. Beernaert, Linearity of calibration curves: use and misuse of the correlation coefficient, *Accredit. Qual. Assur.*, **7** (2002), 281–285.
14. D. L. Massart, B. G. M. Vandeginste, L. M. C. Buydens, S. De Jong, P. J. Lewi, J. Smeyers-Verbeke, *Handbook of Chemometrics and qualimetrics. Part A*, Amsterdam, Elsevier, 1997.
15. P. Vankeerberghen, J. Smeyers-Verbeke, D. Massart, The slope ranking method for homoscedastic calibration lines, *Analusis*, **23** (1995), 247–249.
16. P. Vankeerberghen, J. Smeyers-Verbeke, D. L. Massart, Description and implementation of the system for the run suitability check and explorative method validation, *Lab. Autom. Inform. Mgmt*, **32** (1996), 87–102.
17. W. Xiaoning, J. Smeyers-Verbeke, D. L. Massart, Linearization of atomic absorption calibration curves, *Analusis*, **20** (1992), 209–215.

18. ISO 11843–2, *Capability of Detection – Part 2. Methodology in the Linear Calibration Case*, Geneva, ISO, 2000.
19. J. Van Loco, V. Hanot, G. Huysmans, M. Elskens, J. M. Degroodt, H. Beernaert, Estimation of the minimum detectable value for the determination of PCBs in fatty food samples by GC-ECD: a curvilinear calibration case, *Anal. Chim. Acta*, **483** (2003), 413–418.
20. H. Levene, Robust test for equality of variances, in I. Olkin (ed.), *Contributions to Probalility and Statistics*, California, Stanford University Press, 1960, 278–292.
21. T. S. Breusch, A. R. Pagan, A simple test of heteroskedasticity and random coefficient variation, *Econometrica*, **47** (1979), 1287–1294.
22. H. White, A heteroskedasticity consistent covariance matrix estimator and a direct test for heteroskedasticity, *Econometrica*, **48** (1980), 817–838.
23. L. C. Rodriguez, A. M. G. Campana, F. A. Barrero, C. J. Linares, M. R. Ceba, Validation of an analytical instrumental method by standard addition methodology, *J. AOAC Int.*, **78** (1995), 471–476.
24. M. J. Cardone, New technique in chemical assay calculations. 1. A survey of calculational practices on a model problem, *Anal. Chem.*, **58** (1986), 433–438.
25. M. J. Cardone, New technique in chemical assay calculations 2. Correct solution of the model problem and related concepts, *Anal. Chem.*, **58** (1986), 438–445.
26. F. A. Honorato, R. S. Honorato, M. F. Pimentel, M. C. U. Araujo, Analytical curve or standard addition method: how to elect and design a strategy applied to copper determination in sugarcane spirits using AAS, *Analyst*, **127** (2002), 1520–1525.
27. ISO 3534-1, *Statistics–Vocabulary and Symbols – Part 1: Probability and General Statistical Terms*, Geneva, ISO, 1993.
28. ISO 5725-1 to 6, *Accuracy (Trueness and Precision of Measurement Methods and Results – part 1–6*, Geneva, ISO, 1996.
29. M. Thompson, R. Wood, The international harmonized protocol for the proficiency testing of (chemical) analytical laboratories, *J. AOAC Int.*, **76** (1993), 926–940.
30. W. J. Youden, Graphical diagnosis of interlaboratory test results, *J. Ind. Qual. Contr.*, **15** (1959), 133–137.
31. Eurachem, *The Selection and Use of Reference Materials*, 2002; http://www.eurachem.ul.pt.
32. M. Thompson, S. L. R. Ellison, A. Fajgelj, P. Willetts, R. Wood, Harmonized guidelines for the use of recovery information in analytical measurement, *Pure Appl. Chem.*, **71** (1999), 337–348.
33. NEN 7777, *Environment – Performance Characteristics of Measurement Methods*, Delft, Netherlands Normalisatie-Instituut, 2003.
34. U. Kurfurst, Consideration of the uncertainty in the CRM value with the analysis of solid materials using atomic spectrometry. Part 1. Evaluation and benefit of method uncertainty using certified reference materials, *Accred. Qual. Assur.*, **3** (1998), 406–411.
35. Ph. Quevauviller, J. L. Imbert, P. J. Wagstaffe, G. N. Kramer, B. Griepink, The Certification of the Contents of As, Cd, Cu, Fe, Hg, I, Mn, Pb, Se and Zn in Cod Muscle – CRM 422, Commission of the European Communities, BCR Information, Report EUR 14557 EN, 1993.
36. D. Thornburn Burns, K. Danzer, A. Townshend, Use of the terms "recovery" and "apparent recovery" in analytical procedures, *Pure Appl. Chem.*, **74** (2002), 2201–2205.

37. F. Bolle, J. Van Loco, K. Bouthakrit, J. M. Degroodt, SOP 22/F/1701, *Dosage du plomb et du cadmium dans les produits carnés et végétaux comestibles par spectrométrie d'absorption atomique electrothermique*, Brussels, Scientific Institute of Public Health, 2004.
38. W. Horwitz, L. R. Kamps, K. W. Boyer, Quality Control, *J. AOAC*, **63** (1980), 1344–1354.
39. W. Horwitz, R. Wood, Relationship of (known) control values to (unknown) test values in proficiency studies of pesticide residues, *J. AOAC Int.*, **83** (2000), 2, 399–406.
40. M. Thompson, Recent trends in inter-laboratory precision at ppb and sub-ppb concentrations in relation to fitness for purpose criteria in proficiency testing, *Analyst*, **125** (2000), 385–386.
41. F. E. Grubbs, G. Beck, Extension of sample sizes and percentage points for significance tests of outlying observations, *Technometrics*, **14** (1972), 847–854.
42. R. J. Beckman, R. D. Cook, Outlier......s, *Technometrics*, **25** (1983), 119–254.
43. J. N. Miller, Outliers in Experimental data and their treatment, *Analyst*, **118** (1993), 455–461.
44. Analytical methods committee, Robust statistics - how not to reject outliers. Part 1. Basic concepts, *Analyst*, **114** (1989), 1693–1697.
45. ISO 11843-1, *Capability of Detection. Part 1. Terms and Definitions*, Geneva, ISO, 1997.
46. L. A. Currie, Nomenclature in evaluation of analytical methods including detection and quantification capabilities (IUPAC recommendations 1995), *Anal. Chim. Acta*, **391** (1999), 127–134.
47. L. A. Currie, Detection: international update and some emerging dilemmas involving calibration, the blank and multiple detection decisions, *Chemometrics Intell. Lab. Syst.*, **37** (1997), 151–181.
48. L. A. Currie, Nomenclature in evaluation of analytical methods including detection and quantification capabilities, *Pure Appl. Chem.*, **67** (1995), 1699–1723.
49. F. Torrades, J. G. Raurich, *Anal. Lett.*, **26** (1993), 11, 2503–2512.
50. B. Jülicher, P. Gowik, S. Uhlig, Assesment of detection methods in trace analysis by means of a statistically based in-house validation concept, *Analyst*, **123** (1998), 173–179.

APPENDIX

TABLE 6.A1. Percentiles of Student's *t* Distribution (Single-Sided)

	$1-\alpha$						
d.f.	0.9	0.95	0.975	0.99	0.995	0.9975	0.999
1	3.078	6.314	12.71	31.82	63.66	127.3	318.3
2	1.886	2.920	4.303	6.965	9.925	14.09	22.33
3	1.638	2.353	3.182	4.541	5.841	7.453	10.21
4	1.533	2.132	2.776	3.747	4.604	5.598	7.173
5	1.476	2.015	2.571	3.365	4.032	4.773	5.894
6	1.440	1.943	2.447	3.143	3.707	4.317	5.208
7	1.415	1.895	2.365	2.998	3.499	4.029	4.785
8	1.397	1.860	2.306	2.896	3.355	3.833	4.501
9	1.383	1.833	2.262	2.821	3.250	3.690	4.297
10	1.372	1.812	2.228	2.764	3.169	3.581	4.144
11	1.363	1.796	2.201	2.718	3.106	3.497	4.025
12	1.356	1.782	2.179	2.681	3.055	3.428	3.930
13	1.350	1.771	2.160	2.650	3.012	3.372	3.852
14	1.345	1.761	2.145	2.624	2.977	3.326	3.787
15	1.341	1.753	2.131	2.602	2.947	3.286	3.733
16	1.337	1.746	2.120	2.583	2.921	3.252	3.686
17	1.333	1.740	2.110	2.567	2.898	3.222	3.646
18	1.330	1.734	2.101	2.552	2.878	3.197	3.610
19	1.328	1.729	2.093	2.539	2.861	3.174	3.579
20	1.325	1.725	2.086	2.528	2.845	3.153	3.552
21	1.323	1.721	2.080	2.518	2.831	3.135	3.527
22	1.321	1.717	2.074	2.508	2.819	3.119	3.505
23	1.319	1.714	2.069	2.500	2.807	3.104	3.485
24	1.318	1.711	2.064	2.492	2.797	3.091	3.467
25	1.316	1.708	2.060	2.485	2.787	3.078	3.450
26	1.315	1.706	2.056	2.479	2.779	3.067	3.435
27	1.314	1.703	2.052	2.473	2.771	3.057	3.421
28	1.313	1.701	2.048	2.467	2.763	3.047	3.408
29	1.311	1.699	2.045	2.462	2.756	3.038	3.396
30	1.310	1.697	2.042	2.457	2.750	3.030	3.385
40	1.303	1.684	2.021	2.423	2.704	2.971	3.307
50	1.299	1.676	2.009	2.403	2.678	2.937	3.261
60	1.296	1.671	2.000	2.390	2.660	2.915	3.232
90	1.291	1.662	1.987	2.368	2.632	2.878	3.183
120	1.289	1.658	1.980	2.358	2.617	2.860	3.160
∞	1.282	1.645	1.960	2.326	2.576	2.807	3.090

TABLE 6.A2. Percentiles of the *F*-Distribution (95%)

$f2/f1$	1	2	3	4	5	6	7	8	9	10	12	14	16	20	30	60	120	∞
1	161.4	199.5	215.7	224.6	230.2	234.0	236.8	238.9	240.5	241.9	243.9	245.4	246.5	248.0	250.1	252.2	253.3	254.3
2	18.51	19.00	19.16	19.25	19.30	19.33	19.35	19.37	19.38	19.40	19.41	19.42	19.43	19.45	19.46	19.48	19.49	19.50
3	10.13	9.55	9.28	9.12	9.01	8.94	8.89	8.85	8.81	8.79	8.74	8.71	8.69	8.66	8.62	8.57	8.55	8.53
4	7.71	6.94	6.59	6.39	6.26	6.16	6.09	6.04	6.00	5.96	5.91	5.87	5.84	5.80	5.75	5.69	5.66	5.63
5	6.61	5.79	5.41	5.19	5.05	4.95	4.88	4.82	4.77	4.74	4.68	4.64	4.60	4.56	4.50	4.43	4.40	4.37
6	5.99	5.14	4.76	4.53	4.39	4.28	4.21	4.15	4.10	4.06	4.00	3.96	3.92	3.87	3.81	3.74	3.70	3.67
7	5.59	4.74	4.35	4.12	3.97	3.87	3.79	3.73	3.68	3.64	3.57	3.53	3.49	3.44	3.38	3.30	3.27	3.23
8	5.32	4.46	4.07	3.84	3.69	3.58	3.50	3.44	3.39	3.35	3.28	3.24	3.20	3.15	3.08	3.01	2.97	2.93
9	5.12	4.26	3.86	3.63	3.48	3.37	3.29	3.23	3.18	3.14	3.07	3.03	2.99	2.94	2.86	2.79	2.75	2.71
10	4.96	4.10	3.71	3.48	3.33	3.22	3.14	3.07	3.02	2.98	2.91	2.86	2.83	2.77	2.70	2.62	2.58	2.54
11	4.84	3.98	3.59	3.36	3.20	3.09	3.01	2.95	2.90	2.85	2.79	2.74	2.70	2.65	2.57	2.49	2.45	2.40
12	4.75	3.89	3.49	3.26	3.11	3.00	2.91	2.85	2.80	2.75	2.69	2.64	2.60	2.54	2.47	2.38	2.34	2.30
13	4.67	3.81	3.41	3.18	3.03	2.92	2.83	2.77	2.71	2.67	2.60	2.55	2.51	2.46	2.38	2.30	2.25	2.21
14	4.60	3.74	3.34	3.11	2.96	2.85	2.76	2.70	2.65	2.60	2.53	2.48	2.44	2.39	2.31	2.22	2.18	2.13
15	4.54	3.68	3.29	3.06	2.90	2.79	2.71	2.64	2.59	2.54	2.48	2.42	2.38	2.33	2.25	2.16	2.11	2.07
16	4.49	3.63	3.24	3.01	2.85	2.74	2.66	2.59	2.54	2.49	2.42	2.37	2.33	2.28	2.19	2.11	2.06	2.01
17	4.45	3.59	3.20	2.96	2.81	2.70	2.61	2.55	2.49	2.45	2.38	2.33	2.29	2.23	2.15	2.06	2.01	1.96
18	4.41	3.55	3.16	2.93	2.77	2.66	2.58	2.51	2.46	2.41	2.34	2.29	2.25	2.19	2.11	2.02	1.97	1.92
19	4.38	3.52	3.13	2.90	2.74	2.63	2.54	2.48	2.42	2.38	2.31	2.26	2.21	2.16	2.07	1.98	1.93	1.88
20	4.35	3.49	3.10	2.87	2.71	2.60	2.51	2.45	2.39	2.35	2.28	2.22	2.18	2.12	2.04	1.95	1.90	1.84
21	4.32	3.47	3.07	2.84	2.68	2.57	2.49	2.42	2.37	2.32	2.25	2.20	2.16	2.10	2.01	1.92	1.87	1.81
22	4.30	3.44	3.05	2.82	2.66	2.55	2.46	2.40	2.34	2.30	2.23	2.17	2.13	2.07	1.98	1.89	1.84	1.78
23	4.28	3.42	3.03	2.80	2.64	2.53	2.44	2.37	2.32	2.27	2.20	2.15	2.11	2.05	1.96	1.86	1.81	1.76
24	4.26	3.40	3.01	2.78	2.62	2.51	2.42	2.36	2.30	2.25	2.18	2.13	2.09	2.03	1.94	1.84	1.79	1.73
25	4.24	3.39	2.99	2.76	2.60	2.49	2.40	2.34	2.28	2.24	2.16	2.11	2.07	2.01	1.92	1.82	1.77	1.71
30	4.17	3.32	2.92	2.69	2.53	2.42	2.33	2.27	2.21	2.16	2.09	2.04	1.99	1.93	1.84	1.74	1.68	1.62
40	4.08	3.23	2.84	2.61	2.45	2.34	2.25	2.18	2.12	2.08	2.00	1.95	1.90	1.84	1.74	1.64	1.58	1.51
60	4.00	3.15	2.76	2.53	2.37	2.25	2.17	2.10	2.04	1.99	1.92	1.86	1.82	1.75	1.65	1.53	1.47	1.39
120	3.92	3.07	2.68	2.45	2.29	2.18	2.09	2.02	1.96	1.91	1.83	1.78	1.73	1.66	1.55	1.43	1.35	1.25
∞	3.84	3.00	2.60	2.37	2.21	2.10	2.01	1.94	1.88	1.83	1.75	1.69	1.64	1.57	1.46	1.32	1.22	1.01

TABLE 6.A3. Percentiles of the *F*-Distribution (99%)

f2/f1	1	2	3	4	5	6	7	8	9	10	12	14	16	20	30	60	120	∞
1	4052	4999	5403	5624	5764	5859	5928	5981	6022	6055	6106	6143	6170	6208	6260	6313	6339	6365
2	98.50	99.00	99.16	99.25	99.30	99.33	99.36	99.38	99.39	99.40	99.42	99.43	99.44	99.45	99.47	99.48	99.49	99.50
3	34.12	30.82	29.46	28.71	28.24	27.91	27.67	27.49	27.34	27.23	27.05	26.92	26.83	26.69	26.50	26.32	26.22	26.13
4	21.20	18.00	16.69	15.98	15.52	15.21	14.98	14.80	14.66	14.55	14.37	14.25	14.15	14.02	13.84	13.65	13.56	13.46
5	16.26	13.27	12.06	11.39	10.97	10.67	10.46	10.29	10.16	10.05	9.89	9.77	9.68	9.55	9.38	9.20	9.11	9.02
6	13.75	10.92	9.78	9.15	8.75	8.47	8.26	8.10	7.98	7.87	7.72	7.60	7.52	7.40	7.23	7.06	6.97	6.88
7	12.25	9.55	8.45	7.85	7.46	7.19	6.99	6.84	6.72	6.62	6.47	6.36	6.28	6.16	5.99	5.82	5.74	5.65
8	11.26	8.65	7.59	7.01	6.63	6.37	6.18	6.03	5.91	5.81	5.67	5.56	5.48	5.36	5.20	5.03	4.95	4.86
9	10.56	8.02	6.99	6.42	6.06	5.80	5.61	5.47	5.35	5.26	5.11	5.01	4.92	4.81	4.65	4.48	4.40	4.31
10	10.04	7.56	6.55	5.99	5.64	5.39	5.20	5.06	4.94	4.85	4.71	4.60	4.52	4.41	4.25	4.08	4.00	3.91
11	9.65	7.21	6.22	5.67	5.32	5.07	4.89	4.74	4.63	4.54	4.40	4.29	4.21	4.10	3.94	3.78	3.69	3.60
12	9.33	6.93	5.95	5.41	5.06	4.82	4.64	4.50	4.39	4.30	4.16	4.05	3.97	3.86	3.70	3.54	3.45	3.36
13	9.07	6.70	5.74	5.21	4.86	4.62	4.44	4.30	4.19	4.10	3.96	3.86	3.78	3.66	3.51	3.34	3.25	3.17
14	8.86	6.51	5.56	5.04	4.69	4.46	4.28	4.14	4.03	3.94	3.80	3.70	3.62	3.51	3.35	3.18	3.09	3.00
15	8.68	6.36	5.42	4.89	4.56	4.32	4.14	4.00	3.89	3.80	3.67	3.56	3.49	3.37	3.21	3.05	2.96	2.87
16	8.53	6.23	5.29	4.77	4.44	4.20	4.03	3.89	3.78	3.69	3.55	3.45	3.37	3.26	3.10	2.93	2.84	2.75
17	8.40	6.11	5.19	4.67	4.34	4.10	3.93	3.79	3.68	3.59	3.46	3.35	3.27	3.16	3.00	2.83	2.75	2.65
18	8.29	6.01	5.09	4.58	4.25	4.01	3.84	3.71	3.60	3.51	3.37	3.27	3.19	3.08	2.92	2.75	2.66	2.57
19	8.18	5.93	5.01	4.50	4.17	3.94	3.77	3.63	3.52	3.43	3.30	3.19	3.12	3.00	2.84	2.67	2.58	2.49
20	8.10	5.85	4.94	4.43	4.10	3.87	3.70	3.56	3.46	3.37	3.23	3.13	3.05	2.94	2.78	2.61	2.52	2.42
21	8.02	5.78	4.87	4.37	4.04	3.81	3.64	3.51	3.40	3.31	3.17	3.07	2.99	2.88	2.72	2.55	2.46	2.36
22	7.95	5.72	4.82	4.31	3.99	3.76	3.59	3.45	3.35	3.26	3.12	3.02	2.94	2.83	2.67	2.50	2.40	2.31
23	7.88	5.66	4.76	4.26	3.94	3.71	3.54	3.41	3.30	3.21	3.07	2.97	2.89	2.78	2.62	2.45	2.35	2.26
24	7.82	5.61	4.72	4.22	3.90	3.67	3.50	3.36	3.26	3.17	3.03	2.93	2.85	2.74	2.58	2.40	2.31	2.21
25	7.77	5.57	4.68	4.18	3.85	3.63	3.46	3.32	3.22	3.13	2.99	2.89	2.81	2.70	2.54	2.36	2.27	2.17
30	7.56	5.39	4.51	4.02	3.70	3.47	3.30	3.17	3.07	2.98	2.84	2.74	2.66	2.55	2.39	2.21	2.11	2.01
40	7.31	5.18	4.31	3.83	3.51	3.29	3.12	2.99	2.89	2.80	2.66	2.56	2.48	2.37	2.20	2.02	1.92	1.80
60	7.08	4.98	4.13	3.65	3.34	3.12	2.95	2.82	2.72	2.63	2.50	2.39	2.31	2.20	2.03	1.84	1.73	1.60
120	6.85	4.79	3.95	3.48	3.17	2.96	2.79	2.66	2.56	2.47	2.34	2.23	2.15	2.03	1.86	1.66	1.53	1.38
∞	6.64	4.61	3.78	3.32	3.02	2.80	2.64	2.51	2.41	2.32	2.18	2.08	2.00	1.88	1.70	1.47	1.32	1.01

7

DEMONSTRATION OF MEASUREMENT CAPABILITIES BY MEANS OF INTERLABORATORY COMPARISON SCHEMES FOR TRACE ELEMENT ANALYSIS IN FOOD

YETUNDE AREGBE AND PIOTR ROBOUCH

European Commission, Joint Research Centre, Institute for Reference Materials and Measurements, Retieseweg 111, 2440 Geel, Belgium, E-mail: Yetunde.Aregbe@ec.europa.eu; Piotr.Robouch@ec.europa.eu

THOMAS PROHASKA

University of Natural Resources and Applied Life Sciences, Muthgasse 18, A-1190 Vienna, Austria, E-mail: thomas.prohaska@boku.ac.at

The Determination of Chemical Elements in Food: Applications for Atomic and Mass Spectrometry.
Edited by Sergio Caroli

7.1 THE EUROPEAN AND INTERNATIONAL MEASUREMENT INFRASTRUCTURE

Over the last few years the discussion on the quality of the food consumed in highly industrialized countries has been constantly growing. Rising affluence, lifestyle changes, and accompanying urbanization also appear to be largely responsible for dietary alterations in some regions that until recently were very traditional in their diet. The combination of lack of movement, stress, bad eating habits, and the accessibility of food of minor nutritional quality leads to obesity and, in the long term, results in severe health damage. Cardiovascular diseases, diabetes, and an increased risk of certain forms of cancer are already known as the epidemics of the century. About 20 percent of the European population suffer from obesity, amongst them 6 million children. For the first time ever, nowadays it is likely that children will die earlier than their parents. About 2–8 percent of the total health care costs in Western countries can be ascribed to obesity [1]. The European Union (EU) is already aware of this increasing threat to economy and health and is working on various scenarios to react to the challenge of bringing nutritional habits into line with the requirements for a healthy lifestyle [2]. In contrast, the rates of malnutrition in poor and underdeveloped countries, particularly in Africa, are still rising, resulting in health deficiencies and, in the worst case, in starvation [3]. The world's population is steadily increasing, as is the gap between the average life expectancy of the population in highly industrialized countries and the population in poor, nonindustrialized countries [4]. Thanks to an increased life expectancy, the ageing of part of the world population results in characteristic disease patterns, some of which are also related to nutrition.

A number of trace elements are proven to be essential for the human metabolism and well-being. A sufficient supply of most of the essential trace elements can usually be achieved by a well-balanced, varied diet. Thus, deficiencies do not only depend on the nature of the farmland and the treatment of foodstuffs, but are increasingly linked to bad nutritional habits. It is well known that some trace elements are likely to have anti-carcinogenic effects. Selenium, in particular, is being clinically tested for the treatment of Alzheimer's disease and Parkinson's syndrome [5, 6]. Some trace elements can also cause chronic or acute toxic or even lethal effects in cases of high intake. As an example, Pb is elevated in patients suffering from these latter diseases [7]. More recently, there have been indications that diseases like arthritis and arthrosis are not only related to ageing and the decline of the bone and cartilage structure, but also to the trace element status of the individual as well [8]. A large number of toxic elements are anthropogenic environmental pollutants that finally end up in the food chain. Therefore, trace element analysis in food is a key issue for public health and the quality of chemical measurements in trace element analysis is gaining ever more importance.

Confidence in trace element analysis is also inescapable in economic matters. The mutual recognition principle in the EU guarantees the free movement of goods and services. In general, goods which are lawfully produced in one Member State cannot be banned from sale on the territory of another Member State. Exceptionally, goods and services can only be banned if urgent health or consumer protection issues are at stake. In the EU active networks and organizations envisage competitiveness guaranteeing economic wealth without jeopardizng consumers' health. The European Food Safety Authority (EFSA) is the institution in the EU responsible for food and feed safety. In close collaboration with national authorities, the EFSA provides independent scientific advice on existing and emerging risks related to food and feed safety [9].

The core of the European food measurement infrastructure is the system of European Community Reference Laboratories (CRLs) for residues. They are designed to improve and implement analytical methodologies and the scientific basis of residue control. The CRLs act via networks of National Reference Laboratories (NRLs) for residues in food monitoring activities. The accreditation infrastructure is also an important component of the European *Acquis*

Communautaire. Accreditation aims at the reliability of measurement results and enables laboratories to demonstrate technical competence to their customers. Metrological concepts in food science have become a widely discussed topic over the last few years. Representing the top of the international measurement infrastructure, the National Metrology Institutes (NMIs) have the mandate to disseminate the best practice and measurement capabilities to the field laboratories (FLs) in their countries. The European and international measurement infrastructure needs high-quality tools to assess the reliability and comparability of measurements results in view of the implementation of EU and international policies regarding internal market activities as well as health and consumer protection issues.

7.1.1 EU Food Policies

The EU legislation depends to a large extent, directly or implicitly, on measurements [10]. Transparency and consistency of the rules laid down in regulations and directives is a requirement of modern legislation texts. The food sector in Europe is of major significance for the European economy. A key concern on the European political agenda of the EU and its Member States is food safety. Far from being an isolated concept, the safety of food is becoming an objective to be incorporated into all EU policies. The European market is one of the largest in the world. It has recently grown remarkably because of the accession of new Member States and will grow even more with the opening towards the countries of Southeast Europe. The European Commission's (EC) White Paper on Food Safety of 12 January 2000 set out the plans for a proactive new food policy. Particular attention must be focused on improving quality standards and reinforcing systems of checks throughout the food chain [11]. It is stated in the White Paper that the European food safety policy should be based on scientific advice to the highest standards of independence, excellence, and transparency. The Scientific Committee on Food (SCF) deals with scientific and technical questions concerning consumer health and food safety associated with the consumption of food products and, in particular, questions relating to toxicology and hygiene. On request of the EC, the SCF reviews the upper level of daily intake of individual vitamins, minerals, and trace elements to provide the basis for the establishment of safety factors [12]. Measurements of contaminants in foodstuffs play a key role in reinforcing control and increasing the capability of the scientific advice system, so as to guarantee a high level of human health and consumer protection [13]. It is essential to keep contaminants at levels which are toxicologically acceptable. Maximum levels of trace elements in foodstuffs are laid down in the EC Regulation (466/2001) on upper levels of contaminants in foodstuffs [14]. The EC Directive 2001/22 regulates sampling and methods of analysis for the official control of certain trace elements with a high risk of harmful toxic effects in human beings, such as Cd, Hg, and Pb [15]. Another major Directive was adopted in 2000. The common implementation strategy for the Water Framework Directive (WFD) 2000/60/EC aims at proper incorporation into the national law of all EU Member States and Norway [16, 17]. European waters should all achieve good status by 2015. Associated with the WFD is the EC Directive 98/83/EC on the quality of water intended for human consumption [18]. The EC Directive 98/83/EC is one of the few pieces of legislation that not only lays down upper levels of trace elements in water intended for human consumption, but also specifies analytical performance characteristics. Each Member State has to ensure that the laboratories that analyze the parameters as listed in 98/83/EC have an analytical quality control system in place and apply analytical methods that meet the minimum stated requirements. All the parameters listed in this Directive are to be measured with the specified trueness and precision and at the specified limit of detection. Most of the trace elements in drinking water must be measured with a trueness of the parametric value of 10 percent [18].

7.1.1.1 The Joint Research Centre Food safety is among the priorities under the EU Sixth Framework Programme (FP6) for research and technological developments of the EC-JRC [19]. The Joint Research Centre (JRC) provides customer-driven scientific and technical support for the conception, development, implementation, and monitoring of EU policies. It is a reference center of science and technology for the EU and serves the common interest of the Member States, while being independent of private or national interests of any kind. The JRC is organized in seven different institutes. One of them is the EC-JRC Institute for Reference Materials and Measurements (IRMM) [20]. The IRMM promotes a common European measurement system in support of EU policies. With an increased focus on food science and metrology, IRMM contributes by performing chemical reference measurements to ensure quality systems in the food chain. A core activity in IRMM is to provide reference measurements of the highest metrological quality for the analysis of trace elements in environmental, clinical, and food matrices, focusing on EU policy implementation support. To this end, the results of the measurements on trace elements must be demonstrated to be reliable and internationally comparable, which directly implies the need for quality control (QC) and recognition of internationally accepted standards. Metrological concepts are indispensable in this process. In this respect IRMM provides, with its International Measurement Evaluation Programme (IMEP$^{@}$), a major tool for benchmarking the measurement performance by means of interlaboratory comparisons (ILCs) based on metrological principles in view of relevant policy implementation.

7.1.2 The International Bureau of Weights and Measures

Metrology is the science of measurements and one of today's key sciences. It is of fundamental importance in industry and trade, but also in the environmental, consumer, and health protection field. The *Bureau International des Poids et Mesures* (BIPM) was set up by the *Convention of the Metre* (Fig. 7.1). The task of the BIPM is to provide worldwide uniformity of measurements and their traceability to the International System of Units (SI), the basis for a coherent common measurement system in view of international comparability of measurement standards and references. The meter convention is a diplomatic treaty signed by representatives of the governments. It gives authority to the Conférence Générale des Poids et Mesures (CGPM) and the Comité International des Poids et Mesures (CIPM) to act in matters of world metrology. It endorses new metrological determinations and international scientific resolutions. The CIPM is advised by consultative committees on the best manner to demonstrate equivalence between national measurement standards. The Consultative Committee for Amount of Substance – Metrology in Chemistry (CCQM) was established in 1993 [21].

In 1999 a new international protocol was signed, the CIPM Mutual Recognition Agreement (MRA). The CIPM-MRA is a process that enables international acceptance of measurement standards of known quality and comparability. It meets the requirements of the ISO/IEC 17025 standard in terms of comparability and reliability of measurement results via traceability to internationally accepted stated references together with their stated measurement uncertainties [22]. The CIPM-MRA strives towards the implementation of the principle "once measured, accepted everywhere" by organizing key comparisons to assess the measurement capability of NMIs and laboratories that are designated to provide reference measurements and standards. These calibration and measurement capability (CMC) claims of the NMIs that are signatories to the CIPM-MRA are listed in the BIPM key comparison database [23]. The objective is the international mutual recognition of calibration and measurement certificates by establishing the degree of equivalence of national measurement standards between NMIs. Since the creation of CCQM in 1993, many NMIs have been asked by their governments to support chemical measurements by disseminating traceability to the industries and society in general, for example, via ILCs, training and Certified Reference Materials (CRMs). Some NMIs organize ILCs on a national level in order to

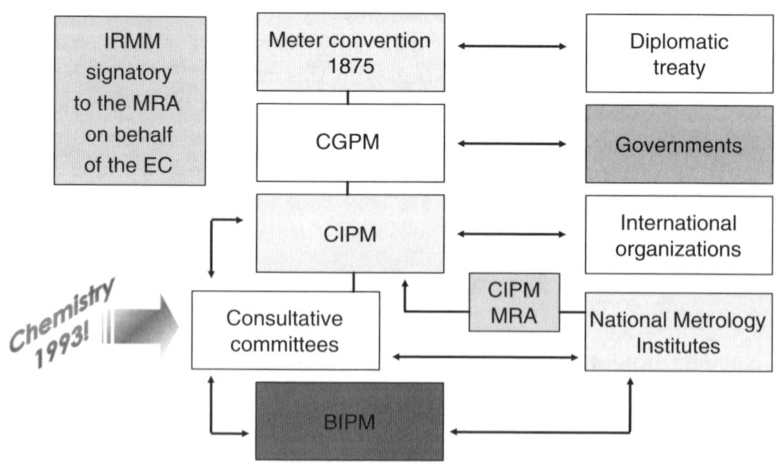

Figure 7.1. The international measurement infrastructure.

monitor the measurement capabilities of laboratories in their country and consequently certify or authorize them for specific tasks. The aim is to provide governments and other parties with a secure technical foundation for wider agreements related to international trade, commerce, and regulatory affairs and to extend the infrastructure to the field of chemical measurements. It is about increased confidence across different countries and national borders. The CIPM-MRA is a completely nonpolitical, transparent, peer review process of measurement capabilities of signatories to the CIPM-MRA. All signatories must have measurement services underpinned by a quality system (ISO/IEC 17025). The IRMM is the signatory to the CIPM-MRA on behalf of the EC. In the chemistry area, the IRMM assists the CIPM-MRA process by providing samples from the same batch as used in IMEP ILCs for CCQM key comparisons and pilot studies [24]. In this way, the IRMM on behalf of the JRC plays a key role in the construction of an internationally structured measurement system.

The Codex Alimentarius Commission has as a mandate to protect the health of consumers and ensure fair practices in the food trade. It was created in 1963 by FAO and WHO to develop food standards, guidelines, and codes of practice under the Food Standards Programme. Food safety standards are developed as reference for the food industry, thereby facilitating the international trade of food commodities. Compliance of food quality laboratories with Codex Alimentarius protects public health and provides consumer confidence in the sector [25]. Very recently, within the scope of cooperation with the Codex Alimentarius Commission, the CCQM has organized group meetings focusing on reference measurement systems for food analysis [26]. Representatives from metrology institutions, the CODEX/AOAC, national food representatives, and proficiency testing providers discussed the approaches of applying metrological concepts, such as comparability, traceability, and uncertainty in food analysis. There where some fields of particular interest for the food sector identified, amongst them trace element analysis in foodstuffs. The elements of major interest are mostly those regulated by international legislation or/and suspected to increase health risks.

7.1.3 Community and National Reference Laboratories

The food measurement infrastructure for residue control in Europe is organized in a network (Fig. 7.2). At the top are the CRLs for residues in living animals and their products [27]. CRLs are designated via a Council Decision and their mandate is laid down in the Council Directive 96/23/CE [28]. On the national level, the NRLs are heading a network of routine food testing laboratories in each of the EU Member States. The NRLs are directly nominated by the relevant Member State. Improvement and implementation of analytical methodology and scientific basis of residue control is part of the functions of a CRL. This also includes method validation and development. The CRLs need to possess the equipment and substances to carry out the analyses for which they are responsible. They must also identify and determine residues in cases where

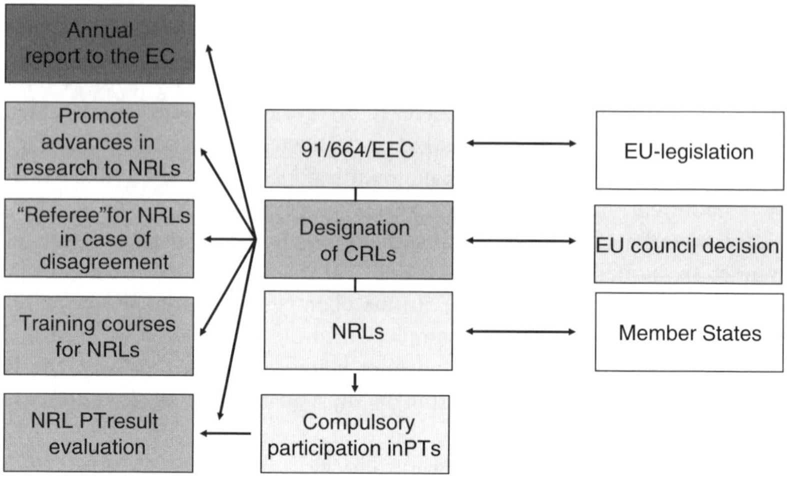

Figure 7.2. The European food measurement infrastructure.

the results of an analysis give rise to a disagreement between Member States. The CRLs provide technical support to the EC in this matter as well as to the NRLs for residues in the Member States. The CRLs must be accredited and must have an efficient quality system in place. They give advice to NRLs regarding the implementation of Good Laboratory Practice (GLP). QC is a main pillar in the CRL–NRL network. The Council Directive 96/23/EC prescribes that the CRLs for residues organize regularly proficiency testing for the NRLs to assess and evaluate their performance. It is mandatory for the NRLs to demonstrate their capabilities to the CRL. In that sense CRLs have similar tasks and competencies in the food measurement infrastructure as NMIs have in the international measurement infrastructure. The CRL has to monitor the capacities of the NRLs and disseminate best practice to the field laboratories via ILCs. Furthermore, the CRLs assist the NRLs to improve their measurement performance by organizing specific training activities. Transparency is provided via an annual report to the EC that summarizes the activity of each CRL. Up until now there has been no protocol for CRLs to demonstrate their measurement capabilities by participating in international comparisons of measurements, to be known as key comparisons [23].

The determination of chemical elements falls within the scope of the CRL for residues at the Istituto Superiore di Sanità (ISS), Rome [29]. By means of proficiency testing, the ISS-CRL constantly monitors the performance of the NRLs as regards the determination of trace elements. Recently, a cooperation was established with the IRMM to organize an IMEP ILC for NRLs on trace elements analysis in tuna fish. As a follow-up of this cooperation specific training activities for NRLs are scheduled to be jointly organized by the IRMM and the ISS-CRL.

7.1.4 The Network of National Accreditation Bodies

The Council Resolution *Enhancing the Implementation of the New Approach Directives* is enforcing accreditation [30]. This piece of legislation lays out a new approach to technical harmonization and standards, the global approach to conformity assessment, and the role of standardization in Europe. Trading partners should be encouraged to adopt standards and regulatory approaches compatible with the EU regulatory framework by the removal of technical barriers to trade. Increasing consistency, transparency, and cooperation of accreditation services within the EU are requested in this legislation to develop more comprehensive policy and guidance for the definition and use of accreditation.

The International Laboratory Accreditation Cooperation (ILAC) promotes laboratory accreditation [31]. Particular emphasis is given to making accredited laboratories attractive not only to industries, regulators, and consumers, but also to governments and governmental institutions as well. Therefore, ILAC aims at developing and harmonizing laboratory and inspection practices to enable assessment and recognition of competency of accredited laboratories. The accreditation bodies organized under ILAC established a mutual recognition arrangement to prove competencies. The ILAC-MRA provides technical underpinning to international trade in order to establish and maintain mutual confidence in the technical competence of ILAC members and their accredited laboratories. The free-trade goal is that a product is tested once and accepted everywhere. One major element of the MRA is that accreditation bodies have to investigate and assess whether their accredited laboratories are compliant with the requirements of the ISO/IEC 17025 Standard [22]. Traceability to the references provided by the NMIs is claimed by accredited laboratories. Therefore, confidence in the NMIs is another cornerstone of the ILAC-MRA. This implies also confidence in the transparent process of measurement capability claims via key comparisons of the BIPM. In 2001 ILAC and the CIPM signed a Memorandum of Understanding striving for the harmonization of future policies on matters of common interest to extend the national and international measurement systems into the broader community of users. On 20 January 2003, ILAC became an Association under the Dutch Law [31]. Full members of ILAC are accreditation bodies that ensure that their accredited laboratories comply with ILAC statutes. These accreditation bodies need to carry out their assessments in compliance with the ISO/IEC 58 Standard and they have been accepted as signatories to the ILAC-MRA [32].

7.1.5 European Cooperation for Accreditation

The process for multilateral recognition in accreditation is very well established in Europe (Fig. 7.3). European Accreditation (EA) is the European network of National Accreditation Bodies (NABs) and a member of ILAC [33]. The main objective of EA is to remove barriers to trade and to ensure the harmonized performance of NABs throughout the European Economic Area. Again the goal is:

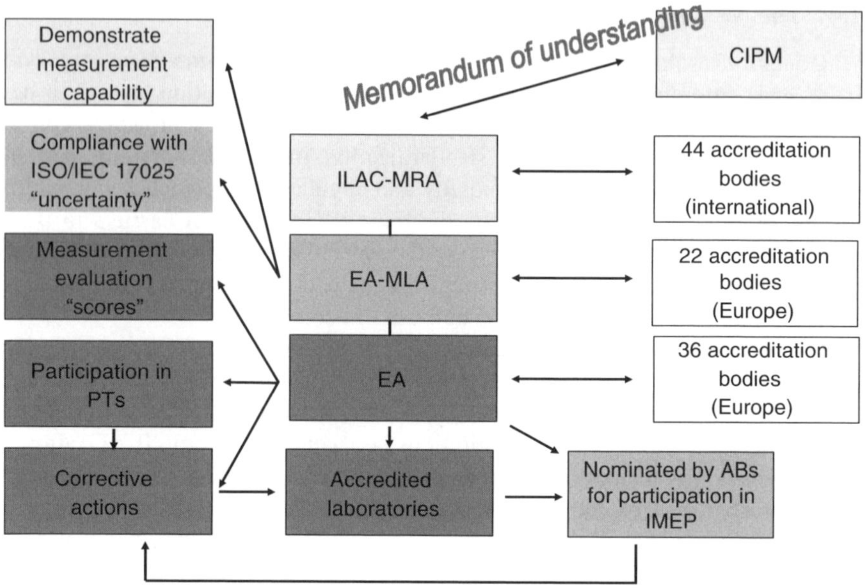

Figure 7.3. Accreditation infrastructure.

once measured, accepted anywhere. The EA Multilateral Agreement (MLA) of equivalence provides recognition and acceptance of certificates and reports issued by organizations accredited by NABs who are signatories to the MLA. EA promotes and develops the principles of conformity assessment. Governments should be convinced to use services of accredited laboratories according to their proven competencies.

In the food sector the majority of laboratories involved in trace element analysis has to be accredited. A very efficient mechanism for those laboratories to demonstrate their measurement capabilities and proficiency is to participate in ILCs. In 2001 a formal letter of intent for cooperation was signed by the Chairman of the EA and the director of IRMM at that time [34]. The cooperation focused mainly in the joint organization of ILCs for accredited laboratories in the frame of IRMM's IMEP programme. In order to further improve the efficiency of accreditation in chemistry with respect to the evaluation and demonstration of the performance of laboratories, in 2004 EA and IRMM agreed to intensify and extend their existing cooperation. Concerning ILCs, an investigation was started towards a more metrology-based approach for performance evaluation of laboratories participating in ILCs. In addition, follow-up training supplementary to the implementation of corrective actions was offered to laboratories after their participation in an IMEP ILC. General training workshops in metrology in chemistry for technical assessors have also been organized [35]. By supporting EA, IRMM assists in intensifying the dialogue with regulators to reinforce confidence in accreditation.

7.2 IMEP®, THE INTERNATIONAL MEASUREMENT EVALUATION PROGRAMME

Laboratories need to prove and improve the quality of their measurements. This is indispensable in view of the demands of consumer protection and competitive contract bidding. Measurements must comply with international standards as set in MRAs. Some 15 years ago the IRMM started to provide the public with a metrological tool to assess global equivalence and reliability of measurement results. The International Measurement Evaluation Programme (IMEP) is organized in support of EU policies in the clinical, environmental and food sectors [36]. The IRMM promotes a common reliable European measurement system in support of EU policies. The selection of test sample material and the analytes under investigation in IMEP ILCs is always related to EU legislation, preferably to regulations or directives that are to be implemented in the Member States or in the status of revision.

The most recent IMEP ILCs on trace elements analysis in wine, rice, and tuna fish were organized in support of the EC Regulation (466/2001) on upper levels of contaminants in foodstuffs. The IMEP ILCs on water were coordinated in support of the EC Directive 98/83/EC on the quality of drinking water intended for human consumption [18]. In general toxic elements that are strictly regulated in the Directive, such as Cd, Hg, and Pb, are elements under investigation in IMEP, but IMEP also focuses on essential trace elements for human beings like Cu, Se, and Zn in food matrices.

7.2.1 Proficiency Testing and IMEP

According to the definitions in the ISO/IEC Guide 43:1997, an ILC is the organization, performance, and evaluation of tests on the same or similar test items by two or more laboratories in accordance with predetermined conditions. In the same guide, Proficiency Testing (PT) is defined as the determination of laboratory testing performance by means of ILCs [37]. Participation in ILCs can serve different purposes. The most frequent is to demonstrate competence and establish degree of equivalence between the results of the participating laboratories, but participation in ILCs can also serve to assign certified values, to validate methods, or to simply monitor and improve laboratory skills. Accredited laboratories are asked by their accreditation bodies to participate in PTs. In that way participation in ILCs enables laboratories to demonstrate competence to third parties. Nevertheless, the main goal remains that participation in ILCs should not be primarily to demonstrate competence to an inspection body or to customers, but mainly to the benefit of the laboratory finding out about the present status and quality of their measurements so that corrective measures can be taken when needed. The IRMM is not a PT provider in the common sense. The EPTIS database lists hundreds of PT schemes operated in Europe and the Americas [38]. EPTIS is a noncommercial, joint publication of 19 international organizations that was established under the auspices of EA, Eurachem, and Eurolab [33, 39]. In 2004 IRMM became a supporting organization to EPTIS.

IMEP has supplementary features to a regular PT scheme; it is rather a whole metrological process that underlines the need of reliability and comparability of measurements. The IRMM is independent of any private or national interest giving scientific support to all EU Member States. For that reason IMEP ILCs are not organized repeatedly on the same matrices or analytes, but they aim to depict the present measurement capability status in various fields where EU policies are relevant, for example, in the food sector.

7.2.2 IMEP Concepts

IMEP® is a publicly accessible metrological ILC scheme. Participation in IMEP is open to all laboratories. Participants in IMEP receive a certified test sample (CTS) with undisclosed reference values. The CTS is a well-characterized material in terms of homogeneity and short-term stability. Participants should treat this test sample according to the procedures that they routinely apply for similar analyses. No analytical methods or instrumental techniques are prescribed. The aim is to shed light on the actual state of practice in chemical measurement performance. Subsequently, participants report their results to the IMEP organizers using a new ILC online IT system. IRMM guarantees full confidentiality concerning the identity of its participants with respect to their reported results. As soon as all participants have reported their results, the online reporting page is closed and the IMEP certificate with the certified reference values (CRVs) is put on the IMEP web site. Furthermore, the IRMM issues individual certificates to each participant. This certificate includes the measurement value for the CTS as reported by the participant, the certified IMEP reference values, the percentage deviation of the reported value from the certified value, and the measurement evaluation score. Finally, the IMEP report to participants is dispatched to each participant and publicly accessible via the IMEP-web site [36].

Participation in IMEP is either direct via IRMM or via the network of IMEP regional/sectional coordinators (RCs). These are contact persons to the IMEP programme in the EU Member States and also outside Europe. Their function is to act as a link between the IMEP ILC cooordinator at the IRMM and the participants (laboratories) in a particular geographic region or within a particular sector of chemistry. Being mostly from metrology institutes or accreditation bodies, RCs have already typically established such an interface. The RCs contact relevant laboratories in their region for participation in IMEP as soon as a new IMEP ILC is launched. They assist in the logistics of the execution, such as the distribution of samples. The complete list of IMEP RCs can be found on the IMEP web site [36]. The IMEP process also includes follow-up training and educational activities for laboratories after participation in an IMEP ILC. The RCs are often involved in identifying needs for follow-up training and future IMEP ILCs. The network of RCs has been very efficient and significantly contributed to the international reputation and acknowledgement of the IMEP programme.

To enable assessment of equivalence of measurements is of major importance on all levels of the measurement infrastructure. IMEP ILCs are linked to internationally recognized measurement capabilities to provide participants with the possibility of demonstrating the quality of their measurements in compliance with national and international requirements as set in MRAs, standards, or quality systems.

7.2.3 IMEP Specific Features

Participants in IMEP can demonstrate the quality of their measurements by comparing their measurement results with an independent certified reference value. This is the main feature of the IMEP programme and the difference from most of the available Pt schemes. It has to be emphasized that IMEP does not compete with regular PT providers' schemes: It is rather a complementary tool that provides, as a service to participants, the comparison of their results with CRVs. These CRVs are completely independent of participants' results and are established in a transparent way via certification campaigns and/or internationally recognized measurement capability claims. Another characteristic feature of IMEP is that right from the start of this programme in 1998 participants were invited to report results with reliable uncertainties. Participants should supply a best estimate of their expanded measurement uncertainty. This became a very important issue for all laboratories that have to meet the requirement of providing reliable measurement results with uncertainties according to the ISO/IEC 17025 Standard [22]. Participants in IMEP have to state whether they evaluate their measurement uncertainty following the *Guide to the Expression of Uncertainty in Measurement* (GUM) or according to similar internationally accepted guidelines [40, 41]. In case they do not follow any guidelines for establishing reliable uncertainties to their measurements, they are asked to report their applied approach for uncertainty estimation.

The two pillars of the IMEP programme, the CRVs and the uncertainty reported by the participants, has recently allowed a new approach to be developed, that is, the result-oriented evaluation of measurement performance relative not only to the reported value, but also to the reported measurement uncertainty. The IMEP process has become more and more attractive to organizations with a European or international mandate for surveillance or inspection of laboratory networks. As a result, participants in IMEP have been increasingly nominated via European or international organizations to benchmark the performance of laboratories from their networks. In the frame of the EA-IRMM collaboration agreement, NABs regularly nominate laboratories that are accredited or in the course of accreditation for participation in IMEP ILCs. In the food sector cooperation was established with the ISS-CRL to organize ILCs and follow-up training activities for NRLs. The International Organization for Wine and Vineyards (IOV) nominated laboratories from their network for participation in the IMEP ILC on wine. Regarding trace elements analysis in water, laboratories were nominated from the Global Atmosphere Watch programme of the World Meteorological Organization of the United Nations for participation in the most recent IMEP ILC on water.

7.2.4 IMEP Certified Reference Values

CRVs are based on primary or other internationally recognized measurement procedures [42]. The underlying philosophy is that the best possible values will serve as reference and these are obtained from well-understood measurement processes rather than via a mere consensus approach. CRVs are required to demonstrate traceability and should have a demonstrated and adequately small uncertainty, as evaluated according to international guidelines. To establish and guarantee the high metrological quality of the IMEP reference values, reference measurements are performed by a network of reference institutes with demonstrated measurement capabilities. The majority of these institutes are NMIs or well-known expert laboratories in a specific field. These reference laboratories prove their claims of measurement capabilities by means of participation in international, mutually recognized CCQM key comparisons [43]. The establishment of the CRVs and the measurement capabilities of the reference institutes are published in a certification report. The characteristics of the sample material and the analytes under investigation are described in detail in this report. For each analyte the reference measurement results of each reference institute are listed and displayed graphically. All reference institutes are identified through their reported reference measurement result. The final calculation of the IMEP reference values is also presented in the certification report. In case the IMEP reference values were not established via a separate certification campaign, but via already demonstrated measurement capability claims, the relevant links are given to CCQM reports and the CMC key comparison database KCDB [23]. It happens sporadically that participants bear skepticism towards the correctness of the IMEP reference value. To overcome this, they are invited to consult all these certification and CCQM reports at the IMEP and the BIPM web sites [23, 24, 36]. As one of the reference institutes, the IRMM also has to demonstrate its measurement capability for the specific analyte in the matrix to be certified. The establishment of IMEP CRVs is therefore a straightforward completely transparent process, independent of any political or commercial interest.

7.2.5 IMEP Participants' Results

In early days IMEP became internationally known because it was one of the few ILCs that was not based on consensus values derived from participants' results. Contrary to common belief, the results of the various IMEP comparisons showed an unexpected large spread of participants' results even for simple measurements, far from the laboratories' own quoted standards of accuracy. To learn more about the origin of this spread, participants in IMEP were also asked to fill in a questionnaire. The questions gather information on the methodological approach, the establishment of traceability via the use of CRMs and the evaluation of uncertainty, the experience of the analyst, the self-declared status of accreditation, and the quality system. Based on general information about the laboratory (e.g., country of origin) and the answers given in the questionnaire, the results obtained for each element are grouped in sets of data. For each set of data, the results are plotted in ascending order against the CRV. The most convenient scale for the graphs is usually ±50 percent deviation from the CRV. No results are excluded! Participants' reported results that are off-scale are shown in textboxes. Recently, results have also been presented graphically, sorted according to their measurement evaluation scores. Several examples of IMEP ILCs organized on water and food matrices are given in the paragraphs which follow. All the participants and certification reports are accessible at the IMEP web site [36].

7.2.5.1 IMEP-12: Water Water is the most essential and precious resource for the well-being of all living species on the planet. The anthropogenic increase of pollutants in drinking water has become a serious threat to part of the world population. To account for this problem four major IMEP-ILCs with increasing numbers of participants, IMEP-3, IMEP-6, IMEP-9, and IMEP-12, on trace element determinations in water were carried out from 1992 to 2000 in support of the Council Directive 98/83/EC on the quality of water intended for human consumption [18]. In this Directive the minimum requirements concerning the regular monitoring of water quality and the parameters to be monitored are stated. In all four water ILCs there were 21 trace elements to be certified [44–48]; among them were B, Cd, Cu, Fe, Ni, and Pb. In the most recent IMEP ILC on water, IMEP-12, 348 laboratories from 46 countries reported measurement results. The matrix was a synthetic water sample prepared by IFA-Tulln according to the IRMM specifications. The amounts of the trace elements certified (As, B, Cd, Cr, Cu, Fe, Mg, Mn, Ni, and Pb) were close to the levels endorsed in the Council Directive 98/83/EC and established via a certification campaign [47]. Figure 7.4 shows a typical IMEP graph of the participants' results for Cd in water. The results are displayed in ascending order. In the caption on top the CRV is given with its expanded uncertainty. This value within its uncertainty is represented in the figure as the gray band reference range. The text box in the top left corner identifies the element. The primary ordinate to the left scales the elemental concentration, whereas the secondary ordinate to the right gives the deviation from the CRV in percentage. As it can be seen from Figure 7.4 there is a large spread of results. The dotted lines indicate the 10 percent trueness of the parametric value for Cd as laid down in the water Directive. Only half of the participants who reported results for Cd in IMEP-12 would meet this requirement. About 20 percent of the participants reported values with a deviation of even more than 50 percent from the CRV.

On special request of IMEP RCs, IMEP results are also presented according to regions or countries. As an example, the results of laboratories from geographic Europe for Cu measurements are given in Figure 7.5. Samples from the same water batch were also used in an EUORMET/METCHEM comparison, representing the European Regional Metrology Organization [49].

7.2.5.2 IMEP-16: Wine Wine is an important commodity appreciated since ancient times. The total volume of wine produced each year exceeds 26 billions liters, thus being one of the major trade goods. QC is a major issue for wine

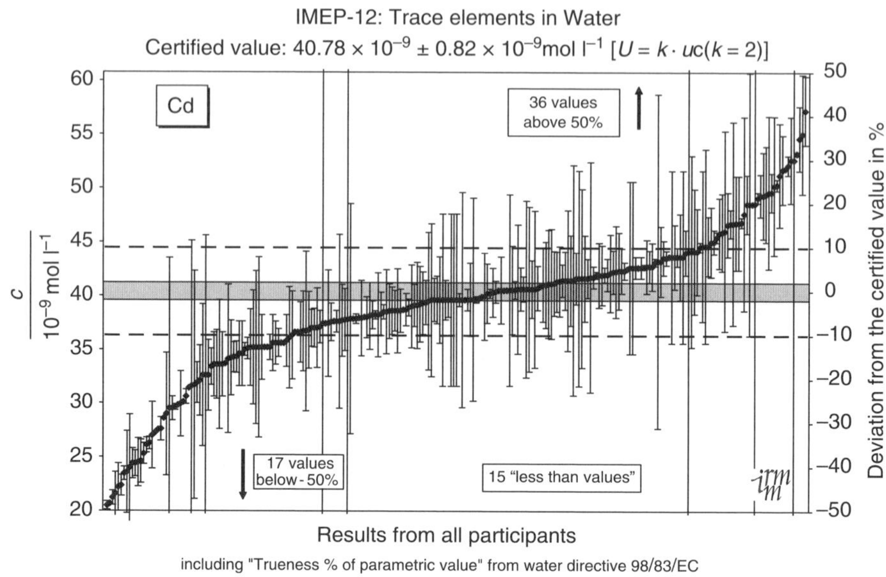

Figure 7.4. IMEP-12. Results for Cd from all the participants.

producers and relies on the results of chemical measurements. The IOV has helped to harmonize regulations on chemical and microbiological parameters [50]. The EC Directive 2001/22 and EC Regulation 2676/90 describe the Community methods for the sampling, sample preparation, and determination of Pb in wine. The official threshold value of Pb in wine is 0.2 mg kg^{-1} [14]. The Pb content in wine depends in the first place on the Pb present in the soil where

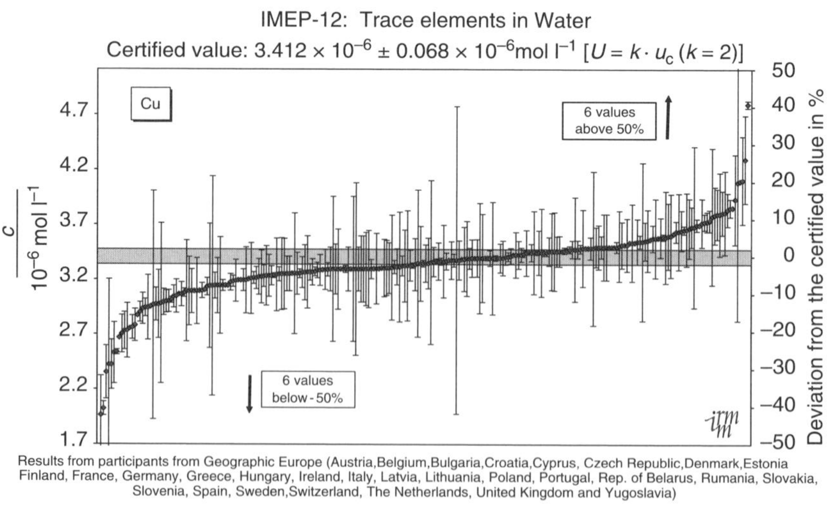

Figure 7.5. IMEP-12. Results for Cu from participants from geographical Europe.

the grapes are grown. Lead contamination can also be due to the wine production process. IMEP-16 focused on the total amount content and the isotopic composition of Pb in red wine. The certified test material was a Bordeaux wine from the shelf. The reference value was established at the IRMM using inductively coupled plasma (ICP) with isotope dilution (ID) – mass spectrometry (MS) [51]. Results were reported by 129 participants from 38 countries, including IOV nominated laboratories. For policy implementation purposes participants were asked to indicate whether they had used the method prescribed in EC regulation 2676/90 or any other method, be it official or not [52]. Figure 7.6 shows the results for the total amount content measurements according to the use of an official method. Furthermore, two-thirds of the results plotted in Figure 7.6 have been reported by laboratories that declared that they were experienced in this kind of analysis. It can be seen that, regardless of the method applied, all the results show a similar spread. This pattern is very often observed in IMEP ILCs. It indicates that reliable results are not necessarily guaranteed by applying a specific, recommended, or officially standardized method, nor by the claimed experience of the analyst.

Participants were also asked about the purpose of their participation in IMEP-16. Seventy percent replied that they participated to monitor internal quality standards. About 30 percent replied that they participated in IMEP-16 to demonstrate measurement capabilities to authorization or accreditation bodies. Samples from the same wine batch were also used as test material in a CCQM pilot study and an EUROMET/METCHEM comparison [53].

7.2.5.3 ***IMEP-19: Rice*** Rice seems to be the oldest cereal cultivated and the main foodstuff for about half of the world population. Asia is the main rice producer and consumer, but also in Latin America and Africa rice is one of the major

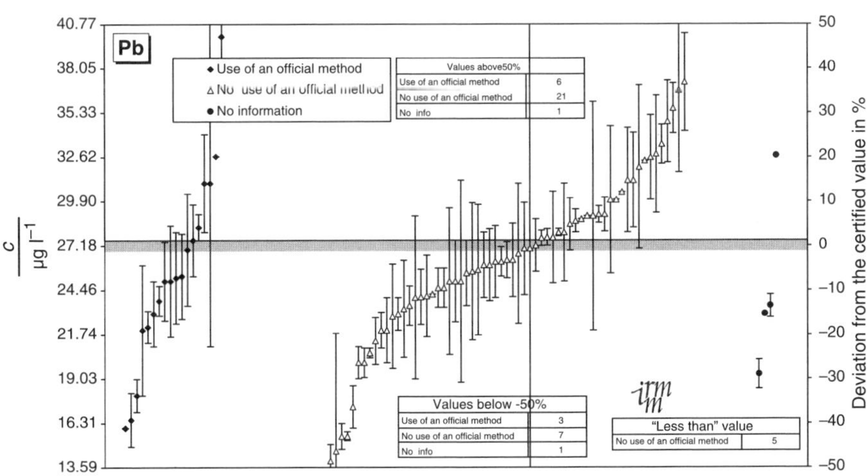

Figure 7.6. IMEP-16. Participants' results for Pb according to the method used, including the one recommended in EC 2676/90.

nutrients. In recent decades rice consumption has been expanding beyond the traditional rice-growing areas, particularly in Europe. It is of common interest to keep contaminants in rice at toxicologically acceptable levels. The rice sample used in IMEP-19 originated from rice grown in Cd-contaminated water. Besides Cd, Pb, Cu, and Zn were also proposed as analytes under investigation to the IMEP-19 participants. Cadmium and Pb are highly toxic. Cadmium has been proven to be carcinogenic and may induce dysfunctions and reproductive deficiencies in humans. Lead may increase cardiovascular diseases in adults. The main human intake of Cd and Pb is via food. Therefore, strict upper levels in rice are set for Cd at a threshold value of 0.2 mg Cd $\cdot$ kg^{-1} in EC 466/2001 [14]. Copper and Zn are well known as essential trace elements for humans. Copper is a constituent of the caeruloplasmin enzyme and is essential for proper cellular metabolism. Zinc is proven to be an essential factor in over 100 enzymes. The EC has recently requested the SCF to review the upper level of the daily intake of Cu and Zn [12]. The reference measurements for Cd, Cu, and Pb were carried out at the IRMM using ID-MS as a primary method of measurement on a Quadrupole (Q) ICP-MS instrument [54]. The reference measurements for Zn were carried out at Bundesanstalt für Materialprüfung (BAM) using IDMS as a primary method of measurement on a multicollector thermal ionization mass spectrometry (TIMS) instrument [55]. A total of 267 laboratories reported results in IMEP-19. Figures 7.7–7.10 shows all the participants' results in IMEP-19 for the four elements according to the analytical technique applied, the quality system, the participation in PTs, and reported uncertainty to customers. Samples from the same rice batch were also used as a test material in a CCQM key comparison and pilot study and an EUROEMT/METCHEM comparison [53, 55].

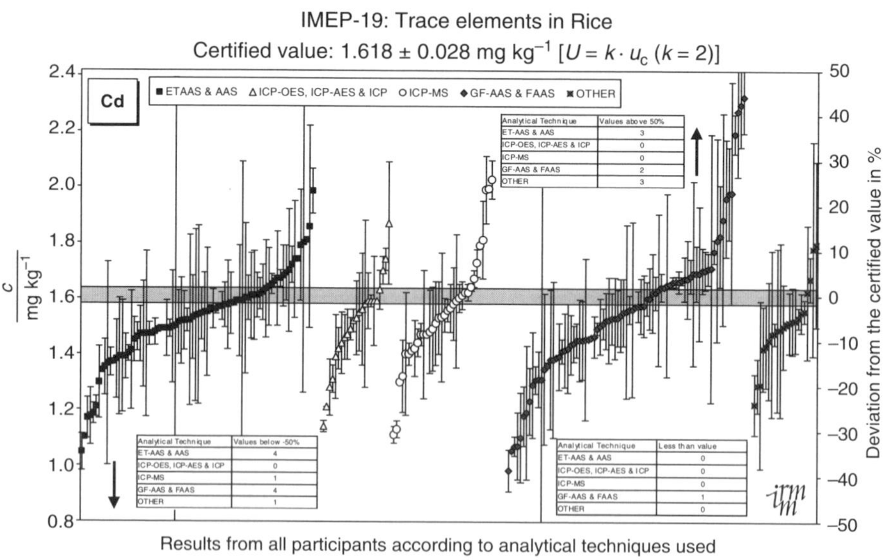

Figure 7.7. IMEP-19. Participants' results for Cd according to the analytical technique.

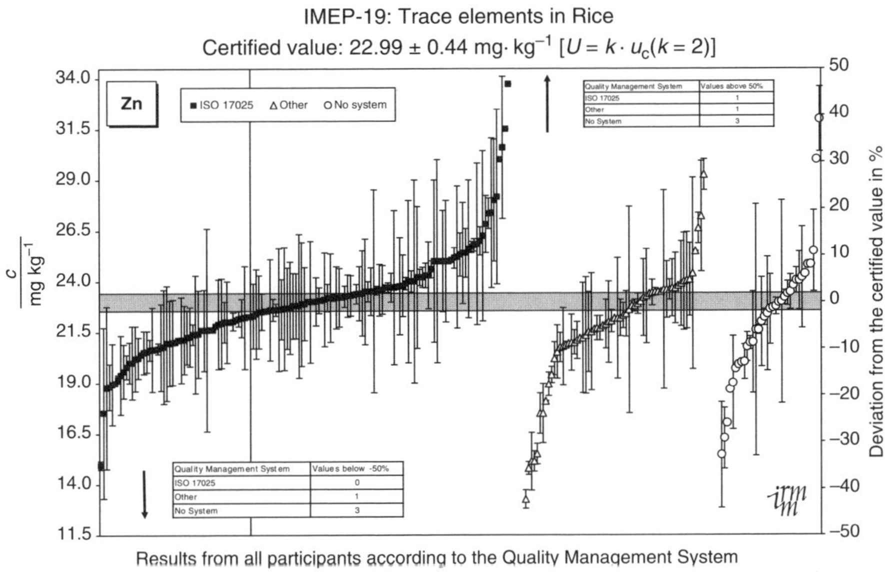

Figure 7.8. IMEP-19. Participants' results for Zn according to the quality system.

7.2.5.4 IMEP-20: Tuna Fish Fish and fishery products are a central element in the diet of a good part of the world population. Recently, concerns about health risks due to elevated contents of Hg in fish were raised by scientists and the media. To address this concern the EU Directorate General for Health and Consumer Protection organized surveys on the consumers' risk perception associated

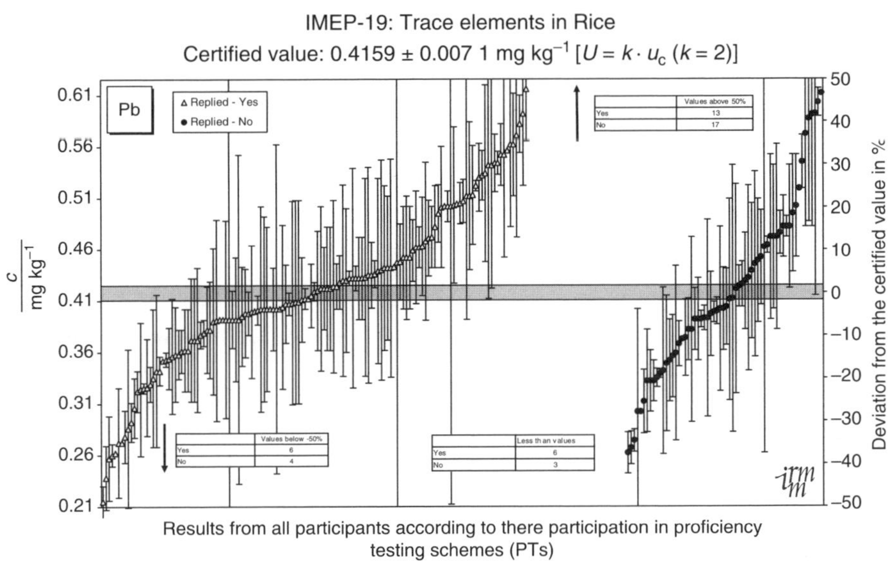

Figure 7.9. IMEP-19. Participants' results for Pb according to participation in PTs.

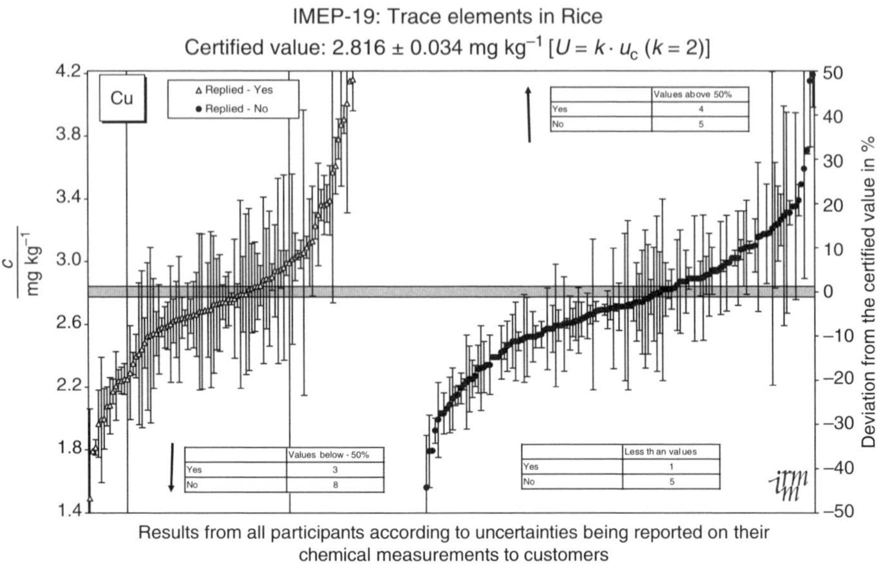

Figure 7.10. IMEP-19. Participants' results for Cu according to the uncertainties reported to customers.

with the presence of Hg in fish [56]. The IRMM asked participants in IMEP-20 to measure the total content of As, Hg, Pb, Se, and methylmercury (CH_3Hg) in a tuna fish product. Mercury is a potential environmental toxicant. The main source of human intake of Hg contaminants originates from CH_3Hg in fish and fishery products. So far, IMEP has focused on the total content measurements of trace elements in different matrices. In IMEP-20 for the first time a species-specific analyte was included in an IMEP ILC. Methylmercury is highly toxic compared to inorganic Hg and is the most abundant organomercury species in the environment. Mercury species may induce alterations in the normal development of the brain of infants and may induce neurological changes in adults. To protect public health, the relevant regulations set maximum levels for Hg and Pb in fishery products. The EC Directive 2001/22 describes the community methods for the sampling, sample preparation, and determination of Hg and Pb in fish [15]. The EC Regulation 466/2001 officially endorses the threshold value of 1 mg Hg kg^{-1} and 0.2 mg Pb kg^{-1} in tuna fish [14]. Another element well known for its high toxicity is As. Arsenic was also offered to participants in IMEP-20 because rules for measurements of As are set in the Commission Decision on implementing Council Directive 96/23/EC concerning the performance of analytical methods and the interpretation of results [57]. Although there is no EU legislation in force regarding Se intake it was decided to also include total Se as an element to be quantified in IMEP-20. Selenium is an essential trace element and certain forms of cancer and cardiovascular diseases have also been associated with Se deficiency. Selenium has the narrowest plateau between concentrations that show deficiency and toxic effects in humans. Recently the EC has requested the SCF to review the upper level of daily intake of individual vitamins and minerals (amongst them Se) and to provide the basis for the establishment of safety factors [12]. Seafood is an important source of Se intake for people in some regions.

In view of the importance of Hg in the food sector, IMEP-20 was organized in collaboration with the ISS-CRL in support of the European network of NRLs. IMEP-20 was also intended as a support to the NRLs and the national networks for food monitoring in the new Member States and food testing laboratories in the candidate countries. The tuna fish sample used as certified test material was a real-life sample withdrawn from the market due to its elevated Hg content. It was perfectly appropriate for the purpose of an ILC, because of its homogeneous distribution of contaminants. In principle, every laboratory involved in food testing or food control could have been offered such a tuna fish sample for analysis by a regular client. In particular, the fact that the Hg content slightly exceeded the upper limits as stated in the EC Regulation 466/2001 made it a very valuable test sample for NRLs and food control laboratories. The reference measurements for Hg and Pb were carried out using ICP-ID-MS [58]. The reference measurements for As and Se were performed by k_0-neutron activation analysis (NAA) [59]. The certified value for CH_3Hg was established by means of species-specific gas chromatography (GC) ID-MS measurements [60]. Measurement results were reported by 235 participants from 24 EU Member States, including all the 10 new Member States and three candidate countries. Only eight participants reported results for CH_3Hg, as shown in Figures 7.11–7.15; some examples of special

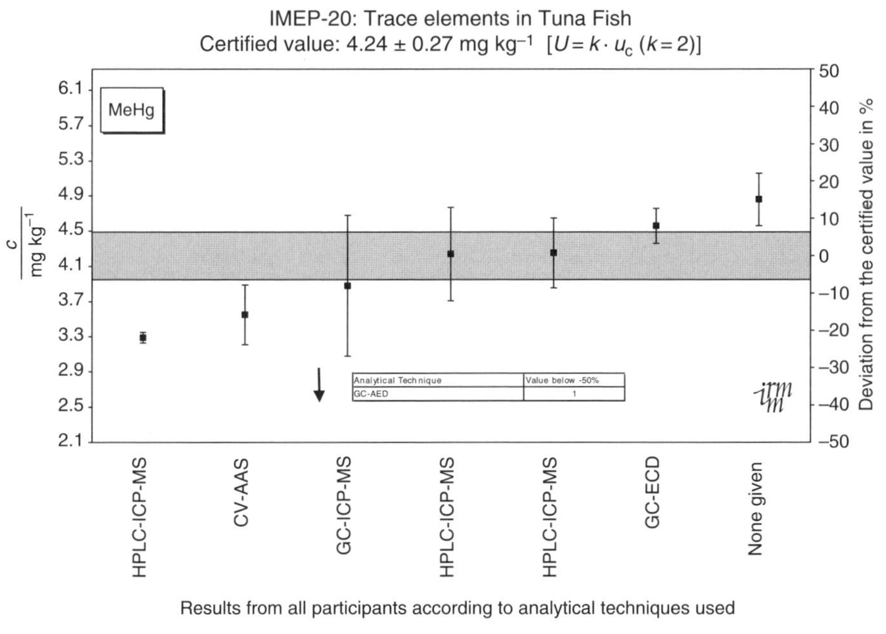

Figure 7.11. IMEP-20. Results of all participants results for methylmercury.

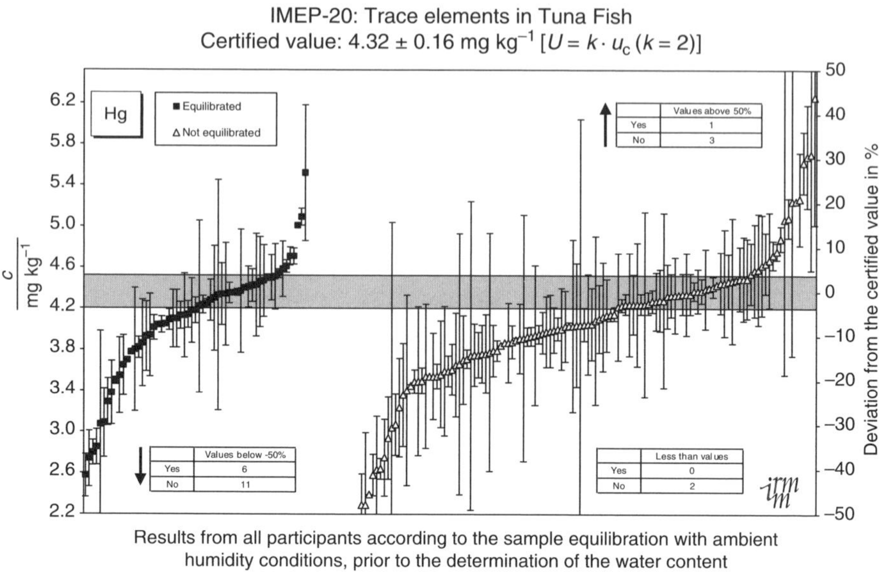

Figure 7.12. IMEP-20. Participants' results for Hg according to sample humidity equilibration.

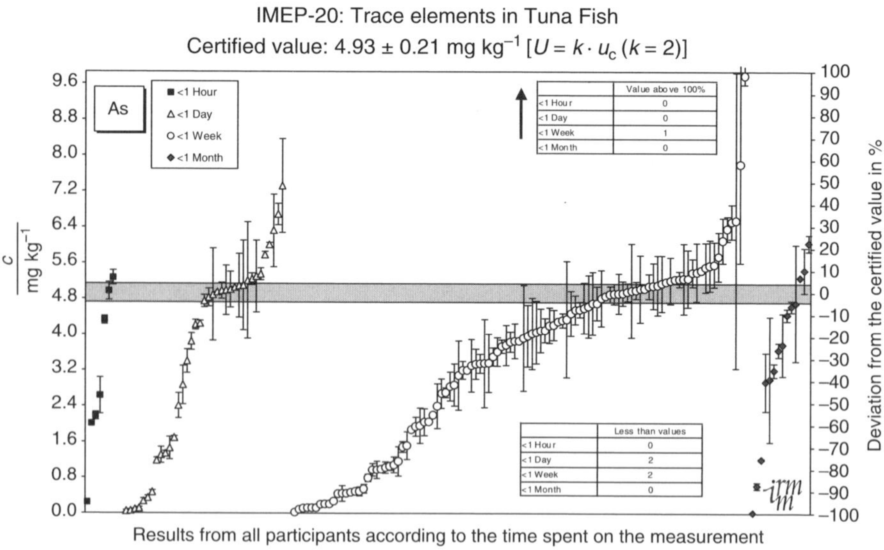

Figure 7.13. IMEP-20. Participants' results for As according to time spent on the measurement.

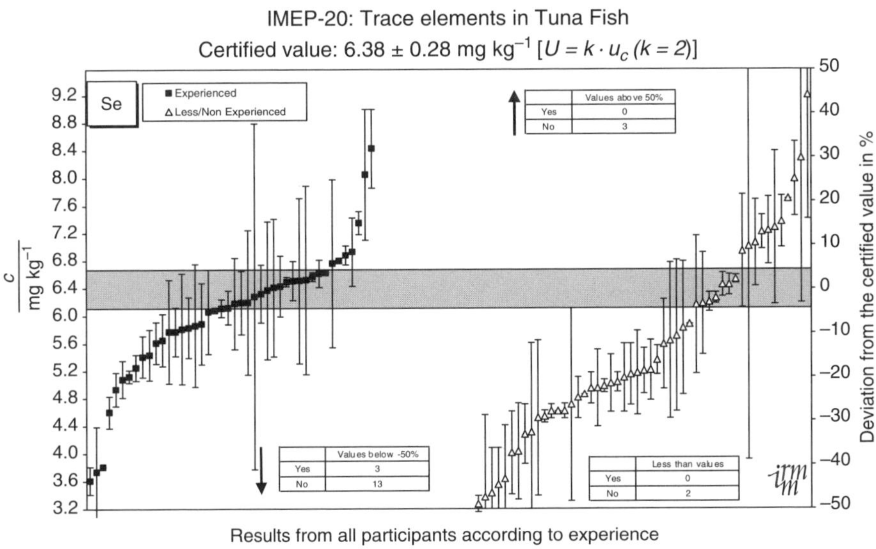

Figure 7.14. IMEP-20. Participants' results for Se according to self-declared level of experience.

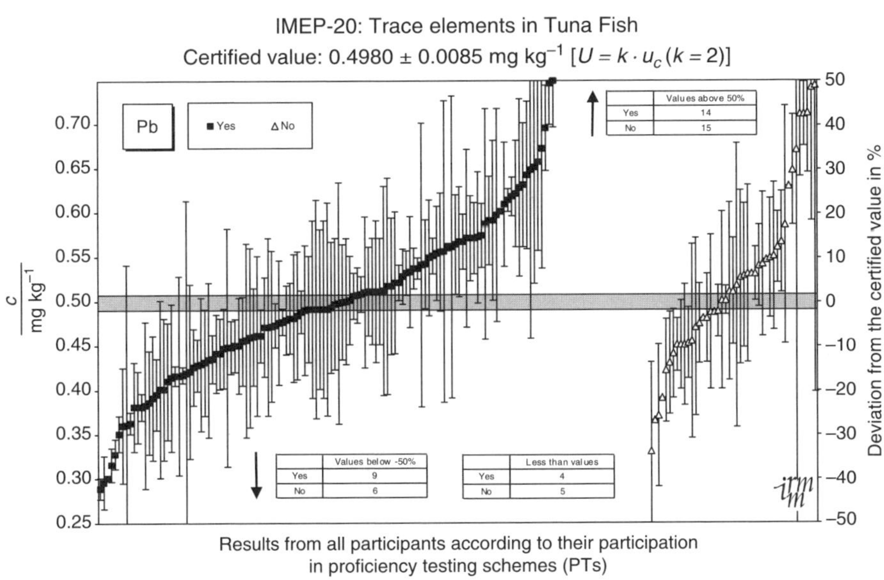

Figure 7.15. IMEP-20. Participants' results for Pb according to participation in PTs.

IMEP graphs are given. These graphs present the information retrieved from the questionnaire. Results in IMEP on solid samples have to be reported as content values corrected for dry mass. There is no protocol prescribed for moisture content determination. Participants are free to use their routine method. The majority of IMEP participants apply the oven-drying method at a certain temperature with or without prior equilibration of the sample at ambient conditions. Other questions regard the time needed for the analysis, the experience of the participants, and their regular participation in PT schemes.

7.3 MEASUREMENT PERFORMANCE EVALUATION IN INTERLABORATORY COMPARISONS

The proficiency of the laboratories involved in trace elements determination is of major importance as their performance is related to the implementation of EU and international legislations, and constitutes crucial information for regulators. Internal QC procedures are required for the majority of laboratories involved in trace element analysis in food. Most of those laboratories must regularly participate in ILC schemes as they have to demonstrate their competence in trace elements determination to regulatory bodies, accreditation bodies, reference laboratories, and customers. Unsatisfactory performance in a series of ILCs can result in the loss of accreditation or mandate as a reference laboratory, or simply in a reduction of the number of customers. Unfortunately, constraints on satisfactory performance are rare in legislation. One of the few exceptions is the EC Directive 98/83/EC on the quality of water intended for human consumption. For each parameter listed in the Directive, the required trueness of parametric value that a laboratory has to achieve is stated in percentages. It is declared in the Directive that "Trueness is the systematic error and is the difference between the mean value of the large number of repeated measurements and the true value" [18]. The water Directive requests that the true value be measured and sets clear criteria for the quality of the measurement. In most cases, where no performance criteria are set by law, a fit for purpose criteria (u_f) has to be used for performance evaluation in ILCs. Fitness for purpose takes into account the end-use the ILC scheme was carried out for. Besides theoretical methods, economic aspects and scientific judgment also have to be taken into account.

7.3.1 Setting Fit-for-Purpose Performance Criteria in IMEP

IMEP provides an independent, CRV of demonstrated quality. This value represents the best estimation of the true value with uncertainty of the analyte under investigation in the matrix. Very often in IMEP the CRVs have a relative expanded uncertainty of less than 2–3 percent. Reference laboratories that contribute to the establishment of the IMEP CRVs, if possible, apply primary analytical methods, like ID-MS, for trace element determination. Routine or field laboratories need not fulfil the terms required for reference laboratories in IMEP. Therefore, the expanded uncertainty of the IMEP reference value is not used as performance criterion, but rather a fit-for-purpose approach is aimed at. In IMEP performance criteria are set in relation to the CRV, either complying with legislation requirements or in agreement with EA or other networks that cooperate with IMEP for specific comparisons, like the ISS-CRL.

7.3.2 *Zeta* Scores and *Zeta*′ Scores

Zeta (z) scores are commonly used as a performance indicator in PT schemes [37, 61, 62]. Sometimes, a consensus value derived from participants' results can serve as reference and the performance criteria, sometimes called target value, equals one standard deviation. This means that 95 percent of the participants would automatically report satisfactory results. In IMEP a z-score is meant as:

$$z = \frac{x - X_{\text{ref}}}{\sigma} \tag{7.1}$$

with X_{ref}, IMEP CRV; x, participant's reported value; and σ, fit-for-purpose criterion (u_{f}), whether chosen or set by law.

The common application of z-score measurement performance evaluation is: $|z| \leq 2$ satisfactory; $2 < |z| \leq 3$ questionable; and $|z| > 3$ unsatisfactory. In IMEP participants are explicitly asked to report their measurement results with uncertainty. Therefore, other scoring approaches also defined in [37] are better applicable to IMEP ILCs. The IRMM developed a slight different approach to calculate E_n-numbers, as described in [63–65], compared to the definition given in [37]. In 2005 the ISO/DIS 13528 on *Statistical methods for use in proficiency testing by ILCs* was issued. According to the definition given in ISO/DIS 13528, the z-score also takes into account the uncertainty reported by the participants [62]. As a consequence, modified *zeta*′ scores are issued to the participants in IMEP rather than modified E_n-numbers [66]:

$$zeta' = \frac{x - X_{\text{ref}}}{\sqrt{u_x^2 + u_{\text{ext}}^2}} \tag{7.2}$$

with X_{ref}, IMEP CRV; x, participant's reported value; u_x, participant's combined uncertainty; and u_{ext}, fit-for-purpose criterion (u_{f}), whether chosen or set by law.

The *zeta*′ scores issued to IMEP participants are based on single performance statistics, accounting for an extended uncertainty u_{ext} of the CRV as well as the reported uncertainty of the participant's measurement result. When calculating *zeta*′-scores according to Equation (7.2), the measurement performance indicators for *zeta*′-scores can be aligned with those of the corresponding z-scores. In case an IMEP participant does not report an uncertainty, the *zeta*′-number equals the z-score: $|zeta'| \leq 2$ satisfactory; $2 < |zeta'| \leq 3$ questionable; $|zeta'| > 3$ unsatisfactory.

Figure 7.16 shows the results for Pb in tuna fish sorted according to ascending *zeta′*-scores for those participants in IMEP-20 who reported measurement results with uncertainties which were calculated according to ISO/GUM or EURACHEM guides for quantifying measurement uncertainty [40, 41]. In the absence of performance characteristics for the uncertainty of the measured value for measurements of metals in tuna fish in the EC Regulation (466/2001), the IRMM, in agreement with the ISS-CRL, selected as performance characteristics an extended uncertainty from the CRV of $X_{\mathrm{ref}} \pm u_{\mathrm{ext}} = X_{\mathrm{ref}} \pm 0.1{\cdot}X_{\mathrm{ref}}$ for the calculation of the *zeta′*-scores.

7.3.3 Setting Criteria for Acceptable Reported Measurement Uncertainties in IMEP

The most frequent objection of using *zeta′*-scores is due to the fact that, according to equation (7.2), a laboratory could improve its performance by artificially increasing the reported uncertainty and thus get a smaller *zeta′*-score. Besides the fact that more and more accreditation bodies carefully investigate nowadays whether accredited laboratories have over- or underestimated their uncertainty, it would not be an advantage on the market for a laboratory to report values on trace elements in a food matrix with a content level that is close to legal limits with a relative uncertainty of, for example, 50 percent or more. Nevertheless, equation (7.2) urges additional criteria of an acceptable uncertainty range. If one looks at the example from IMEP-20 on Pb in tuna fish (Fig. 7.16), a maximum and minimum acceptable level for the combined uncertainty that should be reported by the participants could

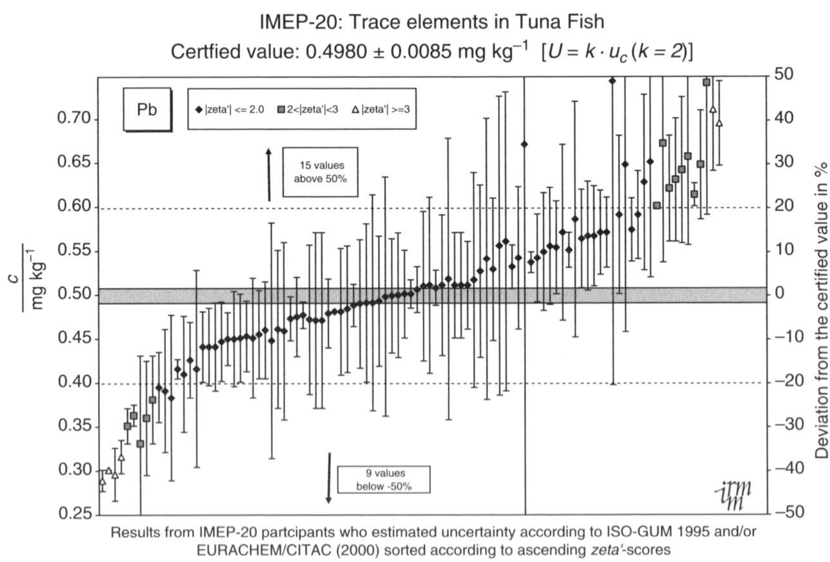

Figure 7.16. IMEP-20. Participants' results for Pb according to measurement performance.

be set – in accordance with the selected performance criteria and the combined uncertainty of the certified reference value – to $u_{ref} \leq u_{lab} \leq 0.1\ X_{ref}$.

As a consequence, satisfactory performance in IMEP-20 would then mean having a result reported with $|zeta'| \leq 2$ and $u_{ref} \leq u_{lab} \leq 0.1{\cdot}X_{ref}$. Laboratories reporting larger uncertainties may not have their experimental procedure under control or may have overestimated some uncertainty components. Laboratories reporting smaller uncertainties are very likely to have either underestimated some of the uncertainty components or not accounted for some uncertainty sources. It has to be emphasized that laboratories with $|zeta'| > 3$ first need to think about the origin of their measurement bias and only subsequently, after corrective measures have been taken, to focus thoroughly on their uncertainty estimation.

7.3.4 The "Naji Plots"

The IRMM developed an easy graphical tool, the so-called Naji plot, to evaluate IMEP participants' results. With the Naji plot approach information about *zeta*′-numbers as well as *z*-scores of single laboratories may be retrieved in a graphical way [63]. Figure 7.17 shows the Hg results of national food testing laboratories that were nominated by their NRLs for participation in IMEP-20. The "Naji plot" provides information concerning *zeta*′-scores while comparing, on the abscissa, laboratory results with the reference value (*z*-score) and, on the ordinate, the ratio of reported uncertainties to the fit-for-purpose criteria $u_{ext} = u_{f.}$. The maximum and minimum acceptable uncertainties are indicated as dashed lines in Figure 7.17. This approach is specifically designed for ILC schemes with independent reference

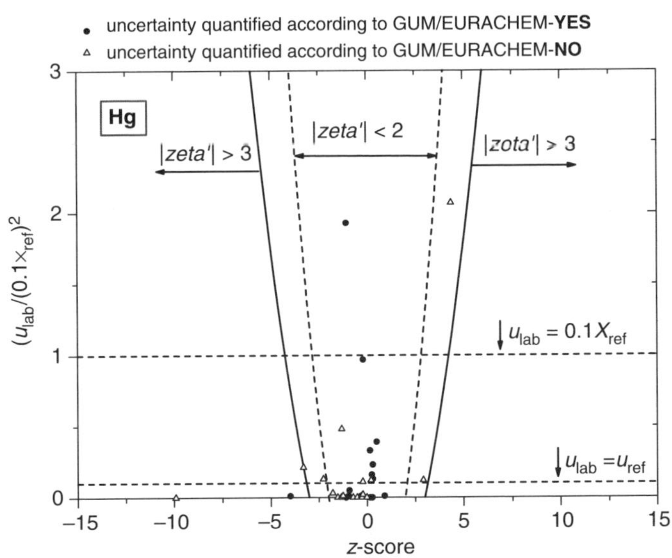

Figure 7.17. IMEP-20. "Naji-plot" of results for Hg from NRL-nominated participants.

values, displaying participants' compliance with performance criteria set relative to the reported value and to the reported measurement uncertainty.

It has to be stressed that, as IMEP coordinator, the IRMM does not evaluate the participants' measurement performance in view of the implementation of corrective actions. It is up to the accreditation or authorization body that nominated the laboratories for participation in IMEP ILCs to assess the laboratories and request corrective measures in order to improve their measurement capabilities.

7.3.5 IMEP Approach Compared to Measurement Evaluation in Routine PT Schemes

While evaluating the laboratory performance according to ISO/DIS 13528, the proficiency test provider computes a normalized bias, which is the difference between the participant's reported measurement result and a reference value divided by a normalization factor [67]. The *z*-score and the *zeta′*-scores are the two scoring equations presented earlier in Section 7.3.1. The ISO/IEC Guide 43 requires that the PT provider clearly states the applied performance criteria and how they were derived [37].

The test sample used in routine PT schemes belongs to one of the following three categories:

- A spike or fortified in-house sample, with a known nominal amount content or concentration of the measurand of interest;
- A reference material (RM) or CRM, with known sample characteristics;
- A sample characterized for homogeneity and stability, but with an unknown content or concentration of the measurand of interest.

In the first two cases measurement performance evaluation scoring can be computed using the reference value and its corresponding uncertainty. However, most of the PT exercises use the third scenario, where the measurand characteristic is derived from the available experimental results of the participants in the PT. In the ideal case where one assumes that no experimental mistakes were made by any participating laboratory, one could indeed derive the reference value from the consensus value as the arithmetic mean of all reported results. When dealing with complex matrices and/or with low analyte concentrations, various effects such as contamination, incomplete extraction/separation, or even human mistakes like the use of wrong units or not taking into account dilution factors, may introduce outlying results. These results may significantly influence the consensus arithmetic mean and the corresponding standard deviation. To account for all this, PT providers often use the robust statistical approach to derive a robust consensus value with a robust standard deviation [68].

As for IMEP, the CRV is derived from measurement results reported by reference laboratories in a certification campaign. These reference laboratories are usually NMIs and expert laboratories that have well-established, demonstrated measurement capabilities and apply, wherever possible, primary methods of measurements. Admittedly, such a strategy is most of the time too expensive for a routine PT provider. In addition, it is difficult for a routine PT provider to select expert laboratories in a specific analytical field. On the other hand, ILCs, with an independent reference value of high metrological quality, provide an incomparable advantage because the performance of all participants is evaluated simply as a normalised bias versus the best estimate of a true value, without using any statistics! Considering all these facts, several questions deserve more detailed reflection.

7.3.5.1 Is the Consensus/Assigned Value Correct? By performing the calculation of any normalized bias the PT provider compares all the participants' results with a given reference. As outcome it can be demonstrated that most of the results are either in good or bad agreement with the reference value. If the robust consensus value was to be computed for the results of As in tuna fish (Fig. 7.13), it would have been about 20 percent lower than the IMEP certified reference value of 4.93 ± 0.21 mg kg^{-1} ($U = k \cdot u_c$, $k = 2$). Based on the IMEP certified range, obtained by different analytical techniques, it is obvious that many laboratories have not yet overcome the experimental difficulties, when dealing with the determination of As in a fish matrix [59].

To give another example it also happened that, having determined Sn in a tomato paste matrix, a laboratory was flagged by the PT provider as an outlyer, when compared with the assigned value obtained by consensus of the participants' results. Further analysis performed using ICP-ID-MS confirmed that the value obtained by the outlyer was correct. It could be proven that the majority of participants had experimental difficulties and that only a few of them had measured the correct Sn content in the respective test material [69]. The PT provider announced, once the laboratory had proven its measurement capability, that it would repeat this PT, but this time using a reference material certified for the Sn content.

7.3.5.2 What is the Uncertainty of the Assigned Value? Many laboratories participate in PTs because they are requested to demonstrate their expertise, in the frame of an ISO/IEC 17025 accreditation. According to this standard, those laboratories should also be able to establish an uncertainty budget and estimate the combined uncertainty of their measurement results. Neither is a straightforward exercise for PT providers to take the uncertainty of an assigned value into account when defining a fit-for-purpose criterion. For the moment there is no guidance available in this matter.

7.3.5.3 Which Assigned Standard Deviation to Choose? When an RM or a CRM is used, the provided uncertainty of the certified analyte – including the uncertainty contribution of homogeneity, characterization, and stability – could be used as assigned standard deviation for proficiency evaluation. In practice, this is seldom done. The common approach is to define a fit-for-purpose criterion as already discussed in Section 7.3.1. Most of the PT schemes are meant to test the performance of laboratories in a specific field of expertise versus requirements defined by a customer or an accreditation body. In this case the value of the assigned standard deviation should be simply determined by professional judgement between laboratory, customer, and/or accreditation body. In some cases performance criteria are laid down in international or national regulations. Even so there is, unfortunately, room for interpretation. When looking at PT schemes organized in support of the WFD much interpretation of the same official text appears [17, 18]. The WFD specifies under paragraph 2.1 "For the following parameters, the specified performance characteristics are that the method of analysis used must, as a minimum, be capable of measuring concentrations equal to the parametric value with a trueness, precision and limit of detection specified." The directive also recalls that: "Trueness is the systematic error and is the difference between the mean value of the large number of repeated measurements and the true value"; "Precision is the random error and is usually expressed as the standard deviation (within and between batch) of the spread of results about the mean. Acceptable precision is twice the relative standard deviation."

In the case of Cd in drinking water the performance criteria are clearly specified as 10 percent for both trueness and precision. Therefore, it is rather surprising to see PT providers stating assigned standard deviations of 5, 7, 10, or even 14 percent for Cd in water, in compliance with the WFD. This is because some PT providers only consider trueness, others, in turn, combine trueness and precision. Some consider precision properly, others take it as a direct estimate of the normalization factor. To improve such situations, the European Union initiated the CoEPT project to study the differences and similarities in the operation of PT schemes and the evaluation of PT results in view of implementing a harmonized approach to provide a basis for the comparability of the PT schemes operated on the market [70].

7.3.5.4 Why is My Reported Uncertainty not Taken into Account? The ISO/IEC 17025 Standard requires a laboratory to be able to estimate the measurement uncertainty. This laboratory may then participate in a PT scheme and report the results with corresponding uncertainties. Assessing its performance as it is commonly done by using the z-score approach can be a disadvantage for this laboratory. It is clear that gross experimental mistakes will not be salvaged by an uncertainty budget calculation, when instead major experimental improvements may be required. However, the laboratory uncertainty may be relevant for questionable results, having $2 < |z| \leq 3$. Combining the assigned standard deviation with the reported uncertainty with the $zeta'$-number approach can result in a satisfactory score. Once more it needs to be emphasized that additional criteria for maximum and minimum acceptable reported uncertainties should be set [64, 65]. IMEP systematically asks participants to report their results with the corresponding estimate of uncertainty, in order to compute $zeta'$-numbers. This is not yet the case for the majority of PT providers (see Table 7.1).

TABLE 7.1. IMEP Versus "Routine" PT Schemes

		Routine PT	IMEP
Assigned value	Nominal value	✓	✓
	By consensus (mean)	✓	
	By consensus (robust statistics)	✓	
	Reference value	✓	✓
Uncertainty of assigned value		Seldom	✓
Assigned standard deviation (SD) for proficiency assessment	Consensus (between lab. st. dev.)	✓	
	Based on customer requirements	✓	✓
	Based on regulatory requirements	✓	✓
Uncertainty of reported results	Computed by the participants	Seldom	✓

7.4 DEMONSTRATION OF MEASUREMENT CAPABILITIES THROUGHOUT THE INTERNATIONAL MEASUREMENT INFRASTRUCTURE

IMEP cannot be understood as solely an ILC or a PT scheme. It is the core of a whole process that enables the assessment of measurement performance throughout the international measurement infrastructure on the basis of high-quality reference measurements. Samples from the same batch are used in parallel to the execution of an IMEP comparison for ILCs to demonstrate equivalence of measurement results on an international, European, or regional level. The organization of all these comparisons in parallel on one sample material needs to be aligned in a strict time planning in order to avoid any exchange of the possible reference value between the participants.

7.4.1 Cooperation with International Organizations

Over recent years fruitful cooperation between the IRMM and different international organizations was established in all fields where chemical measurements are relevant and EU legislation is in force. Besides IMEP being a publicly accessible metrological tool, participants in IMEP are also nominated by international and European organizations from all levels of the measurement infrastructure. EA [33], the International Federation of Clinical Chemistry (IFCC) [71], the WMO [72], the ISS-CRL and NRLs, the Taxation and Customs Union [73], the IOV [50], the European Committee for Standardization [74], and the FOREGS (Geochemical Baseline Programme) [75] use IMEP as a tool to benchmark laboratory performance.

Collaboration can be established in different ways. The most straightforward approach is that international organizations nominate laboratories for participation in one specific IMEP ILC covering the analytes and matrix that is of special interest to the relevant organization. Such cooperation was established between OIV and IRMM in the frame of IMEP-16 *Pb in wine*. Another kind of cooperation is that some samples from a sample batch used in another IMEP ILC are offered to a network of laboratories. In IMEP-15 *Trace Elements in Water*, samples used in IMEP-12 were offered by the IRMM to the WMO, a special agency of the United Nations. The Quality Assurance/Science Activity Center (QA/SAC) coordinated on behalf of the WMO the participation of relevant laboratories. Furthermore, participation in IMEP-15 was of great interest to QA/SAC in view of the intended collaboration between WMO and CIPM [21]. The IRMM could provide IMEP-15 with thorough quality assessment documentation in support to the United Nation's Global Atmosphere Watch (GAW) programme [76]. The third level of cooperation goes beyond the organization of one particular IMEP ILC. It also includes active coordination of the partner institution, issuing of special measurement evaluation reports, and follow-up training for laboratories. IMEP-20 *Trace Elements in Tuna Fish*, in particular, was organized in collaboration with the ISS-CRL in support of the European network of NRLs. The performance criteria in this comparison were set in agreement with the ISS-CRL. The main partner for the IRMM in the organization of IMEP ILCs is EA. The IRMM and EA have revised their collaboration agreement in view of broadening the scope of joint activities in the field of accreditation and metrology in chemistry [34]. The renewed agreement was signed in November 2005 by the EA-chairman and the IRMM Director. The EA-IRMM ILC IMEP-20 on tuna fish in 2004 is the follow-up of a series of ILCs in co-operation with EA. In 2003 the EA-IRMM ILC IMEP-19 *Trace Elements in Rice* was organized in cooperation with EA. In 2002, the first EA-IRMM ILC under the new EA-IRMM collaboration agreement was IMEP-12 *Trace Elements in Water*.

7.4.2 Assistance to the CIPM Mutual Recognition Agreement

The CIPM-MRA is a transparent process to underpin measurement capability claims of signatories to the meter convention. Key comparisons are organized for NMIs or designated laboratories to demonstrate the equivalence of their measurements in view of mutual acceptance of national standards and reference measurements. Their reported results and the equivalence statements of the measurements are included in the Appendix B of the CIPM-MRA, which can support the CMCs in Appendix C of the CIPM-MRA. Once reported there is no possibility of withdrawal of measurement results. Conversely, in a CCQM pilot study results could be withdrawn, but participation does not result in equivalence statements in Appendix B of the CIPM MRA. Therefore, participation in CCQM pilot studies was recently extended to invited expert laboratories. The results of both CCQM key comparisons and pilot studies are published in the Technical Supplement of *Metrologia* [24]. The IRMM assists the CIPM to support the MRA by making available the same samples as used in IMEP for ILCs in Chemistry among NMIs and expert laboratories. The test sample is sent to the NMIs and expert laboratories. Like IMEP ILCs, these institutes are free to measure the analytes under investigation applying their measurement procedures and analytical techniques of choice, but in order to support their measurement capability claims they need to report reliable results of highest metrological quality with demonstrated traceability and a complete uncertainty budget. Figure 7.18 shows the reported result from NMIs together with the IMEP-19 reference value. The IRMM result is circled in red. The same scale of ± 50 percent deviation from the CRV as usually chosen for the display of IMEP results is used for the secondary ordinate to underline the good agreement of results of the NMIs compared to the large spread observed in IMEP ILCs (see also Fig. 7.19). As a signatory to the MRA and as a reference laboratory in IMEP, the IRMM also has to demonstrate its measurement capability for the specific analytes in the matrix to be certified. The IRMM participates preliminarily in the establishment of the Certified Reference Values for IMEP ILCs in CCQM key comparisons or studies where the same sample material is used. This confirms the IRMM's capability to measure trace elements in food matrices by comparison to other NMIs and expert laboratories worldwide.

Furthermore, the IRMM also provides samples from the same batch as used in IMEP to institutes organized in regional metrology organizations like EUROMET or APMP [49, 77]. Figure 7.20 compares the results from the CCQM-K24 key comparison *Cd in rice* on the top of the international measurement infrastructure with the results of EUROMET 565 representing the regional level of the measurement infrastructure with the overall participants' result of IMEP-19, including laboratories nominated by international organizations and

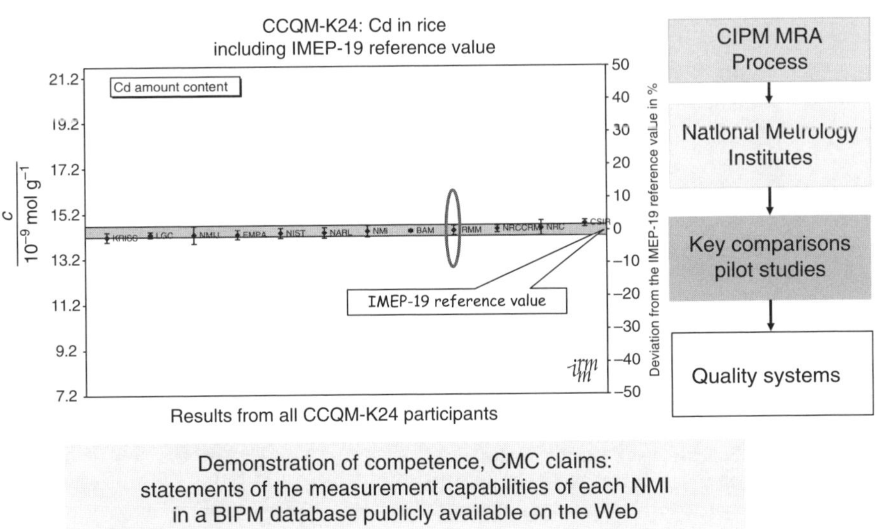

Figure 7.18. The CIPM-MRA process on CQM-K24 "Cd in rice."

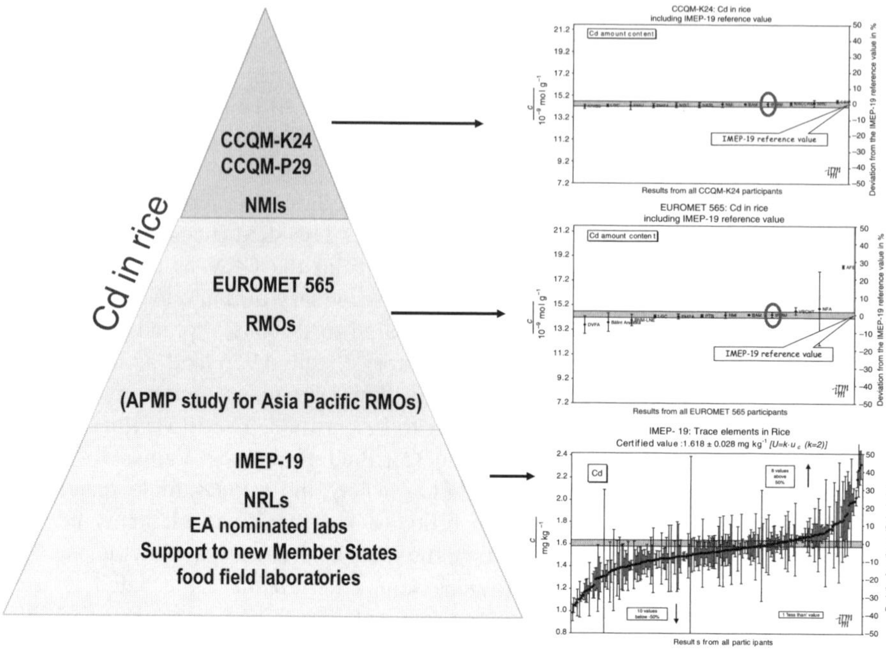

Figure 7.19. International measurement capabilities on Cd in rice.

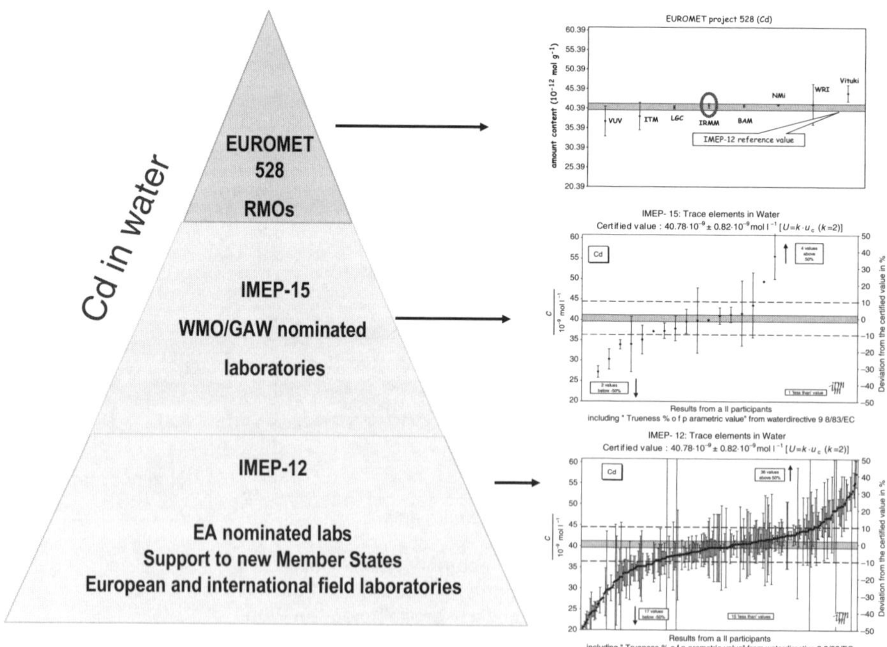

Figure 7.20. International measurement capabilities on Cd in water.

networks. All the graphs are on a scale of ± 50 percent deviation from the certified IMEP-19 reference value. The increase in the spread of the results with respect to the reference value going down the measurement infrastructure is very well displayed in Figure 7.20. The rice material used in IMEP-19 will also become a certified reference material.

Another example is IMEP-12 *Trace Elements in Water*. Samples form the IMEP-12 water batch were also offered to EUROMET [49]. In Figure 7.20 the results of the European NMIs are on the top. In the middle there are the results from the laboratories nominated via the WMO and the GAW programme. Finally, the field laboratories participating in IMEP-12 will be able to compare their results with the results of laboratories that represent their country at EUROMET level and vice versa.

7.4.3 Monitoring the EA Multilateral Arrangement

Accreditation is a very useful tool for laboratories to demonstrate technical competence to their customers. Accredited laboratories need to meet the requirements, according to the ISO/IEC 17025 Standard [22], of providing reliable measurement results with uncertainties. This has recently become a very important aspect in the collaboration agreement between the IRMM and EA because commercial PT providers do not ask participants to report a measurement result with uncertainty. Therefore, IMEP serves as a special tool for the NABs to ensure compliance of their accredited laboratories with the ISO/IEC 17025 Standard, which is a main requirement in the EA multilateral agreement (MLA). One major requirement of the ISO/IEC 17025 Standard is that laboratories have to state their measurement results with uncertainty. Even if they are not regularly reporting uncertainty to their customers, accredited laboratories have to be able to document their uncertainty evaluations before technical assessors. Nowadays assessors look at uncertainty budgets carefully and also openly discuss the problem of over- and underestimating measurement uncertainty. The IRMM is a permanent member of the EA Laboratory Committee Working Group for ILCs in the field of testing. The new IMEP approach of performance evaluation, that is, taking the participant's reported measurement uncertainty into account, is well accepted by the members of EA. The NABs nominate laboratories to participate in IMEP ILCs to evaluate their performance against independent reliable reference values. Unsatisfactory performance in IMEP has as a direct consequence for EA-nominated laboratories: They must take appropriate corrective actions.

In the frame of the EA-IRMM collaboration agreement, 61 laboratories that were nominated via their NABs, reported measurement results in IMEP-20_tuna fish. Furthermore, a number of IMEP-20 participants not nominated by EA indicated that they were accredited or/and authorized for this kind of measurements. Figure 7.21 shows the results for As in tuna fish according to self-declared accreditation or authorization status. It has to be mentioned that due to the overall large spread of As results in IMEP-20, the scale in this graph is ± 100 percent

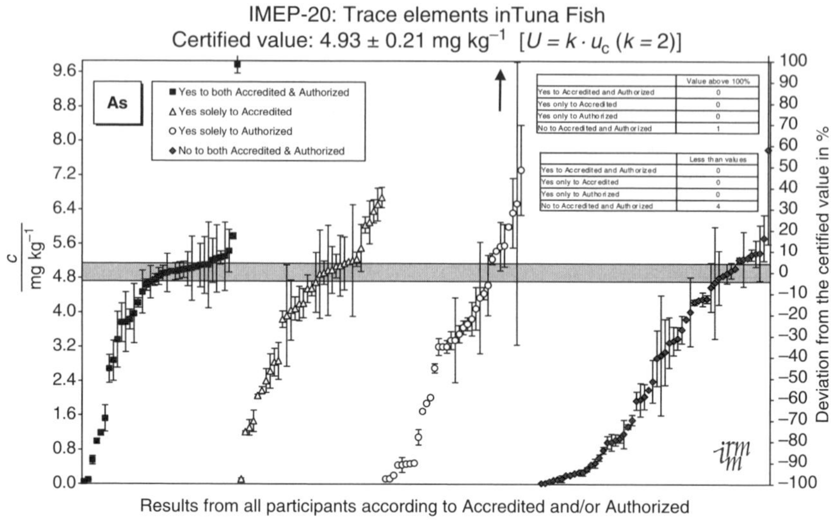

Figure 7.21. IMEP-20. Participants' results for As according to self-declared accreditation status.

deviation of the reference value. From Figure 7.21 it can be seen that there are a number of accredited and authorized laboratories that reported satisfactory measurement results. At the same time Figure 7.22 shows that, even for laboratories with accreditation/authorization status, measurement bias is observed. This kind of IMEP graph can be a very valuable tool for accreditation bodies to draw

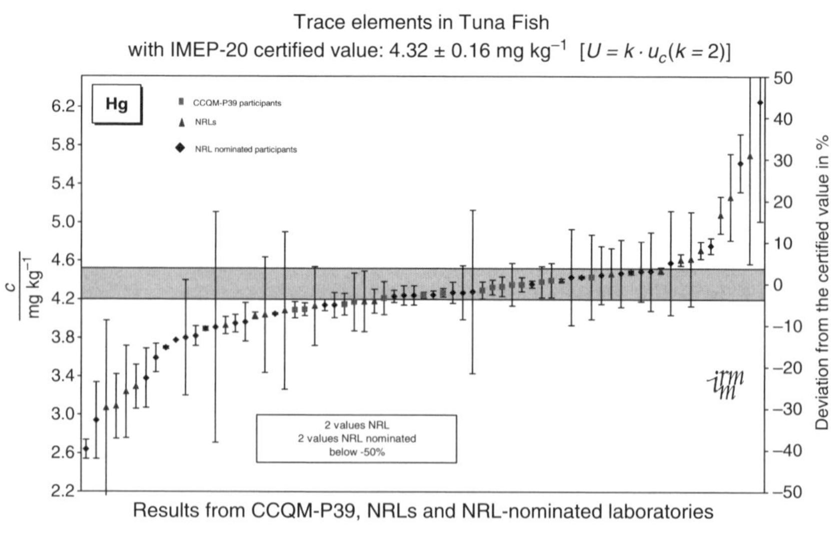

Figure 7.22. IMEP-20 reference results for Hg in tuna fish from the international measurement infrastructure to the field monitoring laboratory.

conclusions on the performance of their accredited laboratories and to decide on the appropriate corrective actions.

The common goal of EA and IRMM is to demonstrate the reliability and comparability of chemical measurements. This will be mainly achieved by the joint organization of ILCs with traceable reference values of high metrological quality, but also by organizing educational and training activities. Therefore, not only does the IRMM provide a tool for NABs to assess the performance of their accredited laboratories, but it also supports EA in monitoring the efficiency of the EA-MLA. In that sense, the IRMM improves the efficiency of accreditation in chemistry.

7.4.4 Strengthening the Network of National Reference Laboratories

IMEP-20 was organized in collaboration with the ISS-CRL in support to the European network of NRLs. The aim was to get a deeper insight into the national approach to the NRL–field laboratory (FL) system in each Member State. Each NRL could nominate five institutes that they usually approach for monitoring and food control activities. Finally, 23 NRLs and 38 NRL-nominated laboratories reported results in IMEP-20. The ISS-CRL cooperation was established to make progress in the overall harmonization of the quantification of residues and contaminants in food of animal origin and, as a consequence, to facilitate the mutual acceptance of experimental data. Cooperation with the ISS-CRL includes coordination of ILCs, but also specific training activities for the NRLs. The tuna fish sample was also offered to CCQM participants for a pilot study. In Figure 7.23 the results from the CCQM-P39 participants, the NRLs and their nominated laboratories are plotted separately. It has to be emphasized that, by means of IMEP-20 measurements on this real-life sample with a Hg content slightly exceeding the upper level of contaminants in tuna fish, measurement capabilities for a measurement close to a legal limit can be compared throughout the whole measurement infrastructure (see Fig. 7.23).

7.4.5 Comparability of Measurement Results

Through the IMEP process the competence of laboratories may be assessed regardless of the methodology they applied or their authorization status. Institutes with different functions (e.g., reference, inspection, surveillance) can be pictured in an objective way and compared with each other. This is what makes IMEP a perfect tool for the international monitoring of the efficiency of networks and multilateral agreements. The purpose is that a regular FL can compare its performance in an IMEP ILC with its NMI or NRL. Likewise, the NMIs and NRLs get a clear picture of the measurement capabilities of laboratories in their countries. This information helps NMIs and NRLs to support routine laboratories in their country with expert advice and calibration services. In addition, NMIs may have the stated responsibility to assure that measurements are traceable.

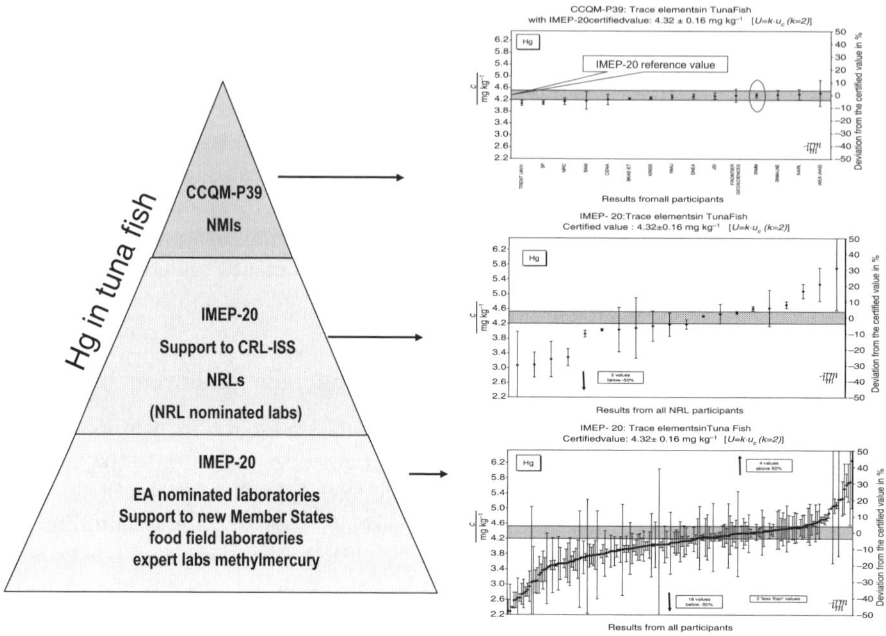

Figure 7.23. International measurement capabilities for Hg in tuna fish.

By using the same sample material for the different ILCs, the quality of carrying out the same analysis can be compared throughout the European and international measurement infrastructure. Therefore, with the IMEP process the equivalent implementation of directives and regulations can be assessed across an enlarged EU.

7.5 TRACEABILITY AND UNCERTAINTY

The need for the determination of different naturally occurring substances in food and feed to assure their quality has gradually shifted to include also a number of anthropogenic environmental pollutants resulting from industrial activity. Food comprises a large number of products of different origin (vegetables, animal products) with a complex composition. In industrialized countries, food production is handled through well-known, controlled processes with numerous input streams and possible sources of contamination. Monitoring all sources of contamination is impossible. Therefore, efforts should concentrate on some of the tissues where contaminants are accumulated in the final food products (e.g., animal fatty tissues). European regulations aim to limit the input of harmful substances in the human food production chain at an early stage. Therefore, an efficient QC has to focus on potential candidates for contamination at the supply levels of the food industry. Besides reference methods, authorities need sound monitoring systems. In particular, for rapid action in case of crisis, systems have to be based on fast screening methods, properly validated towards reference methods for which adapted CRMs exist to demonstrate their reliability. The need for standardized methods and CRMs in this area has been expressed at an international level and by researchers from EU Member States. Thus, it is important that trans-European collaboration addresses these problems. This will assure general usage of the developed methodologies and materials. The main advantages of validated and standardized methods as well as of adequate CRMs of environmental contaminants in food and feed products are improved food safety and increased possibilities for authorities to trace food contamination at early stages. In addition, it will help to eliminate the current economic barriers to trade due to disputes between countries which arise when food contamination accidents take place and no commonly recognized monitoring system exists.

Good measurement science (i.e., metrology) is a prerequisite for the quality of measurement results. The need for accuracy, comparability, and traceability of analytical measurements is significantly increasing. The meaning of traceability and how traceability to the Système International d'Unités can be achieved has been the subject of international debates for a long time. The international system of units is an internationally recognized system based on standards of long-term stability. According to the definition in the International Vocabulary of Basic and General Terms in Metrology (VIM), traceability "is . . . the property of a result of a measurement or the value of a standard whereby it can be related to stated references, usually national or international standards through an unbroken chain of comparisons all having stated uncertainties." If this standard is the "Système international d'Unités via SI quantities and units, an analytical measurement can be stated as traceable to SI [78]. Thus, traceability to SI is ensured via the use of CRMs for element amount content (and also possibly for isotopic composition) and via the assessment of the measurement uncertainty. "Traceable to the SI," therefore, means "traceable to reference values obtained by agreed realizations of the SI units" [41].

It is of main importance to set up a total uncertainty budget following the GUM and EURACHEM guidelines for combined uncertainty to identify the main sources of uncertainty [40, 41]. Results from different laboratories or from the same laboratory at different times have to be comparable with confidence. This is achieved if all laboratories are using the same stated reference. In many cases, this is achieved by establishing a chain of calibrations leading to primary national or international standards, ideally the SI units of measurement [79]. The SI system provides an international infrastructure for realizing comparable measurements by the use of traceable measurements.

EURACHEM is a network of organizations in Europe and has the objective of establishing a system for the international traceability of chemical measurements and the promotion of good quality practices. This network provides a forum for the discussion of common problems and for developing an approach to both technical and policy issues. EURACHEM provides a variety of valuable documents and guides dealing with issues of traceability or quantifying uncertainties in chemical measurements [41].

7.5.1 Compliance with the ISO/IEC 17025 Standard

Measurement results and standards are internationally accepted via their demonstrated comparability and known quality. The test results of a food testing laboratory that is accredited to the ISO/IEC 17025 Standard should be recognized in other countries where laboratory performance is also assessed in accordance with this Standard. This is an aspect of the ILAC MRA [31]. The ISO/IEC 17025 Standard requires traceability to internationally accepted stated references together with their stated measurement uncertainties [22].

7.5.2 Analytical Methods in Trace Elements Analysis

Generally, the analysis of environmental pollutants is considered as a necessary expense that is performed solely if stated by law. With less expensive screening methods and automated modern equipment to analyze suspect samples, the cost of analysis will become much lower. Hence, the attitude towards QA would most probably be more positive and the analytical work much more reliable for the customers. This also strengthens the international competitiveness of European producers. The credibility of the entire monitoring chain (screening methods, reference and standardized methods, as well as CRMs for the quality control of these methods) lies in the adequacy and integration of all three levels of the system. The adequate development and validation of methods is a prerequisite for a harmonized measurement system [80].

The most important analytical techniques which are used in multielement trace analysis are ICP-MS, atomic absorption spectrometry (AAS) and ICP atomic emission spectrometry (AES). NAA is applied as reference method in order to establish certified values. The regular "atomic spectrometry update on clinical and biological materials, foods and beverages" (ASU review) gives an overview of the recent developments in elemental analysis of food and beverages [81].

7.5.3 Certification Measurements in IMEP

The references values in IMEP are measured, if possible, using ID-MS, which is not a routinely applied method in food analysis. ID-MS is considered to be a definitive method of analysis because the chemical manipulations are performed on a direct mass basis and the spectrometric determinations involve only isotope ratio measurements [82]. IDMS was recognized by CCQM as a primary method of measurements, if carried out correctly [83]. CCQM defines a primary method of measurement as a method that is completely described and understood, having the highest metrological qualities and for which the results with a complete uncertainty statement can be given [83]. Primary methods, and hence IDMS, are unique tools in chemical measurements which can lead to results with very small combined uncertainty a priori [84, 85]. Even though ID-MS is limited to the quantification of the spiked element, the possibility of multiple spiking in a single run enables multi-element capabilities even by

ID-MS analysis [86, 87]. ID analysis, particularly by ICP-MS, is therefore known as a powerful strategy capable of an accurate determination of the total amount of an element with small combined uncertainties [88, 89]. Besides the total amount of a toxic element, its chemical form (chemical species) is of increasing interest since the toxicity is mostly dependent on the chemical species. Therefore, elemental speciation is gaining more and more importance and the hyphenation of chromatographic methods to atomic spectrometry becomes a strong tool in food analysis. Nonetheless, the accurate quantification of the species using hyphenated techniques still represents a challenge in analytical sciences. Thus, ICP-ID-MS plays a key role in speciation analysis since it provides accuracy and improved uncertainties over other calibration strategies, including external quantification. The accurate quantification of organometallic species in environmental samples can be hampered by various processes such as incomplete extraction from solid samples, material loss, rearrangement reactions during extraction and/or derivatization, low recoveries, and signal drift [90]. Its application to species-specific determinations has been limited by the nonavailability of commercial species-specific enriched spikes. In the species-specific spiking mode, the original sample is spiked prior to separation, with the advantage that loss of substance has no effect on the analytical result [91]. Due to the lack of standards and CRMs for a variety of elemental species in specific matrices, the application of species-unspecific quantification strategies employing ICP-MS detection has gained significant importance [92]. A constant spike flow provided by a pump is added, after the separation, to the eluted sample of a chromatographic system for on-line ID-MS. Evidently, species-unspecific quantification is the only solution if the identity and structure of the species is unknown. As a main drawback, isotopic equilibration is only obtained directly in the ICP-based approach and the method does not account for elemental losses during sample preparation and separation. A number of speciation applications of on-line ID-MS involved inverse calibration of the spike mass flow. The gravimetric spike flow determination was demonstrated to be a prerequisite for obtaining accurate results [91].

7.5.3.1 IMEP Results According to Analytical Technique or Calibration Strategy Figure 7.24 shows the example of the analysis of Hg in tuna fish in the IMEP-20. The main techniques applied by participants were AAS (with and without hydride generation), ICP-MS and ICP-AES. The results are shown in Figure 7.24 according to the analytical techniques. Unlike the common assumption in the food sector that comparability and reliability of results necessarily has to be method-dependent, it is evident from Figure 7.24 that reliable measurement results can be produced with every method that is appropriate for the analytical problem under investigation.

The analysis of As in tuna fish in Figure 7.25 shows a similar picture. External calibration and standard addition calibration were applied. It is evident that the calibration strategy did not lead to a significant difference in the average laboratory performance.

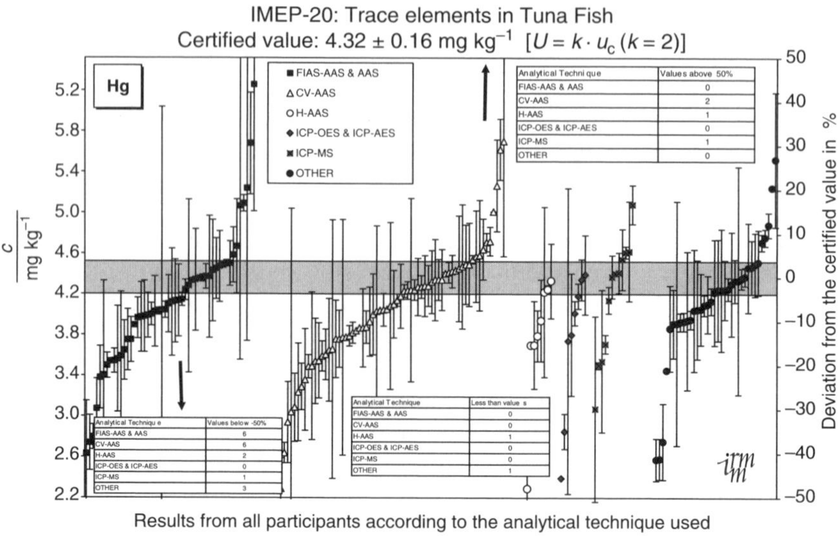

Figure 7.24. IMEP-20. Participants' results for Hg in tuna fish according to the analytical technique.

7.5.4 Certified Reference Materials in Trace Elements Analysis

CRMs are used for the development and validation of analytical methods. More than 200 providers offer a large number of CRMs with different matrices and

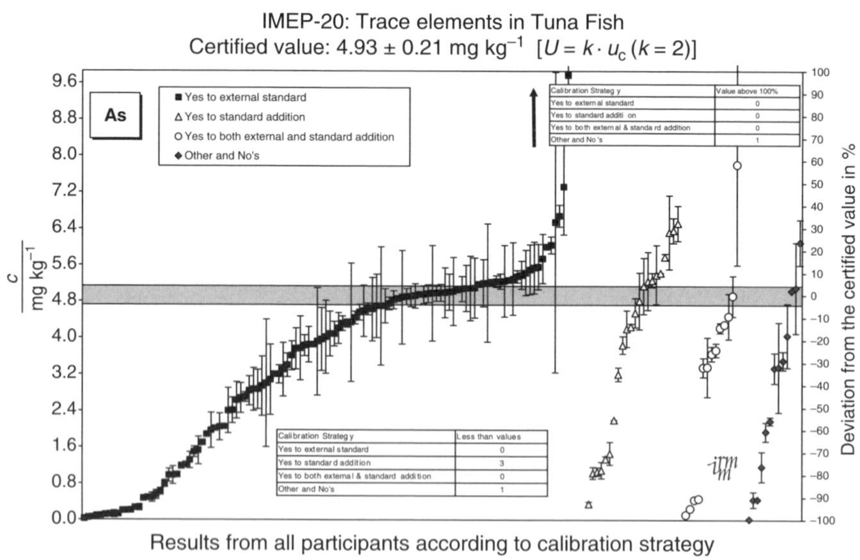

Figure 7.25. IMEP-20. Participants' results for As in tuna fish according to the calibration strategy.

elemental content on the market. They provide information on certified (highest confidence in accuracy), reference (best estimate of the true value), and information values (sufficient information is not available to assign the uncertainty associated with the value). It is often difficult to find the best material for a specific application. The BAM has established a database in order to assist laboratories in finding the adequate reference material. The Code of Reference Materials (COMAR) was established 2002 and has been accessible on the internet since 2003 free of charge [93]. According to COMAR, about 15 percent of the CRMs are produced in the field "Quality of Life," including agricultural products, consumer products, environmental samples, foodstuff, legal control samples, and other CRMs for quality of life. CRMs in the field of food and feed make up not more than 10 percent of the CRMs on the market but with an increasing tendency. Recently, the Virtual Institute on Reference Materials (VIRM) was established as an EU-funded project. VIRM is a nonprofit organization to improve the quality control of products and processes by encouraging, catalyzing and facilitating the increased use of RMs throughout Europe. It should become the meeting place for the European reference material community and a major source of dissemination of information, advice, and training on RMs for producers, distributors, (potential) users, service providers, and research organizations in the EU [94].

Traceability is one major factor that can be achieved via CRMs as main means in the field of chemical metrology. In general, CRMs are applied for the validation of analytical methods. Standard solutions are then used for instrument calibration. Nonetheless, CRMs should not be understood as the solution for all problems in chemical measurement. It goes without saying that the matrix of a CRM should match the analytical problem as exactly as possible. It is clear that there are not CRMs available for all matrices and analytes. Thus, it is important to have the best matrix match.

7.5.4.1 IMEP Results According to the Use of CRMs Figure 7.26 clearly shows that there is no significant difference in the average laboratory performance for Pb in water (IMEP-12) according to the use of water CRMs. This indicates that the use of CRMs does not automatically mean good analytical results.

Figure 7.27 shows that only a few participants indicated that they used a tuna fish CRM for QA. Forty percent of the participants using a CRM for QA, compared to 30 percent not using a CRM, reported Hg results within 5 percent deviation from the certified reference value. Final conclusions on the measurement performance according to the use of CRMs could be drawn if all results were corrected for recovery. Unfortunately this information was not available. It is planned to get information on recovery correction from IMEP participants in future IMEP ILCs.

7.5.5 Uncertainty Evaluation

Uncertainties are a prerequisite for the comparability of analytical measurement results. The ISO/IEC 17025 standard requires traceability to internationally accepted stated references together with their stated measurement uncertainties.

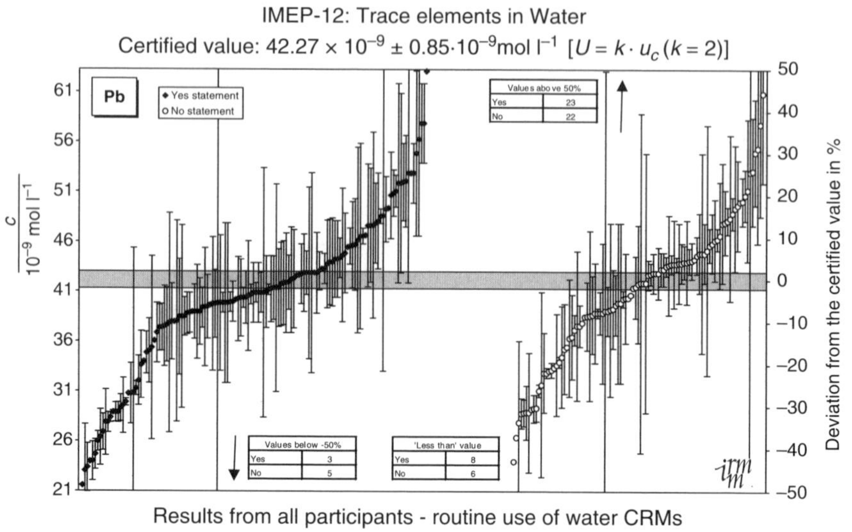

Figure 7.26. IMEP-12. Participants' results for Pb in water according to the use of CRMs.

Uncertainty represents the half-width of the interval, where the measurement results lie. If calculated for an analytical procedure and defined sample type, the estimate of measurement uncertainty may apply to results obtained accordingly. The complete uncertainty budget is a powerful tool to identify main sources of errors. Nonetheless, the value of the uncertainty cannot be used to correct a measurement result. In the food sector the evaluation of measurement uncertainty

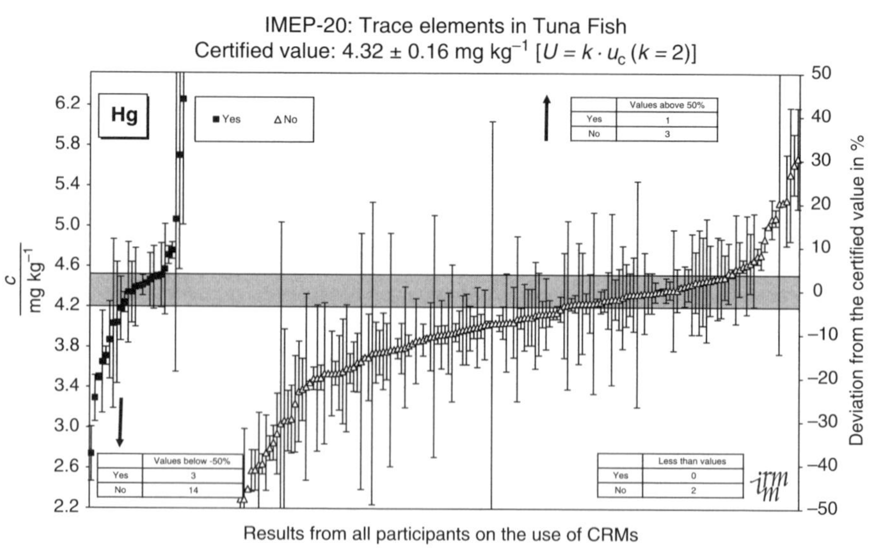

Figure 7.27. IMEP-20. Participants' results for Hg in tuna fish according to the use of CRMs.

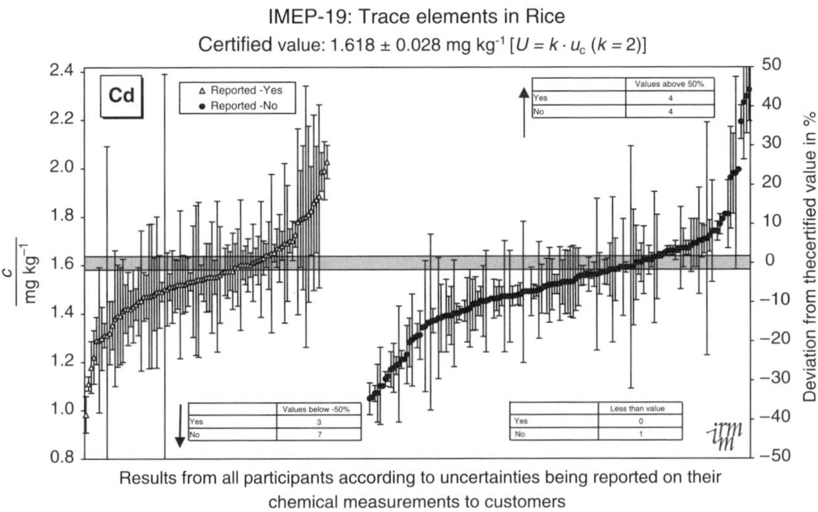

Figure 7.28. IMEP-19. Participants' results for Cd in rice according to uncertainties reported to customers.

is perceived as difficult because of the variety of successive analytical steps (including sample treatment).

7.5.5.1 IMEP Results According to Reported Uncertainties Uncertainty is a tool for harmonization, QC, and validation of a method. Nonetheless, it is evident

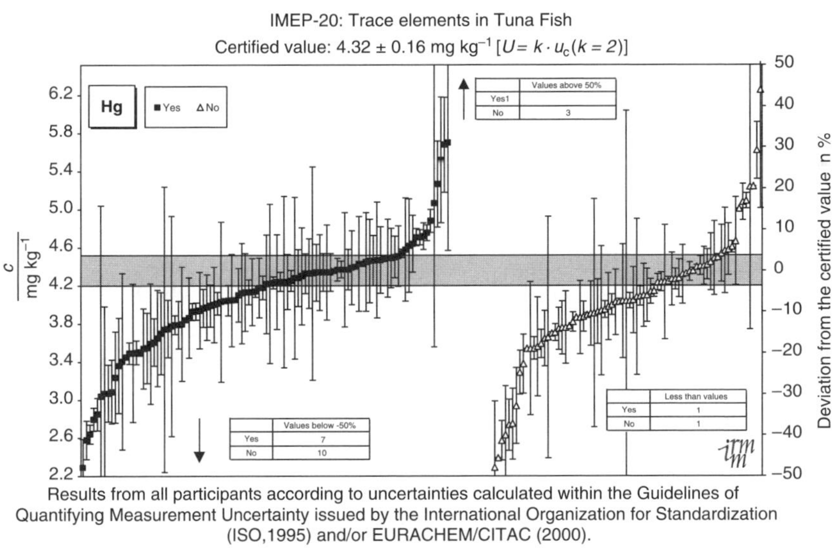

Figure 7.29. IMEP-20. Participants' results for Hg in tuna fish according to reported uncertainties.

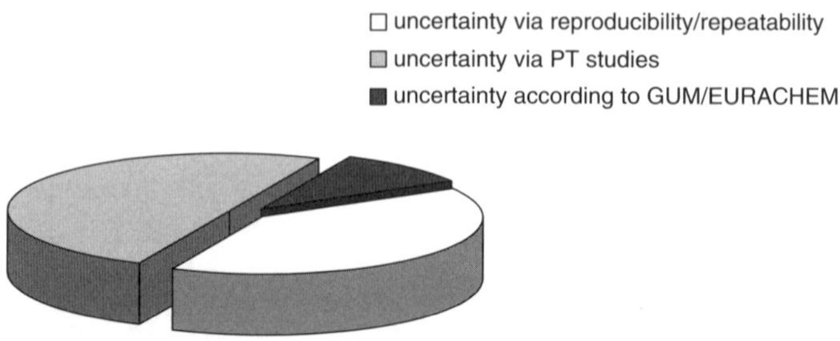

Figure 7.30. Strategies to estimate uncertainty.

that the appropriate application of uncertainty calculation does not automatically lead to correct results. Figures 7.28 and 7.29 shows the participants' results for Cd in rice in IMEP-19. It is clearly visible that the distribution of results is very similar, whether or not an uncertainty according to GUM was reported by the participants. This confirms what was already discussed in Section 7.3.3. Participants reporting unsatisfactory results (i.e., significant deviation from the certified reference value, $|zeta'| > 3$) need to identify their experimental problem and implement the corresponding corrective actions. Only then should the proper uncertainty budget be evaluated.

7.6 TRAINING AS A CONSEQUENCE OF PROPER QUALITY MANAGEMENT IN TRACE ELEMENT ANALYSIS

The need for training is also generated by the requirement set by the ISO/IEC 17025 Standard regarding the competence of laboratories [22]. As laboratories are accredited against this Standard, many of them need training to improve their capacity to demonstrate measurement traceability, estimate uncertainty, and perform validation. With the IMEP-EDUC programme, the IRMM offers training activities to participants after the completion of the respective IMEP ILC. This kind of educational activity after participation in an ILC or PT scheme could serve as an example of follow-up activities offered by regular PT providers on the national level. A training infrastructure starting at a higher education level is a prerequisite to establish a harmonized measurement infrastructure in the future.

7.6.1 IMEP EDUCational

The IMEP EDUCational (IMEP-EDUC) programme is a "learning from mistakes" activity that is linked to the IRMM's Training in Metrology in Chemistry (TrainMiC) programme [35]. Accredited or authorized laboratories with unsatisfactory results in IMEP ILCs need to investigate their entire measurement process in order to trace the origin of the difference between their reported results and the CRVs. The IRMM has launched an educational activity to assist selected IMEP participants in their investigation supplementary to the implementation of corrective actions. The selection of laboratories is preferably done in cooperation with the relevant body or institution that nominated these laboratories for participation in IMEP. The IMEP-EDUC activity gives these selected laboratories the opportunity to discuss the applied measurement approach with an independent experienced scientist (educator). The educator acts as "coach" in order to enable the laboratories to improve their problem-solving strategies. The educator visits each of the laboratories and discusses – in strict confidence with the analyst who performed the measurements – the experiences and problems in carrying out the measurements. A strategy for improving the laboratory performance is set up together with the heads of the laboratory. The general findings of this activity is communicated to all IMEP participants in an anonymous way via the IMEP EDUCational web site to create a communication forum for laboratories with similar problems/experience concerning their measurement performance. The identity of the participating laboratories is not disclosed to the public [36]. Participation in IMEP-EDUC is on a voluntary basis. Furthermore, the aim is to cover, if possible, all measurands under investigation and the main analytical techniques applied to this measurement problem. The IRMM is about to create via the IMEP-EDUC website a communication forum for laboratories to exchange experiences and talk in an undisclosed way about the common problems concerning their measurement performance in IMEP ILCs. The first activity was launched in 2003, following the IMEP-12 ILC *Trace Elements in Water*. As a primary result, specific technical issues were addressed in the IMEP-EDUC homepage [36]:

- General laboratory performance (clean work, sample preparation, handling);
- Organizational perspectives;
- Analytical performance;
- Evaluation of results.

The discussion on uncertainty has become a main issue throughout all contacts with the participating laboratories. Only a minor part of the participating laboratories really applied an uncertainty budget according to GUM [40]. The IMEP-EDUC activity showed clearly that there is the problem of a common language in metrology in Europe. Although all the laboratories were keen to provide a complete uncertainty budget, they encountered several problems when applying the GUM [40]. The major drawback, in their opinion, is that this guide is too complicated and needs "translation" into clear comprehensive guidelines for routine laboratories. Organizations like EA have already reacted to this remark from their accredited laboratories and issued separate EA guidelines (based on GUM) on the expression of uncertainty in quantitative testing [95].

The requirements for a complete uncertainty budget are:

- The complete measurement equation should be reported.
- All input quantities should be stated.

- Correction factors related to sample treatment should be included in the measurement equation.
- Significant sources of uncertainty should be identified.
- The applied evaluation process and type of assumed distribution for the uncertainty estimations should be described.

A common criticism was that it is not feasible to make a complete uncertainty budget for each measured sample in a routine food testing laboratory. During the IMEP-EDUC activity it became quite clear that a proper uncertainty budget is needed once for one analytical protocol. As long as the protocol does not change, the uncertainty budget itself can be used as a "quality tool." A continuous monitoring can be done with a quality control standard. If the repeatability of the measurement is below the uncertainty it can be assumed that the uncertainty established for this analytical protocol applies to similar samples following the same analytical protocol. If the repeatability is larger than the uncertainty, the uncertainty budget has to be revised. In general, laboratories requested a more pronounced training on this specific topic of uncertainty calculation. They also expressed the need for harmonized spreadsheets or software applicable for routine laboratories. A major outcome of IMEP-EDUC from the discussion with the analysts and the heads of laboratories was the urgent need for training on metrology in chemistry issues not only devoted to practitioners, but also to students.

7.6.2 TrainMiC

The TrainMiC of the IRMM has been set up as a training platform for people from all types of organizations. A set of training modules has been constructed that provides understanding in basic measurement matters that apply across sectors (food, environment, clinical). It offers interpretation of existing documents and gives guidance by making concrete examples available. The training material strives towards a congruent and up-to-date view regarding the main topics raised by the ISO/IEC 17025 Standard.

The target audience for the TrainMiC seminars is measurement practitioners from laboratories, technical assessors involved in accreditation, and end users of measurement data from public bodies or enforcement agencies. These seminars are organized in various countries in collaboration with the IRMM by the local partners, such as NMIs, Universities or NABs.

The training material developed by the TrainMiC board was compiled as the EUR report 20841 and is available for download from the TrainMiC web site [96]. At present seven modules are finalized: A general introduction to Metrology in Chemistry; Validation of measurement procedures; Traceability of measurement results; Uncertainty of measurement results; Proper use of certified reference materials (CRM); Interlaboratory comparisons (ILC), and Applied statistics, relevant for the calculation of an uncertainty budget or the performance evaluation. Additional modules are in progress to cover issues such as sampling, QC, and QA.

7.6.2.1 The TrainMiC-ILC Module The aim of this module is to focus on the different kinds of ILCs and/or PTs. The goal is to demonstrate that participating in ILCs or PTs gives evidence of measurement capability and should lead to improved quality of results. The results from ILCs or PTs are of crucial interest for laboratories as these provide clear information on their ability to demonstrate reliable results to their customers. Participation is either voluntary or forced by external requirements (e.g., legal, accreditation, control bodies). Most ILCs and PT schemes involve comparison of participants' results with an assigned value, which has been delivered by a reference laboratory, a subgroup of participants, through consensus from the overall population of test results, or by some other means. Corrective actions after participation to ILCs are also briefly discussed.

7.6.2.2 The TrainMiC-Uncertainty Module Measurement uncertainty is an important ISO/IEC 17025 requirement. This module explains and demystifies the approach of the ISO-GUM to estimate and report the uncertainty of a measurement result obtained following a specific measurement procedure. A clear description of all steps needed for uncertainty evaluation is presented, with the respective examples relevant to the environmental, clinical, and/or food sectors.

7.6.2.3 The TrainMiC-Traceability Module In this module, various issues concerning traceability of chemical measurement results are addressed. As stated in the definition of traceability given above (see Section 7.5) [78], every link in the traceability chain should consist of comparisons that are measurements in accordance with the above-proposed meanings, which include the validation of the measurement procedure and the use of reference materials. Not all chemical measurements are, or should be, traceable to the mole. Other stated references are accepted as well.

7.7 CONCLUSIONS

QC and QA are prerequisites to provide reliable analytical results to customers. Reliability is the key word in view of compliance of laboratories with the ISO/IEC 17025 Standard. Particularly in the field of trace elements analysis, the IMEP process results in a representative picture of the measurement capabilities at a certain point in time on trace elements endorsed in respective EU and international legislation. The general observation is that the majority of laboratories perform very well or satisfactorily in IMEP, but some have serious problems and their claims do not correspond to their actual measurement capability, regardless of whether they are reference, authorized, or accredited laboratories. This emphasizes the fact that the use of IMEP as metrological quality tool is primarily to the benefit of laboratories. The correct application of metrological concepts such as traceability and uncertainty is the basis of providing high-quality and reliable measurement results. There is an increasing demand also in the food sector to establish traceability to the SI in trace elements analysis via CRMs. IMEP also pictures the applied analytical measurement approaches of the participants in view of the main analytical techniques used in trace elements analysis. IMEP shows that the use of standardized methods or RMs does not automatically result in correct measurement results.

A laboratory that performs satisfactorily in an IMEP ILC can claim measurement capabilities towards an independent reference value of highest metrological quality. The successful demonstration of claimed measurement performance is an advantage for each IMEP participant because it can increase the competitiveness of this laboratory on the market, regardless of whether the customer is a private person, an accreditation body, an authorization body, or a governmental institution. On the other hand, participants in IMEP with unsatisfactory results realize that they have some undiscovered mistake in their measurement process. Training and educational activities support the implementation of well-considered corrective measures and the further improvement in QC. The uniqueness of the IMEP process enables comparability of measurement capabilities throughout the international and European measurement infrastructure, from NMIs to the FLs. Participation in relevant IMEP ILCs supports and monitors mutual recognition arrangements and thus increases confidence in measurement results of trace element analysis in food.

ACKNOWLEDGEMENTS

The authors would like to thank P. Taylor, L. Van Nevel, C.Harper, I. Verbist, S. Bynens, E. Garlick, and all other colleagues who made it possible to organize the IMEP programme at The IRMM. The authors also very much acknowledge the efforts of the chairperson of the EA WG for ILCs in testing M. Bednarova and the EA co-ordinator N. Meuree-Vanlaethem for the excellent cooperation. Furthermore, the authors are gratefully indebted to S. Caroli from the ISS-CRL and P. van Houwelingen from DG Enlargement for establishing the contact between the IRMM and the network of NRLs from EU Member States. The good cooperation with the CCQM, especially with the chair of the CCQM-IAWG, M. Sargent, is also very much acknowledged. A special word of thanks goes also to M. Lauwaars for the AOAC related input in the IMEP programme. The authors also wish to acknowledge the members of the TrainMiC board, E. Bułska, S. Duta, M. Lauwaars, I. Leito, N. Majcen, J. Norgaard, M. Suchanek, E. Vassileva, and P. Taylor for their contributions and constructive discussions. Special thanks to all the scientists inside and outside IRMM who contributed to the reprocessing, characterization, and the establishment of the IMEP certified reference values: C. Quétel, J. Snell, E. Vassileva, K. H. Grobecker, M. Bickel, F. Ulberth, S. Yazgan, A. Bernreuther, P. Conneely, G. P. Vermaercke, B. Gawlik, H. Muntau, R. Krska, S. Nelms, J. Diemer, J. Vogl, K. Inagaki, P. Evans, and B. Fairman.

REFERENCES

1. N. Rigby, P. James, *International Obeisity Task Force Position Paper*, September 2003.
2. http://www.nutrition-and-health-2020.net/index.htm.
3. http://www.developmentgoals.org/.
4. http://www.worldpolicy.org/globalrights/econrights/maps-life.html.
5. S. Barath, M. Hsu, D. Kaur, S. Rajagopalan, J. K. Andersen, Glutathione, iron and Parkinson's disease, *Biochem. Pharmac.*, **64** (2002), 1037–1048.
6. C. Smorgon, E. Mari, A. R. Atti, E. Dalla Nora, P. F. Zamboni, F. Calzoni, A. Passaro, R. Fellin, Trace elements and cognitive impairment: an elderly cohort study, *Arch. Gerontol. Geriatr. Suppl.*, **9** (2004), 393–402.
7. B. Wolozin, N. Golts, Iron and Parkinson's disease, *Neuroscientist*, **8** (2002), 22–32.
8. B. Kurz, B. Jost, M. Schunk, Dietary vitamins and selenium diminish the development of mechanically induced osteoarthritis and increase the expression of antioxidative enzymes in the knee joint of STR/1 N mice, *Osteoarthritis Cartilage*, **10** (2002), 119–26.
9. http://www.efsa.eu.int/
10. http://europa.eu.int/eur-lex/en/search/search_lif.html.
11. http://europa.eu.int/scadplus/leg/en/lvb/132041.htm.
12. Scientific Committee on Food, Guidelines of the Scientific Committee on Food for the development of tolerable upper intake levels for vitamins and minerals; *SCF/CS/NUT/UPPLEV/11 FINAL*, 28 November 2000; *http://ec.europa.eu/food/fs/sc/scf/out80a_en.pdf*
13. http://europa.eu.int/comm/food/index.
14. Commission Regulation (EC) No. 466/2001 of 8 March 2001 setting maximum levels for certain contaminants in foodstuff, *Official Journal L 077, 16/03/2001 P. 0014–0021.*
15. Commission Directive 2001/22/EC laying down the sampling methods of analysis for the official control of lead, cadmium, mercury and 3-MCPD in foodstuffs, *Official Journal L 077, 16/03/2001 P. 0014–0021.*
16. Directive 2000/60/EC of the European Parliament and of the Council of 23 October 2000 establishing a framework for Community action in the field of water policy, *Official Journal L 327, 22/12/2000 P. 0001–0073.*
17. *Water Framework Directive Newsletters*, 3rd Edition, October 2004, ISSN: 1725–5589; http://www.eurogeographics.org/eng/documents/WFD_newsletter_3.pdf
18. Council Directive 98/83/EC of 3 November 1998 on the quality of water intended for human consumption, *Official Journal L 330, 05/12/1998 P. 0032–0054.*
19. http://www.jrc.cec.eu.int/welcome.htm.
20. http://www.irmm.jrc.be/.
21. http://www.bipm.org.
22. International Standard, General requirements for the competence of testing and calibration laboratories, *ISO/IEC 17025:2005(E)*
23. http://kcdb.bipm.org/.
24. http://www.bipm.org/metrologia/TechSupp.jsp.

25. FAO/WHO, Codex Alimentaius, www.codexalimentarius.net/standard list.asp.
26. http://www.bipm.org/utils/common/pdf/CCQM10.pdf.
27. http://europa.eu.int/comm/food/food/chemicalsafety/residues/crl_en.htm.
28. Council Directive 96/23/EC of 29 April 1996 on measures to monitor certain substances and residues thereof in live animals and animal products, *Official Journal L 125, 23/05/1996 P. 0010–0032.*
29. http://www.iss.it.
30. Council Resolution of 10 November 2003 on the Communication of the European Commission "Enhancing the Implementation of the New Approach Directives," *O. J. C, 282, 25/11/2003, P. 0003–0004.*
31. http://www.ilac.org/.
32. International Standard, Calibration and testing laboratory accreditation systems – General requirements for operation and recognition, *ISO/IEC Guide 58: 1993.*
33. http://www.european-accreditation.org/.
34. P. Taylor, K. Brinkmann, I. Papadakis, L. Cortez, M. Bednarova, Y. Aregbe, Announcing the Collaboration between the European Co-operation for Accreditation and the Institute for Reference Materials and Measurements: improving the metrological basis of the European Accreditation System for Chemical Measurements, *Accred. Qual. Assur.*, **7** (2002), 168.
35. http://www.trainmic.org/.
36. http://www.imep.ws.
37. International Standard, Proficiency testing by interlaboratory comparisons – Part 1: Development and operation of proficiency testing schemes, *ISO/IEC Guide43:1997.*
38. http://www.eptis.bam.de/.
39. http://www.eurachem.ul.pt/.
40. International Organization for Standardization, *Guide to the Expression of Uncertainty in Measurement*, Geneva, 1995.
41. Eurachem/CITAC. *Guide Quantifying Uncertainty in Analytical Measurement*, 2nd edn, 2000, London, 1995; http://www.p2pays.org/ref/18/17617.pdf.
42. R. Kaarls, T. J. Quinn, The CCQM: a brief review of its origin and present activities, *Metrologia*, **34** (1997), 1–5.
43. http://kcdb.bipm.org//Appendix C.
44. A. Lamberty, L. Van Nevel, R. Moody, P. De Bièvre, IMEP-3: International Comparison of Trace Element Measurements in Synthetic and natural Water, *Accred. Qual. Assur.*, **1** (1996), 71–82.
45. L. Van Nevel, P. D. P. Taylor, U. Örnemark, J. R. Moody, K. G. Heumann, P. De Bièvre, The International Measurement Evaluation Programme (IMEP), *Accred. Qual. Assur.*, **3** (1998), 56–68.
46. I. Papadakis, P. D. P. Taylor, Establishment of SI-traceable reference ranges for the content of various elements in the IMEP-9 water samples, *Accred. Qual. Assur.*, **5** (2000), 331–338.
47. I. Papadakis, Y. Aregbe, J. Nørgaard, L. Van Nevel, P. D. P. Taylor, *IMEP-12, Trace Elements in Water, Certification Report, EUR 19924 EN.* http://www.irmm.jrc.be/html/interlaboratory_comparisons/imep/imep-12/IMEP12_certificationreport.pdf.

48. M. Bednarova, Y. Aregbe, C. Harper, P. D. P. Taylor, Evaluation of laboratory performance in IMEP water interlaboratory comparisons, *Accred. Qual. Assur.*, **10** (2005), 617–626.

49. http://www.EUROMET.org.

50. http://www.oiv.org/.

51. C. R. Quétel,S. M. Nelms, L. Van Nevel, I. Papadakis, P. D. P. Taylor, Certification of the lead mass fraction in wine for comparison 16 of the International Measurement Evaluation Programme, *J. Anal. Atom. Spectrom.*, **16** (2001), 1091–1100.

52. Commission Regulation (EEC) No. 2676/90 of 17 September 1990 determining Community methods for the analysis of wines, *Official Journal L 272, 03/10/1990, P. 0001–0192.*

53. http://www.euromet.org/cgi-bin/projectsearch.pl.

54. E. Vassileva, C. Quétel, Certification of the cadmium, copper and lead amount content in IMEP-19 by Isotope Dilution Inductively Coupled Plasma Mass Spectrometry, Internal Reports GE/R/IM/04/2002/01/31, GE/R/IM/05/2002/01/31, GE/R/IM/06/2002/01/31.

55. Y. Aregbe, P. Taylor, CCQM-P29 pilot study. Cadmium and zinc amount content in rice, *Metrologia Tech. Suppl.* **40** (2003), 08002; http://www.iop.org/EJ/abstract/0026-1394/40/1A/08001/.

56. EC, DG for Health and Consumer Protection, Perceptions of food safety risks associated with the presence of mercury in fish, qualitative study among consumers in four European countries, October 2004. [This study was carried out by Optem, Optem S.A.R.L. au Capital de 30 000 E - R.C.S. Versailles 339.]

57. Commission Decision No. 2002/657/EC of 17 August 2002 implementing Council Directive 96/23/EC concerning performance of analytical methods and the interpretation of results; *Off. J. EC* **L221** (17 August 2002).

58. J. P. Snell, C. R. Quétel, SI-traceable certification of methylmercury amount content in a tuna material by species-specific isotope dilution, *J. Anal. Atom. Spectrom.*, **20** (2005), 447–454.

59. Y. Aregbe, C. Quétel C., P. D. P. Taylor, CCQM-P39: As, Hg, Pb, Se and methylmercury in tuna fish, *Metrologia, Tech. Suppl.*, **41** (2004), 08004.

60. J. P. Snell, C. R. Quétel, L. Lambertson, J. Qvarnstrom, Preparation and certification of ERM-AE670, a ^{202}Hg enriched methylmercury isotopic reference material *J. Anal. Atom. Spectrom.*, **19** (2004), 1315–1324.

61. M. Thompson, R. Wood, The international harmonized protocol for the proficiency testing of (chemical) analytical laboratories, *Pure Appl. Chem.* **65** (1993), 2123–2144 [also published in *J. AOAC Int.* **76** (1993), 926–940].

62. International Standard, Statistical methods for use in proficiency testing by interlaboratory comparisons, *ISO 13528:2005(E).*

63. P. Robouch, N. Younes, P. Vermaercke, The "Naji plot," a simple graphical tool for the evaluation of interlaboratory comparisons, *PTB-IT-10* **149** (2003).

64. Y. Aregbe, P. Robouch, T. Colemean, P. D. P. Taylor, *WMO/IMEP-15 Trace Elements in Water Laboratory Intercomparison*, World Meteorological Organization Global Watch WMO TD No. 1195, 1–21.

65. Y. Aregbe, C. Harper, J. Nørgaard, M. De Smet, P. Smeyers, L. Van Nevel and P. D. P. Taylor, The Interlaboratory Comparison "IMEP-19 Trace Elements in Rice" – a

new approach for measurement performance evaluation, *Accred. Qual. Assur.*, **9** (2004), 323–332.

66. L. Van Nevel, I. Verbist, C. Harper, S. Bynens, P. Smeyers, Y. Aregbe, P. Robouch, P. D. P. Taylor, G. Turk, R. Vocke, W. R. Kelly, IMEP-18, *Sulphur in Diesel fuel (gasoil)*, Report to Participants, EUR 21765 EN, September 2005.
67. International Standard, *Statistical Methods for use in Proficiency Testing by Interlaboratory Comparisons*, ISO 13528:2005(E), 1–66.
68. Royal Society of Chemistry, Analytical Methods Committee, Robust statistics, *Analyst*, **114** (1989), 1693–1697.
69. Y. Aregbe, P. Robouch, Private Communication on PT results for Sn in food matrices with FAPAS and LGC, UK (2005).
70. http://www.eptis.bam.de/coept.htm.
71. http://www.ifcc.org/ifcc.asp.
72. http://www.wmo.ch/index-en.html.
73. http://europa.eu.int/comm/taxation_customs/common/about/welcome/index_en.htm.
74. http://www.cenorm.be/cenorm/index.htm.
75. http://www.gsf.fi/foregs/geochem/.
76. Y. Aregbe, P. Robouch, T. Colemana, P. D. P. Taylor, *WMO/IMEP-15 Trace Elements in Water Laboratory Intercomparison*, WMO TD No. 1195, 1–21.
77. http://www.apmpweb.org/.
78. ISO/VIM, *International Vocabulary of Basic and General Terms in Metrology*. International Organization for Standardization, Geneva, 1993.
79. C. D. Ehrlich, S. D. Rasberry, Traceability timeline, *J. Res. NIST.*, **103** (1998), 93–105.
80. A. J. Mark Green, Practical Guide to Analytical Method Validation, *Anal. Chem.*, **68** (1996), 305A–309A (http://pubs.acs.org/hotartcl/ac/96ac.html).
81. A. Taylor, S. Branch, D. Halls, M. Patriarca, M. White, Atomic spectrometry update. Clinical and biological materials, foods and beverages, *J. Anal. Atom. Spectrom.*, **19** (2004), 505–556.
82. C. Pin, B. Le Fevre, S. Joannon, Isotope dilution and plasma source mass spectrometry: An optimal combination for high precision analysis of trace elements. *Can. J. Anal. Sci. Spectrosc.* **45** (2000), 13.
83. Comité Consultatif pour la Quantité de Matière, *Rapport de la 1er Session*, 1995, Edité par le BIPM, Sèvres, France.
84. J. D. Fassett, P. J. Paulsen, Isotope dilution mass spectrometry for accurate elemental. Analysis, *Anal. Chem.*, **61** (1989), 643A.
85. J. R. Moody, M. S. Epstein, Definitive measurement methods, *Spectrochim. Acta B*, **46B** (1991), 1571.
86. T. Prohaska, C. R. Quétel, C. Hennessy, D. Liesegang, I. Papadakis, P. D. P. Taylor, C. Latkoczy, S. Hann, G. Stingeder, 688 SI-traceable certification of Cu, Cr, Cd and Pb in fly ash candidate reference materials, *J. Environ. Monit.*, **2** (2000), 613–620.
87. E. Vassileva, C. R. Quétel, Certification measurements of the Cd, Cu, and Pb contents in rice using ID-ICPMS, *Anal. Chim. Acta*, **519** (2004), 79–86.

88. K. G. Heumann, *Inorganic Mass Spectrometry*, F. Adams, R. Gijbels, R. J. Grieken, (eds) Wiley, New York, 1988.

89. P. De Bièvre, Stable isotope dilution: an essential tool in metrology, *Fresenius J. Anal. Chem.*, **350** (1994), 277–283.

90. M. Monperrus, E. Krupp, D. Amoiroux, O. F. X. Donard, R. C. Martin-Doimeadios, Potential and limits of speciated isotope-dilution analysis for metrology and assessing environmental reactivity, *Trends Anal. Chem.*, **23** (2004), 261–272.

91. G. Koellensperger, S. Hann, J. Nurmi, T. Prohaska, G. Stingeder, Uncertainty of species unspecific quantification strategies in hyphenated ICP-MS analysis, *J. Anal. Atom. Spectrom.*, **18** (2003), 1047–1055.

92. J. Vogl, K. G. Heumann, Development of an ICP-IDMS method for dissolved Organic carbon determinations and its application to chromatographic fractions of heavy metal complexes with humic substances, *Anal. Chem.*, **70** (1998), 2038–2043.

93. http://www.comar.bam.de/.

94. http://www.virm.net/.

95. EA Expert group on uncertainty of measurement on behalf of the EA Laboratory Committee, EA 4/16 EA Guidelines on the Expression of Uncertainty in Quantitative Testing, Dec. 2003; http://www.european-accreditation.org/n1/doc/EA-4-16.pdf.

96. P. Robouch, E. Bulska, S. Duta, M. Lauwaars, I. Leito, N. Majcen, J. Norgaard, M. Suchanek, E. Vassileva, P. Taylor, TrainMiC – Training in Metrology in Chemistry, EUR 20841 EN, 2003.

SECTION 2

SELECTED APPLICATIONS

8

APPLICATIONS OF INDUCTIVELY COUPLED PLASMA MASS SPECTROMETRY TO TRACE ELEMENT RESEARCH AND CONTROL

FRANCESCO CUBADDA

National Centre for Food Quality and Risk Assessment, Istituto Superiore di Sanità Viale Regina Elena 299, 00161 Rome, Italy, E-mail: francesco.cubadda@iss.it

8.1 INTRODUCTION

In the early 1980s, inductively coupled plasma mass spectrometry (ICP-MS) emerged as a major technique for elemental analysis, an area dominated by atomic spectrometric techniques at that time. Given its features, that is, high

The Determination of Chemical Elements in Food: Applications for Atomic and Mass Spectrometry.
Edited by Sergio Caroli

sensitivity, multielemental capabilities, wide linear dynamic range, high sample throughput, ability to discriminate between isotopes, and suitability as a selective detector in hyphenated methods for elemental speciation, the potential of ICP-MS for the analysis of trace elements and their chemical species in food was soon appreciated. This potential began to be fully realized in the past decade, when successive generations of improved quadrupole instruments were placed on the market and the technique found widespread application in almost every area of food-related research. Sector field (SF) high-resolution instruments, providing both enhanced sensitivity and at least partial relief from spectral interferences, became also available and, mostly in the second half of the 1990s, part of the analytical community turned to this technology for food analysis. Today, ion-molecule chemistry techniques that use collision cells (CC) and dynamic reaction cells (DRC) of various types, are claimed to have opened a new age in ICP-MS history [1]. A variety of CC and DRC instruments have become commercially available, thus enabling, for example, the reliable quantification of P and S at low levels or isotope ratio measurements of essential elements such as Cr, Fe, and Se. The advent of these devices, together with the implementation of different mass analyzers and other technical developments, for example, in the sample introduction systems, provide today's analyst with powerful tools for achieving freedom from interferences and superior detection capability.

The success of ICP-MS for the elemental analysis of food is evidenced by its spread outside analytical chemistry research laboratories, notwithstanding the expensiveness of the instruments and the high running cost in terms of consumables and maintenance. On the one hand, it is being increasingly used for quality control and compliance with food regulations by major food industries or independent analytical laboratories. On the other hand, over recent years many governmental institutions have resorted to the technique for their institutional and control activities and this trend is still continuing. As a result, in the international legislation laying down the analytical methods for the official control of trace elements in food, atomic absorption spectrometry (AAS) is no longer being prescribed as the exclusive detection technique. Rather, either performance criteria are established and any validated method that fulfils these requirements is accepted [2], or ICP-MS is explicitly placed in the list of recommended techniques [3].

Another key factor in the popularity of the technique is its increasingly widespread use in speciation chemistry. The current awareness of the importance of the chemical form in which an element is present in food (e.g., the oxidation state, the nature of the ligands, and the molecular structure), on its possible absorption and fate inside the body, has made elemental speciation a prominent topic of present research in the food science. If speciation critically influences the bioavailability, essentiality, or toxicity of an element, the analytical techniques that enable species discrimination and quantification gather prominence. ICP-MS has the capability for quantifying metal (metalloid)-containing species in a chromatographic or capillary electrophoresis effluent or, as recently demonstrated, in a gel spot after SDS-PAGE via electrothermal vaporization (ETV) or laser ablation (LA) [4, 5]. Hence, it has firmly established as a sensitive, multielement, isotope selective on-line detector in hyphenated setups for speciation analysis.

The aim of this chapter is to provide an overview of the applications of ICP-MS to food research and control. For this purpose, selected examples illustrating various analytical approaches related to specific analytes and matrices are illustrated and the available literature is summarized in easily browsable tables for quick reference. The reported applications cover food safety and food control (including the evaluation of food contact materials and the assessment of radionuclide contamination), dietary intake studies, element metabolism in man, transfer of elements through the food chain, effects of food processing and domestic preparation, and authenticity and origin assessment. The chapter will not treat systematically the analytical challenges of ICP-MS determination of elements and elemental species in food and agricultural matrices, such as spectral and nonspectral interferences, sample preparation, or sample introduction issues. A discussion of these and related topics can be found in a recent review [6] and the references contained therein, as well as in other chapters of this book. The reader is also encouraged to refer to the atomic spectrometry updates on clinical and biological materials, foods, and beverages published yearly by the *Journal of Analytical Atomic Spectrometry* [7–22].

8.2 ELEMENTAL DETERMINATION

8.2.1 Water and Beverages

8.2.1.1 General Aspects The literature concerning liquid matrices, including milk, is summarized in Table 8.1. Depending on the complexity of their nature, beverages can be in many cases directly aspirated into the plasma following acidification or dilution and addition of suitable reagents. Sensitive multielement analysis of drinking water is generally straightforward and allows ultratrace analytes to be directly determined, such as rare earth elements (REEs), even though preconcentration/matrix separation, for example, by means of ion exchange microcolumns [23], has been reported. Arsenic, Sb, Se, and other hydride-forming elements can be determined by the hydride generation (HG) approach, which significantly enhances sensitivity (see Table 8.1). SF instruments offer superior performance for the determination of ultratrace and/or spectrally interfered elements without the need of any preliminary sample treatment [24]. Excellent limits of detection (LoDs) plus complete removal of ArCl overlap have been obtained when the technique was used in conjunction with HG for As determination [25]. The analysis of mineral water can be more challenging, especially in samples with a high mineral or bicarbonate content. In such cases, as well as with more complex liquid matrices, the addition of an appropriate internal standard is recommended.

Fruit juices, which have a more complex organic matrix with a significant carbohydrate content, can be analyzed either following dilution, with subsequent centrifugation/filtration [26] or flow injection (FI) introduction [27], or after full

TABLE 8.1. Literature on Element Determination in Water and Beverages by ICP-MS

Matrix	Elements	Technique[a]	Type of Study[b]	References
Water				
Mineral water	Ag, Al, As, B, Ba, Be, Cd, Co, Cr, Cu, Li, Mn, Mo, Ni, Pb Se, Sb, Sr, Tl, V, Zn	Q	M	[47]
Mineral water	Ag, Al, As, Ba, Be, Bi, Br, Cd, Ce, Co, Cr, Cs, Cu, Dy, Er, Eu, Fe, Ga, Gd, Ge, Hf, Hg, Ho, I, In, La, Li, Lu, Mn, Mo, Nb, Nd, Ni, Pb, Pr, Rb, Sb, Se, Sm	Q	M	[48]
Mineral water	Ag, As, Cd, Cr, Se, V	SF	A	[24]
Mineral water	Cd, Co, Cr, Cu, Mn, Ni, Pb, Se, Zn	Q	M	[165]
Mineral water	Hg	CV-Q (ID)	A, M	[166]
Mineral water	^{226}Ra	SF	A, M	[167]
Mineral water	U	Q	A, M	[168]
Mineral water	U	SF	A, M	[169]
Mineral water	U	Q	I	[170]
Water	Ag, Al, As, Au, B, Ba, Be, Bi, Br, Ca, Cd, Ce, Cl, Co, Cr, Cs, Cu, Dy, Er, Eu, Fe, Ga, Gd, Ge, Hf, Hg, Ho, I, K, La, Li, Lu, Mg, Mn, Mo, Na, Nb, Nd, Ni, P, Pb, Pr, Pt, Rb, S, Sb, Se, Si, Sm, Sn, Sr, Ta, Tb, Te, Th, Ti, Tl, Tm, U, V, Y, Yb, W, Zn, Zr	Q	A	[171]
Water	Ag, Al, As, B, Ba, Be, Bi, Br, Ca, Cd, Ce, Co, Cr, Cs, Cu, Dy, Er, Eu, Fe, Ga, Gd, Ge, Hf, Hg, Ho, I, In, K, La, Li, Lu, Mg, Mn, Mo, Na, Nb, Nd, Ni, Pb, Pr, Rb, Sb, Sc, Se, Sm, Sn, Sr, Ta, Tb, Te, Th, Ti, Tl, Tm, U, V, W, Y, Yb, Zn, Zr	Q	M	[172]
Water	Ag, Al, As, Ba, Be, Cd, Co, Cr, Cu, Fe, Hg, Mn, Mo, Ni, Pb, Se, Sb, Ti, Tl, V, Zn	Q	M	[173]
Water	Ag, Al, As, Ba, Be, Cd, Co, Cr, Cu, Mn, Mo, Ni, Pb, Sb, Se, Th, Tl, U, V, Zn	Q	C	[174]
Water	Al	Q	A	[175]
Water	Al	Q	A	[176]

Water	Al, As, B, Ba, Cd, Cr, Cs, Cu, Fe, Ga, Li, Mn, Mo, Ni, Pb, Rb, Sb, Se, Sr, U, V, Zn	Q	M	[177]
Water	As	SF, Q	A	[178]
Water	As	SF (HG)	A	[25]
Water	As	HG-SF	M	[179]
Water	As, Be, Cd, Co, Cr, Cu, Mn, Ni, Pb, Sb, V, Zn	Q	M	[180]
Water	As, Sb, Se	ETV-Q	A	[181]
Water	As, Sb, Se	HG-Q	A	[182]
Water	Ag, As, B, Ba, Be, Bi, Cd, Co, Cr, Cu, Mn, Mo, Ni, Pb, Se, Sr, Ti, V, Zn	Q	M	[183]
Water	Ca, Co, Cr, Cu, Fe, K, Mn, Na, Ni, Mg, P, V, Zn	Q	M	[184]
Water	Cd, Cr, Cu, Ni, Pb, Zn	Q	M	[185]
Water	Ce, Dy, Er, Eu, Gd, Ho, La, Lu, Nd, Pr, Sc, Sm, Tb, Tm, Y, Yb	Q	M	[186]
Water	Ce, Dy, Er, Eu, Gd, Ho, La, Lu, Nd, Pr, Sc, Sm, Tb, Tm, Y, Yb	Q	A	[23]
Water	Cr	DRC-Q (ID)	A	[187]
Water	F	Q	A	[188]
Water	Hg	Q	A	[189]
Water	Hg	Q	A	[190]
Water	Hg	Q	A	[191]
Water	I	Q	A	[192]
Water	Pb	Q	M	[193]
Water	^{226}Ra	Q	A	[194]
Water	Rb	Q (ID)	A, C, R	[195]
Water	Se	HG-Q	A	[196]
Water	^{99}Tc	Q	A	[197]

(Continued)

TABLE 8.1. *Continued*

Matrix	Elements	Technique[a]	Type of Study[b]	References
Water	Th, U	FI-Q	A	[198]
Water	U	Q	A	[199]
Water RM	Ag, Al, As, B, Ba, Be, Ca, Cd, Co, Cr, Cu, Fe, K, Li, Mg, Mn, Mo, Na, Ni, Pb, Rb, Sb, Se, Sr, Te, Tl, U, V, Zn	Q	A	[200]
Water RM	As, Bi, Ge, Sb, Se, Sn, Te	HG-Q	A	[201]
Water RM	As, Se	HG-Q	A	[202]
Milk and infant formula				
Human milk	Ag, Ce, Co, Cr, Ga, La, Mo, Nb, Ru, Sb, Th, Ti, U	Q	I, O	[66]
Human milk	Al, As, B, Ba, Be, Br, Cd, Co, Cr, Cs, Cu, Fe, Ga, Li, Mn, Mo, Ni, Pb, Rb, Sc, Se, Sr, Th, Ti, Tl, U, V, Zn	Q	A	[203]
Human milk	As, Ca, Cu, Fe, K, Na, Mg, Mn, Pb, Se, Zn	Q	M	[204]
Human milk	Ba, Be, Bi, Cd, Co, Cs, Cu, La, Li, Mn, Mo, Pb, Rb, Sb, Sr, Tl, Zn	Q	O	[63]
Human milk and infant formula	Ag, Al, As, Au, Co, Cr, Cu, Fe, Mn, Ni, Pb, Pt, Sc, Ti, V	SF	A, M	[205]
Human milk and infant formula	Ag, Al, As, Au, Co, Cr, Cu, Fe, Mn, Ni, Pt, Sc, Se, Ti, V	SF	M	[64]
Human milk and infant formula	As, Ca, Cd, Cr, Cu, Mg, Mn, Mo, Ni, Pb, Rb, Zn	Q	A	[206]
Human milk and infant formula	Ba, Ca, Cu, K, Fe, Na, Mg, P, Sr, Zn	Q	I	[207]
Infant formula	Ba, Be, Bi, Cd, Co, Cs, Cu, La, Li, Mn, Mo, Pb, Rb, Sb, Sn, Sr, Tl, Zn	Q	I	[208]
Infant formula	Pb	UN-Q (ID)	A	[209]
Milk	Al, Ca, Cr, Fe, K, Mg, Mn, Na, P, Zn	Q	A	[210]
Milk	Al, Ba, Cu, I, Mn, Mo, Pb, Rb, Se, Sr, Zn	Q	A	[40]
Milk	As, Be, Cd, Co, Cr, Cu, In, Mg, Mn, Ni, Pb, Se, Y, Zn	Q	M	[83]

Milk	Cu, I	FI-Q	A, M	[42]
Milk	I	Q	A	[41]
Milk	I	FI-Q	A, M	[43]
Milk	I	Q	M	[69]
Milk, human milk and infant formula	Al, Ba, Cd, Cr, Cu, Mg, Mo, Ni, Pb, Sn, Zn	Q	I	[68]
Milk, human milk and infant formula	Al, Ca, Cd, Cr, Cu, Fe, Hg, Mg, Mn, Na, Ni, Pb, Se, Sr, Zn	SF	A	[211]
Milk, human milk and infant formula	Al, Ca, Cd, Cr, Cu, Fe, Hg, Mg, Mn, Na, Ni, Pb, Se, Sr, Zn	SF	A	[65]
Milk and infant formula	Al	Q (sl)	M, C	[212]
Milk and infant formula	Ba, Be, Bi, Cd, Co, Cs, Cu, La, Li, Mn, Mo, Pb, Rb, Sb, Sn, Sr, Tl, Zn	Q	A, M	[63]
Milk powder	Ca, Fe, Zn	DRC-Q	A	[213]
Milk powder	Cd, Cr, Pb, Zn	ETV-DRC-Q (ID) (sl)	A	[214]
Milk powder	Pb	Q (ID) (sl)	A, Au	[39]
Milk RMs	I	Q	A	[215]
Other beverages				
Beer	Ag, As, Bi, Cd, Co, Cr, Cs, Cu, Fe, Ga, Hg, In, Mn, Ni, Pb, Rb, Sb, Sn, Th, Tl, U, V, Zn	UN-SF	M	[62]
Fruit juice	Fe, Pb, Sn	Q	A	[26]
Fruit juice	Sn	Q	P	[216]

(*Continued*)

TABLE 8.1. *Continued*

Matrix	Elements	Technique[a]	Type of Study[b]	References
Orange juice	Al, Ag, As, B, Ba, Bi, Ca, Cd, Ce, Co, Cr, Cs, Cu, Eu, Fe, Ga, Gd, I, Li, Mg, Mn, Mo, Na, Nd, Ni, Pb, Pr, Rb, Sb, Sc, Se, Si, Sm, Sn, Sr, Th, Ti, Tl, V, W, Y, Zn, Zr	Q	A, Au	[28]
Orange juice and peel extract	Ba, Co, Cu, Li, Mn, Mo, Ni, Rb, Sn, V, Zn	Q	Au	[60]
Tea infusions	Ag, Al, As, Ba, Be, Bi, Ca, Cd, Co, Cr, Cu, Fe, Li, Mg, Mn, Ni, Pb, Rb, Se, Sn, Sr, Tl, V, Zn	Q	A	[61]
Various alco holic beverages	Ag, Al, As, Au, B, Ba, Be, Bi, Br, Cd, Ce, Co, Cr, Cs, Cu, Fe, Hf, I, Ir, La, Li, Mn, Mo, Nb, Ni, P, Pb, Pd, Pt, Rb, Re, Rh, Ru, Sb, Sc, Sn, Sr, Ta, Te, Th, Ti, Tl, U, V, W, Y, Zn, Zr	SF	A, Au	[219]
Various beverages	I	FI-Q	M	[27]
Various beverages	U	FI-Q	A	[218]
Wine	Ag, Al, As, B, Ba, Be, Bi, Ca, Cd, Ce, Co, Cr, Cs, Cu, Fe, Li, La, Mg, Mn, Mo, Ni, P, Pb, Rb, S, Sb, Si, Se, Sr, Tl, U, V, Zn	Q	Au, O	[33]
Wine	Ag, Al, As, Ba, Be, Bi, Br, Ca, Cd, Ce, Cl, Co, Cs, Cu, Fe, I, La, Li, Mg, Mn, Mo, Ni, P, Pb, Rb, Sb, Se, Sr, Th, Ti, Tl, U, V, Zn	Q	Au	[50]
Wine	Al, As, B, Ba, Be, Bi, Ca, Cd, Ce, Co, Cr, Cs, Cu, Fe, La, Li,.Mg, Mn, Mo, Ni, P, Pb, Rb, S, Sb, Se, Sn, Sr, Ti, Tl, U, V, Zn	Q	Au, O	[51]
Wine	Al, As, B, Ba, Be, Ca, Cd, Ce, Co, Cr, Cs, Cu, Dy, Er, Eu, Fe, Ga, Gd, Hf, Ho, La, Li, Lu, Mn, Mo, Nb, Nd, Ni, Pb, Pr, Rb, Sb, Sc, Sm, Sr, Tb, Ti, Th, Tl, Tm, U, V, W, Y, Yb, Zn, Zr	Q	P, Au	[219]
Wine	Al, As, B, Ba, Ca, Cd, Ce, Cs, Cu, Dy, Er, Eu, Fe, Gd, Ho, La, Li, Lu, Mn, Nd, Ni, Pb, Pr, Rb, Sm, Sn, Sr, Tb, Tm, Yb, Zn	Q	A	[220]
Wine	Al, As, Ba, Ca, Cd, Co, Cs, Cu, Fe, Ga, Hg, K, Li, Mg, Mn, Na, Ni, Pb, Rb, Sr, Tl, Zn	Q	A	[221]

Wine	Al, As, Ba, Cd, Ce, Co, Cr, Cs, Cu, Dy, Er, Eu, Fe, Ga, Gd, Ge, Ho, La, Li, Lu, Mn, Nb, Nd, Ni, Mo, Pb, Pr, Rb, Rh, Ru, Sb, Sm, Sn, Sr, Ta, Tb, Th, Ti, Tl, Tm, U, V, W, Yb, Zn, Zr	FI-Q	Au	[30]
Wine	Al, Ba, Cd, Ce, Co, Cr, Cu, Dy, Er, Eu, Fe, Ga, Gd, Ho, La, Li, Lu, Mn, Nd, Ni, Pb, Pr, Rb, Sm, Sr, Tb, Tm, V, Yb, Zn	Q	A	[222]
Wine	As	FI-Q	A	[223]
Wine	As, Au, Ba, Be, Cd, Ce, Co, Cs, Cu, Dy, Er, Eu, Gd, Ho, La, Lu, Nd, Ni, Pb, Pr, Pt, Rb, Re, Sb, Sm, Sn, Sr, Tb, Te, Th, Ti, Tl, Tm, U, V, W, Yb, Zn, Zr	Q	Au	[224]
Wine	As, B, Ba, Bi, Cd, Ce, Cr, Cs, Cu, Dy, Er, Eu, Fe, Gd, Ho, La, Li, Lu, Mn, Nd, Ni, Pb, Pr, Se, Sm, Sr, Tb, Tm, Rb, V, Yb	Q	A	[32]
Wine	As, Ba, Ce, Co, Cu, Dy, Er, Eu, Ga, Ho, La, Li, Lu, Nd, Ni, Pb, Pr, Rb, Sm, Sr, Tb, Tm, V, Yb, Zn	Q	A	[225]
Wine	As, Cd, Pb	Q	M	[226]
Wine	Ce	FI-Q	A	[227]
Wine	Ce, Dy, Er, Eu, Gd, Ho, La, Lu, Nd, Pr, Sm, Tb, Tm, Yb	Q, SF	A	[34]
Wine	Ce, Dy, Er, Eu, Gd, Ho, La, Lu, Nd, Pr, Sm, Tb, Tm, Yb	Q, SF	A	[52]
Wine	Pb	Q	A, Au	[31]
Wine	Pb	Q	A	[29]
Wine	Pb	FI-Q	A	[228]
Wine	Pb	Q	A	[229]
Wine	Pb	Q	A, M	[57]
Wine	Pb	Q	A	[230]
Wine	Pb	Q	Au, M	[55]
Wine	Pb	TOF, MC, Q	A	[58]

(*Continued*)

TABLE 8.1. *Continued*

Matrix	Elements	Technique[a]	Type of Study[b]	References
Wine	Pb	TOF, SF-MC, Q	A	[59]
Wine	Pb	Q (ID)	A, C	[231]
Wine	Pb	Q	Au	[56]
Wine	Sr	Q	A, Au	[53]
Wine	Sr	SF-MC	A, Au	[54]

[a]Analyzer: Q = quadrupole, CC = collision cell, DRC = dynamic reaction cell, MC = multicollector, SF = sector field. Analytical details: CV = cold vapor, ETV = electro-thermal vaporization, FI = flow injection, HG = hydride generation, ID = isotope dilution, LA = laser ablation, UN = ultrasonic nebulization. Sample introduction: in liquid or slurry (sl) form.

[b]Type of study: A = analytical study, Au = authenticity/fingerprinting/geographical origin assessment, C = collaborative trial, I = dietary intake study, M = monitoring study, P = study on the effect of food processing/preparation, R = reference material certification, O = other studies, including investigations on element uptake/distribution/biotransformation and biological activity.

sample digestion [28]. Dilution is usually applied to reduce the ethanol content of wine before analysis [29]. FI [30], standard addition and isotope dilution (ID) [31], matrix-matching [32], and ethanol removal by heating [33] have also been reported as viable solutions to deal with matrix effects in the analysis of alcoholic beverages. However, depending on the nebulizer used, not only ethanol, but also other organic and inorganic components can produce significant matrix effects that need to be taken care of, especially when red wines are analyzed [34]. The direct analysis of viscous or oily samples requires dilution and addition of surfactants such as Triton X-100, as reported in the analysis of honey [35] or emulsions, for example, of olive oil [36, 37] or vitamin E [38].

In Table 8.1, cow's milk, breast milk, and infant formulae are considered together under the entry "Milk and infant formulae," irrespective of the actual liquid or powdered form of the samples analyzed in each single study, because all these items share a similar matrix and pose analogous analytical problems. Triton X-100 has been used to measure Pb isotope ratios in milk powder samples prepared as slurries [39]. Simple dilution of fresh milk in alkaline media (in order to prevent protein precipitation) has also been successfully applied by other authors [40], especially for I determination [41–43], even though in most studies milk and infant formulae are analyzed following full digestion.

8.2.1.2 Applications Rapid and reliable methods are required for routine monitoring of the quality of the water used for human consumption. Most trace elements are present in drinking water at very low concentrations (few $\mu g\ l^{-1}$ down to $ng\ l^{-1}$). Yet, water can contribute significantly to human exposure to some highly toxic elements due to the amount consumed daily. Owing to its excellent LoDs and multielement capability, ICP-MS has established itself as a superior tool for water analysis and it has found widespread use for drinking water monitoring (Table 8.1). Its features have also been found valuable in studies on the influence that the water used as ingredient in processing or for cooking has on the element content of foods [44, 45]. The determination of long-lived radionuclides is another reported application and several methods have been described for the measurement of natural and artificial radioisotopes in water, including ^{99}Tc, ^{226}Ra, ^{232}Th, and ^{238}U (Table 8.1). In recent times, the technique has been resorted to in some surveys on the inorganic composition of bottled mineral waters, the consumption of which has markedly increased in several countries over the last few years. Two extensive studies carried out in Europe and in Canada reported on the determination of 39 and 21 elements, respectively, by ICP-MS and found that a number of samples did not comply with existing regulations [46, 47].

The use of ICP-MS for elemental characterization in beverages has been covered by a recent review that emphasized the applications in the area of fingerprinting and authenticity assessment [48]. The combination of (multi)elemental/isotopic analyses with powerful multivariate statistical methods can provide a solution for the authentication of food and beverages [49] and there is an increasing awareness of the potential of ICP-MS in this respect. Foods originating from different geographical regions and under different production conditions can be distinguished on the basis of their elemental/isotopic profile. The fingerprinting reflects elemental signatures of the geographical area of origin, related to the underlying geological and climatological background, and incorporates the modifications introduced by processing (if any). Thus, by using chemometric approaches authenticity assessment becomes possible, including evaluation of geographical origin, product classification (e.g., organic versus conventional foods), identification of product adulteration and substitution.

In the case of wine, the main application is to develop a fingerprint technique for the identification of geographical origin that would facilitate the detection of frauds and protect prestigious wineries. Trace elements analysis by ICP-MS followed by a statistical evaluation of the data using discriminant analysis has been shown to be a suitable approach for determining the geographical origin of wine. For example, it allowed Spanish wines from three different regions to be discriminated according to 35 elements, with only seven elements (Cd, Cr, Cs, Er, Ga, Mn, and ^{86}Sr) providing most of the differentiation [30]. It also made it possible to distinguish with 95 percent accuracy between Spanish and English wines (red, rosé, and white) and completely differentiate white wines from the two countries using 48 elements (Fig. 8.1). In similar studies, Taylor et al. distinguished between 59 wines from the Okanagan Valley (British Columbia) and 36 wines from the Niagara Peninsula (Ontario) by means of 10 elements (Al, Co, Mn, Mo, Rb, Sb, Sr, U, V, and Zn), with Sr being the most important discriminating element [50], while Greenough et al. differentiated 25 wines from several vineyards in the Okanagan Valley on the basis of six elements (Al, As, Cd, Co, Cu, and V) [51]. Moreover, they found a strong correlation between some elements

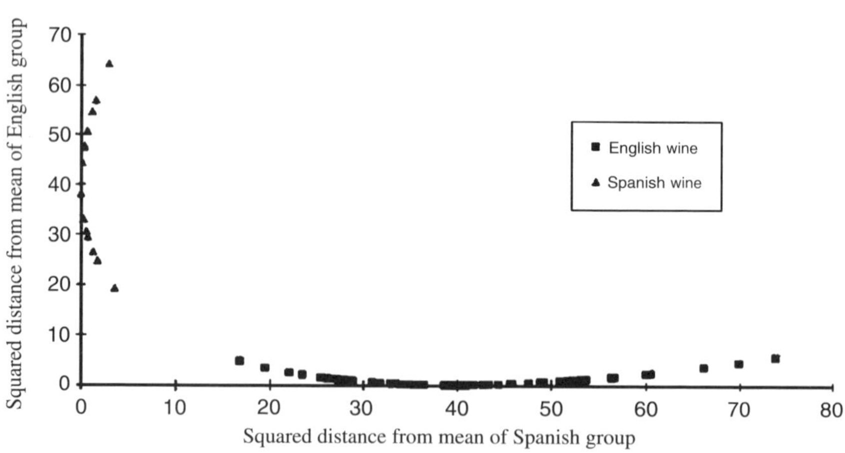

Figure 8.1. Classification of English and Spanish white wines by discriminant analysis using 48 elements determined by FI-ICP-MS. Reprinted from *Food Chem.*, **15**, M. J. Baxter, H. M. Crews, M. J. Dennis, I. Goodall, D. Anderson, The determination of the authenticity of wine from its trace element composition, © 1997, with permission from Elsevier [30].

(Ba, Ca, Mg, Mo, P, and Zn) and wine color. REEs have also been used to identify the geographical location of the vineyard of origin following chondrite-normalization (see Table 8.1), but it has been demonstrated that, if bentonite is used in wine processing for clearing from tarnish components, the REEs pattern will indicate the origin of the bentonite rather than that of the finished wine, as this material is a significant source of REE contamination [52].

Isotopic fingerprinting with stable isotopes has been found useful for wine authentication as well. The use of isotope patterns as biogeochemical tracers is firmly established with the aim of geographical profiling. Thus, the $^{87}Sr:^{86}Sr$ ratio reflects the geological time of the soil on which grape is grown as the ^{87}Rb decays to stable ^{87}Sr, while ^{86}Sr is nonradiogenic. Similarly, relative abundances of Pb isotopes vary with geological age and consequently with geographical location, as three out of four Pb isotopes (^{206}Pb, ^{207}Pb, and ^{208}Pb) originate from decay of U and Th, and exhibit greater variation than the lighter isotope (^{204}Pb), that is, the only stable primordial Pb isotope. Owing to the isobaric overlap of ^{87}Sr and ^{87}Rb, the measurement of the $^{87}Sr:^{86}Sr$ ratio in wine is not an easy task, and has been accomplished either through column separation and use of a mathematical equation to correct for the interference of the residual ^{87}Rb [53] or by resorting to multicollector (MC) SF-ICP-MS by using the $^{88}Sr:^{86}Sr$ ratio for mass bias correction, this leading to a significant improvement in isotope ratio precision [54].

The occurrence of Pb in wine is due to natural (soil-related) and anthropogenic sources, including atmospheric pollution and contamination from the materials utilized for wine-making and storage. In recent years, the contribution from the latter sources has declined and the total Pb content of wine has been found to decrease. Lead isotope ratios have been investigated as a tool to locate wine origin by several authors [31, 55, 56] and also to date the age of port wine [57]. Recently, MC and time-of-flight (TOF) instruments have been shown to substantially improve the precision of Pb isotope ratios with respect to quadrupole (Q) ICP-MS, even though laborious sample treatment and preconcentration are needed in the case of TOF [58, 59]. On the other hand, MC-SF-ICP-MS appears ideal for the task, as more and more detection power is required for precise isotope ratio measurements due to the ever smaller Pb concentration in recent vintages.

Apart from wine, several other works involving multielemental characterization of beverages have been carried out with different scopes, for example, origin assessment of orange juices [28, 60], study of the tea infusion process [61], and investigation of the interdependences among trace elements in beer [62].

The multielement capabilities of ICP-MS have also been extensively employed for the characterization of human milk in studies aiming at verifying the reference ranges (RRs) proposed for both essential and harmful elements, and contribute to the establishment of RRs for new (ultratrace) elements [63–65]. Transfer of selected elements from food into human milk [66], changes in milk composition during lactation [67], and adequacy of the intakes of essential trace elements in breast-fed infants [63, 68] are some of the other topics investigated in related studies. For those infants who are not breast-fed, the nutritional adequacy and safety of substitutive formulae are of major concern. A number of studies on the composition of commercially available formulae and their suitability for infant nutrition from the standpoint of element content have been performed with ICP-MS detection (see Table 8.1). As far as single-element studies are concerned, noteworthy investigations have appeared on the geographical and seasonal variations of the I content of commercial cow milk observed in some countries of Northern Europe [27, 43, 69]. Differences between organic and conventional milk as well as trends in milk-derived products were also addressed in these studies.

8.2.2 Food

8.2.2.1 General Aspects The literature concerning solid foods is summarized in Table 8.2. Total element determination using conventional (i.e., liquid) sample introduction requires a dissolution step for analyte presentation to the instrument. The choice of the digestion method is made on the basis of several considerations, including efficiency in analyte solubilization, potential for loss of volatile elements and contamination of samples, reagent-related interferences and matrix effects, sample throughput, LoD requirements, and safety aspects. Microwave (MW)-assisted digestion with HNO_3, alone or with the addition of H_2O_2, is a widely used method for the dissolution of food samples. In particular, MW digestion with closed vessel systems has gained popularity, being a simple and fast dissolution technique that minimizes acid consumption, risk of sample contamination, and loss of volatile elements. HNO_3 is the acid of choice for ICP-MS analysis as it causes minimal interferences, while small amounts of other acids can be added for specific analytical purposes, for example, HF for the solubilization of Si-bound elements in vegetable matrices containing substantial amounts of Si [70]. Alternative solutions to digestion with concentrated acids have also been reported for a rapid, multielement screening of biological and botanical materials, for example, extraction with water-soluble tertiary amine solutions containing EDTA [71].

Solid foods in powder form can be analyzed directly by means of LA- or ETV-ICP-MS to eliminate time-consuming sample dissolution procedures (see Table 8.2). However, this requires the preparation of homogeneous powdered samples and the subsequent analytical determination is not as straightforward as the one based on liquid sample introduction. Another way to perform direct analysis of solid foods is to grind and suspend them into slurries. The viability of slurry nebulization relies on the ability to prepare samples of fine particle size in a reproducible manner and on the adoption of suitable (e.g., high-solids) nebulizers. Otherwise, slurries can be analyzed by ETV-ICP-MS resorting to the ultrasonic slurry sampling technique [72–74].

For some elements, fit-for-purpose sample preparation is required in order to obtain reliable analytical results. Iodine is a well-known case, its determination being complicated by loss of iodide (as HI) from HNO_3 solutions, memory effects due to volatilization in the sample introduction system and by matrix effects

TABLE 8.2. Literature on Element Determination in Food by ICP-MS

Matrix	Elements	Technique[a]	Type of Study[b]	References
Cereals, pulses, and derived products				
Cereals and cereal products	Cd	Q	I	[232]
Cereals and pulses	Cd, Pb	Q	M	[233]
Cereals, pulses, and cereal products	Cd, Co, Cr, Cu, Mn, Pb, Zn	Q	M	[234]
Cereal RMs	Al, As, B, Ba, Ca, Cd, Co, Cr, Cu, Fe, K, Li, Mg, Mo, Ni, Pb, Sb, Se, Sn, Sr, Th, Ti, Tl, U, V, Zn	Q	A	[235]
Cereal RMs	Al, As, B, Ba, Ca, Cd, Co, Cr, Cu, Fe, Hg, K, Mg, Mn, Mo, Na, Ni, P, Pb, S, Sb. Se, Si, Sr, Th, Ti, Tl, U, V, Zn	Q	A	[70]
Durum wheat and derived products	As, Cd, Ni, Pb	Q (UN)	P	[44]
Durum wheat and derived products	Cd, Cr, Fe, Ni, Pb	Q (UN)	A	[236]
Durum wheat and derived products	Cd, Cr, Fe, Ni, Pb	Q (UN)	P	[45]
Mung beans	As	Q	A, O	[235]

(*Continued*)

TABLE 8.2. *Continued*

Matrix	Elements	Technique[a]	Type of Study[b]	References
Pea	Ag, Al, Au, B, Ba, Be, Bi, Ca, Cd, Ce, Co, Cr, Cs, Cu, Dy, Er, Eu, Fe, Ga, Gd, Ge, Hf, Ho, In, Ir, K, La, Li, Lu, Mg, Mn, Mo, Na, Nb, Nd, P, Pb, Pd, Pr, Pt, Rb, Re, Rh, Ru, S, Sb, Sc, Se, Si, Sm, Sn, Sr, Ta, Tb, Te, Th, Ti, Tl, Tm, U, V, W, Y, Yb, Zn, Zr	SF	O	[96]
Rice	Ag, As, B, Be, Cd, Ce, Co, Cr, Cs, Cu, Dy, Er, Eu, Gd, Ho, La, Li, Lu, Mo, Nd, Ni, Pb, Pd, Pr, Rb, Sb, Sc, Sm, Sn, Sr, Tb, Th, Ti, Tm, U, V, Y, Yb, Zn	Q	O	[238]
Rice	As, Cd, Co, Cr, Cu, Fe, Mn, Pb, V, Zn	UN-Q, -SF	M	[239]
Rice	As, Cd, Mo, Ni	Q	M	[240]
Rice	B	Q	Au	[97]
Rice	B, Gd, Ho, Mg, Rb, Se, W	Q	O	[98]
Rice	Cd	Q	I, O	[241]
Rice	Cd	Q	P	[242]
Rice flour	Cd, Cr, Cu, Hg, Pb	ETV-DRC-Q (ID) (sl)	A	[243]
Rice flour RM	Ag, Cd	DRC-Q	A	[244]
Rice flour RM	Cd, Cu, Pb, Zn	Q (ID)	A	[245]
Rice flour RMs	Cd, Cr, Fe, Pb	Q (ID)	A, R	[246]
Various cereals	Ag, Al, As, Ba, Bi, Cd, Ce, Co, Cr, Cs, Cu, Fe, Hg, In, La, Mn, Mo, N, Pb, Rb, Sb, Se, Sn, Sr, Th, Ti, Tl, U, V, Zn	SF	A	[247]

Various cereals	Ce, Dy, Er, Eu, Gd, Ho, La, Lu, Nd, Pr, Sc, Sm, Tb, Tm, Y, Yb	Q	A	[248]
Various cereals	Pt	Q, SF	A, O	[100]
Wheat (soft and durum)	Al, As, Cd, Co, Cr, Cu, Fe, Mn, Mo, Ni, Pb, Se, Sn, V, Zn	Q	A	[249]
Wheat	Cd, Pb, Se, Sr	SF	O	[99]
Wheat	U	Q	M	[250]
Wheat, buckwheat and quinoa	As, Co, Cr, Cu, Fe, Li, Mn, Mo, Se, Sr, V, Zn	Q	O	[251]
Wheat porridge	Cd	Q	O	[94]
Wheat RM	Ce, Nd, Sm	Q (ID)	A	[252]
Wheat RM	Se	HG-Q (ID)	A	[253]
Fruit, vegetables and other plant food				
Berries	Ag, Al, As, Ba, Bi, Cd, Ce, Co, Cr, Cs, Cu, Fe, Hg, In, La, Mn, Mo, N, Pb, Rb, Sb, Se, Sn, Sr, Th, Ti, Tl, U, V, Zn	SF	M	[254]
Berries	Ce, Cs, Sr	Q	O	[103]
Cabbage	Al, As, B, Ba, Ca, Cd, Co, Cr, Cu, Fe, Hg, Mn, Mo, Ni, Pb, Rb, Se, Sb, Sr, Tl, Zn	Q, SF	R	[255]
Cabbage	^{129}I	Q	A	[256]
Cocoa	Cd, Pb	Q	O	[257]
Fruits, berries, and nuts	As, Cd, Cr, Cu, Pb, Zn	SF	O	[102]

(*Continued*)

TABLE 8.2. *Continued*

Matrix	Elements	Technique[a]	Type of Study[b]	References
Fruit and vegetables	Sn	Q	P	[216]
Onion	Ag, Al, Au, B, Ba, Be, Bi, Ca, Cd, Ce, Co, Cr, Cs, Cu, Dy, Er, Eu, Fe, Ga, Gd, Ge, Hf, Ho, In, Ir, K, La, Li, Lu, Mg, Mn, Mo, Na, Nb, Nd, P, Pb, Pd, Pr, Pt, Rb, Re, Rh, Ru, S, Sb, Sc, Se, Si, Sm, Sn, Sr, Ta, Tb, Te, Th, Ti, Tl, Tm, U, V, W, Y, Yb, Zn, Zr	SF	O	[96]
Potato	Ag, Al, Au, Ba, Bi, Ca, Cd, Co, Cr, Cs, Cu, Dy, Er, Fe, Ga, Gd, Ho, In, Ir, La, Lu, Mn, Mo, Nb, Nd, P, Pb, Pd, Pr, Pt, Rb, Re, Rh, Ru, Sb, Sc, Sm, Sn, Sr, Ta, Tb, Th, Ti, Tl, Tm, U, V, Y, Yb, Zn	SF	O	[95]
Rapeseed	As, Cd, Cr, Cu, Mo, Ni, Pb, Rb, Ti, Tl, V, Zn	Q	O	[101]
Tea	Cd, Co, Cr, Cs, Hg, La, Li, Nd, Ni, Pb, Pr, Rb, Se, Sn, Ti, V, Zr	Q	Au	[258]
Tea	Ce, Dy, Er, Eu, Gd, Ho, La, Lu, Nd, Pr, Sm, Tb, Tm, Y, Yb	Q	A	[259]
Tea	Co, Cr, Cs, Cu, Ni, Pb, Rb, Ti, V	Q	Au	[260]
Various fruit	Sn	Q	M	[85]
Various vegetables	Ce, Dy, Er, Eu, Gd, Ho, La, Lu, Nd, Pr, Sc, Sm, Tb, Tm, Y, Yb	Q	A	[248]
Various vegetables	Pb	FI-Q (ID)	A	[261]

Various vegetables	Pt	Q, SF	A, O	[100]
Vegetable RMs	As, Cd, Co, Cr, Cu, Fe, Mn, Mo, Ni, Pb, Zn	Q	A	[262]
Mushrooms				
Mushrooms	Ag, B, Ca, Cd, Ce, Cs, Cu, Dy, Er, Eu, Fe, Gd, Hg, Ho, In, La, Lu, Mg, Mn, Nd, K, Na, P, Pb, Pr, Rb, Sm, Sr, Tb, Th, Tl, Tm, U, Yb, Zn	UN-SF	M	[263]
Mushrooms	Cd, Pb	Q	M	[264]
Sweeteners				
Honey	As, Be, Cd, Co, Cs, Cu, Li, Mn, Ni, Pd, Pb, Rb, Rh, Sb, Se, Sn, Te, Tl, V	Q (ETV, UN)	A	[265]
Honey	As, Cd, Cr, Cu, Fe, Mn, Ni, Pb, Sn, V, Zn	UN-Q, -SF	A	[266]
Honey	As, Cd, Cr, Cu, Fe, Mn, Ni, Pb, Se, Zn	Q, SF	A	[267]
Honey	As, Cd, Cr, Cu, Fe, Ni, Pb, Pt, Sn, V	UN-Q, -SF	A	[268]
Honey	Ba, Cu, Pb, Zn	Q (ID)	A	[35]
Maple syrup	As, Cd, Co, Cu, Mn, Pb, V, Zn	Q	A	[269]
Yeast and supplements				
Ca supplements	Cd, Pb	Q	C	[270]
Ca supplements	Pb	Q	A	[271]
Ca supplements	Pb	SF	M	[272]
Dietary supplements	As, Cd, Hg, Pb	SF	M	[273]

(*Continued*)

TABLE 8.2. *Continued*

Matrix	Elements	Technique[a]	Type of Study[b]	References
Vitamin E supplements	Cr, Ni, Sn, Pb	Q	A	[38]
Yeast	Se	Q	A	[274]
Fats and oils				
Lard	Cd, Co, Cr, Cu, Ge, Hg, Ni, Pb, Se, Sn, Zn	Q	M, O	[84]
Olive oil	Ba, Cd, Co, Cr, Cu, Mn, Ni, Tl, Pb, U, V, Zn	Q	A	[36]
Olive oil	Al, Ba, Bi, Cd, Co, Cu, Mn, Ni, Pb, Sn, V	FI-Q	A	[37]
Vegetable oil	Cd, Pb, Zn	ETV-Q	A	[275]
Dairy products				
Cream and cheese	I	Q	A, M	[43]
Cheese	I	Q	M	[69]
Meat and offal				
Beef	Pb	Q	P	[105]
Bovine muscle	As, Cd, Cu, Pb	UN-Q, -SF	A	[276]
Bovine liver RM	Ag, Al, As, Cd, Co, Cu, Fe, Hg, Mn, Mo, Pb, Rb, Sb, Se, Sr, Tl, V, Zn	Q	A	[277]
Bovine liver RM	Al, As, B, Ca, Cd, Co, Cr, Cu, Fe, LI, Mg, Mn, Mo, Ni, Pb, Se, Sr, V, Zn	Q (LA) (s)	A	[278]
Bovine liver RM	As, Se	HG-Q	A	[202]
Bovine liver RM	Ca, Cd, Cu, Fe, Mg, Mn, Na, Zn	Q	A	[279]
Bovine liver RM	I	SF (ID)	A	[280]
Bovine liver RM	Se	DRC-Q	A	[illegible]

Cattle meat and offal	As, Be, Co, Cd, Cr, Cu, In, Mg, Mn, Ni, Pb, Se, Y, Zn	Q	M	[83]
Seafood				
Fish	Cd, Hg, Pb	ETV-Q (ID) (sl)	A	[73]
Fish	Hg	Q	A, O	[281]
Fish	Sn	Q (ID)	R	[282]
Fish	Pt	SF	O	[283]
Fish and shellfish	As	Q	M	[284]
Fish RM	Cr	Q (ID)	A	[285]
Fish RM	Hg	Q (ID)	A	[286]
Fish RMs	As, Co, Cd, Cr, Cu, Fe, Hg, Mg, Mn, Na, Ni, Pb, Zn	Q (ID)	A, R	[287]
Fish RMs	Cd, Cu, Pb	ETV-Q (ID)	A	[288]
Fish RMs	Hg	ETV-Q (sl)	A	[72]
Fish RMs	Se	HG-Q (ID)	A	[289]
Lobster	As, Co, Cd, Mo, Se	FI-Q	A	[119]
Mussel	Ba, Cd, Co, Cr, Cu, Ga, Ni, Pb, Rb, V	SF	M	[290]
Mussel	Ca, Cu, Mg, Mn, Fe, Zn	Q	A	[291]
Oyster RM	As, Ca, Cd, Co, Cr, Cu, Fe, Mg, Mn, Pb, Se, Sn, Sr, Th, U, V, Zn	CC-Q (UN)	A	[292]
Oyster RM	Cd, Cr, Ni, Pb	FI-Q	A	[293]
Seafood	As, Cd, Pb	Q	A	[294]
Seafood	As, Cd, Cr, Hg, Pb, Se	Q (ID)	A, R	[295]
Seafood	I	Q	A, M	[80]
Seafood RMs	Ag, As, Cd, Co, Cr, Cu, Fe, Hg, Mn, Ni, P, Pt, Sb, Se, V, Zn	Q	A	[296]
Seafood RMs	Ag, Cd, Co, Cu, In, Mo, Ni, Pb, Sb	FI-TOF (UN)	A	[297]
Seafood RMs	As	Q	A	[298]
Seafood RMs	Cd, Cu, Mn, Pb, V, Zn	FI-Q	A	[299]

(*Continued*)

TABLE 8.2. *Continued*

Matrix	Elements	Technique[a]	Type of Study[b]	References
Seafood RMs	Sn	Q	A	[300]
Shellfish	Al, B, Ba, Cd, Co, Cr, Cu, Fe, Ga, Li, Mn, Mo, Nb, Ni, Pb, Sr, Ta, Ti, Tl, V, Zn, Zr	Q	A	[301]
Shellfish	Al, As, Ba, Bi, Cd, Co Cr, Cu, Hg, Mn, Mo, Ni, Pb, Rb, Se, Sn, Sr, Th, U, V, Y Zn, Zr + lanthanides	Q, SF	A	[302]
Shellfish	U	SF	A, M	[169]
Miscellaneous				
Infant food	Ca, P	DRC-Q	A	[303]
Milk and mixed diet RMs	Al, As, Ca, Cl, Co, Cr, Cu, Fe, K, Mn, Mo, Na, Ni, P, Rb, Zn	LA-Q (s)	A	[304]
Milk and tea leaves RMs	Al, Ba, Ca, Cr, Cu, Fe, K, Mg, Mn, Na, Ni, P, Pb, Rb, Sr, Zn	LA-Q (s)	A	[305]
Milk powder and soy flour	I	FI-Q	C	[306]
Mint, nettle and their infusions	As, Ba, Ca, Cd, Co, Cr, Cu, Fe, I, Li, Mg, Mn, Ni, Pb, Se, Sr, Ti, V, Zn	Q	O	[307]
Mixed diet	Al, As, Ca, Cd, Co, Cr, Cu, Fe, Hg, K, Li, Mg, Mn, Mo, Na, Ni, Pb, Sb, Se, Sn, Zn	Q	I	[308]
Mixed diet	Al, Cd, Cr, Cu, Mn, Ni, Pb, V, Zn	Q	A	[309]
Mixed diet	As	FI-Q	A	[310]
Mixed diet	As, Be, Cd, Cr, Cu, Pb, Sb, Zn	Q	I	[311]
Mixed diet	Ba, Bi, Cd, Co, Cs, Mo, Ni, Pb, Th, Tl, U	Q	I	[312]
Mixed diet	Ba, Ca, Cu, K, Fe, Na, Mg, P, Sr, Zn	Q	I	[207]

Mixed diet	Ca, Fe, P	Q	I	[313]
Mixed diet	Cd	Q	I, O	[241]
Mixed diet	Cd, Pb	Q	I	[314]
Mixed diet	Cd, Pb	Q	I, O	[315]
Mixed diet	Cd, Pb	Q	I	[316]
Mixed diet	Cu	Q	I	[317]
Mixed diet	Cu, K, Mg, Na, Zn	Q	I	[318]
Mixed diet	Pb	Q	I, O	[319]
Mixed diet	Se	Q	I	[320]
Mixed diet	Th	Q	A	[321]
Mixed diet	Th, U	Q	A	[322]
Mixed diet RMs	Cu, Fe, Ni	ETV-Q (ID) (sl)	A	[74]
Oyster and bovine liver RMs, mixed diet	Pb	FI-Q (ID)	A	[323]
Various foods	Al, As, Au, B, Ba, Bi, Ca, Cd, Co, Cr, Cu, Fe, Ge, Hg, Ir, Li, Mn, Mo, Ni, Pb, Pd, Pt, Rh, Ru, Sb, Se, Sn, Sr, Tl, Zn	Q	I	[89]
Various foods	Al, As, B, Ba, Cd, Co, Cr, Mo, Ni, Pb, Sn, Sr, V	FI-Q	A	[324]
Various foods	Al, As, Cd, Cr, Cu, Hg, Ni, Pb, Se, Sn, Zn	Q	I	[325]
Various foods	Al, B, Cd, Co, Cr, Cu, Mn, Mo, Ni, Zn	Q	I	[326]
Various foods	As	Q	I	[327]
Various foods	As	Q	I	[328]
Various foods	As, Cd, Co, Cr, Hg, Mn, Mo, Ni, Pb, Pd, Se, Sn, Tl, V, Zn	Q	A	[329]

(*Continued*)

TABLE 8.2. *Continued*

Matrix	Elements	Technique[a]	Type of Study[b]	References
Various foods	As, Cd, Cu, Hg, Mo, Pb, Se, Sn, Zn	Q	A	[330]
Various foods	As, Cd, Pb	UN-Q	I	[331]
Various foods	Ba, Cd, Co, Cr, Cs, Cs, Cu, Li, Mn, Mo, Ni, Rb, Sc, Th, U, V, Sr	Q	A	[332]
Various food	Br, I	FI-Q	I	[333]
Various foods	Ca, Cu, Fe, K, Mg, Na, P, Zn	Q	I, O	[334]
Various foods	Cd, Pb	Q	A	[335]
Various foods	I	Q	M	[79]
Various foods	I	Q	A	[76]
Various foods	I	Q	O	[78]
Various foods	I	Q	A	[77]
Various foods	Si	Q, SF	C	[336]
Various foods	Sn	Q	I	[337]
Various foods	Pu	SF (UN)	A	[338]
Various foods	Th, U	Q	I	[339]
Various foods	Th, U	Q	I	[340]
Various food and mixed diet	U	Q	I	[170]
Various RMs	Ag, As, Bi, Cd, Co, Cr, Cu, Mn, Ni, Pb, Sb, Se, Sn, Te, V	ETV-Q (sl)	A	[341]
Various RMs	Ag, Cd, Hg, Pb, Tl	ETV-Q (ID)	A	[342]
Various RMs	Al, As, B, Be, Br, Cd, Ce, Co, Cr, Cu, Dy, Er, Eu, Fe, Gd, Hg, Ho, La, Li, Lu, Mg, Mn, Mo, Nd, Ni, Pb, Pr, Se, Sn, Sm, Tb, Th, Tm, U, V, Y, Yb, Zn	Q	A	[343]

Various RMs	Al, As, Ba, Ca, Cd, Ce, Cr, Cs, Cu, Fe, La, Li, Mg, Mn, Mo, Ni, Rb, Sb, Se, Sr, Tl, V, Y, Zn	Q	A	[344]
Various RMs	Al, As, Cd, Co, Cr, Cu, Fe, Mn, Mo, Ni, Pb, Se, Sn, V, Zn	Q	A	[345]
Various RMs	Al, As, Cd, Hg, Pb, Se	Q	A	[346]
Various RMs	Al, Cd, Cr, Mo, Ni, Pb, Rb, Zn	Q	A	[347]
Various RMs	As	Q	A	[348]
Various RMs	As, Ba, Ca, Cr, Cu, Mn, Mo, Pb, Rb, Se, Zn	Q	A	[71]
Various RMs	As, Co, Cr, Cu, Fe, Mn, Ni, Se, Sr, V, Zn	Q	A	[349]
Various RMs	As, Se	Q	A	[350]
Various RMs	As, Se	SF	A	[351]
Various RMs	As, Se, V	Q	A	[352]
Various RMs	B	Q (ID)	A	[353]
Various RMs	B	Q	A	[354]
Various RMs	B	Q	A	[355]
Various RMs	Br, I	Q	A	[356]
Various RMs	Ca, Co, Cr, Cu, Fe, K, Mg, Mn, Na, Ni, P, V, Zn	Q	A	[357]
Various RMs	Cd	Q	A	[358]
Various RMs	Cd, Co, Cu, I, Mn, Ni, Pb, Rb, Zn	Q	A	[359]
Various RMs	Cd, Cr, Cu, Fe, Mn, Ni, Pb, Zn	Q	C	[360]
Various RMs	Ce, Dy, Er, Eu, Gd, Ho, La, Lu, Nd, Pr, Sm, Tb, Tm, Y, Yb	Q	A	[361]
Various RMs	Co, Cr, Cu, Fe, Mn, Mi, Sc, V, Zn	SF	A	[362]
Various RMs	Cr	Q (UN)	A	[363]

(Continued)

TABLE 8.2. *Continued*

Matrix	Elements	Technique[a]	Type of Study[b]	References
Various RMs	Cr, V	Q	A	[364]
Various RMs	Cs, I, Sr, Th, U	Q	C	[365]
Various RMs	Cu, Mn, Ni	ETV-Q (sl)	A	[366]
Various RMs	Hg	Q	A	[367]
Various RMs	Hg	CV-Q (ID)	R	[368]
Various RMs	Hg	Q	A	[369]
Various RMs	Hg, Pb	Q (ID)	R	[370]
Various RMs	I	Q	A	[81]
Various RMs	I	Q	A	[75]
Various RMs	I	Q	A	[82]
Various RMs	I	Q (ID)	A	[371]
Various RMs	I	Q (ID)	A	[372]
Various RMs	Pb	ETV-Q (s)	A	[373]
Various RMs	Pb	SF, Q, CC-MC (ID)	A	[374]
Various RMs	Se	HG-Q (ID)	A	[375]
Various RMs	Se	ETV-Q (ID) (s)	A	[376]
Various RMs	Se	HG-Q (ID)	A	[377]
Various RMs	Se	CC-Q (ID)	A	[378]
Various RMs	Si	SF (ID)	A	[379]
Various RMs	Zn	Q (ID)	A	[380]

[a]Analyzer: Q = quadrupole, CC = collision cell, DRC = dynamic reaction cell, MC = multicollector, SF = sector field, Analytical details: CV = cold vapor, ETV = electro-thermal vaporization, FI = flow injection, HG = hydride generation, ID = isotope dilution; LA = laser ablation, UN = ultrasonic nebulization. Sample introduction: in liquid form unless otherwise indicated [solid (s), slurry [sl]].

[b]Type of study: A = analytical study, Au = authenticity/fingerprinting/geographical origin assessment, C = collaborative trial, I = dietary intake study, M = monitoring study, P = study on the effect of food processing/preparation, R = reference material certification, O = other studies, including investigations on element uptake/distribution/biotransformation and biological activity.

leading to signal suppression. Either the use of a strong oxidizing agent ($HClO_4$) for the conversion of I^- into nonvolatile species (IO_3^-) [75], or the creation of an alkaline environment for the suppression of I volatilization, for example, by dissolution with ammonia or tetramethylammonium hydroxide [76–78], or adding NH_3 after acid digestion [79, 80], has been used to overcome these problems. These methods, as well as the Schöniger combustion technique [81], have been evaluated and compared by Knapp et al. [82].

8.2.2.2 ***Applications*** As already pointed out above, ICP-MS has established itself as a major technique in the field of food safety and food control. The wide dynamic range and speed of multielement measurements makes it a particularly powerful tool for surveillance and legislative work, and proves valuable in dealing with emergencies such as contamination incidents. For instance, in the *Lead in Feed Incident* in the UK (1989) rapid determinations of 11 elements in different matrices allowed the source of toxic effects noticed in beef, dairy cattle, and other animals to be traced back to imported cattle feed, contaminated by Pb and other elements [83]. In the lard poisoning episode in Jiangxi Province, China (1999), in which more than 1000 people were seriously poisoned by industrial lard contamination by organotin compounds, screening of 11 analytes in lard, blood, urine, and organ samples indicated the presence of high levels of several elements and a huge contamination by Sn, suggesting that organotins played a dominant role in poisoning. This hypothesis was confirmed by organotin determination by gas chromatography (GC)-FPD and GC-MS [84].

The latter example points out the potential of the technique as a screening method for the identification of toxic organometallic compounds, for example, in pesticide residue analysis. With an initial screening of total Sn by ICP-MS in a survey on cyhexatin (an organotin acaricide), a limited number of suspect fruit samples (three out of 72) were identified and then submitted to confirmatory analysis with a GC-MS method [85]. This approach resulted in a ~30 percent reduction of the survey cost in terms of staff hours.

Another application of ICP-MS in the area of food safety is the evaluation of food contact materials. Q [86] and TOF [87] mass spectrometers have been employed in the compositional analysis of paper and board material intended for food contact. Studies on the migration of metals from food contact plastics using food simulants and ICP-MS detection have also been reported. One of them showed how to analyze different simulants without the need of any time-consuming preparation, that is, aqueous acetic acid (3 percent w/v) directly, aqueous ethanol (15 percent v/v) after proper dilution, the olive oil simulant following emulsion preparation by means of tetralin and Triton X-100 [88].

The ICP-MS potential has also been exploited in studies on the dietary intake of trace elements, which usually involve a very high sample throughput (see Table 8.2). Taking advantage of the speed of Q instruments, it is possible to deal with a large number of samples within a reasonable time, thus reducing the delay between the execution of the survey and the availability of the results. This allows, if necessary on the basis of the resulting exposure data, for the timely adoption of appropriate measures to protect public health. Cadmium and Pb – among contaminants – and Cu and Zn – among essential elements – are those most widely investigated because of their toxicological/nutritional significance and ease of determination by ICP-MS without major interferences (see Table 8.2). By combining similar food categories into a limited number of major food groups it has been possible to determine up to 30 elements [89] and estimate the relevant intake.

A closely related field of investigation is represented by nutrition studies on mineral and trace element metabolism in man. These studies are performed by means of chemical balance and isotope methods, which use stable isotopes or radioisotopes to determine the absorption and retention/utilization of a nutrient. A main advantage of stable isotope techniques is safety; that is, studies can be performed using intrinsically labeled foods on all population groups including the most sensitive (i.e., infants, pregnant women, and the elderly), to whom administration of a radioisotope is undesirable. Several multiple-isotope labeling options exist and the duration of the experiment is not limited, which may be the case with short-lived radioisotopes. Thus, even though tracers are costly, dose requirements are higher in comparison with radioisotopes, and external analysis of biological samples is required, stable isotopes are valuable tools in nutrition studies and their use appears to be a future growth area. Thermal ionization mass spectrometry (TIMS) is recognized as a reference technique for isotope analysis, but ICP-MS reduces sample preparation and analysis times, offers a precision that is generally adequate for most tracer studies, and is indeed excellent in the case of MC instruments [90, 91]. The fact that most inorganic nutrients have atomic masses $\leq$80 amu poses a particular problem because of the isobaric interferences in this mass region, which limit both the accuracy and the precision of isotope analysis. As previously observed, these drawbacks have been effectively dealt with in the instruments of the last generation, equipped with CC and DRC systems. For instance, the accurate determination of Se at masses 76, 77, and 80 in food, plasma, urine, and feces in an experimental study with ^{77}Se-enriched yeast in humans by DRC-ICP-MS has been reported [92] (Fig. 8.2). Stable isotope techniques employing ICP-MS detection have also been applied in studies regarding harmful elements such as Ni, an element which can stimulate allergic reactions in susceptible individuals [93], or Cd [94].

Table 8.2 reports many applications of ICP-MS to studies concerning the transfer of elements through the food chain. For instance, the multielement capabilities of ICP-MS have been used together with statistical analysis of concentration data with the aim of characterizing nutritional profiles of crops and investigating the effects of soil types and agricultural practices. Bibak et al. [95] and Gundersen et al. [96] determined more than 50 major, trace, and ultratrace elements in vegetables (onions, peas, and potatoes) grown with different fertilization practices and in conventional/organic conditions, respectively. By applying analysis of variance, discriminant partial least-square regression, and principal component analysis, the cultivation method was found to affect the uptake pattern of several elements and the authors claimed that the methodology could be used as authenticity control for organic cultivation after further method development.

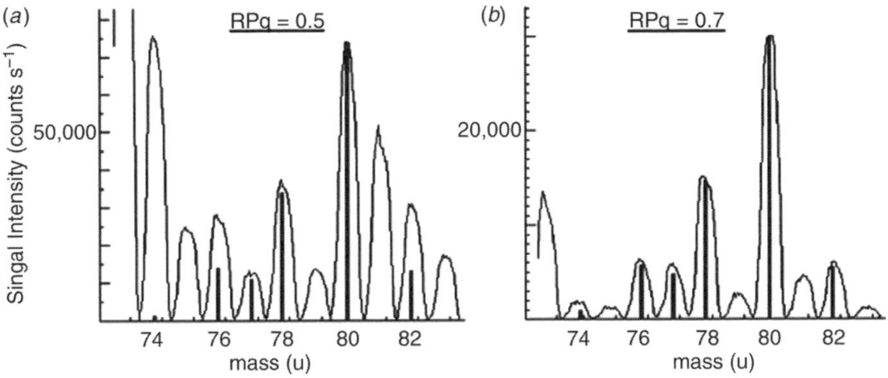

Figure 8.2. Accurate determination of Se in an HNO_3 digest of a nonenriched feces sample from a human study with ^{77}Se-enriched yeast by DRC-ICP-MS. The effect of the settings of the DRC parameter RPq on the Se mass spectrum is shown. The overlaid bars indicate the natural isotopic pattern of selenium, which has been normalized to ^{80}Se. Reproduced from [92] by permission of the Royal Society of Chemistry.

Authenticity studies on cereal products have been carried out by various authors. ICP-MS was used to measure $^{11}B/^{10}B$ in brown rice, soil, and water following sample digestion and separation of matrix elements in a study aiming at identifying the production area of rice [97]. Kelly et al. [98] used isotopic and elemental analysis by isotope ratio mass spectrometry (IRMS) and ICP-MS to determine the geographical origin of premium long-grain rice. Samples cultivated in the USA, Europe, and Basmati regions were analyzed for 50 elements, and nine key variables ($\delta^{13}C$, $\delta^{18}O$, B, Gd, Ho, Mg, Rb, Se, and W) were identified by canonical discriminant analysis as providing the maximum discrimination between rice samples from the above areas. In a similar study, Branch et al. investigated the use of SF-ICP-MS in conjunction with $\delta^{13}C$ and $\delta^{15}N$ determination and multivariate statistics for identifying the origin of soft wheat [99].

Element uptake from soil and transfer into the edible parts of plants have been addressed in several other studies. Soil-to-plant transfer factors in fruit and vegetables grown in various agricultural conditions have been determined for, for example, Pt [100], Tl [101], and various other metal contaminants [102]. In a study on stable isotopes of fission product elements (Ce, Cs, Sr), an in vitro enzymolysis method has been applied to investigate the solubilization of the analytes from fodder in a simulated ruminant digestion [103]. The effect of inhibitors of fission product solubility was also considered and essential elements were determined simultaneously to evaluate potential nutrition problems for the animals from the use of such inhibitors. Selective leaching of individual classes of metal complexes with different ligands and sequential enzymolysis have been recently applied to estimate the potential bioavailability to humans of Cd and Pb in cocoa powder and related products [104].

The final links in the food chain – processing and domestic preparation – may greatly affect element content and bioaccessibility in ready-to-eat foods. For instance, significant variations in the concentrations of several elements as a result of processing of durum wheat grain into semolina and pasta have been assessed in studies which employed multielement ICP-MS detection with ultrasonic nebulization (UN) of low levels of analytes in cereal-based matrices and water [44, 45]. Milling significantly reduced the concentrations of the investigated elements according to an element-specific pattern. Cooking of pasta caused a 50–60 percent reduction of the levels of all elements on a dry weight basis. Another study on the effect of cooking using ICP-MS investigated the possible migration of Pb from the bone of joints into beef meat [105]. High Pb levels can be found in bone because the skeleton is the major repository for the chronic deposition of the metal in mammals, but no significant leaching was observed in the case of beef meat.

As already mentioned, another field where ICP-MS has found advantageous use is the determination of long-lived radionuclides, which are often difficult and time-consuming to measure with radioanalytical techniques. The determination of ^{129}I, ^{232}Th, ^{238}U and other nuclides in various food groups has been reported by several authors, often with the aim of estimating their dietary intake (see Table 8.2). This application also represents a future growth area.

8.3 ELEMENTAL SPECIATION

8.3.1 Water and Beverages

8.3.1.1 General Aspects Several hyphenated techniques with ICP-MS detection are used for elemental speciation. The coupling of high-performance liquid chromatography (HPLC) and ICP-MS has gained high popularity for the purpose of speciation because the effluent flow rate of HPLC columns matches the typical liquid flow rate of common nebulizers and the application range is wide on account of the different chromatographic principles exploited (reversed-phase, ion-exchange, etc.). The low tolerance of ICP-MS for organic modifiers and salts (buffers, ion-pairing reagents) added to HPLC mobile phases to enhance separation is a problem, and thus a compromise between chromatographic separation and detection ability is usually adopted.

Size-exclusion chromatography (SEC) is used for the separation of large molecules according to their effective size in solution. In combination with ICP-MS it allows the molecular weight of biomolecules in nondenaturing conditions to be estimated, and it is frequently employed when dealing with complex samples for fractionation purposes, that is, for the identification of various classes of species of an element and to determine the sum of its concentrations in each class.

The hyphenation with GC is more complicated than with liquid chromatography, but offers several analytical advantages, including fewer spectral interferences, obtainment of a more stable plasma, higher resolving power, and higher sample introduction efficiency (i.e., improved LoDs). Its application, however, is restricted to compounds which are volatile and thermally stable, or can be brought into such a form after derivatization. In recent years, capillary electrophoresis (CE) has also been used as a separation technique for the speciation of metal and metalloid species as it features highly efficient separation, rapid analyses and minute sample size requirements. So far, its major limitations have been complexity of interface, limited reproducibility, and relatively poor LoDs.

The literature on element speciation in water and beverages using ICP-MS detection is summarized in Table 8.3. In most cases, on-line HPLC-ICP-MS is used without the need of any sample treatment, except, for example, ethanol removal under reduced pressure in the case of wine [106]. More details on sample preparation and analytical approaches can be found in a recent review on elemental speciation in beverages [47].

8.3.1.2 Applications Arsenic speciation in drinking and mineral water is important for the assessment of human exposure and potential health risks. Generally, inorganic As largely predominates in water, and thus the focus is on the assessment of the concentrations and the relative amounts of As (III) and As (V), arsenite having a higher chronic toxicity. Ion-exchange chromatography (IEC), reversed-phase ion-pairing chromatography (RPIPC), and CE have been coupled to ICP-MS for As speciation in water (see Table 8.3). To enhance sensitivity when CE has been employed as the separation technique, the use of HG has been reported, also for the simultaneous speciation of As and Se [107]. The simultaneous separation of inorganic Se species has also been achieved by RPIPC [108]. Another species of interest in drinking water analysis is Cr (VI), considerably more toxic than Cr (III) and classified as carcinogenic to humans. Chromium speciation in water has been carried out by HPLC-ICP-MS, using both IEC and RPIPC (see Table 8.3).

Speciation of halogens is important for water analysis. The first to determine inorganic halogen species in drinking water and Miso soup were Salov et al. [109], who quantified six anions (I^-, IO_3^-, Br^-, BrO_3^-, Cl^-, and ClO_3^-) by HPLC-ICP-MS. In recent years, interest in BrO_3^- and IO_3^- determination in drinking water has grown, as they are formed by oxidation of Br^- and I^- during the ozonation treatment for water disinfection. Ozonation has emerged as one of the most promising alternatives to chlorination, challenged for the possible production of carcinogenic disinfection by-products (e.g. trihalomethanes). However, BrO_3^- formation represents a potential health risk because this species is classified as a probable human carcinogen by IARC. High-sensitivity BrO_3^- determination by IEC and ICP-MS has been demonstrated by various authors (see Table 8.3), also with the use of UN and ID (Fig. 8.3). Heumann et al. [110] speciated I by various separation approaches, using both quadrupole and SF intruments for detection, and reported the presence of organo-I species in addition to I^- and IO_3^- in mineral water. Subsequently, Dudoit and Pergantis [111] developed an IEC method for the simultaneous determination of As, Br, and I inorganic species in drinking water.

TABLE 8.3. Literature on Elemental Speciation in Water and Beverages Using ICP-MS as the Detection Technique

Matrix	Elements	Technique[a]	Type of Study[b]	References
Water				
Mineral water	As	RP IP-Q	A	[381]
Mineral water	As	IE-Q	A	[382]
Mineral water	As	CE-Q	A	[383]
Water	As	IE-Q	A	[384]
Water	As	CE-HG-Q	A	[385]
Water	As	IE-Q	A	[386]
Water	As	IE-Q	A	[387]
Water	As	IE-Q	A	[388]
Water	As	IE-Q	A, M	[389]
Water	As	RP IP-Q	M	[390]
Water	As, Ge, Sb	HG-Q	A	[391]
Water	As, Br, I	IE-Q	A	[111]
Water	As, Se	CE-HG-Q	A	[107]
Water	As, Se	RP IP-Q	A	[108]
Water	Br	IE-Q (UN)	A	[392]
Water	Br	IE-Q	A	[393]
Water	Br	IE-Q	A	[394]
Water	Br	IE-Q (ID)	A	[395]
Water	Br	FI-IE-Q	A	[396]
Water	Cr	IE-Q	A	[397]
Water	Cr	IE-Q	A	[398]
Water	Cr	IE-Q	A	[399]
Water	Cr	RP IP-DRC-Q	A	[400]
Water	Cr, V	Q (FI-ASV)	A	[401]
Water	Hg	RP IP-CV-Q	A	[402]
Water	I	IE-Q, RP IP-Q, SF	A	[110]

Water	P	GC-CC-Q	A	[113]
Water RMs	As	FI-HG-Q	A	[403]
Water and miso soup	Br, Cl, I	SE-Q	A	[109]
Milk and infant formula				
Human milk	Ba, Bi, Cd, Cu, Li, Mg, Mn, Mo, Pb, Sr, Tl, Zn	SE-Q	O	[404]
Human milk	Ba, Ca, Cd, Co, Cs, Li, Mo, Ni, Pb, Sr, Tl, V	SE-Q	A	[405]
Human milk	I	CE-Q	A	[117]
Human milk	Se	CE-Q	A	[115]
Human milk	Se	CE-Q	A	[116]
Human milk and infant formula	Br, Cd, Co, Cu, Fe, I, Mn, Mo, P, Pb, Rb, S, Se, Zn	SE-Q	O	[114]
Milk	I	IE-Q	A, M	[406]
Milk, human milk, and infant formula	Br, Ca, Cu, Fe, I, Mg, Mn, Sr, Zn,	SE-Q	A, O	[407]
Milk, human milk, and infant formula	I	SE-Q	A, O	[408]
Other beverages				
Tea infusions	Al	SE-Q	A	[118]
Tea infusions	Al, Ba, Ca, Cu, Fe, Mg, Mn, Ni, Pb, Rb, Sr, Zn	IE-, SE-Q	A, P	[61]
Tea infusions	Cd, Ce, Co, Cs, Dy, Er, Eu, Gd, Ho, La, Lu, Mo, Nd, Ni, Pb, Pr, Sm, Sn, Tb, Th, Tl, Tm, U, W, Y, Yb, Zr	SE-Q	A	[409]

(*Continued*)

TABLE 8.3. *Continued*

Matrix	Elements	Technique[a]	Type of Study[b]	References
Tea infusions	Cd, Cu, Hg, Mn, Mo, Ni, Pb, Sr, Zn	RP-Q	A, P	[410]
Tea infusions	Ce, Dy, Er, Eu, Gd, Ho, La, Lu, Nd, Pr, Sc, Sm, Tb, Tm, Y, Yb	SE-Q	A, O	[411]
Wine	Al, Ba, B, Bi, Ca, Ce, Co, Cs, Cu, Ge, La, Mg, Mn, Nb, Nd, Pb, Pr, Rb, Sn, Sr, Th, U, V, Y, W, Zn, Zr	UF-Q	A	[412]
Wine	As	RP IP-Q	A	[413]
Wine	Pb	SE-Q	A	[106]
Wine and club soda	As	IE-Q	A	[414]

[a]The following acronyms are used for hyphenated separation techniques: liquid chromatography, (i) IE = ion exchange chromatography, (ii) SE = size-exclusion chromatography, (iii) RP = reversed-phase chromatography; IP = ion pair; G = gas chromatography; CE = capillary electrophoresis. Other acronyms: CC = collision cell, CV = cold vapor, DRC = dynamic reaction cell, FI = flow injection, HG = hydride generation, ID = isotope dilution, Q = quadrupole analyzer, SF = sector field analyser, UF = ultrafiltration, UN = ultrasonic nebulization.

[b]Type of study: A = analytical study, M = monitoring study, P = study on the effect of food processing/preparation, O = other studies, including investigations on element uptake/distribution/biotransformation and biological activity.

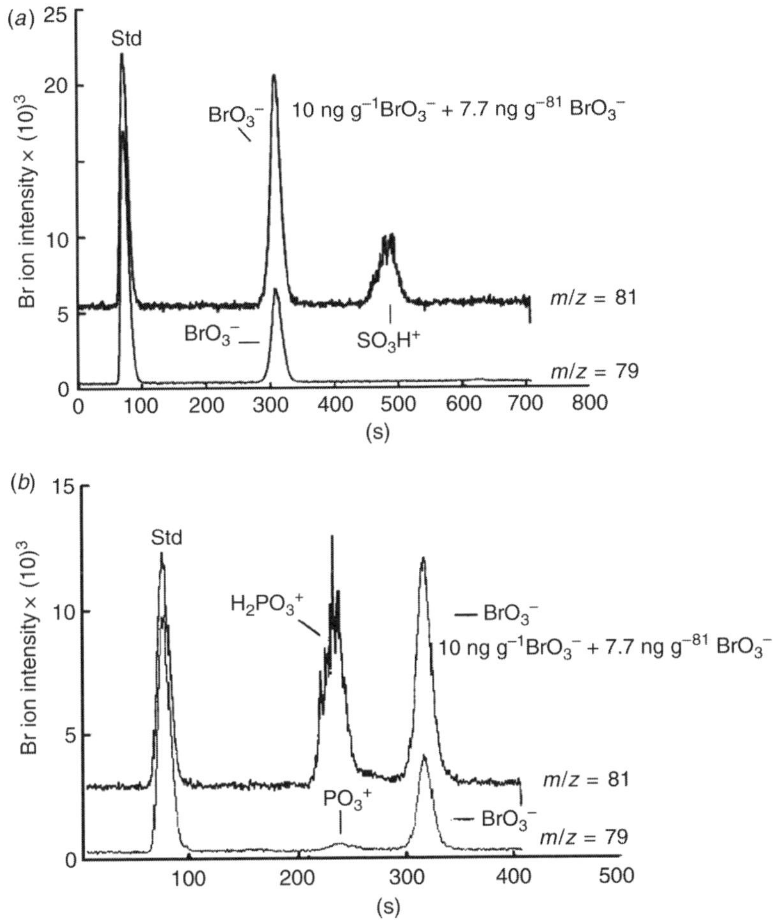

Figure 8.3. ID analysis of BrO_3^- in drinking water by ion chromatography with ICP-MS detection: study of polyatomic ions produced at mass 79 and 81 from a 300 μg g^{-1} sulfate matrix (*a*) and a 100 μg g^{-1} phosphate matrix (*b*). "Std" denotes the peak produced by a 100 μL post-column standard injection of BrO_3^-. Reprinted from [395] with permission Copyright 1999 American Chemical Society.

The combination of GC with low-pressure He ICP-MS, a system that allows fragment spectra to be taken, thus providing structural information about chromatographic eluents, has been reported for the separation of organobromine standards, prefiguring future applications in halogenated pesticides analysis [112]. Recently, Vonderheide et al. [113] coupled GC with a CC-ICP-MS for the determination of organophosphorus pesticides in tap water using He as collision gas.

In view of the role that milk plays in the nutrition and well-being of newborns, elemental speciation of both essential and potentially toxic elements in this matrix is of particular importance and studies have been conducted using SEC for the fractionation of element-binding compounds (see Table 8.3). Brätter and coworkers [114] compared breast milk and infant formula and showed that the protein-binding pattern of trace elements was significantly different. In the formula, the binding pattern depended on the main component (cow milk or soy), its processing (hydrolysis), and the chemical form (inorganic) of the added compounds. To compensate for poor absorption, some elements had been added at higher concentrations than those found in human milk. For example, Fe was up to 20-fold higher and presented a very different binding pattern from that of human milk.

Iodine and Se speciation in breast milk provides an example of the use of CE in hyphenated systems with ICP-MS detection. By employing CE, Michalke and colleagues determined selenoaminocids and identified two chemical forms of iodine, I^- and thyroxine, which were present in comparable amounts in milk [115–117]. Other authors used SEC and IEC for I speciation in various types of milk and infant formulae (see Table 8.3) and found I^- as the main species, with the exception of breast milk and formulae. The latter were found to contain less I than commercial and human milk, and mostly as an unidentified macromolecular compound.

Few applications concerning elemental speciation in beverages have been reported and all deal with the fractionation of metals in tea infusions or the determination of the specific form of As and Pb in wine (see Table 8.3). For instance, Owen et al. [118] applied SEC-ICP-MS to study Al fractions in tea infusions and simulated gastrointestinal digests. Tea (*Camellia sinensis*) belongs to the small group of plants that take up Al in very high quantities in their above-ground tissues and thus Al speciation and bioavailability in tea infusions is of particular interest. After simulated gastrointestinal digestion, it was found that the soluble labile fraction, that is, potentially available for absorption, corresponded to 15 percent of the Al from the tea infusion [118]. Szpunar et al. [106] used SEC-ICP-MS to investigate Pb speciation in wine and found that the metal was mostly bound to a pectic polysaccharide, namely 6-deoxy-L-mannogalacturonan (rhamnogalacturonan II), with a molecular weight of 12 kDa. This compound, which form dimers (dRG-II) cross-linked by a negatively charged borate ester moiety, is present in the primary plant cell wall and released into wine during the vinification process. The dRG-II borate can bind at least 10 times more Pb^{2+} than is typically present and wine could therefore have detoxifying properties due to the strong binding capacity of free dRG-II for inorganic Pb.

8.3.2 Food

8.3.2.1 General Aspects Sample preparation for speciation analyses of solid foods is not as straightforward as previously shown for water and beverages. Extraction, clean-up, and preconcentration can be required and the preservation of the identity and original distribution of the species in the analyzed matrix is no simple matter. In order to obtain acceptable recoveries, extractions are usually repeated several times (each time the supernatant solution is separated from the solid residue by centrifugation), and a number of treatments have been applied to enhance the extraction efficiency and/or reduce the time required. These include sonication, MW-assisted extraction, accelerated solvent extraction, and supercritical fluid extraction [119–122]. Enzymes are also used for leaching species from proteins and other macromolecules. Solid-phase microextraction (SPME) has recently attracted interest as a useful solvent-free extraction procedure [123]. SPME can collect volatile species (directly or following derivatization) from either the sample headspace or the liquid phase, and nonvolatile species from the liquid phase. Thus, it can be used in conjunction with different separation techniques, such as HPLC, GC and CE.

The widely different extraction efficiencies observed for different analytes in relation to the sample matrix and the solvent/extraction procedure utilized call for caution in the interpretation of the results of speciation studies. A compound may be found to largely predominate in an extract containing only a small fraction of the total element concentration and one should be aware that this compound could actually represent a minor species. Moreover, the concentration of the investigated element in the other extracts (if any) and in the residue should be determined to check the accuracy of results. Establishing a mass balance of the analytes in all fractions of the analytical procedure is recommended as it accounts for analyte extraction efficiency and helps evaluating the overall accuracy of speciation results (Fig. 8.4).

As it happens with liquid matrices, species identification is generally carried out by comparing retention times, but, owing to the greater complication of the sample matrix and the wider range of chemical species encountered, the risk for species misidentification is higher. Thus, multidimensional approaches (i.e., combinations of two or more separation methods) or identity confirmation by molecular mass spectrometry, for example, electrospray ionization (tandem) mass spectrometry ESI-(MS)-MS, are resorted to in many studies. The unavailability of the standards of many naturally occurring element species often makes mass spectrometry the only viable solution for species identification. Things are further complicated by the low detection power of molecular mass spectrometry for many practical applications as well as the paucity of Certified Reference Materials (CRMs) for trace element species currently on the market, which makes accuracy assessment of speciation analyses rather difficult.

8.3.2.2 Applications Arsenic speciation has been the subject of an intensive research activity and HPLC-ICP-MS has established as a robust analytical technique for this purpose (see Table 8.4). Interest in As speciation arises from the widely different toxicity of the various species as well as their uneven distribution in food commodities. In particular, seafood contains fairly high As concentrations due to the presence of organo-Ascompounds with a low or negligible toxicity, such as arsenobetaine (AB), the dominant form in fish and crustaceans, and As-containing ribosides. These latter, also known as arsenosugars, are the major As species in seaweeds and, together with AB, in bivalve molluscs.

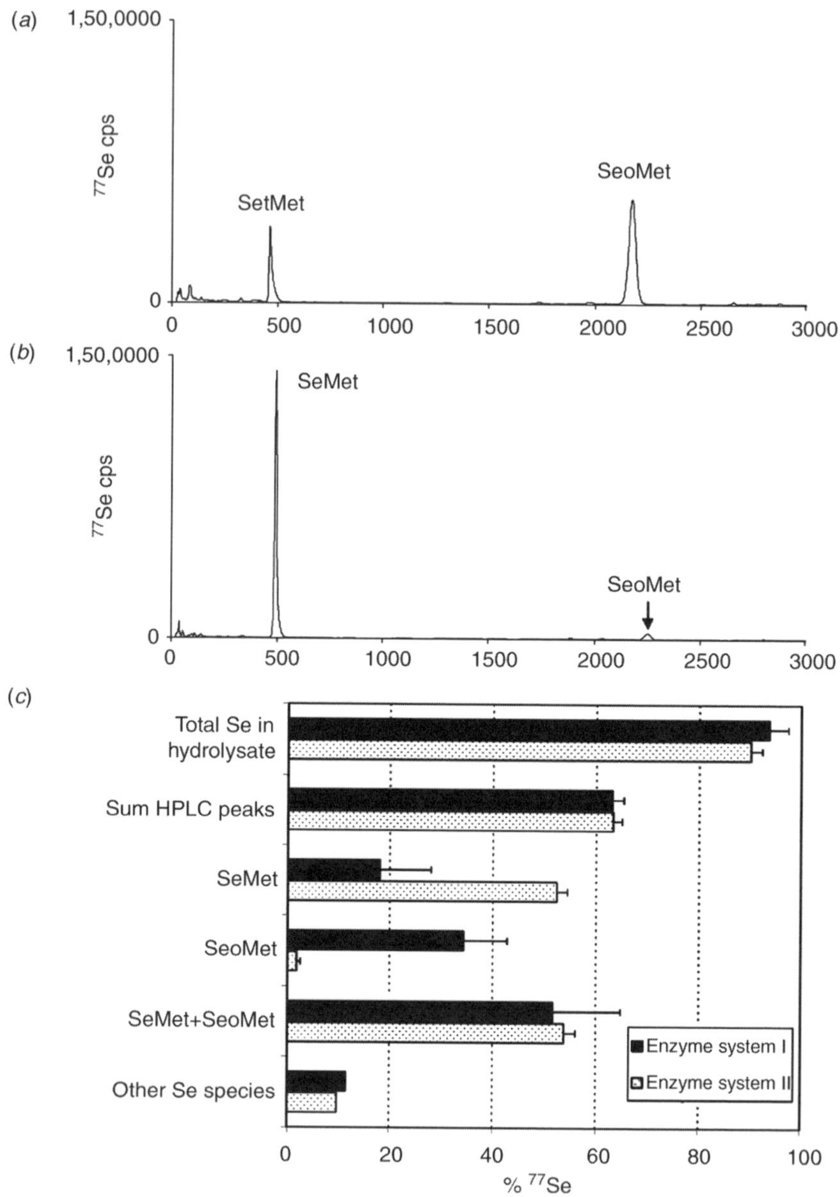

Figure 8.4. Cation exchange HPLC-ICP-MS analysis of ^{77}Se-enriched yeast using two enzymatic sample preparation strategies: (*a*) sequential use of β-glucosidase and a protease mixture; (*b*) protease XIV; and (*c*) mass balance of selenium and selenium species fractionated during the sample preparation and HPLC separation procedure. SeMet = selenomethionine; SeOMet = selenomethionine Se-oxide. Reproduced from [133] by permission of The Royal Society of Chemistry.

TABLE 8.4. Literature on Elemental Speciation in Food Using ICP-MS as the Detection Technique

Matrix	Elements	Technique[c]	Type of Study[b]	References
Cereals, pulses, and derived products				
Bakery products	Br	IE-Q	A	[152]
Bread	Br	RP-Q	A	[151]
Bread	Br	IE-Q	A	[415]
Rice	As	IE-Q	A	[416]
Rice	As	IE-Q	A	[417]
Rice	As	IE-Q	A	[418]
Soybean flour and white bean	Cd, Co, Cu, Fe, Mn, Mo, Ni, P, S, Se, Zn	SE-Q	O	[164]
Various cereals	Se	IE-Q	O	[419]
Fruit, vegetables, and other plant food				
Apple	As	IE-Q	A	[120]
Apple	As	IE-, RP IP-Q	A	[420]
Apple and carrot	B, Ba, Ce, Cu, Mg, Pb, Sr, Zn	SE-Q	A	[162]
Brazil nut	Se	RP IP-Q	A	[421]
Broccoli	Se	RP IP-Q (UN)	A	[422]
Broccoli, garlic and onion	Se	IE-Q	A	[423]
Carrot	As	IE-Q	O	[424]
Carrot	As	IE-Q	A	[425]
Cocoa	Cd, Pb	SE-Q	A	[426]

(*Continued*)

TABLE 8.4. *Continued*

Matrix	Elements	Technique[a]	Type of Study[b]	References
Coffee and coffee beans	Se	GC-CC-Q	M, P	[427]
Garlic	Se	IE-, RP-, RP IP-Q	A, O	[428]
Garlic and onion	Se	RP IP-Q	A	[429]
Garlic	Se	RP IP-Q	A	[430]
Garlic	Se	RP IP-Q	A	[431]
Garlic	Se	RP IP-Q	O	[432]
Garlic	Se	RP-Q	A, O	[433]
Garlic and onion	Se	RP IP-Q	A, M	[434]
Garlic and onion	Se	RP*chir*-Q	A	[138]
Garlic, onion, and others plants	Se	RP IP-Q	A	[435]
Radish	As	IE-Q	O	[436]
Radish	B	SE-Q	A	[163]
Various nuts	As	IE-, RP IP-Q	A, M	[437]
Various nuts	Se	SE-, RP IP-Q	A	[131]
Mushrooms				
Mushrooms	As	IE-Q	O	[138]
Mushrooms	As	IE-Q	O	[439]
Mushrooms	As	IE-, RP IP-Q	O	[440]
Mushrooms	As	IE-Q	A	[441]
Mushrooms	As	IE-Q	A	[442]
Mushrooms	Bi, Co, Cu, Fe, I, Mo, Ni, Se, Zn	SE-Q	A	[443]
Yeast and supplements				
Se supplement	Se	IE-Q	A	[444]

Se supplement	Se	RP IP-Q	A, M	[445]
Se supplement	Se	RP-, RP IP-CC-Q	A	[446]
Se supplement	Se	G-Q	A	[447]
Se supplement	Se	RP*chir*-, IE-Q	A	[139]
Se supplement	Se	RP IP-Q	A	[448]
Se supplement	Se	RP*chir*-Q	A	[140]
Se supplement	Se	G*chir*-Q	A	[142]
Se supplement	Se	IE-DRC-Q	O	[134]
Yeast	Se	RP IP-Q	A	[449]
Yeast	Se	IE-, RP-, RP IP-Q	A, O	[428]
Yeast	Se	RP*chir*-HG-Q	A	[136]
Yeast	Se	RP IP-Q	A	[429]
Yeast	Se	RP-Q	A	[450]
Yeast	Se	SE-, IE-, RP IP-Q	A	[122]
Yeast	Se	RP IP-Q	A	[430]
Yeast	Se	RP IP-Q	A	[431]
Yeast	Se	G*chir*-Q	A	[141]
Yeast	Se	RP IP-Q	O	[432]
Yeast	Se	RP IP-Q	A, M	[434]
Yeast	Se	RP*chir*-Q	A	[138]
Yeast	Se	RP*chir*-Q	A	[135]
Yeast	Se	SE-, RP IP-Q	A	[451]
Yeast	Se	RP*chir*-, M*chir*-, G*chir*-Q	A	[143]
Yeast	Se	RP IP-Q	A	[435]
Yeast	Se	IE-Q, GE-ETV-Q	A	[4]
Yeast	Se	GE-ETV-Q	A	[452]
Yeast	Se	SE-RP-CC-Q	A	[453]
Yeast	Se	SE-IE-Q	A	[454]

(Continued)

TABLE 8.4. *Continued*

Matrix	Elements	Technique[a]	Type of Study[b]	References
Yeast	Se	SE-CE-Q	A	[455]
Yeast	Se	RP*chir*-Q	O	[137]
Yeast	Se	RP IP-Q	A, O	[456]
Yeast	Se	RP IP-Q	O	[457]
Yeast	Se	GE-LA-DRC	A	[5]
Yeast	Se	SE-Q	A	[458]
Yeast	Se	IE-DRC-Q	A	[133]
Yeast and algae	Se	IE-DRC-Q	A	[459]
Yeast and Se supplement	Se	RP IP-Q	A	[460]
Vitamin B_{12} supplement	Co	CE-Q	A	[156]
Meat and offal				
Beef	Fe	SE-SF	A	[154]
Chicken	As	RP-Q	A	[153]
Chicken	As	IE-Q	A	[461]
Chicken	Zn	RP-Q	A	[158]
Pig kidney	Cd	SE-Q	P	[157]
Porcine liver	Cu, Fe, Mn, Zn	SE-, RP-Q	A	[161]
Seafood				
	As	RP IP-Q	O	[462]
Algae and oyster RM				
Bivalves	As	SE-, RP IP-Q	A	[125]
Fish	As	IE-Q	A	[463]
Fish	As	IE-, RP IP-Q	A	[464]
Fish	As	IE-Q	O	[465]

Fish	As	IE-Q	A	[466]
Fish	As	IE-Q	O	[467]
Fish	As	IE-Q	A	[468]
Fish	As	RP IP-Q	A	[469]
Fish	Hg	G-Q	A	[470]
Fish	Hg	G-Q	A	[471]
Fish	Hg	RP IP-CV-Q	A	[472]
Fish	Hg	RP-Q	A	[473]
Fish	Se	IE-Q	A	[474]
Fish	Se	SE-Q	A, O	[132]
Fish	Sn	RP IP-Q	A	[475]
Fish	Sn	RP IP-Q	A	[476]
Fish	Sn	RP IP-Q	A	[477]
Fish RM	As	RP IP-Q	A	[478]
Fish RM	As	SE-, RP IP-Q	A	[124]
Fish RM	As	RP IP-Q	A	[479]
Fish RM	As	M-Q	A	[480]
Fish RM	As	IE-Q	A	[481]
Fish RM	As	n.a.	C	[482]
Fish RM	As	IE-Q	A	[483]
Fish RMs	As	IE-Q	A	[484]
Fish RM	As	IExcl-HG-Q	A	[485]
Fish RM	Hg	RP-CV-Q	A	[486]
Fish RM	Hg	RP-Q	A	[487]
Fish RM	Hg	CE-CV-Q, -SF	A	[488]
Fish RM	Sn	G-LP-Q	A	[112]
Fish RM	Sn	G-Q	A	[147]
Fish RMs	As	IE-Q	A	[489]

(*Continued*)

TABLE 8.4. *Continued*

Matrix	Elements	Technique[a]	Type of Study[b]	References
Fish RMs	Hg	G-Q (ID)	A	[490]
Fish RMs	Hg	RP-Q	A	[491]
Fish RMs	Hg	SPME-TD-Q	A	[144]
Fish RMs	Hg	G-Q (ID)	A	[492]
Fish RMs	Hg	G-Q (ID)	A	[145]
Lobster RM	Hg	G-LP-Q	A	[493]
Mussel	As	RP IP-Q	O	[494]
Mussel	As	IE-Q (HG)	A	[495]
Mussel	As	IE-Q	O	[496]
Mussel	Sn	G-Q	A	[150]
Mussel	Ag, Al, Ca, Cd, Co, Cr, Cu, Fe, Hg, Mn, Mo, Ni, Pb, Sb, Sn, U, V, Zn	SE-SF	O	[160]
Mussel	Cd, Cu, Zn	SE-, FPLC-Q, -SF	A	[159]
Mussel RM	Sn	G-Q	A	[146]
Oyster	Se	IE-Q	A	[497]
Oyster	Se	IE-Q	A	[498]
Oyster RM	As	SE-, RP-, IE-Q	A	[130]
Scallop	As	RP IP, IE-Q	O	[499]
Seafood	As	IE-Q	A, O	[500]
Seafood	As	RP IP-Q	A,O	[501]
Seafood	As	RP IP-Q	A	[502]
Seafood	As	SE-, RP IP-Q	O	[129]
Seafood	As	IE-Q	A, R	[295]
Seafood	As	n.a.	R	[503]
Seafood	As	IE-Q	A	[121]

Seafood	As	IE IP-Q	A	[504]
Seafood	As	IE-Q	M	[505]
Seafood	As	IE-Q	A	[506]
Seafood	As, Se	RP IP-Q	A	[507]
Seafood RMs	As	IE-Q	A, O	[508]
Seafood RMs	As	IE-Q	A	[509]
Seafood RMs	Hg	FI-Q (ID)	A	[510]
Seafood RMs	Hg	G-Q, RP-Q (ID)	A	[511]
Seafood RMs	Hg	ETV-Q	A	[512]
Seafood RMs	Se	RP-Q	A	[513]
Seaweed	As	IE-Q (HG)	A	[514]
Seaweed	As	IE-Q	O	[515]
Seaweed	As	SE-IE-Q	A	[127]
Seaweed	As	SE-IE-Q	A	[128]
Seaweed	As	RP IP-Q	A	[413]
Seaweed	As	SE-IE-RP-Q	A	[129]
Seaweed	As	IE-Q	O	[516]
Seaweed	As	IE-Q	A	[517]
Seaweed	As	IE-Q	A	[518]
Seaweed	As	FI-HG-Q	A	[403]
Seaweed	As	IE-Q	M, P, O	[519]
Shrimp	As	IE-Q	A	[520]
Shrimp	As	IE-Q	O	[521]
Shrimp	As	IE-Q	A	[414]
Shrimp	As	IE-Q	O	[522]
Miscellaneous				
Food composites	As	IE-Q	A	[523]
Infant foods	As	IE-Q	A	[524]

(Continued)

TABLE 8.4. *Continued*

Matrix	Elements	Technique[a]	Type of Study[b]	References
Infant foods	As	IE	A, M	[525]
Meat and seafood	Fe	SE-SF	A	[155]
Mussel and potato	Sn	G-Q	A	[149]
Various foods	As	IE IP-Q	A	[526]
Various foods	As	IE-Q	A	[527]
Various foods	As	IE IP-Q	A	[528]

[a]The following acronyms are used for hyphenated separation techniques: liquid chromatography, (i) IE = ion exchange chromatography, (ii) SE = size-exclusion chromatography, (iii) RP = reversed-phase chromatography, (iv) M = micellar chromatography; IP = ion pair; G = gas chromatography; CE = capillary electrophoresis; GE = gel electrophoresis. *Chir* indicates chiral separation. Other acronyms: CC = collision cell, CV = cold vapor, DRC = dynamic reaction cell, ETV = electro-thermal vaporization, FI = flow injection, HG = hydride generation, ID = isotope dilution; LP = low pressure, Q = quadrupole analyzer, SF = sector field analyzer, SPME = solid-phase microextraction, TD = thermal desorption, UN = ultrasonic nebulization.

[b]Type of study: A = analytical study, C = collaborative trial, M = monitoring study, P = study on the effect of food processing/preparation, R = Reference material certification, O = other studies, including investigations on element uptake/distribution/biotransformation and biological activity.

A range of minor compounds are also found in marine organisms, which has stimulated research work to understand their biogenesis and their position in the metabolic pathways of As in the sea.

Due to the fact that some of the As compounds are permanently cationic (quaternary arsonium compounds such as AB, arsenocholine, and tetramethylarsonium ion), whilst the others are readily and differently ionized depending on the pH, either anion-/cation-exchange chromatography or RPIPC have been used for their separation (see Table 8.4). Another separation mechanism applied was SEC, which in combination with RPIPC enabled the analysis of the full range of As compounds in seafood [124–126]. Even though SEC is typically employed for the separation of large proteins and polysaccharides, secondary adsorption and ion exchange effects can be advantageously used to resolve ions with a high charge-to-mass ratio, such as organo-As compounds. SEC has also been applied in multidimensional approaches for As speciation in seaweeds and an oyster reference material (RM) (Fig. 8.5), which employed IEC and reversed-phase chromatography (RPC) for species identification and identity confirmation/recognition by ES-MS-MS [127–129].

Even though so far most of the speciation work on As has been carried out on seafood, recently a range of food items containing definitely lower levels of As, but at least partly in inorganic form (e.g., vegetables grown in As-rich soils, rice) has been investigated (see Table 8.4). The inorganic species are highly toxic, As (III) being more toxic than As (V), and in foods of terrestrial origin these are found together with dimethylarsinic acid (DMA) and monomethylarsonic acid (MMA), compounds with a lower toxicity, but suspected to be cancer promoters. Mushrooms represent an exception on account of both the high As levels often detected and the presence, in some species, of unexpected compounds such as AB, which is typical of marine animals.

Contrary to As, Se is essential to man and – as shown in Table 8.4 – has attracted the greatest interest as regards speciation. Selenium exists in biological systems in the form of both inorganic species, namely Se (IV) and Se (VI), and, mainly, organo-Se compounds, having a range of molecular masses and charges, from simple methylated species to selenoaminoacids (SeMet in microorganisms and plants, selenocysteine in animals) and selenoproteins containing selenocysteinyl residues. The extraordinary variety of naturally occurring Se species and the recent evidence of the possible role of this element in the prevention of certain forms of cancer and other diseases made Se speciation a most dynamic area of investigation. Applications include the investigation of Se species in Se-enriched yeast and crops (e.g., garlic, onion, broccoli, and cereals) and in naturally Se-rich foods (e.g., some nuts), quality and safety control of Se supplements, identification of the biochemical mechanisms and chemical forms of Se with a disease-preventive effect. Low-molecular-weight compounds have been determined by coupling ICP-MS with IEC, RPC, and RPIPC (see Table 8.4), while SEC has been used for the separation of Se-containing proteins into fractions or the screening of Se compounds on a wide range of molecular masses [122, 131, 132].

Recent trends in Se speciation include the use of multidimensional approaches based on combinations of the above-mentioned techniques, the fractionation of

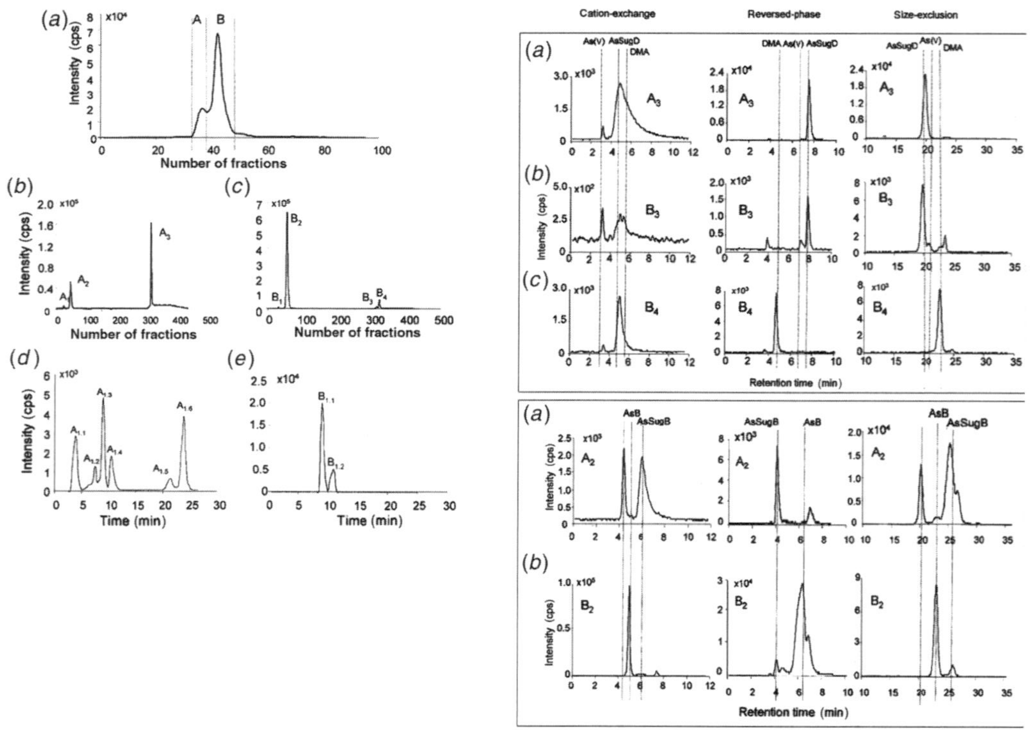

Figure 8.5. On the left: fractionation of As species in a oyster extract by three-dimensional liquid chromatography with ICP-MS detection. (*a*) Size-exclusion LC: A, arseno sugar fraction; B, low molecular mass fraction. (*b*) Anion-exchange LC of the arseno sugar fraction (A). (*c*) Anion-exchange LC of the low molecular mass fraction (B). (*d*) Cation-exchange LC of the void (A1) of the anion-exchange chromatogram shown in (*b*). (*e*) Cation-exchange LC of the void (B1) of the anion-exchange chromatogram shown in (*c*). On the right: verification of the chromatographic purity of fractions in anion-exchange chromatography by three different chromatographic systems (late eluting fractions A3, B3, and B4 above, A2 and B2 fractions below). Reproduced from [130] by permission of the Royal Society of Chemistry.

Se-containing proteins by means of sodium dodecylsulfate polyacrylamide gel electrophoresis (SDS-PAGE) with subsequent analysis of the gel bands containing the proteins by ETV- or LA-ICP-MS, and the adoption of CC and DRC to improve ICP-MS detection of Se (see Table 8.4). Recently, Larsen and coworkers reported the use of HPLC-DRC-ICP-MS for the characterization of a ^{77}Se-labeled and enriched yeast with regard to isotopic composition and content of Se species for later use in human absorption studies [133] and the speciation and bioavailability assessment of Se in yeast-based intervention agents used in cancer chemo-prevention studies [134]. Another topic that has attracted interest is Se chiral speciation. SeMet, other α-selenoaminoacids, and related compounds are chiral, and different enantiomers have different physiological activities. Dietary supplements containing SeMet as the Se source can provide different levels of bioavailable Se depending on the presence of the natural form (L-SeMet) or the synthetic racemate, D-SeMet being poorly metabolized by the human organism. Thus, suitable speciation tools for the enantiomeric analysis of selenoaminoacids has been set up, mainly by coupling HPLC, with a range of chiral stationary phases, and ICP-MS [38, 135–140]. The use of GC-ICP-MS with precolumn derivatization has also been reported [141, 142]. Different alternatives for chiral speciation of SeMet in selenized yeast with ICP-MS Se detection has been compared by Perez Mendez et al. [143] with respect to enantioselectivity, sensitivity, and analysis time, and an HPLC method employing a Chirobiotic T column has been found to be the best choice.

Hyphenated techniques with ICP-MS detection have been used for the speciation of well-known food contaminants, such as Hg and organo-Sn compounds. Mercury speciation primarily focuses on the distinction between the inorganic forms and the main organic compound found in fish, methyl-Hg, the toxicity of which is distinctly higher. HPLC-ICP-MS has been resorted to in some of these studies, but GC-ICP-MS following ethylation with sodium tetraethylborate ($NaBEt_4$) remains the most widely used technique (see Table 8.4), even though SPME for sample introduction has been reported [144] and SPME-GC-ICP-MS has been recently used with ID for RM certification [145]. As regards organo-Sn compounds, RPIPC has been applied by Caruso and coworkers to identify tributyl-Sn and triphenyl-Sn in fish following different extraction procedures (see Table 8.4). However, the technique of choice for organo-Sn speciation is GC-ICP-MS, which has been applied to seafood analysis either by using multicapillary microcolumns [146], or following SPME [147], or even in conjunction with species specific ID [148]. Vercauteren et al. [149] used SPME with GC-ICP-MS for the determination of the organo-Sn pesticide fentin (triphenyl-Sn) in mussels and potatoes and subsequently applied a new extraction technique, stir bar sorptive extraction, to organo-Sn speciation in mussels [150]. The combination of GC with a low-pressure He ICP mass spectrometer has also been reported for organo-Sn speciation in a fish RM [112]. Besides organo-Sn compounds, other nonnaturally occurring substances have been determined in food by ICP-MS hyphenation with suitable separation techniques. Examples are the HPLC-ICP-MS detection of BrO_3^- in bread and bakery products following use of the additive $KBrO_3$ as flour improver [151, 152], or residues of the growth promoter 4-hydroxy-3-nitrophenyl-arsonic acid (roxarsone) in chicken tissues [153].

Finally, other classes of compounds investigated in speciation studies are the complexes of some essential elements with biosynthesized macrocyclic chelating agents (e.g., porphyrins, cobalamins) and the complexes of various elements with major organic components of foods (e.g., peptides, proteins, polysaccharides). Harrington et al. [154, 155] investigated Fe speciation in meat and seafood by SEC with SF-ICP-MS detection. In a study on vitamin B_{12} supplements, Baker and Miller-Ihli [156] performed Co speciation by CE and micellar electrokinetic chromatography and successfully distinguished cyanocobalamin (the form most commonly used in vitamin preparations) from free Co, other cobalamins, and a potentially harmful corrinoid analog (cobinamide dicyanide).

As regards the complexes of elements with other food macromolecules, the investigation of the binding patterns of Cd, Cu, Zn, and other metals to proteins, including metallothioneins (MTs) and MT-like proteins, is a well-established research field, and studies on food matrices (meat, seafood) using SEC-ICP-MS or other ICP-MS coupled techniques have been reported [157–161]. In order to shed further light on the potential bioavailability of protein-bound elements, these studies need to address changes in element solubility following heat-induced protein denaturation during cooking [157]. As for other metallobiomolecules, SEC-ICP-MS has been used to investigate metal–carbohydrate complexes in fruit and vegetable samples [162], the binding of B in radish [163], and of various elements in soybean flour and white bean [164].

8.4 CONCLUSIONS AND FUTURE DIRECTIONS

An examination of the publication trends concerning the use of ICP-MS in food-related areas (Fig. 8.6) shows that after less than one decade from its first application in this field, the number of published works increased by a factor of 10. In recent years, about 50–70 papers on food-related topics have appeared annually and nearly half of them concern speciation topics. ICP-MS represents today an essential tool for inorganic and bioinorganic research on foods and its versatility, that is, the ease of coupling with different sample introduction systems, as well as the hyphenation with devices for species separation, is undoubtedly a key point of its success. Moreover, the progress in instrumentation has overcome most of the past limitations of the technique and with the availability of MC-, CC-, and DRC-ICP-MS instruments a growing number of applications in any area of food and nutrition research are expected.

Food authentication, application of stable isotope tracers, characterization of complex metallobiomolecules, and investigation of their role in biological processes are all areas in which ICP-MS can play a more and more critical role. Speciation will continue to be a very dynamic field of investigation and the integration of speciation and metabolism studies with the use of stable isotopes as species-specific tracers is likely to become a research area of vital importance. On

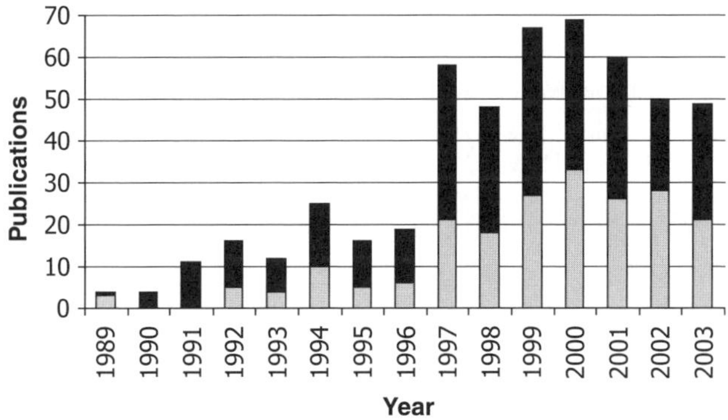

Figure 8.6. Applications of ICP-MS to food-related areas: publication history (1989–2003). Publications concerning speciation are marked in white.

the whole, these technical developments and research trends can greatly contribute to a deeper knowledge of element bioavailability and to understanding how it is influenced by the element chemical forms.

ICP-MS-coupled techniques are expected to increase in popularity, possibly leading to the commercial availability of ready-to-use speciation systems. Among the several approaches so far developed for species identification and quantification with ICP-MS detection in foods, some led to the establishment of sufficiently practicable and reliable analytical protocols for selected analytes/matrices. In a few cases, therefore, there is already the potential for validating routine control methods and this should facilitate the laying-down of species-specific regulations in the international legislation on trace elements in food.

REFERENCES

1. D. W. Koppenaal, G. C. Eiden, C. J. Barinaga, *J. Anal. Atom. Spectrom.*, **19** (2004), 561–570.
2. EC, Commission Directive 2001/22/EC of 8 March 2001 laying down the sampling methods and the methods of analysis for the official control of the levels of lead, cadmium, mercury and 3-MCPD in foodstuffs, *Off. J. Eur. Commun.*, **L77** (16 March 2001), 14–21.
3. EC, Commission Decision 2002/657/EC of 14 August 2002 implementing Council Directive 96/23/EC concerning the performance of analytical methods and the interpretation of results. *Off. J. Eur. Commun.*, **L221** (17 August 2002), 8–36.
4. H. Chassaigne, C. C. Chery, G. Bordin, A. R Rodriguez, *J. Chromatogr. A*, **976** (2002), 409–422.
5. C. C. Chéry, D. Gunther, R. Cornelis, F. Vanhaecke, L. Moens, *Electrophoresis*, **24** (2003), 3305–3313.
6. F. Cubadda, *J. AOAC Int.*, **87** (2004), 173–204.
7. H. M. Crews, D. J. Halls, A. Taylor, *J. Anal. Atom. Spectrom.*, **5** (1990) 75R–149R.

8. S. Branch, H. M. Crews, D. J. Halls, A. Taylor, *J. Anal. Atom. Spectrom.* **6** (1991) 69R–107R.
9. S. Branch, L. Ebdon, M. Ford, M. Foulkes, P. O'Neill, *J. Anal. Atom. Spectrom.*, **6** (1991), 151–154.
10. A. Taylor, S. Branch, H. M. Crews, D. J. Halls, *J. Anal. Atom. Spectrom.*, **7** (1992), 67R–154R.
11. A. Taylor, S. Branch, H. M. Crews, D. J. Halls, *J. Anal. Atom. Spectrom.*, **8** (1993), 79R–149R.
12. A. Taylor, S. Branch, H. M. Crews, D. J. Halls, M. White, *J. Anal. Atom. Spectrom.*, **9** (1994), 87R–146R.
13. A. Taylor, S. Branch, H. M. Crews, D. J. Halls, M. White, *J. Anal. Atom. Spectrom.*, **10** (1995), 61R–125R.
14. A. Taylor, S. Branch, H. M. Crews, D. J. Halls, M. White, *J. Anal. Atom. Spectrom.*, **11** (1996), 103R–186R.
15. A. Taylor, S. Branch, H. M. Crews, D. J. Halls, L. M. W. Owen, M. White, *J. Anal. Atom. Spectrom.*, **12** (1997), 119R–221R.
16. A. Taylor, S. Branch, A. Fisher, D. Halls, M. White, *J. Anal. Atom. Spectrom.*, **16** (2001), 421–446.
17. A. Taylor, S. Branch, D. Halls, L. M. W. Owen, M. White, *J. Anal. Atom. Spectrom.*, **15** (2000), 451–487.
18. A. Taylor, S. Branch, D. Halls, M. Patriarca, M. White, *J. Anal. Atom. Spectrom.*, **18** (2003), 385–427.
19. A. Taylor, S. Branch, D. Halls, M. Patriarca, M. White, *J. Anal. Atom. Spectrom.*, **18** (2004), 505–556.
20. A. Taylor, S. Branch, D. Halls, M. Patriarca, M. White, *J. Anal. Atom. Spectrom.*, **17** (2002), 414–455.
21. A. Taylor, S. Branch, D. Halls, L. M. W. Owen, M. White, *J. Anal. Atom. Spectrom.*, **13** (1998), 57R–106R.
22. A. Taylor, S. Branch, D. Halls, L. M. W. Owen, M. White, M. *J. Anal. Atom. Spectrom.*, **14** (1999), 717–781.
23. W. Zhu, E. W. B. de Leer, M. Kennedy, G. J. F. R. Alaerts, *Fresenius J. Anal. Chem.*, **360** (1998), 74–80.
24. J. Riondato, F. Vanhaecke, L. Moens, R. Dams, *J. Anal. Atom. Spectrom.*, **15** (2000), 341–345.
25. B. Klaue, D. B. Blum, *Anal. Chem.*, **71** (1999), 1408–1414.
26. H. M. Al-Swaidan, *Anal. Lett.*, **21** (1988), 1469–1475.
27. L. B. Rasmussen, E. H. Larsen, L. Ovesen, *Eur. J. Clin. Nutr.*, **54** (2000), 57–60.
28. G. J. Martin, J. B. Fournier, P. Allain, Y. Mauras, *Analusis*, **25** (1997), 7–13.
29. J. Goossens, T. De Smaele, L. Moens, R. Dams, *Fresenius J. Anal. Chem.*, **347** (1993), 119–125.
30. M. J. Baxter, H. M. Crews, M. J. Dennis, I. Goodall, D. Anderson, *Food Chem.*, **60** (1997), 443–450.
31. J. R. Dean, L. Ebdon, R. C. Massey, *Food Addit. Contam.*, **7** (1990), 109–116.
32. A. Stroh, P. Brückner, U. Völlkopf, *Atom. Spectrosc.*, **15** (1994), 100–106.

33. J. D. Greenough, H. P Longerich, S. E. Jackson, *Can. J. Appl. Spectrosc.*, **41** (1996), 76–80.

34. S. Augagneur, B. Medina, J. Szpunar, R. Lobinski, *J. Anal. Atom. Spectrom.*, **11** (1996), 713–721.

35. A. P. Packer, M. F. Giné, *Spectrochim. Acta*, **56B** (2001), 69–75.

36. J. R. Castillo, M. S. Jimenez, L. Ebdon, *J. Anal. Atom. Spectrom.*, **14** (1999), 1515–1518.

37. M. S. Jimenez, R. Velarte, J. R. Castello, *J. Anal. Atom. Spectrom.*, **18** (2003), 1154–1162.

38. C. A. Ponce de Leon, M. Montes Bayon, J. A. Caruso, *Anal. Bioanal. Chem.*, **374** (2002), 230–234.

39. J. R. Dean, L. Ebdon, R. Massey, *J. Anal. Atom. Spectrom.*, **2** (1987), 369–374.

40. J. A. Nobrega, Y. Gelinas, A. Krushevska, R. Barnes, *J. Anal. Atom. Spectrom.*, **12** (1997), 1243–1246.

41. H. Baumann, *Fresenius J. Anal. Chem.*, **338** (1990), 809–812.

42. S. Stürup, A. Buchert, *Fresenius J. Anal. Chem.*, **354** (1996), 323–326.

43. E. H. Larsen, P. Knuthsen, M. Hansen, *J. Anal. Atom. Spectrom.*, **14** (1999), 41–44.

44. F. Cubadda, A. Raggi, F. Zanasi, M. Carcea, *Food Addit. Contam.*, **20** (2003), 353–360.

45. F. Cubadda, A. Raggi, E. Marconi, *Microchem. J.*, **79** (2005), 97–102.

46. A. Misund, B. Frengstad, U. Siewers, C. Reimann, *Sci. Total Environ.*, **244** (1999), 21–41.

47. R. W. Dabeka, H. B. S. Conacher, J. F. Lawrence, W. H. Newsome, A. McKenzie, H. P. Wagner, R. K. H. Chadha, K. Pepper, *Food Addit. Contam.*, **19** (2002), 721–732.

48. J. Meija, S. Mounicou, J. A. Caruso, *J. AOAC Int.*, **87** (2004), 205–224.

49. H. M. Crews, in P. R. Ashurst, M. J. Dennis (Eds), *Analytical Methods of Food Authentication*, Blackie, London, 1998, 270–291.

50. V. F. Taylor, H. P. Longerich, J. D. Greenough, *J. Agric. Food Chem.*, **51** (2003), 856–860.

51. J. D. Greenough, H. P. Longerich, S. E. Jackson, *Aust. J. Grape Wine Res.*, **3** (1997), 75–83.

52. N. Jakubowski, R. Brandt, D. Stuewer, H. R. Eschnauer, S. Görtges, *Fresenius J. Anal. Chem.*, **364** (1999), 424–428.

53. C. M. Almeida, M. T. S. D. Vasconcelos, *J. Anal. Atom. Spectrom.*, **16** (2001), 607–611.

54. M. Barbaste, K. Robinson, S. Guilfoyle, B. Medina, R. Lobinski, *J. Anal. Atom. Spectrom.*, **17** (2002), 135–137.

55. B. Medina, S. Augagneur, M. Barbaste, F. E. Grouset, P. Buat-Meard, *Food Addit. Contam.*, **17** (2000), 435–445.

56. R. Larcher, G. Nicolini, P. Pangrazzi, *J. Agric. Food Chem.*, **51**, (2003), 5956–5961.

57. C. M. R. Almeida, M. T. S. D. Vasconcelos, *Anal. Chim. Acta*, **396** (1999), 45–53.

58. X. D. Tian, H. Emteborg, M. Barbaste, F. C. Adams, *J. Anal. Atom. Spectrom.*, **15** (2000), 829–835.

59. M. Barbaste, L. Halicz, A. Galy, B. Medina, H. Emteborg, F. C. Adams, R. Lobinski, *Talanta*, **54** (2001), 307–317.

60. W. A. Simpkins, H. Louie, M. Wu, M. Harrison, D. Goldberg, *Food Chem.*, **71** (2000), 423–433.

61. K. E. Ødegård, W. Lund, *J. Anal. Atom. Spectrom.*, **12** (1997), 403–408.

62. B. Wyrzykowska, K. Szymczyk, H. Ichichashi, J. Falandysz, B. Skwarzec, S. Yamasaki, *J. Agric. Food Chem.*, **49** (2001), 3425–3431.

63. M. Krachler, F. S. Li, E. Rossipal, K. J. Irgolić, *J. Trace Elem. Med. Biol.*, **12** (1998), 159–176.

64. M. Krachler, T. Prohaska, G. Koellensperger, E. Rossipal, G. Stingeder, *Biol. Trace Elem. Res.*, **76** (2000), 97–112.

65. F. A. Rivero Martino, M. L. Fernandez Sanchez, A. Sanz-Medel, *Anal. Chim. Acta*, **442** (2001), 191–200.

66. O. Wappelhorst, I. Kühn, H. Heidenreich, B. Markert, *Nutrition*, **18** (2002), 316–322.

67. M. Krachler, E. Rossipal, K. J. Irgolić, *Biol. Trace Elem. Res.*, **65** (1998), 53–74.

68. G. H. Biego, M. Joyeux, P. Hartemann, G. Debry, *Food Addit. Contam.*, **15** (1998), 775–781.

69. L. Dahl, J. A. Opsahl, H. M. Meltzer, K. Julshamn, *Br. J. Nutr.*, **90** (2003), 679–685.

70. X. Feng, S. Wu, A. Wharmby, A. Wittmeier, *J. Anal. Atom. Spectrom.*, **14** (1999), 939–946.

71. J. A. Nobrega, Y. Gelinas, A. Krushevska, R. Barnes, *J. Anal. Atom. Spectrom.*, **12** (1997), 1239–1242.

72. M. J. Liaw, S. J. Jiang, Y. C. Li, *Spectrochim. Acta*, **52B** (1997), 779–785.

73. H. C. Liao, S. J. Jiang, *J. Anal. Atom. Spectrom.*, **14** (1999), 1583–1588.

74. N. J. Miller-Ihli, S. A. Baker, *Spectrochim. Acta*, **56B** (2001), 1673–1686.

75. E. H. Larsen, M. B. Ludwigsen, *J. Anal. Atom. Spectrom.*, **12** (1997), 435–439.

76. P. A. Fecher, I. Goldmann, A. Nagengast, *J. Anal. Atom. Spectrom.*, **13** (1998), 977–982.

77. D. Andrey, P. Zbinden, M. K. Wah, W. Lee, *Atom. Spectrosc.*, **22** (2001), 299–305.

78. G. Jahreis, W. Hausmann, G. Kiessling, K. Franke, M. Leiterer, *Exp. Clin. Endocr. Diab.*, **109** (2001), 163–167.

79. K. M. Eckhoff, A. Maage, *J. Food Compos. Anal.*, **10** (1997), 270–282.

80. K. Julshamn, L. Dahl, K. Eckhoff, *J. AOAC Int.*, **84** (2001), 1976–1983.

81. P. Schramel, S. Hasse, *Mikrochim. Acta*, **166** (1994), 205–209.

82. G. Knapp, B. Maichin, P. Fecher, S. Hasse, P. Schramel, *Fresenius J. Anal Chem.*, **362** (1998), 508–513.

83. H. M. Crews, M. J. Baxter, T. Bigwood, J. A. Burrell, L. M. Owen, C. Robinson, C. Wright, R. C. Massey, *Food Addit. Contam.*, **9** (1992), 365–378.

84. G. B. Jiang, Q. F. Zhou, B. He, J. Y. Liu, *Sci. China Ser. B-Chem.*, **43** (2000), 531–539.

85. W. A. C. Anderson, S. A. Thorpe, L. M. Owen, S. E. Anderson, H. M. Crews, S. L. Reynolds, *Food Addit. Contam.*, **15** (1998), 288–292.

86. L. Castle, C. P. Offen, M. J. Baxter, J. Gilbert, *Food Addit. Contam.*, **14** (1997), 35–44.

87. E. Skrzydlewska, M. Balcerzak, F. Vanhaecke, *Anal. Chim. Acta*, **479** (2003), 191–202.
88. P. J. Fordham, J. W. Gramshaw, H. M. Crews, L. Castle, *Food Addit. Contam.*, **12** (1997), 651–669.
89. G. Ysart, P. Miller, H. Crews, P. Robb, M. Baxter, C. de L'Argy, S. Lofthouse, C. Sargent, N. Harrison, *Food Addit. Contam.*, **16** (1999), 391–403.
90. R. M. Barnes, *Adv. Exp. Med. Biol.*, **445** (1998), 379–396.
91. T. Walczyk, *Fresenius J. Anal. Chem.*, **370** (2001), 444–453.
92. J. J. Sloth, E. H. Larsen, S. H. Bügel, S. Moesgaard, *J. Anal. Atom. Spectrom.*, **18** (2003), 317–322.
93. M. Patriarca, T. D. Lyon, G. S. Fell, *Am. J. Clin. Nutr.*, **66** (1997), 616–621.
94. H. M. Crews, L. M. Owen, N. Langford, S. J. Fairweather-Tait, T. E. Fox, L. Hubbard, D. Phillips, *Toxicol. Lett.*, **112–113** (2000), 201–207.
95. A. Bibak, S. Stürup, V. Haahr, P. Gundersen, V. Gundersen, *J. Agric. Food Chem.*, **47** (1999), 2678–2684.
96. V. Gundersen, I. E. Bechmann, A. Behrens, S. Stürup, *J Agric. Food Chem.*, **48** (2000), 6094–6102.
97. A. Kawasaki, H. Watanabe-Oda, in G. Holland, S. D. Tanner (Eds), *Plasma Source Mass Spectrometry. New Developments and Applications*, The Royal Society of Chemistry, Cambridge, Special Publication, 241 (1999), 173–181.
98. S. Kelly, M. Baxter, S. Chapman, C. Rhodes, J. Dennis, P. Brereton, *Eur. Food Res. Technol.*, **214** (2002), 72–78.
99. S. Branch, S. Burke, P. Evans, B. Fairman, C. S. J. Wolff Briche, *J. Anal. Atom. Spectrom.*, **18** (2003), 17–22.
100. S. Lustig, S. Zang, B. Michalke, P. Schramel, W. Beck, *Fresenius J. Anal. Chem.*, **357** (1997), 1157–1163.
101. P. Medek, J. Pavlickova, J. Zbiral, E. Cizmarova, V. Kuban, *Int. J. Environ. Anal. Chem.*, **81** (2001), 207–219.
102. L. Samsøe-Petersen, E. H. Larsen, P. B. Larsen, P. Bruun, *Environ. Sci. Technol.*, **36** (2002), 3057–3063.
103. P. Robb, L. M. W. Owen, H. M. J. Crews, *Anal. Atom. Spectrom.*, **10** (1995), 625–629.
104. S. Mounicou, J. Szpunar, D. Andrey, C. J. Blake, R. Lobinski, *Food Addit. Contam.*, **20** (2003), 343–352.
105. M. J. Baxter, J. A. Burrell, H. M. Crews, A. Smith, R. C. Massey, *Food Addit. Contam.*, **9** (1992), 225–235.
106. J. Szpunar, P. Pellerin, A. Makarov, T. Doco, P. Williams, B. Medina, R. Lobinski, *J. Anal. Atom. Spectrom.*, **13** (1998), 749–754.
107. M. L. Magnuson, J. T. Creed, C. A. Brockhoff, *Analyst*, **122** (1997), 1057–1061.
108. P. Thomas, K. Pereira, D. Koller, *Analyst*, **25** (1997), 19–23.
109. V. V. Salov, J. Yoshinaga, Y. Shibata, M. Morita, *Anal. Chem.*, **64** (1992), 2425–2428.
110. K. G. Heumann, L. Rottmann, J. Vogl, *J. Anal. Atom. Spectrom.*, **9** (1994), 1351–1355.
111. A. Dudoit, S. A. Pergantis, *J. Anal. Atom. Spectrom.*, **16** (2001), 575–580.
112. J. W. Waggoner, L. S. Milstein, M. Belkin, K. L. Sutton, J. A. Caruso, H. B. Fannin, *J. Anal. Atom. Spectrom.*, **15** (2000), 13–18.

113. A. P. Vonderheide, J. Meija, M. Montes-Bayón, J. A. Caruso, *J. Anal. Atom. Spectrom.*, **18** (2003), 1097–1102.

114. P. Brätter, I. Navarro Blasco, V. E. Negretti De Brätter, A. Raab, *Analyst*, **123** (1998), 821–826.

115. B. Michalke, P. Schramel, *Fresenius J. Anal. Chem.*, **357** (1997), 594–599.

116. B. Michalke, P. Schramel, *J. Chromatogr. A*, **807** (1998), 71–80.

117. B. Michalke, *J. Anal. Atom. Spectrom.*, **14** (1999), 1297–1302.

118. L. M. W. Owen, H. M. Crews, R. C. Massey, *Chem. Spec. Bioavailab.*, **4** (1992), 89–96.

119. J. A. Brisbin, J. A. Caruso, *Analyst*, **127** (2002), 921–929.

120. J. A. Caruso, D. T. Heitkemper, C. B. Hymer, *Analyst*, **126** (2001), 136–140.

121. D. Kuehnelt, K. J. Irgolić, W. Goessler, *Appl. Organomet. Chem.*, **15** (2001), 445–456.

122. C. Casiot, J. Szpunar, R. Lobinski, M. Potin-Gautier, *J. Anal. Atom. Spectrom.*, **14** (1999), 645–650.

123. D. Beauchemin, *Anal. Chem.*, **74** (2002), 2873–2894.

124. Y. Shibata, M. Morita, *Anal. Chem.*, **61** (1989), 2116–2118.

125. Y. Shibata, M. Morita, *Appl. Organomet. Chem.*, **6** (1992), 343–349.

126. J. S. Edmonds, Y. Shibata, K. A. Francesconi, R. J. Rippingale, M. Morita, *Appl. Organomet. Chem.*, **11** (1997), 281–287.

127. S. McSheehy, J. Szpunar, *J. Anal. Atom. Spectrom.*, **15** (2000), 79–87.

128. S. McSheehy, M. Marcinek, H. Chassaigne, J. Szpunar, *Anal. Chim. Acta*, **410** (2000), 71–84.

129. S. McSheehy, P. Pohl, R. Lobinski, J. Szpunar, *Anal. Chim. Acta*, **440** (2001), 3–16.

130. S. McSheehy, P. Pohl, R. Lobinski, J. Szpunar, *Analyst*, **126** (2001), 1055–1062.

131. S. S. Kannamkumarath, Katarzyna Wrobel, Kazimierz Wrobel, A. Vonderheide, J. A. Caruso, *Anal. Bioanal. Chem.*, **373** (2002), 454–460.

132. G. Önning, I. A. Bergdahl, *Analyst*, **124** (1999), 1435–1438.

133. E. H. Larsen, J. Sloth, M. Hansen, S. Moesgaard, *J. Anal. Atom. Spectrom.*, **18** (2003), 310–316.

134. E. H. Larsen, M. Hansen, H. Paulin, S. Moesgaard, M. Reid, M.Rayman, *J. AOAC Int.*, **87** (2004), 225–232.

135. S. Perez Mendez, E. Blanco Gonzalez, A. Sanz-Medel, *J. Anal. Atom. Spectrom.*, **15** (2000), 1109–1114.

136. S. Perez Mendez, E. Blanco Gonzalez, M. L. Fernandez Sanchez, A. Sanz-Medel, *J. Anal. Atom. Spectrom.*, **13** (1998), 893–898.

137. C. A. Ponce de Leon, M. M. Bayon, C. Paquin, J. A. Caruso, *J. Appl. Microbiol.*, **92** (2002), 602–610.

138. C. A. Ponce de Leon, K. L. Sutton, J. A. Caruso, P. C. Uden, *J. Anal. Atom. Spectrom.*, **15** (2000), 1103–1107.

139. K. L. Sutton, C. A. Ponce de Leon, K. L. Ackley, R. M. C. Sutton, A. M. Stalcup, J. A Caruso, *Analyst*, **125** (2000), 281–286.

140. M. Montes-Bayon, C. B'Hymer, C. A. Ponce de Leon, J. A. Caruso, *J. Anal. Atom. Spectrom.*, **16** (2001), 945–950.

141. S. Perez Mendez, M. Montes Bayon, E. Blanco Gonzalez, A. Sanz-Medel, *J. Anal. Atom. Spectrom.*, **14** (1999), 1333–1337.

142. C. Devos, K. Sandra, P. Sandra, *J. Pharm. Biomed.*, **27** (2002), 507–514.

143. S. Perez Mendez, E. Blanco Gonzalez, A. Sanz-Medel, *Biomed. Chromatogr.*, **15** (2001), 181–188.

144. Z. Mester, J. Lam, R. Sturgeon, J. Pawliszyn, *J. Anal. Atom. Spectrom.*, **15** (2000), 837–842.

145. L. Yang, Z. Mester, R. E. Sturgeon, *J. Anal. Atom. Spectrom.*, **18** (2003), 431–436.

146. J. Szpunar, S. McSheehy, K. Polec, V. Vacchina, S. Mounicou, I. Rodriguez, R. Lobinski, *Spectrochim. Acta*, **55B** (2000), 779–793.

147. S. Aguerre, G. Lespes, V. Desauziersb, M. Potin-Gautier, *J. Anal. Atom. Spectrom.*, **16** (2001), 263–269.

148. M. Monperrus, O. Zuloaga, E. Krupp, D. Amouroux, R. Wahlen, B. Fairman, O. F. X. Donard, *J. Anal. Atom. Spectrom.*, **18** (2003), 247–253.

149. J. Vercauteren, A. De Meester, T. De Smaele, F. Vanhaecke, L. Moens, R. Dams, P. Sandra, *J. Anal. Atom. Spectrom.*, **15** (2000), 651–656.

150. J. Vercauteren, D. Peres, C. Devos, P. Sandra, F. Vanhaecke, *Anal. Chem.*, **73** (2001), 1509–1514.

151. M. J. Dennis, A. Burrell, K. Mathieson, P. Willetts, R. C. Massey, *Food Addit. Contam.*, **11** (1994), 633–639.

152. D. T. Heitkemper, L. A. Kaine D. S. Jackson, K. A. Wolnik, *J. Chromatogr. A*, **671** (1994), 101–108.

153. J. R. Dean, L. Ebdon, M. E. Foulkes, H. M. Crews, R. C. Massey, *J. Anal. Atom. Spectrom.*, **9** (1994), 615–618.

154. C. F. Harrington, S. Elahi, S. A. Merson, P. Ponnampalavanar, *Anal. Chem.*, **73** (2001), 4422–4427.

155. C. F. Harrington, S. Elahi, S. A. Merson, P. Ponnampalavanar, *J. AOAC Int.*, **87** (2004), 253–258.

156. S. A. Baker, N. J. Miller-Ihli, *Spectrochim. Acta*, **55B** (2000), 1823–1832.

157. H. M. Crews, J. R. Dean, L. Ebdon, R. C. Massey, *Analyst*, **114** (1989), 895–899.

158. L. M. W. Owen, H. M. Crews, R. C. Hutton, A. Walsh, *Analyst*, **117** (1992), 649–655.

159. C. N. Ferrarello, M. R. F. de la Campa, J. F. Carrasco, A. Sanz-Medel, *Anal. Chem.*, **72** (2000), 5874–5880.

160. C. N. Ferrarello, M .R. F. de la Campa, C. Sariego-Muniz, A. Sanz-Medel, *Analyst*, **125** (2000), 2223–2229.

161. V. Nischwitz, B. Michalke, A. Kettrup, *J. Anal. Atom. Spectrom.*, **18** (2003), 444–451.

162. J. Szpunar, P. Pellerin, A. Makarov, T. Doco, P. Williams, R. Lobinski, *J. Anal. Atom. Spectrom.*, **14** (1999), 639–644.

163. T. Matsunaga, T. Ishii, H. Watanabe, *Anal. Sci.*, **12** (1997), 673–675.

164. R. Koplik, H. Pavelkova, J. Cincibuchova, O. Mestek, F. Kvasnika, M. Suchanek, *J. Chromatogr. B*, **770** (2002), 261–273.

165. N. A. Nkono, O. I. Asubiojo, *Sci. Total Environ.*, **208** (1997), 161–163.

166. J. L. Mann, S. E. Long, W. R. Kelly, *J. Anal. Atom. Spectrom.*, **18** (2003), 1293–1296.

167. C. J. Park, P. J. Oh, H. Y. Kim, D. S. Lee, *J. Anal. Atom. Spectrom.*, **14** (1999), 223–227.
168. M. El Himri, A. Pastor, M. de la Guardia, *Fresenius J. Anal. Chem.*, **367** (2000), 151–156.
169. M. Forte, R. Rusconi, C. Margini, G. Abbate, S. Maltese, P. Badalamenti, S. Bellinzona, *Radiat. Prot. Dosim.*, **97** (2001), 325–328.
170. M. Galletti, L. D'Annibale, V. Pinto, C. Cremisini, *Health Phys.*, **85** (2003), 228–235.
171. J. L. Fernandez-Turiel, J. F. Llorens, F. Lopez-Vera, C. Gomez-Artola, I. Morell, D. Gimeno, *Fresenius J. Anal. Chem.*, **368** (2000), 601–606.
172. C. Reimann, K. Bjorvatn, B. Frengstad, Z. Melaku, R. Tekle-Haimanot, U. Siewers, *Sci. Total Environ.*, **311** (2003), 65–80.
173. M. A. Saleh, E. Ewane, J. Jones, B. L. Wilson, *J. Food Compos. Anal.*, **14** (2001), 127–152.
174. J. E. Longbottom, T. D. Martin, K. W. Edgell, S. E. Long, M. R. Plantz, B. E. Warden, *J. AOAC Int.*, **77** (1994), 1004–1023.
175. P. L. Trentini, M. Ascanelli, B. Zanforlini, F. Venturini, G. Bucci, F. Fagioli, *J. Anal. Atom. Spectrom.*, **8** (1993), 905–909.
176. A. Martin-Esteban, P. Fernandez, C. Perez-Conde, A. Gutierrez, C. Camara, *Anal. Chim. Acta*, **304** (1995), 121–126.
177. J. Suzuki, Y. Katsuki, H. Ogawa, K. Suzuki, H. Matsumoto, K. Yasuda, *J. Food Hyg. Soc. Jpn.*, **41** (2000), 387–396.
178. Y. Takaku, K. Masuda, T. Kobayashi, T. Shimamura, *Bunseki Kagaku*, **43** (1994), 897–903.
179. S. C. Peters, J. D. Blum, B. Klaue, M. R. Karagas, *Environ. Sci. Technol.*, **33** (1999), 1328–1333.
180. M. J. Duane, S. Facchetti, *Sci. Total Environ.*, **172** (1995), 133–144.
181. B. Fairman, T. Catterick, *J. Anal. Atom. Spectrom.*, **12** (1997), 863–866.
182. J. Bowman, B. Fairman, T. Catterick, *J. Anal. Atom. Spectrom.*, **12** (1997), 313–316.
183. K. Shiraishi, T. Nakajima, Y. Takaku, A. Tsumura, S. Yamasaki, I. P. Los, I. Y. Kamarikov, M. G. Buzinny, A. V. Zelensky, *J. Radioanal. Nucl. Chem.*, **173** (1993), 313–321.
184. N. J. Miller-Ihli, S. A. Baker, *J. Food Compos. Anal.*, **14** (2001), 619–629.
185. P. Fecher, A. Nagengast, R. Sperk in G. Holland, S. D. Tanner (Eds), *Plasma Source Mass Spectrometry – Developments and Applications*, The Royal Society of Chemistry, Special Publication, Cambridge, 202 (1997), 240–252.
186. J. L. M. de Boer, W. Verweij, T. van der Velde-Koerts, W. Mennes, *Water Res.*, **30** (1996), 190–198.
187. Y. L. Chang, S. J. Jiang, *J. Anal. Atom. Spectrom.*, **16** (2001), 1434–1438.
188. M. Montes-Bayon, A. Rodriguez-Garcia, J. I. Garcia-Alonso, A. Sanz-Medel, *Analyst*, **124** (1999), 27–31.
189. M. J. Powell, E. S. K. Quan, D. W. Boomer, D. R. Wiederin, *Anal. Chem.*, **64** (1992), 2253–2257.
190. J. Allibone, E. Fatemian, P. J. Walker, *J. Anal. Atom. Spectrom.*, **14** (1999), 235–239.

191. E. Fatemian, J. Allibone, P. J. Walker, *Analyst*, **124** (1999), 1233–1236.

192. Y. Takaku, T. Shimamura, K. Masuda, Y. Igaraschi, *Anal. Sci.*, **11** (1995), 823–827.

193. E. S. Hall, E. Murphy, *J. Radioanal. Nucl. Chem.*, **175** (1993), 129–138.

194. V. F. Hodge, G. A. Laing, *Radiochim. Acta*, **64** (1994), 211–215.

195. U. Örnemark, P. D. P. Taylor, P. De Bievre, *J. Anal. Atom. Spectrom.*, **12** (1997), 567–572.

196. M. A. Quijano, A. M. Gutierrez, M. C. Perez-Conde, C. Camara, *J. Anal. Atom. Spectrom.*, **10** (1995), 871–874.

197. R. C. Richter, S. R. Koirtyohann, S. S. Jurisson, *J. Anal. Atom. Spectrom.*, **12** (1997), 557–562.

198. E. H. Evans, J. B. Truscott, L. Bromley, P. Jones, J. Turner, B. E. Fairman, in R. W. Morrow, J. S. Crain (Eds), *Applications of Inductively Coupled Plasma–Mass Spectrometry to Radionuclide Determinations*, ASTM Special Technical Publication, 1344 (1998), 79–88.

199. M. E. Ketterer, C. J. Khourey, in R. W. Morrow, J. S. Crain (Eds), *Applications of Inductively Coupled Plasma–Mass Spectrometry to Radionuclide Determinations*, ASTM Special Technical Publication, 1344 (1998), 120–136.

200. J. Soldevila, M. El Himri, A. Pastor, M. de la Guardia, *J. Anal. Atom. Spectrom.*, **13** (1998), 803–807.

201. L. S. Zhang, S. M. Combs, *J. Anal. Atom. Spectrom.*, **11** (1996), 1043–1048.

202. L. S. Zhang, S. M. Combs, *J. Anal. Atom. Spectrom.*, **11** (1996), 1049–1054.

203. D. Amarasiriwardena, M. Kotrebai, A. Krushevska, R. M. Barnes, *Can. J. Anal. Sci. Spectrom.*, **42** (1997), 69–78.

204. L. C. Yang, M. Zhang, S. Lin, D. Y. Chen, M. H. Zheng, *Mikrochim. Acta*, **142** (2003), 85–88.

205. T. Prohaska, G. Köllensperger, M. Krachler, K. De Winne, G. Stingeder, L. Moens, *J. Anal. Atom. Spectrom.*, **15** (2000), 335–340.

206. T. Alkanani, J. K. Friel, S. E. Jackson, H. P. Longerich, *J. Agric. Food Chem.*, **42** (1994), 1965–1970.

207. B. L. Gulson, K. J. Mizon, M. J. Korsch, K. R. Mahaffey, A. Taylor, *J. Environ. Res. A*, **87** (2001), 160–174.

208. M. Krachler, E. Rossipal, *Ann. Nutr. Metab.*, **44** (2000), 68–74.

209. J. J. Thompson, *J. AOAC Int.*, **76** (1993), 1378–1384.

210. S. E. Emmett, *J. Anal. Atom. Spectrom.*, **3** (1988), 1145–1146.

211. F. A. Rivero Martino, M. L. Fernandez Sanchez, A. Sanz-Medel, *J. Anal. Atom. Spectrom.*, **15** (2000), 163–168.

212. M. J. Baxter, J. A. Burrell, H. M. Crews, R. C. Massey, *Food Addit. Contam.*, **8** (1991), 653–660.

213. K. L. Chen, S. Jiang, *Anal. Chim. Acta*, **470** (2002), 223–228.

214. C. Y. Ho, S. J. Jiang, *J. Anal. Atom. Spectrom.*, **17** (2002), 688–692.

215. H. Vanhoe, F. Van Allemeersch, J. Versieck, R. Dams, *Analyst*, **118** (1993), 1015–1019.

216. G. H. Biego, H. Biaudet, M. Joyeux, G. Debry, *Sci. Aliment.*, **16** (1996), 623–630.

217. I. Rodushkin, F. Ödman, P. K. Appelblad, *J. Food Compos. Anal.*, **12** (1999), 243–257.

218. Z. Karpas, A. Lorber, L. Halicz, I. Gavrieli, in R. W. Morrow, J. S. Crain (Eds), *Applications of Inductively Coupled Plasma-Mass Spectrometry to Radionuclide Determinations*, ASTM Special Technical Publication, 1344 (1998), 89–98.

219. C. M. R. Almeida, M. Vasconcelos, *J. Agric. Food Chem.*, **51** (2003), 4788–4798.

220. M. M. Castiñeira, R. Brandt, von A. Bohlen, N. Jakubowski, *Fresenius J. Anal. Chem.*, **370** (2000), 553–558.

221. M. Y. Perez-Jordan, J. Soldevila, A. Salvador, A. Pastor, M. de la Guardia, *J. Anal. Atom. Spectrom.*, **13** (1998), 33–39.

222. C. M. Almeida, M. Teresa, S. D. Vasconcelos, *Anal. Chim. Acta*, **463** (2002), 165–175.

223. S. Wangkarn, S. A. Pergantis, *J. Anal. Atom. Spectrom.*, **14** (1999), 657–662.

224. J. P. Perez-Trujillo, M. Barbaste, B. Medina, *Anal. Lett.*, **36** (2003), 679–697.

225. C. M. R. Almeida, M. Vasconcelos, M. Barbaste, B. Medina, *Anal. Bioanal. Chem.*, **374** (2002), 314–322.

226. M. Barbaste, B. Medina, J. P. Perez-Trujillo, *Food Addit. Contam.*, **20** (2003), 141–148.

227. J. F. Tyson, H. Ge, E. R. Denoyer, *J. Anal. Atom. Spectrom.*, **12** (1997), 1163–1167.

228. J. Goossens, L. Moens, R. Dams, *Anal. Chim. Acta*, **293** (1994), 171–181.

229. S. Augagneur, B. Medina, F. E. Grousset, *Fresenius J. Anal. Chem.*, **357** (1997), 1149–1152.

230. C. M. R. Almeida, M. T. S. D. Vasconcelos, *J. Anal. Atom. Spectrom.*, **14** (1999), 1815–1821.

231. C. R. Quetel, S. M. Nelms, L. Van Nevel, I. Papadakis, P. D. P. Taylor, *J. Anal. Atom. Spectrom.*, **16** (2001), 1091–1100.

232. S. Shimbo Z. W. Zhang, T. Watanabe, H. Nakatsuka, N. Matsuda-Inoguchi, K. Higashikawa, M. Ikeda, *Sci. Total Environ.*, **281** (2001), 165–175.

233. Z. W. Zhang, T. Watanabe, S. Shimbo, K. Higashikawa, M. Ikeda, *Sci. Total Environ.*, **220** (1998), 137–145.

234. C. P. Sanchez-Castillo, P. J. S. Dewey, M. D. Reid, M. De Lourdes Solano, W. P. T. James, *J. Food Compos. Anal.*, **10** (1997), 312–333.

235. S. Wu, X. Feng, A. Wittmeier, *J. Anal. Atom. Spectrom.*, **12** (1997), 797–806.

236. F. Cubadda, A. Raggi, *Microchem. J.*, **79** (2005), 91–96.

237. K. Van Den Broeck, C. Vandecasteele, J. M. C. Geuns, *J. Anal. Atom. Spectrom.*, **12** (1997), 987–991.

238. M. Pham Thi Huynh, F. Carrot, S. Chu Pham Ngoc, M. Dang Vu, G. Revel, *J. Radioanal. Nucl. Ch.*, **217** (1997), 95–99.

239. S. D'Ilio, M. Alessandrelli, R. Cresti, G. Forte, S. Caroli, *Microchem. J.*, **73** (2002), 195–201.

240. T. D. Phuong, P. V. Chuong, D. T. Khiem, S. Kobot, *Analyst*, **124** (1999), 553–560.

241. T. Watanabe, Z. W. Zhang, C. S. Moon, S. Shimbo, H. Nakatsuka, N. Matsuda-Inoguchi, K. Higashikawa, M. Ikeda, *Int. Arch. Occ. Env. Health*, **73** (2000), 26–34.

242. T. Moriyama, K. Shindoh, Y. Taguchi, H. Watanabe, A. Yasui, T. Joh, *J. Food Hyg. Soc. Jpn*, **44** (2003), 145–149.

243. P. C. Li, S. J. Jiang, *Anal. Chim. Acta*, **495** (2003), 143–150.

244. C. C. Chang, H. T. Liu, S. Jiang, *Anal. Chim. Acta*, **493** (2003), 213–218.

245. C. J. Park, J. K. Suh, K. W. Lee, S. H. Lee, *Anal. Sci.*, **13** (1997), 429–432.

246. C. J. Park, J. K. Suh, *J. Anal. Atom. Spectrom.*, **12** (1997), 573–577.

247. I. Rodushkin, T. Ruth, A. Huhtasaari, *Anal. Chim. Acta*, **378** (1999), 191–200.

248. H. Zhou, J. Liu, *Atom. Spectrosc.*, **18** (1997), 92–194.

249. F. Cubadda, M. Baldini, M. Carcea, L. A. Pasqui, A. Raggi, P. Stacchini, *Food Addit. Contam.*, **18** (2001), 778–787.

250. O. A. F. Al-Dayel, *Asian J. Chem.*, **14** (2002), 1307–1310.

251. J. Lintschinger, N. Fuchs, H. Moser, R. Jäger, T. Hlebeina, G. Markolin, W. Gössler, *Plant Food Hum. Nutr.*, **50** (1997), 223–237.

252. B. Li, Y. Sun, M. Yin, *J. Anal. Atom. Spectrom.*, **14** (1999), 1843–1848.

253. M. Haldimann, B. Zimmerli, *Mitt. Geb. Lebensm. Hyg.*, **85** (1994), 111–131.

254. I. Rodushkin, F. Ödman, H. Holmström, *Sci. Total Environ.*, **231** (1999), 53–65.

255. R. Herzig, A. Rehnert, S. Korhammer, J. Kumpulainen, P. Schramel, H. Muntau, T. Linsinger, P. Quevauviller, *Trends Anal. Chem.*, **21** (2002), 746–761.

256. R. J. Cox, C. J. Pickford, M. Thompson, *J. Anal. Atom. Spectrom.*, **7** (1992), 635–640.

257. S. Mounicou, J. Szpunar, D. Andrey, C. Blake, R. Lobinski, *Analyst*, **127** (2002), 1638–1641.

258. A. Marcos, A. Fisher, G. Rea, S. J. Hill, *J. Anal. Atom. Spectrom.*, **13** (1998), 521–525.

259. X. Cao, G. Zhao, M. Yin, J. Li, *Analyst*, **123** (1998), 1115–1119.

260. A. Moreda-Pineiro, A. Fisher, S. J. Hill, *J. Food Compos. Anal.*, **16** (2003), 195–211.

261. M. J. Baxter, J. A. Burrell, H. M. Crews, R. C. Massey, D. J. McWeeny, *Food Addit. Contam.*, **6** (1989), 341–349.

262. R. Koplik, E. Curdova, M. Suchanek, *Fresenius J. Anal Chem.*, **360** (1998), 449–451.

263. J. Falandysz, K. Szymczyk, H. Ichihashi, L. Bielawski, M. Gucia, A. Frankowska, S. I. Yamasaki, *Food Addit. Contam.*, **18** (2001), 503–513.

264. P. Mattila, K. Könkö, M. Eurola, J. M. Pihlava, J. Astola, L. Vahteristo, J. Hietaniemi, J. Kumpulainen, M. Valtonen, V. Piironen, *J. Agric. Food Chem.*, **49** (2001), 2343–2348.

265. M. Bettinelli, U. Baroni, S. Spezia, C. Terni, *Atom. Spectrosc.*, **21** (2000), 195–204.

266. S. Caroli, G. Forte, M. Alessandrelli, R. Cresti, M. Spagnoli, S. D'Ilio, J. Pauwels, G. N. Kramer, *Microchem. J.*, **67** (2000), 227–233.

267. G. Forte, S. D'Ilio, S. Caroli, *J. AOAC Int.* **84** (1), (2001), 972–1975.

268. S. Caroli, G. Forte, A. L. Iamiceli, B. Galoppi, *Talanta*, **50** (1999), 327–336.

269. S. A. Bhandari, D. Amarasiriwardena, *Microchem. J.*, **64** (2000), 73–84.

270. B. P. Bourgoin, D. Boomer, M. J. Powell, S. Willie, D. Edgar, D. Evans, *Analyst*, **117** (1992), 19–22.

271. R. E. Wolf, *Atom. Spectrosc.*, **18** (1997), 169–174.

272. G. M. Scelfo, A. R. Flegal, *Environ. Health. Perspect.*, **108** (2000), 309–313.

273. S. P. Dolan, D. A. Nortrup, P. M. Bolger, S. G. Capar, *J. Agric. Food Chem.*, **51** (2003), 1307–1312.

274. N. E. Renard, T. A. Tompkins, *Biol. Trace Elem. Res.*, **88** (2002), 185–191.

275. S. J. Huang, S. J. Jiang, *J. Anal. Atom. Spectrom.*, **16** (2001), 664–668.

276. S. Caroli, M. Spagnoli, G. Forte, M. Alessandrelli, R. Cresti, S. D'Ilio, J. Pauwels, G. N. Kramer, G. Záray, *Microchem. J.*, **67** (2000), 235–243.

277. J. P. Schmit, M. Youla, Y. Gelinas, *Anal. Chim. Acta*, **249** (1991), 495–501.

278. N. I. Ward, F. R. Abou-Shakra, S. F. Durrant, *Biol. Trace Elem. Res.*, **26–27** (1990), 177–187.

279. J. Dombovari, J. S. Becker, H. J. Dietze, *Fresenius J. Anal. Chem.*, **367** (2000), 407–413.

280. W. Kerl, J. S. Becker, H. J. Dietze, W. Dannecker, *J. Anal. Atom. Spectrom.*, **11** (1996), 723–726.

281. J. A. Moreton, H. T. Delves, *J. Anal. Atom. Spectrom.*, **13** (1998), 659–665.

282. K. Okamoto, *Spectrochim. Acta*, **46B** (1991), 1615–1622.

283. S. Zimmermann, C. M. Menzel, Z. Berner, J. D. Eckhardt, D. Stüben, F. Alt, J. Messerschmidt, H. Taraschewski, B. Sures, *Anal. Chim. Acta*, **439** (2001), 203–209.

284. M. De Gieter, M. Leermakers, R. Van Ryssen, J. Noyen, L. Goeyens, W. Baeyens, *Arch. Environ. Con. Tox.*, **43** (2002), 406–417.

285. J. W. H. Lam, J. W. McLaren, B. A. J. Methven, *J. Anal. Atom. Spectrom.*, **10** (1995), 551–554.

286. M. J. Campbell, G. Vermeir, R. Dams, P. Quevauviller, *J. Anal. Atom. Spectrom.*, **7** (1992), 617–621.

287. D. Beauchemin, J. W. McLaren, S. N. Willie, S. S. Berman, *Anal. Chem.*, **60** (1988), 687–691.

288. C. C. Chang, S. J. Jiang, *J. Anal. Atom. Spectrom.*, **12** (1997), 75–80.

289. C. Moor, J. W. H. Lam, R. E. Sturgeon, *J. Anal. Atom. Spectrom.*, **15** (2000), 143–149.

290. I. E. Bechmann, S. Stürup, L. V. Kristensen, *Fresenius J. Anal. Chem.*, **368** (2000), 708–714.

291. H. El Azouzi, M. L. Cervera, M. de La Guardia, *J. Anal. Atom. Spectrom.*, **13** (1998), 533–538.

292. S. F. Boulyga, H.-J. Dietze, J. S. Becker, *Mikrochim. Acta.*, **137** (2001), 93–103.

293. C. F. Harrington, B. E. Fairman, T. Catterick, *Spectrosc. Eur.*, **9** (1997), 10–16.

294. B. S. Sheppard, D. T. Heitkemper, C. M. Gaston, *Analyst*, **119** (1994), 1683–1686.

295. E. H. Larsen, G. A. Pedersen, J. W. McLaren, *J. Anal. Atom. Spectrom.*, **12** (1997), 963–968.

296. W. Maher, S. Forster, F. Krikowa, P. Snitch, G. Chapple, P. Craig, *Atom. Spectrosc.*, **22** (2001), 361–370.

297. K. Benkhedda, H. Goenaga Infante, E. Ivanova, F. C. Adams, *J. Anal. Atom. Spectrom.*, **15** (2000), 1349–1356.

298. M. J. Campbell, C. Demesmay, M. Ollé, *J. Anal. Atom. Spectrom.*, **9** (1994), 1379–1384.

299. L. Ebdon, A. S. Fisher, P. J. Worsfold, H. Crews, M. Baxter, *J. Anal. Atom. Spectrom.*, **8** (1993), 691–695.

300. K. Saeki, N. Nakatani, T. H. Le, S. Tanabe, *Bunseki Kagaku*, **47** (1998), 135–139.

301. E. S. de Lima, M. F. Costa, A. Pastor, M. de la Guardia, *Atom. Spectrosc.*, **22** (2001), 405–413.

302. S. Sakao, H. Uchida, *Anal. Chim. Acta*, **382** (1999), 215–223.

303. M. C. Wu, S. J. Jiang, T. S. Hsi, *Anal. Bioanal. Chem.*, **377** (2003), 154–158.

304. S. F. Durrant, N. I. Ward, *Food Chem.*, **49** (1994), 317–323.

305. S. F. Durrant, *Analyst*, **117** (1992), 1585–1592.

306. P. Fecher, C. Walther, J. Sondermann, *Deut. Lebensm.-Rundsch.*, **95** (1999), 133–142.

307. A. Lozak, K. Soltyk, P. Ostapczuk, Z. Fijalek, *Sci. Total Environ.*, **289** (2002), 33–40.

308. L. Noël, J. C. Leblanc, T. Guerin, *Food Add. Contam.*, **20** (2003), 44–56.

309. L. J. Melnyk, J. N. Morgan, R. Fernando, E. D. Pellizzari, O. Akinbo, *J. AOAC Int.*, **86** (2003), 439–447.

310. A. Lasztity, A. Krushevska, M. Kotrebai, R. M. Barnes, D. Amarasiriwardena, *J. Anal. Atom. Spectrom.*, **10** (1995), 505–510.

311. H. Muto, T. Abe, Y. Takizawa, K. Kawabata, K. Yamaguchi, K. Saitoh, *Sci. Total Environ.*, **144** (1994), 231–239.

312. K. Shiraishi, J. F. McInroy, Y. Igarashi, *J. Nutr. Sci. Vitaminol.*, **36** (1991), 81–86.

313. Z. W. Zhang, S. Shimbo, K. Miyake, T. Watanabe, H. Nakatsuka, N. Matsuda-Inoguchi, C. S. Moon, K. Higashikawa, M. Ikeda, *Eur. J. Clin. Nutr.*, **53** (1999), 226–232.

314. Z. W. Zhang, S. Shimbo, N. Ochi, M. Eguchi, T. Watanabe, C. S. Moon, M. Ikeda, *Sci. Total Environ.*, **205** (1997), 179–187.

315. M. Ikeda, Z. W. Zhang, S. Shimbo, T. Watanabe, H. Nakatsuka, C. S. Moon, N. Matsuda-Inoguchi, K. Higashikawa, *Sci. Total Environ.*, **249** (2000), 373–384.

316. S. Shimbo, Z. W. Zhang, C. S. Moon, T. Watanabe, H. Nakatsuka, N. Matsuda-Inoguchi, K. Higashikawa, M. Ikeda, *Int. Arch. Occ. Env. Health*, **73** (2000), 163–170.

317. Y. Pang, D. L. MacIntosh, P. B. Ryan, *J. Nutr.*, **131** (2001), 2171–2176.

318. S. Shimbo, Z. W. Zhang, K. Miyake, T. Watanabe, H. Nakatsuka, N. Matsuda-Inoguchi, C. S. Moon, K. Higashikawa, M. Ikeda, *Eur. J. Clin. Nutr.*, **53** (1999), 233–238.

319. M. Ikeda, Z. W. Zhang, S. Shimbo, T. Watanabe, H. Nakatsuka, C. S. Moon, N. Matsuda-Inoguchi, K. Higashikawa, *Am. J. Ind. Med.*, **38** (2000), 271–280.

320. Z. W. Zhang, S. Shimbo, J. B. Qu, T. Watanabe, H. Nakatsuka, N. Matsuda-Inoguchi, K. Higashikawa, M. Ikeda, *Biol. Trace Elem. Res.*, **80** (2001), 125–138.

321. P. Akhter, S. D. Orfi, H. Kawamura, N. Ahmad, M. J. Khaleequr-Rahman, *Environ. Radioactiv.*, **62** (2002), 123–127.

322. K. Shiraishi, Y. Takaku, K. Yoshimizu, Y. Igarashi, K. Masuda, J. McInroy, G. Tanaka, *J. Anal. Atom. Spectrom.*, **6** (1991), 335–338.

323. J. R. Dean, L. Ebdon, H. M. Crews, R. C. Massey, *J. Anal. Atom. Spectrom.*, **3** (1988), 349–354.

324. C. F. Harrington, S. Elahi, P. Ponnampalavanar, T. M. D'Silva, *Atom. Spectrosc.*, **20** (1999), 174–179.

325. G. Ysart, P. Miller, M. Croasdale, H. M. Crews, P. Robb, M. Baxter, C. de L'Argy, N. Harrison, *Food Add. Contam.*, **17** (2000), 775–786.

326. G. H. Biego, M. Joyeux, P. Hartemann, G. Debry, *Sci. Total Environ.*, **217** (1998), 27–36.

327. R. A. Schoof, L. J. Yost, J. Eickhoff, E. A. Crecelius, D. W. Cragin, D. M. Meacher, D. B. Menzel, *Food Chem. Toxicol.*, **37** (1999), 839–846.

328. T. Roychowdhury, T. Uchino, H. Tokunaga, M. Ando, *Food Chem. Toxicol.*, **40** (2002), 1611–1621.

329. H. Zhou, J. Liu, *Atom. Spectrosc.*, **18** (1997), 115–118.

330. J. J. Fardy, I. M. Warner, *J. Radioanal. Nucl. Chem.*, **157** (1992), 239–244.

331. J. M. Llobet, G. Falcó, C. Casas, A. Teixidó, J. L. Domingo, *J. Agr. Food Chem.*, **51** (2003), 838–842.

332. K. Shiraishi, *J. Radioanal. Nucl. Chem.*, **238** (1998), 67–73.

333. M. Rose, P. Miller, M. Baxter, G. Appleton, H. M. Crews, M. Croasdale, *J. Environ. Monitor.*, **3** (2001), 361–365.

334. N. Matsuda-Inoguchi, S. Shimbo, Z. W. Zhang, S. Srianujata, O. Banjong, C. Chitchumroonchokchai, T. Watanabe, H. Nakatsuka, K. Higashikawa, M. Ikeda, *Eur. J. Clin. Nutr.*, **54** (2000), 187–194.

335. N. Salvato, *Ind. Aliment. Italy*, **34** (1995), 504–515.

336. K. Van Dyck, H. Robberecht, R. Van Cauwenbergh, H. Deelstra, J. Arnaud, L. Willemyns, F. Benijts, J. A. Centeno, H. Taylor, M. E. Soares, M. L. Bastos, M. A. Ferreira, P. C. D'Haese, L. V. Lamberts, M. Hoenig, G. Knapp, S. J. Lugowski, L. Moens, J. Riondato, R. Van Grieken, M. Claes, R. Verheyen, L. Clement, M. Uytterhoeven, *J. Anal. Atom. Spectrom.*, **15** (2000), 735–741.

337. G. H. Biego, M. Joyeux, P. Hartemann, G. Debry, *Arch. Environ. Con. Toxicol*, **36** (1999), 227–232.

338. P. Evans, S. Elahi, K. Lee, B. Fairman, *J. Environ. Monitor.*, **5** (2003), 175–179.

339. K. Shiraishi, M. Yamamoto, *J. Radioanal. Nucl. Chem.*, **196** (1995), 89–96.

340. M. Yukawa, Y. Watanabe, Y. Nishimura, Y. Guo, Z. Yongru, H. Lu, W. Zhang, L. Wei, Z. Tao, *Fresenius J. Anal. Chem.*, **363** (1999), 760–766.

341. D. Pozebon, V. L Dressler, A. J. Curtius, *J. Anal. Atom. Spectrom.*, **13** (1998), 1101–1105.

342. D. Pozebon, V. L. Dressler, A. J. Curtius, *Talanta*, **51** (2000), 903–911.

343. F. Laborda, M. J. Baxter, H. M. Crews, J. Dennis, *J. Anal. Atom. Spectrom.*, **9** (1994), 727–736.

344. J. K. Friel, C. S. Skinner, S. E. Jackson, H. P. Longerich, *Analyst*, **115** (1990), 269–273.

345. F. Cubadda, A. Raggi, A. Testoni, F. Zanasi, *J. AOAC Int.*, **85** (2002), 113–121.

346. P. Zbinden, D. Andrey, *Atom. Spectrosc.*, **19** (1998), 214–219.

347. R. D. Satzger, *Anal. Chem.*, **49** (1988), 2500–2504.

348. S. Branch, L. Ebdon, M. Ford, M. Foulkes, O'Neill, *J. Anal. Atom. Spectrom.*, **6** (1991), 151–154.

349. B. L. Sharp, A. S. Bashammakh, C. M. Thung, J. Skilling, M. Baxter, *J. Anal. Atom. Spectrom.*, **17**, (2002), 459–468.

350. A. Krushevska, M. Kotrebai, A. Lasztity, R. M. Barnes, D. Amarasiriwardena, *Fresenius J. Anal. Chem.*, **355** (1996), 793–800.

351. A. T. Townsend, *Fresenius J. Anal. Chem.*, **364** (1999), 521–526.

352. S. J. Hill, M. J. Ford, L. Ebdon, *J. Anal. Atom. Spectrom.*, **7** (1992), 719–725.

353. F. G. Smith, D. R. Wiederin, R. S. Houk, C. B. Egan, R. E. Serfass, *Anal. Chim. Acta*, **248** (1991), 229–234.

354. S. Evans, U. Krähenbühl, *Fresenius J. Anal. Chem.*, **349** (1994), 454–459.

355. A. M. S. Nyomora, R. N. Sah, P. H. Brown, R. O. Miller, *Fresenius J. Anal. Chem.*, **357** (1997), 1185–1191.

356. B. Schnetger, Y. Muramatsu, *Analyst*, **121** (1996), 1627–1631.

357. S. A. Baker, D. K. Bradshaw, N. J. Miller-Ihli, *Atom. Spectrosc.*, **20** (1999), 167–173.

358. D. Karunasagar, J. Arunachalam, *Anal. Chim. Acta*, **441** (2001), 291–296.

359. Y. Gelinas, A. Krushevska, R. M. Barnes, *Anal. Chem.*, **70** (1998), 1021–1025.

360. M. Haldimann, B. Zimmerli, K. Rieder, U. Mueller, P. Lischer, *Mitt. Lebensm. Hyg.*, **90** (1999), 697–713.

361. M. Yin, B. Li, *Spectrochim. Acta*, **53B** (1998), 1447–1454.

362. A. T. Townsend, *J. Anal. Atom. Spectrom.*, **15** (2000), 307–314.

363. F. Cubadda, S. Giovannangeli, F. Iosi, A. Raggi, P. Stacchini, *Food Chem.*, **81** (2003), 463–468.

364. A. Krushevska, S. Waheed, J. Nóbrega, D. Amarisiriwardena, R. M. Barnes, *Appl. Spectrosc.*, **52** (1998), 205–211.

365. R. M. Parr, H. Kawamura, G. V. Iyengar, *Biol. Trace Elem. Res.*, **71–72** (1999), 5–13.

366. R. W. Fonseca, N. J. Miller-Ihli, *Spectrochim. Acta*, 51B (1996), 1591–1599.

367. S. Cao, H. Chen, X. Zeng, *J. Anal. Atom. Spectrom.*, **14** (1999), 1183–1186.

368. S. J. Christopher, S. E. Long, M. S. Rearick, J. D. Fassett, *Anal. Chem.*, **73** (2001), 2190–2199.

369. L. Perring, D. Andrey, *Atom. Spectrosc.*, **22** (2001), 371–378.

370. L. Moens, H. Vanhoe, F. Vanhaecke, J. Goossens, M. Campbell, R. Dams, *J. Anal. Atom. Spectrom.*, **9** (1994), 187–191.

371. G. Rädlinger, K. G. Heumann, *Anal. Chem.*, **70** (1998), 2221–2224.

372. M. Haldimann, A. Eastgate, B. Zimmerli, *Analyst*, **125** (2000), 1977–1982.

373. Y. Okamoto, *Fresenius J. Anal. Chem.*, **367** (2000), 300–305.

374. J. Ruiz-Encinar, J. I. Garcia-Alonso, A. Sanz-Medel, S. Main, P. J. Turner, *J. Anal. Atom. Spectrom.*, **16** (2001), 322–326.

375. W. T. Buckley, J. J. Budac, D. V. Godfrey, K. M. Koenig, *Anal. Chem.*, **64** (1992), 724–729.

376. F. Vanhaecke, S. Boonen, L. Moens, R. Dams, *J. Anal. Atom. Spectrom.*, **12** (1997), 125–130.

377. S. Boulyga, J. Dombovari, J. S. Becker, H. J. Dietze, *Atom. Spectrosc.*, **21** (2000), 149–155.

378. L. Hinojosa Reyes, J. M. Marchante Gayon, J. I. Garcia-Alonso, A. Sanz-Medel, *J. Anal. Atom. Spectrom.*, **18** (2003), 11–16.

379. P. Klemens, K. G. Heumann, *Fresenius J. Anal. Chem.*, **371** (2001), 758–763.

380. K. Y. Patterson, C. Veillon, P. B. Moser-Veillon, G. F. Wallace, *Anal. Chim. Acta.*, **258** (1992), 317–324.

381. P. Thomas, K. Sniatecki, *J. Anal. Atom. Spectrom.*, **10** (1995), 615–618.

382. S. Saverwyns, X. Zhang, F. Vanhaecke, R. Cornelis, L. Moens, R. Dams, *J. Anal. Atom. Spectrom.*, **12** (1997), 1047–1052.

383. M. Van Holderbeke, Y. Zhao, F. Vanhaecke, L. Moens, R. Dams, P. Sandra, *J. Anal. Atom. Spectrom.*, **14** (1999), 229–234.

384. A. G. Cox, C. W. McLeod, M. Nishikawa, D. Wiederin, in G. Holland, S. D. Tanner (Eds), *Plasma Source Mass Spectrometry – Developments and Applications*, The Royal Society of Chemistry, Cambridge, Special Publication, 202 (1997), 145–151.

385. M. L. Magnuson, J. T. Creed, C. A. Brockhoff, *J. Anal. Atom. Spectrom.*, **12** (1997), 689–695.

386. M. Moldovan, M. M. Gomez, M. A. Palacios, C. Camara, *Microchem. J.*, **59** (1998), 89–99.

387. P. A. Gallagher, C. A. Schwegel, X. Y. Wei, J. T. Creed, *J. Environ. Monitor.*, **3** (2001), 371–376.

388. J. A. Day, M. Montes-Bayon, A. P. Vonderheide, J. A. Caruso, *Anal. Bioanal. Chem.*, **373** (2002), 664–668.

389. L. S. Milstein, A. Essader, E. D. Pellizzari, R. A. Fernando, O. Akinbo, *Environ. Int.*, **28** (2002), 277–283.

390. A. Shraim, N. C. Sekaran, C. D. Anuradha, S. Hirano, *Appl. Organomet. Chem.*, **16** (2002), 202–209.

391. M. J. Ellwood, W. A. Maher, *J. Anal. Atom. Spectrom.*, **17** (2002), 197–203.

392. J. T. Creed, M. L. Magnuson, J. D. Pfaff, C. Brockhoff, *J. Chromatogr. A*, **753** (1996), 261–267.

393. A. Seubert, M. Nowak, *Fresenius. J. Anal. Chem.*, **360** (1998), 777–780.

394. M. Nowak, A. Seubert, *Anal. Chim. Acta*, **359** (1998), 193–204.

395. J. T. Creed, C. A. Brockhoff, *Anal. Chem.*, **71** (1999), 722–726.

396. A. R. Elwaer, C. W. McLeod, K. C. Thompson, *Anal. Chem.*, **72** (2000), 5725–5730.

397. C. Barnowski, N. Jakubowski, D. Stuewer, J. A. C. Broekaert, *J. Anal. Atom. Spectrom.*, **12**, (1997), 1155–1161.

398. S. Saverwyns, K. Van Hacke, F. Vanhaecke, L. Moens, R. Dams, *Fresenius J. Anal. Chem.*, **363** (1999), 490–494.

399. F. Sacher, B. Raue, J. Klinger, H. J. Brauch, *Int. J. Environ. Anal. Chem.*, **74** (1999), 191–201.

400. Y. L. Chang, S. J. Jiang, *J. Anal. Atom. Spectrom.*, **16** (2001), 858–862.

401. J. R. Pretty, E. A. Blubaugh, J. A. Caruso, T. M. Davidson, *Anal. Chem.*, **66** (1994), 1540–1547.

402. C. C. Wan, C. S. Chen, S. J. Jiang, *J. Anal. Atom. Spectrom.*, **12** (1997), 683–687.

403. S. L. Anderson, S. A. Pergantis, *Talanta*, **60** (2003), 821–830.

404. E. Coni, B. Bocca, B. Galoppi, A. Alimonti, S. Caroli, *Microchem. J.*, **67** (2000), 187–194.

405. E. Coni, A. Alimonti, A. Bocca, F. La Torre, E. Menghetti, E. Miraglia, S. Caroli, *Trace Elem. Electrol.*, **13** (1996), 26–32.

406. M. Leiterer, D. Truckenbrodt, K. Franke, *Eur. Food Res. Technol.*, **213** (2001), 150–153.

407. F. A. R. Martino, M. L. F. Sanchez, A. Sanz-Medel, *J. Anal. Atom. Spectrom.*, **17** (2002), 1271–1277.

408. M. L. Fernandez Sanchez, J. Szpunar, *J. Anal. Atom. Spectrom.*, **14** (1999), 1697–1702.

409. H. Matsuura, A. Hokura, F. Katsuki, A. Itoh, H. Haraguchi, *Anal. Sci.*, **17** (2001), 391–398.

410. W. R. L. Cairns, S. J. Hill, L. Ebdon, *Microchem. J.*, **54** (1996), 88–110.

411. D. Wang, C. Wang, G. Zhao, Z. Wei, Y. Tao, X. Liang, *Biosci. Biotechnol. Biochem.*, **65** (2001), 1987–1992.

412. M. M. Castiñeira, P. Burba, N. Jakubowski, J. Andersson, *Anal. Bioanal. Chem.*, **376** (2003), 174–181.

413. S. Wangkarn, S. A. Pergantis, *J. Anal. Atom. Spectrom.*, **15** (2000), 627–633.

414. B. S. Sheppard, J. A. Caruso, D. T. Heitkemper, K. A. Wolnik, *Analyst*, **117** (1992), 971–975.

415. T. Akiyama, M. Yamanaka, Y. Date, H. Kubota, M. H. Nagaoka, Y. Kawasaki, T. Yamazaki, C. Yomota, T. Maitani, *J. Food Hyg. Soc. Jpn*, **43** (2002), 348–351.

416. D. T. Heitkemper, N. P. Vela, K. R. Stewart, C. S. Westphal, *J. Anal. Atom. Spectrom.*, **16** (2001), 299–306.

417. I. Pizarro, M. Gomez, M. A. Palacios, C. Camara, *Anal. Bioanal. Chem.*, **376** (2003), 102–109.

418. M. D'Amato, G. Forte, S. Caroli, *J. AOAC Int.*, **87** (2004), 238–243.

419. M. Stadlober, M. Sager, K. J. Irgolić, *Food Chem.*, **73** (2001), 57–366.

420. C. B'Hymer, J. A. Caruso, *J. Liq. Chromatogr. R. T.*, **25** (2002), 639–653.

421. A. P. Vonderheide, K. Wrobel, S. S. Kannamkumarath, C. B'Hymer, M. Montes-Bayon, C. Ponce de Leon, J. A. Caruso, *J. Agric. Food Chem.*, **50** (2002), 5722–5728.

422. M. T. Roberge, A. J. Borgerding, J. W. Finley, *J. Agric. Food Chem.*, **51** (2003), 4191–4197.

423. H. Ge, X.-J. Cai, J. F. Tyson, P. C. Uden, E. R. Denoyer, E. Block, *Anal. Commun.*, **33** (1996), 279–281.

424. H. Helgesen, E. H. Larsen, *Analyst*, **123** (1998), 791–796.

425. N. P. Vela, D. T. Heitkemper, K. R. Stewart, *Analyst*, **126** (2001), 1011–1017.

426. S. Mounicou, J. Szpunar, R. Lobinski, D. Andrey, C. J. Blake, *J. Anal. Atom. Spectrom.*, **17** (2002), 880–886.

427. J. Meija, J. M. Bryson, A. P. Vonderheide, M. Montes-Bayon, J. A. Caruso, *J. Agric. Food Chem.*, **51** (2003), 5116–5122.

428. S. M. Bird, H. Ge, P. C. Uden, J. F. Tyson, E. Block, E. J. Denoyer, *J. Chromatogr. A.*, **789** (1997), 5116–5122.

429. P. C. Uden, S. M. Bird, M. Kotrebai, P. Nolibos, J. F. Tyson, E. Block, E. Denoyer, *Fresenius J. Anal. Chem.*, **362** (1998), 447–456.

430. M. Kotrebai, S. M. Bird, J. F. Tyson, E. Block, P. C. Uden, *J. Chromatogr. A*, **866** (1999), 51–63.

431. M. Kotrebai, M. Birringer, J. F. Tyson, E. Block, P. C.Uden, *Anal. Commun.*, **36** (1999), 249–252.

432. C. Ip, M. Birringer, E. Block, M. Kotrebai, J. F. Tyson, P. C. Uden, D. J. Lisk, *J. Agric. Food Chem.*, **48** (2000), 2062–2070.

433. S. McSheehy, W. Yang, F. Pannier, J. Szpunar, R. Lobinski, J. Auger, M. Potin-Gautier, *Anal. Chim. Acta*, **421** (2000), 47–153.

434. M. Kotrebai, M. Birringer, J. F. Tyson, E. Block, P. C. Uden, *Analyst*, **125** (2000), 71–78.

435. P. C. Uden, R. Hafezi, M. Kotrebai, P. Nolibos, J. F. Tyson, E. Block, *Phosphorus, Sulfur Silicon Relat. Elem.*, **171** (2001), 31–56.

436. P. Tlustoš, W. Goessler, J. Szakova, J. Balik, *Appl. Organomet. Chem.*, **16** (2002), 216–220.

437. S. S. Kannamkumarath, K. Wrobel, K. Wrobel, J. A. Caruso, *J. Agric. Food Chem.*, **52** (2004), 1458–63.

438. A. R. Byrne, Z. Šlejkovec, T. Stijve, L. Fay, W. Goessler, J. Gailer, K. J. Irgolić, *Appl. Organomet. Chem.*, **9** (1995), 305–313.

439. Z. Slejkovec, A. R. Byrne, T. Stijve, W. Goessler, K. J. Irgolić, *Appl. Organomet. Chem.*, **11** (1997), 673–682.

440. D. Kuehnelt, W. Goessler, K. J. Irgolić, *Appl. Organomet. Chem.*, **11** (1997), 289–296.

441. E. H. Larsen, *Spectrochim. Acta*, **53B** (1998), 253–265.

442. M. Slekovec, W. Goessler, K. J. Irgolić, *Chem. Spec. Bioavailab.*, **11** (1999), 115–123.

443. R. G. Wuilloud, S. S. Kannamkumarath, J. A. Caruso, *J. Agric. Food Chem.*, **52** (2004), 1315–1322.

444. J. Zheng, H. Greschonig, F. Liu, W. Kosmus, *Trace Elem. Electrol.*, **15** (1998), 70–75.

445. C. B'Hymer, J. A. Caruso, *J. Anal. Atom. Spectrom.*, **15** (2000), 1531–1539.

446. J. M. Marchante-Gayon, C. Thomas, I. Feldmann, N. Jakubowski, *J. Anal. Atom. Spectrom.*, **15** (2000), 1093–1102.

447. M. V. Pelaez, M. M. Bayon, J. Alonso, A. Sanz-Medel, *J. Anal. Atom. Spectrom.*, **15** (2000), 1217–1222.

448. J. Zheng, M. Ohata, N. Furuta, W. Kosmus, *J. Chromatogr. A*, **874** (2000), 55–64.

449. S. M. Bird, P. C. Uden, J. F. Tyson, E. Block, E. J. Denoyer, *J. Anal. Atom. Spectrom.*, **12** (1997), 785–788.

450. C. Casiot, V. Vacchina, H. Chassaigne, J. Szpunar, M. Potin-Gautier, R. Lobinski, *Anal. Commun.*, **36** (1999), 77–80.

451. S. McSheehy, P. Pohl, J. Szpunar, M. Potin-Gautier, R. Lobinski, *J. Anal. Atom. Spectrom.*, **16** (2001), 68–73.

452. C. C. Chéry, H. Chassaigne, L. Verbeeck, R. Cornelis, F. Vanhaecke, L. Moens, *J. Anal. Atom. Spectrom.*, **17** (2002), 576–580.

453. T. Lindemann, H. Hintelmann, *Anal. Chem.*, **74** (2002), 4602–4610.

454. S. McSheehy, F. Pannier, J. Szpunar, M. Potin-Gautier, R. Lobinski, *Analyst*, **127** (2002), 223–229.

455. S. Mounicou, S. McSheehy, J. Szpunar, M. Potin-Gautier, R. Lobinski, *J. Anal. Atom. Spectrom.*, **17** (2002), 15–20.

456. K. Wrobel, K. Wrobel, Caruso, J. A. *J. Anal. Atom. Spectrom.*, **17** (2002), 1048–1054.

457. M. Yoshida, S. Sugihara, T. Suenaga, C. Naito, K. Fukunaga, H. Tsuchita, *J. Nutr. Sci. Vitaminol.*, **48** (2002), 401–404.

458. J. R. Encinar, L. Ouerdane, W. Buchmann, J. Tortajada, R. Lobinski, J. Szpunar, *Anal. Chem.*, **75** (2003), 3765–3774.

459. E. H. Larsen, M. Hansen, T. Fan, M. Vahl, *J. Anal. Atom. Spectrom.*, **16** (2001), 1403–1408.

460. M. Kotrebai, S. M. Bird, J. F. Tyson, E. Block, P. C. Uden, *Spectrochim. Acta*, **54B** (1999), 1573–1591.

461. A. Polatajko, J. Szpunar, *J. AOAC Int.*, **87** (2004), 233–237.

462. V. W. M. Lai, W. R. Cullen, C. F. Harrington, K. J Reimer, *Appl. Organomet. Chem.*, **11** (1997), 797–803.

463. S. Branch, L. Ebdon, P. O'Neill, *J. Anal. Atom. Spectrom.*, **9** (1994), 33–37.

464. K. L. Ackley, C. B'Hymer, K. L. Sutton, J. A. Caruso, *J. Anal. Atom. Spectrom.*, **14** (1999), 845–850.

465. K. Hanaoka, W. Goessler, K. Yoshida, Y. Fujitaka, T. Kaise, K. J. Irgolić, *Appl. Organomet. Chem.*, **13** (1999), 765–770.

466. J. W. McKierna n, J. T. Creed, C. A. Brockhoff, J. A. Caruso, R. M. Lorenzana, *J. Anal. Atom. Spectrom.*, **14** (1999), 607–613.

467. K. A. Francesconi, S. Khokiattiwong, W. Goessler, S. N. Pedersen, M. Pavkov, *Chem. Commun.*, **2000**, 1083–1084.

468. J. Kirby, W. Maher, *Appl. Organomet. Chem.*, **15** (2001), 108–115.

469. K. Wrobel, K. Wrobel, B. Parker, S. S. Kannamkumarath, J. A. Caruso, *Talanta*, **58** (2002), 899–907.

470. S. Slaets, F. Adams, I. R. Pereiro, R. Lobinski, *J. Anal. Atom. Spectrom.*, **14** (1999), 851–857.

471. Q. Tu, J. Qian, W. Frech, *J. Anal. Atom. Spectrom.*, **15** (2000), 1583–1588.

472. C. S. Chiou, S. J. Jiang, K. S. K. Danadurai, *Spectrochim. Acta*, **56B** (2001), 1133–1142.

473. R. Rai, W. Maher, F. Kirkowa, *J. Anal. Atom. Spectrom.*, **17** (2002), 1560–1563.

474. H. M. Crews, P. A. Clarke, D. J. Lewis, L. M. Owen, P. R. Strutt, A. J. Izquierdo, *J. Anal. Atom. Spectrom.*, **11** (1996), 1177–1182.

475. U. T. Kumar, J. G. Dorsey, J. A. Caruso, E. H. Evans, *J. Chromatogr. A*, **654** (1993), 261–268.

476. U. T. Kumar, N. P. Vela, J. G. Dorsey, J. A. Caruso, *J. Chromatogr. A*, **655** (1993), 340–345.

477. N. P. Vela, J. A. Caruso, *J. Anal. Atom. Spectrom.*, **11** (1996), 1129–1135.

478. D. Beauchemin, M. E. Bednas, S. S. Berman, J. W. McLaren, K. W. M. Siu, R. E. Sturgeon, *Anal. Chem.*, **60** (1988), 2209–2212.

479. D. Beauchemin, K. W. M. Siu, J. W. McLaren, S. S. Berman, *J. Anal. Atom. Spectrom.*, **4** (1989), 285–289.

480. S. L. Cleland, L. K. Olson, J. A. Caruso, J. M. Carey, *J. Anal. Atom. Spectrom.*, **9** (1994), 975–979.

481. J. Mattusch, R. Wennrich, *Anal. Chem.*, **70** (1998), 3649–3655.

482. F. Lagarde, M. B. Amran, M. J. F. Leroy, C. Demesmay, M. Olle, A. Lamotte, H. Muntau, P.Michel, P. Thomas, S. Caroli, E. Larsen, P. Bonner, G. Rauret, M. Foulkes, A. Howard, B. Griepink, E. A. Maier, *Fresenius J. Anal. Chem.*, **363** (1999), 18–22.

483. T. Lindemann, A. Prange, W. Dannecker, B. Neidhart, *Fresenius J. Anal. Chem.*, **368** (2000), 214–220.

484. K. Falk, H. Emons, *J. Anal. Atom. Spectrom.*, **15** (2000), 643–649.

485. T. Nakazato, T. Taniguchi, H. Tao, M. Tominaga, A. Miyazaki, *J. Anal. Atom. Spectrom.*, **15**, (2000), 1546–1552.

486. D. S. Bushee, *Analyst*, **113** (1988), 1167–1170.

487. C. W. Huang, S. J. Jiang, *J. Anal. Atom. Spectrom.*, **8** (1993), 681–686.

488. M. S. da Rocha, A. B. Soldado, E. Blanco, A. Sanz-Medel, *J. Anal. Atom. Spectrom.*, **16** (2001), 951–956.

489. W. Goessler, D. Kuehnelt, C. Schlagenhaufen, Z. Slejkovec, K. J. Irgolić, *J. Anal. Atom. Spectrom.*, **13** (1998), 183–187.

490. H. Hintelmann, R. D. Evans, *Fresenius J. Anal. Chem.*, **358** (1997), 378–385.

491. R. Falter, G. Ilgen, *Fresenius J. Anal. Chem.*, **358** (1997), 401–406.

492. L. Yang, V. Colombini, P. Maxwell, Z. Mester, R. E. Sturgeon, *J. Chromatogr. A*, **1011** (2003), 135–142.

493. B. Rosenkranz, G. O'Connor, E. H. Evans, *J. Anal. Atom. Spectrom.*, **15** (2000), 7–12.

494. J. Gailer, K. A. Francesconi, J. S. Edmonds, K. J. Irgolić, *Appl. Organomet. Chem.*, **9** (1995), 341–355.

495. T. Dagnac, A. Padró, R. Rubio, G. Rauret, *Talenta*, **48** (1999), 763–772.

496. K. A. Francesconi, J. Gailer, J. S. Edmonds, W. Goessler, K. J. Irgolić, *Comp. Biochem. Phys. C*, **122** (1999), 131–137.

497. P. Moreno, M. A. Quijano, A. M. Gutierrez, M. C. Perez-Conde, C. Camara, *J. Anal. Atom. Spectrom.*, **16**, (2001), 1044–1050.

498. P. Moreno, M. A. Quijano, A. M. Gutierrez, M. C. Perez-Conde, C. Camara, *Anal. Bioanal. Chem.* **374** (2002), 466–476.

499. V. W. M. Lai, W. R. Cullen, S. Ray, *Mar. Chem.*, **66** (1999), 81–89.

500. E. H. Larsen, G. Pritzi, S. H. Hansen, *J. Anal. Atom. Spectrom.*, **8** (1993), 1075–1083.

501. S. X. C. Le, W. R. Cullen, K. J. Reimer, *Environ. Sci. Technol.*, **28** (1994), 1598–1604.

502. P. Thomas, K. Sniatecki, *Fresenius J. Anal. Chem.*, **351** (1995), 410–414.

503. F. Lagarde, M. B. Amran, M. J. F. Leroy, C. Demesmay, M. Olle, A. Lamotte, H. Muntau, P. Michel, P. Thomas, S. Caroli, E. Larsen, P. Bonner, G. Rauret, M. Foulkes, A. Howard, B. Griepink, E. A. Maier, *Fresenius J. Anal. Chem.*, **363** (1999), 5–11.

504. U. Kohlmeyer, J. Kuballa, E. Jantzen, *Rapid Commun. Mass Spectrom.*, **16** (2002), 965–974.

505. W. H. Li, C. Wei, C. Zhang, M. Van Hulle, R. Cornelis, X. R. Zhang, *Food Chem. Toxicol.*, **41** (2003), 1103–1110.

506. J. J. Sloth, E. H. Larsen, K. Julshamn, *J. Anal. Atom. Spectrom.*, **18** (2003), 452–459.

507. X. C. Le, X. F. Li, V. Lai, M. Ma, S. Yalcin, J. Feldmann, *Spectrochim. Acta*, **53B** (1998), 899–909.

508. E. H. Larsen, *Fresenius J. Anal. Chem.*, **352** (1995), 582–588.

509. A. Chatterjee, *Talanta*, **51**, (2000), 303–314.

510. D. Beauchemin, K. W. M. Siu, S. S. Berman, *Anal. Chem.*, **60** (1988), 2587–2590.

511. H. Hintelmann, R. Falter, G. Ilgen, R. D. Evans, *Fresenius J. Anal. Chem.*, **358** (1997), 363–370.

512. S. N. Willie, D. C. Gregoire, R. E. Sturgeon, *Analyst*, **122** (1997), 751–754.

513. M. A. Quijano, P. Moreno, A. M. Gutierrez, M. C. Perez Conde, C. Camara, *J. Mass Spectrom.*, **35** (2000), 878–884.

514. P. A. Gallagher, X. Wei, J. A. Shoemaker, C. A. Brockhoff, J. T. Creed, *J. Anal. Atom. Spectrom.*, **14** (1999), 1829–1834.

515. J. Feldman, K. John, P. Pengprecha, *Fresenius J. Anal. Chem.*, **368** (2000), 116–121.

516. B. M. Gamble, P. A. Gallagher, J. A. Shoemaker, X. Wei, C. A. Schwegel, J. T. Creed, *Analyst*, **127** (2002), 781–785.

517. R. Tukai, W. A. Maher, I. J. McNaught, M. J. Ellwood, *Anal. Chim. Acta*, **457** (2002), 173–185.

518. M. Van Hulle, C. Zhang, X. R. Zhang, R. Cornelis, *Analyst*, **127** (2002), 634–640.

519. C. Wei, W. H. Li, C. Zhang, M. Van Hulle, R. Cornelis, X. R. Zhang, *J. Agric. Food Chem.*, **51** (2003), 5176–5182.

520. E. H. Larsen, S. Stürup, *J. Anal. Atom. Spectrom.*, **9** (1994), 1099–1105.

521. D. A. Hunter, W. Goessler, K. A. Francesconi, *Mar. Biol.*, **131** (1998), 543–552.

522. K. A. Francesconi, D. A. Hunter, B. Bachmann, G. Raber, W. Goessler, *Appl. Organometal. Chem.*, **13** (1999), 669–679.

523. L. S. Milstein, A. Essader, C. Murrell, E. D. Pellizzari, R. A. Fernando, J. H. Raymer, O. Akinbo, *J. Agric. Food Chem.*, **51** (2003), 4180–4184.

524. M. Pardo-Martinez, P. Vinas, A. Fisher, S. J. Hill, *Anal. Chim. Acta*, **441** (2001), 29–36.

525. N. P. Vela, D. T. Heitkemper, *J. AOAC Int.*, **87** (2004), 244–252.

526. S. Londesborough, J. Mattusch, R. Wennrich, *Fresenius J. Anal. Chem.*, **363** (1999), 577–581.

527. P. Zbinden, D. Andrey, C. Blake, *Atom. Spectrosc.*, **21** (2000), 205–215.

528. U. Kohlmeyer, E. Jantzen, J. Kuballa, S. Jakubik, *Anal. Bioanal. Chem.*, **377** (2003), 6–13.

9

DANISH MONITORING SYSTEM FOR FOODS 1998–2003. CONTENT OF As, Cd, Hg, Ni, Pb, AND Se AND DIETARY INTAKE BY CHILDREN AND ADULTS

ERIK H. LARSEN

National Food Institute, Technical University of Denmark, Department of Food Chemistry, Mørkhøj Bygade 19, DK-2860 Søborg, Denmark, E-mail: ehl@dfvf.dk

INGE ROKKJÆR

Danish Veterinary and Food Administration, Regional Laboratory, Sønderskovvej 5, DK-8520 Lystrup, Denmark

TUE CHRISTENSEN

National Food Institute, Technical University of Denmark, Department of Nutrition, Mørkhøj Bygade 19, DK-2860 Søborg, Denmark

The Determination of Chemical Elements in Food: Applications for Atomic and Mass Spectrometry.
Edited by Sergio Caroli

9.1 INTRODUCTION

The content and dietary intake of trace elements in foodstuffs sold on the Danish market has been repeatedly investigated in the Danish Food Monitoring Programmeme since its initiation in 1983 [1]. In this chapter the results of the fourth monitoring period, 1998–2003, are reported and compared with earlier monitoring periods. Twenty years of continued monitoring of toxic trace elements in food, namely As, Cd, Hg, Ni, Pb, and Se, have generated results of great importance for the assessment and handling of risks associated with the consumption of food sold on the Danish market. The data obtained represent a comprehensive benchmark of the concentration level and variation of the five toxic trace elements in commercially available foods. Based on this knowledge, specific cases of geographically localized food contamination caused by environmental contamination of soil or ambient air can be evaluated and any necessary action taken [2]. Finally, the monitoring data obtained for Danish foods provide a scientifically sound background for establishing legislation at the national and European level. The samples taken of each of the food items included in the monitoring programme were analyzed individually and in this way detailed information on the variation of trace elements in foods eaten by the Danes could be gained. The results were combined with food consumption data in order to estimate the Danes' dietary intake of trace elements [3]. In each 5 year period the same food items were sampled in order to monitor changes in the content of the trace elements of interest over time. The Danes' food consumption pattern was surveyed independently with some overlapping in time with the repeated monitoring cycles. This allowed for an estimation of the Danes' current dietary intake of trace elements.

The aims of the food-monitoring programme 1998–2003 were: (i) to monitor the content and changes over time of trace elements in food sold on the Danish market; and (ii) to estimate the intake of trace elements with the entire diet and also to evaluate if any associated health consequences were likely to occur.

9.2 METHODS OF SAMPLING, ANALYSIS, AND QUALITY ASSURANCE

The samples taken during the 1998–2003 period and the chemical analyses were organized in eight surveys, each covering food items from the same main category of foods. The number of samples of each food item was first of all dimensioned on the basis of the cost of the laboratory work. Secondly, the number of samples was decided on the basis of earlier experience regarding variation in trace element content, expected rate of change in content over time, and the expected contribution of the food item to the total trace element intake. The Public Food Inspection Services in various parts of Denmark were in charge of the nationwide sampling scheme. The types and numbers of food samples taken are given in Appendices 1–6.

9.2.1 Chemical Analyses and Quality Assurance

The regional laboratories located in Århus and Odense analyzed the content of As, Cd, Hg, Ni, Pb, and Se. The samples were prepared according to common household practice, but none of the foods were cooked prior to analysis. Only edible parts of the foods were used and adhered soil was removed by brushing under clean water. The methods of analysis involved isolation of the relevant tissue or part of sample using utensils that did not contaminate the samples. The homogenized food sample was dissolved by microwave (MW)-assisted wet-ashing in quartz vessels (Multiwave, Anton Paar, Austria) with HNO_3. Following further dilution to 100 mL volume by water ($<0.2\ \mu S\ cm^{-1}$), the trace element content was determined by inductively coupled plasma mass spectrometry (ICP-MS) using an Agilent 7500 instrument (Agilent Technologies, Waldbronn, Germany). A summary of the analytical settings used is given in Table 9.1. The analytical work was organized and run in batches comprising 15–20 unknown samples, one blank, one double determination for each 10 unknown samples, and one Certified Reference Material (CRM). Based on these controls, the Limit of Detection (LoD), precision, and accuracy were estimated for each of the eight surveys. In case of deviations from a set of criteria for tolerable variation of blanks and for values obtained for CRMs (x-charts) and for double determinations (R-charts), all

TABLE 9.1. Summary of ICP-MS Method of Analysis and Figures of Merit

Method Performance	Unit	As	Cd	Hg	Ni	Pb	Se
Mean blank concentration[a]	ng l^{-1} in solution	1.3	0.30	3.2	7.1	6.20	1.8
Limit of detection[b]	μg kg^{-1} wet weight	0.2	0.05	0.48	1.1	0.92	0.27
Relative standard deviation, s_r^c	%	1.6	3.0	2.3	3.9	3.4	2.9
Recovery[d]	%	100	103	97	101	101	107
		Analytical Method					
Isotopes detected	amu	75	114	202	60	208	82
Integration time per mass	ms	300	300	300	300	300	300
Repetitions		3	3	3	3	3	3
Internal standard		Te	Rh	Rh	Ge	Bi	Te

[a]Expressed as element concentration in solution.
[b]Expressed for 2 g (wet weight) sample intake and 100 mL final volume following the three sigma criterion.
[c]Mean value calculated from double determinations of element concentrations >10 LoD.
[d]Mean value for standard added to sample digest diluted to 100 mL volume.

the analyses in that batch were repeated. The LoDs, which were calculated according to the three-sigma criterion, were estimated from the variance of the analytical blank values. Results indicated by "less than" in Appendices 1–6 were below the LoD value for the analytical survey in which the result was produced. The repeatability has been summarized for each trace element in Table 9.1. The values correspond to the within-day random error for the analysis of trace element concentrations above 10 times the LoD.

9.2.2 Handling of Low Results and Data Analysis

Analytical results close to the LoD are associated with relative standard deviations larger than those given in Table 9.1; for values below the LoD, the confidence interval exceeds 100 percent of the value. Their mean value, which was the best approximation to the true concentration, was used as the best available figure for the true content. In contrast, using zero or the LoD value for these low concentrations would under- or overestimate, respectively, the trace element intake.

9.2.3 Estimation of the Dietary Trace Element Intake

The intake estimates are based on dietary intake data collected in the Danish nationwide food consumption survey of 2000–2002 [3]. The food consumption data were sampled throughout the 3 year period in order to account for possible seasonal variations in dietary habits. The representative sample of Danes included a total of 4120 respondents (2167 female and 1953 male) aged 4–75 years. The Danish nationwide food consumption survey used a 7 day prospective food record with a precoded (semi-closed) questionnaire with answering categories for the most commonly eaten foods and dishes in the Danish diet. The questionnaire was organized according to the normal meal pattern in a day. For food items not found in the precoded categories the participant was asked to write in the missing food. Amounts were given in household measures, for example, pieces, glasses, cups, spoons. Standard portion sizes were used in the conversion of the reported amounts to weight (grams). For many food items/dishes the participant had a choice of several portion sizes and was aided by pictures of portioned foods. Composite foods (e.g., dishes) were disaggregated into ingredients by means of 443 standard recipes. Due to the simplified design of the questionnaire, it was possible to represent the total diet by the intake of 314 food items. The final result of these conversions was then recalculated and expressed as the daily mean intake of the 7 day food registration for each participant in the survey.

The dietary trace element intake and its variation among all respondents or in specific age groups was estimated by summation of the food products consumed by each of the respondents, and the mean content of the six trace elements for each of the 314 reported food items. Since the monitoring system provided data for the trace element content only in 97 food items, additional data were gathered from other surveys of trace elements in food sold in Denmark. Furthermore, for food items that were not included in the analytical surveys, concentration data from similar foods were transferred. This was the case for several diary products and types of bread. No corrections for losses or gains during preparation of foods were applied, as the current knowledge about this is insufficient. The mean values of trace element content in the foods were used because they were assumed to reflect the consumer's average long-term exposure to a trace element in marketed foods. Individuals who mainly consume food originating from geographically localized areas, such as home-grown products, were not included in the intake estimations. For the purpose of comparing the intake estimates (micrograms per day) with the Provisional Tolerable Weekly Intake (PTWI) given as micrograms per body weight per week (WHO), the intake was expressed as micrograms per person per day by ratioing to the reported body mass for each respondent.

9.3 RESULTS

9.3.1 Chemical Analyses

The 7585 results for the content of As, Cd, Hg, Ni, Pb, and Se in 96 food items sampled during the 1998–2003 period are summarized in Appendices 1–6. The ICP-MS-based analytical method has an LoD, which for Hg and Pb is about 10-fold lower than the LoDs obtained by atomic absorption spectrometry (AAS) used in the former 1993–1997 monitoring period [4]. This improved detection capability has provided a better analytical accuracy for the above trace elements present in the low micrograms per kilogram concentration range. Since the concentrations of Hg and Pb in many food items are particularly low, the contributions to the total trace element intake from such foods have become more accurate. Furthermore, a set of results for trace elements in drinking water, which is an important source of trace element intake, has been generated in 2004 (unpublished data).

Although not originally planned as a part of the monitoring system, these comprehensive results have provided an improved and updated knowledge on drinking water (tap water) as a contributor to the total trace element intake. In summary, the change of method of analysis and the availability of newly generated data for trace element content in certain food groups have made possible a more comprehensive intake estimation of the six trace elements.

9.3.2 Cadmium

The data for Cd offer many modestly lower mean concentration values in the fourth period as compared with the third. Cadmium concentration in potatoes and carrots has been monitored since the beginning of the Danish food monitoring system in 1983. The results for these marker foods are shown in Figure 9.1*a*,*b*. The mean Cd content has diminished over the 20-year monitoring period by a factor of approximately 2–2.5 in these vegetables. The results for Cd in the kidneys of slaughter animals – one of the marker organs – is lower in comparison with the third monitoring period. Both sets of results for these marker foods of vegetable and animal origin indicate that the burden from Cd has decreased.

9.3.3 Lead

The Pb concentration in curly kale, a marker food for atmospheric deposition of Pb, has decreased by a factor of approximately 8 in 2003 in comparison with the 1980s [1]. The reduction of the Pb content found in kale as well as in spinach (not shown) coincides with the reduction in Pb emissions from combustion of leaded petrol in Denmark and other European countries. The remaining

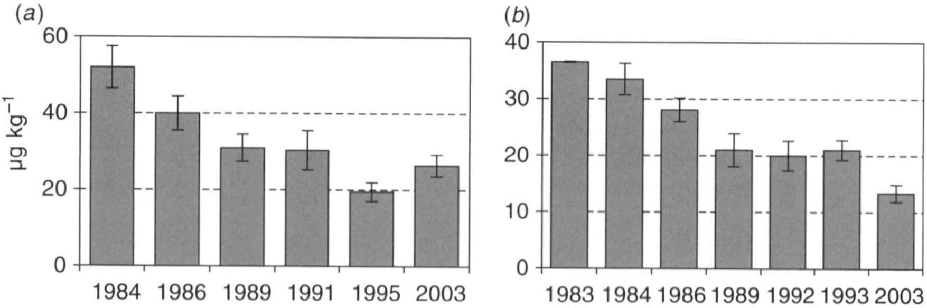

Figure 9.1. Mean concentration of cadmium in marker foods for the 1984–2003 period. (*a*) Carrots; (*b*) potatoes.

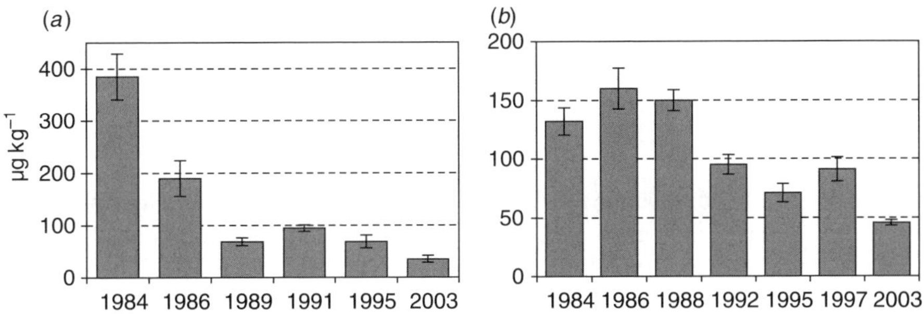

Figure 9.2. Mean concentration of Pb in marker foods for the 1984–2003 period. (*a*) Curly kale; (*b*) Ox kidneys.

contamination of leafy vegetables by this metal is likely to originate from a combination of atmospheric deposition of a variety of emission sources and from Pb-contaminated dust in the top-soil. The direct root uptake from contaminated soil is, however, very modest [2].

Similarly to Cd, Pb is taken up by animals' kidneys, which may be used to monitor the exposure of the animal to Pb via its fodder. A marked decline by approximately a factor of 3 in the mean Pb content has occurred for ox kidney, as shown in Figure 9.2*b*, but in pig kidney the Pb concentration remained unchanged between the third and the fourth monitoring period (data not shown).

9.3.4 Nickel

In general, the content of Ni in the food items listed in Appendix 3 ranges widely within the fourth monitoring period and also varies between periods [4]. The wide range of results prevents the finding of any significant changes in the Ni content, and the Ni content in rye bread, which was tested as a marker food for this metal, has not changed from the third monitoring period [5]. As in the case of Cd in drinking water, newly generated data (Table 9.2) has provided updated knowledge on the occurrence of Ni in tap water.

TABLE 9.2. Content (μg l^{-1}) of Four Trace Elements in Tap Water Sampled Throughout Denmark in the Spring of 2004. The LoD is the Limit of Detection of the ICP-MS Analytical Method Used

	N	Mean	Minimum	Maximum	LoD
Cd	208	0.11	<0.01	1.91	0.01
Hg	208	<0.01	<0.01	<0.01	<0.01
Ni	208	2.21	0.35	21.3	0.04
Pb	208	0.87	0.03	9.76	0.01

9.3.5 Mercury

The content of Hg in fish, contributing substantially to the dietary intake of this element, is present at a stable concentration level in comparison with the third monitoring period. Using plaice and cod as marker foods, the concentration of Hg in these fish species remains stable over the four monitoring periods. In geographically localized areas of the Danish seas, Hg emissions from chlorine-alkali and other chemical industries earlier led to an extremely elevated content of this element in marine biota including fish, particularly in the Sound [6]. These emission sources have been discontinued and for over a decade the Hg content in locally caught fish has decreased [7].

9.3.6 Selenium

The Se content in food is generally stable over the years, although large differences exist between foods of animal and plant origin. This can be explained by the conventional farming practice of supplementing the animals' fodder with Se. In contrast, the Se content of arable soil is not fortified by this element and consequently the content in cereals and vegetables is low. This is caused by the low natural Se content in Danish arable soil. The highest Se content was found in cabbage species such as curly kale, Brussels sprouts and cauliflower, which naturally concentrate Se.

9.3.7 Arsenic

Arsenic mainly occurs in marine foods as a result of natural processes in the sea. The content found in fish vary largely for the same fish species. Part of the variation in the As content in flounder, herring, and cod could be explained by salinity differences between the seas where the fish was caught [8]. In general, the As content was high in fish caught in waters with a high degree of salinity (the North Sea and the Kattegat) and low in more brackish waters (the Belt Sea and The Baltic).

9.3.8 Food Consumption Data

The reported dietary intake data for each of the 4120 respondents were transformed into a daily intake in grams of the 314 foods by the database system General Intake Estimation System (GIES), an in-house software system developed by the National Food Institute. The database can be queried for food intake data, intake of energy, nutrients and trace elements for an individual respondent or for any grouping of respondents. Table 9.3 shows the average intake of foods aggregated by food group in grams per day for all 4120 respondents.

TABLE 9.3. Consumption of Food (g day^{-1}) in Denmark Aggregated into Food Groups

Name of Food Group	Number of Foods per Group	Consumption (g day^{-1})
Milk and milk products	19	350.5
Cheese and cheese products	15	27.0
Ice cream	1	8.5
Cereals and cereal products	49	215.7
Vegetables and vegetable products	52	267.6
Fruit and fruit products	51	262.5
Meat and meat products	34	108.5
Fish and fish products	22	17.4
Poultry and poultry products	8	25.3
Eggs and egg products	4	15.7
Fatty foods	18	35.5
Sweets	14	32.5
Beverages	16	1916.6
Condiments and spices	10	4.9
Snacks	1	0.6
All food groups	314	3289

9.4 DISCUSSION

9.4.1 Trace Element Intake

The calculated dietary intake of As, Cd, Hg, Ni, and Se for Danes between 15 and 75 years of age is given in Table 9.4.

In general, the estimated intakes of Cd, Hg, Ni, and Pb for the fourth monitoring period are lower than those calculated for the three previous monitoring periods. This reduction may have been caused by a combination of changes in dietary habits as reported in the 2000–2002 national dietary survey, and by a reduced content of trace elements in the food items investigated. The dietary intake of the toxic trace elements by infants is listed in Table 9.5. Young children ingest more food than adults in relation to their body weight, and therefore the associated dietary intake of trace elements per kilogram of body weight amounts to higher values.

9.4.2 Cadmium

Cadmium may accumulate in the body, primarily in kidneys and liver, and has a half-life of several decades. The toxic effect occurs in the kidneys and may lead to proteinuria. A PTWI value has been established at 7 μg kg^{-1} body weight equivalent to 72 μg person^{-1} day^{-1} [9]. Cadmium has been classified as a

TABLE 9.4. Estimated Intakes by Adults (15–75 Years Old) of Six Trace Elements via the Danish Diet During Four Five-Year Monitoring Periods Covering 1983–2003

Monitoring Period		As		Cd		Hg		Ni	Pb		Se	
		μg day^{-1}	%PTWI[a]	μg day^{-1}	%PTWI	μg day^{-1}	%PTWI	μg day^{-1}	μg day^{-1}	%PTWI	μg day^{-1}	%NNR
1998–2003	Mean	62[b]	<2%	10	14%	2	4%	109	19	7%	43	87%
	95th percentile	227[b]		17	24%	4	8%	197	31	13%		
	5th percentile										23	
1993–1997	Mean			16	22%	4	7%	167	18	7%	48	96%
	95th percentile			24	33%	6	11%	278	28	11%		
	5th percentile										27	
1988–1992	Mean			17	24%	5	10%	157	27	11%	49	98%
	95th percentile			28	40%	9	18%	281	46	18%		
	5th percentile										25	
1983–1987	Mean	118	<4%	20	29%	7	14%	199	42	17%	51	102%
	95th percentile			32	46%	15	30%	252	76	30%		
	5th percentile										28	

The intakes are given as the mean and 95th percentile values or, for Se, as the mean and 5th percentile values. The PTWI and the Nordic Nutrition Recommendations (NNR) indicate upper tolerable and target intakes, respectively. For Ni neither recommended intake nor PTWI value exists.

[a]The assumption is made that less than 5% of the total content in seafood is inorganic arsenic [16] for which a PTWI value has been set.

[b]The estimated intake of arsenic is based on data for vegetables, meats, poultry, fish, and beverages.

TABLE 9.5. Estimated Intakes by Infants (4–6 Years) of Cd, Hg, and Pb Using Age-Specific Consumption Data from the Danish Dietary Survey for the 2000–2002 Period

		Cd		Hg		Pb	
	N	μg day^{-1}	% PTWI	μg day^{-1}	% PTWI	μg day^{-1}	% PTWI
Mean	230	7.7	36%	1.3	16%	9.7	13%
95th percentile		11.9	57%	2.9	45%	15.4	20%

carcinogen when inhaled, but such an effect was neither substantiated nor disproved when the exposure was oral [10]. The mean and 95th percentile of Cd intakes (Table 9.4), which have been estimated at 10 and 17 μg day^{-1} for the 1998–2003 period, are equivalent to 14 and 24 percent of the PTWI value, respectively. The median Cd intake of 9.6 μg day is in good accordance with the mean value. The food groups that contribute the most to the intake are bread and cereals followed by vegetables, as shown in Figure 9.3. The decrease in Cd intake in comparison with the previous monitoring period mainly regards bread and cereals, vegetables and beverages including drinking water [4].

For children of 4–6 years of age the mean and 95th percentile intake of cadmium (Table 9.5) at 36 and 57 percent of the PTWI, respectively, are relatively high. The margin is modest between the young child's intake of Cd via food in Denmark and the PTWI value that corresponds to the adverse effect level among the most sensitive individuals. When setting the PTWI value an average

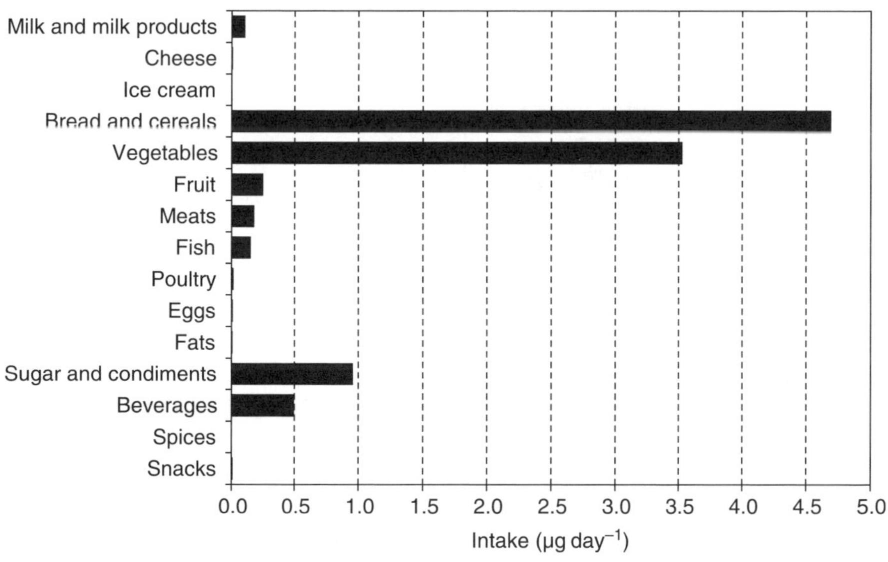

Figure 9.3. Intake of Cd from main food groups by Danes aged 15–75 years.

intestinal absorption rate of 5 percent was assumed. This value may vary depending on factors such as food matrix and the chemical species of Cd present in the food. For a more detailed risk assessment, studies are needed on Cd absorption rate and speciation in the foods that contribute most to the intake.

9.4.3 Lead

Ingested Pb is accumulated in the body and the most sensitive adverse effect is associated with the maldevelopment of the central nervous system in the fetus and newborn child. A possible association between increased Pb content in blood and reduced intelligence quotient has been substantiated and a lower threshold value could not be set. A PTWI value has been established at 25 $\mu g\ kg^{-1}$ body weight, which is equivalent to 257 $\mu g\ person^{-1}\ day^{-1}$ [11]. The mean and 95th percentile intake of Pb (Table 9.4), which have been estimated at 19 and 31 $\mu g\ day^{-1}$ for the 1998–2003 period, are equivalent to 7 and 13 percent of the PTWI value, respectively. The median intake value, which is 16 $\mu g\ day^{-1}$, is in good agreement with the mean value. The food groups that mostly contribute to the Pb intake are beverages followed by vegetables, bread and cereals, fruit, and sugars, as shown in Figure 9.4.

The dietary intake of Pb is unchanged in comparison with the third monitoring period, but much lower than those estimated for the first and second monitoring periods (Table 9.4). This is consistent with the general decrease in the Pb content in foods in the same time period. The dietary Pb intake for adults is not a matter of concern.

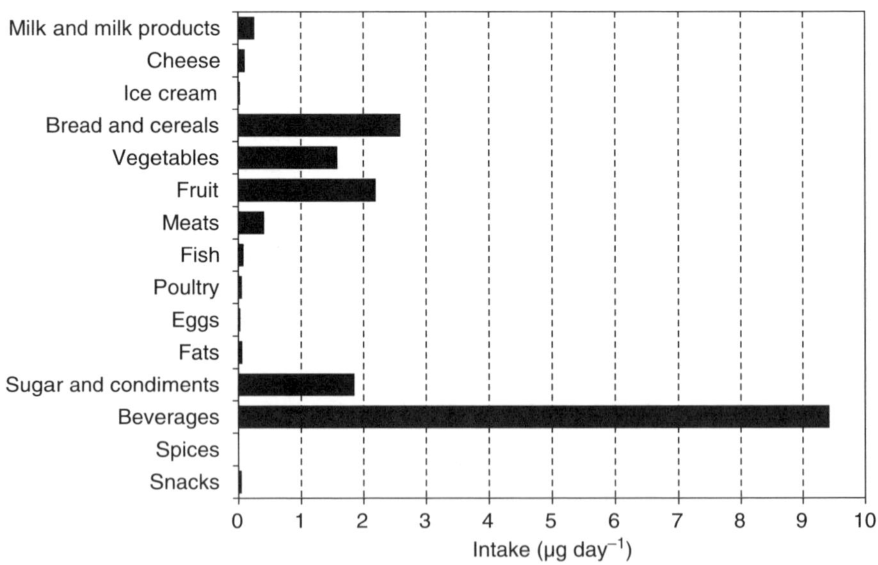

Figure 9.4. Intake of Pb from main food groups by Danes aged 15–75 years.

The food consumption of a 2-year-old child with a body weight of 15 kg averages 59 percent of that of an adult [3]. The assumption that the child's diet is composed of the same foods as that of the adult is, however, an approximation because of known differences regarding certain Pb-containing food items, such as wine and offal. Therefore, a conservative estimate of the mean and 95th percentile Pb intake for the 2-year-old is equal to or less than 11 μg day^{-1} (20 percent of the PTWI) and 17 μg day^{-1} (31 percent of the PTWI), respectively. The tendency of small children to ingest soil or inhale Pb-containing dust from, for example, play grounds or house dust represents an additional source of Pb exposure that should be kept to a minimum. If not strictly observed, the PTWI value may be exceeded when these environmental Pb sources of exposure are accounted for as well.

For a 4–6-year-old child however, the mean and 95th percentile of Pb intakes estimated on the basis of age-specific consumption data from the 2000–2002 dietary survey (Table 9.5) correspond to 13 and 20 percent of the PTWI value, respectively.

9.4.4 Nickel

The dietary intake of Ni does not represent any health risk to the general population. Although troublesome to some sensitized individuals, a tolerable oral intake of this metal has not been established. The mean and 95th percentile intakes of Ni estimated at 104 and 190 μg day^{-1}, respectively (Table 9.4) are lower than for the three previous monitoring periods.

The variation in Ni intake, as shown in the histogram in Figure 9.5, has a median value of 95 μg day^{-1}, but the upper tail indicates that some individuals are exposed to up to 350 μg day^{-1}. The food groups that contribute the most to the intake of Ni are beverages followed by cereals and milk, as shown in Figure 9.6. Considering the risk of additional Ni exposure from metallic surfaces in direct contact with food, for example, water kettles or Ni-containing kitchen utensils, adverse effects among sensitized individuals may occur. More importantly, the Ni content in tap water (Table 9.2), which might vary due to its

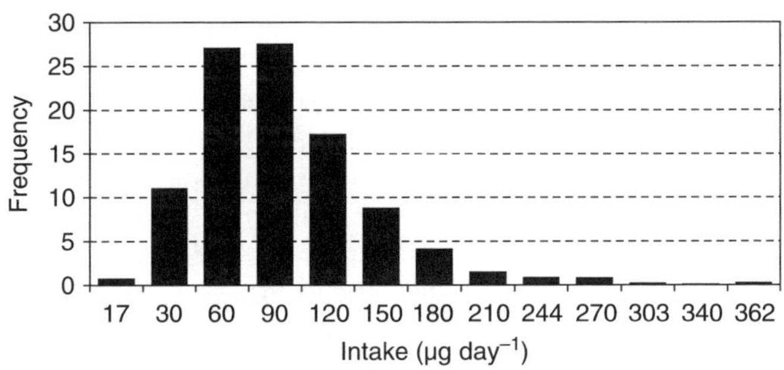

Figure 9.5. Distribution in Ni intake in the Danes aged 15–75 years.

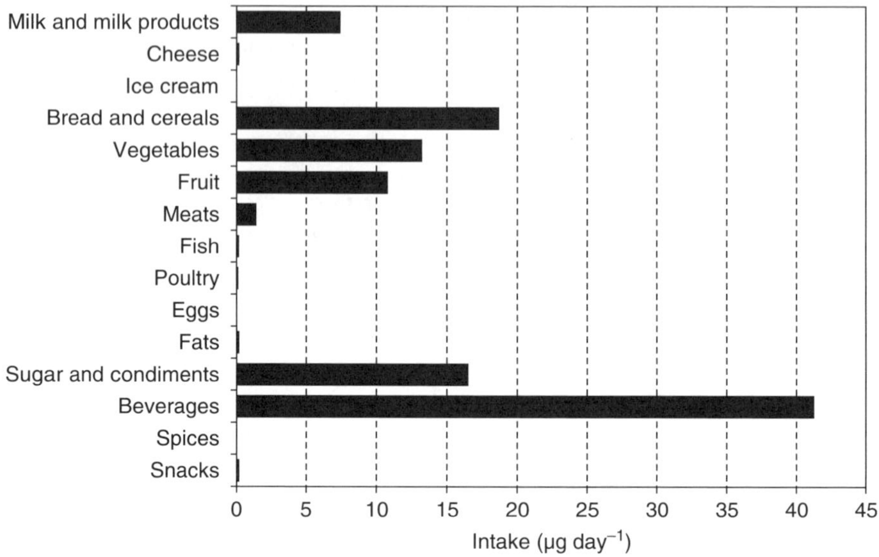

Figure 9.6. Intake of Ni from main food groups by Danes aged 15–75 years.

migration from water installations, may result in an intake not reflected by the present mean estimate. In cases of high Ni migration from drinking water installations, the intake via tap water may, for certain individuals, increase by 20–30 μg day^{-1}. The food groups that contribute the most to the intake of Ni are beverages, followed by bread, cereals and sugars.

9.4.5 Mercury

Ingested Hg accumulates in the body. The most toxic species is methyl-mercury, which occurs in fish. The adverse effect of inorganic Hg first shows in the kidneys while methyl-mercury may affect the central nervous system. A PTWI value has been established at 5 $\mu g\ kg^{-1}$ body weight per week for Hg in general, and 1.6 $\mu g\ kg^{-1}$ body weight per week specifically for methyl-mercury [12]. The mean and 95th percentile dietary intakes of Hg, which have been estimated at 1.9 and 4.1 $\mu g\ day^{-1}$ for the 1998–2003 period (Table 9.4), have been evaluated in the following way. About 60 percent of the mean mercury intake (1.1 $\mu g\ day^{-1}$) originates from fish (Fig. 9.7). It is assumed that all mercury contained in fish is present as methyl-mercury, whereas the mercury in all other foods occurs as inorganic Hg. Since fish is the main contributor to the total Hg intake and the PTWI value for methyl-mercury is lower than that for Hg, the estimated intake of this metal as methyl-mercury from fish will lead to the most restrictive risk assessment. As indicated in Table 9.4, the methyl-mercury fraction of the total Hg intake corresponds to 12 percent of the PTWI for the mean and to 40 percent of the PTWI for the 95th percentile intakes.

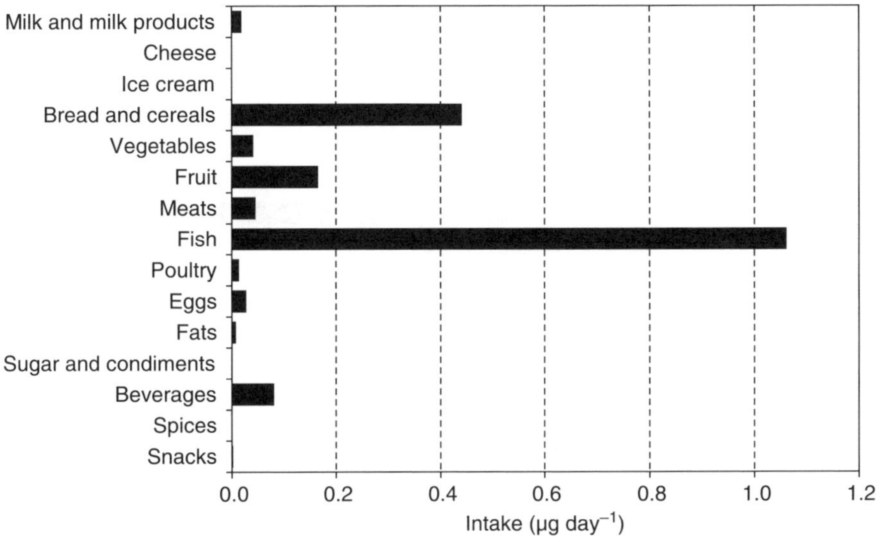

Figure 9.7. Intake of Hg from main food groups by Danes aged 15–75 years.

In comparison with previous monitoring periods, the Hg intake was evaluated more seriously during the 1998–2003 monitoring period. This is because of the newly issued, lower tolerable intake for methyl-mercury which reflects the greater toxicological concern about the Hg species. For a 4–6 year-old child the mean and 95th percentile Hg intake have been estimated at 16 and 45 percent of the PTWI value, respectively.

9.4.6 Selenium

Selenium is an essential element for humans and is a constituent of selenoamino acids contained in selenoenzymes, such as glutathione peroxidases. This group of enzymes catalyzes the reduction of peroxides and thereby takes part in the anti-oxidative defense of the body. A severely low Se intake may cause heart disease (Keshan's disease). Therefore, possible adverse health effects are associated with a low intake of this element and justify the focus on the low percentile Se intake in the population. The food groups that contribute the most to the Se intake are meat (including offal) followed by bread and cereals, and fish, as shown in Figure 9.8.

The estimated mean and 5th percentile intake of Se in the 1998–2003 period were 43 and 23 $\mu g\ day^{-1}$, respectively, and the Se intake have shown only small changes with a decreasing tendency over that past 20 years (Table 9.4). Selenium intake is in good agreement with the recommended level of 40 and 50 $\mu g\ day^{-1}$ for women and men, respectively. Few individuals have an intake below the lower intake of 20 $\mu g\ day^{-1}$ [13]. The recommendations do not, however, take into consideration the possible cancer-preventive effect of Se at higher doses [14].

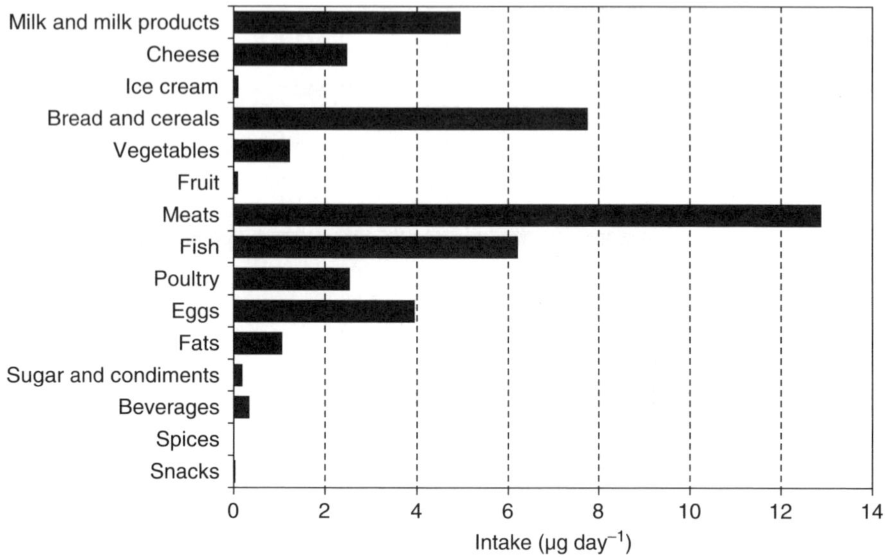

Figure 9.8. Intake of Se from main food groups by Danes aged 15–75 years.

9.4.7 Arsenic

Being a potentially toxic element in its inorganic chemical forms, As occurs in seafood as the nontoxic arsenobetaine and other minor organoarsenicals [9, 15]. Inorganic As, however, may lead to the development of skin cancer. From the total diet given in Table 9.4, the intake of As was estimated in the first monitoring period at 118 $\mu g\ day^{-1}$ [1]. In the two following monitoring periods (second and

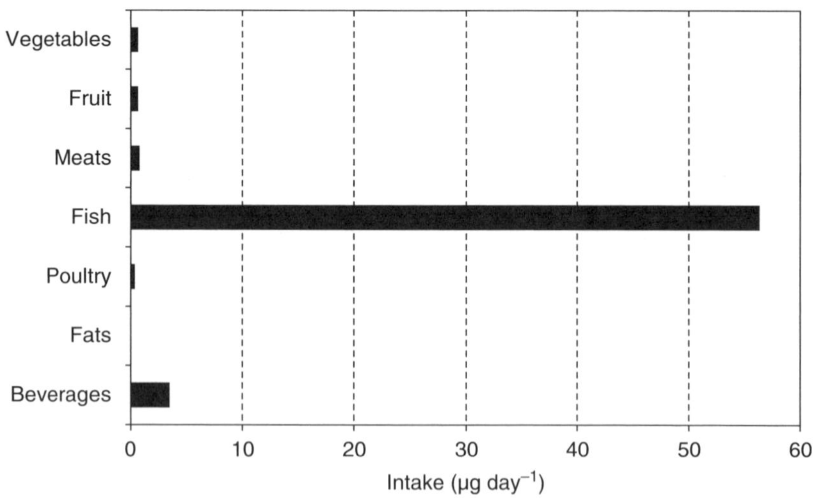

Figure 9.9. Intake of As from main food groups by Danes aged 15–75 years.

third) only the As content in fish was followed and consequently a re-evaluation of the total dietary As intake was not possible. In the fourth period, however, a wider range of foods were included in the programme (Appendix 6) and the As intake from these food groups was estimated at 62 μg day^{-1} (Table 9.4). Most of the intake (91 percent of the total intake) originates from fish, as shown in Figure 9.9.

Assuming the inorganic As occurring in fish and other seafood products as 5 percent of the total As, the intake of the inorganic forms via seafood corresponds to 2 percent of the PTWI value for inorganic As, which is 15 μg kg^{-1} body weight per week or 154 μg per person per day [16, 17]. A more relevant method for the monitoring of As in seafood in relation to human health risk assessment should therefore focus on the speciation of the inorganic As species, rather than on the total As content [18].

9.5 CONCLUSIONS

The monitoring programme comprises the trace elements As, Cd, Hg, Ni, Pb, and Se. A total of 7585 samples of the 98 foods most frequently consumed by the Danes were sampled on the Danish market and analyzed. The results of the monitoring programme for the 1998–2003 period showed that the content of Cd, Hg, and Pb in foods marketed in Denmark remains at a stable concentration level in comparison with the previous monitoring period. As opposed to these trace elements – which to some extent are present in food due to environmental contamination – the content of As, Ni, and Se is mainly the result of natural biogeochemical processes and supplementation. The estimated adult Cd and Hg intakes amount to a relatively high fraction of the respective PTWI values. Keeping in mind that an intake equal to the PTWI value corresponds to the adverse effect intake level, and that the tolerable intake of Cd might be reduced in the future, further monitoring of those food items that contribute the most to the dietary intake is warranted.

The trace element intake per kg of body weight by children, however, is even higher than that for adults. For this reason it is important to continue the study of the dietary habits of children, including children 2–4 years old. Such consumption data will allow for an important future estimation of the intake of those trace elements that pose a particular risk to young children, that is, Hg and Pb.

REFERENCES

1. National Food Agency. Food monitoring in Denmark. Nutrients and contaminants 1983–1987. Publication No. 195, 1990. Available from: Danish Veterinary and Food Administration, Søborg.
2. L. Samsøe-Petersen, E. H. Larsen, P. B. Larsen, P. Bruun. Uptake of trace elements and PAHs by fruit and vegetables from contaminated soils, *Env. Sci. Technol.*, **36** (2002), 3057–3063.

3. Danish Institute for Food and Veterinary Research. The Danes' dietary habits 2000–2002. Main results, 2005. Available from: Danish Institute for Food and Veterinary Research, Søborg.
4. E. H. Larsen, N. L. Andersen, A. Møller, A. Petersen, G. K. Mortensen, J. Petersen, Monitoring the content and intake of trace elements from food in Denmark, *Food. Addit. Contam.*, **19** (2002), 33–46.
5. Danish Veterinary and Food Administration. Monitoring system for foods, 1993–1997, Part 2, Chemical contaminants, 2001. Available from: Danish Veterinary and Food Administration, Søborg.
6. L. A. Jørgensen, B. Pedersen, Trace metals in fish used for time trend analysis and as environmental indicators, *Mar. Pollut. Bull.*, **28** (1994), 235–243.
7. National Food Agency. Contamination of Fish from Øresund. Mercury in cod and flounder, Publication No. ILF 1994.10, 1994. Available from: Danish Institute for Food and Veterinary Research, Søborg.
8. E. H. Larsen, K. A. Francesconi, Arsenic concentrations correlate with salinity for fish taken from the North Sea and Baltic waters, *J. Mar. Biol. Ass. UK*, **83** (2003), 283–284.
9. FAO/WHO, Toxicological evaluation of certain food additives and contaminants. WHO Food Additive Series, 24, 1989. Available from: World Health Organization, Geneva, Switzerland.
10. Scientific Committee for Food. Reports from the Scientific Committee for Food, Thirty-sixth Series, 1995. Available from: The European Commission, Luxembourg.
11. FAO/WHO, Evaluation of certain food additives and contaminants, WHO Technical Report Series, 837, 1993. Available from: World Health Organization, Geneva.
12. http://www.inchem.org/documents/jecfa/jeceval/jec_1307.htm.
13. Nordic Council of Ministers. Nordic Nutrition Recommendations NNR 2004-integrating nutrition and physical activity. Nord 2004:013, 2004. Available from Nordic Council of Ministers, Copenhagen.
14. L. C. Clark, G. F. Combs, B. W. Turnbull, E. H. Slate, D. K. Chalker, J. Chow, L. S. Davis, R. A. Glover, G. F. Graham E. G. Gross, A. Krongrad, J. L. Lesher, H. K. Park, B. B. Sanders, C. L. Smith, J. R. Taylor, Nutritional Prevention of Cancer Study Group, Effects of selenium supplementation for cancer prevention in patients with carcinoma of the skin. A randomized controlled trial. *J. Am. Med. Assoc.*, **276** (1996), 1957–1963.
15. E. H. Larsen, G. Pritzl, S. H. Hansen, Arsenic speciation in seafood samples with emphasis on minor constituents: an investigation using high-performance liquid chromatography with detection by inductively coupled plasma mass spectrometry, *J. Anal. Atom. Spectrom.*, **8** (1993), 1075–1084.
16. J. S. Edmonds, K. J. Francesconi, Arsenic in seafood: human health aspects and regulations, *Mar. Pollut. Bull.*, **26** (1993), 665–674.
17. FAO/WHO. Toxicological evaluation of certain food additives and contaminants, WHO Food Additive Series, 24, 1989. Available from: World Health Organization, Geneva.
18. Larsen EH, Berg T. Trace element speciation and international food legislation – A Codex Alimentarius position paper on arsenic as a contaminant. In L. Ebdon, H. Crews, R. Cornelis, O. F. X. Donard, Ph. Quevauviller (eds), *Trace Element Speciation for Environment, Food and Health*, Cambridge, Royal Chemical Society, 2001, 251–260.

APPENDIX 1

TABLE 9.A1. Content of Pb (μg kg^{-1} Fresh Weight) in Foods Sampled in the Fourth Monitoring Period (1998–2003)

Foodstuff	Number	Number <LoD	Average	Minimum	Maximum	Median	0.90 Quantile
			Vegetables				
Aubergines	10	2	6.2	<0.4	48.2	1.1	9.8
Beans	9	0	3.1	0.6	8.7	2.7	4.9
Beetroots	10	0	5.6	3.5	8.1	5.4	7.8
Broccoli	10	1	5.3	<0.4	23.5	2.3	9.5
Brussel sprouts	9	0	5.8	2.2	20.5	3.8	8.7
Cabbage	11	0	1.2	0.5	5.3	0.8	1.5
Carrots	39	0	8.8	0.5	34.0	6.1	22.5
Cauliflower	10	4	1.1	<0.8	2.2	1.0	2.1
Celeriac	9	0	3.2	0.8	5.8	3.5	4.1
Celery	12	0	3.0	1.0	7.4	2.3	6.3
Chinese cabbage	10	0	1.6	0.4	4.0	1.3	2.8
Cucumber	10	0	1.4	0.6	3.4	1.3	2.4
Curly kale	15	0	35.4	5.1	79.5	32.7	77.7
Leeks	11	0	2.1	0.6	3.6	2.2	3.4
Lettuce	31	0	16.3	0.5	139	8.1	35.9
Mushrooms	10	0	2.0	0.8	3.9	1.9	3.1
Onions	20	0	1.9	0.7	8.0	1.3	2.9
Peas	10	0	3.6	1.1	9.0	3.1	6.4
Pepper	13	0	1.9	1.1	4.5	1.7	2.8
Potatoes	30	8	1.7	<0.8	7.2	1.5	2.9
Rhubarb	8	0	47.0	23.5	87.6	33.9	85.6
Spinach	28	0	22.4	2.4	92.7	12.4	52.5
Squash	10	0	1.5	0.6	2.9	1.2	2.7
Tomato	20	0	1.6	0.5	5.7	0.9	3.0
			Offal				
Liver, calf	20	0	22.6	6.0	60.9	18.5	45.6
Liver, chicken	12	10	<6.0	<6.0	6.2	<6.0	<6.0
Liver, ox	14	0	19.0	9.7	35.6	18.0	26.7
Liver, pig	19	9	6.8	<6.0	20.0	6.0	9.8
Kidney, calf	23	0	46.0	22.0	80.9	42.6	71.5
Kidney, ox	43	0	45.7	20.0	84.9	43.8	66.5
Kidney, pig	32	10	9.8	<6.0	56.8	7.3	14.0
			Meat Products				
Pate, meat	10	0	6.5	2.5	12.0	6.4	8.3
Pork, flank, cooked	10	2	3.9	<2.0	13.0	3.3	5.3
Pork, liver paste	10	0	5.1	2.7	8.4	5.0	7.9

(Continued)

TABLE 9.A1. *Continued*

Foodstuff	Number	Number <LoD	Average	Minimum	Maximum	Median	0.90 Quantile
Pork, mettwurst, smoked	11	2	3.6	<2.0	7.3	3.1	6.8
Pork, saddle, smoked	10	8	<2.0	<2.0	3.4	<2.0	2.7
Pork, sausage, frankfurter	13	2	9.7	<2.0	70.7	3.8	9.3
Pork, sausage, saveloy	9	1	6.3	<2.0	27.0	3.8	11.1
Sausage, chicken	11	1	4.7	<2.0	9.2	4.1	7.4
Sausage, salami	10	2	4.1	<2.0	11.0	3.5	7.2
Sausage, turkey	8	2	3.0	<2.0	4.3	3.2	4.0
			Meat				
Beef	41	18	<5.0	<5.0	29.0	<5.0	9.5
Chicken	30	23	<5.0	<5.0	15.0	<5.0	6.8
Lamb	7	6	<5.0	<5.0	5.5	<5.0	<5.0
Mutton	2	2	<5.0	<5.0	<5.0	<5.0	<5.0
Pork	43	30	<5.0	<5.0	24.6	<5.0	<5.0
Turkey	12	11	<5.0	<5.0	5.0	<5.0	<5.0
Veal	18	10	<5.0	<5.0	18.0	<5.0	<5.0
			Fish				
Cod	30	19	3.7	<2.0	37.0	<2.0	7.0
Cod liver	10	4	5.0	<2.0	14.0	3.5	10.0
Eel	8	1	6.1	<2.0	12.0	4.2	12.0
Flounder	22	12	2.6	<2.0	13.0	<2.0	5.9
Garfish	15	5	3.8	<2.0	23.6	2.6	4.3
Herring	18	2	8.5	<2.0	30.3	5.4	20.7
Mackerel	22	17	<2.0	<2.0	10.0	<2.0	5.6
Mussel	15	0	142	31	383	144	180
Plaice	21	13	<2.0	<2.0	6.5	<2.0	3.1
Shrimp	10	3	4.3	<2.0	7.0	4.6	6.7
Tuna	12	8	<2.0	<2.0	4.8	<2.0	3.1
			Beverages				
Apple juice	6	0	5.7	1.3	10.9	5.2	10.2
Fruit juice	6	0	39.5	4.0	207	6.9	108
Orange juice	11	4	4.2	<0.6	15.2	1.9	13.3
Red wine	15	0	28.4	12.9	43.8	26.8	43.0
Soft drinks	21	12	1.4	<0.6	11.3	<0.6	1.8
Water, carbonated	4	2	0.8	<0.6	2.1	<0.6	1.7
White wine	10	0	30.5	16.5	58.0	27.8	47.6

(*Continued*)

TABLE 9.A1. *Continued*

Foodstuff	Number	Number <LoD	Average	Minimum	Maximum	Median	0.90 Quantile
			Dairy Products				
Blended spread	5	4	<4.0	<4.0	5.0	<4.0	<4.0
Butter	5	5	<4.0	<4.0	<4.0	<4.0	<4.0
Cheese, brie	10	4	3.5	<2.0	10.6	2.5	7.6
Cheese, firm, Danbo	10	3	3.4	<3.0	4.2	3.4	4.1
Cocoa, instant	9	0	2.7	1.5	5.8	2.2	4.5
Cream, whipping	10	6	<2.0	<2.0	3.9	<2.0	3.7
Ice cream, dairy	11	5	2.9	<2.0	11.6	2.3	3.6
Milk, buttermilk	10	5	<0.4	<0.4	0.6	<0.4	0.5
Milk, partly skimmed	10	5	0.5	<0.4	1.8	<0.4	0.8
Milk, skimmed	10	4	0.8	<0.4	3.1	0.5	1.3
Milk, skimmed with chocolate	2	0	2.3	2.3	2.4	2.3	2.3
Milk, whole	10	8	<0.5	<0.5	1.0	<0.5	0.6
Ymer	10	4	1.7	<0.6	7.8	0.6	3.9
Yogurt	11	7	<0.6	<0.6	1.9	<0.6	1.1
			Bread and Cereals				
Bread, white	10	7	<20.0	<20.0	42.0	<20.0	29.4
Bread, white coarse grain	10	8	<20.0	<20.0	28.0	<20.0	27.1
Bread, white roll	10	7	<20.0	<20.0	27.0	<20.0	20.7
Bread, wholemeal	10	10	<20.0	<20.0	<20.0	<20.0	<20.0
Breakfast cereal	10	7	<20.0	<20.0	41.0	<20.0	26.6
Corn flakes	10	8	<20.0	<20.0	61.0	<20.0	25.0
Crispbread, rye	7	7	<20.0	<20.0	<20.0	<20.0	<20.0
Crispbread, wheat	3	3	<20.0	<20.0	<20.0	<20.0	<20.0
Oats, rolled	10	10	<20.0	<20.0	59.0	<20.0	<20.0
Pasta	10	6	21.0	<20.0	59.0	<20.0	55.4
Rice, polished	6	6	<20.0	<20.0	<20.0	<20.0	<20.0
Rye bread	10	9	<20.0	<20.0	27.0	<20.0	<20.0
Rye bread, wholemeal	20	19	<20.0	<20.0	24.0	<20.0	<20.0
Rye flour, whole meal	10	7	<20.0	<20.0	33.0	<20.0	25.8
Wheat flour	10	8	<20.0	<20.0	20.0	<20.0	20.0

APPENDIX 2

TABLE 9.A2. Content of Cd (μg kg^{-1} Fresh Weight) in Foods Sampled in the Fourth Monitoring Period (1998–2003)

Foodstuff	Number	Number <LoD	Average	Minimum	Maximum	Median	0.90 Quantile
			Vegetables				
Aubergines	10	1	2.9	<0.3	16.9	0.9	4.4
Beans	9	1	1.0	<0.3	2.2	1.0	1.8
Beetroots	10	0	28.0	12.5	42.5	28.0	37.0
Broccoli	10	0	6.0	2.6	10.6	5.7	9.8
Brussel sprouts	9	0	7.3	5.0	11.7	6.7	10.2
Cabbage	11	0	4.1	1.4	9.3	3.8	6.9
Carrots	39	0	26.4	2.1	62.8	23.5	56.3
Cauliflower	10	0	4.4	2.1	7.6	3.9	7.0
Celeriac	9	1	63.5	<0.3	116	58.9	108
Celery	12	0	21.5	1.3	49.4	20.8	38.4
Chinese cabbage	10	0	13.9	6.2	36.1	10.9	20.6
Cucumber	10	8	<0.3	<0.3	1.7	<0.3	0.5
Curly kale	15	0	16.7	3.9	50.2	11.7	29.5
Leeks	11	0	19.8	1.8	49.2	19.2	37.5
Lettuce	31	0	21.3	3.5	69.6	15.3	40.6
Mushrooms	10	0	13.5	3.9	38.2	9.4	25.9
Onions	20	0	16.4	5.3	40.6	15.1	25.3
Peas	10	0	3.3	1.4	8.5	2.4	5.1
Pepper	13	4	2.8	<0.3	13.6	0.5	9.4
Potatoes	30	0	13.4	0.8	37.6	12.7	20.8
Rhubarb	8	0	24.2	8.1	44.7	20.5	40.7
Spinach	28	0	63.8	<0.3	278	58.4	92.5
Squash	10	3	1.7	<0.3	9.4	0.9	2.3
Tomato	20	9	3.1	<0.3	21.8	0.6	7.7
			Offal				
Liver, calf	20	0	40.8	11.3	239	24.6	52.7
Liver, chicken	12	0	15.6	6.7	46.9	14.2	18.2
Liver, ox	14	0	64.2	11.7	174	59.9	91.8
Liver, pig	19	0	25.1	13.6	47.5	23.8	33.9
Kidney, calf	23	0	126	44.0	448	105	195
Kidney, ox	43	0	434	29.0	2710	345	643
Kidney, pig	32	0	206	61.7	461	176	349
			Meat Products				
Pate, meat	10	0	9.4	3.9	18.1	7.7	17.2
Pork, flank, cooked	10	9	1.1	<1.0	2.8	<1.0	2.0

(Continued)

TABLE 9.A2. *Continued*

Foodstuff	Number	Number <LoD	Average	Minimum	Maximum	Median	0.90 Quantile
Pork, liver paste	10	0	12.7	10.6	17.5	11.9	15.3
Pork, mettwurst, smoked	11	2	2.7	<1.0	4.0	3.2	3.8
Pork, saddle, smoked	10	10	<1.0	<1.0	1.3	<1.0	1.0
Pork, sausage, frankfurter	13	8	2.1	<1.0	5.1	1.8	4.9
Pork, sausage, saveloy	9	5	2.0	<1.0	3.8	1.9	3.2
Sausage, chicken	11	4	2.8	<1.0	6.6	2.9	5.3
Sausage, salami	10	9	<1.0	<1.0	2.5	<1.0	1.8
Sausage, turkey	8	6	1.2	<1.0	2.6	1.2	2.2
Meat							
Beef	41	40	<3.0	<3.0	<3.0	<3.0	<3.0
Chicken	30	30	<3.0	<3.0	<3.0	<3.0	<3.0
Lamb	7	7	<3.0	<3.0	<3.0	<3.0	<3.0
Mutton	2	2	<3.0	<3.0	<3.0	<3.0	<3.0
Pork	43	43	<3.0	<3.0	<3.0	<3.0	<3.0
Turkey	12	12	<3.0	<3.0	<3.0	<3.0	<3.0
Veal	18	18	<3.0	<3.0	<3.0	<3.0	<3.0
Fish							
Cod	30	29	<0.6	<0.6	0.8	<0.6	<0.6
Cod liver	10	0	24.8	2.9	75.0	18.0	43.0
Eel	8	4	0.6	<0.6	1.5	0.6	1.2
Flounder	22	9	1.1	<0.6	7.4	0.6	1.8
Garfish	15	7	1.4	<0.6	5.0	0.7	3.1
Herring	18	0	4.3	0.7	13.8	4.0	7.6
Mackerel	22	0	3.9	1.2	13.1	3.2	4.3
Mussel	15	0	101	54.8	169	95.8	140
Plaice	21	16	<0.6	<0.6	1.4	<0.6	1.1
Shrimp	10	1	25.8	<0.6	56.1	26.3	50.6
Tuna	12	0	11.6	5.4	32.4	7.6	20.6
Beverages							
Apple juice	6	0	0.4	0.3	0.5	0.4	0.5
Beer, lager	15	13	<0.1	<0.1	0.3	<0.1	<0.1
Fruit juice	6	0	2.1	0.1	8.9	0.6	5.7
Orange juice	11	6	0.1	<0.1	0.4	<0.1	0.4

(*Continued*)

TABLE 9.A2. *Continued*

Foodstuff	Number	Number <LoD	Average	Minimum	Maximum	Median	0.90 Quantile
Red wine	15	1	0.3	<0.1	0.6	0.3	0.4
Soft drinks	21	21	<0.1	<0.1	<0.1	<0.1	<0.1
Water, carbonated	4	4	<0.1	<0.1	<0.1	<0.1	<0.1
White wine	10	0	0.3	0.1	0.5	0.3	0.5
			Dairy Products				
Blended spread	5	5	<3.0	<3.0	<3.0	<3.0	<3.0
Butter	5	5	<3.0	<3.0	<3.0	<3.0	<3.0
Cheese, brie	10	10	<1.0	<1.0	<1.0	<1.0	<1.0
Cheese, firm, Danbo	10	9	<1.0	<1.0	1.1	<1.0	<1.0
Cocoa, instant	9	0	2.9	2.2	3.4	2.9	3.3
Cream, whipping	10	10	<1.0	<1.0	<1.0	<1.0	<1.0
Ice cream, dairy	11	9	<1.0	<1.0	2.4	<1.0	1.6
Milk, buttermilk	10	7	0.2	<0.1	0.8	<0.1	0.7
Milk, partly skimmed	10	10	<0.3	<0.3	<0.3	<0.3	<0.3
Milk, skimmed	10	10	<0.2	<0.2	<0.2	<0.2	<0.2
Milk, skimmed with chocolate	2	0	3.3	3.3	3.4	3.3	3.4
Milk, whole	10	10	<0.3	<0.3	<0.3	<0.3	<0.3
Ymer	10	8	<0.4	<0.4	0.5	<0.4	0.4
Yoghurt	11	10	<0.4	<0.4	1.0	<0.4	<0.4
			Bread and Cereals				
Bread, white	10	0	20.4	15.0	24.0	21.0	24.0
Bread, white coarse grain	10	0	42.5	14.0	81.0	40.5	71.1
Bread, white roll	10	0	22.7	17.0	34.0	21.5	28.6
Bread, wholemeal	10	0	24.6	20.0	29.0	24.0	28.1
Breakfast cereal	10	1	27.8	<1.5	66.0	27.0	54.3
Corn flakes	10	2	3.0	<1.5	8.2	2.3	5.6
Crispbread, rye	7	0	7.5	5.4	10.0	6.7	9.5
Crispbread, wheat	3	0	22.3	18.0	30.0	19.0	27.8
Oats, rolled	10	0	23.2	13.0	110.0	23.5	29.8
Pasta	10	0	21.7	8.9	33.0	22.0	29.4
Rice, polished	6	0	17.9	8.2	27.0	17.0	26.5
Rye bread	10	10	7.2	5.1	12.0	7.0	9.7
Rye bread, wholemeal	20	12	26.8	4.5	110.0	14.5	61.6
Rye flour, whole meal	10	10	8.1	4.3	9.9	8.7	9.7
Wheat flour	10	1	32.1	15.0	58.0	30.0	43.6

APPENDIX 3

TABLE 9.A3. Content of Ni (μg kg^{-1} Fresh Weight) in Foods Sampled in the Fourth Monitoring Period (1998–2003)

Foodstuff	Number	Number <LoD	Average	Minimum	Maximum	Median	0.90 Quantile
			Vegetables				
Aubergines	10	0	6	1	16	3	12
Beans	9	0	250	78	434	251	346
Beetroots	10	0	26	15	39	27	34
Broccoli	10	0	97	5	329	35	273
Brussel sprouts	9	0	38	28	47	39	47
Cabbage	11	0	23	5	53	15	49
Carrots	39	0	44	13	145	31	105
Cauliflower	10	0	22	9	38	21	37
Celeriac	9	0	81	2	137	76	135
Celery	12	0	27	8	62	22	46
Chinese cabbage	10	0	20	7	41	19	31
Cucumber	10	1	5	<1	14	3	11
Curly kale	15	0	87	28	190	80	151
Leeks	11	0	28	8	64	18	50
Lettuce	31	0	35	4	173	25	78
Mushrooms	10	0	4	1	10	3	7
Onions	20	0	34	7	73	35	52
Peas	10	0	449	130	1070	281	1043
Pepper	13	1	26	<1	96	6	74
Potatoes	30	0	33	4	365	17	56
Rhubarb	8	0	78	38	133	68	113
Spinach	28	0	60	9	341	28	137
Squash	10	0	50	3	156	33	107
Tomato	20	4	16	<1	142	4	29
			Offal				
Liver, calf	20.0	19	<9	<9	42	<9	<9
Liver, chicken	12.0	7	<9	<9	24	<9	13
Liver, ox	14.0	13	<9	<9	73	<9	<9
Liver, pig	19.0	15	<9	<9	22	<9	11
Kidney, calf	23.0	11	<9	<9	16	<9	10
Kidney, ox	43.0	10	14	<9	153	10	19
Kidney, pig	32.0	3	43	<6	217	18	111
			Meat Products				
Pate, meat	10	0	52	14	213	34	79
Pork, flank, cooked	10	0	18	<9	58	15	32

(Continued)

TABLE 9.A3. *Continued*

Foodstuff	Number	Number <LoD	Average	Minimum	Maximum	Median	0.90 Quantile
Pork, liver paste	10	0	26	9	84	16	50
Pork, mettwurst, smoked	11	0	47	13	144	26	117
Pork, saddle, smoked	10	3	<9	<9	<9	<9	<9
Pork, sausage, frankfurter	13	0	88	<9	524	39	215
Pork, sausage, saveloy	9	0	37	19	58	32	55
Sausage, chicken	11	1	39	<9	154	21	105
Sausage, salami	10	1	14	<9	40	14	20
Sausage, turkey	8	0	16	<9	25	16	24
			Meat				
Beef	41	25	<7	<7	35	<7	<7
Chicken	30	16	<7	<7	44	<7	<7
Lamb	7	5	<7	<7	24	<7	11
Mutton	2	1	<7	<7	<7	<7	<7
Pork	43	34	<7	<7	65	<7	<7
Turkey	12	11	<7	<7	<7	<7	<7
Veal	18	10	<7	<7	37	<7	10
			Fish				
Cod	30	18	8	<5	71	<5	20
Cod liver	10	2	17	<5	70	11	25
Eel	8	6	<5	<5	5	<5	5
Flounder	22	13	7	<5	36	<5	16
Garfish	15	5	9	<5	21	7	17
Herring	17	12	<5	<5	21	<5	8
Mackerel	22	21	<5	<5	7	<5	<5
Mussel	15	0	197	80	506	145	375
Plaice	21	2	20	<5	70	15	42
Shrimp	10	5	20	<5	143	<5	38
Tuna	12	10	14	<5	143	<5	9
			Beverages				
Apple juice	6	0	9	5	12	10	12
Beer, lager	15	0	9	2	58	4	4
Fruit juice	6	0	113	5	578	20	311
Orange juice	11	0	14	4	21	14	19

(*Continued*)

TABLE 9.A3. *Continued*

Foodstuff	Number	Number <LoD	Average	Minimum	Maximum	Median	0.90 Quantile
Red wine	15	0	21	11	30	22	26
Soft drinks	21	3	4	<1	11	2	10
Water, carbonated	4	2	1	<1	2	1	2
White wine	10	0	23	13	39	21	33
			Dairy Products				
Blended spread	5	5	<7	<7	<7	<7	<7
Butter	5	5	<7	<7	<7	<7	<7
Cheese, brie	10	0	49	39	56	48	55
Cheese, firm, Danbo	10	0	67	57	76	67	75
Cocoa, instant	9	0	172	112	216	164	211
Cream, whipping	10	10	<3	<3	<3	<3	<3
Ice cream, dairy	11	0	24	15	74	19	24
Milk, buttermilk	10	0	10	9	13	9	12
Milk, partly skimmed	10	0	7	6	8	7	7
Milk, skimmed	10	0	8	7	9	8	9
Milk, skimmed with chocolate	2	0	181	176	186	181	185
Milk, whole	10	0	10	5	13	10	11
Ymer	10	0	18	16	19	18	18
Yogurt	11	0	12	7	16	13	15
			Bread and Cereals				
Bread, white	10	7	<60	<60	176	<60	112
Bread, white coarse grain	10	2	154	<60	308	150	295
Bread, white roll	10	9	<60	<60	62	<60	<60
Bread, wholemeal	10	8	<60	<60	107	<60	77
Breakfast cereal	10	2	392	<60	1200	190	939
Corn flakes	10	7	<60	<60	156	<60	129
Crispbread, rye	7	5	<60	<60	95	<60	76
Crispbread, wheat	3	2	<60	<60	72	<60	64
Oats, rolled	10	0	1411	380	1680	1195	2462
Pasta	10	0	94	67	124	99	116
Rice, polished	6	0	286	187	368	275	352
Rye bread	10	11	<60	<60	119	<60	80
Rye bread, wholemeal	20	9	123	<60	414	84	318
Rye flour, whole meal	10	3	68	<60	105	69	99
Wheat flour	10	9	<60	<60	74	<60	<60

APPENDIX 4

TABLE 9.A4. Content of Hg (μg kg^{-1} Fresh Weight) in Foods Sampled in the Fourth Monitoring Period (1998–2003)

Foodstuff	Number	Number <LoD	Average	Minimum	Maximum	Median	0.90 Quantile
			Vegetables				
Aubergines	10	10	<0.2	<0.2	<0.2	<0.2	<0.2
Beans	9	9	<0.2	<0.2	<0.2	<0.2	<0.2
Beetroots	10	10	<0.4	<0.4	<0.4	<0.4	<0.4
Broccoli	10	10	<0.2	<0.2	<0.2	<0.2	<0.2
Brussel sprouts	9	9	<0.4	<0.4	<0.4	<0.4	<0.4
Cabbage	11	11	<0.2	<0.2	<0.2	<0.2	<0.2
Carrots	39	22	<0.2	<0.2	0.5	<0.2	0.4
Cauliflower	10	10	<0.4	<0.4	<0.4	<0.4	<0.4
Celeriac	9	3	0.3	<0.2	0.5	0.3	0.4
Celery	12	7	0.2	<0.2	0.4	<0.2	0.3
Chinese cabbage	10	10	<0.2	<0.2	<0.2	<0.2	<0.2
Cucumber	10	10	<0.2	<0.2	<0.2	<0.2	<0.2
Curly kale	15	0	5.4	1.9	11.1	4.7	9.2
Leeks	11	11	<0.2	<0.2	<0.2	<0.2	<0.2
Lettuce	31	22	0.2	<0.2	2.0	<0.2	0.6
Mushrooms	10	0	3.6	1.3	10.5	2.6	7.0
Onions	20	19	<0.2	<0.2	0.6	<0.2	<0.2
Peas	10	10	<0.4	<0.4	<0.4	<0.4	<0.4
Pepper	13	13	<0.2	<0.2	<0.2	<0.2	<0.2
Potatoes	30	29	<0.4	<0.4	0.8	<0.4	<0.4
Rhubarb	8	8	<0.2	<0.2	<0.2	<0.2	<0.2
Spinach	28	0	0.8	0.2	2.6	0.6	1.5
Squash	10	10	<0.2	<0.2	<0.2	<0.2	<0.2
Tomato	20	20	<0.2	<0.2	<0.2	<0.2	<0.2
			Offal				
Liver, calf	20	20	<3.0	<3.0	4.0	<3.0	<3.0
Liver, chicken	12	11	<3.0	<3.0	6.1	<3.0	3.6
Liver, ox	14	14	<3.0	<3.0	3.5	<3.0	<3.0
Liver, pig	19	18	<3.0	<3.0	13.3	<3.0	<3.0
Kidney, calf	23	19	<3.0	<3.0	6.3	<3.0	5.3
Kidney, ox	43	25	5.6	<3.0	14.9	4.9	10.9
Kidney, pig	32	25	5.0	<3.0	24.5	<3.0	8.1
			Meat Products				
Pate, meat	10	10	<1.0	<1.0	<1.0	<1.0	<1.0
Pork, flank, cooked	10	10	<1.0	<1.0	<1.0	<1.0	<1.0

(*Continued*)

TABLE 9.A4. *Continued*

Foodstuff	Number	Number <LoD	Average	Minimum	Maximum	Median	0.90 Quantile
Pork, liver paste	10	10	<1.0	<1.0	1.2	<1.0	1.1
Pork, mettwurst, smoked	11	11	<1.0	<1.0	<1.0	<1.0	<1.0
Pork, saddle, smoked	10	10	<1.0	<1.0	<1.0	<1.0	<1.0
Pork, sausage, frankfurter	13	13	<1.0	<1.0	<1.0	<1.0	<1.0
Pork, sausage, saveloy	9	9	<1.0	<1.0	<1.0	<1.0	<1.0
Sausage, chicken	11	10	<1.0	<1.0	4.7	<1.0	1.2
Sausage, salami	10	9	<1.0	<1.0	2.0	<1.0	1.3
Sausage, turkey	8	8	<1.0	<1.0	<1.0	<1.0	<1.0
			Meat				
Beef	41	41	<2.0	<2.0	<2.0	<2.0	<2.0
Chicken	30	28	<2.0	<2.0	3.1	<2.0	<2.0
Lamb	7	7	<2.0	<2.0	<2.0	<2.0	<2.0
Mutton	2	2	<2.0	<2.0	<2.0	<2.0	<2.0
Pork	43	40	<2.0	<2.0	6.5	<2.0	<2.0
Turkey	12	12	<2.0	<2.0	<2.0	<2.0	<2.0
Veal	18	18	<2.0	<2.0	<2.0	<2.0	<2.0
			Fish				
Cod	30	0	57.9	21.1	160	48.9	94.7
Cod liver	10	1	17.2	<2.6	47.9	13.1	41.7
Eel	8	0	178	20.5	488	94.8	459
Flounder	22	0	40.0	2.9	102	36.8	60.1
Garfish	15	0	71.6	29.9	184	55.9	130
Herring	18	0	41.2	18.1	77.1	34.6	64.5
Mackerel	22	0	34.1	22.2	44.7	34.2	42.2
Mussel	15	0	10.3	2.6	27.7	8.0	22.2
Plaice	21	0	35.3	2.9	80.4	28.7	72.4
Shrimp	10	0	20.0	5.9	57.7	13.9	36.3
Tuna	12	0	474	154	1900	218	968
			Beverages				
Apple juice	6	6	<0.3	<0.3	<0.3	<0.3	<0.3
Beer, lager	15	15	<0.3	<0.3	<0.3	<0.3	<0.3
Fruit juice	6	6	<0.3	<0.3	<0.3	<0.3	<0.3
Orange juice	11	11	<0.3	<0.3	<0.3	<0.3	<0.3

(*Continued*)

TABLE 9.A4. *Continued*

Foodstuff	Number	Number <LoD	Average	Minimum	Maximum	Median	0.90 Quantile
Red wine	15	15	<0.3	<0.3	<0.3	<0.3	<0.3
Soft drinks	21	21	<0.3	<0.3	<0.3	<0.3	<0.3
Water, carbonated	4	4	<0.3	<0.3	<0.3	<0.3	<0.3
White wine	10	10	<0.3	<0.3	<0.3	<0.3	<0.3
			Dairy Products				
Blended spread	5	5	<0.9	<0.9	<0.9	<0.9	<0.9
Butter	5	5	<0.9	<0.9	<0.9	<0.9	<0.9
Cheese, brie	10	0.4	<0.4	<0.4	<0.4	<0.4	<0.4
Cheese, firm, Danbo	10	10	<0.5	<0.5	<0.5	<0.5	<0.5
Cocoa, instant	9	0	0.4	0.1	1.3	0.4	0.6
Cream, whipping	10	10	<0.4	<0.4	<0.4	<0.4	<0.4
Ice cream, dairy	11	11	<0.4	<0.4	<0.4	<0.4	<0.4
Milk, buttermilk	10	7	<0.1	<0.1	0.2	<0.1	0.1
Milk, partly skimmed	10	10	<0.1	<0.1	<0.1	<0.1	<0.1
Milk, skimmed	10	9	<0.1	<0.1	0.1	<0.1	<0.1
Milk, skimmed with chocolate	2	0	0.3	0.3	0.3	0.3	0.3
Milk, whole	10	10	<0.1	<0.1	<0.1	<0.1	<0.1
Ymer	10	5	0.1	<0.1	0.3	<0.1	0.1
Yogurt	11	10	<0.1	<0.1	0.1	<0.1	<0.1

APPENDIX 5

TABLE 9.A5. Content of Se (μg kg^{-1} Fresh Weight) in Foods Sampled in the Fourth Monitoring Period (1998–2003)

Foodstuff	Number	Number <LoD	Average	Minimum	Maximum	Median	0.90 Quantile
			Vegetables				
Aubergines	10	5	0.7	<0.4	1.7	0.5	1.4
Beans	9	1	3.9	<0.4	7.2	4.5	7.0
Beetroots	10	6	<0.8	<0.8	1.4	<0.8	1.2
Broccoli	10	1	7.1	0.4	20.2	6.0	14.3
Brussel sprouts	9	0	25.5	6.0	54.2	29.1	48.5
Cabbage	11	2	5.1	<0.4	29.3	2.5	5.7
Carrots	39	10	2.5	<0.4	13.0	1.2	7.8

(*Continued*)

TABLE 9.A5. *Continued*

Foodstuff	Number	Number <LoD	Average	Minimum	Maximum	Median	0.90 Quantile
Cauliflower	10	1	22.8	<0.8	166.0	6.1	31.9
Celeriac	9	1	4.3	0.4	12.0	2.5	7.6
Celery	12	4	2.1	<0.4	7.2	2.4	3.5
Chinese cabbage	10	4	2.5	<0.4	14.1	1.0	5.3
Cucumber	10	7	<0.4	<0.4	1.9	<0.4	0.8
Curly kale	15	4	29.6	<0.8	95.5	15.4	83.8
Leeks	11	1	7.0	<0.4	23.6	3.5	20.7
Lettuce	31	17	1.1	<0.4	7.3	<0.4	2.8
Mushrooms	10	0	129	35.4	473	92.7	181
Onions	20	2	7.2	<0.4	49.7	2.0	21.1
Peas	10	0	8.9	4.0	24.0	6.4	14.1
Pepper	13	6	5.0	<0.4	53.9	0.7	2.7
Potatoes	30	9	1.4	<0.8	3.4	1.1	2.8
Rhubarb	8	8	<0.4	<0.4	<0.4	<0.4	<0.4
Spinach	28	2	6.0	<0.4	88.8	1.1	11.2
Squash	10	4	1.4	<0.4	4.1	1.2	2.9
Tomato	20	8	2.7	<0.4	22.7	0.9	5.8
			Offal				
Liver, calf	20	0	367	62	1050	307	572
Liver, chicken	12	0	461	315	538	469	514
Liver, ox	14	0	337	119	717	322	501
Liver, pig	19	0	457	371	618	449	559
Kidney, calf	23	0	1229	767	1600	1220	1457
Kidney, ox	43	0	1135	391	1530	1140	1427
Kidney, pig	32	0	1639	1150	2326	1610	1938
			Meat Products				
Pate, meat	10	0	188	118	294	145	290
Pork, flank, cooked	10	0	104	66.0	141	105	137
Pork, liver paste	10	0	196	161	220	196	212
Pork, mettwurst, smoked	11	0	74.4	52.0	93.0	80.0	93.0
Pork, saddle, smoked	10	0	116	82.0	164	113	144
Pork, sausage, frankfurter	13	0	71.1	57.0	93.0	71.0	76.8
Pork, sausage, saveloy	9	0	66.9	58.0	82.0	65.0	76.4
Sausage, chicken	11	0	103	63.0	203	99.0	121
Sausage, salami	10	0	94.7	75.0	113	96.0	109

(*Continued*)

TABLE 9.A5. *Continued*

Foodstuff	Number	Number <LoD	Average	Minimum	Maximum	Median	0.90 Quantile
Sausage, turkey	8	0	78.1	61.0	88.0	82.0	85.9
			Meat				
Beef	41	0	101	61.0	161	99.4	130
Chicken	30	0	94.6	64.0	149	94.4	107
Lamb	7	0	23.6	12.0	37.0	23.0	31.0
Mutton	2	0	22.5	21.0	24.0	22.5	23.7
Pork	43	0	126	95.3	259	123	138
Turkey	12	0	89.0	32.7	137	95.6	126
Veal	18	0	76.4	13.0	106	81.1	99.1
			Fish				
Cod	30	0	269	167	377	272	339
Cod liver	10	0	635	296	1742	539	811
Eel	8	0	335	238	677	295	473
Flounder	22	0	233	170	358	230	324
Garfish	15	0	262	222	297	258	296
Herring	18	0	331	218	535	292	501
Mussel	15	0	294	158	565	293	393
Mackerel	22	0	250	201	375	245	292
Plaice	21	0	325	177	1223	302	394
Shrimp	10	0	187	134	237	179	222
Tuna	12	0	772	488	1141	791	1040
			Beverages				
Apple juice	6	4	0.3	<0.3	0.6	<0.3	0.6
Beer, lager	15	2	1.2	<0.3	2.5	1.3	2.5
Fruit juice	6	3	1.7	<0.3	9.5	<0.3	5.0
Orange juice	11	4	0.4	<0.3	0.8	0.4	0.7
Red wine	15	3	0.7	<0.3	2.2	0.6	1.5
Soft drinks	21	16	0.4	<0.3	2.8	<0.3	1.9
Water, carbonated	4	1	0.5	<0.3	1.0	0.5	0.8
White wine	10	4	0.4	<0.3	1.3	0.3	0.9
			Dairy Products				
Blended spread	5	5	<8.0	<8.0	<8.0	<8.0	<8.0
Butter	5	5	<8.0	<8.0	<8.0	<8.0	<8.0
Cheese, brie	10	0	58.4	5.7	105	50.7	94.7
Cheese, firm, Danbo	10	0	87.5	69.1	94.8	90.3	94.8
Cocoa, instant	9	0	13.9	11.8	15.3	13.6	15.1

(*Continued*)

TABLE 9.A5. *Continued*

Foodstuff	Number	Number <LoD	Average	Minimum	Maximum	Median	0.90 Quantile
Cream, whipping	10	0	12.9	10.1	16.0	12.7	15.6
Ice cream, dairy	11	2	8.4	<4.0	16.0	7.9	12.1
Milk, buttermilk	10	0	15.2	11.8	17.0	15.7	16.6
Milk, partly skimmed	10	0	15.4	11.3	20.1	15.0	19.4
Milk, skimmed	10	0	16.6	14.0	19.3	16.4	19.0
Milk, skimmed with chocolate	2	0	13.1	12.1	14.1	13.1	13.9
Milk, whole	10	0	16.1	12.8	19.0	15.9	18.3
Ymer	10	0	26.2	23.2	28.4	26.8	27.8
Yogurt	11	0	18.3	15.5	20.2	18.9	19.7
			Bread and Cereals				
Bread, white	10	5	44.3	<40	98	44	75.5
Bread, white coarse grain	10	4	56.7	<40	154	55	112.6
Bread, white roll	10	2	61.8	<40	99	63	87.3
Bread, wholemeal	10	9	<40	<40	41	<40	<40
Breakfast cereal	10	1	85.1	<40	207	71	151.2
Corn flakes	10	5	49.5	<40	96	49.5	80.7
Crispbread, rye	7	5	<40	<40	101	<40	69.8
Crispbread, wheat	3	3	<40	<40	<40	<40	<40
Oats, rolled	10	2	49.7	<40	208	51	88.8
Pasta	10	2	63.7	<40	208	52	97.3
Rice, polished	6	2	87	<40	149	87.5	147.5
Rye bread	10	7	<40	<40	54	<40	<40
Rye bread, wholemeal	20	19	<40	<40	113	<40	<62.2
Rye flour, whole meal	10	8	<40	<40	50	<40	43.7
Wheat flour	10	9	<40	<40	52	<40	<40

APPENDIX 6

TABLE 9.A6. Content of As (μg kg^{-1} Fresh Weight) in Foods Sampled in the Fourth Monitoring Period (1998–2003)

Foodstuff	Number	Number <LoD	Average	Minimum	Maximum	Median	0.90 Quantile
			Vegetables				
Aubergines	10	4	2.2	<0.4	8.5	0.5	6.8
Beans	9	1	1.5	<0.4	2.7	1.3	2.6
Beetroots	10	4	1.1	<0.7	2.7	0.9	2.0
Broccoli	10	0	1.9	0.7	4.2	1.0	3.8
Brussel sprouts	9	0	4.8	1.1	8.9	4.9	7.6
Cabbage	11	1	0.9	<0.4	3.1	0.7	1.2
Carrots	39	0	2.6	0.7	6.3	2.2	4.9
Cauliflower	10	2	2.0	<0.7	5.8	1.4	4.2
Celeriac	9	0	2.7	0.6	7.3	2.1	4.4
Celery	12	0	2.8	1.3	6.8	2.3	4.1
Chinese cabbage	10	0	2.2	0.6	4.3	1.6	3.9
Cucumber	10	3	2.0	<0.4	6.1	0.9	5.0
Curly kale	15	0	5.0	2.1	8.2	5.1	7.4
Leeks	11	0	2.1	0.7	4.8	1.7	3.6
Lettuce	31	1	3.8	<0.4	16.1	2.3	9.4
Mushrooms	10	0	20.1	9.0	38.9	17.5	30.4
Onions	20	0	5.5	0.8	28.6	2.7	11.1
Peas	10	1	1.6	<0.7	2.8	1.4	2.7
Pepper	13	5	0.7	<0.4	2.3	0.5	1.4
Potatoes	30	14	1.4	<0.7	10.5	1.0	2.7
Rhubarb	8	0	1.9	0.7	5.3	1.4	3.4
Spinach	28	0	8.1	1.5	24.1	6.2	17.9
Squash	10	3	1.5	<0.4	5.4	0.8	3.4
Tomato	20	9	0.5	0.4	1.0	0.4	0.9
			Offal				
Liver, calf	20	1	19.0	<4.0	112	13.0	26.8
Liver, chicken	12	1	22.7	5.7	62.1	19.5	40.4
Liver, ox	14	2	12.8	4.2	32.0	11.5	22.1
Liver, pig	19	0	14.5	6.0	23.0	14.0	22.2
Kidney, calf	23	0	43.1	14.0	63.3	49.0	61.5
Kidney, ox	43	0	47.2	14.0	136	46.0	63.3
Kidney, pig	32	0	34.0	13.0	79.6	35.0	61.3
			Meat Products				
Pate, meat	10	0	13.3	8.5	19.0	13.0	17.2
Pork, flank, cooked	10	0	7.7	3.4	13.0	7.1	11.2

(*Continued*)

TABLE 9.A6. *Continued*

Foodstuff	Number	Number <LoD	Average	Minimum	Maximum	Median	0.90 Quantile
Pork, liver paste	10	0	12.6	11.0	15.0	12.0	15.0
Pork, mettwurst, smoked	11	1	6.4	<3.0	11.0	6.5	9.7
Pork, saddle, smoked	10	0	7.0	4.3	12.0	6.8	8.5
Pork, sausage, frankfurter	13	0	5.4	<3.0	11.0	4.8	8.7
Pork, sausage, saveloy	9	0	8.0	4.5	12.0	7.8	11.2
Sausage, chicken	11	0	56.1	7.3	261	18.0	206
Sausage, salami	10	0	7.2	4.5	11.0	6.6	9.8
Sausage, turkey	8	0	9.9	4.8	16.0	9.7	14.6
Meat							
Beef	41	3	5.1	<5.0	14.0	<5.0	8.3
Chicken	30	1	14.2	<5.0	61.4	9.2	31.2
Lamb	7	5	<5.0	<5.0	<5.0	<5.0	<5.0
Mutton	2	2	<5.0	<5.0	<5.0	<5.0	<5.0
Pork	43	3	7.1	<5.0	81.1	<5.0	8.1
Turkey	12	1	5.5	<5.0	11.0	5.8	7.3
Veal	18	3	<5.0	<5.0	8.0	5.1	6.4
Fish							
Cod	30	0	2409	316	12179	1931	4619
Cod liver	10	0	2773	1565	4003	2673	3830
Eel	8	0	477	243	654	505	608
Flounder	22	0	2000	197	6667	1061	4520
Garfish	15	0	352	120	771	343	476
Herring	18	0	1514	334	2648	1470	1999
Mackerel	22	0	2027	1458	2612	2084	2435
Mussel	15	0	1077	413	2024	972	1858
Plaice	21	0	10708	3268	23989	8569	19807
Shrimp	10	0	4099	271	20924	2555	5554
Tuna	12	0	1179	677	2327	1132	1453
Beverages							
Apple juice	6	5	2.1	<2.0	6.3	<2.0	4.0
Beer, lager	15	6	4.2	<2.0	10.0	3.5	6.0
Fruit juice	6	2	9.3	<2.0	43.8	2.4	24.3
Orange juice	11	5	3.4	<2.0	12.0	2.0	9.3

(*Continued*)

TABLE 9.A6. *Continued*

Foodstuff	Number	Number <LoD	Average	Minimum	Maximum	Median	0.90 Quantile
Red wine	15	3	4.4	<2.0	14.0	2.7	9.7
Soft drinks	21	19	<2.0	<2.0	2.3	<2.0	<2.0
Water, carbonated	4	3	3.5	<2.0	8.9	<2.0	6.8
White wine	10	0	8.7	2.8	21.3	7.7	12.9
Dairy products							
Blended spread	5	5	<8.0	<8.0	<8.0	<8.0	<8.0
Butter	5	5	<8.0	<8.0	<8.0	<8.0	<8.0
Cheese, brie	10	5	7.4	<4.0	13.9	6.3	13.9
Cheese, firm, Danbo	10	2	6.6	<5.0	9.9	6.2	9.2
Cocoa, instant	9	3	1.1	<1.0	1.5	1.2	1.5
Cream, whipping	10	9	<4.0	<4.0	4.2	<4.0	<4.0
Ice cream, dairy	11	6	<3.0	<3.0	5.5	<3.0	4.9
Milk, buttermilk	10	2	0.8	<0.7	1.2	0.8	1.1
Milk, partly skimmed	10	2	1.4	<0.8	2.9	1.1	2.6
Milk, skimmed	10	0	1.1	0.9	1.4	1.0	1.4
Milk, skimmed with chocolate	2	1	1.1	<1.0	1.4	1.1	1.3
Milk, whole	10	1	1.6	<0.9	2.1	1.7	2.0
Ymer	10	2	1.4	<1.0	1.7	1.5	1.7
Yoghurt	11	4	1.1	<1.0	1.9	1.1	1.6

10

TRACE ELEMENTS IN THE TOTAL DIET TYPICAL OF NORTHERN ITALY

M. Bettinelli

C.M.B. Sistemi integrati per i laboratori, la qualità e l'ambiente, Via G. Landi 77, 29100 Piacenza, Italy, E-mail: maurizio.bettinelli@libero.it

S. Spezia, A. Gatti, A. Ronchi, and C. Minoia

Laboratory for Environmental and Toxicological Measurements, S. Maugeri Foundation, Via Ferrata 8, 27100 Pavia, Italy

C. Roggi and G. Turconi

Department of Health Applied Sciences, Section of Human Nutrition, University of Pavia, Pavia, Italy

The Determination of Chemical Elements in Food: Applications for Atomic and Mass Spectrometry.
Edited by Sergio Caroli

10.1 INTRODUCTION

The significance of essential trace elements and minerals for human health is well documented [1]. The deficiency of a particular trace element is not a common condition in industrialized countries; however, the intake of some minerals or trace elements might be inadequate in specific segments of the population [2]. On the other hand, it has been recognized that the intake of food contaminated with elements can give rise to acute toxicity. During the last decade it has become more and more evident that certain chemical contaminants can produce pathologies even after years of latency because the substances accumulate in the body over a long period of time and can exert adverse effects only when their levels in certain tissues reach critical values. Under the same conditions, many factors, such as state of nutrition, gender, age, element retention percentage, chemical forms and bond sites can influence the concentration of trace elements in the human body.

Several paths of environmental exposure can significantly affect the uptake of these substances, but food is the matrix mostly involved in the trace element passage from the environment to the human body [3]. The entry of trace elements into the food chain depends on the environmental matrix they are contained in, for example, soil, water, and atmosphere. Trace elements can pass, through different ways, from these matrices to intermediate links and then into the human body [4]. Cadmium, Cr, Hg, Ni, and Pb are the trace elements mostly involved in the exposure of humans to chemical contaminants [5]. Many studies have shown that leafy vegetables easily accumulate air-borne contaminants (Cd, Cr, and Pb), while vegetables growing underground, such as potatoes and carrots, accumulate Cd more than other trace elements [6]. High levels of potentially toxic elements are also found in cultivated mushrooms (e.g., the cap is rich in Ni, while Cd is more concentrated in the stalk) [7].

A study carried out by Alberti-Fidanza et al. [8] on heavy metal dietary exposure of a population of students in central Italy showed that bread contains amounts (in $\mu g\ 100\ g^{-1}$) of Cd (5.6), Cr (66.9), Ni (55.6), and Pb (21.5) that are greater than in other investigated foods. In freshly produced milk, many differences in trace element levels have been found depending on the season, the kind of fodder, environmental conditions and the production system [9]. Winter milk is rich in Cd and Pb, and summer milk is rich in Al [10]. As for cheese, apart from the above factors, the different production phases modify the trace element content as well: curdling increases Al, Cd, and Pb content, while salting reduces all element content, except Al [11, 12]. Maturation is a very delicate phase because some toxic elements such as Al and Cd can contaminate cheese [12]. In human milk, trace element content changes during breast-feeding [13]. There is a significant reduction of all trace elements other than Co, which increases by 100 percent. Generally, the concentrations of toxic elements such as Cd and Hg, and other potentially interesting elements such as Ba, Be, La, Li, and Sb, decrease in mature milk from 20 to 11 percent as compared with colostrum, except for Pb, the reduction of which proceeds at a slower pace. A study carried out in 2000 in

Lombardy on 231 breast milk samples taken on the third or fourth day after delivery showed extremely different concentration values for Pb ($17.1 \pm 31.8\ \mu g\ l^{-1}$, range 2–216 $\mu g\ l^{-1}$) and Cd ($1.7 \pm 2.4\ \mu g\ l^{-1}$, range 1–20 $\mu g\ l^{-1}$). Some 9.5 percent of breast-feds on day 6 of breast-feeding ingested quantities of Pb higher than the Provisional Tolerable Weekly Intake (PTWI) (25 $\mu g\ kg^{-1}$) [14].

Mussels (*Mytilus edulis*) have a content of trace elements, both essential and toxic (As, Cd, Hg, Ni, and Pb), that is tightly correlated to the geographic zone, water salinity, and degree of sea contamination [15].

Many studies have evaluated the daily dietary intake of toxic trace elements by the general population. Data collected in the 1990s by a Finalized Project of the Italian National Research Council show that toxic elements consumed by the Italian population were below the tolerable levels; however, the intake of Cr and Hg approached the limit values [16, 17]. Conversely, a study carried out in Oltrepò Pavese showed Hg dietary intakes definitely under PTWI (5 $\mu g\ kg^{-1}$) for the whole investigated population, while Cd and Pb intake values were over PTWI for subjects at the 90th and 95th percentile, that is, those more exposed [18–20]. Also for Ni, subjects at the 90th and 95th percentiles presented intake values over 600 $\mu g\ day^{-1}$, that is, the value set by WHO as the toxicity level over which particularly sensitive subjects could have allergic reactions. Foods that mostly contributed to the daily intake were cereals for Cd (59 percent), fruits (34 percent) and cereals (26 percent) for Hg, fruits (53 percent) and cereals (26 percent) for Pb, and vegetables (42 percent) and cereals (24 percent) for Ni. It is interesting to note that, even if fish is the product mostly contaminated by Hg, the contribution of fish to the total Hg intake is only 4 percent, because of the low frequency of its consumption.

A study carried out in France revealed an intake values of 98 $\mu g\ day^{-1}$ for Cr, 4.2 mg day^{-1} for Al, 27 $\mu g\ day^{-1}$ for Cd, and 231 $\mu g\ day^{-1}$ for Ni [21]. The Cr dietary intake was thus calculated to be 25 percent from the group of meat, eggs, and poultry, 20 percent from dairy products, and 20 percent from cereals and beverages [21]. Lead levels were high in meat cooked on a hot plate because of the release of this metal by the surface itself. The highest value of Pb in meat was found in raw sausages [8].

Studies on the dietary intake of Cd and Pb carried out in the 1990s in Italy using different methods showed that values were higher than Predicted Tolerable Daily Intake (PTDI) for Pb in two surveys in Modena (641 $\mu g\ day^{-1}$, 220–456 $\mu g\ day^{-1}$) and for Cd in Catania (44–183 $\mu g\ day^{-1}$), Marche (52–138 $\mu g\ day^{-1}$) and Modena (63–151 $\mu g\ day^{-1}$) [22]. Cadmium and Pb daily intake in some European countries have values lower than the PTWI in all European countries, ranging from 17 to 50 percent, of PTWI for Cd and from 8 to 83.5 percent of PTWI for Pb [23].

The present investigation was undertaken to estimate the intake of trace elements from the total diet of the population in northern Italy; elements such as Al, B, Cd, Co, Li, Mo, Mn, Ni, Pb, Rb, Sb, Sn, Sr, and V are not frequently reported in the open literature. The contribution of the most representative food groups (cereals and cereal products, vegetables, fruit, milk and dairy products, meat and meat products, fish, sweet products, and beverages) was evaluated in a stepwise manner, that is, through:

- Identification of the predominant foods and prepared meals;
- Measurement of toxic and essential trace element concentrations in foods and prepared meals; and
- Estimation of the intakes by the population.

Data on the northern Italian population dietary consumptions, as published in the literature, were used for the Total Diet Study (TDS) [24]. As for food samplings, two areas were identified for each of the three regions considered in the study (Piemonte, Lombardia, and Veneto) for a total of six areas. In the diet of these regions rice is preferred to regular pasta and pasta made with eggs, filled or unfilled, which in turn prevail in central and southern Italy. The rice intake is, for example, higher in the northwest where it is produced, while the northeast shows the highest intake of grain and flour, probably because of the tradition of preparing cornmeal mash (*polenta*). Yogurt, cream, butter, and tea are more appreciated in these regions than in other parts of Italy, while the intake of cheese and milk is similar across regions. People in the north consume less olive oil than in the center or south; conversely, the intake of butter was higher in the north with a high consumption of margarine in the northwest only.

Tomatoes are preferred in the south, while the north (the northeast in particular) has the highest intake of preserved vegetables. People living in northern Italy eat higher quantities of sweets. In particular, the northwest has the highest intakes of honey, candies, jam, and chocolate, while the northeast shows the highest intake of cakes, biscuits, pastries, and ice-cream. There are differences between regions as regards mutton and horse meat consumption; the former is preferred in the south and the latter in the northeast. Fish and seafood – fresh and frozen – and eggs are more frequently eaten in the south, while fish and preserved seafood are more popular in the north. Finally, pork and legumes – dry and canned – show the lowest values in the northeast.

Samples of all the mentioned foodstuffs were collected in different shops of the small/medium/large scale retail trade. Concerning the largely-scale retail trade, foods were bought in the best known supermarkets operating in northern Italy.

The study was also carried out taking into account the recipes used in the mentioned areas to prepare different meals according to the local cooking traditions. Foods and prepared meals were homogenized using a blender equipped with titanium blades, packed in acid-washed polypropylene containers, and deep-frozen at −40°C.

10.2 EXPERIMENTAL

10.2.1 Apparatus

Digestions were carried out in a microwave (MW) sample preparation system CEM (Corp., Matthews, NC, USA) model MARS-Xpress equipped with a 1400 W magnetron with adjustable power down to 1 percent increments and operating at a

MW frequency of 2455 MHz. The operating conditions for the sample digestion procedures are given below. The system also has a rotating turntable and a cavity exhaust fan with programmable speed to convey corrosive fumes to an exhaust hood. Temperature measurement was recorded at 1 s intervals with an IR thermometer sensor. An in-built digital computer controls and monitors MW oven operations including time, temperature, and power settings.

The elements Al, Mn, and Sr were determined by means of a Perkin–Elmer Optima 4300DV inductively coupled plasma emission spectrometry (ICP-AES) instrument (axial mode), equipped with an AS-90 Plus autosampler, a cross-flow nebulizer, and a Scott-type spray chamber in Ryton. The instrumental operating parameters are listed in Table 10.1.

Other elements were measured by a Perkin–Elmer SCIEX ELAN 6100 DRCII Inductively Coupled Plasma Mass Spectrometry (ICP-MS) instrument equipped with a cyclonic spray chamber, a concentric nebulizer and a dynamic reaction cell (DRC). In the vented (standard) mode, no reaction gas is present in the cell and the instrument shows the typical characteristics of a quadrupole-based ICP-MS apparatus. When the gas is introduced into the cell an ion-molecule reaction takes place that can be tailored so as to eliminate spectroscopic interferences. Experimental conditions are summarized in Table 10.2.

Boron, Li, Mo, Pb, and Sb were determined in the standard mode, while Al, Cd, Co, Ni, Mn, Rb, Sb, Sn, and V were determined in the DRC mode. The determination of Ni was done with a gas flow of 0.15 ml min^{-1} of CH_4, while for the other elements NH_3 was used as cell gas at 0.4 ml min^{-1}. The determination of Se by flow injection hydride generation atomic absorption spectrometry (FI-HG-AAS) was carried out by means of the Perkin–Elmer FIAS 200 system, equipped with the Perkin–Elmer autosampler AS-90, and connected to an electrically heated quartz cell installed on a PerkinElmer absorption spectrometer AAS 4100. The analytical conditions are given in Table 10.3.

TABLE 10.1. Instrumental Parameters and Working Conditions for ICP-AES Analysis

Model	Perkin–Elmer Optima 4300DV (Simultaneous Spectrometer)
Incident power (W)	1300
Reflected power (W)	<5
Ar gas flow rate (l min^{-1})	
Outer gas	15
Auxiliary gas	0.5
Aerosol gas	0.8
Nebulizer	
Model	Cross-flow gem-tip
Sample flow rate (ml min^{-1})	1.0
Pump of sample injection	Perkin–Elmer
Autosampler	AS90 Plus
Viewing mode	Axial

TABLE 10.2. Instrumental Parameters and Working Conditions for DRC-ICP-MS Determinations

RF power (W)	1250		
Ar gas flow rate (l min^{-1})			
Plasma argon	15		
Nebulizer flow	1.0		
Auxiliary flow	1.3		
Sample flow rate	1		
Nebulizer	Concentric cyclonic spray chamber		
Interface cones	Platinum		
Scan mode	Peak hopping		
Resolution	Normal		
Reading time (ms)	150		
Dwell time (ms)	25		
Sweeps/readings	2		
Readings/replicates	3		
Number of replicates	5		
Sample time (s)	110		
Sample read delay (s)	50		
Autosampler wash delay (s)	45		
Calibration mode	External calibration		
Calibration standard	1, 5, 10, 50, 100 μg l^{-1}		
Curve fit	Linear through zero		
DRC Parameters	*DRC Vented (Standard Mode)*	*DCR Pressurized (Enhanced Mode)*	
Cell *q* parameter	0.25	0.65	0.65
Cell *a* parameter	0	0	0
Cell gas	No gas	NH_3 at 0.4 ml min^{-1}	CH_4 at 0.15 ml min^{-1}

TABLE 10.3. Instrumental Parameters and Working Conditions for Se Determination by FI-HG-AAS

Element	Se
Wavelength (nm)	196
Slit (nm)	2
Sample volume injected (μl)	200
Linearity (μg l^{-1})	10
Cell temperature (°C)	900
Carrier solution	HCl 3% (w/v)
Reductant solution	$NaBH_4$ 0.2% (w/v) in NaOH 0.05% (w/v)
Carrier gas	
Flow carrier gas (100 ml min^{-1})	Argon

10.2.2 Certified Reference Materials

To check the accuracy of the analytical procedure, the following Certified Reference Materials (CRMs) were used: Peach Leaves, NIST SRM 1547; Tomato Leaves, NIST SRM 1573; Pine Needles, NIST SRM 1575; and Wheat Flour, NIST SRM 1567. All of them are produced by the National Institute of Standards and Technology (NIST Gaithersburg, MD, USA).

10.2.3 Reagents and Calibrants

A number of Suprapur reagents (E. Merck, Darmstadt, Germany) were used, namely, HNO_3 (65 percent m/v), HCl (37 percent m/v), and HF (40 percent m/v). The $NaBH_4$ solution (0.2 percent m/v) was prepared by dissolving $NaBH_4$ powder in deionized water (2 g in 1000 mL water) stabilized with NaOH 0.05 percnt (m/v) (500 mg NaOH in the same 1000 mL solution).

Multielemental standard solutions were prepared from 10 mg l^{-1} multielemental AAS standards and ICP-MS calibration standard (CPI International, Amsterdam, The Netherlands n. 3, n. 4, MS3, and MS1), by dilution with water containing the same amount of acids as the samples.

High-purity water was produced starting from distilled water using a Milli-Q^{TM} deionizing system (Millipore, Bedford, MA, USA). Glassware was cleaned by soaking overnight in a 10 percent (w/v) HNO_3 solution and then rinsing with deionized water.

10.2.4 Sample Preparation

CRMs were handled according to NIST's specifications. Aliquotes of 250 mg of each CRM were added with 4 mL of HNO_3, 4 mL of HCl, and 200 μL of HF in the vessels of the oven where also the blanks were also run under the working conditions summarized in Table 10.4.

After cooling, the digests were filtered and brought to a final volume of 50 ml. Such solutions are ready to be analyzed by ICP-AES, ICP-MS, and FI-HG-AAS. Five replicate samples and the relative blanks were prepared on different working days to obtain the limits of detection (LoDs) of the method.

About 700–800 mg of the homogenized sample (corresponding to about 500 mg of cooked food) were treated as previously described for the CRMs. Volumes of 10 mL of liquid samples (beverages) were treated with 4 mL of HNO_3 (65 percent,

TABLE 10.4. Working Conditions for Microwave Digestion

STEP	1	2
Power (W)	1000	1000
Ramp time (min)	10	4
Hold time (min)	1	25
Temperature (°C)	160	180

m/v) and 4 mL of HCl (37 percent, m/v), and heated at 90°C for approximately 1 h. After cooling, the samples were diluted to 50 mL with deionized water.

10.3 RESULTS AND DISCUSSION

Quality control (QC) and quality assurance (QA) programs were developed during the present investigation. The main components of the program include: (i) development of standard operating procedures complying with all the essential aspects of good analytical practice; and (ii) validation of the analytical methods using the above CRMs. Pivotal aspects of the preanalytical phases include the following: (i) precautions taken to minimize external contamination during sampling, sample preparation, and chemical analysis, including appropriate checking of the purity of standards and reagents; (ii) the use of HF in the acid mixture in order to obtain the total dissolution of the food samples [25–29]. For liquid samples, the LoDs (in $\mu g\, l^{-1}$) for the elements determined by ICP-MS were, as an average, 0.05, 5 for those determined by ICP-AES, and 0.5 for Se as determined by FI-HGAAS.

Regarding the samples digested as previously reported, the estimated LoDs (in $\mu g\, g^{-1}$) were 0.001 for ICP-MS, 0.1 for ICP-AES, and 0.01 for FI-HGAAS. The entire analytical procedure was tested for both measurement accuracy and precision in order to assess the degree of reliability to be assigned to the data generated in this investigation. The level of accuracy was monitored by adding one of the above CRMs to each series of samples. Precision was ascertained by replicating the entire analytical procedure an adequate number of times (five in all cases). The results obtained for the CRMs tested are summarized in Table 10.5.

Analytical data obtained for the said CRMs were acceptable, with recovery percentages varying between 80 and 120 percent depending on the element. Only for elements for which the measured concentrations were very close to the LoDs, namely V in the NIST 1547, Sb in the NIST 1567, and B in the NIST 1547 and 1573, did the recovery worsen to 140, 170 and 70 percent, respectively. Precision was also found to be satisfactory, with relative standard deviations (RSDs) always lower than 20 percent, with the only exception being Sr in the NIST 1567 (28 percent). Higher RSD values were found for elements present at concentrations lower than 0.1 $\mu g\, g^{-1}$ as reported for the Wheat Flour, NIST 1567.

Duplicate analyses were performed on a representative selection of all the actual samples collected. Different types of foods and cooked meals were analyzed by two different techniques, that is, ICP-AES and ICP-MS. In order to evaluate the accuracy of the adopted instrumental procedure correlation analysis with an LR method was carried out on the data. For the comparison R^2, slope b and intercept a were evaluated, which in a theoretical model take on the values $R^2 = 1$ (maximum accordance), $b = 1$ (absence of additive or multiplicative effects), and $a = 0$ (absence of bias). Typical results obtained are those illustrated for Al ($R^2 = 0.94$, $a = -0.92$, $b = 1.02$) and Mn ($R^2 = 0.83$, $a = -0.13$, $b = 0.92$) in Figures 10.1 and 10.2, respectively.

TABLE 10.5. Concentrations (μg g^{-1}) of Elements in NIST SRMs After Microwave Digestion and ICP-MS Analysis (n = 5)

	NIST 1547		NIST 1573		NIST 1567		NIST 1575	
Element	Certified	Found	Certified	Found	Certified	Found	Certified	Found
Al	249 ± 8	230 ± 14	(1200)[a]	—	—	—	545 ± 30	490 ± 27
B	29 ± 2	22 ± 2	(30)[a]	23 ± 4	(1.5)[b]	—	(17 ± 3)[b]	20 ± 2
Cd	0.026 ± 0.003	0.024 ± 0.002	(2.5 ± 0.2)[b]	1.9 ± 0.3	0.032 ± 0.007	0.027 ± 0.008	< 0.5	0.03 ± 0.01
Co	(0.07)[a]	0.16 ± 0.03	(0.53 ± 0.05)[b]	0.48 ± 0.05	(0.021)[b]	0.025 ± 0.006	(0.1)[a]	0.14 ± 0.06
Mn	98 ± 3	96 ± 4	238 ± 7	225 ± 10	8.5 ± 0.5	8.8 ± 0.8	675 ± 15	690 ± 11
Ni	0.69 ± 0.09	0.79 ± 0.10	(1.17 ± 0.08)[b]	1.18 ± 0.05	(0.180)[a]	0.170 ± 0.11	(3.5)[a]	2.9 ± 0.6
Pb	0.87 ± 0.03	0.77 ± 0.05	6.3 ± 0.3	5.2 ± 0.5	0.02 ± 0.01	—	10.8 ± 0.5	11.2 ± 1.1
Sb	(0.02)[a]	0.04 ± 0.01	(0.036 ± 0.007)[b]	0.034 ± 0.006	(0.020)[b]	0.034 ± 0.005	(0.2)[a]	0.21 ± 0.02
Sr	53 ± 4	45 ± 5	44.9 ± 0.3	44 ± 10	(0.97 ± 0.11)[b]	1.25 ± 0.35	4.8 ± 0.2	4.5 ± 0.3
V	0.37 ± 0.03	0.52 ± 0.05	(1.3 ± 0.3)[b]	1.1 ± 0.2	(11.25)[b]	11.3 ± 1.1	(0.40 ± 0.05)[b]	—

[a]Values reported, but not certified.

[b]Consensus values reported by Gladney et al. [30].

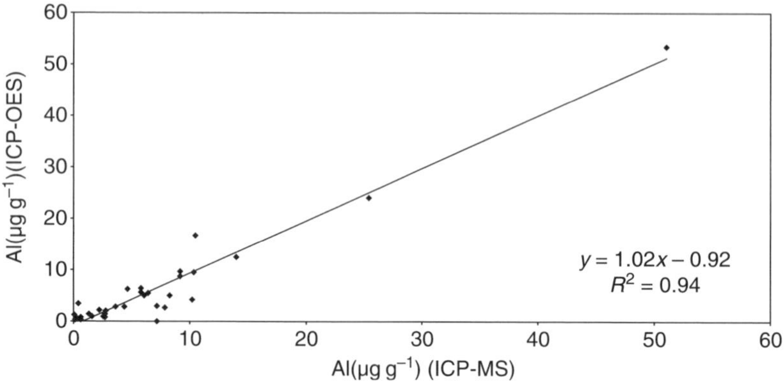

Figure 10.1. Comparison between Al concentrations (μg g^{-1}) in foods and cooked meals as determined by ICP-AES and ICP-MS.

Toxic and essential trace elements intakes were computed from food consumption data of the population in northern Italy and element concentration measurements. The results (Table 10.6) were compared with the Italian Recommended Dietary Allowance (RDA) for essential trace elements and with the Provisional Tolerable Daily/Weekly Intake (PTDI/PTWI) for toxic elements. Finally, an assessment was made of the extent to which each food contributes to the total intake of trace elements by the population under study and the most contaminated foods were also detected. On the whole, mean intake values satisfied the Italian RDA for all the essential trace elements [31], while the potentially toxic trace elements were below the relevant threshold limits.

The Al intake is 3.6 mg day^{-1} (range, 0.3–5.2 mg day^{-1}), which is well below the PTWI (7 mg week^{-1} per kg body weight for a 65 kg adult male) [1], and falls within the lowest range of intake reported for Western diets (3–14 mg day^{-1}) [32].

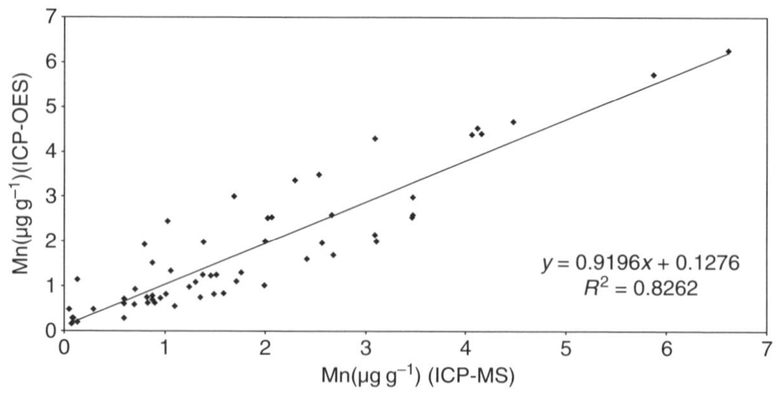

Figure 10.2. Comparison between Mn concentrations (μg g^{-1}) in foods and cooked meals as determined by ICP-AES and ICP-MS.

TABLE 10.6. Essential and Toxic Element Daily Intakes in the Population of Northern Italy

Trace Elements	Mean Value	Minimum Value	Maximum Value	PTDI[a] or RDA[a]	% of PTDI
Al[b] (mg)	3.58	0.28	5.19	65	6
B[c] (mg)	1.65	0.54	3.19	13	13
Cd[b] (μg)	13.6	1.4	32.7	65	21
Co[b] (μg)	28.9	15.4	57.3	—	—
Li[b] (μg)	29.9	8.2	42.4	100	30
Mo[d] (μg)	79.6	32.6	106.2	50–100	—
Mn[d] (mg)	1.38	0.45	2.19	1–10	—
Ni[c] (μg)	361.1	45.0	764.2	600	60
Pb[b] (μg)	55.2	25.8	66.6	232	24
Rb (mg)	1.19	0.66	2.27	—	—
Sb[b] (μg)	8.5	4.1	19.6	—	—
Se[d] (μg)	228	52.0	524.8	—	—
Sn[b] (mg)	1.57	0.006	4.44	140	1
Sr (mg)	1.14	0.67	1.88	—	—
V[c] (μg)	12.2	3.3	17.9	—	—

[a]From WHO [1] for PTDI and from LARN [31] for RDA.
[b]Potentially toxic elements.
[c]Probably essential elements.
[d]Essential elements.

Food categories that mostly contribute to the Al total daily intake (see Table 10.7) are cereals and tubers (32 percent), and milk and dairy products (33 percent), which together make up 65 percent of total daily metal intake as reported in the literature [33].

The B intake is 1.65 mg day^{-1} (range 0.54–3.19 mg day^{-1}), which is slightly higher than the mean basal requirements (range 0.5–1.35 mg day^{-1}), but below the toxic level of 13 mg day^{-1} [1]. The intake is similar to that reported for 22 Canadian women, 1.33 $\pm$ 0.13 mg day^{-1} [34], but lower than that reported for Britons, 2.8 $\pm$ 1.5 mg day^{-1} [34]. The food category that mostly contributes to the B total daily intake (see Table 10.7) is vegetables and fruit (47 percent) [1].

The Cd intake is 14 μg day^{-1} (range 1.4–33 μg day^{-1}), which is well below the PTDI of 70 μg day^{-1} for a 65 kg adult male [1] and close to the values reported in various European countries, ranging from 10 to 30.2 μg day^{-1} [35]. The food categories that mostly contribute to the Cd total daily intake (see Table 10.7) are cereals and tubers (55 percent) followed by vegetables and fruit (30 percent).

As regards the Co intake, this is 29 μg day^{-1} (range 15–57 μg day^{-1}), about one order of magnitude higher than the mean reported by Noel et al. [36], 4 μg day^{-1} for the French population, and that reported for Canadians, 11 μg day^{-1} [37]. Food categories that most contribute to the Co total daily intake (see Table 10.7) are cereals and tubers (41 percent), followed by vegetables and fruit (27 percent).

TABLE 10.7. Percentage of Trace Element Intake According to Food Categories

	Trace Element Intake (%)														
Food Categories	Al	B	Cd	Co	Li	Mo	Mn	Ni	Pb	Rb	Sb	Se	Sn	Sr	V
Cereals and tubers (bread, pizza, pasta, rice, and potatoes)	32	12	55	41	31	45	62	34	51	26	30	34	97	30	37
Meat, fish and seafood, eggs (beef, pork, poultry, ham, salami, fish, seafood, eggs)	13	10	15	22	20	16	10	20	17	26	19	20	2	15	18
Milk and dairy products (milk, yogurt, cheese)	33	16	0	5	6	8	3	21	4	7	14	21	0	17	5
Vegetables and fruit (vegetables, legumes, citrus fruit, fresh fruit)	18	47	30	27	33	25	20	15	17	30	22	15	1	30	23
Sweet foods (dessert, ice cream)	2	0	0	2	3	3	2	0	4	4	1	0	0	3	2
Beverages (mineral water, soft drinks, wine, beer)	2	15	0	3	7	3	3	10	7	7	14	10	0	5	15

Furthermore, the Li intake is 30 μg day^{-1} (range 8–42 μg day^{-1}), which is below the values reported for American diets (60–70 μg day^{-1}) and Turkish diets (102 μg day^{-1}), but similar to Finnish intake values (35 μg day^{-1}) [38] and French values (11 μg day^{-1}) [35]. Food categories that mostly contribute to the Li total daily intake (see Table 10.7) are vegetables and fruit (33 percent) followed by cereals and tubers (31 percent).

In the case of Mo, the intake is 79.6 μg day^{-1} (range 32.6–106.2 μg day^{-1}), which falls within the Italian recommended range of 50–100 μg day^{-1} [31] and is the lowest value of the range of 80–250 μg day^{-1} reported in four studies by IAEA [39]. The estimated mean daily intakes of French consumers is 112 μg day^{-1} [35]. The food categories to which most of the Mo total daily intake can be ascribed are cereals and tubers (45 percent) and vegetables and fruit (25 percent; see Table 10.7).

The Mn intake is 1.38 mg day^{-1} (range 0.45–2.19 mg day^{-1}), which hardly meets the Italian recommended values of 1–10 mg day^{-1}. On the other hand, it is worth noting that most data in the literature concerning the habitual diets all over the world range from 0.36 to 1.78 mg day^{-1} [40]. Higher Mn intake values were found for the adult population (3.1 mg day^{-1} in Canada, 2.7 mg day^{-1} in New Zealand, and 2.9 mg day^{-1} in the United States) [41–43]. The food category that mostly contributes to the Mn total daily intake is cereals and tubers (see Table 10.7; 62 percent).

In turn, the Ni intake is 361 μg day^{-1} (range 45–764 μg day^{-1}). The maximum intake is around 600 μg day^{-1}, the highest acceptable value [1] above which sensitized individuals may show allergic symptoms. The value obtained in this study exceeds the ranges allowed in the UK (140–150 μg day^{-1}) [44, 45], the United States (69–162 μg day^{-1}) [46], Denmark (60–260 μg day^{-1}), [47], and France (74 μg day^{-1}) [35]. The food category that mostly contributes to the Ni total daily intake is cereals and tubers (see Table 10.7; 34 percent).

The Pb intake is 55.2 μg day^{-1} (range 25.8–66.6 μg day^{-1}), which is below the PTDI of 232 μg day^{-1} for a 65 kg adult male (25 μg kg^{-1} of body weight weekly), but within the ranges reported for various European countries, from 16 to 179 μg day^{-1} [23]. The median value obtained in the present study (50 μg day^{-1}) is lower than that reported for the 24 h median intakes in three areas of Belgium (95 μg day^{-1}) [48], but higher than that reported for the French population (34 μg day^{-1}) [36]. The food category that gives place to the largest Pb total daily intake (see Table 10.7) is cereals and tubers (51 percent).

In the case of Rb, the intake is 1.37 mg day^{-1} (range 0.66–2.27 mg day^{-1}), slightly lower than the value reported in the USA (2.44 ± 0.54 mg day^{-1}) [41]. Cereals and tubers are the food category that mostly contributes to the Rb total daily intake (34 percent; see Table 10.7).

The Sb intake is 8.5 μg day^{-1} (range 4.1–19.6 μg day^{-1}), that is, twice as much as the value of 4 ± 0.5 μg day^{-1} reported for the United States [41]. The food category that most contributes to Sb total daily intake is cereals and tubers (30 percent), and vegetables (20 percent; see Table 10.7).

The Se intake is 228 μg day^{-1} (range 52–525 μg day^{-1}); it is higher than the value of 110 ± 8 μg day^{-1} reported in the USA [41], but still in the range ascertained for other countries ranging from 6 to 500 μg day^{-1} [49]. Almost only cereals and tubers (34 percent) contribute significantly to the Se total daily intake (see Table 10.7).

The intake of Sn is 1.57 mg day^{-1} (range 0.006–4.44 mg day^{-1}), that is, below the PTWI of 14 mg kg^{-1} body weight [50]. It is slightly higher than the value of 0.65 mg day^{-1} reported in The Netherlands [51], but similar to that reported in a 1997 UK Total Diet Study [44]. The food category that almost alone contributes to the Sn total daily intake is cereals and tubers (97 percent; see Table 10.7). The use of canned tomatoes for cooking both pasta and pizza may well play a significant role.

The Sr intake is 1.14 mg day^{-1} (range 0.67–1.88 mg day^{-1}), that is, slightly lower than the value of 1.62 $\pm$ 0.24 mg day^{-1} reported in the USA [45]. This can be traced back to food categories such as cereals and tubers (34 percent), milk and dairy products (21 percent), and meat, fish and seafood (20 percent; see Table 10.7).

As regards the V intake, a value of 12.2 μg day^{-1} (range 3.3–17.9 μg day^{-1}) has been ascertained. This meets the recommended daily intake of 10 μg day^{-1} [1] and is largely below 10 mg day^{-1}, the highest acceptable value above which toxic symptoms are shown [52]. The intake value obtained in this study is close to the range of 6.2–18.3 μg day^{-1} reported in the USA total diet study of 1987 [32] and to that reported by Myron et al. of 12.4–30.1 μg day^{-1} [53] and compares well with the value of 13 μg day^{-1} reported in the UK total diet study of 1985 [54]. The food category that mostly contributes to the V total daily intake is cereals and tubers (37 percent; see Table 10.7).

According to the food categories mentioned above, the percentage distribution of trace element intakes for the population of northern Italy could be estimated. As an example, the typical intakes (percent) of Cd, Co, Se, and Sn are shown in Figures 10.3–10.6.

The results obtained in this study show that sweets and beverages poorly contribute to trace element intake, while cereals (bread, pasta, pizza, rice) and tubers (potatoes), followed by vegetables and fruit, contribute primarily to total daily intakes. The mean intake through cereals and tubers is between 30 and 55 percent

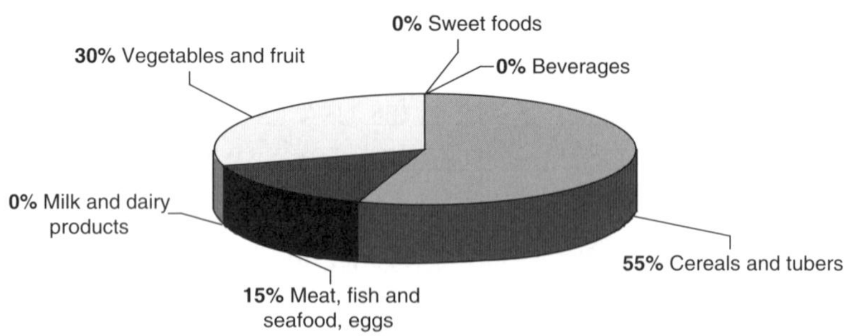

Figure 10.3. Percentage of Cd intake in the population of northern Italy according to food categories.

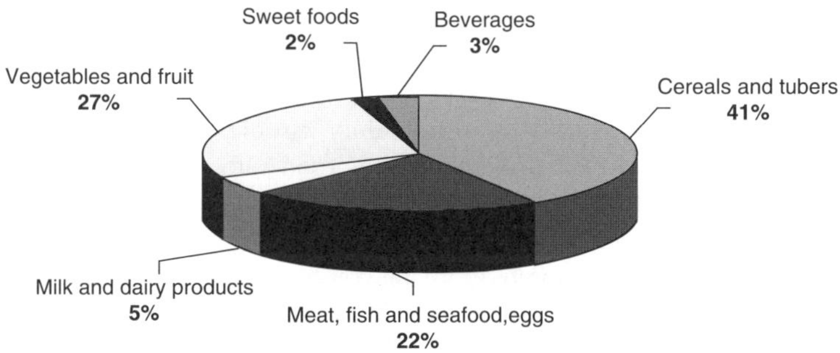

Figure 10.4. Percentage of Co intake in the population of northern Italy according to food categories.

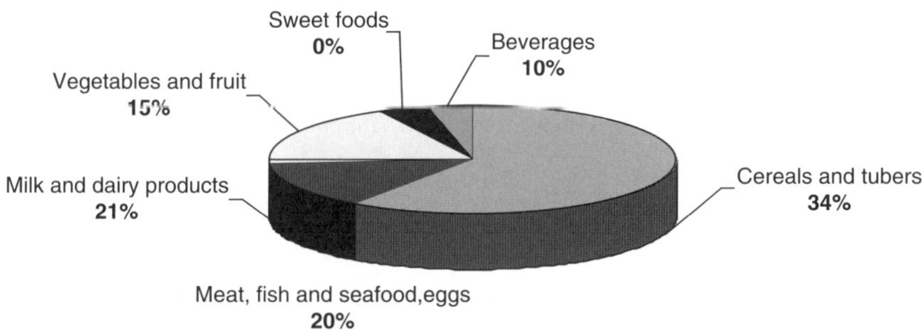

Figure 10.5. Percentage of Se intake in the population of northern Italy according to food categories.

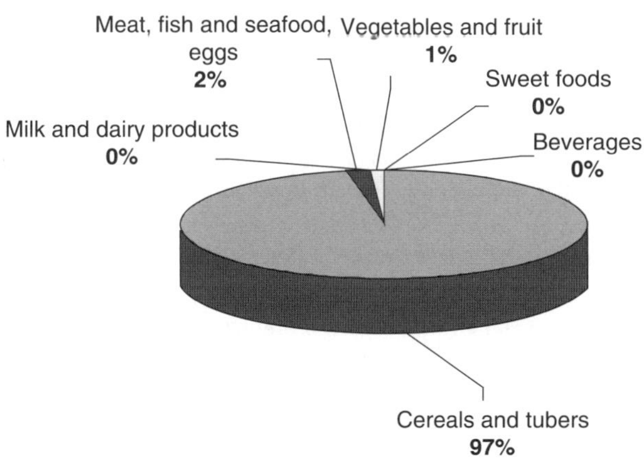

Figure 10.6. Percentage of Sn intake in the population of northern Italy according to food categories.

for Al, Cd, Co, Li, Mo, Ni, Pb, Rb, Sb, Sr, and V, 62 percent for Mn, and 97 percent for Sn. The exceedingly high value of Sn can be explained by the canned tomatoes used for cooking pasta and pizza. The lowest intake through cereals is that of B (12 percent), which is mainly associated with vegetables and fruit (47 percent). For all the other elements, this food category provides an intake ranging from 15 to 30 percent.

There is still much ongoing debate on the validity of the different approaches to perform dietary analysis (single food analysis, duplicate diet analysis, total diet, and total mixed diet analysis) to provide an adequate answer to nutritional questions. The approach followed in this study has proven useful for the assessment of the actual dietary intake of trace metals that may be lost and/or added during the preparation of meals. The measurements done on cooked samples prepared according to the Italian local traditions provides an assessment of the actual concentrations of trace elements uptaken by the population through the diet.

REFERENCES

1. World Health Organization, *Trace Elements in Human Nutrition and Health*, Geneva 1996.
2. K. Van Dokkum, Trace element intake in Europe: safe and adequate? In B. Sandstrom, P. Walter (Eds), *Role of Trace Elements for Health Promotion and Disease Prevention*, 1998, 8–17.
3. Y. Miyamoto, A. Kajikawa, J. H. Zaidi, T. Nakanishi, K. Sakamoto, Minor and trace element determination of food spices and pules of different origins by NAA and PAA, *J. Radioanal. Nucl. Chem.*, **234** (2000), 747–765.
4. C. P. Sanchez-Castillo, P. J. S. Dewey, M. D. M. Solano de Lourdes, W. P. T. James, The mineral and trace element content of Mexican cereals, cereals products, pulses and snacks: preliminary data. *J. Food Comp. Anal.*, **10** (1997), 312–333.
5. E. L. Gunderson, FDA total diet study, July 1986–April 1991. Dietary intakes of pesticides, selected elements and other chemicals, *J. AOAC Int.*, **78** (1995), 1353–1363.
6. D. Voutsa, A. Grimanis, C. Samara, Trace element in vegetables grown in an industrial area in relation to soil and air particulate matter, *Environ. Pollut.*, **94** (1996), 325–335.
7. L. Ràcz, L. Papp, B. Prokai, Zs. Kovács, Trace element determination in cultivated mushrooms: an investigation of manganese, nickel and cadmium intake in cultivated mushrooms using ICP atomic emission, *Microchem. J.*, **54** (1996), 444–451.
8. A. Alberti-Fidanza, G. Burini, G. Perriello, Trace elements in food and meals consumed by students attending the faculty cafeteria, *Sci. Total Environm.*, **287** (2002), 133–140.
9. E. Coni, A. Bocca, D. Ianni, S. Caroli, Preliminary evaluation of the factors influencing the trace element content of milk and diary products, *Food Chem.*, **52** (1995), 123–130.
10. S. J. Joos, M. Betts, Copper intakes and consumption patterns of chocolate foods as sources of copper for individuals in the 1987–88 nationwide food consumption survey, *Nutr. Res.*, **16** (1996), 41–52.

11. E. Coni, A. Bocca, P. Coppolelli, S. Caroli, C. Cavallucci, M. Trabalza Marinucci, Minor trace element content in sheep and goat milk and diary products, *Food Chem.*, **57** (1996), 253–260.

12. R. M. Tripathi, R. Raghunath, V. N. Sastry, T. M. Krishnamoorthy, Daily intake of heavy metals by infants through milk and milk products, *Sci. Total Environ.*, **227** (1999), 229–235.

13. E. Rossipal, M. Krachler, Pattern of trace elements in human milk during the course of lactation, *Nutr. Res.*, **18** (1998), 11–24.

14. G. Turconi, M. Guarcello, C. Livieri, S. Comizzoli, L. Maccarini, A. M. Castellazzi, A. Pietri, G. F. Piva, C. Roggi, Evaluation of xenobiotics in human milk and ingestion by the newborn: an epidemiological survey in Lombardy (Northern Italy), *Eur. J. Nutr.*, **43** (2004), 191–197.

15. B. D. Struck, R. Pelzer, P. Ostapczuck, H. Emons, C. Mohl, Statistical evaluation of ecosystem properties influencing the uptake of As, Cd, Co, Cu, Hg, Mn, Ni, Pb, and Zn in seaweed (*Fucus vesiculosus*) and common mussel (*Mytilus edulis*), *Sci. Total Environ.*, **207** (1997), 29–42.

16. G. P. Santaroni, G. Ingrao, P. Belloni, L. Gabelli, Levels of some trace elements in Italian diets. In: RAISA (Ed.), *Agrobiotechnologies in Agricultural Products and By-products Improvement Processes*, Volterra, CNR, 1992, 1087–1088.

17. L. Gambelli, P. Belloni, G. Ingrao, G. P. Santaroni, Some heavy metal levels in Italian diets. In: RAISA (Ed.), *Agrobiotechnologies in Agricultural Products and By-products Improvement Processes*, Sarteano, CNR, 1994, 187–192.

18. C. Roggi, E. Merlo, C. Minoia, Dietary lead exposure, *Ann. Ig.*, **9** (1997), 59–65.

19. C. Roggi, E. Merlo, C. Minoia, Dietary nickel exposure, *Ann. Ig.*, **9** (1997), 221–227.

20. C. Roggi, L. Maccarini, M. C. Mancini, A. Ronchi, Dietary cadmium and mercury exposure, *Ann. Ig.*, **10** (1998), 95–101.

21. G. H. Biego, M. Joyeux, P. Hartemann, G. Debry, Daily intake of essential minerals and metallic micropollutants from foods in France, *Sci. Total Environ.*, **217** (1998), 27–36.

22. G. Casalicchio, *I Microelementi nell'Ecosistema Terrestre. Micronutrienti, Microutili e Microtossici nel Suolo, nei Vegetali, negli Animali e nell'Uomo*, Pitagora Editrice, Bologna, 2000.

23. L. Nasreddine, D. Parent-Massin, Food contamination by metals and pesticides in the European Union. Should we worry? *Toxicol. Lett.*, **127** (2002), 29–41.

24. A. Turrini, A. Saba, D. Perrone, E. Cialfa, A. D'Amicis, Food consumption patterns in Italy: the INN-CA Study 1994–1996, *Eur. J. Clin. Nutr.*, **55** (2001), 571–588.

25. H. J. Reid, S. Greenfield, T. E. Edmonds, Investigation of decomposition products of microwave digestion of food samples, *Analyst*, **120** (1995), 1543–1548.

26. I. Novozamsky, R. van Eck, V. J. G. Houba, J. J. van der Lee, Solubilization of plant tissue with nitric acid, hydrofluoric acid, hydrogen peroxide in a closed-system microwave digestor, *Commun. Soil Sci. Plant Anal.*, **27** (1996), 867–875.

27. Da-Hai Sun, J. K. Waters, T. P. Mawhinney, Microwave digestion for determination of aluminium, boron and 13 other elements in plants by inductively coupled plasma atomic emission spectrometry, *J. AOAC Int.*, **80** (1997), 647–650.

28. E. J. Gawalko, T. W. Nowicki, J. Babb, R. Tkachuk, Comparison of closed-vessel and focused open-vessel microwave dissolution for determination of cadmium, copper,

lead, and selenium in wheat, wheat products, corn bran, and rice flour by transverse-heated graphite furnace atomic absorption spectrometry, *J. AOAC Int.*, **80** (1997), 379–387.

29. M. Hoenig, H. Baeten, H. Vanhentenrijk, E. Vassileva, Ph. Quevauviller, Critical discussion on the need for an efficient mineralization procedure for the analysis of plant material by atomic spectrometric methods, *Anal. Chim. Acta*, **358** (1998), 85–94.
30. E. S. Gladney, C. E. Burns, D. R. Perrin, I. Roelandts, T. E. Gill, 1982 *Compilation of Elemental Concentration Data for NBS Biological, Geological, and Environmental Standard Reference Materials*, NBS Special Publication, 1984, 260–288.
31. SINU, Società Italiana di Nutrixzione Umana, *Recommended intake values of energy and nutrients for the Italian population*, 1996, http://www.sinu.it/larn.asp.
32. J. L. Greger, N. J. Baier, Excretion and retention of low or moderate levels of aluminium by human subjects, *Food Chem. Toxicol.*, **21** (1983), 473–476.
33. J. A. T. Pennington, Aluminium content of foods and diets, *Food Addit. Cont.*, **5** (1988), 119–128.
34. W. B. Clarke, R. S. Gibson, Lithium, boron and nitrogen in 1-day composites and a mixed-diet standard, *J. Food Comp. Anal.*, **1** (1988), 209–220.
35. E. I. Hamilton, M. J. Minski, Abundance of the chemical elements in man's diet and possible relations with environmental factors, *Sci. Total Environ.*, **1** (1972/1973), 375–394.
36. L. Noel, J. C. Leblanc, T. Guerin, Determination of several elements in duplicate meals from catering establishments using closed vessel microwave digestion with inductively coupled plasma mass spectrometry detection estimation of daily dietary intake, *Food Addit. Contam.*, **20** (2003), 44–56.
37. R. W. Dabeka, A. D. McKenzie, Survey of lead, cadmium, fluoride, nickel, and cobalt in food composites and estimation of dietary intakes of these elements by Canadians in 1986–1988, *J. AOAC Int.*, **78**(4), (1995), 897–909.
38. G. V. Iyengar, W. B. Clarke, R. G. Dawing, Determination of boron and lithium in diverse biological matrices using neutron activation – mass spectrometry (NA-MS), *Fresenius J. Anal. Chem.*, **338** (1990), 562–566.
39. R. M. Parr, Human dietary intakes of trace elements: a global literature survey mainly for the period 1970–1991: data listings and references, Report IAEA-NAHRES-12, Vienna, International Atomic Energy Agency, 1992.
40. A. Gormican, Inorganic elements in foods used in hospital menus, *J. Am. Diet. Assoc.*, **56** (1970), 397–403.
41. G. V. Iyengar, W. R. Wolf, J. T. Tanner, E. R. Morris, Content of minor and trace elements, and organic nutrients in representative mixed total diet composites from USA, *Sci. Total Environ.*, **256** (2000), 215–226.
42. R. S. Gibson, C. A. Scythes, Trace element intakes of women, *Br. J. Nutr.*, **48** (1982), 241–248.
43. B. E. Guthrie, M. F. Robinson, Daily intakes of manganese, copper, zinc and cadmium by New Zealand women, *Br. J. Nutr.*, **38** (1977), 55–63.
44. G. A. Smart, J. C. Sherlock, Nickel in foods and the diet, *Food Addit. Contam.*, **4** (1987), 61–71.

45. G. Ysart, P. Miller, M. Croasdale, H. Crews, P. Robb, M. Baxter, C. de L'Argy, N. Harrison, 1997 UK Total Diet Study. Dietary exposure to aluminium, arsenic, cadmium, chromium, copper, lead, mercury, nickel, selenium, tin, and zinc, *Food Addit. Contam.*, **17** (2000), 775–786.

46. J. A. T. Pennington, J. W. Jones, Molybdenum, nickel, cobalt, vanadium, and strontium in total diets, *J. Am. Diet. Assoc.*, **87** (1987), 1644–1650.

47. N. K. Veien, M. R. Andersen, Nickel in Danish food, *Acta Dermato-venereologica*, **66** (1986), 502–509.

48. J. P. Buchet, R. Lauwerys, A. Vandevoorde, J. M. Pycke, Oral daily intake of cadmium, lead, manganese, copper, chromium, mercury, calcium, zinc, and arsenic in Belgium: duplicate meal study, *Food Chem. Toxicol.*, **21** (1983), 19–24.

49. C. Reilly, Selenium: a new entrant into the functional food arena, *Trends Food Sci. Technol.*, **9** (1998), 114–118.

50. World Health Organization, *Evaluation of Certain Food Additives and Contaminants*, Thirty-third report of the Joint WHO Expert Committee on Food Additives, WHO Technical Report Series, 776 (1989).

51. W. Van Dokkum, Minerals and trace elements in total diets in the Netherlands, *Br. J. Nutr.*, **61** (1989), 7–15.

52. E. Dimond, J. Caravaca, A. Benchimol, Vanadium: excretion, toxicity, lipid effect in man, *Am. J. Clin. Nutr.*, **12** (1963), 49–53.

53. D. R. Myron, T. J. Zimmermann, T. R. Shuler, I. M. Klevay, D. E. Lee, F. H. Nielsen, Intake of nickel and vanadium by humans. A survey of selected diets, *Am. J. Clin. Nutr.*, **31** (1978), 527–531.

54. W. W. Evans, J. I. Read, D. Caughlin, Quantification of results for estimating elemental dietary intakes of lithium, rubidium, strontium, molybdenum, vanadium, and silver, *Analyst*, **110** (1985), 873–877.

11

CAR CATALYTIC CONVERTERS AND THE CONTAMINATION OF FOOD BY PLATINUM-GROUP ELEMENTS

CHIARA FRAZZOLI, ROBERTA CAMMARONE, AND SERGIO CAROLI

Istituto Superiore di Sanità, Rome, Viale Regina Elena 299, 00161, Italy,
E-mail: frazzoli@iss.it, caroli@iss.it

11.1 INTRODUCTION

As prescribed by the European Directive 94/12/EEC, the US EPA 1970 Clean Air Act, and other legal provisions, by the year 2000 pre-established emission standards must be complied with for all new cars. Cars registered as from 1993 are fitted with catalytic converters for treating exhaust gases and thus significantly

The Determination of Chemical Elements in Food: Applications for Atomic and Mass Spectrometry.
Edited by Sergio Caroli

reduce (up to 90 percent) the hazardous emission of hydrocarbons, CO and NO_x. Over the years various catalytic control approaches have been developed in order to achieve more stringent toxic emission standards.

The noble metals of the platinum-proup elements (PGEs), basically Pd, Pt, and Rh, were introduced as active components of catalytic converters in 1975. These devices use an alumina or cordierite monolith treated with an Al_2O_3 washcoat, which in turn is doped with rare earth oxides and up to 0.10–0.15 percent (w/w) of the said PGEs. At an early stage, Pt was the active metal at the highest concentration in the catalyst, which typically contained 0.08 percent Pt, 0.04 percent Pd, and 0.005–0.007 percent Rh. Progress in the catalytic converter technology and the need to achieve the performance required by car manufacturers prompted a dramatic change in the relative amounts of PGEs, and Pt was partly replaced by Pd and Rh [1]. The role of Pt and Pd is the high-temperature catalytic oxidation of hydrocarbons and CO into the less toxic CO_2, whereas Rh reduces NO_x primarily to N_2.

During normal vehicle operation, both the thermal cracking of the catalyst structure and/or the mechanical surface abrasion of the catalytic converters cause PGEs to be released via the exhaust pipe into the surrounding environment [2]. A range of 0.5–1.4 kg $year^{-1}$ for the annual amount of Pt emitted from automobile catalytic converters was reported [3], with an average daily emission factor in the magnitude order of nanograms per kilometer along roadsides. PGE environmental concentration is positively correlated to high speed [4, 5] and performance of the catalytic devices [6], even if the frequent "stop–start" pattern typical of urban settings seems to most severely contribute to the breakdown process by subjecting the converter to very large and rapid temperature gradients.

It is widely accepted [7, 8] that most of the PGEs from catalytic converters spread into the atmosphere as nanocrystalline particles (mainly in the zero oxidation state, but also as oxides) attached to alumina particles. The amount of PGE released into the environment by the catalyst can be measured by determining PGE content in the exhaust fumes collected directly from the exhaust pipe. Palacios et al. [6] reported that most PGEs are released in particulate form ($Pt > 95$ percent, $Pd > 85$ percent, and $Rh > 90$ percent) and described the dependence of particle size distribution on the production process of the catalytic converter. At a simulated speed of 80 km h^{-1} 70 percent of the particles are inhalable [9], and mostly deposit in the upper respiratory tract (diameter larger than 10.2 μm). Particles with a diameter less than 10 μm (thoracic particle fraction or PM_{10}) can reach the tracheobronchial region, whereas a minor fraction has a diameter small enough (<4 μm) to pass the upper respiratory system, and deposits in the alveolar region of the human lung [10, 11].

11.2 SOURCES OF ENVIRONMENTAL POLLUTION BY PGEs

11.2.1 Car Catalytic Converters

Since PGE-based car catalytic converters were introduced, the increasing amounts of polluting PGEs have prompted extensive research and interest in determining their concentration in environmental samples (such as soil, grass, airborne particles, sludge, water, and road dust) to make a realistic assessment of the risk that this ongoing phenomenon represents for man and the environment.

Although a large number of parameters affect airborne PGEs content (e.g., particle distribution, traffic type, meteorological conditions, and different driving conditions), engine test experiments under city traffic simulating conditions have been performed. The determination of PGEs in exhaust fumes shows that they are released in the nanograms per kilometer range [2, 4, 10, 12].

Gómez et al. [13] as well as Caroli et al. [14] reported that PGE levels at traffic sites were significantly higher than in rural or nonpolluted zones. As in the case of airborne particulate matter, the adoption of car catalysts seems to have caused the increase in PGEs road dust up to the nanograms per gram level. Additional information supporting the assumption of PGEs enrichment in environmental samples as a consequence of traffic emissions, for example, soil, sediments, and tunnel and road dust, has been gained along with data on the ability of grass and other plants to take up and accumulate PGEs under controlled experiments. Gómez et al. [13] and Caroli's team [15] quote abundant literature on the results obtained in different European and non-European cities. A study was also conducted in Ghana, were the levels of Pt were found to be similar to those in Europe, probably as a consequence of the increasing number of old cars equipped with catalytic converters [16].

Palladium and Pt contents in surface soil were found to decrease exponentially with increasing distance from highways [17]. On the other hand, these metals can also be transported to water bodies through run-off [18]. Medium- and long-term investigations have provided evidence of a constant increase in environmental PGEs [19, 20]. In fact, in their monitoring study of Pt concentration in sediments from the Upper Mystic Lake (Boston), Rauch and Hemond [19] recorded a level of Pt relatively constant until 1975, when car catalytic converters were first introduced in the United States. The subsequent steep increase, followed by a plateau from 1992 onward, is deemed to be probably due, respectively, to the high rate of car catalytic converters adopted in the early years and then to the relatively stable number of cars equipped with these devices as well as to improvements in catalysts technology and composition. As for road sediments, it was found that the Pt concentration in samples collected at a car park in Göteborg increased by 3–6 orders of magnitude over 10 years [3].

11.2.2 Other Anthropogenic Sources

In 1999 Schäfer et al. [20] observed a significant increase in Pt in sewage sludge incineration ashes from municipal sewage plants over the period 1993–1997. These findings substantiated the assumption that urban sludge not only receives emissions from traffic, but also from other anthropogenic sources. From the comparison of data available on PGEs concentrations in exhaust fumes and in environmental matrices and the subsequent modeling with traffic statistics, an inconsistency was spotted [2] suggesting that, even if the major input is attributable to the wide use of PGE-based car catalytic converters, PGE emission is not limited to these devices, and other sources must be taken into account, such as street materials, metallurgical plants, fertilizers or pesticides, rubber tyres, and fuel products. The contribution of sources other than catalysts was, however, rather small with respect to that of catalytic converters.

Because of its properties and resistance to oxidation and corrosion, Pt is a good catalyst widely used in different industrial processes, such as polymerization of olefins, gasoline refining, or ammonia oxidation as well as catalytic purification of industrial exhausts. It is also used as a reinforcing agent for plastic and concrete, in insulation for buildings, in the hulls of ships, fuselages of aircraft and petrol tankers, and in everyday products such as ladders and fishing rods. Many new technologies, from electronic industries, mobile telephones, or glass fiber optics for telecommunications to computer screens, make use of PGEs. In 1998 very high Pt levels (1.1 mg kg^{-1}) were found in sludge from cities with high numbers of jewellery manufacturing producers [21]. Taking into account other everyday uses such as in fountain pens, razors and jewellery, it is clear that the extent of use of PGEs will increasingly be an issue of concern in the years to come. So far, however, there is little information available about the contribution of these sources to the overall load of PGEs in the environment [22, 23]. Figure 11.1 depicts the possible pathway of PGEs from car catalytic converters to the environment and food.

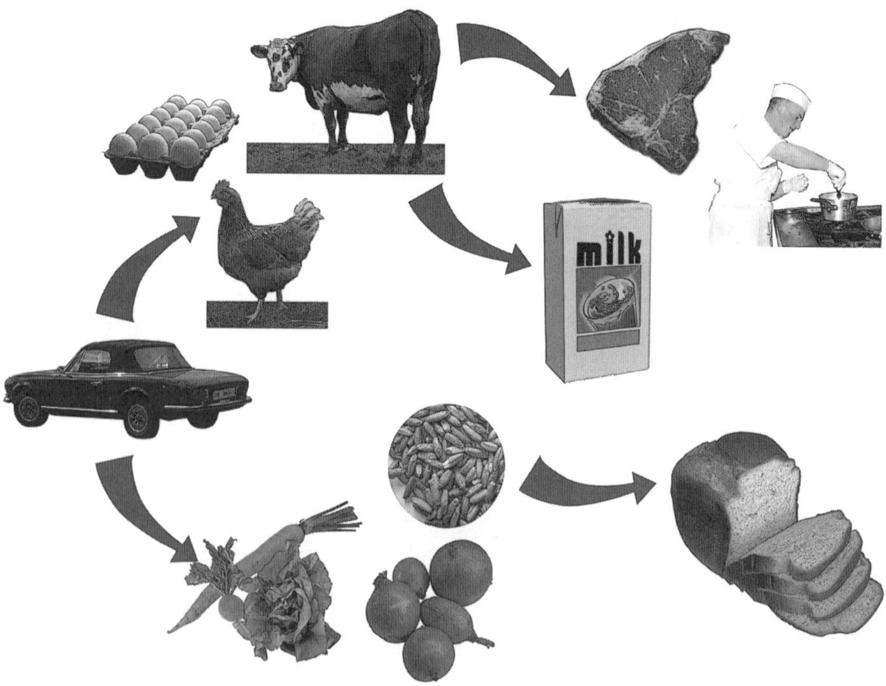

Figure 11.1. Possible pathway of PGEs from car catalytic converters to the environment and food.

As with Pt, the levels of Pd and Rh in the environment are also increasing more quickly than would be expected if catalytic converters were the only source. A reason for this was given by Palacios et al. [6], who inferred that other automobile-related sources of Pd and Rh (such as fuel, oil, spark plugs, or engine abrasion) could be playing a nonnegligible role. Like Pt, these elements primarily come from catalytic converters, but have a higher mobility in the environment than Pt. This possibility was substantiated Rauch et al. [24], who reported that, when analyzing road and urban river sediments, Pt turned out to be attached to particles containing Ce (another element released by car catalytic converters, where it acts as a stabilizer), while Pd and Rh were not and were therefore more easily mobilized.

PGEs are also widely used in medicine, for example, as active components of antitumoral drugs and in a range of surgical implants and prosthetics. Since the late 1970s, Pd has been used worldwide by the dental industry in the development of Au–Pd and affordable Pd-based restorative dentistry. A number of properties make it fit for the manufacture of dental crowns and bridge alloys [25], that is, good solubility with several metals, low specific gravity, ability to impart good mechanical properties, excellent tarnish and corrosion resistance, and a melting point high enough to sustain the firing temperatures of most porcelains (950–1020°C). Palladium-based alloys are also biocompatible with the oral environment and cheaper than those of either Au or Pt. Dental alloys can contain up to 18 percent Pt and up 80 percent Pd. For instance, in 1975 Pd and Pt concentrations in urine and saliva of subjects with dental prostheses of this kind were reported to be higher by factor of 10–800 than in the case of individuals without such prostheses [25].

Pollution from medical applications of PGEs has also raised some concern. In fact, more than 30 years ago Pt coordination complexes were introduced in the clinical practice as very active antitumoral agents. After Rosenberg's discovery in 1969 [26], clinical trials began of *cis*-2-6 diamminedichloroplatinum(II) (*cis*-DDP, also called *cis*-platin) as an anticancer drug, and its effective activity in the clinical treatment of a variety of human cancers was subsequently shown. The introduction of Pt complexes as cytotoxic agents into chemotherapy in 1978 gave rise to a host of studies on the chemistry and biochemistry of these compounds. Pharmacokinetic studies on the one hand and the need to minimize toxic side effects on the other hand stimulated the interest in the determination of ultratrace amounts of Pt in biological fluids and tissues. *cis*-Platin was originally licensed for its antineoplastic activity against a wide range of human tumors. Its proven antiproliferative activity drove the oncologic community to prepare and test many new Pt derivatives. Thus, in an attempt to improve the efficaciousness of chemotherapy, reduce the severe side-effects of *cis*-platin (e.g., the dose-limiting nephrotoxicity), and increase the doses that can be administered, second- and third-generation platinum complexes were developed. As neither hospital sewage nor urine and excretions from patients administered with Pt-based drugs undergo any special treatment, effluents from hospitals have become another, nonnegligible, source of Pt-polluting waste water and sewage sludge [27]. As reported by Kümmerer and Helmers [28], only a small part (10–20 percent) of carboplatin (another Pt-based antiblastic drug) is bound to protein and is thus bioavailable.

In the first 24 h after administration, 50–70 percent of the administered carboplatin is excreted with urine and thus discharged with the hospital effluents, whereas 51 days are required for clearing 31–85 percent of the administered *cis*-platin dose. It is noteworthy that the half-life of the excretion of the remainder is very long, as evidenced by the findings of Schierl et al. [29] on the long-term pharmacokinetics of this drug, with a urinary Pt concentration a factor of 40 higher than normal 8 years after the last administration. On the other hand, Kümmerer and Helmers demonstrated that pollution due to Pt-based anticancer drugs caused by hospitals is not particularly significant even though nonnegligible (in the range of 3.3–12.6 percent) when compared with Pt emissions from cars equipped with catalytic converters. In 1999 Kümmerer et al. estimated a total release of 14.2 kg of Pt in 1996 in Germany for this kind of source versus a total amount of 187.2 kg of Pt ascribable to cars [30].

11.3 BIOAVAILABILITY AND TOXICITY OF PLATINUM-GROUP ELEMENTS

The emission of noble metals and the possible consequences for man and the environment are still being debated. Although PGEs belong to the group of elements with the lowest abundance in the earth's crust, with a mean concentration of 5 $\mu g\ kg^{-1}$ [31], their increasing use and their potential for polluting the environment are threatening human health. These precious metals have long been considered harmless, because of their inertness, nontoxicity and nonallergenicity in the metallic state [32]. In contrast, some compounds, especially the soluble salts, were found to be toxic. In particular, chronic industrial exposure to them is responsible for a syndrome known as platinosis which is characterized by respiratory and cutaneous hypersensitivity [33]. On the other hand, metallic Pd is known to cause contact dermatitis.

In 1989 the increasing awareness of the importance of the noxious effects induced by Pt anticancer drugs, that is, a certain level of mutagenicity, carcinogenicity, and/or teratogenicity, prompted a survey on the occupational hazard caused to sanitary personal by the manipulation and exposure to such antiproliferative drugs [34]. Some Pt coordination complexes used in the treatment of tumors have been shown to cause extensive DNA damage when delivered at high doses [35]. Adverse effects of industrially used coordination complexes were also observed by Büngerr et al. in 1996 for Pd and Rh [36]. In particular, cytotoxic effects of Pd salts were three times lower than those found for analogous Pt compounds, while the Rh salts proved to be 30 times less toxic. The authors also reported that the levels of toxicity of different complexes of the same metal do not differ significantly from one another, indicating that the metal itself is toxic. As regards mutagenicity, Rh salts turned out to be considerably less mutagenic than the analogous Pt compounds, and the tested Pd salts gave no evidence of any mutagenic property. The Pt and Pd genotoxicity was also studied by Gebel et al. in 1997 [37].

In several occupational cases and epidemiological studies, it was proved that exposure to halogenated Pt salts could cause respiratory sensitization. The relatively high sensitizing effects of some Pt complexes is well documented [22], for example, through the studies of Rosner and Merget on haxachloroplatinate and tetrachloroplatinate salts [38]. The persistence of Pt salts sensitization and the long-lasting adverse health consequences (such as rhinitis, asthma conjunctivitis, and contact urticaria) among precious metals refinery workers have been thoroughly investigated [39–41]. Skin reactivity was significantly associated with average air concentrations of Pt salts in employees' work area. On the other hand, Venables et al. [42] confirmed cigarette smoking as a risk factor for the sensitization to Pt salts. Some Pt complexes may interfere with essential enzymatic activity by binding the N and S atoms in proteins [43, 44].

As regards occupational exposure, Pt compounds such as hexachloroplatinic (IV) acid have been studied with regard to respiratory allergic and skin reactions [45, 46]. The occupational exposure to PGEs of employees, including workmen, in Pt refineries or in Pt catalyst plants (for the assembly or recycling of motor vehicle catalysts), were also investigated and a number of papers report the significant increase (up to 100 times) of the Pt concentrations in body fluids of exposed workers [47–49]. Analogously, despite standard safety precautions and good work practices, elevated urinary Pt levels were found in pharmacy technicians regularly involved in the preparation of antineoplastic drugs [50].

As few data are available on human toxicity of PGEs, several methods have been developed to quantify extremely low levels of these elements in body fluids so as to monitor their intake via the inhalation route. Occupationally related uptake of Pt by inhalation in urban settings in the case of individuals heavily exposed to traffic, such as road workers, taxi and bus drivers, and police officers, showed a progressive increase in urinary Pt and Pd excretion and sensitization [51–54]. Even more worrying seem to be the findings by Farago et al. [55] regarding urine of workers in contact with precious metal salts. Moreover, Caroli et al. [56] investigated the exposure to Pd, Pt, and Rh of schoolchildren from the urban and suburban areas of Rome, and urine concentrations of Pd and Rh were found to be definitely associated with traffic density of the relevant areas of residence. A similar study on adults with no occupational exposure was conducted in Budapest and Vienna by Záray et al. [57]. The low values found by Caroli et al. in the case of children aged between 6 and 10 years can be explained by the data of Salma et al. [58], according to whom the deposition fractions in the alveolar compartment of the adult lung is ca. three times higher than in children, who have as a consequence an internal exposure lower than that of adults. The influence of different types and age of Pt- and Pd-containing dental alloys was found to be of great importance in patients [48, 52] and the exposure of dental technicians was also ascertained. Very little information has been published on the bioavailability and toxicity of Pd, Pt, and Rh after exposure via inhalation. The level of Pt in nasal lavage fluid, related to inflammation in the upper respiratory tract, was proposed as biomarker for traffic-related airborne particulate matter (PM) [59], whereas after 90 days of intratracheal inhalation by laboratory animals (rats) of particles simulating emission from car catalytic converters, Pt was found in blood, urine, faeces, and all important organs, and up to 30 percent of the finely dispersed Pt was bioavailable [10, 60].

After emission, the PGEs undergo various physical and chemical transformations, potentially resulting in migration and deposition into different environmental compartments [61–63]. Although the toxicology of PGEs is still an open question, effects on animal and plant organisms living in contaminated areas may well be of some concern and strongly linked to the bioavailability of these metals. With regards to the increase in PGEs emissions and to their health effects (from allergic to cytotoxic consequences), the final chemical forms of PGEs and their mobility, transformation, and bioavailability in the environment are of key importance. PGEs are located primarily in the vicinity of the emission sources, on roads, near industries, and so on. They can enter the food chain and accumulate in the human body through different pathways, for example, aerosol deposition, uptake from urban contaminated waste water, sludges, or hospital effluents, or direct input in soil via artificial fertilizers. The soluble fraction of the emitted PGEs, estimated to be less than 10 percent of the total amount [10], represents the amount immediately available for various organisms (from microorganisms and acquatic macroinvertebrates to plants) and for agriculture-related matrices (soils, animal feeds, and zootechnical products). It should not be overlooked, however, that some transformation reactions under environmental conditions, such as oxidation and complexation, can further increase the bioavailability of PGEs.

Bioaccumulation and bioconcentration routes of environmental PGEs were investigated in river water and sediments. Aquatic organisms feeding on sediments, such as the freshwater isopod *Asellus aquaticus*, were subjected to different studies [18, 62]. This isopod was selected for pollution monitoring studies because it lives on sediments and takes up metals through ingestion of polluted food particles and adsorption of metal dissolved in water. Moreover, it plays a key role in the food chain, being an important food source for fish and predators. Toxicity effects observed in the *Asellus aquaticus* exposed to Pt suggest that an increase in PGEs emission may affect the aquatic ecosystem as well as humans.

Some studies [64–66] report that the emission of Pt as microparticles makes this element rather bioavailable, allowing one to assume that it can be taken up by plants growing near roads, enter the food chain, and thus contribute to human internal exposure. In this regard, Ballach studied some plants exposed to dissolved Pt salts [67]. When exposed to Pt, cloned poplars were found to exhibit a reduced uptake of nutrient solutions and an impaired water uptake by the roots with the ensuing water stress. The author also showed a high accumulation rate of water-soluble Pt^{4+} and Ce in the plant roots along with the low Pt transport rate to the upper plant parts. Other studies [68, 69] substantiated the fact that Pt in the vegetative plant parts is mainly taken up by adsorption of airborne particles or dust deposition.

On the other hand, reactivity, bioavailability, and toxicity of Pt are not necessarily correlated with its total content, but depend also upon factors other than the soil properties and the metal-uptake kinetics of plants. Proteins are known to play an important role in the transport and metabolism of metals in both plants and animals, and studies of Pt speciation in plants have shown that under environmental conditions Pt binds to proteins of high molecular mass [70]. The pathways and the related toxicity of Pt species may therefore be closely correlated to the protein binding. In this regard, the assessment of the actual risk posed by PGEs to living organisms requires chemical speciation studies to identify their more bioavailable and toxic forms. In fact, the identification of the different chemical species and their physical properties (such as oxidation state, chemical bonds, and possible association with other components of the matrix) is ever more acknowledged to be a key aspect for proper risk assessment studies.

As particle size decreases, the solubility of the emitted particles can be expected to increase. Organometallic compounds can be formed under environmental conditions from organic precursors, most commonly through biological methylation related to the presence of microorganisms. The biotransformation of toxic substances in the environment is an important issue since lipid solubility, volatility, and persistence may be increased by the combination of methyl groups with the inorganic moieties. Rauch and Morrison [71] reported that the transport of Pt in soil is possible only if it is present as a complex. On the basis of this assumption, a new investigative approach has been developed, aimed at generating information concerning the transformation behavior of Pt in soils and plants, and thus at identifying and quantifying not only the total concentration of the metals, but also their various potentially toxic forms. Because of the expected low metal concentration (especially in native materials), the separation method of the metal species prior to detection, from centrifugation to chromatographic or electrophoretic strategies [72], should preserve the type of binding between the metal and the organic moieties.

11.4 QUANTIFICATION OF PLATINUM-GROUP ELEMENTS

Although PGEs concentration is still low in most biological, environmental and food samples, their possible accumulation [61] pose an important public health problem due to their toxic potential. In order to investigate the occupational, environmental and dietary exposure to PGEs, and estimate the potential danger for human health, during the recent years several analytical techniques have been used and developed to tackle the various problems posed by different kinds of matrices (Table 11.1). These techniques include electrochemical techniques, for example, adsorptive cathodic stripping voltammetry (CSV), instrumental neutron activation analysis (INAA), total reflection X-Ray fluorescence (TXRF), laser-excited atomic fluorescence spectrometry (LEAFS), electrothermal atomic absorption spectrometry (ET-AAS), inductively coupled plasma atomic emission spectrometry (ICP-AES), and inductively coupled plasma mass spectrometry (ICP-MS). The extremely low concentrations of PGEs put severe constraints on the determination of these elements, and special attention has to be paid to prevent any contamination or loss during sampling, treatment and analyses [73].

TABLE 11.1. Most Frequently Used Analytical Techniques for the Determination of PGEs in Various Matrices

Technique	Element	Matrix	Reference
CSV	Pt	Food groups and body fluids	[63]
		Airborne particulate matter	[77]
		Isopods	[18]
		Biotic material, grass	[72, 74]
		Road dust, urban river sediments	[76]
		Body fluids	[65, 75]
ICP-MS	Pd, Pt, Rh	—	[5, 51, 93, 105, 111]
		Food groups	[115]
		Airborne particulate matter, road dust, tunnel dust, grass	[15, 94, 99, 108, 100]
		Road and urban river sediments	[19, 97]
		Volcanic fluids	[107]
		Urine	[14, 106, 110]
	Pt, Rh	Airborne particulate matter	[109]
	Pd, Pt	Food groups	[116]
		Airborne particulate matter, road dust	[101]
		Water	[98]
		Urine	[32]
	Pd	Food groups	[113]
	Pt	Food groups and biological samples	[114, 117]
		Plants, beans	[69, 70, 95]
		Road dust, soil	[16]
		Water, sediments	[92]
		Urine and airborne particulate matter	[57, 91]
ICP-AES	Pd, Pt	Geological samples	[89]
	Pt	Biological samples	[87, 88]
TXRF	Pd	Biological and environmental samples	[90]
AAS	Pd, Pt	Blood, urine	[79, 86]
	Pd	Food groups	[118]
		Airborne particulate matter, road dust	[83, 84]
		[*Method development and validation*]	[78]
	Pt	Wine	[80]
		Biotic and environmental materials	[81]
		Blood, plasma, urine	[82, 85]

Highly sensitive (picogram range) voltammetric methods have also been set up for the analysis of Pt in biotic, environmental, and biological materials [65, 74–77]. These methods are based on the formation and adsorption of Pt formazone or ethylenediamine complexes. These complexes are formed in situ on a hanging Hg drop electrode and can catalyze the reduction of protons to H. The associated current reduction is correlated to Pt concentration. Since this technique is affected by even very low residual levels of organic matrix in the solution, dry-ashing is considered a more suitable digestion procedure for the total disruption of organics in road dust and river sediments [76]. Alternatively, digestion with *aqua regia* was also proposed for the determination of airborne Pt [77]. In their turn, neither ET-AAS nor INAA are sensitive enough for direct measurements in the low ng kg^{-1} or ng l^{-1} range, although some approaches have been suggested which exploit preliminary separation or enrichment steps [78]. On the other hand, if INAA is a highly expensive method rarely employed for PGEs detection, the accuracy of data obtained by means of ET-AAS for very low levels strongly depends on background absorption correction. To this end, optimization studies for both analyte modifiers and furnace conditions have been carried out [79–81]. Other teams have developed methods for PGEs by ET-AAS with Zeeman-effect background correction, with or without the flow injection procedure for on-line matrix separation and preconcentration [82–85]. The separation of PGEs complexes from the matrix by liquid–liquid extraction for analyte enrichment prior to ET-AAS determination was also investigated [86]. Some efforts have been made using solid sampling techniques after preconcentration on an ion exchange resin. A fit-for-purpose combined procedure was developed using a solid-phase extraction in a flow injection (FI) system for subsequent detection by ET-AAS, electrothermal vaporization laser atomic fluorescence spectrometry (ETV-LAFS), ICP-MS, and INAA. Inductively coupled plasma atomic emission spectrometry (ICP-AES) was used to measure the concentration of PGEs in biological samples (urine, serum, tissues) [87, 88] and geological preconcentrated samples [89], whereas classical dc arc atomic emission spectrometry (AAES) was used for the determination of Pt in plants [96].

A separation and enrichment procedure prior to Pd determination by TXRF was implemented by Messerschmidt et al. [90]. The authors found that the reductive coprecipitation of Pd along with Hg followed by evaporation of this metal enables the Pd detection in the picogram range. In some cases, the low concentration levels of PGEs make the direct measurement of these analytes difficult, and the removal of major concomitants becomes inescapable. For the reliable determination of ultratrace levels of PGEs, the best performance and convenience are offered by mass spectrometry (MS) instrumentation, because of its multielement capability, reliability in terms of sensitivity, accuracy, and precision, and outstanding detection power. Platinum can be separated from matrix elements (in particular, Hf, the oxide of which adversely affects the Pt signal) by adsorptive chromatography before introduction into the plasma. To compensate for the decrease in sample volume after enrichment, FI was used to introduce the sample in the nebulizer [91, 92]. Methods for the determination of trace PGEs by ICP-MS coupled with NiS fire assay collection and Te co-precipitation have also been reported [93, 94], whereas isotope dilution (ID) is also resorted to for quantification [17, 92]. Thus, better limits of detection are attainable by separating and/or preconcentrating the analyte and removing interfering matrix components.

Several efficient sample introduction systems have been conceived to achieve higher precision, accuracy, and sensitivity, such as thermospray nebulization (TN) [95], ETV [88, 99], and hydraulic high pressure nebulization. The use of desolvation devices, for example, ultrasonic nebulization (UN), in combination with

ICP-MS, instead of the more conventional pneumatic nebulization, increases the detection power by an order of magnitude [51, 97]. These matrix elimination modes, or rather the plasma power reduction to achieve the so-called cold plasma condition or even the polyatomic cluster disruption by means of a collision cell (CC), can solve the most common interference problems, with the additional advantage that the detection power is enhanced, primarily thanks to the improved ion transmission and the lower detection noise. The severe mass interferences caused by polyatomic ions in the determination by ICP-MS of ultra-low levels of PGEs actually calls for their minimization. Many attempts have been made to this end in environmental and biological matrices by separating elements which might lead to mass interferences and/or preconcentrating the analyte [98]. Various chemical approaches during the sample preparation procedures were developed. These go from the removal of the analyte from the original matrix by ion-exchange [100], to the hyphenation of quadrupole (Q)-ICP-MS instrumentation with ion chromatography [17, 101], and the reduction of the interfering elements through chelation [102, 103]. Other studies aimed at demonstrating the advantages of on-line reduction or removal of interfering species through the use of highly selective catalytic reactions [104], fused-silica capillaries modified with ion-exchange reagents prior to nebulization [105], or with the assistance of dynamic reaction cell (DRC) or CC devices. At the time of writing and to the best of the authors' knowledge, only one work on the application of the DRC for PGEs determinations has been published [111]. DRC and CC devices are post-plasma cells which do not interfere with the plasma operating conditions or the plasma chemistry. The processes taking place between the reaction gas and the ion beam in a DRC reduce interferences by converting interfering species into polyatomic species with a different m/z or, conversely, by reacting solely with the analyte of interest producing a new polyatomic species at a m/z free from interferences.

At present, the most fit-for-purpose instrumental approach to overcome mass interferences in ICP-MS is sector field (SF)-ICP-MS, although still with some serious constraints. At present, SF-ICP-MS is acknowledged to be the best technique to quantify traces of PGEs in complexes matrices. Owing to the low background noise and the high sensitivity obtained with this instrument, extremely good limits of detection (LoDs) are attained, allowing the direct determination without any chemical preconcentration steps. In this way, the risk of contamination due to sample handling is reduced.

The overall influence of interferences depends on their relative abundance in the sample matrix as compared with that of the analyte. Provided that the analytical signal is sufficiently strong and plasma conditions are stable during the analytical run, interferences may be mathematically corrected by estimating and controlling daily the contribution of the interfering species on the PGEs signals through the analysis of matrices spiked with standard solutions. Corrections can be calculated through equations elaborated daily, in which the corrected analytical signal is obtained from the apparent one, the signal due to the element present in the interfering molecule, and the rate of interference formation. Mass interferences can cause several difficulties in the correct determination of PGEs,

TABLE 11.2. Major Potential Mass Interferences on Pd Isotopes

Analyte		Interferent		Minimum Required Resolution ($m/\Delta m$)
Isotope	Abundance[a] (%)	Species	Abundance[a] (%)	
$^{104}Pd^+$	11.14	$^{40}Ar^{64}Zn^+$	48.44	8500
		$^{38}Ar^{66}Zn^+$	0.02	6200
		$^{36}Ar^{68}Zn^+$	0.06	9000
		$^{40}Ar^{64}Ni^+$	0.90	7900
		$^{88}Sr^{16}O^+$	82.38	33,500
		$^{87}Sr^{17}O^+$	0.003	23,600
		$^{86}Sr^{18}O^+$	0.02	21,200
		$^{87}Rb^{17}O^+$	0.01	22,100
		$^{208}Pb^{2+}$	52.35	1200
		$^{104}Ru^+$	18.7	54,700
$^{105}Pd^+$	22.33	$^{40}Ar^{65}Cu^+$	0.003	7300
		$^{36}Ar^{69}Ga^+$	0.20	92,000
		$^{89}Y^{16}O^+$	99.76	27,600
		$^{88}Sr^{17}O^+$	0.03	1,000,000
		$^{87}Sr^{18}O^+$	0.014	30,900
		$^{87}Rb^{18}O^+$	0.056	28,400
$^{106}Pd^+$	27.33	$^{40}Ar^{66}Zn^+$	27.79	7200
		$^{38}Ar^{68}Zn^+$	0.013	6800
		$^{36}Ar^{70}Ge^+$	0.002	9300
		$^{90}Zr^{16}O^+$	51.3	26,500
		$^{88}Sr^{18}O^+$	0.16	70,600
		$^{89}Y^{17}O^+$	0.037	58,800
		$^{106}Cd^+$	1.25	27,900
$^{108}Pd^+$	26.46	$^{38}Ar^{70}Ge^+$	0.014	8600
		$^{36}Ar^{72}Ge^+$	0.005	7300
		$^{40}Ar^{68}Zn^+$	18.7	6500
		$^{92}Zr^{16}O^+$	17.1	540,000
		$^{91}Zr^{17}O^+$	0.004	216,000
		$^{92}Mo^{16}O^+$	14.8	40,000
		$^{108}Cd^+$	0.89	1,080,000
$^{110}Pd^+$	11.72	$^{40}Ar^{70}Ge^+$	20.4	6000
		$^{38}Ar^{72}Ge^+$	1.9	5300
		$^{36}Ar^{74}Ge^+$	0.12	7000
		$^{40}Ar^{70}Zn^+$	0.62	6300
		$^{36}Ar^{74}Se^+$	0.003	7300
		$^{94}Mo^{16}O^+$	9.23	20,000
		$^{92}Mo^{18}O^+$	0.03	366,000
		$^{92}Zr^{18}O^+$	0.034	78,500
		$^{94}Zr^{16}O^+$	17.3	26,800
		$^{93}Nb^{17}O^+$	0.037	
		$^{110}Cd^+$	12.49	52,300

[a]Monoatomic interferents: natural abundance; polyatomic species: maximum possible abundance calculated as the product of the natural abundances of the two isotopes forming the molecular ion divided by 100.

TABLE 11.3. Major Potential Mass Interferences on Pt Isotopes

Analyte		Interferent		Minimum Required Resolution ($m/\Delta m$)
Isotope	Abundance[a] (%)	Species	Abundance[a] (%)	
$^{194}Pt^+$	32.90	$^{178}Hf^{16}O^+$	27.2	8100
		$^{177}Hf^{17}O^+$	0.007	9600
		$^{176}Hf^{18}O^+$	0.01	8800
		$^{176}Yb^{18}O^+$	0.02	9200
		$^{176}Lu^{18}O^+$	0.005	9200
$^{195}Pt^+$	33.80	$^{179}Hf^{16}O^+$	13.6	8200
		$^{178}Hf^{17}O^+$	0.01	6900
		$^{177}Hf^{18}O^+$	0.04	8800

[a]Maximum possible abundance calculated as the product of the natural abundances of the two isotopes forming the molecular ion divided by 100.

especially when using a simple Q mass filter [2, 17, 51, 94]. As shown in Tables 11.2–11.4, the determination of Pt mainly suffers from the presence of Hf oxides, whereas the determination of Pd and Rh is severely crippled by other oxides. Argides can also exert significant isobaric interferences [106–108].

Mathematical corrections may be impractical when the ratio of apparent analyte signal to real analyte signal is too large or when its high variability from sample to sample makes the correction ineffective [109, 110]. As such, neither Pd nor Rh could be determined directly in natural samples without prior matrix removal. Recent studies conducted by Simpson et al. [111] with a DRC coupled to an ICP-MS indicate the possibility of removing oxide-based interferences (ZrO on Pd, HfO on Pt) from PGEs isotopes in natural samples, such as an *aqua regia* soil leach solution, using O_2 as the reactive gas. As the PGEs are relatively inert, O_2 promotes oxidation and association reactions of interfering elements from monoxides to multiple-oxide species, for example, HfO_4^+.

TABLE 11.4. Major Potential Mass Interferences on Monoisotopic Rh

Analyte		Interferent		Minimum Required Resolution ($m/\Delta m$)
Isotope	Abundance[a] (%)	Species	Abundance[a] (%)	
$^{103}Rh^+$	100	$^{38}Ar^{65}Cu^+$	0.02	7200
		$^{40}Ar^{63}Cu^+$	68.89	8040
		$^{36}Ar^{67}Zn^+$	0.01	10,100
		$^{87}Sr^{16}O^+$	6.99	102,900
		$^{87}Rb^{16}O^+$	27.76	147,000
		$^{85}Rb^{18}O^+$	0.14	17,200
		$^{206}Pb^{2+}$	24.14	1248

[a]Monoatomic interferents: natural abundance; polyatomic species: maximum possible abundance calculated as the product of the natural abundances of the two isotopes forming the molecular ion divided by 100.

As regards the occurrence of other signal interferences (such as instrument drift, cone deposition, or matrix signal suppression), most authors report the feasibility of their compensation by correcting the analyte intensity for the intensity of an internal standard (IS). Isotope 115 of In is frequently used as an IS because it is monoisotopic, free from isobaric interferences (except for ^{115}Sn, 0.36 percent abundance, generally present at negligible levels in samples), as a rule absent in the sample matrix, not interfering with the analytes, and with a ionization potential similar to those of PGEs.

11.5 LEVELS OF PGEs IN FOOD

Only a few studies so far have dealt with the analysis of PGEs in foods, as shown by the information given in Tables 11.5–11.7. An investigation of 1992 concluded that the diet can contribute to the total exposure to these metals by approximately one-half [63]. In that study, the authors used adsorptive voltammetry after preconcentration and matrix removal to determine Pt in a number of cooked and freeze-dried foodstuffs. In 1994 the UK surveillance programme for chemicals in food issued one of the periodic Total Diet Studies for the assessment of the dietary exposure of consumers to various potentially toxic elements. The design of the study involved 20 groups of similar foods and for the first time Pd, Pt, and Rh were included in the list of contaminants to be assessed [112]. Samples were prepared and cooked, when appropriate, according to usual practices. In most cases, the element contents was less than the relevant LoDs. In 1997 Zhou and Liu determined Pd in various foods and food materials such as beef, mussels, rice, corn flour, celery, shaddock, raw sugar, tea, orange juice, peanut oil, and red wine (Table 11.6), as well as offal, i.e., chicken stomach and pork liver, by means of Q-ICP-MS [113]. They measured the Pd isotope at mass 104 and applied the mathematical correction equation for interfering Ru. In 2005 Fragnière et al. [114] used SF-ICP-MS in the low-resolution (LR) mode for the determination of Pt in various freeze-dried food samples. A desolvation system was employed to minimize the formation of the interfering Hf oxide so as to avoid the need for mathematical corrections in the majority of cases. As regards vegetables, higher Pt levels have been found in the vegetative parts than in the regenerative ones. The relatively high Pt concentrations found in plant leaves and leafy vegetables suggest that they can be ascribed to exogenous contamination [114, 115].

As already discussed above, Q-ICP-MS may still require matrix removal and preconcentration because the Q mass analyzer is unable to resolve most of the spectral interferences affecting the analytical masses of PGEs. Moreover, the detection power might be inadequate for the performance of reliable measurements in some samples. PGEs determination is hampered by mass interferences and requires high analytical selectivity and sensitivity. Double-focussing SF-ICP-MS with mass resolution capabilities higher than those afforded by Q instruments (currently up to 10,000) can properly deal with most mass interferences that have

TABLE 11.5. Concentrations of Pt in Foods

	Mean Concentration of Pt (ng kg^{-1})					
Reference	[63]		[112]	[117]	[115, 116]	[114]
Year of sampling	1989		1994	1998	2005	2005
Country of sampling	AUS[a]		UK[a]		IT[a]	CH[a]
Food groups	(DW)[b]	RSD %	(FW)[b]		(FDS)[b]	(FDS)[b]
Full cream milk	1030	14	<100		74	
Skimmed milk	1450	14			73	
Milk derivatives	700	4				
Wholemeal flour	1090	12	100		87	64[c]
White flour	900	11			93	
Miscellaneous cereals	4980	17	100			
Rice flour	6500	8				
Sugar and preserves			100			
Edible animal tissues (beef, pork, veal)					23	
Offal	17,900	15	<100			
Beef	10,400	5	<100		19	73[c]
Pork	5840	12			45	
Lamb	1550	14				
Chicken	12,900	9	100		71	
Fish	3730	14	<100			
Canned salmon	7700	4				
Oils and fats			200			
Eggs	14,300	9	<100		33	
Potatoes	3880	6	<100		18	
Carrots	2740	9			16	20[c]
Onions	6620	9			LoQ	
Beans	8060	17				
Mushrooms				20		
Green vegetables (external leaves)			100		41	170[c]
Other vegetables (internal leaves, tubers, bulbs)			100		18	43[c]
Fruit						
Blackberries			<100	6		
Pumpkin	5430	19				
Apple	2720	19				
Grape	700	9				
Banana	3530	22				
Peanuts/nuts	1110	16	100			
Beverages (tea, wine, coffee, soft drinks)			<100			10[c]

[a]AUS = Australia; IT = Italy; CH = Switzerland.
[b]FW = fresh weight; DW = dry weight; FDS = freeze-dried sample.
[c]Data obtained from bar graphs.

TABLE 11.6. Concentration of Pd Values in Foods

	Mean Concentration of Pd (ng kg^{-1})			
Reference	[118]	[112]	[113]	[115, 116]
Year of sampling	1962	1994	1997	2005
Country of sampling		UK[a]	PRC[a]	IT[a]
Food groups	(FW)[b]	(FW)[b]	(ns)[b]	(FDS)[b]
Full cream milk		<300		3040
Skimmed milk				12,700
Wholemeal flour		2000	1000	3280
White flour				13,600
Miscellaneous cereals		900		
Rice flour			2000	
Raw sugar			3000	
Sugar and preserves		500		
Edible animal tissues (beef, pork, veal)				11,200
Offal		2000		
Ham	2000	600		
Beef	700	400	3000	
Pork	500		3000	
Turkey				3460
Chicken	600	<300	3000	
Fish		2000	9000	
Mussels			46,000	
Oils and fats		400	900	
Eggs		400		2830
Vegetables (mean value on leaves, tubers, and bulbs)		600	9000	47,800
Potatoes		500		
Fruit		400	2000[c]	
Nuts		3000		
Beverages (mean value on tea, coffee, soft drinks)		400		
Tea			80,000	
Red wine			6000	

[a]IT = Italy; PRC = People's Republic of China; UK = United Kingdom.
[b]FW = fresh weight; FDS = freeze-dried sample; ns = not specified.
[c]Orange juice.

made Pd analysis difficult. The resolution power offered by this technique is, however, still inadequate to satisfactorily deal with some strong interferences on the Pd and Rh signals (Tables 11.3–11.4). PGEs contamination of foods was investigated by Caroli's team in 2005–2006 [115, 116] on freeze-dried food aliquots by means of the above second generation of sector field ICP-MS.

The influence of $^{179}Hf^{16}O^{+}$ on $^{195}Pt^{+}$ was found to be mathematically corrigible due to the low concentration of Hf in the investigated samples when working

TABLE 11.7. Concentrations of Rh Values in Foods

	Mean Concentration of Rh (ng kg^{-1})	
Reference	[112]	[115]
Year of sampling	1994	2005
Country of sampling	UK[a]	IT[a]
Food groups	(FW)[b]	(FDS)[b]
Full cream milk	300	1680
Skimmed milk		1090
Wholemeal flour	200	139
White flour		2230
Sugar and preserves	100	
Animal tissues (beef, pork, veal, offal) and meat products	700	
Poultry	100	
Fish	200	
Oils and fats	100	
Eggs	100	
Vegetables (sum of leaves, tubers, bulbs)	200	
Fruit	100	
Nuts	4000	
Beverages (sum of tea, coffee, soft drinks)	100	

[a]UK = United Kingdom; IT = Italy.
[b]FW = fresh weight; FDS = freeze-dried sample.

in LR mode ($m/\Delta m = 300$). While Pt isotopes are primarily affected by Hf oxides, the isotopes of Pd and Rh are influenced by several species and are therefore more difficult to accurately compensate for. Mass interferences on Pd include argides and oxides. Different approaches for mass interference correction for Pd have been proposed in the literature (Table 11.1). Palladium possesses five stable isotopes. The most abundant isotopes are ^{106}Pd and ^{108}Pd, respectively, with 27.3 and 26.5 percent of the total Pd. When these isotopes are selected for quantification and Cd is expected in the sample, the contribution of the isobaric interference of the monoatomic Cd ion cannot be overlooked. The Cd contribution to the signals of ^{106}Pd and ^{108}Pd can be corrected by monitoring the $^{111}Cd^+$ signal, which, in turn, must be mathematically corrected for the influence of $^{95}Mo^{16}O^+$. This last correction cannot be disregarded if a significant amount of Mo is detected during a preliminary semiquantitative investigation of the sample. The semiquantitative study of the sample matrix to be analyzed provides useful information on the major interference problems to be faced and corrected. On the basis of the expected levels of the analytes and other potentially interfering elements, a proper selection can be done of the analytical isotope to be quantified and of the interfering ions to be corrected for. In the case of ^{106}Pd and ^{108}Pd, the polyatomic interferences due to argides turned out to be of minor importance, whereas the most severe problems were posed by formation of oxides, generally overlooked when the oxygen isotope involved is scarcely abundant (^{17}O, ^{18}O) and the elements forming the oxides are only minor constituents. It is generally accepted that the most severe interferences affecting Pd accurate quantification arise from mass overlap of oxide ions of Rb and Sr. In some works [115, 116], the low counts HR mode ($m/\Delta m = 10{,}000$) allowed for the physical resolution of most interferences, thus avoiding the risk of underestimating the analyte signal as a result of the mathematical subtraction of the combined effects of the various overlapping signals on that of the analyte.

The need for additional research to better assess mathematical, physical, and/or chemical approaches in the analyses of PGEs is also substantiated by the available literature data on Rh. In fact, oxides were ascertained to heavily affect the signal of $^{103}Rh^{+}$ to such an extent that the mathematical corrections might be completely inadequate. Thus, in the already quoted UK surveillance estimate, only provisional results obtained by SF-ICP-MS in the medium resolution (MR) mode are given (Table 11.7).

Trace metals including Pt in wild fungi and wild blackberries sampled in 1998 were determined by Q ICP MS in UK [117]. On a fresh weight basis, the levels of Pt were 6 ng kg^{-1} in blackberries and 20 ng kg^{-1} in mushroom. Concentrations of Pt in wild fungi were significantly higher at urban sites than at rural sites. Some results are available for Pd by INAA analysis of meat [118]. A method based on ET-AAS was developed for Pt in wine. Pt levels in most French wines were less than 10 μg l^{-1} [80].

Cucumber plants were subjected to controlled cultivation experiments to investigate the process through which Pt can enter the food chain by accumulation in plants. The said vegetable strongly accumulated bioavailable Pt forms, with accumulation factors much higher in the roots than in the shoots [119]. At present, it is difficult to assess the reliability of all these measurements, because there are no food-based certified reference materials (CRMs) with figures on these metals, and only a few comparable measurements have been made on the same materials by different techniques. Along with preliminary information indicating that Pd ends up in wastes or incineration ashes during recycling of cars and dental prostheses [120], the preliminary data (i.e., Pd shows higher mobility in the environment than Pt) make most of the aspects related to Pd emissions, including its environmental fate and adverse effects on human health, a matter of growing concern [121].

Quite obviously, the quality of analytical results strongly depends on the representativeness of the sample examined, the appropriateness of the pretreatment procedures applied for the quantitative conversion of analytes into detectable forms, and the fitness for purpose of the laboratory setting. Because of the low concentration and the numerous interferences which affect the various detection systems, measurements methods must be thoroughly validated. Even if values obtained by several laboratories for PGEs concentrations in environmental and biological samples were compliant with the basic requirements of quality assurance (QA) and internal quality control (QC), striking differences have been sometimes observed among results obtained from different research groups. Hence, routine applications must be based on robust QC and QA approaches. Needless to say, the availability of ad hoc CRMs will play an ever more important role in the quantification of PGEs in food matrices [122].

11.6 CONCLUSIONS

Natural processes, deliberate or accidental contamination during agro-zootechnical production, processing, and transportation can actually affect PGE levels in foods. Although environmental concentrations of PGEs are still relatively low, a number of reports clearly show that they are progressively increasing their environmental concentrations with ensuing concern on their possible adverse effects on human health.

At present, only limited information is available about the toxic effects of long-term exposure to low levels of Pt on human health, and even more scanty data are reported in the case of Pd and Rh, which require a more in-depth human risk assessment. The number and quality of studies performed over the last two decades in this field reveal beyond any doubt that the environmental concentrations of PGEs are detectable both at significantly higher levels than before and at higher levels in intense traffic areas than in rural or nonpolluted zones. In this framework, the preliminary results on the ability of PGEs to enter the food chain shed some new light on the global risk posed by these elements to the consumer health. The concentrations detected in food samples, in fact, support the assumption that a significant exposure to these elements may occur through the diet. Hence, carefully planned and scrupulously conducted dietary surveys like those carried out in the UK in 1994 [112] and in Germany a few years later [123] are urgently needed. By the same token, it is hoped that further bioavailability and toxicological studies, much along the same line of thought as, for example, those recently completed by Gagnon et al. [124], can be undertaken as well as studies on target tissues [125, 126] and breastfeeding [127] to achieve a sound and realistic risk assessment of the exposure to these contaminants through food commodities.

REFERENCES

1. W. B. Williamson, J. C. Summers, J. F. Skowron, Catalyst technologies for future automotive emission systems, *Soc. Autom. Engng*, 1989, 3.41–3.51.
2. M. Moldovan, M. M. Gómez, M. A. Palacios, Determination of platinum, rhodium and palladium in car exhaust fumes, *J. Anal. Atom. Spectrom.*, **8** (1999), 1163–1169.
3. S. Rauch, On the environmental relevance of platinum group elements, Report ISBN 91-7291-029-1, Chalmers University of Technology, Göteborg, Sweden, 2001.
4. E. Helmers, Platinum emission rate of automobiles with catalytic converters – comparison and assessment of results from different approaches, *Environ. Sci. Pollut. Res.*, **4** (1997), 100–103.

5. M. M. Hidalgo, M. M. Gómez, M. A. Palacios, Trace enrichment and measurement of flow injection inductively coupled plasma mass spectrometry, *Fresenius J. Anal. Chem.*, **354** (1990), 420–423.

6. M. A. Palacios, M. M. Gómez, M. Moldovan, G. M. Morrison, S. Rauch, C. McLeod, R. Li, J. Laserna, P. Lucena, S. Caroli, S. Alimonti, P. Schramel, S. Lustig, U. Wass, B. Stenbom, M. Luna, J. C. Saenz, J. Santamaría, J. M. Torrents, Platinum-group elements: quantification in collected exhaust fumes and studies of catalyst surfaces, *Sci. Total Environ.*, **257** (2000), 1–15.

7. M. Balcerzak, Analytical methods for the determination of platinum in biological and environmental materials. A review, *Analyst*, **122** (1997), 67R–74R.

8. V. R. Schölg, G. Indlekofer, P. Oelhafen, Mikropartikelemissionen von Verbrennungsmotoren mit Abgasreinigung-Röntgen-Photoelektronenspektroscopie in der Umweltanalytik, *Angew. Chem.*, **99** (1987), 312–322.

9. S. Knoloch, H. König, ICP-MS Determinations in Automotive Catalyst Exhaust, in G. Holland, A. N. Eaton (eds), *Applications of Plasma Source Mass Spectrometry. II*, The Royal Society of Chemistry, London, 1993, pp. 108–114.

10. M. Moldovan, M. A. Palacios, M. M. Gómez, G. Morrison, S. Rauch, C. McLeod, R. Ma, S. Caroli, A. Alimonti, F. Petrucci, B. Bocca, P. Schramel, M. Zischka, C. Pettersson, U. Wass, M. Luna, J. C. Saenz, J. Santamaría, Environmental risk of particulate and soluble platinum group elements released from gasoline and diesel engine catalytic converters, *Sci. Total Environ.*, **296** (2002), 199–208.

11. R. Merget and G. Rosner, Evaluation of the health risk of platinum group metals emitted from automotive catalytic converters, *Sci. Total Environ.*, **270** (2001), 165–173.

12. H. P. König, R. F. Hertel, W. Koch, G. Rosner, Determination of platinum emissions from a three-way catalyst-equipped gasoline engine, *Atmos. Environ.*, **26A** (1992), 741–745.

13. B. Gómez, M. A. Palacios, M. Gómez, J. L. Sanchez, G. Morrison, S. Rauch, C. McLeod, R. Ma, S. Caroli, A. Alimonti, F. Petrucci, B. Bocca, P. Schramel, M. Zischka, C. Pettersson, U. Wass, Levels and risk assessment for humans and ecosystems of platinum group elements in the airborne particles of some European cities, *Sci. Total Environ.*, **299** (2002), 1–19.

14. S. Caroli, F. Petrucci, B. Bocca, M. Krachler, F. Forastiere, A. Alimonti, Exposure to platinum-group metals released by automotive catalytic converters: the case of urban youngsters, in A. M. Roussel, R. A. Anderson, A. E. Favier (eds), *Trace Elements in Man and Animal*, Kluwer Academic/Plenum Publishers, New York, vol. 10, 2000, pp. 667–670.

15. B. A. Leśniewska, B. Godlewska-Żyłkiewicz, B. Bocca, S. Caimi, S. Caroli, A. Hulanicki, Platinum, palladium and rhodium content in road dust, tunnel dust and common grass in Białystok area (Poland), A pilot study, *Sci. Total Environ.*, **321** (2004), 93–104.

16. M. E. Kylander, S. Rauch, G. M. Morrison, K. Andam, Impact of automobile emissions on the levels of platinum and lead in Accra, Ghana, *J. Environ. Monit.*, **5** (2003), 91–95.

17. M. Müller, K. G. Heumann, Isotope dilution inductively coupled plasma quadrupole mass spectrometry in connection with a chromatographic separation for ultra-trace determinations of platinum group elements (Pt, Pd, Ru, Ir) in environmental samples, *Fresenius J. Anal. Chem.*, **368** (2000), 109–115.

18. S. Rauch and G. M. Morrison, Platinum uptake by the freshwater isopod *Asellus aquaticus* in urban rivers, *Sci. Total Environ.*, **235** (1999), 261–268.
19. S. Rauch, H. Hemond, Sediment-based evidence of platinum concentration changes in an urban lake near Boston, Massachusetts, *Environ. Sci. Technol.*, **37** (2003), 3283–3288.
20. J. Schäfer, J. D. Eckhardt, Z. A. Berner, D. Stüben, Time-dependent increase of traffic-emitted platinum group elements (PGE) in different environmental compartments, *Environ. Sci. Technol.*, **33** (1999), 3166–3170.
21. E. Helmers, H. Schwarzer, M. Schuster, Comparison of palladium and platinum in environmental matrices: palladium pollution by automobile emission?, *Environ. Sci. Pollut. Res.*, **5** (1998), 44–50.
22. International Programme of Chemical Safety, Environmental Health Criteria, no. 125, *Platinum*, World Health Organization, Geneva, 1991.
23. International Programme of Chemical Safety, Environmental Health Criteria, no. 226, *Palladium*, World Health Organization, Geneva, 2002.
24. S. Rauch, G. M. Morrison, M. Motelica Heino, O. Donard, Evaluation of speciation, transport and ecological risk of palladium, platinum and rhodium in urban stormwater systems. *Proceedings of the 8th International Conference on Urban Storm Drainage*, the Institution of Engineers, Australia, 1999.
25. R. Rushforth, Superior physical properties make palladium an ideal dental metal, *Platinum Met. Rev.*, **48** (2004), 30–31.
26. B. Rosenberg, L. VanCamp, J. E. Trosko, V. H. Mansour, Platinum compounds: a new class of potent antitumour agents, *Nature*, **222** (1969), 385–386.
27. T. Hees, B. Wenclawiak, S. Lustig, P. Schramel, M. Schwarzer, M. Schuster, D. Verstraete, R. Dams, E. Helmers, Distribution of Platinum Group Elements (Pt, Pd, Rh) in environmental and clinical matrices: composition, analytical techniques and scientific outlook, *Environ. Sci. Pollut. Res.*, **5** (1998), 105–111.
28. K. Kümmerer, E. Helmers, Hospital effluents as a source for platinum in the environment, *Sci. Total Environ.*, **193** (1997), 179–184.
29. R. Schierl, B. Rohrew, J. Hohnloser, Long-term platinum excretion in patients treated with cisplatin, *Cancer Chemother. Pharmac.*, **36** (1995), 75–78.
30. K. Kümmerer, E. Helmers, P. Hubner, G. Mascart, M. Milandri, F. Reinthaler, M. Zwakenberg, European hospitals as a source for platinum in comparison with other sources, *Sci. Total Environ.*, **225** (1999), 155–165.
31. B. Mason, *Principles of Geochemistry*, 2nd edn, Wiley, New York, 1958.
32. J. Begerow, M. Turfeld, L. Dunemann, Determination of physiological palladium and platinum levels in urine using double focusing magnetic sector field ICP-MS, *Fresenius J. Anal. Chem.*, **359** (1997), 427–429.
33. P. E. Brubaker, J. P. Moran, K. Bridbord, F. G. Heuter, Noble metals: a toxicological appraisal of potential new environmental contaminants, *Environ. Health Perspect.*, **10** (1975), 39–56.
34. C. Dominici, M. Börzsönyi, S. Caroli, F. Petrucci, M. A. Castello, Occupational exposure to antiproliferative drugs in health care workers. Proceedings of the Italian-Hungarian Meeting on Genotoxicity of Environmental Chemicals: Scientific and Regulatory Aspects, *Ann. Ist. Super. Sanità*, **25** (1988), 601–604.

35. W. R. Leopold, E. C. Miller, J. A. Miller, Carcinogenicity of antitumor *cis*-platinum(II) coordination complexes in the mouse and rat, *Cancer Res.*, **39** (1979), 913–918.

36. J. Bünger, J. Stork, K. Stalder, Cyto- and genotoxic effects of coordination complexes of platinum, palladium and rhodium *in vitro*, *Int. Arch. Occup. Environ. Health*, **69** (1996), 33–38.

37. T. Gebel, H. Lantzsch, K. Plebow, H. Dunkelberg, Genotoxicity of platinum and palladium compounds in human and bacterial cells, *Mutat. Res. Gen. Toxicol. Environ. Mutag.*, **389** (1997), 183–190.

38. G. Rosner, R. Merget, Allergenic potential of platinum compounds, in A. D. Dayan, R. F. Hertel, E. Heseltine, G. Kazantzis, E. M. Smith, M. T. Van der Venne (eds), *Immunotoxicity of Metals and Immunotoxicology*, Plenum Press, New York, 1990, pp. 93–102.

39. R. Schierl, A. Pethran, G. Fruhmann, Urinary platinum in hospital personnel after platinum-drug exposure, *Proceedings of the International Symposium on Biological Monitoring in Occupational and Environmental Health*, Espoo, Finland, 1996.

40. D. B. Baker, P. H. Gann, S. M. Brooks, J. Gallagher, I. Bernstein, Cross-sectional study of platinum salts sensitization among precious metals refinery workers, *Am. J. Indust. Med.*, **18** (1990), 653–664.

41. R. Merget, G. S. Werninghaus, F. Bode, E. M. Bergmann, W. Zachgo, J. Meier-Sydow, Quantitative skin prick and bronchial provocation tests with platinum salt, *Br. J. Indust. Med.*, **48** (1991), 830–837.

42. K. M. Venables, M. B. Dally, A. J. Nunn, J. F. Stevens, R. Stephens, N. Farrel, J. V. Hunter, M. Stewart, E. G. Hunghes, A. J. Newman Taylor, Smoking and occupational allergy in workers in a platinum refinery, *Br. Med. J.*, **299** (1989), 939–942.

43. R. J. Belt, K. J. Himmelstein, T. F. Patton, S. J. Bannister, L. A. Sternson, A. J. Repta, Pharmacokinetics of non-protein-bound platinum species following administration of *cis*-dichlorodiammineplatinum(II), *Cancer Treat. Rep.*, **63** (1979), 1515–1521.

44. P. Melius, M. E. Friedman, Complexes of platinum with polypeptides and proteins, *Inorg. Perspect. Biol. Med.*, **1** (1977), 1–18.

45. B. Lindell, *Platinum, DECOS and NEG Basis for an Occupational Standard*, Nordic Council of Ministers, **14** (1997), 10–11.

46. R. Merget, R. Kulzer, A. Kniffka, F. Alt, R. Breitstadt, T. Bruening, Platinum concentrations in sera of catalyst production workers are not predictive of platinum allergy, *Int. J. Hyg. Environ. Health*, **205** (2002), 347–351.

47. J. Messerschmidt, F. Alt, U. Jerono, G. Tölg, J. Angerer, K. H. Schaller, Investigations of platinum concentration within body fluids of non-exposed and occupationally exposed persons, *6th International Workshop of Trace Element Analytical Chemistry in Medicine and Biology*, Neuherberg, 1991, pp. 24–26.

48. C. E. W. Herr, M. Jankofsky, J. Angerer, W. Kuster, N. I. Stilianakis, U. Gieler, T. Eikmann, Influences on human internal exposure to environmental platinum, *J. Exp. Anal. Environ. Epidemiol.*, **13** (2003), 24–30.

49. K. H. Schaller, J. Angerer, F. Alt, J. Messerschmidt, A. Weber, The determination of platinum in blood and urine as a tool for the biological monitoring of internal

exposure, *Proceedings of the SPIE – Int. Soc. Opt. Eng. 1993, International Conference on Monitoring of Toxic Chemicals and Biomarkers*, **1716** (1992), 498–504.

50. I. Ensslin, A. Pethran, R. Schieler, G. Frumann, Urinary platinum in hospital pharmacy personnel occupationally exposed to platinum-containing drugs, *Int. Arch. Occup. Environ. Health*, **65** (1994), 339–342.
51. M. Krachler, A. Alimonti, F. Petrucci, K. J. Irgolić, F. Forestiere, S. Caroli, Analytical problems in the determination of platinum-group metals in urine by quadrupole and magnetic sector field inductively coupled plasma mass spectrometry, *Anal. Chim. Acta*, **363** (1998), 1–10.
52. J. Begerow, L. Dunemann, Internal platinum and palladium exposure of the general population with emphasis on the exposure from automobile exhaust and dental restorative alloys, in F. Zereini, F. Alt (eds), *Anthropogenic Platinum-group Element emissions. Their Impact on Man and Environment*, Springer, Berlin, 2000, pp. 227–236.
53. I. Iavicoli, B. Bocca, F. Petrucci, O. Senofonte, G. Carelli, A. Alimonti, S. Caroli, Biomonitoring of traffic police officers exposed to airborne platinum, *Occup. Environ. Med.*, **61** (2004), 636–639.
54. R. Schierl, Urinary biomonitoring of platinum in occupational medicine, in F. Zereini, F. Alt (eds), *Anthropogenic Platinum-group Element Emissions. Their Impact on Man and Environment*, Springer, Berlin, 2000, pp. 237–242.
55. M. E. Farago, P. Kavanagh, R. Blanks, J. Kelly, G. Kazantis, I. Thornton, P. R. Simpson, J. M. Cook, H. T. Delves, G. E. M. Hall, Platinum concentration in urban soil, and in blood and urine in the United Kingdom, *Analyst*, **123** (1998), 451–454.
56. S. Caroli, A. Alimonti, F. Petrucci, B. Bocca, M. Krachler, F. Forestiere, M. T. Sacerdote, S. Mallone, Assessment of the exposure of urban youngsters to platinum-group metals released by automotive catalytic converters, *Spectrochim. Acta*, **56B** (2001), 1241–1248.
57. Gy. Záray, M. Óvári, I. Salma, I. Steffan, M. Zeiner, S. Caroli, Determination of platinum in urine and airborne particulate matter from Budapest and Vienna, *Microchem. J.*, **76** (2004), 31–34.
58. I. Salma, I. Balásházy, R. Winkler-Heil, W. Hofmann, Gy. Záray, Effect of particle mass size distribution on the deposition of aerosols in the human respiratory system, *J. Aerosol Sci.*, **33** (2002), 119–132.
59. R. P. F. Schins, D. Polat, J. Begerow, M. Turfeld, A. Becker, P. J. A. Borm, Platinum levels in nasal lavage fluid as a biomarker for traffic-related exposure and inflammation in children, *Sci. Total Environ.*, **334–335** (2004), 447–445.
60. W. E. Pepelko, J. G. Orthoefer, Y. Y. Yang, Effects of 90 days exposure to catalytically treated automobile exhaust in rats, *Environ. Res.*, **19** (1979), 91–101.
61. S. Rauch, G. M. Morrison, Routes for bioaccumulation and transformation of platinum in the urban environment, in F. Alt, F. Zereini (eds), *Anthropogenic Platinum-group Metals Emission and Their Effect on Man and Environment*, Springer, Berlin, 2000, pp. 85–93.
62. M. Moldovan, S. Rauch, M. M. Gómez, M. A. Palacios, G. M. Morrison, Bioaccumulation of palladium, platinum, and rhodium from urban particulates and sediments by the freshwater isopod *Asellus aquaticus*, *Water Res.*, **35** (2001), 4175–4183.
63. G. T. Vaughan, T. M. Florence, Platinum in human diet, blood and excreta, *Sci. Total Environ.*, **111** (1992), 47–58.

64. G. Rosner, H. P. König, W. Koch, H. Kock, R. F. Hertel, H. Windt, Engine test experiments to assessing platinum uptake by plants, *Angew. Botanik*, **65** (1991), 127–132.

65. J. Messerschmidt, F. Alt, G. Tölg, K. H. Schaller, J. Angerer, Adsorptive voltammetric procedure for the determination of platinum base-line levels in human body fluid, *Fresenius J. Anal. Chem.*, **343** (1992), 391–394.

66. F. Alt, A. Bambauer, K. Hoppstock, B. Mergler, G. Tölg, Platinum traces in airborne particulate matter. Determination of whole content, particle size distribution and soluble platinum, *Fresenius J. Anal. Chem.*, **346** (1993), 693–696.

67. H. J. Ballach, Impact studies on cloned poplars, *Environ. Sci. Pollut. Res.*, **4** (1997), 131–139.

68. E. Helmers, Elements accompanying platinum emitted from automobile catalysts, *Chemosphere*, **33** (1996), 883–885.

69. S. Lustig, S. Zang, B. Michalke, P. Schramel, W. Beck, Platinum determination in nutrient plants by inductively coupled plasma mass spectrometry with special respect to the hafnium oxide interference, *Fresenius J. Anal. Chem.*, **357** (1997), 1157–1163.

70. D. Klueppel, N. Jakubowski, J. Messerschmidt, D. Stuewer, D. Klockow, Speciation of platinum metabolites in plants by size-exclusion chromatography and inductively coupled plasma mass spectrometry, *J. Anal. Atom. Spectrom.*, **13** (1998), 255–262.

71. S. Rauch, G. M. Morrison, Routes for bioaccumulation and transformation of platinum in the urban environment, in F. Alt, F. Zereini (eds), *Anthropogenic Platinum-group Element Emissions*, Springer, Berlin, 1999, pp. 245–255.

72. J. Messerschmidt, F. Alt, G. Tölg, Detection of platinum species in plant material, *Electrophoresis*, **16** (1995), 800–803.

73. S. Caroli, E. Coni, N. Violante, F. Petrucci, S. Caimi, Alcuni aspetti del trattamento di materiale biologico prima della determinazione di elementi in traccia, *Ann. Ist. Super. Sanità*, **31** (1995), 219–224.

74. K. Hoppstock, F. Alt, K. Cammann, G. Weber, Determination of platinum in biotic and environmental materials. Part II: a sensitive voltammetric method, *Fresenius J. Anal. Chem.*, **335** (1989), 813–816.

75. O. Nygren, G. T. Vaughan, T. M. Florence, G. M. P. Morrison, I. M. Warner, L. S. Dale, Determination of platinum in blood by adsorptive voltammetry, *Anal. Chem.*, **62** (1990), 1637–1640.

76. C. Wei, G. M. Morrison, Platinum in road dusts and urban river sediments, *Sci. Total Environ.*, **146/147** (1994), 169–174.

77. R. Schierl, G. Fruhmann, Airborne platinum concentrations in Munich city buses, *Sci. Total Environ.*, **182** (1996), 21–23.

78. M. Schuster, M. Schwarzer, Selective determination of palladium by on-line column preconcentration and graphite furnace atomic absorption spectrometry, *Anal. Chim. Acta*, **328** (1996), 1–11.

79. H. Jones, Determination of platinum and palladium in blood and urine by flameless atomic absorption spectrometry, *Anal. Chem.*, **48** (1976), 1472–1474.

80. C. Cabrera-Vique, P. L. Teissedre, M. T. Cabanis, J. C. Cabanis, Determination of platinum in wine by graphite furnace atomic absorption spectrometry, *J. AOAC Int.*, **80** (1997), 57–61.

81. F. Alt, U. Jerono, J. Messerschmidt, G. Tölg, The determination of platinum in biotic and environmental materials, I. μg/kg- to g/kg-range, *Mikrochim. Acta*, **III** (1988), 299–304.

82. S. M. Hopfer, L. Ziebka, F. W. Sunderman, J. R. Sporn, B. R. Greenberg, Direct analysis of platinum in plasma and urine by electrothermal atomic absorption spectrophotometry, *Ann. Clin. Lab. Sci.*, **19** (1989), 389–396.

83. M. Schuster, M. Schwarzer, G. Risse, Determination of palladium in environmental samples, in F. Alt, F. Zereini (eds), *Anthropogenic Platinum-group Element Emissions. Their Impact on Man and Environment*, Springer, Berlin, 2000, pp. 173–182.

84. A. Limbeck, J. Rendl, H. Puxbaum, ETA-AAS determination of palladium in environmental samples with on-line preconcentration and matrix separation, *J. Anal. Atom. Spectrom.*, **18** (2003), 161–165.

85. E. Ivanova, F. Adams, Flow injection on-line sorption preconcentration of platinum in a knotted reactor coupled with electothermal atomic absorption spectrometry, *Fresenius J. Anal. Chem.*, **361** (1998), 445–450.

86. J. Begerow, M. Turfeld, L. Dunemann, Determination of physiological noble metals in human urine using liquid-liquid extraction and Zeeman electrothermal atomic absorption spectrometry, *Anal. Chim. Acta*, **340** (1997), 277–283.

87. F. B. Lo, D. K. Arai, M. A. Nazar, Determination of platinum in urine by inductively coupled plasma atomic emission spectrometry, *J. Anal. Toxicol.*, **11** (1987), 242–246.

88. A. Alimonti, F. Petrucci, C. Dominici, S. Caroli, Determination of Pt in biological samples by inductively coupled plasma atomic emission spectrometry (ICP-AES) with electrothermal vaporization (ETV), *J. Trace Elem. Electrolytes Health Dis.*, **1** (1987), 79–83.

89. Y. S. Chung, R. M. Barnes, Determination of gold, patinum, palladium and silver in geological samples by inductively coupled plasma atomic emission spectrometry after poly(dithiocarbamate) resin pre-treatment, *J. Anal. Atom. Spectrom.*, **3** (1988), 1079–1082.

90. J. Messerschmidt, A. Von Bohlen, F. Alt, R. Klockenkämper, Separation and enrichment of palladium and gold in biological and environmental samples, adapted to the determination by total reflection X-ray fluorescence, *Analyst*, **125** (2000), 397–399.

91. H. Mukai, Y. Ambe, M. Morita, Flow injection inductively coupled plasma mass spectrometry for the determination of platinum in airborne particulate matter, *J. Anal. Atom. Spectrom.*, **5** (1990), 75–80.

92. D. Colodner, E. A. Boyle, J. M. Edmond, Determination of rhenium and platinum in natural waters and sediments by flow injection isotope dilution inductively coupled plasma mass spectrometry, *Anal. Chem.*, **65** (1993), 1419–1425.

93. S. Yali, G. Xiyun, D. Andao, Determination of platinum group elements by inductively coupled plasma-mass spectrometry combined with nickel sulphide fire assay and tellurium coprecipitation, *Spectrochim. Acta*, **53B** (1998), 1463–1467.

94. M. B. Gómez, M. M. Gómez, M. A. Palacios, ICP-MS determination of Pt, Pd and Rh in airborne and road dust after tellurium coprecipitation, *J. Anal. Atom. Spectrom.*, **18** (2003), 80–83.

95. M. Parent, H. Vanhoe, L. Moens, R. Dams, Determination of low amounts of platinum in environmental and biological materials using thermospray nebulization inductively coupled plasma-mass spectrometry, *Fresenius J. Anal. Chem.*, **354** (1996), 664–667.

96. V. Otruba, M. Strnadová, B. Skalníková, Determination of platinum in plants by emission spectrometry after preconcentration on modified silica gel, *Talanta*, **40** (1993), 221–224.

97. S. Rauch, M. Motelica-Heino, G. M. Morrison, O. F. X. Donald, Critical assessment of platinum group element determination in road and urban river sediments using ultrasonic nebulization and high resolution ICP-MS, *J. Anal. Atom. Spectrom.*, **15** (2000), 329–334.

98. G. E. M. Hall, J. C. Pelchat, Determination of palladium and platinum in fresh water by inductively coupled plasma mass spectrometry and activated charcoal preconcentration, *J. Anal. Atom. Spectrom.*, **8** (1993), 1059–1065.

99. C. Lüdke, E. Hoffmann, J. Skole, S. Artelt, Particle analysis of car exhaust by ETV-ICP-MS, *Fresenius J. Anal. Chem.*, **355** (1996), 261–263.

100. K. Kanitsar, G. Koellensperger, S. Hann, A. Limbeck, H. Puxbaum, G. Stingeder, Determination of Pt, Pd and Rh by inductively coupled plasma sector field mass spectrometry (ICP-SFMS) in size-classified urban aerosol samples, *J. Anal. Atom. Spectrom.*, **18** (2003), 239–246.

101. M. C. Bruzzoniti, S. Cavalli, A. Mangia, C. Mucchino, S. Sarzanini, E. Marasco, Ion chromatography with inductively coupled plasma mass spectrometry, a powerful analytical tool for complex matrices estimation of Pt and Pd in environmental samples, *J. Chromatogr.*, **997**(A) (2003), 51–63.

102. D. B. Taylor, H. M. Kingston, D. J. Nogay, D. Koller, R. Hutton, On-line solid-phase chelation for the determination of eight metals in environmental waters by ICP-MS, *J. Anal. Atom. Chem.*, **11** (1996), 187–191.

103. L. Ebdon, A. S. Fisher, P. J. Worsfold, H. Crews, M. Baxter, On-line removal of interferences in the analysis of biological materials by flow injection inductively coupled plasma mass spectrometry, *J. Anal. Atom. Chem.*, **8** (1993), 691–695.

104. A. A. Ensafi, K. Zarei, Highly selective catalytic determination of ultra trace amounts of rhodium by linear sweep voltammetry, *Fresenius J. Anal. Chem.*, **361** (1998), 103–105.

105. H. G. Riepe, V. Loreti, R. Garcia-Sanchez, C. Cámara, J. Bettmer, Removal of interfering elements in ICP-QMS for the determination of Pt, Rh, and Pd by chemically modified sample introduction capillaries, *Fresenius J. Anal. Chem.*, **370** (2001), 488–491.

106. F. Petrucci, N. Violante, O. Senofonte, M. De Gregorio, A. Alimonti, S. Caroli, G. Forte, A. Cristaudo, Development of an analytical method for monitoring worker populations exposed to platinum-group element, *Microchem. J.*, **76** (2004), 131–140.

107. T. P. Fischer, S. Shuttleworth, P. A. O'Day, Determination of trace and platinum-group elements in high ionic-strength volcanic fluids by sector field inductively coupled plasma mass spectrometry (ICP-MS), *Fresenius J. Anal. Chem.*, **362** (1998), 457–464.

108. F. Petrucci, B. Bocca, A. Alimonti, S. Caroli, Determination of Pd, Pt and Rh in airborne particulate and road dust by high-resolution ICP-MS: a preliminary investigation of the emission from automotive catalysts in the urban area of Rome, *J. Anal. Atom. Spectrom.*, **15** (2000), 525–528.

109. B. Bocca, F. Petrucci, A. Alimonti, S. Caroli, Traffic-related platinum and rhodium concentrations in the atmosphere of Rome, *J. Environ. Monit.*, **5** (2003), 563–568.

110. B. Bocca, A. Alimonti, A. Cristaudo, E. Cristallini, F. Petrucci, S. Caroli, Monitoring of the exposure to platinum-group elements for two Italian population groups through urine analysis, *Anal. Chim. Acta*, **512** (2004), 19–25.

111. L. A. Simpson, M. Thomsen, B. J. Alloway, A. Parker, A dynamic reaction cell (DRC) solution to oxide based interferences in inductively coupled plasma mass spectrometry (ICP-MS) analysis of the noble metals, *J. Anal. Atom. Spectrom.*, **16** (2001), 1375–1380.

112. G. Ysart, P. Miller, H. Crews, P. Robb, M. Baxter, C. De L'Argy, S. Lofthouse, C. Sargent, N. Harrison, Dietary exposure estimates of 30 elements from the UK total diet study, *Food Addit. Contam.*, **16** (1999), 391–403.

113. H. Zhou, J. Liu, The simultaneous determination of 15 toxic elements in foods by ICP-MS, *Atom. Spectrosc.*, **18** (1997), 115–118.

114. C. Fragnière, M. Haldimann, A. Eastgate, U. Krähenbühl, A direct ultratrace determination of platinum in environmental food and biological samples by ICP-SFMS using a desolvation system, *J. Anal. Atom. Spectrom.*, **20** (2005), 626–630.

115. C. Frazzoli, R. Cammarone, S. Caroli, Uptake of platinum-group elements with the diet: a preliminary investigation, *Pure Appl. Chem.*, **78** (2006), 69–78.

116. C. Frazzoli, R. Cammarone, S. Caroli, Investigation of palladium and platinum contamination in food by means of sector field inductively coupled plasma mass spectrometry, 2006, *Food Addit. Contam.*, **23** (2007), 1–7.

117. C. A. Weeks, M. Croasdale, M. A. Osborne, L. Hewitt, P. F. Miller, P. Robb, M. J. Baxter, P. D. Warriss, T. G. Knowles, Multi-element survey of wild edible fungi and blackberries in the UK, *Food Addit. Contam.*, **23** (2006), 140–147.

118. R. C. Koch, J. Roesmer, Application of activation analysis to the determination of trace element concentrations in meat, *J. Food Sci.*, **27** (1962), 309–320.

119. D. Verstraete, J. Riondato, J. Vercauteren, F. Vanhaecke, L. Moens, R. Dams, M. Verloo, Determination of the uptake of $[Pt(NH_3)_4)(NO_3)]_2$ by grass cultivated on a sandy loam soil and by cucumber plants, grown hydroponically, *Sci. Total Environ.*, **218** (1998), 153–160.

120. A. Cowley, Palladium, in *Platinum*, Johnson Matthey, London, 1999, pp. 31–37, 50–51.

121. F. Zereini, F. Alt (eds), *Palladium Emission in the Environment: Analytical Methods, Environmental Assessment and Health Effects*, Springer, New York, 2006, pp. 640.

122. M. Zischka, P. Schramel, H. Muntau, A. Rehnert, M. G. Gómez, B. Stojanik, G. Wannemaker, R. Dams, P. Quevauviller, E. A. Maier, A new certified reference material for the quality control of palladium, platinum and rhodium in road dust, BCR-723, *Trends Anal. Chem.*, **21** (2002), 851–868.

123. J. Wittsiepe, P, Schrey, M. Wilhelm, J. Begerow, L. Dunemann, Dietary intake of platinum and gold by children from Germany using duplicate portion sampling, *J. Trace Elem. Med. Biol.*, **17** (2003), 117–122.

124. Z. E. Gagnon, C. Newkirk, S. Hicks, Impact of platinum group metals on the environment: A toxicological, genotoxic and analytical chemistry study, *J. Environ. Sci. Health Part A*, **41** (2006), 397–414.

125. S. Zimmermann, A. von Bohlen, J. Messerschmidt, B. Sures, Accumulation of the precious metals platinum, palladium and rhodium from automobile catalytic converters in Paratenuisentis ambiguus as compared with its fish host, Anguilla Anguilla, *J. Helminthol.*, **79**(1) (2005), 85–89.

126. B. A. Lesniewska, J. Messerschmidt, N. Jakubowski, A. Hulanicki, Bioaccumulation of platinum group elements and characterization of their species in Lolium multiflorum by size-exclusion chromatography coupled with ICP-MS, *Sci. Total Environ.*, **322**(1–3) (2004), 95–108.

127. M. Krachler, T. Prohaska, G. Koellensperger, E. Rossipal, G. Stingeder, Concentrations of selected trace elements in human milk and in infant formulas determined by magnetic sector field inductively coupled plasma–mass spectrometry, *Biol. Trace Elem. Res.*, **76**(2) (2000), 97–112.

12

ARSENIC AND OTHER POTENTIALLY TOXIC TRACE ELEMENTS IN RICE

CHIARA FRAZZOLI, MARILENA D'AMATO, AND SERGIO CAROLI

Istituto Superiore di Sanità, Viale Regina Elena 299, 00161 Rome, Italy, E-mail: caroli@iss.it, frazzoli@iss.it, damato@iss.it

GYULA ZÁRAY

Department of Inorganic and Analytical Chemistry, Eötvös University, Budapest, Hungary, E-mail: zaray@ludens.elte.hu

12.1 INTRODUCTION

Rice is an annual grass that belongs to the *Gramineae* family. The cultivated species of rice are basically two, namely, the Asian *Oryza sativa* (common rice), growing worldwide, and the African *Oryza glaberrima*, growing on a limited scale in West

The Determination of Chemical Elements in Food: Applications for Atomic and Mass Spectrometry.
Edited by Sergio Caroli

Africa. Asian cultivated rice evolved into three eco-geographical types, *Indica, Japonica*, and *Javanica*. It is estimated that about 120,000 distinct rice varieties exist in the world, which differ from one another in terms of growth duration, endosperm traits, level of tolerance to coldness, or other abiotic stresses (e.g., drought, salinity, and submergence), as well as resistance to diseases and pests. In some countries, varieties are classified also according to the season in which they grow [1].

The nutritive characteristics of rice and its relative low cost make of this cereal a staple food in many countries, meeting almost completely (up to 80 percent) the food needs of half of the world population, and contributing over 50 percent of protein to the daily food intake [2]. From a general viewpoint, the major rice-producing countries are also large rice consumers and exporters. In fact, China and India, because of their hot and humid climates, favorable to the cultivation of rice, cover more than 50 percent of the world rice production and account for about 25 percent of the worldwide export of rice [3, 4].

Such a large consumption of this foodstuff may pose a health risk to the consumer as regards the content of potentially toxic elements as a consequence of cultivation in contaminated groundwater. In fact, in several Asian countries, for example, Bangladesh, China, India (West Bengal), and Taiwan, there has been substantial evidence of As groundwater contamination with direct consequences as regards the As content in vegetables and rice grown in those areas [5–10]. In 1995, the Codex Alimentarius Commission adopted safety and quality criteria for the rice produced for human consumption and other Codex working groups set maximum limits for pesticides residues and discussed possible limits for heavy metals such as Cd and Pb [11, 12].

Rice is, and will be for many years to come, a key parameter in the world economy. Hence, to better protect the health of the consumer, it is of crucial importance that the content of essential and potentially noxious elements in this foodstuff be constantly monitored. This is all the more true in the case of As since this element is often present at significant levels in the water where rice is cultivated. In this context, spectrometric techniques are pivotal to gain reliable quantitative information. The sections which follow, on the one hand, summarize the most significant findings in this field over the past few years and, on the other hand, report the so far unpublished results of an investigation recently undertaken by the authors of this chapter.

12.2 SELECTED STUDIES ON TRACE ELEMENTS IN RICE

Needless to say, rice is of paramount importance in the diet of Asian populations. It is not surprising, then, that the vast majority of the investigations undertaken to date come from countries in the Far East.

In a study conducted in Japan in 1989 and 1990, a number of food commodities (including rice) were analyzed for their contents of As, Be, Cd, Ce, Cr, Hg, Pb, Sb, and Sn [13]. Food samples were collected at the Akita University Hospital to mirror hospital diets, which, in turn, were similar to those of the population at large. The ratio of rice to the total intake of the various foodstuffs (primarily fish, shellfish, meat, milk, eggs, vegetables, and fruits) was in most cases between 55 and 60 percent. Samples were analyzed as a pool of the various foods consumed with the diet. After freeze-drying, the composite samples were wet-ashed with HNO_3 and H_2O_2 and diluted as necessary with distilled water prior to analysis with inductively coupled plasma mass spectrometry (ICP-MS). The accuracy of experimental data was checked by means of a Certified Reference Material (CRM), namely the NIST Standard Reference Material 1577 Bovine Liver, with satisfactory results. The average concentrations detected for the elements investigated were found to be as follows (in $\mu g\ g^{-1}$): As, 1.75 ± 0.67; Be, 0.18 ± 0.02; Cd, 0.11 ± 0.28; Ce, 0.11 ± 0.41; Cr, 0.96 ± 0.33; Cu, 2.68 ± 0.96; Hg, 0.003 ± 0.010; Pb, 0.07 ± 0.03; Sb, 0.015 ± 0.006; Sn, 0.11 ± 0.28; and Zn, 16.6 ± 1.98. The figures obtained for As were relatively higher than those typically reported for the same food categories.

One of the largest exporters of rice in the world is Vietnam. In the late 1990s a project was launched by a Vietnamese and an Australian University to gain reliable information on the quality of Vietnamese rice from the viewpoint of its elemental content at major, minor, and trace levels, and to provide a better understanding of the agricultural, geographic, and environmental effects in order to optimize rice quality [14]. Thirteen Vietnamese brown and white rice varieties were collected in 16 localities, mostly from the Red River region (one of the primary rice production areas in Vietnam) so as to represent different growth localities, climatic seasons, varieties, and production types. Some Australian samples were also taken into account for comparative and reference purposes. Rice grains were dried and de-hulled prior to closed-vessel microwave (MW)-assisted digestion by means of HNO_3 and H_2O_2. Analytical determinations of 14 elements (Al, As, Ca, Cd, Cu, Fe, K, Mg, Mn, Mo, Na, Ni, P, and Zn) were performed using flame atomic absorption (FAAS) for Na, quadrupole (Q)-ICP-MS for As, Cd, Mo, and Ni, and inductively coupled plasma atomic emission spectrometry (ICP-AES) for all the other elements. Thorough validation of the analytical methods was accomplished. In particular, accuracy was checked by resorting to the CRM NIES 10 Rice Flour, with generally good agreement with the certified values. The average concentrations of all the elements investigated in Vietnamese rice fell into the ranges reported for Australian and other overseas rices. The concentration ranges obtained for Vietnamese and Australian rice samples were, respectively (in $\mu g\ g^{-1}$): Al, 0.8–6.9 and 1.0–5.1; As, 0.032–0.465 and 0.120–0.776; Ca, 37–113 and 40–84; Cd, <0.003–0.048 and <0.003; Cu, 1.1–5.8 and 1.8–2.8; Fe, 4.6–11.2 and 2.1–6.0; K, 497–1796 and 633–1254; Mg, 137–887 and 164–421; Mn, 5.9–16.3 and 8.3 and 17.3; Mo, 0.108–1.030; Na, 2.5–15.1 and 2.8–6.4; Ni, <0.100–2.022 and <0.100–0.204; P, 819–2589 and 789–1440; and Zn, 14.6–30.1 and 13.8–22.8. It could be concluded that Vietnamese rice has elemental concentration ranges largely overlapping those of Australian rice. Another fact worth mentioning is that Vietnamese rice has rather high concentrations of nutritional elements such as K, Mg, and P, whereas the levels of potentially toxic trace elements (e.g., Cd) are very low. Finally, the concentration of Ni was found to be higher in white than in brown rice, which is probably ascribable to the translocation behavior of this metal in the rice plant.

A simple but effective method was set up by Qiao-Yun Ye et al. [15] to determine Cd in rice by flow injection (FI) on-line precipitation preconcentration coupled with FAAS, thus overcoming the drawbacks of batchwise operation and further enhancing preconcentration and separation. The key steps of the method consisted in the on-line precipitation preconcentration obtained by merging the sample with ammonia solution, collection of the ensuing precipitates by a knotted reactor without filtration, dissolution of the precipitates with diluted HNO_3, and on-line detection by FAAS. Accuracy was estimated by means of the CRM GBW 08511 Rice Flour. Good agreement was observed between the experimental and certified values of Cd. The method was applied to a number of rice brands from the region of Tianjin (China). Cadmium concentrations spanned the range $<0.002-0.91\ \mu g\, g^{-1}$, the lower end being the limit of detection (LoD) (3σ criterion). The developed method turned out to be easy to use, sensitive, environmentally friendly, and cost-effective, and is thought to be a promising alternative to electrothermal atomization atomic absorption spectrometry (ET-AAS).

Rice dominates the Taiwanese diet [16]. A total of 407 rice samples and 83 shellfish samples were collected from various areas of Taiwan. The elements taken into consideration were As, Cd, Cr, Cu, Hg, Ni, Pb, and Zn. To quantify As, the hydride generation (HG) approach was chosen. To this end, samples were refluxed with HNO_3 and H_2SO_4, added with formic acid, diluted with deionized water, and finally reacted with $NaBH_4$ and NaOH to generate AsH_3. For the quantification of all the other elements, the rice samples were refluxed with HNO_3 and then added with $HClO_4$. Arsenic, Cd, Cr, Cu, Hg, Ni, Pb, and Zn were then determined by means of ICP-AES. As regards Hg, the cold vapor (CV) mode was resorted to after digesting the samples with concentrated HNO_3 in a water bath at 80°C. The analytical technique chosen for Hg was FAAS. In-house validation of the methods was done using the CRMs NIST SRM 1568a Rice Flour, GBW 08510 Rice Flour, and GBW 08511 Rice Flour, with satisfactory results in terms of measurements accuracy. The concentrations ascertained for the elements under test in rice samples (pool of barn and packed types) were the following (in $\mu g\, g^{-1}$): As, $<0.10-0.63$; Cd, $<0.05-0.20$; Cr, $<0.10-5{,}45$; Cu, $<0.05-7.60$; Hg, $<0.01-0.04$; Ni, $<0.10-3.75$; Pb, $<0.10-0.20$; and Zn, $<0.05-70.1$. These levels were considered to be of no concern. For all rice varieties, the concentrations of, for example, Cd and Hg were below Taiwanese food sanitary standards (0.50 and 0.05 $\mu g\, g^{-1}$, respectively). On the basis of the weekly intake of the said elements from rice and shellfish, it could be concluded that the provisional tolerable weekly intakes recommended by WHO/FAO and USNAS were not exceeded.

The trace elements content of ancient rices (red and black) and modern rice (e.g., the Koshihikari brand) was investigated by Kaneda et al. [17]. Red and black rices were brought from China into Japan more than 2000 years ago and gave way to present-day rices. The nutritious, tonic, and hematopoietic properties of black rice have been known and exploited for a long time. On the other hand, modern varieties have been bred to improve the flavor and enhance environmental resistance. In order to evaluate the relevant nutritional value in terms of concentration of trace elements in such rices, five elements (Ca, Fe, K, Mg, and P) were quantified by neutron activation analysis (NAA) and a further eight elements (Al, Co, Cr, Cu, Fe, Mn, Mo, and Zn) were determined by ICP-MS. For all rice varieties, sample preparation consisted of the subsampling of three different parts (hull, bran, and white core), extraction with ethanol, concentration by rotary evaporation, and freeze-drying. In the case of NAA measurements, aliquots of the freeze-dried powder were sealed in double-layered chlorine-free polyethylene bags prior to irradiation. For ICP-MS, the rice extracts were treated repeatedly with HNO_3, H_2O_2, and $HClO_4$ and diluted as necessary before presentation to the spectrometer. No explicit mention is made in this study of the validation of the analytical methods used, in particular as regards the use of CRMs. As a rule, the highest content of trace elements was found in the bran portions for the three rice varieties under test. The black rice hull from Toyama Prefecture turned out to be an excellent source of Ca, Cu, Fe, Mn, and Zn, whereas the Koshihikari rice hull from Niigata Prefecture was found to be a good supplier of K, Mg, and Na. The findings of this study were deemed beneficial for the sound development of trace element supplementation regimens.

In order to estimate the dietary intake of essential and trace elements via regular consumption of rice by the Korean population, background levels were investigated [18]. A total of 63 samples were collected in eight administrative areas in South Korea and analyzed by ICP-AES for their content of 16 minor and trace elements. Dried, finely milled rice grains were treated first with HNO_3 and then with $HClO_4$. After dilution with deionized water the solutions were ready for analysis. A rigorous quality control (QC) program was implemented with CRM NIST SRM 1568a Rice Flour. Accuracy of results was always found to be within the 95 percent confidence interval with the only exception of Al. The concentration ranges found were, in order (in $\mu g\ g^{-1}$): Al, 3.03–15.9; As, 0.043–1.99; Ba, 0.37–1.52; Ca, 74.3–117; Cd, 0.010–0.032; Co, 0.002–0.010; Cr, 0.068–0.127; Cu, 1.29–2.53; Fe, 5.41–18.0; K, 1480–2790; Mg, 680–1390; Na, 2.94–23.7; Ni, 0.014–0.179; Pb, 0.081–0.374; Si, 14.2–88.9; and Zn, 14.8–19.7. Differences in element content among the eight administrative areas were, in most cases, not significant. On the basis of the above figures and assuming an average rice consumption of ca. 250 g per day, the daily intakes of As, Cd, Cu, Pb, and Zn via Korean white rice might, in some cases, pose potential health problems.

Jayasekera and Freitas were the first to investigate the content of 24 chemical elements in rice grain and flour from Sri Lanka [19]. The purpose of their study was to compare the data obtained with the maximum permissible levels prescribed by international guidelines. Determinations were made by means of INAA with k_0 standardization. The rice samples were collected from local markets and were representative of two major varieties, raw polished and parboiled rice. Prior to analysis, samples were freeze-dried and ground to fine powder. The CRM IAEA V-8 rye flour was used to check the accuracy of experimental measurements. The following mean concentrations (averages of various brands and producers) were obtained (in $\mu g\ g^{-1}$): Al, 7.6; As, 0.06; Br, 0.25; Ca, 94; Cd, $<$0.5; Cl, 180; Co, 54; Cr, 0.77; Cs, 0.027; Cu, 3.0; Fe, 168; Hg, $<$0.09; K, 2046; La, 0.004; Mg, 895; Mn, 15.2; Mo, 0.83; Na, 12.3; P, 1625; Rb, 33.6; S, 1250; Sc, 1.6; Se, 0.17; and Zn, 23.8. Calcium, P, and S were also quantified by energy-dispersive X-ray fluorescence (EDXRF) analysis. The variations observed in different rice samples could be well explained by factors such as rice brand, soil composition, use of agricultural chemicals, postharvest processing, and environmental causes. From a general viewpoint, however, the levels of essential elements were found to be adequate, whereas potentially toxic elements were present at concentrations raising no concern for human health in spite of the fact that rice in Sri Lanka accounts for more than 50 percent of the average daily intake of food.

Multielemental analysis of rice is ever more challenging and improved accuracy is constantly sought, as testified by the increasing number of papers published over the past decade. In this context, the development of novel methods and their validation are crucial. Belarra and coworkers, for example, described the direct determination of Fe and Zn in solid rice by means of ET-AAS [20]. Rice of two different varieties (long and short grain) was analyzed. The grains were chopped, suspended in water drops, and directly placed onto the platform of the pyrolytic graphite tube. The rice matrix was removed during the calcination step (600–1000°C) with no appreciable analyte loss. Results compared well with those obtained by conventional approaches, ET-AAS analysis of samples after dissolution by acid digestion with HNO_3 or calcination at 800°C and dissolution in *aqua regia*. The direct method allowed the two elements to be quantified down to 10 $\mu g\ g^{-1}$ with a relative standard deviation of 6–10 percent and a throughput of four samples per hour, quite convenient where expeditiousness is favored over precision. In the samples investigated the mean concentrations of Fe and Zn were (in $\mu g\ g^{-1}$) 10.7 $\pm$ 1.1 and 11.0 $\pm$ 0.8, respectively.

Tsukada et al. [21] compared the performance of NAA and Q-ICP-MS in determining major and trace elements (Ag, Al, Ba, Ca, Cd, Ce, Co, Cr, Cs, Cu, Fe, K, La, Mg, Mn, Mo, Na, Ni, Pb, Rb, Sb, Sc, Se, Sr, Th, U, V, Zn, and Zr) in soil and plants (potato petiole and potato tuber). The accuracy afforded by the two methods in the analysis of plants was tested using the CRMs NIST SRM 1568a and NIES 10a, 10b, and 10c Rice Flour, thus providing good evidence of the applicability of these methods to rice as well.

Digestion of samples is in itself a challenging task. Cabanis et al. quantified trace Cd in food samples (rice, beef, and cheese cream) by four different digestion procedures based on: H_2SO_4–HNO_3 in an open flask; HNO_3 under pressure; H_2SO_4–HNO_3 with refluxing; and HNO_3–$HClO_4$–H_2O_2 with refluxing [22]. Dry ashing procedures were not considered because of the risk of volatilization of the analyte at temperatures exceeding 450°C. The study aimed at assessing the effects of digestion reagents and food matrix components on the ET-AAS determination of Cd. In this respect, the procedure based on H_2SO_4–HNO_3 with refluxing was found to be the best. The rice under test contained Cd at a concentration of 0.065 $\mu g\ g^{-1}$.

Gawalko and coworkers elucidated the characteristics of closed-vessel and focused open-vessel MW sample preparation techniques for the subsequent determination of Cd, Cu, and Pb, and Se in wheat, wheat products, corn bran, and rice flour by ET-AAS with transverse-heated graphite furnace [23]. Both techniques were evaluated using a total of 15 CRMs for the three food types. In the case of rice, these were the NIES 10a, 10b, and 10c Rice Flour and the NRC Rice Flour. The two techniques were equivalent in terms of agreement with the certified figures. Up to 12 samples could be processed with the closed system with minimal amounts of acid and reduced risk of contamination. Larger masses of samples could be treated by the open-vessel system.

A few years later Momen et al. also investigated the issue of optimal sample preparation of wheat, corn, barley, rye, oats, millets, and rice for the subsequent determination of Ag, Al, Ba, Bi, Cd, Cr, Cu, Fe, Ga, In, Mg, Mn, Pb, and Zn by ICP-AES [24]. Samples were carefully selected so as to represent major commercially available categories. The cereals were subjected to grinding and sieving down to 100 μm. The fine powders were digested using the following procedures: (i) wet digestion with HNO_3–H_2SO_4; (ii) wet digestion with HNO_3–H_2O_2–H_2SO_4; (iii) dry ashing after addition of an ethanol solution of $Mg(NO_3)_2$ as an ashing aid, subsequent evaporation of moisture at 90°C and ashing in an electric muffle furnace at 500°C; (iv) same as (iii) except that the methanolic solution also contained HNO_3. The average concentration values found for the various elements in rice and rice flour (pool of the four digestion procedures) were, respectively (in $\mu g\,g^{-1}$): Ag, 0.015 ± 0.001 and 0.012 ± 0.001; Al, 4.4 ± 0.1 and 4.3 ± 0.01; Ba, 0.11 ± 0.01 and 0.10 ± 0.01; Bi, 0.22 ± 0.01 and $0.21 \pm 0.\ 01$; Cd, 0.024 ± 0.001 and 0.020 ± 0.002; Cr, 0.20 ± 0.01 and 0.16 ± 0.01; Cu, 2.3 ± 0.2 and 2.3 ± 0.1; Fe, 13.0 ± 0.5 and 9.1 ± 0.3; Ga, 0.056 ± 0.003 and 0.056 ± 0.003; In, 0.34 ± 0.02 and $0.27 \pm 0.\ 01$; Mg, 557 ± 15 and 510 ± 14; Mn, 11.9 ± 0.6 and 10.5 ± 0.4; Pb, 0.044 ± 0.003 and 0.032 ± 0.002; and Zn, 23 ± 2 and 19 ± 1. The accuracy of the investigated methods was verified by means of the CRM NIST SRM 1586a Rice Flour and found to be satisfactory in all cases, with the only exception of Bi, Fe, and Cd for dry-ashed sample, probably as a consequence of analyte volatilization and/or adsorption on the inner wall of the crucible. Reproducibility was also found to be sufficient. Results showed that there were no statistically significant differences in efficiency between the two wet digestion procedures nor between the two dry ashing procedures. Hence, it was concluded that the use of H_2O_2 for wet digestion and of HNO_3 for dry ashing was not critical. Although both approaches turned out to be acceptable, wet digestion was recommended for its better analytical recovery.

The National Institute for Environmental Studies (NIES) of the Environment Agency of Japan is a well-established producer of biological and environmental CRMs. In this context, in 1991 Okamoto described the preparation of the multielemental CRM NIES 10 Unpolished Rice Flour [25]. Unpolished rice, instead of polished rice, was selected as the raw material for this CRM since the former was used as a test material for environmental contamination. In fact, unpolished rice can contain Cd at a level of 1.0 μg g^{-1}, which exceeds the threshold of 0.4 μg g^{-1} set by Japanese authorities (the consumption of Cd-contaminated rice is the main cause of the Itai-Itai disease). Three different levels of Cd were chosen (low, medium, and high). The batch of unpolished rice was pulverized with a rotor speed-mill, sieved through a 0.5 mm screen, and dried in an air-oven at 80°C to reduce the moisture content to about 5 percent. The homogenized fine powder was packaged into 1000 acid-washed glass bottles, containing 60 g of candidate CRM each. FAAS, ET-AAS, ICP-AES, ID-ICP-MS, thermal ionization mass spectrometry (TIMS), NAA, and proton-induced X-ray emission (PIXE) analysis were the techniques applied in the certification project in which 18 qualified laboratories participated. The three CRMs thus obtained were labeled NIES 10a, NIES 10b, and NIES 10c, with a content of 0.023 ± 0.003, 0.32 ± 0.02, and 1.82 ± 0.06 μg g^{-1}, respectively. Also Ca, Cu, Fe, K, Mg, Mn, Mo, Na, Ni, P, Rb, and Zn were certified along with Cd. For other elements (namely, Al, As, Br, Cl, Co, Cr, Hg, Se, and Sr) only indicative values were achieved.

A few years later, the lack of certified concentrations for some crucial elements in the commercially available rice-based CRMs prompted the Korean Research Institute of Standards and Science (KRISS) to undertake the preparation of rice flour CRMs at elevated (A series) and normal (B series) levels of trace metals [26]. The preparation of the candidate CRMs included crushing of the rice grains, mixing, sieving down to less than 0.5 mm, distribution and sealing in Pyrex glass, and ^{60}Co γ-irradiation for sterilization. Pilot analyses of the candidate CRMs were performed by INAA, ET-AAS, ICP-AES, and ICP-MS, and a number of elements could be quantified (As, Ca, Cd, Cr, Cu, Fe, K, Mg, Mn, Mo, Na, P, Pb, and Zn). As a part of these activities, Chang et al. applied isotope dilution (ID)-ICP-MS to the determination of Cd, Cu, Pb, and Zn in the CRM KRISS B Rice Flour [27]. The analytical method was validated by using the CRMs NIST SRM 1568a Rice Flour and NIES 10a Rice Flour. The concentrations were: Cd, 33.9 ± 0.2 ng g^{-1}; Cu, 3.00 ± 0.1 μg g^{-1}; Pb, 29.9 ± 0.2 ng g^{-1}; and Zn 22.6 ± 0.7 μg g^{-1}.

Cadmium concentration of the CRMs GBW 08510, GBW 08511, and GBW 08512 Rice Flour (high, medium, and low Cd content, respectively) was certified by the National Research Center for Certified Reference Materials (Beijing, China) [28]. The raw material was obtained from Cd- contaminated rice grown in the Songjang county (Shangai). The CRMs were prepared through blowing to remove dust and husk, grinding, sieving down to a 42 mesh (GBW 08510 and GBW 08511) or a 60 mesh (GBW 085129), packaging in glass bottles (40 g each), and ^{60}Co irradiating for long-term preservation. After HNO_3–$HClO_4$ dissolution of samples, ID-SF-ICP-MS was used to quantify Cd concentrations. These were found to be, in order (in μg g^{-1}): 2.6019 ± 0.0097; 0.5044 ± 0.0034; and 0.0069 ± 0.0006.

The CRM NIST SRM 1568a Rice Flour, along with the CRMs NIST SRM 1566 Oyster Tissue, 1567 Wheat Flour, and 1570 Trace Elements in Spinach, was revalidated nearly two decades after the original certification was issued [29].

INAA and thermal neutron capture prompt γ-ray activation analysis (PGAA) were used to this end. Unopened units were shown to have stable dry weight bases and stable mass fractions. The concentrations measured in this study for Al, As, B, Br, Ca, Cl, Co, Cu, Fe, K, Mg, Mn, Mo, Na, Rb, S, Se, and Zn were in very good agreement with the certified or consensus values, which justifies their continued use for QC purposes.

The issues of method validation and assessment of measurement uncertainty in the determination of potentially toxic trace elements in rice are of permanent interest for the scientific community. In this context, the sources of uncertainty associated with the determination of Cd, Cu, Pb, and Zn have been recently estimated in rice through an interlaboratory comparison [30]. Four Brazilian laboratories participated in the proficiency test. The analytical technique used were FAAS, ET-AAS, and ICP-AES. The rice samples were supplied by the Institute for Reference Materials and Measurements (IRMM), Joint Research Center of the European Commission, within the scope of the interlaboratory comparison International Measurement Evaluation Programme (IMEP) 19 Trace Elements in Rice (see also Chapter 7 in this book). Three out of the four laboratories reported values close to the reference values. It was emphasized that, in order to establish a reliable uncertainty budget, all significant sources of uncertainty should be identified.

Over the past decade several studies ascertained the content of several trace elements, primarily Cd, in rice cultivated in different countries (see, e.g., [31–40]). A recent study reported a total As concentration in various types of Italian rice as ranging from 80 to 280 ng g^{-1}, and compared these figures with those obtained for other countries [41]. The ability of As to accumulate in rice can be ascribed to the fact that As is similar to P in terms of biogeochemical behavior [8, 42–44].

Speciation analysis, on the other hand, plays a pivotal role in a more realistic assessment of the impact on human health of the intake of potentially toxic elements through the diet (see also Chapters 16–22 in this book). Recent studies on chemical speciation of As in rice showed that the percentage of the inorganic As (III) and As (V) species, i.e., the most toxic ones, range from 11 to 91 percent, a much higher percentage than in other foods [45, 46].

To shed further light on the actual content of trace elements in rice from various countries, an investigation was recently undertaken which reached completion at the moment of writing this chapter. The study is reported in detail in the following section.

12.3 TRACE ELEMENTS IN ITALIAN AND ASIAN RICE: A COMPARATIVE INVESTIGATION

12.3.1 Introductory Aspects

In spite of the wealth of information summarized in Section 12.2, there is still a keen need for systematic investigations aimed at quantifying the presence of trace elements in food. The purpose of this study was to compare the contents of some trace elements in different types of rice cultivated in Italy as well as in some Asian countries.

12.3.2 Materials and Methods

12.3.2.1 ***Sample Collection*** Ten rice samples of Italian origin (five of which were parboiled) and 11 samples from Asian countries were obtained from Italian retailers. The Asian samples were from China (three), Thailand (two), Bangladesh (two), Sri Lanka (one), India (one), Pakistan (one), and Philippines (one).

Aliquots were subsampled from the total amounts available for each type of rice, put into polyethylene bags, coded for unequivocal identification, and stored in a desiccator until analysis.

12.3.2.2 ***Reagents*** The chemicals employed in this work were as follows: single-element calibrant stock solutions (1000 mg l^{-1}) of As, Cd, Co, Cu, Fe, Mn, Pb, Rh, and Y in 2 percent HNO_3 (Spex Industry, Edison, NJ, USA), 65 percent HNO_3 Ultrapur and 30 percent H_2O_2 Suprapur (both supplied by Merck, Darmstadt, Germany). All solutions were prepared with 18 MΩ deionized water (EASY pure UV, Barnstead/Thermolyne, Dubuque, IA, USA).

12.3.3 Analysis

Manipulations of samples were always performed in a Class-100 clean room (Tamco, Rome, Italy) to minimize the risk of any possible exogenous contamination by the elements under test. Prior to analysis all samples under test were subjected to MW-assisted acid digestion. To this end, about 0.5 g of material were placed in a PTFE vessel and added with a mixture of 5 mL HNO_3, 1 mL H_2O_2, and 1 mL high-purity deionized water. Water was added to the digestion mixture to minimize the evolution of gases caused by the high content of carbohydrates in the rice matrix. The vessels were sealed, placed in an MW oven (MLS-1200 Mega, FKV, Bergamo, Italy), and then subjected twice to the following MW digestion cycle: 2 min at 250 W, 2 min at 0 W, 5 min at 250 W, 5 min at 400 W and 5 min at 600 W. After cooling, the digests were gravimetrically diluted to 20 g with high-purity deionized water.

To quantify the trace elements of interest plasma-based techniques were used, namely: (i) ICP-AES using an Optima 3100 instrument (Perkin–Elmer, Norwalk, CT, USA) equipped with a cross-flow nebulizer and a Ryton Scott spray chamber; (ii) Dynamic Reaction Cell (DRC) Q-ICP-MS using an Elan 6100 spectrometer (PerkinElmer, Norwalk, CT, USA) equipped with a quartz cross-flow Meinhard nebulizer and a cyclonic spray chamber; (iii) SF-ICP-MS using an Element1 (ThermoElectron, Bremen, Germany) with a pneumatic nebulizer and a Ryton Scott spray chamber.

Copper, Fe, Mn, and Zn were quantified by ICP-AES as the expected high levels of concentration of these elements in the matrix under study were well above the detection limits characteristic of this technique. Arsenic, Cd, Co, and Pb required more powerful techniques such as Q-ICP-MS and SF-ICP-MS.

The standard addition calibration mode was adopted for all determinations since matrix effects were found to be significant, as expected for complex samples as those under test. Yttrium was used as the internal standard at the concentration of 100 ng g^{-1} for ICP-AES and Rh at the concentration of 5 ng g^{-1} for both Q-ICP-MS and SF-ICP-MS. The accuracy of the whole procedure was tested using the CRM NIST SRM 1568a Rice Flour, as set out in Table 12.1.

It is well known that the signal of $^{75}As^+$ may be significantly altered by the formation of the polyatomic species $^{35}Cl^{40}Ar^+$. To minimize this interference, a mathematical correction equation was applied to estimate its magnitude through the signal of $^{37}Cl^{40}Ar^+$ at mass 77 amu. After comparing the data obtained with and without the mathematical correction in the case of the CRM it was concluded that the interference was negligible. Hence, only uncorrected data for As have been considered, this being confirmed by determinations performed by SF-ICP-MS in the medium resolution (MR) mode ($m/\Delta m = 4000$). Moreover, in the SF-ICP-MS determinations, also the small interferences caused by other argides and oxides ($^{59}Co^{16}O^+$, $^{39}K^{36}Ar^+$) on the $^{75}As^+$ signal could be disregarded by integrating only the central part of the peak. Again, it turned out that the correction was unnecessary.

For the quantification of Cd, Co, and Pb the masses at 114, 59, 202, and 208 amu were chosen. As for $^{114}Cd^+$ determinations by Q-ICP-MS, a mathematical equation was applied to correct for $^{114}Sn^+$ interference through the Sn signal at mass 118 amu. The determination by SF-ICP-MS in the MR mode might cause the influence of $^{98}Mo^{16}O^+$, $^{98}Ru^{16}O^+$ and $^{114}Sn^+$ on the $^{114}Cd^+$ signal, while solving the interference of $^{74}Ge^{40}Ar^+$ and $^{74}Se^{40}Ar^+$ on the analyte. All in all, the accuracy of $^{114}Cd^+$ results gives evidence of the suitability of SF-ICP-MS in the MR mode for the determination of Cd in the rice matrix.

The interferences due to oxides formation, for example, $^{98}Mo^{16}O^+$ and $^{43}Ca^{16}O^+$ on masses $^{114}Cd^+$ and $^{59}Co^+$, respectively, were minimized with Q-ICP-MS by optimizing the nebulizer gas flow so as to obtain an intensity ratio for CeO^+/Ce^+ lower

TABLE 12.1. Accuracy As Tested by Means of the CRM NIST SRM 1568a Rice Flour. Data are Expressed as Mean Concentration in ng $g^{-1} \pm$ SD (n = 10)

	Technique			
Element	ICP-AES	Q-ICP-MS	SF-ICP-MS	Certified Values[a]
As	—	330 ± 20	320 ± 30	290 ± 30
Cd	—	26 ± 3	26 ± 2	22 ± 2
Co	—	20 ± 1	18± 1	(18)
Cu	2300 ± 300	—	—	2400 ± 300
Fe	7700 ± 500	—	—	7400 ± 900
Mn	19,400 ± 1500	—	—	20,000 ± 1600
Pb	—	8 ± 1	9 ± 2	<10
Zn	18,500 ± 1900	—	—	19,400 ± 500

[a]Figures within parentheses are informative values.

than 3 percent. The determination of $^{59}Co^+$ in the MR mode by SF-ICP-MS was not affected by the main interferences ($^{43}Ca^{16}O^+$ and $^{118}Sn^{++}$). In the case of $^{208}Pb^+$, the integration windows were optimized so as to eliminate counts due to oxides; the signals arising from argide formation were, in turn, fully separated.

12.3.4 Results and Discussion

The results obtained are summarized in Tables 12.2–12.6. In general, no striking differences in element concentrations were observed between Italian and Asian rice samples. The results obtained from SF-ICP-MS closely match those of Q-ICP-MS, as reported in Tables 12.4 and 12.5. The reliability of the analytical methods was confirmed by the data obtained by the CRM (Table 12.1).

As regards the potentially toxic elements, namely As and Cd, their concentrations were found to be slightly lower in Asian than in the Italian rice varieties, with ranges of 34–246 and 130–377 ng g^{-1} for As and of 5–49 and 16–67 ng g^{-1} for Cd, respectively. It is worth noting that the present results are in agreement with previous literature data [9, 10, 41].

In the case of Pb, no differences in concentration were observed between the two sets of samples. The results for Cd and Pb were indicative of a low-level pollution of the areas under study. In this regard, none of the sample exceeds the relevant FAO/WHO limit of 200 ng g^{-1} level for both analytes.

For the essential elements, higher concentrations were found in Asian rices, even though, in both cases, their levels are comparable with those obtained in previous studies [10, 42]. The observed ranges of concentration for Asian and Italian samples, expressed in nanograms per gram, are respectively as follows: Co, 4–26 and 4–14; Cu, 1015–3005 and 961–2291; Fe, 1893–9949 and 1363–3673; Mn, 5505–17,612 and 3544–11,371; Zn, 6762–20,337 and 5977–18,574. In more detail Mn and Zn levels in parboiled Italian rice (see Table 12.2, rice varieties 5 through 9) were lower than in Asian rice, which may be a consequence of

TABLE 12.2. Concentrations of Cu, Fe, Mn, and Zn in Italian Rice Varieties As Obtained by ICP-AES. Data are Expressed as Mean Concentration in ng $g^{-1} \pm$ SD (n = 3)

Rice Variety	Cu	Fe	Mn	Zn
1	2118 ± 232	2885 ± 320	9056 ± 1068	17,209 ± 106
2	1664 ± 50	1363 ± 106	7745 ± 685	12,167 ± 1213
3	2088 ± 139	2653 ± 110	8052 ± 622	11,961 ± 1030
4	1395 ± 69	2110 ± 226	9255 ± 442	12,219 ± 1043
5	1959 ± 74	3673 ± 347	6364 ± 491	9382 ± 1004
6	961 ± 100	1856 ± 183	3544 ± 320	6700 ± 461
7	1823 ± 36	2552 ± 238	4474 ± 70	8196 ± 585
8	1661 ± 168	2745 ± 115	5879 ± 636	7066 ± 533
9	2291 ± 134	2096 ± 330	5442 ± 321	5977 ± 76
10	1344 ± 260	2689 ± 76	11,371 ± 629	18,574 ± 1128

TABLE 12.3. Concentrations of Cu, Fe, Mn, and Zn in Asian Rice Types as Obtained by ICP-AES. Data are Expressed as Mean Concentration in ng g^{-1} ± SD (n = 3)

Country of Origin	Cu	Fe	Mn	Zn
1[a]	1358 ± 42	2426 ± 37	17,612 ± 1140	11,686 ± 1340
2[a]	2558 ± 224	9949 ± 383	4664 ± 597	6762 ± 207
3[b]	2017 ± 301	2688 ± 157	9670 ± 54	15,567 ± 825
4[b]	3005 ± 73	2391 ± 450	5505 ± 176	20,337 ± 682
5[b]	1015 ± 62	2864 ± 474	12,693 ± 594	13,386 ± 740
6[c]	2284 ± 120	5627 ± 366	6177 ± 510	14,198 ± 675
7[d]	2158 ± 27	2854 ± 164	7989 ± 265	15,294 ± 383
8[e]	1015 ± 41	2505 ± 133	7639 ± 706	12,563 ± 600
9[f]	2263 ± 260	3027 ± 220	4173 ± 463	8555 ± 919
10[g]	2377 ± 180	3153 ± 126	9213 ± 772	14,317 ± 901
11[g]	1205 ± 105	1893 ± 210	9097 ± 649	17,098 ± 1652

[a]Bangladesh, [b]China, [c]India, [d]Pakistan, [e]Philippines, [f]Sri Lanka, [g]Thailand.

TABLE 12.4. Concentrations of As, Cd, Co, and Pb in Italian Rice Types as Obtained by Plasma-Mass Techniques. Data are Expressed as Mean Concentration in ng g^{-1} ± SD (n = 3)

Rice Variety	Technique	As	Cd	Co	Pb
1	Q-ICP-MS	183 ± 9	32.9 ± 4.1	4.23 ± 0.61	3.34 ± 0.14
	SF-ICP-MS	169 ± 24	33.8 ± 3.5	3.25 ± 0.26	2.47 ± 0.27
2	Q-ICP-MS	130 ± 17	47.0 ± 5.3	7.98 ± 0.70	3.21 ± 0.48
	SF-ICP-MS	127 ± 12	46.0 ± 0.4	6.53 ± 0.44	2.44 ± 0.67
3	Q-ICP-MS	211 ± 14	45.6 ± 4.7	6.67 ± 1.9	1.69 ± 0.44
	SF-ICP-MS	229 ± 15	47.6 ± 3.9	3.9 ± 0.4	2.19 ± 0.39
4	Q-ICP-MS	165 ± 23	27.1 ± 2.9	4.44 ± 0.67	4.09 ± 0.50
	SF-ICP-MS	148 ± 25	25.9 ± 1.7	5.01 ± 0.84	3.92 ± 0.32
5	Q-ICP-MS	377 ± 14	67.2 ± 5.8	5.74 ± 0.41	18.6 ± 3.3
	SF-ICP-MS	390 ± 17	69.2 ± 4.7	6.89 ± 0.55	17.0 ± 2.8
6	Q-ICP-MS	143 ± 10	31.5 ± 2.5	13.9 ± 1.5	7.50 ± 0.93
	SF-ICP-MS	140 ± 4	30.4 ± 1.9	15.9 ± 1.7	8.94 ± 0.73
7	Q-ICP-MS	377 ± 10	19.4 ± 1.6	6.40 ± 0.57	5.69 ± 0.77
	SF-ICP-MS	374 ± 13	19.0 ± 1.5	7.89 ± 0.91	6.07 ± 0.81
8	Q-ICP-MS	296 ± 15	15.7 ± 2.1	5.79 ± 0.71	4.79 ± 0.21
	SF-ICP-MS	297 ± 11	22.4 ± 0.7	7.00 ± 0.68	3.63 ± 0.13
9	Q-ICP-MS	233 ± 32	50.6 ± 5.7	10.7 ± 0.9	3.11 ± 0.45
	SF-ICP-MS	233 ± 38	49.5 ± 5.1	11.9 ± 1.2	3.56 ± 0.65
10	Q-ICP-MS	136 ± 15	38.8 ± 3.6	5.17 ± 0.83	3.21 ± 0.78
	SF-ICP-MS	138 ± 17	40.1 ± 4.2	4.65 ± 0.79	3.93 ± 0.69

TABLE 12.5. Concentrations of As, Cd, Co, and Pb in Asian Rice Types as Obtained by Plasma-Mass Techniques. Data are Expressed as Mean Concentration in ng g^{-1} ± SD (n = 3)

Country of Origin	Technique	As	Cd	Co	Pb
1[a]	Q-ICP-MS	97.2 ± 10.0	34.2 ± 4.2	10.3 ± 0.7	8.04 ± 1.38
	SF-ICP-MS	87.1 ± 11.3	32.6 ± 3.4	11.0 ± 0.7	5.09 ± 1.07
2[a]	Q-ICP-MS	162 ± 30	48.8 ± 5.3	14.0 ± 1.2	21.6 ± 3.2
	SF-ICP-MS	164 ± 30	47.2 ± 4.8	14.6 ± 1.9	21.5 ± 3.3
3[b]	Q-ICP-MS	177 ± 10	17.9 ± 2.3	15.4 ± 2.2	3.33 ± 0.45
	SF-ICP-MS	157 ± 7	16.5 ± 1.5	13.2 ± 2.7	3.91 ± 0.52
4[b]	Q-ICP-MS	246 ± 7	5.92 ± 0.67	4.08 ± 0.53	5.18 ± 0.40
	SF-ICP-MS	204 ± 2	5.34 ± 0.82	3.03 ± 0.49	5.69 ± 0.28
5[b]	Q-ICP-MS	119 ± 9	23.3 ± 2.4	7.83 ± 1.14	1.00 ± 0.04
	SF-ICP-MS	124 ± 12	22.7 ± 1.9	8.21 ± 0.95	1.32 ± 0.19
6[c]	Q-ICP-MS	123 ± 5	20.9 ± 2.1	7.76 ± 0.87	20.8 ± 2.3
	SF-ICP-MS	129 ± 6	18.8 ± 1.8	7.96 ± 1.08	22.1 ± 2.8
7[d]	Q-ICP-MS	59.1 ± 2.5	32.1 ± 2.9	6.86 ± 0.65	2.48 ± 0.33
	SF-ICP-MS	61.6 ± 4.3	33.1 ± 2.7	6.29 ± 0.62	2.28 ± 0.38
8[e]	Q-ICP-MS	112 ± 10	15.3 ± 2.3	4.24 ± 0.29	2.92 ± 0.41
	SF-ICP-MS	113 ± 7	13.2 ± 1.8	3.94 ± 0.23	3.43 ± 0.39
9[f]	Q-ICP-MS	33.8 ± 3.2	4.53 ± 0.62	25.6 ± 1.4	5.48 ± 0.82
	SF-ICP-MS	33.8 ± 2.7	4.14 ± 0.11	22.3 ± 2.7	5.85 ± 0.85
10[g]	Q-ICP-MS	95.8 ± 12.3	24.2 ± 3.6	6.24 ± 0.72	3.27 ± 0.61
	SF-ICP-MS	92.3 ± 9.8	27.3 ± 3.9	5.37 ± 0.64	2.84 ± 0.41
11[g]	Q-ICP-MS	140 ± 19	17.3 ± 1.5	17.4 ± 0.2	6.77 ± 0.53
	SF-ICP-MS	140 ± 17	17.3 ± 1.0	19.0 ± 0.4	8.57 ± 1.17

[a]Bangladesh, [b]China, [c]India, [d]Pakistan, [e]Philippines, [f]Sri Lanka, [g]Thailand.

TABLE 12.6. Average Concentrations and Concentration Ranges of Some Elements in Italian and Asian Rices

	Italian Samples			Asian Samples		
Element	n	Mean (ng g^{-1})	Range (ng g^{-1})	n	Mean (ng g^{-1})	Range (ng g^{-1})
As	10	224	130–377	11	124	34–246
Cd	10	38	16–67	11	22	5–49
Co	10	7	4–14	11	11	4–26
Cu	10	1730	961–2291	11	1932	1015–3005
Fe	10	2462	1363–3673	11	3579	1893–9949
Mn	10	7118	3544–11,371	11	8585	5505–17,612
Pb	10	6	2–19	11	7	1–22
Zn	10	10,945	5977–18,574	11	13,615	6762–20,337

parboiling. This result is consistent with data from previous studies [32, 36]. A study conducted on the nutrient composition of commercial rice from Brazil provided some evidence of loss of Mn and Zn as a consequence of the parboiling process, perhaps caused by their inhomogeneous distribution in the grain or from the different retention behavior in terms of solubility, migration rate, and variations in the hydrothermal process [43]. For Cu and Zn, the variations among samples can be primarily ascribed to their different availability from soil, strongly depending on soil pH and organic matter [44].

The varieties of rice examined in this study show only modest differences in trace element content between the Italian and Asian samples. The slight variations observed can be ascribed to specific characteristics of the botanical varieties, soil composition, and availability for each element, and local environmental contamination. The concentration of potentially toxic elements in rice ascertained in this study raise no specific concerns. On the other hand, the key role played by rice consumption in Asia diets calls for careful evaluation of the average daily intake for elements such as As, Cd, and Pb, which might exceed the established tolerance level.

12.4 CONCLUSIONS

Further progress in the knowledge of the intake of trace elements with rice depends primarily on the achievement of three major goals, namely: (i) systematic and internationally coordinated action to gain information on the presence of trace elements in this staple food as a function of the geographic origin, soil nature, environmental conditions of growth, and processing by the producer, which is all the more important in the age of globalization of the market; (ii) ascertainment (wherever applicable) of the chemical forms under which trace elements are present in rice so as to better assess the effects (be they beneficial or adverse) of their intake; in this case synergism among highly reputed expert laboratories is also becoming more and more the rule in order to adequately face this challenge; and (iii) full implementation of quality systems in all phases of the experimental work to attach credibility, validity, and comparability to the data obtained. The role of CRMs in this context cannot be exaggerated. Only fully reliable information can pave the way to a sound evaluation of the benefits and risks associated with the intake of trace elements with rice, this being obviously of the greatest importance to the scientist, the decision maker, and, ultimately, to the layman. Needles to say, these three facets are closely interwoven and only when they are complied with in a holistic manner will substantial steps forward become possible.

REFERENCES

1. G. S. Khush, Origin, dispersal, cultivation and variation of rice, *Plant Mol. Biol.*, **35** (1997), 25–34.
2. Rice Information, Non-series Titles, 3 FAO, Rome, 2002.
3. *Concept Paper: the International Year of Rice 2004*, Secretariat FAO, 2003.

4. *International Year of Rice, Economics and the International Year of Rice*, FAO, Rome, 2004; available at www.rice2004.org.
5. B. K. Mandal, K. T. Suzuki, Arsenic round the world: a review, *Talanta* **58** (2002), 201–235.
6. A. A. Meharg, M. Rahman, Arsenic contamination of Bangladesh paddy field soils: implications for rice contribution to arsenic consumption, *Environ. Sci. Technol.*, **37** (2003), 229–234.
7. T. Roychowdhury, H. Tokunaga, M. Ando, Survey of arsenic and other heavy metals in food composites and drinking water and estimation of dietary intake by the villagers from an arsenic affected area of West Bengal, India, *Sci. Total Environ.*, **308** (2003), 15–35.
8. M. J. Abedin, J. Feldmann, A. A. Meharg, Uptake of arsenic species in rice (*Oryza sativa L.*) plants, *Plant Physiol.*, **128** (2002), 1120–1128.
9. J. M. Duxbury, A. B. Mayer, J. G. Lauren, N. Hassan, Food chain aspects of arsenic contamination in Bangladesh, *J. Environ. Sci. Health*, **A38** (2003), 61–69.
10. M. G. M. Alam, E. T. Snow, A. Tanaka, Arsenic and heavy metal contamination of vegetables grown in Samta village, Bangladesh, *Sci. Total Environ.*, **308** (2003), 83–96.
11. Codex alimentarius, Codex standard for rice, *Codex Stan.*, 198, FAO-UNO WHO, 1995.
12. Codex alimentarius, *Codex Stan.*, 210, FAO-UNO, WHO, 2001.
13. H. Muto, T. Abe, Y. Takizawa, K. Kawabata, K. Yamaguchi, K. Saitoh, Simultaneous multielemental analysis of daily food samples by inductively coupled plasma mass spectrometry, *Sci. Total Environ.*, **144** (1994), 231–239.
14. Tran Dong Phuong, Pham Van Chuong, D. Tong, Khiem, S. Kokot, Elemental content of Vietnamese rice. Part 1. Sampling, analysis and comparison with previous studies, *Analyst*, **124** (1999), 553–560.
15. Qiao-Yun Ye, Yan Li Jiang, Xiu-Ping Yan, Determination of trace cadmium in rice by flows injection on-line filterless precipitation-dissolution preconcentration coupled with flame atomic absorption spectrometry, *J. Agric. Food Chem.*, **51** (2003), 2111–2114.
16. Haw-Tarn Lin, Sue-Sun Wong, Gwo-Chen Li, Heavy metal content of rice and shellfish in Taiwan, *J. Food Drug Anal.*, **12** (2004), 167–174.
17. I. Kaneda, H. Yasui, Y. Adachi, J. Takada, H. Sakurai, Determination of trace element concentrations in ancient rice (red and black rices) and pressent-day rice (Koshihikari): relationship among the trace element concentrations, species, harvest site and rice parts, *Biomed. Res. Trace Elem.*, **16** (2005), 241–249.
18. Myung Chae Jung, Seong-Taek Yun, Jin-Soo Lee, Jong-Un Lee, Baseline study on essential and trace elements in polished rice from South Korea, *Environ. Geochem. Health*, **27** (2005), 455–464.
19. R. Jayasekera, M. C. Freitas, Concentrations levels of major and trace elements in rice from Sri Lanka as determined by the k_0 standardization method, *Biol. Trace Elem. Res.*, **103** (2005), 83–96.
20. M. A. Belarra, M. P. Martinez-Garbayo, J. M. Anzano, J. R. Castello, Graphite-furnace atomic absorption spectrometric method for direct determination of iron and zinc in solid rice samples, *Anal. Sci.*, **12** (1996), 483–488.
21. H. Tsukada, A. Takeda, H. Hasegawa, S. Ueda, T. Iyogi, Comparison of NAA and ICP-MS for the determination of major and trace elements in environmental samples, *J. Radioanal. Nucl. Chem.*, **263** (2005), 773.

22. M. T. Cabanis, G. Cassanas, J. C. Cabanis, S. Brun, Comparison of four methods for digesting food samples for determination of trace levels of cadmium by flameless atomic absorption spectrophotometry, *J. Assoc. Off. Anal. Chem.*, **71** (1988), 1033–1037.

23. E. J. Gawalko, T. W. Nowicki, J. Babb, R. Tkachuk, Comparison of closed-vessel and focused open-vessel microwave dissolution for determination of cadmium, copper, lead, and selenium in wheat, wheat products, corn bran, and rice flour by transverse-heated graphite furnace atom, *Int. J. Assoc. Off. Anal. Chem.*, **80** (1997), 379–387.

24. A. A. Momen, G. A. Zachariadis, A. N. Anthemidis, J. A. Stratis, Development and validation of routine analysis methods for the determination of essential, nonessential, and toxic minor and trace elements in cereal and cereal flour samples by inductively coupled plasma atomic emission spectrometry, *Int. J. Assoc. Off. Anal. Chem.*, **88** (2005), 1797–1810.

25. K. Okamoto, Preparation and certification of rice flour-unpolished reference material, *Sci. Total Environ.*, **107** (1991), 29–44.

26. Yong Sam Chung, Young Yu Chung, Kyung Haeng Cho, Joung Hae Lee, Determination of trace and toxic elements in Korean rice CRM by INAA, ICP and AAS, *J. Radioanal. Nucl. Chem.*, **215** (1997), 129–134.

27. Chang Joon Park, Jung Kee Suh, Kwang Woo Lee, Sang Hak Lee, Determination of Cd, Cu, Zn and Pb in rice flour reference materials by isotope dilution inductively coupled plasma mass spectrometry, *Anal. Sci.*, **13** (1997), 429–432.

28. Mo-Tian Zhao, Jun Wang, Baikeng Lu, Hai Lu, Certification of the cadmium content in certified reference materials for Cd rice flour, *Rap. Commun. Mass Spectrom.*, **19** (2005), 910–914.

29. D. L. Anderson, W. C. Cunningham, Revalidation and long-term stability of National Institute of Standards and Technology standard reference materials 1566, 1567, 1568a, and 1570, *Int. J. Assoc. Off. Anal. Chem.*, **83** (2000), 1121–1134.

30. A. K. Avila, T. O. Araujo, R. M. H. Borges, P. R. G. Couto, M. F. G. Rosias, O. G. F. Rocha, D. S. Vaitsman, V. Dias, Evaluation of uncertainty sources associated with toxic trace element concentrations in rice, *Accred. Qual. Assoc.*, **10** (2005), 277–282.

31. T. Watanabe, S. Shimbo, C.-S. Moon, Z.-W. Zhang, M. Ikeda, Cadmium contents in rice samples from various areas in the world, *Sci. Total Environ.*, **184** (1996), 191–196.

32. S. Shimbo, Z.-W. Zhang, T. Watanabe, H. Nakatsuka, N. Matsuda-Inoguchi, K. Higashikawa, M. Ikeda, Cadmium and lead contents in rice and other cereal products in Japan in 1998–2000, *Sci. Total Environ.*, **281**(2001), 165–175.

33. E. Cortes Toro, H. A. Das, J. Fardy, Z. Bin Hamzah, R. K. Iyer, Sun Laiyan, N. Leelhaphunt, Y. Muramatsu, R. M. Parr, I. H. Qureshi, S. M. Resnizki, S. Surtipanti, S. A. Tarafdar, M. B. A. Vasconcellos, Toxic heavy metals and other trace elements in foodstuffs from 12 different countries. An IAEA coordinated research program, *Biol. Trace Elem. Res.*, **43–45** (1994), 415–422.

34. M. Schuhmacher, J. L. Domingo, J. M. Llobet, J. Corbella, Cadmium, chromium, copper, and zinc in rice and rice field soil from southern Catalonia, Spain, *Bull. Environ. Contam. Toxicol.*, **53** (1994), 54–60.

35. I. F. Rivai, H. Koyama, S. Suzuki, Cadmium content in rice and its daily intake in various countries, *Bull. Environ. Contam. Toxicol.*, **44** (1990), 910–916.

36. J. O. Nriagu, T.-S. Lin, Trace metals in wild rice sold in the United States, *Sci. Total Environ.*, **172** (1995), 223–228.
37. H. K. Das, A. K. Mitra, P. K. Sengupta, A. Hossain, F. Islam, G. H. Rabbani, Arsenic concentrations in rice, vegetables, and fish in Bangladesh: a preliminary study, *Environ. Int.*, **30** (2004), 383–387.
38. S. W. Al Rmalli, P. I. Haris, C. F. Harrington, M. Ayub, A survey of arsenic in foodstuffs on sale in the United Kingdom and imported from Bangladesh, *Sci. Total Environ.*, **337** (2005), 23–30.
39. T. Roychowdhury, T. Uchino, H. Tokunaga, M. Ando, Survey of arsenic in food composites from an arsenic-affected area of West Bengal, India, *Food Chem. Toxicol.*, **40** (2002), 1611–1621.
40. R. A. Schoof, L. J. Yost, J. Eickhoff, E. A. Crecelius, D. W. Cragin, D. M. Meacher, D. B. Menzel, A market basket survey of inorganic arsenic in food, *Food Chem. Toxicol.*, **37** (1999), 839–846.
41. S. D'Ilio, M. Alessandrelli, R. Cresti, G. Forte, S. Caroli, Arsenic content of various types of rice as determined by plasma-based techniques, *Microchem. J.* **73** (2002), 195–201.
42. A. A. Meharg, J. Hartley-Whitaker, Arsenic uptake and metabolism in arsenic resistant and non-resistant plant species, *New Phytol.*, **154** (2002), 29–43.
43. R. J. B. Heinemann, P. L. Fagundes, E. A. Pinto, M. V. C. Penteado, U. M. Lanfer-Marquez, Comparative study of nutrient composition of commercial brown, parboiled and milled rice from Brazil, *J. Food Comp. Anal.*, **18** (2005), 287–296.
44. Chrycanthus Chukwuma, Evaluating baseline data for copper, manganese, nickel and zinc in rice, yam, cassava and guinea grass from cultivated soils in Nigeria, *Sr., Agr. Ecosyst. Environ.*, **53** (1995), 47–61.
45. M. D'Amato, G. Forte, S. Caroli, Identification and quantification of the major species of arsenic in rice, *Int. J. Assoc. Off. Anal. Chem.*, **87** (2004), 238–243.
46. D. T. Heitkemper, N. P. Vela, K. R. Stewart, C. S. Westphal, Determination of total and speciated arsenic in rice by ion chromatography and inductively coupled plasma mass spectrometry, *J. Anal. At. Spectrom.*, **16** (2001), 299–306.

13

TOTAL ANALYSIS AND DISTRIBUTION OF TRACE ELEMENTS IN HUMAN, COW, AND FORMULA MILK

RAFAEL R. DE LA FLOR ST. RÈMY, MARÍA LUISA FERNÁNDEZ SÁNCHEZ, AND ALFREDO SANZ-MEDEL

Department of Physical and Analytical Chemistry, Faculty of Chemistry, University of Oviedo, C/ Julián Clavería 8, Oviedo, E-33006 Spain, E-mail: asm@uniovi.es

The Determination of Chemical Elements in Food: Applications for Atomic and Mass Spectrometry.
Edited by Sergio Caroli

13.1 INTRODUCTION

13.1.1 Definition and Physical Properties

Milk is defined by the International Dairy Federation (IDF) as "The product obtained by the normal secretion of the mammary gland without any addition or subtraction" [1]. This biological fluid is characterized by physical and chemical properties which make milk so essential for the correct development and growth of mammals as set forth below.

Milk is an opaque white liquid, more or less yellowish depending on the β-carotene content in its fatty constituents. Milk odour is soft although characteristic. Its taste depends on the animal species, but it is generally pleasant and sweet. Some of the more relevant physical properties used for the quality control (QC) of milk [1] are detailed in Table 13.1 (in the case of cow's milk, as a typical example).

13.1.2 Chemical Composition of Milk

The chemical composition determines the nutritive properties of milk and its quality. The major constituents are water, fatty matter, proteins, and sugar; however, they are not necessarily more important than minor constituents, such as vitamins, enzymes, and minerals, as the latter also play an important role in nutrition. The determination of such minor constituents is therefore of great significance [2–4].

As it can be seen in Table 13.2, human milk composition is quite different from that of cow's milk. Casein and mineral contents are lower in human milk than in cow's milk, whilst the lactose content is higher in the former. With regards to fatty matter, both types of milk present similar contents, but the total protein is over three-fold higher in cow's milk.

13.1.2.1 Carbohydrates Nearly all the carbohydrates of milk consist of lactose, although there are also other lower concentration carbohydrates (lactose is the disaccharide constituted by two monosaccharide molecules, one of glucose

TABLE 13.1. More Relevant Physical Constants and Features of Cow Milk

Physical Constants	Value
pH (20°C)	6.5–7.0
Acidity[a]	15–18 °D[a]
Density	1.028–1.036 g ml^{-1}
Viscosity	2.0–2.1 cp
Freezing point	0.51–0.55°C
Boiling point	100.17°C

[a]°D: Dornic grade; 1°D corresponds to 0.1 g of lactic acid per liter of milk.

TABLE 13.2. Cow's and Human Milk Compositions (w/w% contents)

	Cow's Milk	Human Milk
Water	87.5	87.7
Lactose	4.8	7.0
Fatty matter	3.7	3.8
Total proteins	3.25	0.9
Caseins	2.6	0.25
Minerals	0.7	0.2
Vitamins	Trace level	Trace level
Enzymes	Trace level	Trace level

and the other of galactose). In cow's milk the content of lactose is 4.8 percent. In cow colostrum, this content is more than 3 percent. The lactose content in human milk is higher than in cow's milk (7.1 percent) and also contains other oligosugars such as L-fucose, *N*-acetylglucosamine, and neuraminic acid; these substances have an important biological activity, particularly for the synthesis of a growth factor for *Lactobacillus bifidus*, a bacteria almost unique to the intestines of breast-fed infants. This is one of the reasons why mother's milk is to be preferred to cow milk. On the other hand galactose is essential for the synthesis of cerebrosides, which are mostly found in the myelin sheath of nerves. In this regard, a relationship between the lactose content in the milk of certain species and the development of the nervous system has been suggested [5].

13.1.2.2 Lipids As the capacity of the stomach of newborns is very limited, lipids and fat are necessary to provide the large amounts of energy required by the newborn's high body growth rate. Milk fat is the major source of lipid that mammalian newborns use for accumulating body adipose tissue and for the development of their nervous system. The major lipids in milk fat are triglycerides, which represent more than 98 percent of the neutral lipids. From a quantitative point of view, there is no difference in the fat content in cow's (3.8 percent) and human milk (3.7 percent), but they do differ in their qualitative composition: in fact, in cow's milk the predominant fatty acids are saturated acids, while polyunsaturated acids are the main fatty acids in human milk, as shown in Table 13.3.

This different qualitative fat composition is very important as linoleic (omega-6) and linolenic (omega-3) acids are essential, especially for the young infant. In

TABLE 13.3. Fatty Acids in Cow and Human Milk (%)

Fatty Acids	Human Milk	Cow's Milk
Saturated	46	64
Monounsaturated	39	32
Polyunsaturated	15	4

fact, they cannot be formed in the human organism. Omega-3 and omega-6 acids are the precursors of prostaglandins, very important compounds in physiological functions such as inflammatory, coagulation, and immunological processes. The linoleic acid content in cow milk varies from 1.2 to 2.0 percent, while this value is much higher in human milk (8–9 percent), another reason for which human milk is to be preferred for the newborn.

13.1.2.3 Proteins The total amount of proteins in milk is much higher during the colostrum period than in that of mature milk (once milk composition is already stable). The protein content is 12 and 35 g l^{-1} in human and cow's milk, respectively. This higher protein content is directly related to the higher growth rate of calves. The chemical contents of the principal proteins present in human and cow's milk are summarized in Table 13.4.

Milk proteins can be classified into two groups, caseins and whey proteins. Their proportion differs widely in human and cow's milk. The whey proteins/caseins ratio is 20 : 80 in cow's milk and 60 : 40 in human milk. On the other hand, water-soluble whey proteins are much less abundant than caseins in cow's milk.

It is interesting to note that β-lactoglobulin is the major protein in cow's milk whey, but it is absent from human milk whey. However, the content of α-lactoalbumin and immunoglobulins (Ig) is higher in human than in cow's milk whey. Serum albumin and lactoferrins are found in similar concentrations in human and cow's milk whey, and the amount of lactoferrin is higher in colostrum than in mature milk. In any case, lactoferrin is present in milk throughout the lactation period.

TABLE 13.4. Protein Content in Cow's and Human Milk

	Human Milk (g l^{-1})	Cow Milk (g l^{-1})
Caseins	3–5	26
Casein-αs1	—	10
Casein-αs2	—	2.6
Casein-β	3–5	9.3
Casein-γ	—	0.8
Casein-κ	1–3	3.3
Whey proteins	4–6	5–7
β-Lactoglobulin	—	2–3
α-lactoalbumin	1–3	0.8–1.2
Serum albumin	0.3	0.4
IgA secretory	0.5–1	—
IgA	0.1	0.03
IgG	0.01	0.6
IgM	0.02	0.05
Lactoferrin	1–3	0.01–0.1

TABLE 13.5. Vitamins (mg l^{-1}) in Cow's and Human Milk

	Cow's Milk	Human Milk
Vitamin A (retinol)	0.33	0.57
Provitamin A (carotene)	0.2	0.4
Vitamin D	0.0006	0.0006
Vitamin E	0.98	6.64
Vitamin B1 (thiamine)	0.44	0.16
Vitamin B2 (riboflavin)	1.8	0.4
Vitamin B3 (niacin)	0.94	1.47
Vitamin B6 (pyridoxine)	0.64	0.1
Vitamin B12	0.0043	0.0003
Vitamin C	21.1	43
Vitamin K	0.04	0.03
Biotin	0.031	0.008
Folic acid	0.05	0.05
Pantothenic acid	3.46	1.84

13.1.2.4 Vitamins Vitamins are organic compounds essential for the correct development of organisms. Therefore, foods, and especially milk, must supply them at the needed amounts. Normal values of vitamins in human and cow milk are summarized in Table 13.5. In general terms, vitamins can be classified as hydrosoluble and liposoluble. According to this feature, the liposoluble A, D, E, and K vitamins are in fat globules, while the remaining vitamins are in milk plasma.

13.1.2.5 Minerals The main mineral constituents in milk are calcium and chlorine, magnesium chloride, phosphate, and citrate. Minerals in milk are mainly present as soluble salts or in colloidal form associated with caseins. Their concentrations may vary enormously. Thus, the minerals present in milk can be classified according to their concentration level as major and minor elements, with small quantitative contributions from trace and ultratrace elements. The total content of minerals in mammalian milks should correspond to the growth requirements of each biological species. Accordingly, the mineral total content in cow milk is four times higher than in human milk.

The most significant differences are found for all major elements (e.g., Ca, K, Mg, Na, and P). However, the ratio Ca : P, which is related to structural functions extremely important for the correct development of the newborn, is higher in human milk. Despite their low levels in milk, minor and trace elements play a central role in the organism, since their deficiency or excess may generate undesirable effects in the newborn. This fact and the centrality of milk in the human diet explain the growing interest in the determination of trace elements in milk [6–8]. Table 13.6 summarizes the concentration levels for major, trace and ultratrace elements in human and cow milk.

TABLE 13.6. Element Normal Values (mg ml^{-1}) Expected for Cow and Human Milk

Element	Cow's Milk	Human Milk
Calcium (major)	1200	310
Cobalt (ultratrace)	0.009	0.002
Copper (minor)	0.034	0.402
Iron (minor)	0.394	0.661
Magnesium (major)	120	40
Manganese (ultratrace)	0.016	0.010
Nickel (ultratrace)	0.010	0.010
Potassium (major)	1570	530
Phosphorus (major)	920	150
Selenium (ultratrace)	0.014	0.015
Sodium (major)	480	150
Strontium (minor)	0.236	0.105
Zinc (minor)	3.40	3.42

13.1.3 Role of Milk in the Nutrition of Infants and Adults

Human milk is the best food for babies as it provides all the nutrients needed for about the first 6 months (26 weeks) of life. Moreover, it contains nutrients that are fit for the unique needs of the human infant, such as certain essential polyunsaturated fatty acids and certain milk proteins, as well as Fe in a readily absorbable form. Human milk also contains immunological and bioactive substances, absent from commercial infant formulas, which confer protection from bacterial and viral infections and may aid gut adaptation and development of the newborn.

Human milk is reliably superior to all substitutes, including commercial infant formulas. The composition of human milk changes during feeds, according to the time of day and even during the course of lactation, supplying the adequate contents of macro- and micronutrients to the newborn all along the lactation period. Adequate nutrition is particularly important in newborns, because they are characterized by a high body growth rate and, consequently, they are more prone to suffer deficiencies in essential elements [9, 10]. Special attention must be paid to the nutrition of premature infants, as certain essential elements are not stored by the fetus during its development into the uterus. Nevertheless, there are situations where it may be preferable, or necessary, to substitute breast milk with an alternative feeding; then, appropriate milk formulas may take the place of human milk. Most commercial infant formulas are based on cow's milk and have been designed to mimic the nutrient composition of human milk. As noted before, the concentrations of protein and electrolytes (e.g., Cl, K, and Na) in human milk are lower than in cow's milk, whereas the opposite is true for certain minerals, primarily Fe and to a lesser extent Zn. In addition, commercial infant formulas lack the nonnutritional, bioactive components of human milk (protective and trophic factors), and the quality of their proteins and lipids (amino acid and fatty acid profiles) may not be optimal for the needs of the baby. Nevertheless, commercial infant formulas provide a satisfactory alternative (and sole) source of nutrition to young infants up to about 6 months of age. Even after the introduction of complementary foods, formulas continue to make a major contribution to infant energy and nutrient requirements and, in the absence of breastfeeding, it should be the main fluid in the diet for the first nine months and possibly beyond.

On the other hand, milk is not only an essential food for infants, but for children and adults as well. Children need sufficient nutrients and energy to meet the demands of growth and development. Demands for nutrients such as protein, Ca, Fe, and Zn are relatively high, and teenagers require quite large amounts of B vitamins – thiamine, riboflavin and niacin. In addition, approximately 45 percent of the adult skeleton is laid down during adolescence.

Although no food is a complete diet in itself (except breast milk for the first few months of life), cow's milk contains nearly all the constituents of nutritional importance to man. Besides providing calories which are essential for growth, milk also supplies a wide range of nutrients which are well utilized by the body and are essential for the growth. In particular, milk provides energy (calories) and is a good source of protein, Ca, Zn, and vitamins A (whole milk only), B2 (riboflavin), and B12. Milk also makes a valuable contribution to intakes of I, niacin (a B vitamin) and vitamin B6. Other minerals present in moderate amounts include Se. As vitamins A and D are contained in the cream fraction, they are present at higher concentrations in high-fat milks.

Milk is a "nutrient-dense" food, which means that the amount of nutrients supplied by milk is high in relation to its caloric content. Milk is therefore a particularly valuable food for children, whose nutritional needs are high in relation to their caloric needs. It is one of the richest sources of Ca, (as are also other dairy products, such as cheese and yogurt), which is better absorbed by the body than Ca from some other foods. The reason for this is that dairy products contain a favorable balance of Ca, Mg, and phosphate, plus lactose (milk sugar) which, together with digested milk proteins, make absorption of Ca easier. Calcium is an essential nutrient which must be provided regularly in the diet. It is vital for the structural integrity of the skeleton and of teeth. There is considerable medical evidence to indicate that the achievement of peak bone density at the end of the growth period is greatly assisted by a diet that provides adequate amounts of Ca. It has been shown that a 1 percent increase in peak bone density results in a 12 percent reduction of osteoporosis risk.

Milk is also kind to teeth because it contains a type of sugar (lactose) which significantly reduces the risk of decay. Lactose is, compared with table sugar (sucrose), less cariogenic and relatively free from sweetness.

13.1.4 Milk as a Source of Essential Elements

Essential elements (e.g., Cu, Fe, Zn, and major elements such as Ca, Mg, K, and Na) are involved in the form of metalloproteins and enzymes in many bio chemical processes, including enzymatic, hormonal, transport, and structural functions, of assistance in keeping the ionic equilibria and osmotic pressure. These factors are all the more important in the case of infants, as during growth and development the human body is extremely sensitive to the deficiency or the excess of essential elements. Thus, the intake of essential elements from milk has to be adequate in order to assure human health. On the other hand, milk can also be a source of exposure to potentially toxic elements, such as Al, Cd, Hg, and Pb, which might be responsible for long-term health risks, especially for young children.

Major elements of milk (e.g., C, H, N, O, P, and S) are also the constituents of biomolecules such as proteins, carbohydrates, and lipids. Major chemical elements in milk (e.g., Ca, Cl, K, Mg, and Na), which play an important electrolytic role in cells, should be included in this group.

13.1.4.1 Iron Normal adults are estimated to contain 4–5 g of the metal. The Recommended Dietary Allowances (RDAs) for Fe are 10 mg per day for an adult male and 14 mg per day for an adult female. Most Fe exists in complex forms bound to protein as heme compounds (hemoglobin and myoglobin) or nonheme compounds as transferrin and ferritin. It is noteworthy that 80 percent of total Fe is used for oxygen transport [11–13]. Iron content in milk varies with the biological species and stage of lactation and is resistant to changes in the level of dietary Fe. The average metal concentrations in cow's and human milk are similar, but the level of Fe is three times higher in colostrum than in mature milk. Iron in milk occurs primarily in combination with several proteins, especially lactoferrin and transferrin.

13.1.4.2 Zinc RDAs are 12–15 mg per day for adults and 19 mg per day for lactating females. Zinc is an essential component of the enzyme systems that are involved in carbohydrate and protein metabolism, and nucleic acid synthesis. Also, the key role of this element in boosting the natural immune system against diseases such as malaria and pneumonia has been shown [11, 12].

13.1.4.3 Copper The content of Cu in the human body is estimated to range between 50 and 80 μg. The RDA is 1.5–3.0 mg per day for adults. Copper is an essential component of several enzymes and is required in bone formation, cellular respiration, cardiac function, connective tissue development, and myelination of the spinal cord [11–14]. This metal is also necessary for Fe absorption and mobilization. Again, Cu content in milk differs with the biological species, stage of lactation, and diet intake. In all species colostrum is substantially richer in Cu than mature milk is.

13.1.4.4 Selenium This metal is an essential trace element that functions as a component of enzymes involved in antioxidant protection and thyroid hormone metabolism. The existence of a number of selenoproteins has been demonstrated. In several intra- and extracellular glutathione peroxidases and iodothyronine

5′-deiodinases, Se is located at the active centers as the selenoamino acid selenocysteine (SeCys) [13, 15]. Glutathione peroxidase catalyzes the breakdown of H_2O_2, phospholipids hydroperoxides and other free hydroperoxides. Erythrocyte glutathione peroxidase contains four Se atoms, in the form of selenocysteine, which is essential for its biological activity.

13.1.4.5 Manganese The human body is estimated to contain a total of 10–20 mg Mn. The RDA is between 2 and 5 mg per day. Manganese is involved in many enzymatic processes as a metalloenzyme or as a cofactor. In contrast to the Mn-activated enzymes, there are few Mn metalloenzymes (arginase, pyruvate carboxilase, and Mn-superoxide dismutase) [14]. The content of Mn in milk responds rapidly to changes in dietary intakes of this metal.

13.1.4.6 Iodine The major function of I in the human body is the formation of the thyroid hormone (thyroxin). It is also involved in muscle function and growth. The RDAs are 150 mg (adult male) and 110 mg (adult female) per day [15].

13.1.4.7 Chromium This metal is essential for normal insulin and glucose function. Chromium is involved in the glucose tolerance factor affecting insulin and the clearance of glucose from the body [14, 15].

13.1.4.8 Nickel Nickel is an essential element for certain animals, but not in humans. However, nickel may be involved in enzyme activity in humans [15].

13.1.4.9 Silicon This element is essential for growth and skeletal development in certain animal species. Silicon is currently not considered to be essential for humans, but it may be involved in the formation of connective tissue, bone matrix, atherosclerosis, hypertension, and in the aging process [14, 15].

13.1.4.10 Vanadium Vanadium is believed to be essential in certain animals, although there is still no evidence of this in humans. A V deficiency may alter Fe metabolism, impair heme synthesis, and affect lipid metabolism [15].

13.2 SAMPLING AND SAMPLE PRETREATMENT

13.2.1 Collection and Storage of Milk Samples

The overall analytical performance of a given determination will strongly depend on the quality of the preanalytical steps such as sampling, sample handling, and storage of the samples. When analyzing total concentrations of an element in milk, factors such as possible contamination and adequate storage are considered to be very important. It is therefore necessary to define all the conditions under which the components of the sample will remain unaltered; existing species should be kept as stable as possible, while contaminations or losses (e.g., due to interactions of sample constituents with containers and other material), which could affect the reliability of the data, should be carefully minimized and checked.

Selection of the material for sampling and storage is critical and an appropriate choice will reduce the risk of contamination. Therefore, milk samples should be collected in an inert and clean container. Unfortunately, no completely contamination-free containers exist. As an example, collection tubes or instruments made of glass cannot be used for Al, Mn, and Pb sampling, storage, or measurements. Plastics with a Cd-based softener or Zn-doped stoppers are to be avoided when analyzing Cd and Zn, respectively. Other metals such as Co, Cu, Fe, and Se present much less risk of contamination. Needless to say, the risk of contamination increases as analyte concentrations in the sample decrease. For this reason it will be always necessary to chemically clean all the materials and instruments in order to avoid trace element contamination and check for the absence of such phenomena. Containers may be thoroughly rinsed with a solution which releases the elements of interest from the material (e.g., using 5 percent HNO_3 and then washing with milli-Q water) [16, 17].

Once the material has been chosen for sampling and storage, it is important to establish a protocol for the collection of milk samples. The extraction will be achieved manually or by means of a plastic milk extractor, a new one for each sampling. The nipple and the skin around it should be cleaned with soap and distilled water before collection. The milk thus collected should then be stored in plastic containers previously decontaminated, as explained above. The containers should be labeled and kept at $-20^{\circ}C$ until analysis.

13.2.2 Sample Treatment for Total Trace Element Determination in Milk Samples

The method of sample preparation to be used for a given analysis is governed by the nature and concentration of the analyte, the nature (solid or liquid) and type of matrix, the available sample amount, and also by the instrumental technique employed. Freeze-dried samples will require some form of digestion or dissolution in order to be analyzed by a classic atomic technique (i.e., using nebulization). Liquids might be analyzed by direct nebulization, but this is not always possible due to matrix interferences. Milk pretreatment may be necessary under such circumstances.

The main problem in carrying out total multielemental determinations in milk (as in other biological samples) is the nature of the matrix, which may interfere with the analytical technique employed for the measurement. In this sense, pretreatment of the samples becomes necessary so as to minimize matrix effects as much as possible (e.g., by destroying the organic matrix). An alternative to such destructive acid attacks is the direct analysis in milk whey samples by simply diluting the sample previously obtained by centrifugation. The main preparation procedures for milk samples (whole, skimmed, or freeze-dried) can be classified as follows: (a) use of diluted solutions in order to minimize matrix and molecular absorption interferences; (b) protein precipitation and separation; (c) complete oxidation procedures (dry ashing and wet digestions); and (d) slurries.

13.2.2.1 Dilution As milk is a viscous liquid, it is usually necessary, or at least advisable, to dilute the milk sample with an appropriate solvent. Deionized water or dilute HNO_3 are usually employed to this end. Also surfactants, such as Triton X-100, are added to milk samples to lower surface tension and promote thorough mixing of the diluted sample. The concentration of Triton X-100 varies from 0.05 to 2 percent [18, 19]. Final dilution factors as high as 50 or 100 have been chosen for metals such as Cu, Fe, and Zn (because of their high levels in milk) when using high sensitivity atomic techniques, for example, electrothermal atomization atomic absorption spectrometry (ET-AAS) and inductively coupled plasma mass spectrometry (ICP-MS). At such high dilutions, matrix effects are insignificant and the majority of the elements that have limits of detection (LoDs) below 1 $\mu g\, l^{-1}$ can be determined. It is important to note that, using ICP-MS for the final measurements, simple direct dilution methods have been found to be inadequate for the measurement of elements in the 40–80 amu range in whole milk: Apart from the possible clogging of the nebulizer and sampler cone, the spectral interferences caused by the formation of polyatomic species are very strong in this region. However, elemental determinations in milk whey can be carried out by ICP-MS just by simple dilution of the whey with redistilled water.

13.2.2.2 Protein Precipitation Procedures involving protein precipitation with trichloroacetic acid for Cu, Fe, and Zn have been developed for flame atomic absorption spectrometry (FAAS) determinations in body fluids [20, 21]. The physical separation of caseins and fat from milk whey can also be achieved using centrifugation at 30,000***g***.

13.2.2.3 Dry Ashing and Wet Ashing Dry ashing procedures for whole milk are carried out at relatively high temperatures (450–500°C) in air at atmospheric pressure (sometimes using ashing aids to improve efficiency and/or to minimize losses). The biggest advantage of this procedure is that large amounts of sample can be ashed and the ashed material can be dissolved in a small volume of acid. In this fashion trace elements can be preconcentrated, a decisive factor when very low concentrations are to be determined. This procedure is also recommended when a sample solution completely free of organic matter is required [22–24]. Conversely, these procedures entail disadvantages that may not be disregarded, such as contamination phenomena, analyte losses, and possible matrix effects resulting from the high levels of the ashing aids used.

Oxidative decomposition by wet ashing is the most widely used sample preparation method for milk samples. Acid digestion involves the use of mineral acids and oxidizing agents to dissolve the sample. Some of the acids commonly used in these procedures are HCl, HNO_3, H_2SO_4, and $HClO_4$. Wet decomposition can be performed with either open or closed systems. Open systems entail

Teflon beakers or test tubes on a hot plate [25–27]. Closed digestion systems produce pressures above atmospheric which provide higher temperatures and facilitate the complete oxidation of the samples; digestion time and reagent consumption can be also reduced in this way. In addition, loss of volatile elements is minimized and the rate of digestion is increased, yielding low analytical blanks.

On the other hand, microwave (MW) irradiation is currently a popular source of heating. It has been shown that MW digestion can be even more efficient than conventional heating [28]. Using MW irradiation for wet ashing, both speed and efficiency of decomposition are accomplished and the use of $HClO_4$ can be avoided [29].

In an exhaustive paper [30], several procedures, including dry ashing and wet digestion, were compared for the determination of Zn by Inductively coupled plasma atomic emission spectrometry (ICP-AES) and the residual C content after digestion was evaluated as an indicator of the corresponding decomposition efficiency. High-pressure digestions were shown to completely destroy milk organic C by just adding HNO_3.

The choice of the appropriate mineral acid and oxidant depends on the final analytical technique to be used. For instance, HCl is not recommended for furnace analysis as it can cause Cl interferences, while H_2SO_4 is not desirable with ICP-AES or ICP-MS because of transport interferences derived from its viscosity. With ICP-MS HNO_3 and H_2O_2 are preferred since the effect of polyatomic interferences is minimal as compared with HCl, $HClO_4$ and H_2SO_4, which introduce polyatomic ions such as ClO^+ and SO^+ [28–32]. In any case, in order to minimize corrosion of the metal sampler and skimmer cones with ICP-MS, final sample solutions should not contain high acid concentrations (e.g., above 10 percent for HNO_3).

13.2.2.4 Slurries An alternative to dry and wet decompositions is the preparation of slurries, which have been shown to provide a convenient way to introduce solid material into the plasma torch. The solid sample is not digested; rather, it is finely ground and suspended in a liquid to be then introduced as an aerosol of fine, hydrated, solid particles. Thus, the direct analysis of solid samples as slurries reduces both the risk of sample contamination and the time required for sample preparation.

The preparation of milk samples as slurries to be analyzed by ICP-AES was addressed by Carrión et al. [33]. The samples were emulsified with *n*-ethoxy nonylphenol and then diluted 10-fold with 1 percent HNO_3. Aqueous solutions with the same amount of emulsifier and acid were used as calibration standards. In another work samples of milk and formulae were prepared as slurries and were then analyzed using aqueous standards [7]. Dean et al. [34] used ICP-MS to determine Pb isotope ratios in milk powders just suspended in a diluted solution of Triton X-100.

13.3 METHODS FOR TOTAL ELEMENTAL ANALYSIS IN WHOLE MILK

13.3.1 Atomic Spectrometry Techniques

Atomic spectrometry techniques are the most widely used for the total elemental analysis in milk samples. The choice of FAAS, ET-AAS, or ICP-AES depends on the concentration of the element sought, the required accuracy and precision, the number of elements to be determined, and so on.

For instance, those elements present at relatively high concentration levels in milk are usually determined by FAAS or flame atomic emission spectrometry (FAES). On the other hand, if better LoDs are needed (e.g., ng g^{-1}), the technique of choice would be ET-AAS. For multielemental analysis plasma-based techniques are recommended, such as AES for major elements and trace elements, or, as described below, mass spectrometry (MS) for major, trace, and ultratrace elements. The most relevant applications of these atomic techniques for elemental analysis in milk samples are summarized in Table 13.7.

13.3.1.1 Flame Atomic Emission/Absorption Spectrometry FAES has been applied in milk samples for the determination of the major elements K and Na [35, 36]. Other elements such as Sr can be also determined by FAES [37].

FAAS is a well established method for the determination of the major milk elements Ca, K, Mg, and Na after dissolution or mineralization of the samples [38–41]. In fact, the practical advantages of FAAS, that is, speed, simplicity, precision, cost, and ease of operation, are ideal when large sample throughput is required. Thus, many practical milk measurements in clinical, occupational health, toxicology, and nutrition laboratories are still performed by FAAS.

Determinations of Cu, Fe, and Zn have also been accomplished using FAAS [42–47]. For instance, Cu, Fe, and Zn have been determined by directly aspirating diluted milk containing an aqueous detergent (0.1 percent Meriten, 0.1 percent Saponin or 0.01 percent Triton X-100) [48, 49]. It has to be pointed out, however, that milk is a difficult matrix for direct analysis as a consequence of its high fat content and the presence of inorganic salts (especially of Ca, K, and P).

When analyte concentrations fall below the LoD, or the LoD is not good enough to achieve a precise direct determination, preconcentration is used to concentrate the analyte, to remove the matrix, or both. In particular, solid-phase extraction has been found to be very effective in improving LoDs [50–53], thus allowing even potentially toxic elements such as Al and Pb to be determined.

Conventional FAAS is chracterized by poor detection power. Serious interferences from hydride-forming elements such as As, Sb, and Se are well known. Hydride generation techniques may circumvent these problems, providing an excellent tool to determine those elements at trace and ultratrace levels; this is particularly useful for the determination of Se in milk samples [54–56]. Other

TABLE 13.7. Most Relevant Applications of Atomic Techniques for the Total Elemental Analysis of Milk Samples

Reference	Elements Determined	Technique	Sample	Pretreatment
[38]	Ca, K, Mg, Na	FAAS	Milk powder	Direct dispersion
[39]	Ca	FIA-FAAS	Whole milk	
[42]	Cu, Fe	FIA-FAAS	Whole milk	MW digestion HNO_3/ Triton X-100
[18]	Ca, Fe, K, Mg, Na, Zn	FAAS	Powder whole milk	Dispersion in Triton X-100
[22]	Ca, Mg	FAAS	Powder milk	Dry ashing
[23]	Cd, Cr, Cu, Mn, Zn	FAAS	Whole milk	Dry ashing
[51]	Co, Cu, Fe, Zn	FAAS	Whole milk	Solid-phase extraction
[25]	Ca, Cu, Fe, K, Mg, Na, Zn	FAAS	Whole milk	Open digestion HNO_3–$HClO_4$
[68]	Fe, Se	ET-AAS	Whole milk	Dilution water/ tertiary amines
[19]	Zn	ETAAS	Powder and whole milk	Dilution Triton X-100
[69]	Cd	ET-AAS	Whole milk	Dilution Triton X-100
[74]	Pb	ET-AAS	Whole milk	Dilution Triton X-100
[70]	Cd	ET-AAS	Powder milk	Slurry
[67]	Al, Cr, Mn, Mo	ET-AAS	Milk powder	Slurry
[71]	Al, Cd, Cr, Cu, Mn, Ni, Pb, Se, Zn	ET-AAS	Whole milk, skimmed milk and milk powder	Slurries – Triton X-100
[107]	Cu, Fe, Ni, Pb,	ET-AAS	Milk powder	Direct injection

[73]	V	ET-AAS		Dilution Triton X-100
[72]	Si	ET-AAS	Whole milk	Dilution Triton X-100
[108]	[Se]	ET-AAS	Whole milk	Dilution Triton X-100
[109]	[Cr]	ET-AAS	Whole milk	Dilution Triton X-100
[24]	Cd, Cu, Hg, Pb, Zn	ET-AAS	Milk powder	Dry ashing
[110]	Cd, Cr, Cu, Fe, Mn, Mo, Ni, Pb, Zn	ET-AAS	Whole milk	Dry ashing
[75]	I	ET-AAS	Whole and powdered milk	MW distillation
[111]	Cd, Pb	ET-AAS	Whole milk	MW digestion HNO_3
[112]	[Se]	ET-AAS	Milk powder	MW digestion HNO_3
[64]	Al	ET-AAS	Milk powder	MW digestion HNO_3
[113]	[Cr]	ET-AAS		Digestion HNO_3 and H_2SO_4
[114]	[Cr]	ET-AAS	Whole milk	Digestion HNO_3 and V_2O_5
[33]	Ca, Cu, Fe, Mg, P	ICP-AES	Powder and whole milk	Dilution with Zn HNO_3
[7]	Ca, Cu, Fe, K, Mg, Mn, Na, P, Zn	ICP-AES	Milk powder	Dilution Triton X-100
[115]	[P]	ICP-AES	Milk powder	Dilution, glacial acetic acid.
[116]	Cu, Fe, Mn, Zn	ICP-AES	Milk powder	Digestion H_2SO_4
[117]	[Sn]	ICP-AES	Condensed milk	MW digestion and high pressure asher
[118]	Al, As, B, Ba, Cd, Ca, Cr, Co, Cu, Fe, Pb, Mg, Mn, Mo, Ni, P, K, Se, Na, Sr, Tl, V, Zn	ICP-AES	Evaporated milk	MW digestion/HNO_3

(*Continued*)

TABLE 13.7. *Continued*

Reference	Elements Determined	Technique	Sample	Pretreatment
[76]	Al, Ba, Cd, Cr, Cu, Fe, Mg, Mn, Zn	ICP-AES	Whole milk	Dry ashing
[78]	Al, Ba, Cu, I, Mn, Mo, Pb, Rb, Se, Sr, Zn	ICP-AES	Powdered, skimmed and whole milk	Dilution
[26]	Ca, Cu, Fe, K, Mg, Mn, Na, P, Zn	ICP-AES	Whole milk	Open digestion HNO_3–$HClO_4$
[27]	Ca, Cu, Fe, Mg, Mn, P, K, Na, Zn	ICP-AES	Milk powder	Open digestion HNO_3–$HClO_4$
[63]	Cd, Co, Cr, Cu, Fe, Mn, Ni, Pb, Zn	ICP-AES/ ET-AAS	Milk powder	Digestion HNO_3–$HClO_4$–HCl
[80]	Ag, Al, As, Ba, Be, Ca, Cd, Co, Cr, Cu, Fe, Hg, Mg, Mn, Mo, Na, Ni, Pb, Sb, Sn, Sr, Ti, Tl, U, V, Zn	ICP-AES	Whole milk	MW digestion HNO_3
[88]	Ca, Cu, Fe, K, Mg, Mn, Na, P, Zn	ICP-AES	Milk powder	Slurries
[82]	Al, Ba, Cd, Co, Cr, Cu, Fe, Mg, Mn, Ni, Pb, Pt, Sr, Zn	ICP-AES	Whole milk	Dry ashing
[84]	Ca, Cu, Fe, Mg, Mn, Zn	ICP-AES	Whole milk	MW digestion HNO_3–H_2O_2

[6]	Al, As, Ca, Cd, Cl, Co, Cr, Cu, F, Fe, Hg, I, K, Mg, Mn, Mo, Na, Ni, P, Pb, Sb, Se, Sn, V, Zn	ICP-AES	Whole milk	
[91]	Ba, Be, Bi, Cd, Co, Cs, Cu, La, Li, Mn, Mo, Pb, Rb, Sb, Sn, Sr, Tl, Zn.	ICP-QMS	Milk powder	MW digestion $HNO_3-H_2O_2$
[90]	Ba, Cd, Ca, Cs, Co, Cu, Ce, La, Mg, Mn, Mo, Ni, Pb, Rb, Sn, Sr, Zn	ICP-QMS	Whole milk	Digestion HNO_3
[103]	[I]	ICP-QMS	Milk powder	High pressure asher autoclave
[119]	Al, As, Li, B, Ba, Be, Cd, Co, Cr, Cs Cu, Fe, Ga, Mn, Mo, Ni, Pb, Rb, Sc, Se, Sr, Th, Ti, U, V, Zn	ICP-QMS	Whole milk and milk powder	High pressure asher autoclave
[120]	I, Cu	ICP-QMS	Whole milk	Dilution KOH/TMAH
[100]	Cd, Cr, Pb, Zn	RC-ICP-MS	Milk powder	Slurries
[99]	Ca, Fe, Zn.	RC-ICP-MS	Milk powder	MW digestion HNO_3
[101]	Al, Cd, Co, Cr, Cu, Fe, I, Mn, Pb, Se, Zn	ICP-RC-MS	Whole milk	MW digestion $HNO_3-H_2O_2$
[96]	Al, Ca, Cd, Cr, Cu, Fe, Hg, Mg, Mn, Na, Ni, Pb, Se, Sr, Zn	SF-ICP-MS	Milk whey	Dilution

(*Continued*)

TABLE 13.7. *Continued*

Reference	Elements Determined	Technique	Sample	Pretreatment
[106]	Al, Ca, Cd, Cr, Cu, Fe, Hg, Mg, Mn, Na, Ni, Pb, Se, Sr, Zn	SF-ICP-MS	Whole milk, skimmed milk and milk whey	MW digestion $HNO_3-H_2O_2$
[98]	Ag, Al, As, Au, Co, Cr, Cu, Fe, Mn, Ni, Pt, Sc, Se, Ti, V	SF-ICP-MS	Whole milk	MW digestion $HNO_3-H_2O_2$
[8]	Ag, Al, As, Au, Co, Cr, Cu, Fe, Mn, Ni, Pb, Pt, Sc, Ti, V	SF-ICP-MS	Whole milk	MW digestion $HNO_3-H_2O_2$

approaches for the generation of volatile species are also possible, such as cold vapor (CV) generation, although it is restricted to Cd and Hg. In fact, CV generation of Hg can be most useful [57, 58] and this technique could be extended to ultratrace determinations of Cd [59]. Recently, solid-phase preconcentration and CV generation have been proposed for the determination of Hg [60].

13.3.1.2 Graphite Furnace Atomic Absorption Spectrometry Graphite furnace atomic absorption spectrometry (GFAAS), the most popular form of ET-AAS, is today a common technique widely used in routine laboratories and has become a powerful tool for the analysis of trace and ultratrace elements in clinical and biological samples [61]. The main advantages of this technique are low cost, simplicity, excellent detection power, and the fact that it allows very low sample volumes to be used (5–20 μL). In this sense, this technique allows LoDs for many elements in the order of 0.01 $\mu g\,l^{-1}$ in solution or 1 $\mu g\,g^{-1}$ in solid samples to be achieved [62]. However, the technique is prone to spectral and matrix interferences.

Direct milk analysis methods seem ideal for routine analysis laboratories: They are simple, rapid, involve little sample pretreatment, reduce the risks of contamination and should not require considerable operator skills. In principle, ET-AAS could be suited to milk analysis. Unfortunately, the results obtained so far are not accurate because of excessive background absorption, carbonaceous residue build-up in the graphite tube, and foaming and splattering of the samples during the drying stage. As a result, direct milk analysis methods with this technique have not become popular. Sample pretreatment [63–65], background correction using D_2 lamp [19, 66, 67] or Zeeman effect [68, 69], standard additions [19], matrix modifiers [63, 69, 70], pyrolytic tubes, and atomization on a L'vov platform [66, 67, 70] have all been widely used in order to avoid or overcome these interferences.

ET-AAS determination of several metals, including Al, Cd, Cu, Cr, Mn, Ni, Pb, Se, and Zn, was carried out by Garcia et al. [71]. They studied the contents of these elements in whole, low-fat, skimmed, condensed, and evaporated milk. The total elemental concentration ranges found were in good agreement with those reported in the literature, although significant differences in the levels of Al, Cd, and Pb were detected in the different types of milk analyzed (probably due to contamination phenomena in the production process). Copper and Mn levels were reported to be slightly higher in condensed milk, possibly due to the high concentration of added sugar in this matrix.

ET-AAS and ICP-AES techniques for the determination of Cd, Co, Cr, Cu, Fe, Mn, Ni, and Pb in milk powder were compared by Vuchkova et al. [63]. Samples were digested both by an acidic treatment and by dry ashing. A universal temperature programme was reported for ET-AAS, without chemical modifier addition, because the high Mg and P content of the sample can act this way. Elements in the concentration range of ng g^{-1} (Cd, Co, Cr, Ni, and Pb) were determined by ET-AAS with good accuracy and precision. The detection power of ICP-AES did not allow them to be directly determined, whilst elements in the range of $\mu g\ g^{-1}$ (e.g., Cu, Fe, and Zn) were easily determined using this approach.

The determination of Zn in human milk collected during the first days of lactation was carried out by simple dilution of the samples (1 + 99 v/v) with 0.1 percent Triton X 100 [19]. Normal values for Zn in colostrum and transitional milk varied from 45 to 318 μM and from 30 to 146 μM, respectively. The major factor influencing Zn concentration in human milk seems to be the stage of lactation.

Bermejo Barrera et al. determined Si [72] and V [73] in infant formulas, human milk and cow's milk. As for Si, the highest levels were found in the infant formulas, while no significant differences were observed between human and cow's milk. Human milk exhibited the highest V levels. The concentration range was reported to be very narrow for these samples.

The direct determination of Fe and Se by ET-AAS was proposed without any digestion, that is, just by simple dilution of the milk samples [68, 69, 74]. The elements were determined in bovine milk after dilution of the sample with water-soluble tertiary amines (10 percent). Pyrolytic graphite tubes, Pd as the chemical modifier, and Zeeman background correction were used. The concentration of Fe varied from 0.61 to 1.17 $mg\ l^{-1}$ and Se concentrations were in the 17–122 $\mu g\ l^{-1}$ range.

Powder milk samples have also been introduced directly as slurries for analysis by ET-AAS [67, 70, 71]. MW-assisted distillation of I for indirect determination of the element in commercial and formula milks has also been described [75].

*13.3.1.3 **Inductively Coupled Plasma Atomic Emission Spectrometry*** ICP-AES is a technique half-way between FAAS and ET-AAS in terms of detection power. Among all ICP-AES features its robustness against matrix effects and its ability to carry out multielemental analysis predominate as the most advantageous [76–80]. Multielemental analysis has also been successfully used to establish reference values [6, 76, 81–84] for many major and trace essential elements in different matrices of biological and nutritional interest, particularly in milk samples [81–83]. Reference values for minor and trace element in human milk are collected in Table 13.8.

Bocca et al. [84] applied an ICP-AES method to assess the concentration range of Ca, Cu, Fe, Mg, Mn, and Zn in the milk of 60 lactating mothers (19–40 years of age) of different areas of Italy. The total concentration ranges found were in good agreement with those reported in the literature. The influence of age and residence area of the mothers on the total concentration of the elements was evaluated. Although, their findings indicate that there are no significant differences associated with residence area, some small discrepancies were found in the case of Ca, Mg, Mn, and Zn contents. For Cu a significant difference between urban and rural groups was reported.

Some 20 years ago Schramel [85] reported the analysis of milk by ICP-AES just by diluting the samples in water and using direct nebulization. It was eventually proved, though, that in ICP-AES the direct dilution of milk samples with

TABLE 13.8. Overall Reference Concentration Ranges for Major and Trace Elements in Human Milk

Element	Concentration	Element	Concentration
Al	39–250 μg l^{-1}	Mg	29–38 mg l^{-1}
As	0.2–36 μg l^{-1}	Mn	3–40 μg l^{-1}
Ca	220–300 mg l^{-1}	Mo	0.3–19.5 μg l^{-1}
Cd	0.7–4.6 μg l^{-1}	Na	90–130 mg l^{-1}
Cl	320–410 mg l^{-1}	Ni	1.5–39 μg l^{-1}
Co	0.15–3 μg l^{-1}	P	135 mg l^{-1}
Cr	0.4–5.1 μg l^{-1}	Pb	2–30 μg l^{-1}
Cu	180–751 μg l^{-1}	Sb	1–4 μg l^{-1}
F	7–17 μg l^{-1}	Se	10–62 μg l^{-1}
Fe	202–1710 μg l^{-1}	Sn	1 μg l^{-1}
Hg	0.2–13 μg l^{-1}	V	0.1–0.3 μg l^{-1}
I	41–168 μg l^{-1}	Zn	0.7–4.0 mg l^{-1}
K	410–550 mg l^{-1}		

Reproduced by permission of CRC Press [6].

water leads to poor analytical performance in terms of sensitivity, accuracy, and precision. This fact has been documented by Emmet [86] and related to the fatty nature of milk. On the other hand, Coni et al. [83] have demonstrated that over-large drops form during the nebulization of milk samples, which leads to the incorrect atomization of the different components of milk while they stay in the plasma. Accordingly, these investigators agreed that, owing to the different values in surface tension and viscosity of milk samples with respect to an aqueous solution, analysis using calibration with aqueous standards is inadequate.

In order to eliminate as much as possible the above drawbacks for direct milk sample introduction into the plasma torch, the most common methodology is now based on the elimination of the organic matrix of the samples (digestion), as mentioned in the previous section. However, the digestion of samples increases the duration of the analysis and the risk of contamination due to the more laborious handling of the sample. Again, slurry methods using direct nebulization from milk slurries or dispersions have been developed to avoid such limitations, employing EDTA mixed with ternary amines [87, 88], etoxynonylphenol in acidic medium [89] or Triton X-100 [7].

13.3.2 Inductively Coupled Plasma Mass Spectrometry

ICP-MS is today the most powerful detector in atomic spectrometry, mainly as a consequence of: (i) its extremely high sensitivity; (ii) its multielemental analysis capability; (iii) its unique capability to measure isotope ratios. Despite these advantages the implementation of this technique in routine laboratories is comparatively low mainly because of its high purchase cost.

From the sensitivity viewpoint this technique presents LoDs even lower than ET-AAS. Comparing quadrupole (Q) ICP-MS versus sector field (SF) ICP-MS, the LoDs obtained with the latter in its low resolving power mode are almost one order of magnitude lower than those afforded by Q-ICP-MS. However, with SF-ICP-MS, the increase in resolving power means a loss of sensitivity (about 16 times by changing from 300 to 3000 nominal resolving power). In any case, it is interesting to note that, for some limited cases (e.g., blanks from the H_2O_2 used in the oxidative digestion), the LoDs observed were similar with both techniques (i.e., Cr, Mo, and Pb).

Milk is a particularly difficult matrix for ICP-MS as it contains large amounts of proteins and inorganic salts plus small organic molecules. This high concentration of dissolved matter causes serious matrix effects in ICP-MS. Therefore, digestion or dilution prior to the analysis may be desirable. In addition, polyatomic species as oxide and argide ions are always present and they add to isobaric and polyatomic interferences coming from the major elements in milk. For elemental analysis of milk samples these polyatomic spectral interferences may constitute one of the most important limitations of Q-ICP-MS systems. Despite this drawback, many studies involving milk analysis have been carried out using Q-ICP-MS.

13.3.2.1 Quadrupole Mass Analyzers A study to examine longitudinal and gestational mineral content (Ba, Cd, Ca, Cs, Co, Cu, Ce, La, Mg, Mn, Mo, Ni, Pb, Rb, Sn, Sr, and Zn) in human milk was carried out by Friel et al. [90]. Human milk from mothers of premature and full-term infants during the first three months of lactation were analyzed and trace element determinations were carried out by Q-ICP-MS, after the digestion of the milk samples with HNO_3. The results indicated lower Ca and Pb in premature milk. Calcium was the only nutritionally significant element to differ between groups. Molybdenum showed a definite decrease with time, in both premature and full-term milk, suggesting that the milk Mo content is homeostatically regulated. The rest of the elements did not appear to display any specific pattern.

Krachler et al. [91] carried out total determinations of trace elements (Ba, Be, Bi, Cd, Co, Cs, Cu, La, Li, Mn, Mo, Pb, Rb, Sb, Sn, Sr, Tl, and Zn) by Q-ICP-MS in 13 formulas based on cow's and soy milk and also in Austrian cow's milk. Some essential elements, such as As, Cr, Se, and V, suffered from serious interferences from polyatomic ions and could not be determined under routine conditions by Q-ICP-MS. In this study, the authors determined only those elements for which isotopes free from serious interferences were available. Concentrations of trace elements such as Cu, Mn, and Zn in the formula powders varied considerably. As pre-formulas and follow-up formulas are partly supplemented with these essential trace elements, their concentrations proved to vary considerably depending on the commercial brand. Also, concentrations of all trace elements quantified were lower in cow's milk than in formulas, which did not meet the dietary requirements of infants.

To overcome polyatomic interferences in Q-ICP-MS a host of physical approaches (i.e., optimization of plasma parameters, addition of supplementary plasma gases, the use of alternative plasma gases or collision/dynamic reaction cell systems with high mass resolution spectrometers) and chemical means (based on separations) are available. Separations of the analyte from the interfering matrix, either during sample preparation or by using an appropriate sample pre-treatment device (e.g., an electrothermal vaporizer) are in use today. Alternatively, adequate corrections involving mathematical algorithms implemented within the instrument software may be used.

13.3.2.2 High Mass Resolution Spectrometers (sector Field Mass Analyzers)

The more general method to overcome limitations coming from spectral interferences from polyatomic ions is the use of high mass resolution spectrometers which combine a magnetic and an electric sector field analyzer [92, 93]. Many molecular interferences in ICP-MS can be eliminated by using these mass analyzers at a resolving power of 2000–5000, sufficient to overcome most molecular ion interferences in biological material [94, 95].

Several publications have addressed multielemental analysis in milk whey, cow's milk, human milk and infant formulas using SF-ICP-MS (see Table 13.9). Rivero Martino et al. [96] reported the multielemental analysis of a wide range of elements in human, cow's, and formula milk whey samples. Milk whey was obtained by centrifugation and then diluted 1 : 4 (v/v) with milliQ water for minor and trace elemental analysis (Al, Cd, Cr, Cu, Fe, Hg, Mn, Ni, Pb, Se, Sr, and Zn) and 1 : 1999 (v/v) for major elements (Ca, Mg, and Na). Gallium, Rh,

TABLE 13.9. Reported Multielemental Determinations in Human, Formula and Cow's Milk by SF-ICP-MS

Element	Human Milk ($ng\ g^{-1}$)	Infant Formulas ($ng\ g^{-1}$)	Cow's Milk ($ng\ g^{-1}$)
Ag	<0.13–42	<0.13–0.67	nd
Al	<10–380	50–500 26 ± 3	26 ± 3
As	<1.3–30	nd	nd
Au	0.1–2.1	<0.05–0.20	nd
Cd	0.53 ± 0.17	0.71 ± 0.18	0.47 ± 0.13
Co	<0.07–1.5	0.70–1.1	nd
Cr	<0.8–163	<0.8–15	1.1 ± 0.4
Cu	90–1400	920–2550	60 ± 5
Fe	110–1570	1400–12500	237 ± 17
Mn	1.8–25	29–55	53 ± 4
Ni	<0.13–6.4	11–18	9.5 ± 1.3
Pb	0.5 ± 0.4	5.6 ± 0.4	1.8 ± 0.2
Pt	<0.01–0.04	<0.01	nd
Sc	<0.13–1.2	<0.13	nd
Se	14.1 ± 2.1	17.5 ± 2.0	12.8 ± 1.3
Ti	<2.7–114	<2.7–5.1	nd
V	<0.07–3.7	0.25–0.75	nd
Zn	3800 ± 200	4200 ± 100	3800 ± 0.3

Data from References [97, 98, 106].

and Tl were added as internal standards for the ICP-MS final measurement. Again, the results obtained showed substantial differences in trace element content in human, cow's, and formula milks. The major components (Ca, Mg, and Na) were at higher levels in cow's milk whey than in human milk whey, while the content of some essential elements (Cu, Fe, and Se) was proved to be higher in human milk whey. Comparatively higher levels of essential elements were found in formula milk whey as a result of an excessive addition of essential elements to the formula raw material.

Prohaska et al. [97] evaluated the potential of SF-ICP-MS for reliably quantifying various potentially toxic and essential elements (Ag, Al, As, Au, Co, Cr, Cu, Fe, Mn, Ni, Pb, Pt, Sc, Ti, and V) in human milk and infant formulas. Whole milk was digested using MW irradiation, HNO_3, and H_2O_2. The observed concentration ranges in human milk samples agreed well with the ranges reported in the literature for those elements.

Krachler et al. [98] first reported the concentration of Ag, Au, Pt, Sc, Ti, and V in human milk and infant formulas. Concentrations of Ag and Au showed large variations in human milk that might be associated with dental fillings and jewellery, according to the authors of this study. Platinum concentrations were very low, below the LoDs of the method, for most of the samples. The rest of the elements were at the lower end of the reference range except for Cr, which was five times higher than the higher end of the reference range. All elemental concentrations in infant formulas were approximately one order of magnitude higher than in human milk.

In all these SF-ICP-MS works, the use of medium resolving power ($R = 3000$) demonstrated that polyatomic interferences in whey and milk after the digestion were well separated, with the exception of $^{1}H^{81}Br^{+}$ and $^{40}Ar^{35}Cl^{+}$ interferences on ^{82}Se and ^{75}As, respectively (they would require higher resolving powers of about 10,000).

13.3.2.3 Collision/Dynamic Reaction Cell Systems The development of Q-ICP-QMS equipped with collision/dynamic reaction cells is offering an interesting alternative to high-resolution systems for the removal of spectral interferences. The cells (quadrupoles, hexapoles, or octopoles) are pressurized with a gas or a mixture of gases in order to reduce or eliminate the interfering polyatomic species by collisional dissociation and, mainly, by ion-molecule reactions. Helium is usually used as collision gas while, H_2, NH_3, Xe, and CH_4 are employed as possible reaction gases. The collision/dynamic reaction cell can also increase the ion transmission efficiencies by collisional focusing.

Determination of Ca, Fe, and Zn in milk powder has been carried out by Chen et al. [99] using a dynamic reaction cell (DRC) system. The sample dissolution was accomplished by MW digestion using HNO_3 as the oxidant. Various reaction gases (including NH_3, CH_4, He, O_2, and H_2) were tested and the best results were obtained with CH_4. The LoDs reported were 1, 0.01, and 0.001 ng mL^{-1} for ^{40}Ca, ^{56}Fe, and ^{64}Zn, respectively. This method was applied to the determination of Ca, Fe, and Zn in a reference skimmed milk powder and two milk formulas.

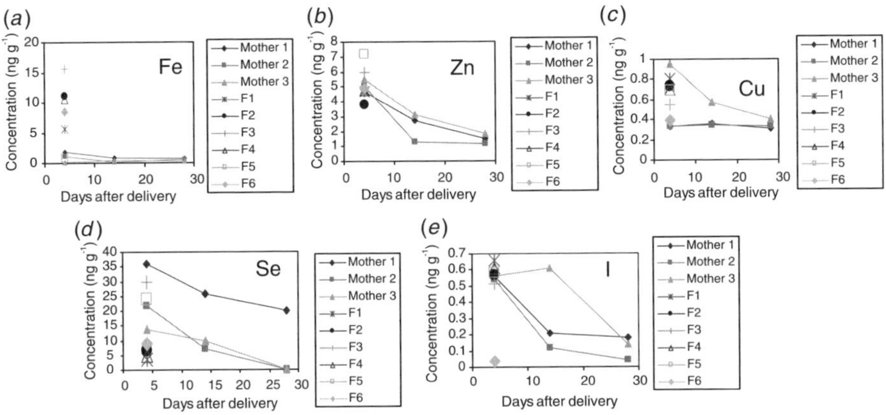

Figure 13.1. Comparative concentrations of Cu, Fe, I, Se, and Zn in human milk at three different lactation periods and in formula milk for premature infants. Reproduced by permission of The Royal Society of Chemistry (RSC) [101].

Ho et al. [100] carried out the determination of Cd, Cr, Pb, and Zn in infant formulas by electrothermal vaporization coupled to a DRC-ICP-MS. The reaction gas used in the reaction cell was NH_3, which proved to be effective for reducing background count rates and lower the LoDs for ^{52}Cr and ^{68}Zn.

Most recently an octopole reaction system (ORS) has been applied by de la Flor St. Remy et al. [101] for multielemental analysis of essential and toxic elements in a comparative study of human and formula milks. In this work human and formula milk samples were digested in an MW oven using HNO_3 and H_2O_2 as oxidants, while H_2 was used as reaction gas in the octopole. The proposed method was applied to the multielemental analysis of premature human milk samples at three different lactating stages (during the first month after birth), and formula milks commercially available for premature newborns. Total concentrations in the two sample types were critically compared. The elemental content in human milk turned out to be higher in the samples corresponding to the first weeks of feeding as shown in the plots of Figure 13.1. As expected, different mothers showed different elemental compositions. However, all mothers' milk tended to a similar elemental concentration level approaching the end of the month. Significant were the differences between human and formula milk (see Fig. 13.1*a–d*). Also Fe content in formula samples was always much higher than in human milk. The artificial fortification of cow's milk-based formulas could easily explain these differences.

13.3.2.4 Isotope Ratio Measurements The analysis of milk samples for the accurate determination of its constituent elements has also been done by isotope dilution analysis [34, 102, 103]. For that purpose, an enriched isotope of the element or elements to be determined is spiked to the sample, altering the natural isotopic composition of the element to be analyzed. By measuring the altered isotope ratios the concentration of the element in the sample can be quantified with high precision and accuracy. In other words, the use of an isotope of the same element represents the ideal internal standard for that particular element determination, and, therefore, isotope dilution results are expected to be highly accurate and precise even if sample (analytes) loss during the preparation or pretreatment processes has taken place. It is not surprising that isotopic dilution analysis (IDA) has been proposed as a reference method [104].

IDA techniques have been used for I and Pb determinations in milk powder. The determination of I in food by isotope dilution has been reported using isotope ^{129}I [102, 103]. IDA has also been applied to the accurate measurement of the Pb content in milk powder [34]. As expected, in all three studies referred to above IDA provided much more accurate and precise values for the sought elements than more conventional calibration approaches.

13.4 ELEMENTAL DISTRIBUTIONS IN THE DIFFERENT FRACTIONS OF MILK: CASEINS, FAT, AND MILK WHEY

The distribution of essential and potentially toxic elements in different gross fractions is the preliminary step to study the particular chemical forms in which they are present in milk [105]. Multielemental distribution patterns in human, cow, and formula milks have been reported by the Sanz-Medel group [106]. Total elemental distributions in skimmed milk and/or milk whey samples were obtained after removing fat and/or caseins by ultracentrifugation. Then, in order to destroy the organic matter, each fraction (whole milk, skimmed milk, and milk whey) was digested by MW-assisted digestion for their minor and major elemental analysis.

Results in skimmed milk and whey showed also that essential elements, including Cu, Fe, Mg, Mn, Se, and Zn, could be associated with fat and caseins depending on the element (see Figs 13.2–13.4). The percentage value of a given element

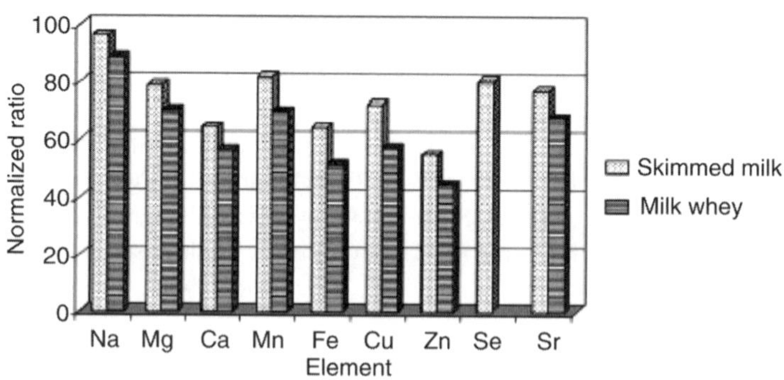

Figure 13.2. Elemental distribution patterns relative to whole milk in different fractions of human milk. Reprinted from *Analytica Chimica Acta*, **442**, F. A. Rivero, M. L. Fernández, A. Sanz-Medel, The potential of double focusing ICP-MS, pp. 191–200, © 2001, with permission from Elsevier [106].

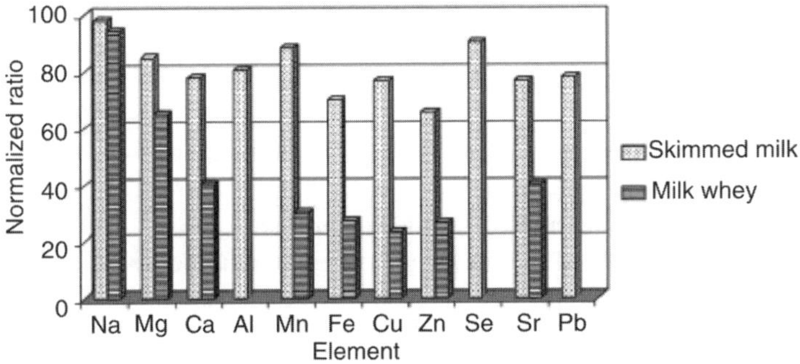

Figure 13.3. Elemental distribution patterns relative to whole milk in different fractions of cow's milk. Reprinted from *Analytica Chimica Acta*, **442**, F. A. Rivero, M. L. Fernández, A. Sanz-Medel, The potential of double focusing ICP-MS, pp. 191–200, © 2001, with permission from Elsevier [106].

contained in skimmed milk ranged from 60 to 90 percent of that in whole milk. For a given element, its distribution in skimmed milk was similar in all types of milk under study. On the other hand, the percentage values of trace elements in milk whey ranged from 20 to 40 percent of the whole cow's and formula milks. These values were quite similar again, in both cases, when the same element was measured. However, the values obtained in human milk whey were higher (between 50 and 70 percent of the whole human milk content), probably due to the comparatively low casein content of this kind of milk. In other words, casein precipitation seems to reduce substantially the level of essential inorganic nutrients.

However, Mg and Na showed a different behavior; in fact, Na showed percentage values in skimmed milk and milk whey close to 90 percent of those present in whole milk, for all types of milk under study. The percentage values obtained for Mg distribution were always close to 85 percent in the case of skimmed milk

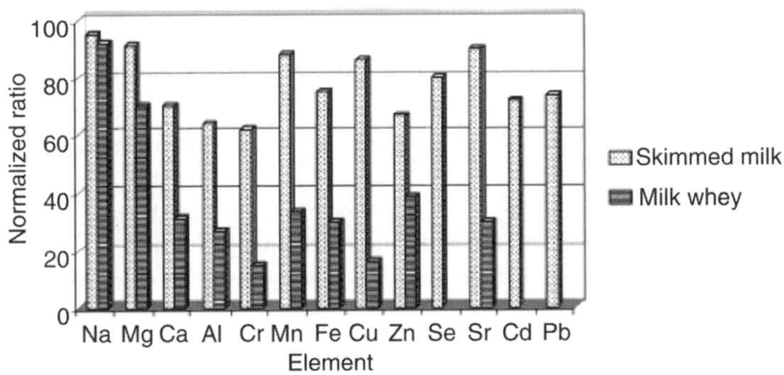

Figure 13.4. Elemental distribution patterns relative to whole milk in different fractions of formula milk. Reprinted from *Analytica Chimica Acta*, **442**, F. A. Rivero, M. L. Fernández, A. Sanz-Medel, The potential of double focusing ICP-MS, pp. 191–200, © 2001, with permission from Elsevier [106].

and close to 60 percent in the case of milk whey as compared with whole milk, for all the four types of milk investigated.

13.5 CONCLUSIONS

An overview of the application of atomic spectrometric techniques to the elemental analysis of milk samples has been given. Elemental composition of milk, its nutritional role, sample preparation methods for analysis and measurement techniques have been described in detail. It appears that ICP-MS and ICP-AES are the most reliable techniques for the multielemental analysis of major, minor, and trace elements in milk samples.

From a nutritional viewpoint, it is necessary to stress the current importance of carrying out the multielemental analysis of milk samples (either human, cow's, or formula milk) in order to establish the reference values of essential elements and quantify the levels of potentially toxic elements. This fact is more relevant to formula milk production for premature babies as some essential elements are not stored by the fetus during its development in the uterus. Attention has been already paid to the qualitative and quantitative composition (analysis) of proteins, lipids, carbohydrates, and, of course, essential elements. However, in the case of human nutrition, knowledge on the particular species (compounds) in which a given element is present (chemical speciation) is now urgently needed, because the absorption and bioavailability of the essential element will strongly depend on that particular chemical form. Thus, although only the total element daily requirements have been considered here, it is important to stress that more attention must be paid to the chemical form in which essential and potentially toxic elements are present in milk. Such aspects are dealt with in detail in Chapter 13 by B. Michalke et al.

REFERENCES

1. F. M. Luquet, Leche y productos lácteos. Vaca – Oveja – Cabra. 1. La leche. De la Mama a la Lechería, *Acribia, Technique et Documentation*, Lavoisier, Paris, 1991.
2. J. C. Collin, P. Compagnone, I. Ryba, A. Baer, Determination of plasmin (alkaline milk proteinase) and chymosin in milk products by the ELISA assay, *Lait*, **68** (1998), 235–239.
3. M. M. Delgado Zamarreño, A. Sánchez Pérez, M. C. Gómez Pérez, M. A. Fernández Moro, J. Hernández Méndez, Determination of vitamins A, E and K1 in milk by high-performance liquid chromatography with dual amperometric detection, *Analyst*, **120** (1995), 2489–2492.
4. M. Ake, H. Fabre, A. K. Malan, B. Mandrou, Column liquid chromatography determination of vitamins A and E in powered milk and local flour: a validation procedure, *J. Chromatogr. A*, **826** (1998), 183–189.
5. C. Alais, G. Linden, *Food Biochemistry*, Ellis Horwood Series in Food Science and Technology, Ellis Horwood, Chichester 1991.

6. S. Caroli, A. Alimonti., E. Coni, F. Petrucci, O. Senofonte, N. Violante, The assessment of reference values for elements in human biological tissues and fluids: A systematic review, *Crit. Rev. Anal. Chem.*, **24** (1994), 363–398.
7. P. J. Mckinstry, H. E. Indyk, N. D. Kim, The determination of major and minor elements in milk and infant formula by slurry nebulization and inductively coupled plasma-optical emission spectrometry (ICP-OES), *Food Chem.*, **65** (1999), 245–252.
8. T. Prohaska, G. Köllespenger, M. Krachler, K. De Winne, G. Stingeder, L. Moens, Determination of trace elements in human milk by inductively coupled plasma sector field mass spectrometry (ICP-SFMS), *J. Anal. Atom. Spectrom.*, **15** (2000), 335–340.
9. E. Coni, P. Falconieri, E. Ferrante, P. Semerano, E. Beccaloni, A. Stacchini, S. Caroli, Reference values for essential and toxic elements in human milk, *Ann. Ist. Super. Sanità*, **26** (1990), 119–130.
10. J. A. Blaymore Bier, A. E. Ferguson, Y. Morales, J. A. Liebling, W. Oh, B. R. Bohd, Breast-feeding infants who were extremely low birth weight, *Pediatrics*, **100** (1997), E3.
11. W. Mertz, *Trace Elements in Human and Animal Nutrition*, 5th edn, Academic Press, New York, 1987.
12. A. Taylor, Detection and monitoring of disorders of essential trace elements, *Ann. Clin. Biochem.*, **33** (1996), 486–510.
13. D. B. Milne, C. A. Burtis, E. R. Ashwood (eds), *Tietz Textbook of Clinical Chemistry*, 2nd edn, W. B. Saunders, Philadelphia, PA, 1994, 1317–1353.
14. G. M. Salido Ruiz, Curso de Elementos Traza, Servicio de Publicaciones de la Universidad de Extremadura, 1993.
15. E. Frieden, *Biochemistry of the Essential Ultratrace Elements*, E. Frieden (ed.), Plenum Press, New York, 1984, pp. 1–15.
16. L. Kosta, Contamination as a limiting parameter in trace analysis, *Talanta* **29** (1982), 985.
17. J. Versiek, R. Cornelis, *Trace Elements in Human Plasma or Serum*, CRC Press, Boca Raton, FL, 1989.
18. I. L. García, P. Vinas, N. Campillo, M. H. Córdoba, Extending the dynamic range of flame atomic absorption spectrometry: a comparison of procedures for the determination of several elements in milk and mineral waters using online dilution, *Fresenius J. Anal. Chem.*, **355** (1996), 57–64.
19. J. Arnaud, A. Favier, J. Alary, Determination of zinc in human milk by eletrothermal atomica absorption spectrometry, *J. Anal. Atom. Spectrom.*, **6** (1991), 647–652.
20. S. S. Ali, G. H. Kazi, T. G. Kazi, Use of comparative sample preparation methods for the extraction of metal ions in milk as determined by atomic absorption spectroscopy, *ACGC Chem. Res. Commun.*, **6** (1997), 9–12.
21. H. T. Delves, Analysis of biological and clinical materials, *Prog. Anal. Atom. Spectrosc.*, **4** (1981), 1–48.
22. R. M. Pollman, Atomic absorption spectrophotometric determination of calcium and magnesium and colorimetric determination of phosphorus in cheese: collaborative study, *J. Assoc. Off. Anal. Chem.*, **74** (1991), 27–31.
23. B. E. Guthrie, Chromium, manganese, copper, zinc and cadmium content, of New Zealand foods, *New Zealand Med. J.*, **82** (1975), 418–424.

24. R. C. Campos, A. J, Curtius, H. Berndt, Combustion and volatilisation of solid samples for direct atomic absorption spectrometry using silica or nickel tube furnace atomisers, *J. Anal. Atom. Spectrom.*, **5** (1990), 669–673.
25. E. M. R. Rodríguez, M. S. Alaejos, C. D. Romero, Mineral concentrations in cow's milk from the Canary Islands, *J. Food Comp. Anal.*, **14** (2001), 419–430.
26. R. L. Sims, L. M. Mullen, D. B. Milne, Application of inductively coupled plasma emission spectroscopy to multielement analysis of foodstuffs used in metabolic studies, *J. Food. Comp. Anal.*, **3** (1990), 27–37.
27. R. F. Suddendorf, K. K. Cook, Inductively coupled plasma emission spectroscopic determination of nine elements in infant formula: collaborative study, *J. Assoc. Off. Anal. Chem.*, **67** (1984), 985–992.
28. M. Burguera, J. L. Burguera, Microwave sample pretreatment in analytical systems. A review, *Quím. Anal.*, **15** (1996), 112.
29. T. Alkanani, J. K. Friel, S. E. Jackson, P. Longerich, Comparison between digestion procedures for the multielemental analysis of milk by inductively coupled plasma mass spectrometry, *J. Agric. Food. Chem.*, **42** (1994), 1965–1970.
30. A. Krushevska, R. M. Barnes, C. J. Amarasiriwaradena, H. Forner, L. Martines., Determination of the residual carbon content by inductively coupled plasma atomic emission spectrometry after decomposition of biological samples, *J. Anal. Atom. Spectrom.*, **7** (1992), 845–850.
31. L. Stewart, J. Martines, R. M. Barnes, Flow-through, microwave-heated digestion chamber for automated sample preparation prior to inductively coupled plasma spectrochemical analysis, *Analyst*, **119** (1994), 1003.
32. M. Krachler, T. Prohaska, G. Koellensperger, E. Rossipal, G. Stingeder, Concentrations of selected trace elements in human milk and in infant formulas determined by magnetic sector field inductively coupled plasma-mass spectrometry, *Biol. Trace Elem. Res.*, **76** (2000), 97–112.
33. N. Carrión, A. Itriago, M. Murillo, E. Eljuri, A, Fernández, Determination of calcium, potassium, magnesium, iron, copper and zinc in maternal milk by inductively coupled plasma atomic emission spectrometry, *J. Anal. Atom. Spectrom.*, **9** (1994), 205–207.
34. J. R. Dean, L. Ebdon, R. Massey, Selection of mode for the measurement of lead isotope ratios by inductively coupled plasma mass spectrometry and its application to milk powder analysis, *J. Anal. Atom. Spectrom.*, **2** (1987), 369–374.
35. M. A. de la Fuente, M. Juárez, Rapid determination of calcium, magnesium, sodium and potassium in milk by flame atomic spectrometry after microwave oven digestion, *Analyst*, **120** (1995), 107–111.
36. J. L. F. C. Lima, C. Delerue Matos, C. M. C. F. V. Vaz, Determination of Ca, Mg, Na, and K in milk by AAS and flame emission spectroscopy using high dilution FIA manifolds based on stream splitting or a dialysis unit, *Atom. Spectrosc.*, **17** (1996), 196–201.
37. B. S. Mohite, S. M. Khopkar, Solvent extraction separation of strontium as 18-crown-6 complex with picrate ion, *Anal. Chem.*, **59** (1987), 1200–1203.
38. M. de la Guardia, A. Salvador, P. Bayarri, R. Farre, Rapid atomic spectrometric determination of sodium, potassium, calcium and magnesium in powdered milk by direct dispersion, *Analyst*, **111** (1987), 1375–1377.

39. J. F. van Staden, A. van Rensburg, Simultaneous determination of total and free calcium in milk by flow injection, *Analyst*, **115** (1990), 605–608.

40. M. C. Yebra Biurrun, M. L. Mella Louzao, A. Bermejo Barrera, M. P. Bermejo Barrera, Comparative study of the determination of calcium in different types of milk by atomic absorption spectrophotometry after different sample treatments, *Alimentaria*, **28** (1991), 23–25.

41. F. Salinas, M. C. Mahedero, M. Jiménez Arrabal, Determinación de calcio y de magnesio en leche por inyección en flujo mediante absorción atómica, *Quím. Anal.*, **11** (1992), 11–16.

42. M. Burguera, J. L. Burguera, A. M. Garaboto, O. M. Alarcón, Determination of iron and copper in infant formula powdered milks by flow injection atomic absorption spectrometry, *Quím. Anal.*, **6** (1987), 427–435.

43. J. Arnaud, A. Favier, J. Alary, Méthodes de dosage du zinc dans le lait humain, *Lait*, **71** (1991), 87–97.

44. M. D. Mingorance, M. L. Pérez-Vázquez, M. Lachica, Microwave digestion methods for the atomic spectrometric determination of some elements in biological samples, *J. Anal. Atom. Spectrom.*, **8** (1993), 853–858.

45. P. Bermejo Barrera, R. Domínguez Gónzalez, R. Soto Ferreiro, A. Bermejo Barrera, Use of surfactant agents in the direct determination of zinc in milk samples by flame atomic absorption, *Analusis*, **23** (1995), 135–136.

46. P. Bermejo, R. Domínguez, A. Reboiro, A. Bermejo, The use of emulsions in the direct determination of zinc in milk fat by FAAS, *Atom. Spectrosc.*, **20** (1999), 161–165.

47. M. D. Silvestre, M. J. Lagarda, R. Farré, C. Martínez-Costa, J. Brines, Copper, iron and zinc determinations in human milk using FAAS with microwave digestion, *Food Chem.*, **68** (1999), 95–99.

48. S. Arpadjan, D. Stojanova, Application of detergents to the direct determination of iron, zinc, and copper in milk by means of flame atomic-absorption spectrophotometry, *Fresenius J. Anal. Chem.*, **302** (1980), 206.

49. S. Arpadzhyan, D. Nakova, Direct determination of iron, zinc, copper and manganese in milk powder by atomic absorption, *Nahrung*, **25** (1981), 359–363.

50. Y. Petit de Peña, M. Gallego, M. Valcárcel, On-line sorbent extraction, preconcentration and determination of lead by atomic absorption spectrometry, *Talanta*, **42** (1995), 211.

51. B. S. Garg, J. S. Bist, R. K. Sharma, N. Bhojak, Solid-phase extraction of metal ions and their estimation in vitamins, steel and milk using 3-hydroxy-1,4-naphthoquinone-inmobilized silica gel, *Talanta*, **43** (1996), 2093–2099.

52. M. Yaman, Simultaneous enrichment of aluminium and lead with cupferron on activated carbon for determination in milk and fruit juices by atomic absorption spectrometry, *Mikrochim. Acta*, **129** (1998), 115–119.

53. B. S. Garg, R. K. Sharma, J. S. Bist, N. Bhojak, S. Mittal, Separation and preconcentration of metal ions and their estimation in vitamin, steel and milk samples using *o*-vanillin-immobilized silica gel, *Talanta*, **48** (1999), 49–55.

54. M. A. Palacios, I. Arévalo, C. Cámara, Determinación de Se en muestras biológicas por absorción atómica mediante generación de hidruros, *Quím. Anal.*, **4** (1985), 320–325.

55. V. E. Negretti de Brätter, P. Brätter, A. Tomiak, An automated microtechnique for selenium determination in human body fluids by flow injection hydride atomic absorption spectrometry (FI-HAAS), *J. Trace Elem. Electrolytes Health Dis.*, **4** (1990), 41–48.
56. F. Li, E. Rossipal, K. J. Irgolić, Determination of selenium in human milk by hydride cold-trapping atomic absorption spectrometry and calculation of daily selenium intake, *J. Agric. Food Chem.*, **47** (1999), 3265–3268.
57. D. C. Wigfield, S. A. Eatock, The matrix effect in the cold-vapor atomic absorption analysis of mercury in various biological tissues, *J. Anal. Toxicol.*, **11** (1987), 137–139.
58. G. Schnitzer, A. Soubelet, C. Testu, C. Chafey, Comparison of open and closed focussed microwave digestions in view of total mercury determination by cold vapor atomic absorption spectrometry, *Mikrochim. Acta*, **119** (1995), 199–209.
59. A. Sanz-Medel, M. C. Valdés-Hevia y Temprano, M. C. Bordel García, M. R. Fernández de la Campa, Generation of cadmium atoms at room temperature using vesicles and its application to cadmium determination by cold vapor atomic spectrometry, *Anal. Chem.*, **67** (1995), 2216–2223.
60. M. A. H. Hafez, I. M. M. Kenawy, M. A. Akl, R. R. Lashein, Preconcentration and separation of total mercury in environmental samples using chemically modified chloromethylated polystyrene-PAN (ion-exchanger) and its determination by cold vapor atomic absorption spectrometry, *Talanta*, **53** (2001), 749–760.
61. A. Taylor, Detection and monitoring of disorders of essential trace elements, *Ann. Clin. Biochem.*, **33** (1996), 486–510.
62. L. De Galan, Atomic spectrometry: a user's view, *Phil. Trans. R. Soc. Lond. A*, **333** (1990), 5–18.
63. L. Vuchkova, L. Margitova, S. Arpadjan, Comparative study on the determination of trace elements in milk powder and cheese products using modifier-free ETAAS and ICP-AES, *Anal. Lab.*, **5** (1996), 41–45.
64. N. Motkosky, B. Kratochvil, Characterization of trace amounts of Al in biological reference materials by ETAAS, *Analyst*, **118** (1993), 1313–1316.
65. P. Martins, D. Pozebon, V. L. Dressler, G. A. Kemieciki, Determination of trace elements in biological materials using tetramethylammonium hydroxide for sample preparation, *Anal. Chim. Acta*, **470** (2002), 195–204.
66. N. Campillo, P. Viñas, I. López-García, M. Hernández-Córdoba, Direct determination of copper and zinc in cow milk, human milk and infant formula samples using ETAAS, *Talanta*, **46** (1998), 615–622.
67. P. Viñas, N. Campillo, I. López-García, M. Hernández-Córdoba, Electrothermal atomic absorption spectrometric determination of molybdenum, aluminium, chromium and manganese in milk, *Anal. Chim. Acta*, **356** (1997), 267–276.
68. P. C. Aleixo, J. A. Lóbrega, Direct determination of iron and selenium in bovine milk by graphite furnace atomic absorption spectrometry, *Food Chem.*, **83** (2003), 457–462.
69. J. Smeyers Verbeke, Q. Yang, W. Penninckx, F. Vandervoort, Effectiveness of palladium chemical modification for the determination of cadmium by graphite furnace atomic absorption spectrometry, *J. Anal. Atom. Spectrom.*, **5** (1990), 393–398.
70. S. Lynch, D. Littlejohn, Development of a slurry atomization method for the determination od cadmium in food samples by eletrothermal atomization atomic absorption spectrometry, *Talanta*, **37** (1990), 825–830.

71. E. M. García, M. L. Lorenzo, C. Cabrera, M. C. López, J. Sánchez, Trace element determination in diferent milk slurries, *J. Dairy Res.*, **66** (1999), 569–579.

72. P. Bermejo Barrera, M. C. Barciela Alonso, R. Domínguez González, A. Bermejo Barrera, J. A. Cocho de Juan, J. M. Fraga Bermúdez, Silicon determination in milk by electrothermal atomic absorption spectrometry using Pd as chemical modifier, *Anal. Bioanal. Chem.*, **374** (2002), 1290–1293.

73. P. Bermejo Barrera, T. Pardiñas Alvite, M. C. Barciela Alonso, A. Bermejo Barrera, J. A. Cocho de Juan, J. M. Fraga Bermúdez, Vanadium determination in milk by atomic absorption spectrometry with electrothermal atomization using hot injection and preconcentration on the graphite tube, *J. Anal. Atom. Spectrom.*, **15** (2000), 435–439.

74. P. Falomir, A. Alegría, R. Barberá, R. Farré, M. J. Lagarda, Direct determination of lead in human milk by ETAAS, *Food Chem.*, **64** (1999), 111–113.

75. P. Bermejo Barrera, M. Aboal Somoza, A. Bermejo Barrera, M. L. Cervera, M. de la Guardia, Microwave assisted distillation of iodine for the indirect atomic absorption spectrometric determination of iodide in milk samples, *J. Anal. Atom. Spectrom.*, **16** (2001), 382–389.

76. S. Caroli, A. Alimonti, P. Delle Femmine, F. Petrucci, O. Senofonte, N. Violante, A. Menditto, G. Morisi, A. Menotti, P. Falconieri, Role of inductively coupled plasma atomic emission spectrometry in the assessment of reference values for trace elements in biological matrices, *J. Anal. Atom. Spectrom.*, **7** (1992), 859–864.

77. V. E. Negretti de Brätter, P. Brätter, A. Reinicke, G. Schulz, W. O. L. Alvarez, N. Alvarez, Determination of mineral and trace elements in total diet by inductively coupled plasma atomic emission spectrometry: comparison of microwave-based digestion and pressurized ashing systems using different acid mixtures, *J. Anal. Atom. Spectrom.*, **10** (1995), 487–491.

78. J. A. Nóbrega, Y. Gélinas, A. Krushevska, R. M. Barnes, Direct determination of major and trace elements in milk by inductively coupled plasma atomic emission and mass spectrometry, *J. Anal. Atom. Spectrom.*, **12** (1997), 1239–1242.

79. M. De Wit, R. Blust, Determination of metals in saline and biological matrices by axial inductively coupled plasma atomic emission spectrometry using microconcentric nebulization, *J. Anal. Atom. Spectrom.*, **13** (1998), 515–520.

80. A. Ikem, A. Nwankwoala, S. Odueyungbo, K. Nyavor, N. Egiebor, Levels of 26 elements in infant formula from USA, UK and Nigeria by microwave digestion and ICP–OES, *Food Chem.*, **77** (2002), 439–447.

81. E. Coni, P. Falconieri, E. Ferrante, P. Semeraro, E. Beccaloni, A. Stacchini, S. Caroli, Reference values for essential and toxic elements in human milk, *Ann. Ist. Super. Sanitá*, **26** (1990), 119–130.

82. E. Coni, S. Caroli, D. Ianni, A. Bocca, A methodological approach to the assessment of trace elements in milk and dairy products, *Food Chem.*, **50** (1994), 203–210.

83. E. Coni, A. Stacchini, S. Caroli, P. Falconieri, Analytical approach to obtaining reference values for minor and trace elements in human milk, *J. Anal. Atom. Spectrom.*, **5** (1990), 581–586.

84. B. Bocca, A. Alimonti, E. Coni, M. Di Pasquale, L. Giglio, A. Piccioli Bocca, S. Caroli, Determination of the total content and binding pattern of elements in human

milk by high performance liquid chromatography–inductively coupled plasma atomic emission spectrometry, *Talanta*, **53** (2000), 295–303.

85. P. Schramel, Improvement of ICP-measurements by using a water-cooled spray chamber, *Fresenius' Z. Anal. Chem.*, **320** (1985), 233–236.
86. S. E. Emmett, Inter-laboratory note. Analysis of liquid milk by inductively coupled plasma mass spectrometry, *J. Anal. Atom. Spectrom.*, **3** (1988), 1145–1146.
87. J. A. Nóbrega, Y. Gélinas, A. Krushevska, R. M. Barnes, Direct determination of major and trace elements in milk by inductively coupled plasma atomic emission and mass spectrometry, *J. Anal. Atom. Spectrom.*, **12** (1997), 1243–1246.
88. K. M. Hua, M. Kay, H. E. Indyk, Nutritional element analysis in infant formulas by direct dispersion and inductively coupled plasma-optical emission spectrometry, *Food Chem.*, **68** (2000), 463–470.
89. N. Carrión, A. Itriago, M. Murillo, E. Eljuri, A. Fernández, Determination of calcium, potassium, magnesium, iron, copper and zinc in maternal milk by inductively coupled plasma atomic emission spectrometry, *J. Anal. Atom. Spectrom.*, **9** (1994), 205–207.
90. J. K. Friel, W. L. Andrews, S. E. Jackson, H. P. Longerich, C. Mercer, A. McDonald, B. Dawson, B. Sutradhar, Elemental composition of human milk from mothers of premature and full-term infants during the first three months of lactation, *Biol. Trace Elem. Res.*, **67** (1999), 225–247.
91. M. Krachler, E. Rossipal, K. J. Irgolić, Trace elements in formulas based on cow and soy milk and in Austrian cow milk determined by inductively coupled plasma mass spectrometry, *Biol. Trace Elem. Res.*, **65** (1998), 53–74.
92. N. Jakubowski, L. Moens, F. Vanhaecke, Sector field mass spectrometers in ICP-MS, *Spectrochim. Acta*, **53B** (1998), 1739–1763.
93. L. Moens, F. Vanhaecke, J. Riondato, R. Dams, Some figures of merit of a new double focusing inductively coupled plasma mass spectrometry, *J. Anal. Atom. Spectrom.*, **10** (1995), 569–574.
94. J. M. Marchante-Gayón, C. Sariego Muñiz, J. I. García Alonso, A. Sanz-Medel, Multi-elemental trace analysis of biological materials using double focusing inductively coupled plasma mass detection, *Anal. Chim. Acta*, **400** (1999), 307–320.
95. J. M. Marchante-Gayón, Double-focusing ICP-MS for the analysis of biological materials, *Anal. Bioanal. Chem.*, **379** (2004), 335–337.
96. F. A. Rivero Martino, M. L. Fernández Sánchez, A. Sanz-Medel, Total determination of essential and toxic elements in milk whey by double focusing ICP-MS, *J. Anal. Atom. Spectrom.*, **15** (2000), 163–168.
97. T. Prohaska, G. Köllensperger, M. Krachler, K. De Winne, G. Stingeder, L. Moens, Determination of trace elements in human milk by inductively coupled plasma sector field mass spectrometry (ICP-SFMS), *J. Anal. Atom. Spectrom.*, **15** (2000), 335–340.
98. M. Krachler, T. Prohaska, G. Köllensperger, E. Rossipal, G. Stingeder, Concentrations of selected trace elements in human milk and in infant formulas determined by magnetic sector field inductively coupled plasma–mass spectrometry, *Biol. Trace Elem. Res.*, **76** (2000), 97–112.
99. K. L. Chen, S. J. Jiang, Determination of Ca, Fe and Zn in milk powder by reaction cell inductively coupled plasma mass spectrometry, *Anal. Chim. Acta*, **470** (2002), 223–228.

100. C.-Y. Ho, S.-J. Jiang, Determination of Cr, Zn, Cd and Pb in milk powder by slurry sampling electrothermal vaporization inductively coupled plasma mass spectrometry, *J. Anal. Atom. Spectrom.*, **17** (2002), 688–692.

101. R. de la Flor St. Remy, M. L. Fernández Sánchez, A. Sanz-Medel, Determination of essential and toxic total elements in premature human milk by inductively coupled plasma mass spectrometry (ICP-ORC-MS), using an octapole reaction cell, *J. Anal. Atom. Spectrom.*, **19** (2004), 616–622.

102. G. Rädlinger, K. G. Heumann, Iodine determination in food samples using inductively coupled plasma isotope dilution mass spectrometry, *Anal. Chem.*, **70** (1998), 2221–2224.

103. M. Haldimann, A. Eastgate, B. Zimerli, Improved measurement of iodine in food samples using inductively coupled plasma isotope dilution mass spectrometry, *Analyst*, **125** (2000), 1977–1982.

104. K. G. Heumann, Isotope dilution mass spectrometry, *Int. J. Mass Spectrom. Ion Proc.*, **118/119** (1992), 575–592.

105. P. Bermejo, R. Domínguez, A. Bermejo, Direct determination of Fe and Zn in different components of cow milk by FAAS with a high performance nebulizer, *Talanta*, **45** (1997), 325–330.

106. F. A. Rivero Martino, M. L. Fernández Sánchez, A. Sanz-Medel, The potential of double focusing ICP-MS for studying elemental distribution patterns in whole milk, skimmed milk and milk whey of different milks, *Anal. Chim. Acta*, **442** (2001), 191–200.

14

USE OF SPECTROCHEMICAL METHODS FOR THE DETERMINATION OF METALS IN FISH AND OTHER SEAFOOD IN LOUISIANA

JOSEPH SNEDDON

Department of Chemistry, McNeese State University, Lake Charles, LA 70609, USA,
E-mail: jsneddon@mcneese.edu

14.1 INTRODUCTION

A map of Louisiana is shown in Figure 14.1. The coastal zone of Louisiana comprises an area south of Interstate 10 (a major highway stretching from Los Angels, California, to Jacksonsville, Florida, of approximately 3000 miles), approximately 30–40 miles wide. It can be roughly divided into two main regions: The region east of the Atchafalaya River is open marshland that reaches

The Determination of Chemical Elements in Food: Applications for Atomic and Mass Spectrometry.
Edited by Sergio Caroli

Figure 14.1. Map of Louisiana. Shown is an insert of USA.

as far north as the Mississippi River. To the west, the Chenier Plains, bound by the Atchafalaya River on the east and the Sabine River on the west, are the dominant feature. The Lousiana coast is bounded on the west by Texas, on the east by Mississippi, and to the south by the Gulf of Mexico. The coastal region in the west is bounded by the Mississippi River. The coast comprises a distance of approximately 400 miles, but also has large open marshland, swamps, lakes, and numerous bayous (bodies of water which are not noticeably moving). They are tide-dependent from the Gulf of Mexico. Due to this, it is not uncommon to have large fluctations in the salinity of the marshes and bayous, ranging from fresh water (0 parts per thousand) through brackish water (approximately 22 parts per thousand), and strictly salt water (approximately 35 parts per thousand).

This area of Louisiana is one of the most productive estuaries in the continental United States in the commercial production of shrimps, menhaden (fish), oysters, crayfish (also known as crawfish and crawdads), crab, mussels, and other marine species. Louisiana is also known as a sportsman's paradise because of the recreational fishing, as well as significant commercial fishing. For example, the shrimp industry is worth over US$350 million per year in Louisiana.

The discovery of oil and gas in the coastal zone has led to large-scale drilling both on shore and in the Gulf of Mexico region. During the 1940s, the need to refine fossil fuels and produce chemicals became a national emergency and a number of locations in Louisiana (as well as other geographical regions in this general area) were chosen to develop this industry. This area was chosen because of its abundant fresh water, the deep-water ports, and remoteness from attack. The need for these products and the absence of environmental regulations led to widespread metal (and organic) pollution. Since the early 1970s stronger regulations have been in place, although accidental spills of chemicals, refined chemicals, and oil still occur. These spills can pollute the waters, soils, and sediments, which are the main feeding grounds of the wildlife and seafood, and can therefore enter the food cycle.

This chapter presents the results of a number of selected studies of elements in various fish and seafood in the Louisiana coastal region. Some results from nearby areas, particularly in the Gulf region and Mississippi delta region, are presented for completeness and comparison. In some cases the studies were performed for toxicological purposes and the prediction of metal concentrations. Some studies also involved organic chemicals and these are included for completeness.

14.2 SPECTROCHEMICAL METHODS

The most widely used spectrochemical methods are flame atomic absorption spectrometry (FAAS), electrothermal atomization atomic absorption spectrometry (ETA-AAS), and inductively coupled plasma atomic emission spectrometry (ICP-AES). Some work has been performed using inductively coupled plasma mass spectrometry (ICP-MS) and the unique properties of Hg have allowed the use of cold vapor (CV) AAS. It is beyond the scope of this chapter to describe these well-established and well-accepted spectrochemical techniques. The reader is referred to several excellent texts which describe in detail the basic principles, instrumentation, and method development of these analytical techniques [1–4]. The most toxic elements, such as As, Cd, Cr, Pb, and particularly Hg have been the most widely studied. Other metals, such as Ba, Cu, Fe, Mn, V, and Zn, have also been investigated.

14.3 SAMPLE COLLECTION AND PREPARATION METHODS

Fish and seafood collected for element determination can be sampled through a variety of techniques such as hook and line, electro-shocking and trapping. Once collected the fish should be placed in a plastic bag (if size permits) and stored on ice for 24 h. Once the fish reaches the laboratory, it should be frozen either before or after dissecting tissue samples. If the samples are dissected prior to freezing, they should be collected in duplication. Any instruments used for dissecting should be stainless steel or (preferably) plastic or polymer and rinsed with de-ionized or distilled water prior to use. Prior to metal determination, the fish needs to be digested, ideally into a solution. Spectrochemical methods on solutions (usually) produce the best accuracy and precision. Several methods have been developed, including HNO_3–H_2SO_4 and addition of $HClO_4$ should some fat tissue remain. H_2O_2 has been used to digest the fatty acid tissue. These traditional digestion methods, often call wet digestion procedures, have been replaced by microwave (MW) digestion procedures. MW digestion procedures allow greater numbers of samples to be processed in less time with less contamination. For an exhaustive discussion of sample procedures the reader is referred to several texts and papers on this subject [5, 6] as well as to other chapters of this book.

Results obtained using these digestion methods are reported on a wet basis. To report a dry-weight basis, a subset of the tissue or fish to be analyzed can be oven-dried and a wet-to-dry weight ratio calculated. The ratio can be used to report metal concentrations in terms of dry weight.

14.4 A SELECTION OF STUDIES ON ELEMENTS IN SEAFOOD

Many studies have been performed on soils and sediments [7–9]. Numerous investigations have involved monitoring metal concentrations in soils and sediments as well as reclaimed marshes over a period of several decades. While this chapter focuses on elemental determination, several studies have simultaneously determined organics. These results are presented in these studies for completeness in order to understand the suspected pollution or contamination of the fish and seafood. Uptake of elements can be affected by organics in the lakes, bayous, and reclaimed marshes. Recently, work has focused on determining metal concentrations (from crude oil and products) in the soils, sediments, and selected plants after oil spills that have been remediated by controlled burning [7]. This information is frequently used to extrapolate the pollution extent in wildlife, fish, and seafood from the area.

A study of the abdominal muscle tissue of crayfish collected from five commercial farms in Southwest Louisiana found average Pb concentrations ranging from 0.25 ± 0.16 to 0.40 ± 0.14 $\mu g\ g^{-1}$ (dry weight) [10]. These authors found that Pb tended to accumulate in the abdominal tissue of the crayfish.

Mercury is (arguably) the most toxic metal in the periodic table. It can exist in several forms, including inorganic Hg and organomercury forms. Methyl-Hg is considered to be the most toxic form or species of Hg. It is considered to be a potent neurotoxin in adults. The most widely form of exposure in adults is caused by the consumption of contaminated fish and “action levels” have been established. If the concentration exceeds the action level, then action is taken to protect the public either by notice, closing fishing ways or by restricting the sale of fish. In the USA, two agencies can protect the public from Hg-contaminated fish: the Environmental Protection Agency (EPA), which has an action level of 0.5 $\mu g\ g^{-1}$ and the Food and Drug Administration (FDA), which has an action level of 1.0 $\mu g\ g^{-1}$. These levels are continuously being monitored and could be reduced in the near future. In addition, young children and pregnant and nursing women are potentially at risk. Organo-Hg compounds in oysters are a particular problem in Louisiana. Cooking of the oysters at elevated temperatures could remove the relatively low-boiling-point organo-Hg compounds. However, oysters are frequently eaten raw. Seldom does a season go by without a local or visitor to the Louisiana area dying from eating contaminated raw oysters, although organo-Hg is not always the problem. To complicate the problem, in Louisiana there is an economic factor. Where privately owned and commercial companies have a relatively short season of (perhaps) a few months in which to harvest their catch, a delay of even a few days can have a severe economic impact. Table 14.1 shows the results of a number of studies of Hg concentrations in fish and seafood found in Louisiana [11]. It should be noted that all the studies were not performed in Louisiana, but are included for reference and comparison. The results, for the most part, are in sub microgram per gram concentrations for all types of seafood. The most widely used methods for determination are those based on CV combined with FAAS as outlined in the introduction, although neutron activation analysis (NAA) and ETA-AAS have also been used. The extremely low concentrations of the elements in the fish and seafood dictate that spectrochemical methods must be very sensitive and capable of determining sub-microgram levels with good accuracy and high precision. Organo-Hg levels are frequently determined by combining gas chromatography (GC) with spectrochemical methods (AAS or AES) as the detector.

In a laboratory-controlled environment and using tidewater silverside (*Menidia beryllina*), Jackson et al. [12] showed that the salinity of water affects the accumulation of Cd. The tidewater silverside is found from Massachusetts to Vera Cruz, Mexico and is an important forage fish for various species of larger organisms including sport fish in Louisiana. In general, the higher the salinity, the less Cd is accumulated in the fish. At low or no salinity, higher concentrations of Cd are found in the fish. Elimination of Cd occurred at a faster rate in fish exposed to higher levels of salinity.

Aquatic heavy metal contamination in Louisiana is a very serious health problem. This issue first came into national media attention in the early 1970s and late 1980s, when several outbreaks of As, Cr, Hg, Pb, and Zn contamination occurred in northern, southeastern, and southwestern Louisiana in the streams, rivers, and lakes where game and commercial fishing was popular. This problem of aquatic heavy metal pollution continued right through the 1990s and was even more of a concern in the New Orleans area. It was in 1995 that sizable concentrations of As contamination were found in the bodies of *Micropterus salmoides* (small-mouth bass fishes) and in *Procambarus clarkii* (red crawfishes) in several commercial lakes and streams in New Orleans [13]. In this study, the effects of As were analyzed as well as the biomagnification effects on the aquatic food chain. Additionally, the study showed the process by which aquatic organisms (fish and invertebrates) biochemically convert inorganic elements to organometallics, which are more easily absorbed into their tissue and muscle cells. Moreover, the study makes a detailed analysis of how this problem continues to grow and how it is causing serious aquatic pollution problems in the mid-south area and for the Mississippi River area in general.

Trace metals (Cr, Cu, Pb, and Zn) were determined in fish and sediment samples from Lake Boeuf, Southeastern Louisiana [14]. Lake Boeuf is a popular

TABLE 14.1 Mercury Concentrations in Fish and Seafood [11]

Species of Fish/Seafood	Tissue Type	Mercury Concentrations
Rainbow trout (*Salmo gaindheri*)	Lateral muscle	1.1–1.4 pg
Cod (*Gadius morhua L.*)	Tissue	0.8 μg g^{-1}
Flounder (*Platichthys flesus L.*)	Tissue	0.15 μg g^{-1}
Northern pike (*Esox lucias*)	Muscle	0.33–1.21 μg g^{-1} wet weight
Large lake trout (*Salvelinus namaycush*)	Muscle	0.12–1.21 μg g^{-1} wet weight
Yellow walleye (*Stizostedion v. vitreum*)	Muscle	0.78 μg g^{-1} wet weight
Brook trout (*Salvelinus fontinalis*)	Muscle	0.12–0.57 μg g^{-1} wet weight
Lake whitefish (*Corregonus clupeaformis*)	Muscle	0.06–0.22 μg g^{-1} dry weight
Gray mullet (*Magil cephalus*)	Muscle	0.03 ± 0.12 μg g^{-1}
Shrimp (*Penaeus aztecus*)		0.06 ± 0.12 μg g^{-1}
Shrimp (*Penaeus stylirostris*)		0.09 ± 0.12 μg g^{-1}
Shrimp (*Penaeus californiensis*)		0.12 ± 0.10 μg g^{-1}
Oyster (*Crassostrea virgininica*)		0.09 ± 0.01 μg g^{-1}
Crab (*Callinectes sapidas*)	Tissue from leg and abdominal region	0.005 μg wet weight
Oyster (*Crassastrea virgina*)	Meat	0.06 μg g wet weight
Clam (*Rangia cuneato*)	Meat	0.11 μg g wet weight
Mussel (*Mytilus edulis*)		0.11 ± 0.6 μg g^{-1} wet weight
Crayfish (*Pacifastacus sp.*)	Tail muscle	0.57–5.72 μg g^{-1} wet weight
Oyster (*Saccostrea curcullata*)	Whole body	0.01–0.834 μ/g^{-1} wet weight
Oyster (*Saccostrea echinato*)	Whole body	0.016–0.125 μg g^{-1} wet weight

recreational fishing lake from which the following sunfish are caught and consumed: bluegill sunfish (*Lepomis macrochirus*); large-mouth bass (*Micropterus salmoides*); and black crappie (*Poxomis nigromaculatus*). After extraction from fish and sediment using MW digestion techniques, metals were quantified using flame atomic absorption spectroscopic methods. Trace metal concentrations in fish did not vary significantly between species or weight of fish; however, a seasonal trend in metal concentration was observed. Chromium concentrations were 2–15 $\mu g\ g^{-1}$, Cu 0.2–4 $\mu g\ g^{-1}$, Pb 0.5–9 $\mu g\ g^{-1}$, and Zn 10–20 $\mu g\ g^{-1}$. Concentrations of the metals studied in core samples of sediment did not vary with depth in the core, suggesting no pollutant metals have been input to the lake.

A study investigated the occurrence and adaptation to Pb and Zn in the western mosquitofish (*Gambusia affinis*) inhabiting contaminated Bayou Trepagnier, Louisiana [15]. This bayou has a long history of industrial contamination. Present levels of Pb and Zn in water and sediment were considerably higher in Bayou Trepagnier than in a nearby control stream. Tissue metal levels of mosquitofish were highly elevated for Pb and (to a lesser extent) for Zn. Fish collected from Bayou Trepagnier and then exposed to Zn in a 96 h laboratory bioassay did not differ in their sensitivity to Zn from those collected from a control site. In contrast, Bayou Trepagnier fish did exhibit an increased resistance to Pb. This difference in resistance to Pb between Bayou Trepagnier fish and control fish was no longer evident when both groups of fish were kept for 34 days in the laboratory under identical conditions (in clean water). While a genetic basis for the difference in resistance between the two populations cannot be fully excluded, it did appear that the elevated Pb resistance in Bayou Trepagnier mosquitofish was due to acclimation (physiological, individually based) rather than adaptation at the population level.

Research on evaluating the liver and brain esterases in the spotted gar fish (*Lepisosteus oculatus*) as biomarkers of the effect of multiple contaminants in the lower Mississippi River basin was undertaken [16]. The study sites were Devil's Swamp (DS), which is situated on the east bank of the Mississippi River just northwest of Baton Rouge, LA, and a pristine control site, Tunica Swamp (TS), which is approximately 30 miles up river from DS near St Francisville, LA. DS is an ecosystem contaminated with chlorinated hydrocarbons, such as hexachlorobenzene (HCB) and hexachlorobutadiene (HCBD), and a variety of heavy metals. Combined pollutants adversely affect the hepatic and neurological systems. Gar with significant body burdens of organics and metals have liver injury at the tissue, cellular, and enzymic (biochemical) levels. In gar from sites with mixed pollution, concurrent liver injury included large melanin-rich macrophage centers, increased Fe pigments, ductal proliferation in the exocrine pancreas, and depression of liver esterase activity. The findings of esterase depression may be directly or indirectly linked to liver lesions due to chemical contamination.

Two species of buffalo fish, smallmouth buffalo (*Ictiobus bubalus*) and bigmouth buffalo (*Ictiobus cyprinellus*), were collected from a contaminated (multiple metals and organic chemicals) Mississippi River Basin ecosystem, DS, and a control site, Tunica Swamp, both near Baton Rouge, LA [17]. The buffalo fish were examined for general health and histopathology effects. This research consisted of the analysis of the prevalence of infectious and noninfectious disease from the contaminated versus the relatively uncontaminated control site. Tissue burdens were also determined by chemical analysis of buffalo fish muscle (fillets) samples. Both bigmouth buffalo and smallmouth buffalo from DS accumulated metals in muscle tissue and generally had a higher prevalence of gill lesions than fish collected from the TS control site. Smallmouth buffalo from DS had higher prevalences of lesions than bigmouth buffalo from DS. The primary types of infectious gill lesions observed in the buffalo fish were protozoan infections in the gill filament cartilage and gill lamellae. Chondrolysis, chondritis, and branchitis were associated with parasitic infection. Eosinophils were the predominant cell type involved with branchitis and chondritis. The proliferative lesions observed were chondroplasia, chloride cell hyperplasia, mucus cell hyperplasia, and epithelial cell hyperplasia and fusion. Based on these preliminary findings, smallmouth buffalo appeared to be a potentially sensitive biomarker of exposure.

Field and laboratory studies were conducted to determine the levels of Cd, Pb, and HCBD in various samples collected from a swamp environment in Louisiana and to assess the toxicities of As, Cd, and Hg to two species of aquatic organisms (bluegills and crawfish) indigenous to this swamp [18]. Cadmium and Pb were present in almost all collected samples. Their concentrations in fish were below the action levels set by EPA for the protection of human health. However, hazardous levels of these two elements were found in samples of crawfish, vegetation, soil, sediment, and water to some extent. Low levels of HCBD were recorded in water and sediment samples. In bioassay studies, Hg was found to be the most toxic metal, while As was the least toxic. Between the two test organisms, bluegills appeared to be more sensitive than crawfish. Mixtures of Cd with As or Hg resulted in a combined toxic effect which was simply additive. However, a synergistic effect was recorded with the mixtures of As and Hg.

Two commercially important species of Louisiana crayfish, *Procambarus clarkii* and *P. acutus*, from the Atchafalaya River Basin, from open ponds, and from the sediment and water of these environments, were sampled three times during two consecutive fishing (production) seasons [19]. The abdominal muscle and hepatopancreatic tissue were analyzed separately. Lead, Hg, and Cd, if present, were in concentrations below the limit of detection (LoD). The heptatopancreatic tissue contained (in $\mu g\ kg^{-1}$) Ba <8, Cu 11–15, and Fe <640. Abdominal muscle samples had $<3\ \mu g\ kg^{-1}$ of most metals. Locations with the highest levels of metal residues in sediment were not necessarily locations where crayfish had the highest levels in their tissues.

The edible parts of several species of fish and other marine organisms inhabiting the Calcasieu River/Lake, in Southwest Louisiana, were determined for As, Cd, Cr, Cu, Pb, Hg, Ag, and Zn [20]. Concentrations of all elements measured in both intra and interspecies showed no significant variation with sample location. Differences in metal concentrations were related to organism mobility and not to site-specific variations in metal loadings. Different species of finfish contained only trace amounts (in $\mu g\ kg^{-1}$) of Ag (<0.01–0.3), As (<0.1–0.3), Cd (0.02–0.08), and Pb (<0.2–0.5).

Concentrations of Cu and Cr were highly variable, in contrast to Zn, which was present in relatively constant amounts across all species (28 ± 7 μg kg^{-1}). Sessile organisms such as oysters had the highest concentrations of heavy metals, including Cd. Periphyton and zooplankton were the only groups that showed differences in metal concentrations with sampling location.

The objectives of a study were to ascertain current contaminant residue concentrations in fish and sediment from lakes in the Atchafalaya River Basin and to determine the influence of overflow and agricultural land use on contaminant levels [21]. Fish and sediment samples were collected from eight lakes that differed in overflow and land-use characteristics. Residue analysis of 24 sediment and 82 fish samples collected from the study lakes showed that contaminant residues were uniformly low. Total dichlorodiphenyltrichloroethane (DDT) concentrations consisting mainly of dichlorodiphenyldichloroethylene (DDE) were found in 98.8 percent of the fish samples and generally were the highest residues, though they rarely exceeded 0.50 μg g^{-1}. Dieldrin, α-BHC, and γ-BHC, in concentrations rarely exceeding 0.05 μg g^{-1}, were detected in approximately 50 percent of the fish samples. Total organochlorine pesticide and polychlorobenzene (PCB) residues were <2.00 μg g^{-1} in all fish from all lakes, except for a 3.46 μg g^{-1} total in spotted gar (*Lepisosteus oculatus*) from one lake. Sediments were free from pesticide residues except for occasional traces (0.01 μg g^{-1}) of α-BHC. Metal residue concentrations in fish and sediment samples were typical of uncontaminated areas. Lake systems in the Atchafalaya River Basin were relatively uncontaminated by agricultural pesticides, and metal and organic residues in fish were similar to those of fish from lakes with the same land-use and overflow characteristics.

Concentration gradients of Ba, Cd, Cr, Cu, Pb, and Zn in surficial sediments of the central Gulf of Mexico (Louisiana, USA) decrease with distance from some petroleum drilling platforms [22]. These gradients arose from platform corrosion, drilling fluids, and cuttings, and other platform-related activities. Although the gradients are somewhat masked by Mississippi river sediment input, they are not explained by sediment variability. Trace metal concentration in four species of seafood, that is, *Penaeus aztecus* (brown shrimp), *Micropogon undulatus* (Atlantic croaker), *Archosargus probatocephalus* (sheepshead), and *Chaetodipterus faber* (spadefish), were not significantly higher than in other areas of the Gulf.

Research examined the pattern of sediment contamination of an urban bayou of New Orleans (formerly a natural waterway) and the potential for human exposure from consumption of fish caught in the bayou [23]. Sediments and soils of Bayou Saint John were evaluated for Cd, Pb, and Zn. Sediment cores were collected at bridges ($n = 130$) and sites between the bridges ($n = 303$) of the bayou. In addition, soil samples ($n = 66$) were collected along the banks. Sediments below the bridges contain significantly more (p-value approximately 10^{-7}) Pb and Zn (medians of 241 and 230 μg kg^{-1}, respectively) than bayou sediments located between bridges (medians of 64 and 77 μg kg^{-1}, respectively). Sediments below bridges of the upper reaches of the bayou contain significantly larger amounts of metals ($p < 10^{-14}$ for Pb and Zn and approximately 10^{-8} for Cd;

medians of 329, 383 and 1.5 μg kg^{-1}, respectively) than sediments below bridges in the lower reaches of the bayou (medians of 0.5, 43, and 31 μg kg^{-1}, for Cd, Pb, and Zn respectively). Likewise, medians for sediments located between bridges contain significantly [$p < 10(-14)$] higher quantities of Cd, Pb, and Zn (1.8, 170, and 203 μg kg^{-1}, respectively) in the upper bayou than Cd, Pb, and Zn (0.8, 48, and 32 μg kg^{-1}, respectively) in the lower reaches of the bayou. The potential risk for human exposure may be magnified by the fact that fishing generally occurs from the numerous bridges that cross the bayou. Poor and minority people do most of the fishing. Most people (87 percent) indicated they ate fish they caught from the bayou.

Significant adverse effects on environmental quality, ecosystem integrity, and human health have often been associated with improper disposal of hazardous materials. A study ascertained the levels of eight potentially toxic elements in various fish species that were collected from a local hazardous-waste-contaminated wetland and estimated the potential health risk associated with consuming such fish [24]. A total of 53 fish samples representing 12 different species were examined. The respective mean concentrations (in ng g^{-1}) of As, Cd, Cr, Cu, Fe, Pb, Hg, and Ni found in edible tissues were as follows: 72.5 ± 103.1, 55.4 ± 34.6, 97.4 ± 111.7, 177.3 ± 328.3, 2711.0 ± 4469.6, 26.0 ± 119.0, 32.7 ± 75.3, and 81.5 ± 178.9. For a 10 kg child eating 6.5 g of fish per day, the computed combined hazard index was 0.475, whereas for 70 kg adults with a daily consumption of 6.5 g (general population), 30 g (sport fishermen), and 140 g (subsistence fishermen), the respective computed hazard indices were 0.068, 0.313, and 1.462. The results indicate that subsistence fishermen had the highest risk for systemic effects, with an exposure exceeding the EPA Reference Dose value. In a 10 kg child, such excess exposure would be reached (even if only for elements such as As and Hg) when applying the EPA-approved maximum fish consumption rate of 54 g day^{-1} to the general population. The cancer risk for As, the only element with an established cancer potency factor from oral exposure, varied from 8×10^{-6} to 253×10^{-6}, indicating an exposure exceeding the widely accepted risk level of 1×10^{-6} (one excess cancer per 10^6 individuals).

Between 1990 and 1994, samples of three shellfish species, that is, blue crab (*Callinectes sapidus*), crayfish (*Procambarus acutis*), and river shrimp (*Macrobrachium ohionii*) and 16 fish species were collected at six sites along the lower Mississippi River by the Louisiana Department of Environmental Quality, Office of Water Resources, in coordination with the EPA [25]. The fish species included bigmouth buffalo (*Ictiobus cyanellus*), blue catfish (*Ictalurus furcatus*), carp (*Cyprinus carpio*), channel catfish (*Ictalurus punctatus*), cobia (*Rachycentron canadum*), flathead catfish (*Pylodictis olivaris*), freshwater drum (*Aplodinotus grunniens*), largemouth bass (*Micropterus salmoides*), long nose gar (*Lepisosteus osseus*), red drum (*Sciaenops ocellatus*), red snapper (*Lutjanus campechanus*), smallmouth buffalo (*Ictiobus bubalus*), spotted gar (*Lepisosteus oculatus*), striped bass (*Morone saxatilis*), white bass (*Morone chrysops*), and white crappie (*Pomoxis annularis*). Organic compounds and heavy metal concentrations were measured in 161 composite fish tissue samples where each composite included 3–10 individual fish. Nineteen chemicals found at measurable levels in sample tissues were used in calculations of lifetime excess cancer and noncancer risks due to fish consumption. On the basis of a consumption rate of one 8 oz meal per week (0.032 kg day^{-1}) and a body weight of 70 kg, it was estimated that there might be 574 chemical-specific cancer risks, 41 total cancer risks, and 697 margins of exposure. The authors identified nine species of concern (blue catfish, carp, channel catfish, cobia, crayfish, flathead catfish, red drum, spotted gar, and striped bass) based on total cancer risk greater than 10^{-4} or margin of exposure greater than 1 and whether or not samples collected in subsequent years resulted in lower risks. The compounds primarily responsible for the elevated risks were aldrin, dieldrin, α-benzene hexachloride (BHC), γ-benzene hexachloride, heptachlorepoxide, As, and Hg.

By examining metal concentrations of biota collected from reference and restored sites and comparing them, the authors would be able to determine whether the dredge material is introducing metal contaminants to the biota present on the restored sites [26]. If the metal concentrations of biota from the restored sites were greater than the concentrations of biota from the reference sites, the authors would then be able to compare the individual restored sites with the reference sites to determine differences between the sites. The authors also considered that there was no difference between the metal concentrations present in the different species of biota collected from the restoration sites. Significant differences in Pb concentrations were found between fiddler crabs (*Uca longisignalis*) collected from the restored sites and reference sites. The authors also found a significant difference in V concentrations of the biota between reference sites and restored sites for *Spartina patens*, *U. longisignalis*, and *Littoria irrorata*.

The objective of a study were to identify lesions and diseases in fish as biomarkers of exposure to environmental contaminants from polluted wetlands typical of the Mississippi River Basin [27]. DS, located northwest of Baton Rouge, LA, is an ecosystem contaminated with a variety of metals and organic chemicals. The nearby control site was TS. Gar were collected from the study sites by electroshocking and crab trap. The fish were necropsied and evaluated for disease and histopathology and tissues were analyzed for metals and organics typical of DS contaminants. Liver lesions occurred in three species of gar (*Lepisosteus osseus*, *L. oculatus*, and *L. platostomus*). The striking liver lesions observed in gar from DS were abundant melanin-rich macrophage centers, pancreatitis, and perivascular cuffing and ductal proliferations in the exocrine pancreas. Other lesions observed were parasitism, inflammation, and multifocal necrosis. Image and pigment analyses of melanomacrophage centers (MMC) in spotted gar (*L. oculatus*) indicated increased levels of hepatocellular and MMC Fe. In turn, larger MMC may be due to significant body burdens of pollutants.

High concentrations of As, Ba, Cd, Cu, Fe Hg, Mn, Ni, Pb, V, and Zn were detected in clams (*Chama macreophylla*), oysters (*Crassostrea virginica*), red snapper fish (*Lutjanus campechanus*), and gray triggerfish (*Balistes capriscus*) collected near discharging and nondischarging platforms [28]. The results were described as a function of the concentrations of elements in produced water discharges versus ambient seawater, concentrations of elements in organisms from reference versus discharge sites, and accumulation of elements in sediments adjacent to oil and gas platforms.

As part of long-term biological effects monitoring of sludge dumping at the Deepwater Municipal Dump Site, bottom fish were collected near the site and analyzed for contaminant body burdens [29]. Bottom fish were collected within 10 miles of the site in September 1989 at approximately 2400 m depth using a Gulf of Mexico shrimp trawl. Muscle and liver contaminant levels were determined as well as data on total body length, weight and volume and liver volume and weight. A total of six fish, three blue hake (*Antimora rostrata)*, two grenadier (*Coryphaenoides aramatus)*, and one cutthroat eel (*Synaphobrachus kaupi*) were selected for organic and metal contaminant analysis. In liver, As (ranging from 5.4 to 26.4 $\mu g\ g^{-1}$), Cu (ranging from 4.4 to 23.9 $\mu g\ g^{-1}$), and Zn (ranging from 37.1 to 51.9 $\mu g\ g^{-1}$) were all elevated compared to the other elements. In muscle tissue, As (ranging from 47.8 to 122.4 $\mu g\ g^{-1}$) and Zn (ranging from 9.5 to 31.8 $\mu g\ g^{-1}$) were found at much higher concentrations than other elements determined (Ag, Cd, Cu, Hg, and Pb). Polychlorinated biphenyls (PCBs) and DDT (as well as DDE and DDD) were found in all samples. Individual pesticide concentrations ranged from undetected to 0.02 to 1.2 $\mu g\ g^{-1}$ in muscle and from 0.8 to 14.4 $\mu g\ g^{-1}$ in the liver. Total PCB concentrations ranged from 0.005 to 0.5 $\mu g\ g^{-1}$ in muscle to 0.6 to 9.4 $\mu g\ g^{-1}$ in the liver. The highest pesticide and PCB concentrations in muscle were found in the cutthroat eel which had similar lipid concentrations in the muscle and liver. Polycyclic aromatic hydrocarbon (PAHs) concentrations were negligible in all samples. Elevated levels of hepatic cytochrome P 4501A were detected in grenadiers and in the cutthroat eel but not in the blue hake. None of the fish examined had any histopathological lesions of the type normally associated with chronic or severe exposure to chemical contaminants. Based on this study, the bottom fish appear to be good indicators of exposure to chlorinated hydrocarbons and trace elements. However, before bottom fish can be used to assess the effects of sewage-sludge dumping at the 106-Mile site, more information is needed on the body burdens in fish not impacted by dumping at the site.

The objective of a study was to determine inorganic and organic contaminant concentrations in edible tissue of fish collected from eight coastal areas receiving waste water discharges and from two reference locations [30]. Trace metal residues were statistically similar regardless of the collection site. Zinc (100 percent detection in all samples), total Hg (100 percent), total As (92 percent), Cu (92 percent), and Se (88 percent) were the more commonly detected trace elements. Mercury concentrations exceeded the Florida health-based standard of 0.5 $\mu g\ g^{-1}$ for limited fish consumption in 30 percent of the total samples and averaged 0.40 (± 1 SD $= 0.22$, range ≤ 0.08–0.85) $\mu g\ g^{-1}$ wet weight. The average total PAH concentrations were 1.79 (± 1.60) $ng\ g^{-1}$ (reference areas) and 2.17 (± 3.29) $ng\ g^{-1}$ (waste water-impacted areas). Pyrene was detected most frequently (63 percent of the total samples) and averaged 0.74 (± 0.35) $ng\ g^{-1}$ wet weight. The average total PCB concentrations were 4.8 (± 7.1) $ng\ g^{-1}$ (reference areas) and 31.6 (±31.3) ng g^{-1} (waste water-impacted areas). Concentrations of dieldrin and *cis*-chlordane were approximately eight times greater, respectively, in fish collected from waste water receiving waters, whereas total DDT and total pesticide concentrations were not elevated in the same areas. Concentrations of total PCBs and all chlorinated pesticides were below US health-based standards. The lack of a published reference data base for fish tissue quality in near-coastal areas of the Gulf of Mexico restricts an assessment of the environmental significance of results from this and similar studies investigating the fate of point source contaminants.

A study was conducted of six Louisiana lakes to examine the relationship between sediment properties including Hg content and health advisories associated with Hg in fish [31]. Comparison was made between three lakes with health advisories (Black Lake, Chicot Lake, and Henderson Lake) and three lakes where the levels of Hg in fish are below health advisory levels (False River, Lake St John, and Miller Lake). Three sediment core samples were collected from each lake and sectioned into 2 cm increments to a depth of 20 cm. Sediment properties measured in each depth increment of the sediment profile included total Hg, ^{137}Cs activity (for sedimentation rate), and sediment organic matter content. Of the lakes studied, those lakes that have health advisories for Hg tended to have higher total Hg contents, usually higher sediment organic matter contents, and higher sedimentation rates than sediments in lakes where health advisories for Hg are not issued.

Meat samples from wild harvested subadult and adult alligators in south Louisiana had low Hg levels, averaging 0.131 μg g^{-1} [32]. The highest level detected (0.386 μg g^{-1}) is much lower than the FDA action level for safe human consumption of fish flesh (Louisiana Department of Environmental Quality, 1998). Samples of both sexes from southwest Louisiana were significantly higher (0.176 μg g^{-1}) than those from southeast Louisiana (0.098 μg g^{-1}). No difference was seen between males and females pooled from all sites. The average alligator meat Hg level in Louisiana (0.131 μg g^{-1}) was lower than values reported in other southeastern states and presents no public health concern.

The concentration of Hg was determined in a variety of fish and sediment cores taken from Lake Boeuf located in southeastern Louisiana [33]. Lake Boeuf is an important recreational and game fishing lake from which the following sunfish are caught and consumed: bluegill sunfish (*Lepomis macrochirus*); redear sunfish (*Lepomis microlophus*); largemouth bass (*Micropterus salmoides*); and black crappie (*Pomoxis nigromaculatus*). These species of fish were collected and Hg levels were determined using the EPA CV analytical method 7471. Mercury concentrations exceeding the FDA action level of 1.0 μg g^{-1} in fish were found in four of 15 bluegill sunfish and two of 15 black crappie. Bluegill sunfish contained the highest concentrations of Hg, ranging in value from <0.02 to 0.92 μg g^{-1} and an average of 0.22 ± 0.31 μg g^{-1}. Redear sunfish had the lowest concentration, averaging only 0.03 ± 0.03 μg g^{-1}. Sediment cores revealed a history of Hg inputs to the lake with concentrations as high as 0.25 μg g^{-1} at depths of about 15 cm.

Trace element concentrations were determined in livers of canvasbacks (*A. valisineria*) collected at Catahoula Lake and the Mississippi River Delta, LA, during the winter of 1987–88 [34]. Forty percent of canvasbacks wintering at Lake Catahoula had elevated concentrations of Pb ($>6.7\ \mu g\ g^{-1}$ dry weight) in the liver and 33 percent had concentrations consistent with Pb intoxication ($>26.7\ \mu g\ g^{-1}$). Based on the number of canvasbacks that winter at Lake Catahoula and the frequency of Pb exposure there, >5 percent of the continental population of canvasbacks may be exposed to Pb at Lake Catahoula alone. Lead concentrations in livers differed among months and were higher in males than females. On the other hand, no differences were noted between adults and immatures. Concentrations of Hg and Se in livers of females differed among months not by age or location. Cadmium concentrations in livers differed by age, location, and month of collection, not by sex. Frequencies and concentrations of trace elements not commonly associated with adverse effects on avian species (Ag, Al, As, Cu, Fe, Mg, Mn, Ni, V, and Zn) were presented. Except for the elevated concentrations of Pb at Catahoula Lake, all trace elements were at background concentrations.

An author noted that Louisiana's vital seafood industry provides nearly 40 percent of domestic seafood production. Unlike Louisianans, most Americans do not enjoy fresh seafood, yet manage to eat over 4 billion pounds of seafood annually, most of it foreign, frozen, and fried! America's commercial fisheries now face crippling economic and environmental pressures from seafood imports, over-fishing, urban, and agricultural waste water runoff, harmful algal blooms, and coastal wetlands loss. As a result of these ecosystem stresses, seafood-borne disease now causes 37 percent of all foodborne illness in the United States. Despite recent advisories on high Hg-containing finfish consumption, seafood provides more than a third of food with essential dietary protein, rich in cardiovascular-protective omega-3 polyunsaturated fatty acids. Louisiana and other coastal-state physicians can effectively curtail the rising threat of local seafood-borne disease outbreaks by supporting responsible coastal restoration and regulation of commercial fishing and by recommending careful inspection, selection, and preparation of seafood [35].

The primary source of nonoccupational exposure to Hg is through the consumption of contaminated fish. Since 1994, the Louisiana Department of Environmental Quality has reported Hg contamination in fish obtained from bodies of water throughout the state and has issued fish consumption advisories accordingly. To determine the extent of Hg intoxication in Louisiana, screening for blood Hg levels was offered to volunteers residing near selected advisory areas [36]. A total of 313 residents participated in the screening, six of which were found to have elevated levels. No level was detected in 48 of the participants, while the remaining participants had normal levels. Significantly higher levels were found in those individuals associated with commercial fishing and those reporting increased fish consumption. For most people, ordinary consumption of fish contaminated with Hg does not currently appear to pose a public health hazard in Louisiana; however, educational efforts regarding the risks of fish consumption in large quantities should be continued.

Oysters (*Crassostrea virginica*) from 50–69 locations (sites) along the Gulf of Mexico coastline, collected annually in 1986, 1987, and 1988, were assayed for 13 trace elements, including most of the metals of concern from an environment quality perspective [37]. Essentially, the entire U.S. Gulf coastline was sampled, from far south Texas to far south Florida. Pooled samples of 20 oysters from three different stations at each site were analyzed by AAS. The concentrations found were generally less than or equal to literature values from other parts of the world thought to be uncontaminated by anthropogenic trace element inputs. A few sites did, however, show apparent trace element pollution and other sites gave anomalous values that cannot readily be explained by either known anthropogenic or natural causes. The range of values for the overall data set (maximum/minimum ratio) varied from 15-fold for Mn to 624-fold for Pb, whereas the coefficient of variation (standard deviation/mean ratio) was generally in the 50–60 $\mu g\ g^{-1}$ range for most elements. Variations were much greater between stations than between years at a given station. Enrichments usually occurred in suites of three to four metals with Ag, Cd, Cu, and Zn being the most common suite. Thus several strong inter-metal correlations were found. There was, however, little correlation between metal levels in oysters and in sediments from the collection sites even when sediment data were ratioed to Al (sediment data are not given here). There was likewise little correlation between oyster metal levels and size, sex, or reproductive stage of the oysters. Geographically, appreciably elevated (>3 times average) metal levels were generally restricted to single sites within bays or estuaries, implying local control. On the other hand, regionally, Ag, Cd, and Se levels were somewhat higher in Texas oysters than in those from Florida, whereas the reverse was true for As and Hg. Concentrations were lower than average for several elements in oysters from central Louisiana, especially Ag, Cd, and Cu. Thus, the Mississippi River outflow and extensive offshore oil development do not seem to enrich oysters in trace elements.

Three channels from Lake Pontchartrain, Louisiana, to the Gulf of Mexico were studied as sample (oysters, clams, and sediments) sources to determine trace element (Ag, As, Be, Cd, Cr, Cu, Hg, Ni, Pb, Se, Tl, and Zn) uptake; only Tl was not detected in the tissue and it was minimally present in the three sediments [38]. Oysters in the Inner Harbor Navigation Canal contained high concentrations of Zn, whereas clams from the other two channels (i.e., the Rigolets and the Chef Menteur Pass) contained high levels of Cu. Sediment metal concentrations were reflected in the tissue levels.

Investigations were made at the West Hackberry and Weeks Island (Louisana) proposed brine diffuser sites during four consecutive seasons (1978–1979) [39]. Samples of suspended particulate matter, epibenthic organisms, and macrocrustaceans were collected during all four seasons. The spatial and temporal distributions of 13 trace metals were determined in each sample matrix. Higher metal concentrations in sediments from the West Hackberry site are in agreement with the higher amounts of silt, clay, and organic C reported in these sediments. There are indications of seasonal variations of Ba, Cd, and Hg in the white shrimp (*Penaeus setiferus*) muscle tissues.

14.5 CONCLUSIONS AND FUTURE WORK

Element concentrations in fish and seafood will continue to be a concern in Louisiana. Long-term studies will be required to determine whether there is an increase, a decrease, or a constant change in element concentrations. Changes in concentration will most likely be monitored by continuing studies on soil, sediments, and plants. Chemical speciation to identify the various forms of the elements could play an important role in fish and seafood safety. Chemical speciation is not limited to Hg compounds, although chronic Hg poisoning by residents in Minamata Bay, Japan, are well documented. In this case, brain damage to new-born babies was attributed to inorganic Hg. Other cases include a spill of tetraalkyl-Pb in the Mediterranean Sea and the decimation of the oyster fields of Arcachon Bay in France. Clough et al. [40] provided an interesting general discussion on chemical speciation, although the emphasis is on isotopic dilution ICP-MS as the technique for speciation.

ACKNOWLEDGMENT

The author acknowledges, with thanks, the financial support of the Environmental Protection Agency (EPA), grant number R-82958401-1.

REFERENCES

1. D. J. Butcher, J. Sneddon, *A Practical Guide to Graphite Furnace Atomic Absorption Spectrometry*, Wiley-VCH, New York, 1998.
2. J. Sneddon, Atomic absorption spectrometry, in Frank Settle (ed.), *Handbook of Instrumental Techniques for Analytical Chemistry*, Prentice Hall, Upper Saddle River, NJ, 1997, 373–393.
3. R. Thomas, *A Practical Guide to ICP-MS*, Marcel Dekker, New York, 2003.
4. J. Nolte, *ICP-Emission Spectrometry, a Practical Guide*, Wiley-VCH, Hoboken, NJ, 2003.
5. M. G. Heagler, A. G. Lindow, J. N. Beck, C. S. Jackson, J. Sneddon, The effect of water chemistry on the bioaccumulation of cadmium in *Menidia beryllina* (cope), the tidewater silverside. Part I. A simple and rapid method for the digestion of fish, *Microchem. J.*, **53** (1996), 472–477.
6. H. M. Kingston, S. J. Haswell (eds), *Microwave Enhanced Chemistry: Fundamentals, Sample Preparation and Application*, ACS Professional Reference Book Series, American Chemical Society, Washington DC, 1991.
7. C. Hardaway, J. Sneddon, J. N. Beck, Determination of metals in crude oil by atomic spectroscopy, A mini-review, *Anal. Lett.*, **37** (2004), 1–19.
8. J. N. Beck, J. Sneddon, Metal concentrations in soils and sediments in southwest Louisiana, *Anal. Lett.*, **33** (2000), 1913–1959.
9. J. N. Beck, J. Sneddon, Use of atomic absorption spectrometry for the determination of metals in sediments in southwest Louisiana, *Microchem. J.*, **66** (2000), 73–113.

10. L. M. Briggs-Reed, M. B. G. Heagler, A comparative analysis of lead concentrations in purged and unpurged crayfish (*Procambarus clarkii*): the significance of digestive tract removal prior to consumption by humans, *Microchem. J.*, **55** (1997), 122–128.

11. J. Sneddon, M. G. Heagler, Determination of mercury by atomic spectroscopy: application to fish, *Advances in Atomic Spectroscopy*, **4** (1998), 213–229.

12. C. S. Jackson, J. Sneddon, M. G. Heagler, A. G. Lindow, J. N. Beck, Use of flame atomic absorption spectrometry and the effect of water chemistry for the study of the bioaccumulation of cadmium in *Menidia beryllina* (cope), the tidewater silverside *Microchem. J.*, **75** (2003), 23–28 .

13. L. R. Brown III, Q. C. Brown, R. H. Miller, Effects of arsenic on *Poecilia reticulata*: ecotoxicity and biomagnification in the aquatic food chain in southeast New Orleans, Abstracts of Papers, *American Chemical Society, 220th Meeting*, 2000.

14. J. Aucoin, R. Blanchard, C. Billiot, C. Partridge, D. Schultz, K. Mandhare, M. J. Beck, J. N. Beck, Trace metals in fish and sediments from Lake Boeuf, southeastern Louisiana, *Microchem. J.*, **62** (1999), 299–307.

15. P. L. Klerks, S. A. Lentz, Resistance to lead and zinc in the western mosquitofish *Gambusia affinis* inhabiting contaminated Bayou Trepagnier, *Ecotoxicology*, **7** (1998), 11–17.

16. T. L. Huang, P. O. Obih, R. Jaiswal, W. R. Hartley, A. Thiyagarajah, Evaluation of liver and brain esterases in the spotted gar fish (*Lepisosteus oculatus*) as biomarkers of effect in the lower Mississippi river basin, *Bull. Environ. Contamin. Toxicol.*, **58** (1997), 688–695.

17. A. Thiyagarajah, W. R. Hartley, S. E. Major, M. W. Broxson, Gill histopathology of two species of buffalo fish from a contaminated swamp, *Mar. Environ. Res.*, **42** (1996), 113–117.

18. A. A. Abdelghani, Y. V. Pramar, T. K. Mandal, P. B. Tchouwou, L. Heyer, Levels and toxicities of selected inorganic and organic contaminants in a swamp environment, *J. Environ. Sci. Health, Part B: Pesticides, Food Contam. Agric. Wastes*, **B30** (1995), 717–731 .

19. M. W. Finerty, J. D. Madden, S. E. Feagley, R. M. Grodner, Effect of environment and seasonality on metal residues in tissues of wild and pond-raised crayfish in southern Louisiana, *Arch. Environ. Contam. Toxico.*, **19** (1990), 94–100.

20. G. J. Ramelow, C. L. Webre, C. S. Mueller, J. N. Beck, J. C. Young, M. P. Langley, Variations of heavy metals and arsenic in fish and other organisms from the Calcasieu river and lake, Louisiana, *Arch. Environ. Contam. Toxicol.*, **18** (1989), 804–818.

21. P. V. Winger, J. K. Andreasen, Contaminant residues in fish and sediments from lakes in the Atchafalaya river basin (Louisiana), *Arch. Environ. Contam. Toxicol.*, **14** (1985), 579–586.

22. J. B. Tillery, R. E. Thomas, Heavy metal contamination from petroleum production platforms in the central Gulf of Mexico, *Proceedings of the Symp./Research into the environmental fate of effluent drilling fluids and cuttings*, Washington, DC, 1980, 562–587.

23. M. Welt, H. W. Mielke, C. Gonzales, K. M. Batist, G. Corey, L. H. Cresswell 3rd, P. W. Mielke, Metal contamination of sediments and soils of Bayou Saint John: a potential health impact to local fishermen?, *Environ. Geochem. Health*, **25** (2003), 387–396.

24. P. B. Tchounwou, A. A. Abdelghani, Y. V. Pramar, L. R. Heyer, C. M. Steward, Assessment of potential health risks associated with ingesting heavy metals in fish collected from a hazardous-waste contaminated wetland in Louisiana, USA, *Rev. Environ. Health*, **11** (1996), 191–203.

25. K. H. Watanabe, F. W. Desimone, A. Thiyagarajah, W. R. Hartley, A. E. Hindrichs, Fish tissue quality in the lower Mississippi river and health riisks from fish consumption, *Sci. Total Environ.*, **302** (2003), 109–126.

26. J. J. Darbonne, M. G. Heagler, Concentrations of cadmium, lead, nickel and vanadium in biota from marsh restoration sites in southwest Louisiana, *Microchem. J.*, **60** (1998), 242–248.

27. W. R. Hartley, A. Thiyagarajah, A. M. Treinies, Liver lesions in the gar fish (*Lepisosteidae*) as biomarkers of exposure, *Mar. Environ. Res.*, **42** (1996), 217–221.

28. J. H. Tregry, R. P. Trocine, K. L. Naito, S. Metz, Assessing the potential for enhanced bioaccumulation of heavy metals from produced water discharges to the Gulf of Mexico, *Environ. Sci. Res.*, **52** (1996), 339–354.

29. W. G. Steinhauer, C. D. Hunt, C. S. Peven, R. E. Hillman, M. J. Moore, J. J. Stegeman, D. Redford, Contaminants and cytochrome P4501A induction in bottom fish collected near the 106-Mile Site, *J. Mar. Environ. Engng*, **3** (1996), 173–187.

30. M. A. Lewis, G. I. Scott, D. W. Bearden, R. L. Quarles, J. Moore, E. D. Strozier, S. K. Sivesrtsen, A. R. Dias, M. Sander, Fish tissues in near-coastel area of the Gulf of Mexico receiving point source discharges, *Sci. Total Environ.*, **284** (2002), 249–261.

31. R. P. Gambrell, R. D. DeLaune, W. H. Patrick Jr., A. Jugsujinda, Mercury distribution in sediment profiles of six Louisiana lakes, *J. Environ. Sci. Health, Pt A: Toxic/Hazard. Substances Environ. Engng*, **A36** (2001), 661–676.

32. R. M. Elsey, V. A. Lance, L. Campbell, Mercury levels in alligator meat in south Louisiana, Louisiana department of wildlife and fisheries, *Bull. Environ. Contam. Toxic.*, **63** (1999), 598–603.

33. T. P. Dupre, T. J. Granier, J. Tamerage, S. Keife, R. Marino, S. O'Rourke, C. Partridge, D. L. Schultz, K. Mandhare, J. N. Beck, Variation of mercury concentration in fish taken from Lake Boeuf, southeastern Louisiana, *Microchem. J.*, **61** (1999), 156–164.

34. T. W. Custer, W. L. Hohman, Trace elements in canvasbacks (*Aythya valisineria*) wintering in Louisiana, USA, 1987–1988, *Env. Pollut.*, **84** (1994), 253–259.

35. J. H. Diaz, Is fish consumption safe?, *J. Louisiana State Med. Soc. (Official Organ of the Louisiana State Medical Society)*, **156** (2004), 44–49.

36. T. M. Bellanger, E. M. Caesar, L. Trachtman, Blood mercury levels and fish consumption in Louisiana, *J. Louisiana State Med. Soc. (Official Organ of the Louisiana State Medical Society)*, **152** (2000), 64–73.

37. B. J. Presley, R. J. Taylor, P. N. Boothe, Trace metals in Gulf of Mexico oysters, *Sci. Total Environ.*, **97–98** (1990), 551–593.

38. C. J. Byrne, L. R. DeLeon, Trace metal residues in biota and sediments from Lake Pontchartrain, Louisiana, *Bull. Environ. Contamin. Toxicol.*, **37** (1986), 151–158 .

39. J. B. Tillery, *Biological/Chemical Survey of Texoma and Capline Sector Salt Dome Brine Disposal Sites off Louisiana, 1978–1979. Volume VII. Determine Trace Metal Composition and Concentration in Major Components of the Ecosystem, Avail.* NTIS Report (1980) (NOAA-TM-NMFS-SEFC-31, NOAA-81020207; order no. PB81-174989), Government Report Announcement, Index (U. S.), **81** (1981), 3275.

40. R. Clough, S. T. Belt, E. H. Evans, B. Fairman, T. Catterick, Isotope dilution ICP-MS for speciation studies, *Appl. Spectrosc. Rev.*, **38** (2003), 101–132.

15

ESSENTIAL AND POTENTIALLY TOXIC CHEMICAL ELEMENTS IN BEVERAGES

PATRICIA SMICHOWSKI AND DANIEL A. BATISTONI†

Comisión Nacional de Energía Atómica, Centro Atómico Constituyentes, Unidad de Actividad Química, Avenida General Paz 1499, B1650KNA San Martín, Provincia de Buenos Aires, Argentina, E-mail: smichows@cnea.gov.ar

†This chapter was almost finished when Daniel A. Batistoni unexpectedly and in a very short time passed away. His friends and colleagues will always remember his honesty and kindness.

The Determination of Chemical Elements in Food: Applications for Atomic and Mass Spectrometry.
Edited by Sergio Caroli

15.1 INTRODUCTION

The importance of the intake of metallic and many nonmetallic elements by living organisms is at present recognized both because of the toxic properties of some elements and their beneficial effects as essential participants in numerous specific processes that occur at the subcellular level. Oxygen transport, photosynthesis, respiration, and N fixation are largely dependent upon elements as ubiquitous as Fe [1]. Copper has been found to be critical in regulating cascade electron transfer processes in cells and Zn is known to behave as a transynaptic messenger. The dissemination of Mn in the environment – owing to its addition to gasoline – has prompted researchers to reevaluate its properties as an essential dietary element in comparison to observed subtle signs of toxicity [2]. The involvement of metals such as Al, Mn, and Zn in neurodegenerative processes is presently well documented [3]. Solubility, complexation capacity, and interchange of oxidation states are among the physical–chemical properties that make transition metals and many semi-metals of the groups 14–16 usual components of liquid foods, thus allowing them to be uptaken through a regular diet. In addition, the wide diffusion of common beverages (such as cocoa and instant tea and coffee) and natural infusions of vegetal origin, with demonstrated or assumed medicinal properties, have expanded the sources of intake of both toxic and essential elements.

As in many other fields of human activity, modern analytical technology plays an increasingly important role in studies on the dissemination and pathways of metals through the food chain. Trace and ultratrace concentration levels that could not be attained with conventional instrumental procedures are now routinely measured. The potential influence of elements that used to be neglected or at least not properly taken into account is nowadays carefully considered in the standards that regulate the suitability of bottled waters, fruit juices, and alcoholic beverages for human consumption. The observed distribution of metals in wines and other alcoholic beverages is employed in many cases to obtain information about the geographical origin of different varieties.

Since some form of liquid sample presentation is common to most atomic spectrometric techniques, these may be considered as the method of choice for the identification and quantification of trace metals in liquid foods. In spectrometric techniques, after conversion of the sample into microspray, chemical flames as in flame atomic absorption spectrometry (FAAS) and atomic fluorescence spectrometry (AFS) [4], and high-temperature plasmas, as in direct current plasma atomic emission spectrometry (DCP-AES) or inductively coupled plasma atomic emission spectrometry (ICP-AES) [5, 6], are generally employed as atomization cells, aimed at generating a relatively stable population of free atoms and ions. These species are thermally or radiationally excited and then sensed through the emission or absorption of the characteristic radiation associated with electronic transitions at given wavelengths. Electrothermal atomizers, as typically in electrothermal atomization (ET) AAS, constitute an alternative way of producing an atomic vapor for the subsequent absorption of photons [7]. On the other hand, plasma sources may also be highly efficient systems to produce ions that are detectable after appropriate selection based on the mass-to-charge ratio, as in inductively coupled plasma mass spectrometry (ICP-MS) [8]. Obviously, the characterization of the isotopic composition of the element is a direct consequence. On the other hand, one of the key operational differences between atomic and mass spectrometric techniques is the direct interaction of sample components with the detector that takes place in the latter (i.e., only photons get into an atomic spectrometric system). The intrinsic properties of the different signal generation–transduction environment, together with the particular characteristics of the available instrumentation determine the advantages and limitations of each experimental setup regarding figures of merit, inter-element interferences, matrix effects, and economy and simplicity of operation.

In this chapter, aspects of the atomic spectrochemical methods relevant to the analysis of food beverages are considered and discussed in the context of recently reported applications of the methodologies. It should be noted that, as a rule, only the most recent literature data are taken into account so as to keep the length of this chapter within reasonable limits. This in no way precludes the existence of valuable earlier papers, as exemplified by the material included in previously published reviews [9, 10] addressing many of the topics dealt with here. It is also worth mentioning that a considerable number of approaches for spectrochemical analysis of natural matrices other than foods can be, after proper adaptation, applied to trace analysis of food and beverages. In this context the subfield of sample preparation and pre-treatment deserves a special mention, owing to its remarkable development in the last years, resulting from the popularity gained by microwave (MW) digestors, preconcentration devices, and automatic liquid transport methods applicable on- and off-line. Sample presentation to the nebulizer-atomizer devices – mostly based on pneumatic operation and regarded as the critical step in atomic spectrometric methods – has greatly benefited from liquid sample transport strategies, such as nonsegmented flow injection analysis (FIA) that allows for the sequential, on-line pretreatment of the sample and its subsequent injection into the spray generation interface. These procedures, as well as on-line analyte derivatization techniques that improve the power of detection of some metalloids and the coupling of separation methods to atomic detectors, are given particular consideration in this chapter.

15.2 BOTTLED AND MINERAL WATERS

15.2.1 General Considerations

Drinking waters other than tap water include mineral as well as enriched water. The latter is manufactured by purifying tap or well water and adding mineral salts in a controlled way to simulate the composition of natural products. Mineral waters reflect in many instances the geochemical characteristics of the regions where they are collected and bottled. Bottled water is one of the most regulated food products and multielement analytical atomic spectrometric methods have helped in establishing the permitted contents of potentially toxic metals and metalloids, as well as of naturally occurring elements, including some essential ones. Along with a considerable number of trace organic compounds, about 10 elements – earmarked below as inorganic chemicals – and six additional elements earmarked as secondary inorganic parameters are usually investigated in connection to quality standards of commercial samples. Tables 15.1 and 15.2 show the limits of detection (LoDs) of inorganic components and secondary inorganic parameters for commercial control purposes along with the corresponding values of standards of quality as defined by the U.S. Food and Drug Administration (FDA).

Local regulations may vary among the different countries. In the United States all bottled water products must comply with the FDA Standard of Quality [Section 165.110(b) of Title 21 of the Code of Federal Regulations (CFR)], aiming at ensuing the aptitude and safety of the products from production and packaging through consumption. Furthermore, FDA labeling rules for bottled water involve a Standard of Identity. This defines, in a standardized fashion, the different terms usually added to bottle labels, such as artesian, drinking, mineral, purified, sparkling, and spring. Soft drinks such as soda or tonic water are excluded from these norms. Numerous literature references discuss aspects of interest regarding regulations for European and other countries [11–13]. More recently, information related to exposure effects from several elements due to human consumption has been reported [14–16], including comments on the effect of dietary metals as carcinogens [17].

In Europe the span in element concentrations covered by 56 brands of bottled mineral waters randomly collected all over the continent has been considered in great detail by Misund et al. [18]. Determinations of 66 elements were carried out by ICP-AES and ICP-MS, complemented with ion chromatography (IC). Variations for single elements as large as six orders of magnitude were found, which are attributable to ample regional variations in the composition of natural waters. The authors stress that many elements known to be extremely toxic can reach relatively high levels, even though no maximum allowable concentrations have been established (at the time of writing) in the EU. Apart from the intrinsic importance of the report, it represents a typical application of ICP with atomic or mass spectrometric detection for the trace analysis of water. In this case, about 82 percent of the results were obtained by ICP-MS.

TABLE 15.1. Typical Limits of Detection and Standards of Quality, as Defined by U.S. FDA, of Inorganic Chemicals for Commercial Bottled Waters

Inorganic Chemicals	Limit of Detection (mg l^{-1})	FDA SOQ (mg l^{-1})
Antimony	0.001	0.006
Arsenic	0.002	0.05
Barium	0.1	2
Beryllium	0.001	0.004
Cadmium	0.001	0.005
Chromium	0.05	0.1
Cyanide	0.05	0.1
Fluoride	0.5	2/1.3[a]
Lead	0.001	0.005
Mercury	0.0002	0.002
Nickel	0.05	0.1
Nitrate (N)	0.1	10
Nitrite (N)	0.01	1
Total nitrate + nitrite	0.1	10
Perchlorate	5	—[b]
Selenium	0.01	0.05
Thalium	0.0005	0.002

[a]Highest value: if naturally occurring. Lowest value: if added.
[b]Not available.

15.2.2 Applications of Individual Techniques

Among spectrometric methods of analysis, the U.S. Environmental Protection Agency (EPA) analytical protocols corresponding, for example, to Methods 200.8 (ICP-MS), 200.7 (ICP-AES) and 200 Series (FAAS) are well established standard methodologies for the elemental analysis of waters. Although at present superseded by plasma techniques, FAAS remains a very simple and cost-effective measurement technique for soluble metals. The equipment essentially allows

TABLE 15.2. Typical Limits of Detection and Standards of Quality, as Defined by U.S. FDA, of Secondary Inorganic Parameters for Commercial Bottled Waters

Secondary Inorganic Parameters	Limit of Detection (mg l^{-1})	FDA SOQ (mg l^{-1})
Aluminum	0.05	0.2
Chloride	0.5	250
Copper	0.01	1
Iron	0.01	0.3
Manganese	0.001	0.05
Silver	0.005	0.1
Sulphate	2	250
Zinc	0.1	5

elements to be sequentially determined, but a considerable degree of automation for the determination of several elements in the same sample is currently possible through computer control of the instrumentation. Minimum determinable concentrations are limited by the atomization characteristics of the analytical flames. For that reason an analyte enrichment step is often necessary, particularly for refractory elements (Cr, Mo, Ti, V). A typical approach, off-line separation–preconcentration by ion-exchange resins is described in an earlier paper by Korkisch and Krivanec [19] for the quantification of Mo and V in tap and bottled mineral water. After acidification of the sample with citric acid and heating to remove CO_2, the medium is adjusted at pH 3.0 with Na citrate and ascorbic acid. A column of strongly basic anionic resin (Dowex 1-X8) is employed to adsorb the (anionic) citrate complexes of the analyte metals. Elution of V is accomplished with 6 M HCl, and subsequently Mo is recovered by elution with a mixture of 2 M $HClO_4$ and 1 M HCl. Analysis of the eluates is performed, after suitable evaporation and redissolution to reduce the solution volumes, by FAAS with an N_2O–acetylene flame. Sodium sulfate is used as an additive to sample and calibrant solutions for the determination of Mo. Measured concentration ranges were 0.1–0.9 $\mu g\ l^{-1}$ for V and 0.2–13 $\mu g\ l^{-1}$ for Mo.

Collection of metal complexes of the analytes on suitable adsorbing materials is often employed as an enrichment step in combination with flame methods. In a procedure proposed by Solyak et al. [20], five metals [Co(II), Cu(II), Cr(III), Fe(III), and Pb(II)] were complexed with calmagite {3-hydroxy-4-[(6-hydroxy-*m*-tolyl)azo]-naphthalenesulfonic acid} and subsequently collected on a soluble cellulose nitrate membrane filter. In this way an effective separation from alkaline and alkaline earth metals was achieved, based on the differences in their complex formation constants and those of the transition elements. The experimental parameters were optimized for the quantitative recovery of the elements. After hot dissolution of the filter with HNO_3, the analytes were determined by FAAS. Minimum detectable concentrations ranged from 0.06 $\mu g\ l^{-1}$ for Cu to 2.5 $\mu g\ l^{-1}$ for Cr.

The advantage of the selective adsorption of a particular element oxidation state has been exploited for on-line element preconcentration and speciation analysis of Cr by FAAS. Cespón Romero et al. [21] described an FIA system employing a minicolumn made of a chelating resin containing poly(aminophosphonic) acid groups, able to selectively retain Cr(III) ions. An FIA manifold was employed for efficient preconcentration and subsequent elution of Cr(III) with a small volume of 0.5 M HCl. The original sample was also treated with ascorbic acid to reduce Cr(VI) to Cr(III) and total Cr is determined as Cr(III) after appropriate retention and elution. Eluates are introduced into an N_2O–acetylene flame connected to the column outlet. The concentration of Cr (VI) is obtained by difference. Employing a sample volume of 6.6 mL, LoD for total Cr is 0.2 $\mu g\ l^{-1}$. A study of FI operational variables, interferences, and precision is reported for the analysis of tap, mineral, and river waters.

When ultimate LoDs are not a stringent requirement, as in the case of the determination of Ca and Mg, the robustness and versatility of routine FAAS measurements may be enhanced by employing variations of FI analysis as methods of sample presentation. The application of a methodology called sequential injection analysis (SIA) has been described by Araújo and coworkers [22]. Briefly, time control of the injections and the use of a fluid selecting valve on a one-line manifold allows measurements to be performed under controlled dispersion and the sample dilution strategies and the consumption of the releasing agent (Al) added to suppress interferences to be optimized. Analysis of 15 mineral waters for Ca and Mg gave reproducibilities better than 3 percent, with sampling rates higher than 110 h^{-1}.

Even a technique of higher detection power as ET-AAS may require some sort of previous analyte enrichment for difficult elements. In the determination of As and Se in mineral waters described by Hudnik and Gomiscek [23], coprecipitation of both elements on hydrated Fe(III) oxide was employed to improve LoDs, otherwise impaired by matrix effects. A graphite tube furnace was the atomization cell, with the atomic vapor sampled with element electrodeless discharge lamps. After treatment of the sample with Fe(III) solution at the appropriate pH, the oxide precipitate was filtered and dissolved and the solution volume reduced to 5 mL of 0.2 M H_2SO_4. Ten-microliter volume aliquots of sample and standard solutions were injected into the furnace. Reported LoDs were 0.2 and 0.5 μg l^{-1} for As and Se, respectively.

Preconcentration of Mo from mineral and produced waters has been performed by Ferreira et al. [24], employing polyurethane foam as sorbent, prior to the determination by ET-AAS. The thiocyanate complex of Mo(VI) was adsorbed onto a minicolumn filled with the sorbent, followed by elution with a 3 M HNO_3 solution. With an enrichment factor of 10, an LoD of 0.08 μg l^{-1} was achieved. Good agreement in the analysis of mineral waters was found between the results of the proposed method and those obtained by ICP-MS.

Aluminum has deserved particular attention among the studied metals because of its connection with neurovegetative disorders and its controversial status as a risk factor for the Alzheimer's disease [25]. Data on the content of Al in more than 100 bottled water samples, comprising 43 Italian and foreign brands, have been reported by Plessi and Monzani [26]. ET-AAS enabled these authors to reach a characteristic mass (i.e., the mass of analyte that produces an absorbance value of 0.0044) of 38 pg Al from 20 μL sample injections. It was found that all the concentration values were below the 50 μg l^{-1} limit allowed by the EU guideline.

A detailed study of the occurrence of Al in drinking waters (tap, still mineral, and sparkling mineral water) as well as in fruit juices and soft drinks was done by López and collaborators [16]. ET-AAS employing pyrolitically coated graphite tubes provided with a Lvov platform was chosen as determination technique. Water samples were injected into the furnace without previous treatment. Fruit juices and soft drinks were mineralized with HNO_3 in the presence of V_2O_5 prior to dilution and subsequent injection. Magnesium nitrate was selected as matrix modifier in order to improve reproducibility. The authors also evaluated the incidence of the type of containers. The attained (absolute) LoD of 4 pg, with a characteristic mass of 10 pg, was adequate for the determinations. The ranges of measured Al concentration values were (in ng ml^{-1}) 4.2–165.3 for drinking water, 49.3–1144.6 for fruit juices, and 44.6–1053.3 for soft drinks. The study was useful to estimate the daily dietary intake of Al coming from those types of sources in Spain. In a related paper, García et al. [27] employed the same digestion process for sample preparation to the determination of Cr in various beverages, including potable water by ET-AAS. Sensitivity, accuracy, and precison of the method were verified. Concentration values ranged from nondetectable to 11.80 μg l^{-1} Cr in potable water.

Some refractory elements cannot be determined by ET-AAS at the levels usually present in waters. That is the case with M. El Himri et al. [28] developed a fast and accurate procedure, without any prior treatment, to analyze tap and mineral waters from Spain and Morocco for this highly toxic element. ICP-MS was employed. The analytical isotope selected was ^{238}U, with Rh as internal standard. An LoD of 2 ng l^{-1} was obtained. The estimated repeatability was 3 percent at the concentration level of 73 ng l^{-1}. The method was validated by comparison with a radiochemical procedure devised for natural samples and by analysis of a Certified Reference Material (CRM). Multi-element capabilities of ICP-AES have also been employed for surveys of trace elements. Al-Saleh and Al-Doush [29] reported the concentrations of dissolved Be, Cd, Cr, Cu, Fe, Mg, Mn, Hg, Ni, Se, Sr, V, and Zn in 21 samples of retail bottled waters from Riyadh, Saudi Arabia. It was found that Cd, Fe, Hg, Ni, and Zn were present at concentrations higher than the limits recommended by the EU and World Health Organization (WHO) guidelines.

Methods of speciation analysis, based on the coupling of high resolution separation techniques with selective spectroscopic detection, have had a great impact on the characterization of mineral waters in terms of toxic metal and nonmetal content. Detailed surveys of different approaches are included in several recent specialized monographs and reviews [30–33]. Contamination with different forms of As is of major concern, particularly when the element is present in the trivalent state. Food can be contaminated with As(V) from natural sources as well as from human activities (use of weed killers, pesticides, phosphate fertilizers, and wood preservatives). Organic forms of As, such as monomethylarsonic (MMA) and dimethylarsinic (DMA) acids, arsenobetaine (AB) and arsenocholine (AC), are in general considered of lower or negligible toxicity. Derivatization by hydride generation (HG) is commonly employed as sample presentation procedure for AAS, ICP-AES, and ICP-MS. However, the efficiency of formation of the gaseous covalent hydrides varies with the As form, requiring an intermediate step designed to destroy the organic part of the molecule. This situation is clearly exemplified in a work by López et al. [34] describing the on-line determination of six As species in waters and other environmental samples. Separation of As(III), As(V), MMA, DMA, AB and AC was achieved by high-performance liquid chromatography (HPLC). Before injecting the eluent into the hydride generator and phase separator, a thermoreactor was interposed to oxydize the organic arsenicals with persulfate solutions at 110°C. The gaseous hydrides are generated from As in inorganic form and then swept by an Ar current into an AAS instrument and excited with an electrodeless discharge lamp. Absolute LoDs ranged from 0.4 ng for AC to 0.8 ng for As(V). As expected, the mineral water tested only showed the presence of As(V) at the 50 ng l^{-1} concentration level, due to the prevailing oxidic conditions of the samples.

The impact of highly selective and sensitive mass spectrometric techniques in the field of food analysis, as well as recent advances in the analysis of trace elements of nutritional interest by ICP-MS, have been highlighted in a very comprehensive review authored by Careri and coworkers [35]. To attain ultimate LoDs required in speciation analysis in the case of toxic contaminants, the specificity offered by ICP-MS may be further enhanced by hyphenation. As result of a thorough study carried out by Van Holderbeke et al. [36], a method for fast, baseline separation of four anionic As species [As(II), As(V), MMA, and DMA] and two cationic arsenicals (AB and AC) from mineral water was developed by employing the ICP-MS instrument as detector for capillary electrophoresis (CE). The interface consisted of a stainless steel tee tube that permitted the introduction of liquid from the CE tubing through the horizontal arm and a forced flow of sheath liquid through the lower arm of the tee. The stainless steel device also allowed the CE outlet electrode to be grounded. A steel tube connected the tee with a microconcentric nebulizer. Two different nebulizers were tested, in both cases mounted to a standard expansion chamber. For the electrophoretic separation with ICP-MS detection the buffer was 20 mM borate at pH 9.4 added with an osmotic flow modifier. A similarly prepared solution, but containing 50 mM borate, was employed with UV detection, but this method was suitable only for relatively high concentrations of the analytes. Sample stacking conditions were investigated, that may improve the sensitivity and the effects of the sheath liquid (0.14 M HNO_3) flow rate and of a laminar flow induced on the CE capillary by the natural suction of the nebulizers. The sensitivity of UV detection was insufficient to detect the As species at the concentration levels required ($\mu g\ l^{-1}$). Reported LoDs for ICP-MS detection, defined for a signal-to-noise (S/N) ratio of 10 : 1, were in the 1–2 $\mu g\ l^{-1}$ As range, with migration times from 1.73 min for As(III) to 2.53 min for MMA. Excellent agreement between external calibration and single internal standard addition was found for the As (V) concentration (near 28 $\mu g\ l^{-1}$) measured on a Volvic mineral water sample.

To attain the ultimate LoDs required for speciation analysis in the case of toxic contaminants the specificity offered by ICP-MS may be further enhanced by hyphenation. In addition, the application of a calibration method based on isotopic dilution (ID) is also possible. Heumann and coworkers [31] evaluated the coupling of ID-ICP-MS with IC, reversed-phase (RP) HPLC and size-exclusion (SE) HPLC, as well as capillary column gas chromatography (CCGC) for speciation analysis of elements in mineral, drinking, and environmental waters. The methodology was intended to guarantee the accuracy of results, because variations in the concomitant compounds present in the eluted chromatographic peaks may affect the ionization efficiency of the plasma, impeding the safe utilization of external chromatographic calibration. The authors describe in detail two different ID calibration modes that use a species-specific or species-unspecific spike solution. The second mode was used combined with IC separation to analyze four samples of commercial mineral water and one of drinking water to assess the distribution of iodate, iodide, organic organoiodine, and total I (the latter quantified by direct ID-ICP-MS). Interestingly, the relative abundance of the IO_4^- and I^- species differs among the samples of mineral waters, a result that is ascribed by the authors to differences in ozone treatment during water processing. Anionic organo-I species were detected at low concentrations (sub-microgram per liter) in two samples. The total I content varied between 9.0 and 18.0 μg l^{-1}. Different combinations of chromatography with ID-ICP-MS were also proposed for the speciation analysis of Se.

Numerous methods of determination of As species in beverages employ AAS as the detection technique. However, it does not show sufficient detection power for low levels of As and a preconcentration step is often necessary. On the other hand, while ET-AAS fulfills the requirements in terms of detection power, it suffers from serious interference effects in the analysis of complex matrices, making the use of chemical modifiers necessary. HG in conjunction with AFS constitutes a good alternative for beverage analysis because of its low LoDs and affordability in terms of cost and freedom from maintenance. Speciation analysis of inorganic As in Slovenian bottled mineral waters was addressed by van Elteren and coworkers with two instrumental approaches [37]. In one of the methods, HPLC separation was coupled to HG-AFS. An anion-exchange column was employed with 50 mM KH_2PO_4 at pH 6.1 as the eluent. Formation of gaseous hydrides from the eluted AsO_2^- and AsO_3^- ions was performed with $NaBH_4$ in acid medium as reductant. Spectroscopic detection was carried out with an AFS instrument that employs a hydrogen–air flame excited with a boosted hollow cathode lamp. In the second method, used for comparison, As(III) was determined after selective coprecipitation with dibenzildithiocarbamate. Total As was determined in a similar fashion after a prereduction step to As(III). The same FI configuration employed for HPLC was used, but without including the chromatographic column. The concentration of pentavalent As was then calculated by difference. The trivalent As concentration in waters was found to be lower than the LoD in both methods. The authors also report thermodynamic calculations of the protonation and distribution of As inorganic species as function of pH and E_h.

Finally, a combination of bidirectional electrostacking (a technique related to isotacophoresis and CE) with ET-AAS has been investigated by He et al. [38]. Although it cannot be considered a routine analytical technique as yet, the method offers interesting potentialities for speciation analysis of Cr(III) and Cr(VI) in waters, with respective LoDs of 6 and 5 ng ml^{-1}, arising from preconcentration factors of around 10.

15.3 SOFT BEVERAGES AND FRUIT JUICES

15.3.1 General Considerations

Trace elements levels in soft beverages, fruit juices, and nectars are usually controlled by manufacturers for different reasons, not only related to toxicity. For example, in addition to the possible toxic effects associated with Al, this metal may produce unfavorable flavor. Furthermore, Al can cause some beverage products to lose their color. Fruit soft drinks containing azo dyes may fade in color due to combination with Al. Fruit and vegetables are poor sources of Se. The concentration of this element in fruits is low and highly variable depending on the Se content in soils, the type of fruit and the season in which the fruits are harvested. However, Se levels may be increased in processed products. A safe and adequate daily dietary intake has been established at 10–200 μg per day depending on age [39]. Specifically, the Food and Nutrition Board of the US National Research Council [40] has recommended a dietary Se allowance of 0.87 μg kg^{-1} or, with rounding, 55 and 70 μg kg^{-1} for the reference North American adult female and male, respectively. The main exposure to inorganic Sn for humans arises from food, because trace amounts are present in most natural fruits and other foods. Total Sn content in foods varies over a wide range and differences exist between canned or packaged and fresh foods. Considerably larger amounts of Sn may be present in packaged juices because this metal can migrate from plastic bottles and cans into food and beverages. The presence of oxidizing agents (nitrates), pH, and storage temperature greatly favors the leaching of Sn into the canned material. Since nowadays most cans are lacquered, Sn contamination by leaching has been substantially reduced. Biological effects of absorbed Sn have been reported not only at real toxic levels, but also at sub-toxic levels (e.g., some mg kg^{-1} of food) and may include interactions with other essential elements or with enzyme systems [41].

15.3.2 Applications of Individual Techniques

A study of the influence of ordered media on the determination of Pb by HG-ICP AES in soft drinks was conducted by Valdés Hevia y Temprano et al. [42]. Different organized molecular assemblies (micelles and vesicles) were tested for PbH_4 generation using the ammonium peroxidisulfate–HNO_3 system as the oxidant and Na tetrahydroborate as the reducing agent. This method was characterized by better LoDs (9 ng ml^{-1}) and precision (1.4 at the 50 ng ml^{-1} level) than conventional HG. Lead was quantified directly in commercial fruit juices. It was found in the range 77.3 ± 3.1 to 91.7 ± 4.2 ng ml^{-1}. Further investigations of the same group have focused on the evaluation of other reaction media that may improve LoDs. Commercial fruit juices were analyzed directly for their Pb content by HG-ICP-AES without any sample pretreatment and the results were compared with those obtained by ET-AAS [43]. The generation of PbH_4 was achieved using $K_2Cr_2O_7$–lactic acid as the reaction medium and $NaBH_4$ as the reductant. The LoDs was lowered to 2 ng ml^{-1}. The mixture selected provided high detection power, good precision, and low interference level, and proved to be adequate for the determination of Pb in apple, pineapple, and orange juices at concentrations from 13.0 ± 5.2 to 66.4 ± 7.3 ng ml^{-1}.

Four acid oxidant mixtures were tested and compared for Pb determination in food and beverages by HG-ICP-AES [44]. Apple juice samples were digested with

5 mL of HNO_3 and 5 mL of H_2O_2. Lead is among the analytes hardest to convert to hydride and the chemical yield of the reaction is rather low. For this reason, treatment of the sample with a strong oxidant as ammonium persulfate was necessary. The presence of an oxidizing agent is related to the formation of metastable Pb(IV) compounds before the formation of PbH_4 in the reaction of Pb with $NaBH_4$. The study demonstrated that PbH_4 was formed with 85 percent efficiency in tartaric acid, with a better control of interferences. After hydride conversion, a concentration as low as 4.4 ng ml^{-1} was detected with a precision of 5.4 percent. Lead content in commercial apple juice was found to be 26.3 ± 1.2 ng ml^{-1}.

Aluminum levels in fruit juices and soft drinks were determined by ET-AAS after digestion of samples with an HNO_3–V_2O_5 mixture [15]. Concentrations ranged from 49.3 to 1144.6 μg l^{-1} in fruit juices ($n = 47$) and from 44.6 to 1053.3 μg l^{-1} in soft drinks ($n = 88$). An absolute LoD of 40 pg was reached. According to several authors [45], Al dietary intake must not exceed 6 mg per day to avoid potentially toxic levels. Taking into account estimated data for the consumption of fruit juice (0.036 L per person per day) and soft drinks (0.041 L per person per day) reported by Spanish authorities [46], and considering the mean values of Al detected, the authors estimated a contribution of these beverages to the total dietary Al intake of 156 μg Al per day, an amount that is not considered to represent an imminent toxicological risk.

Interest in procedures that avoid complete sample digestion has grown in recent years. Suspensions of finely powdered samples (slurries) have been used by Zezzi Arruda et al. [47] for the determination of Al in fruit and tomato juices using a semi-on-line approach to introduce samples into a graphite furnace. The automatic slurry sample FIA system designed is claimed to work without the addition of a thixotropic agent or magnetic stirring bar, vortex mixing, gas bubbling, or ultrasonic agitation prior to injection into the atomization cell. Direct slurry measurements on juice samples were performed by simply diluting 1 mL of juice to 20 mL with 0.2 percent (v/v) HNO_3. Aluminium concentration in slurry and liquid phase (filtrate) of pineapple, pineapple and grape, orange, apple, and tomato juices varied from 91 to 769 μg l^{-1}. The metal content in natural orange juice (241 μg l^{-1}) was similar to the level found in processed orange juice (227–242 μg l^{-1}), but there were significant differences as for pH among different fruit juices (orange, lemon, pineapple). While the content of Al in some commercial products was fairly high, it was much lower than that reported for foodstuffs.

García et al. [27] determined Cr by ET-AAS in 90 samples of some of the most common beverages in the regular diet. To prevent the formation of refractory carbides, the graphite tubes and associated Lvov platforms were pretreated with a saturated ammonium molybdate solution and heated several times in accordance with the temperature program used. Samples were treated with an HNO_3–V_2O_5 acid digestion mixture at 120°C for 90 min. Chromium levels in peach, peach and grape, orange, pineapple, apple, and tomato juices ranged from 0.08 to 11.85 μg l^{-1}. In other soft drinks (orange, lemon, soda water, cola, mix of tropical fruits, and tea), Cr was found in the range of 6.04–59.26 μg l^{-1}.

Levels of Mn (15–22 μg ml^{-1}) in commercial pineapple juice were found to be much higher than those of other elements such as Cr, Cu, Fe, and Ni, and much higher than those detected in other juices [48]. FAAS was used for metal determination. A major part of Mn was found in the filtered juice, with only a small fraction in the pulp retained by a 0.4 μm filter. Electron paramagnetic resonance spectroscopy (EPRS) showed that Mn was mostly present as the $Mn(H_2O)_6^{2+}$ species. Concentrations of Mn determined in various pineapple juice samples after different treatment conditions (digested whole juice, digested filtrate, and plain filtrate) ranged from 1.2 to 23 μg ml^{-1}. Chemical characterization of pineapple juices and nectars was achieved by principal component analysis [49] using 23 parameters. Trace metals included as descriptors and determined by FAAS were Ca, Cu, Fe, K, Mg, Mn, Na, and Zn. The study also revealed that pineapple appears to concentrate Mn to a much larger extent than other fruit juices.

Selenium was determined in a juice made from the wild fruit Lantingguo (*Vuccinium uliginosam*) that is characterized for its content of amino acids, vitamins, and essential elements. For the determination of Se by ET-AAS [50], five matrix modifier combinations (Cu–Ni, Pd–Mg, Pt–Cu, Pt–Ni, and Pt–Mg) were tested and compared for suppressing the interfering effects of seven co-existing elements (Ca, Fe, Mg, Mn, K, P, and Zn). According to the study results, Pd–Ni was found to give the best performance. The characteristic mass was estimated to be 28 pg. Speciation analysis of Se in juices has been also reported. Burguera et al. [51] developed a closed system for the determination of Se(IV) and Se(VI) by FIA-HG-AAS with focused MW-assisted on-line prereduction of Se(VI) to Se(IV). The samples and the prereductant solutions [4 M HCl for Se(IV) and 12 M HCl for Se(VI)] which circulated in a closed-flow circuit were injected by means of a time-based devise. Sample preparation was very simple: juice samples were obtained by squeezing the citric fruits and, after filtration through a filter paper. They were kept under refrigeration until analysis. The LoDs were estimated to be 1.0 μg l^{-1} for Se(IV) and 1.5 μg l^{-1} for Se(VI), and the sample throughput was about 50 measurements per hour. Selenium (IV) was the predominant species in all the samples analyzed with a relative content of 71, 100, and 68 percent in lemon, mandarin, and orange juice, respectively.

In an attempt to use milder acidic conditions the prereduction of Se(IV) to Se(VI) was carried out with a mixture of HCl and HBr (10 percent v/v each) instead of HCl alone ($\geq$50 percent v/v) [52]. Experimental parameters were selected by a univariate optimization method. The main advantage of the MW heating was that it allowed for a strict control over the heating power as well as over the time the heating was applied. Seven samples of orange juice were analyzed. Selenium was present in five of them as a mixture of Se(IV) and Se(VI), Se(IV) being the predominant species with concentrations ranging from 5.20 ± 0.08 to 9.50 ± 0.09 μg l^{-1}.

Antimony species, namely Sb(III), Sb(V), and trimethylstiboxide (TMeSbO), were selectively generated and measured in orange juice samples by FIA-HG-ICP-AES with fluoride as the modifier [53]. The influence of different fluoride concentrations on the reduction and prereduction of the species was investigated.

The study showed that at concentrations of 100 mg l^{-1} fluoride and 1.2 percent KI (as the prereducing agent), the formation of stibine from Sb(V) was suppressed, while TMeSbO showed no signal depression. On the other hand, the use of 100 mg l^{-1} fluoride in the absence of KI led to complete signal suppression for Sb(V) with apparent no influence on the signal intensity of Sb(III) and TMeSbO. While the concentration of TMeSbO was measured directly, Sb(III) and Sb(V) concentrations were calculated on the basis of the three analyzing steps, giving LoDs of 1.1, 1.2, and 1.4 $\mu g\ l^{-1}$, respectively

A scheme was proposed by Weber [54] for the separation and identification of different Sn species in lemon juice. Soluble Sn species were separated from other constituents by means of column switching liquid chromatography and ET-AAS was used for detection. Calcium, Cu, Fe, Mg, and Zn were measured on-line in the same fraction by FAAS. Gas chromatography (GC) with MS detection of the organic ligands after derivatization confirmed that citric acid was the relevant ligand for binding metals in lemon juice. Investigation of six commercial lemon juices (with different packing material and origins) showed large variations of total Sn, but few differences among Sn species. Total Sn varied from 0.1 to 142 ng ml^{-1} and the sample with the lowest Sn concentration (not detectable) was the only one delivered in a glass bottle, without any food additive.

Target trace element levels in orange juices have been used to classify and monitor the quality, authenticity, adulteration, and country of origin of the products. A relationship between elemental concentration and region of origin was first reported by McHard et al. [55]. These authors found that Ba concentration was related to the country of origin. In a comprehensive study, the concentrations of 22 trace elements were determined in 482 samples of Australian and Brazilian orange juices and Australian peel extracts using ICP-AES [56]. Samples were treated with HNO_3 in a microwave oven until a clear solution was obtained. Means and ranges of elements were established over a five-year survey. The authors reported that regional differences were apparent for the level of one or more trace elements in the Australian juices, probably related to differences in soils and rootstock. According to the multivariate analysis of trace elements in juices from both countries, a clear differentiation was evident.

15.4 WINES AND OTHER ALCOHOLIC BEVERAGES

15.4.1 General Considerations

Wine is a widely consumed beverage worldwide, with thousands of years of tradition and a remarkable commercial and social value. The evaluation of the quality of wine is a permanent concern for manufactures, merchants, and consumers. The presence of potentially toxic elements in wines is associated with soil contamination and also related to atmospheric precipitation, pesticides, and materials used in production, transportation, and storage. Although at the end of alcoholic fermentation there is a significant reduction of the mineral content, this may not be enough to prevent some problems related to wine stability, namely, precipitation and changes in wine organoleptic properties. Trace elements may exist in wines as free metal ions, as complexes with organic acids, and as complexes with proteins, polyphenols, and polysaccharides. Among the several aspects that must be accounted for in the analysis of wine, the diversity of analytes, their wide range of concentrations depending on the type of wine, and the complexity of the matrix are perhaps the most relevant ones. Wine has been over the years one of the beverages most extensively consumed by man, and is the alcoholic beverage that presents the relatively highest level of Pb. Lead may cause health problems such as encephalopathy and kidney damage. It is not by chance that many of the papers reported in the literature on wine analysis are devoted to Pb determination. The International Office of Vine and Wine (IOVW) has progressively reduced the threshold limit value of Pb in wines down to 200 $\mu g\ l^{-1}$ [57].

Iron is other of the major metallic constituents of wines. According to the EU legislation, Fe content in commercial wines is in the range 0.5–5.0 mg l^{-1} [58], while for domestic and homemade wines these values are higher, reaching 20 mg l^{-1} [59]. Iron content in wines is ascribed either to direct uptake of grapes from the soil they are grown in or to the migration of metallic species from the equipment used for harvesting, conveying, and storing of grapes and the extracted wine [60]. The latter is the most important source of Fe contamination of wine, while natural uptake is limited [61]. At trace levels, Fe is involved in metabolism and fermentation processes, as an enzyme activator, stabilizer, and functional component of proteins. At higher levels, Fe may alter the redox systems of the product in favor of oxidation, affecting sensory characteristics and participating in the formation of complexes with tannins and phosphates that result in product instabilities [60].

The presence of As in wines is attributed to several different factors such as type of soil, use of fertilizers and pesticides, climatic conditions, and manufacturing processes. Legislation concerning As content exists in some countries. For example, in the specific case of Spain, the legislation sets a limit of 1 mg l^{-1} for wines [62]. Furthermore, from an enological perspective, the presence of other elements such as Al may produce bad tastes and the clouding of wines [63].

15.4.2 Applications of Individual Techniques

Since Pb is present in beverages at very low concentrations, there is a need for methods sensitive enough to detect it at microgram per liter levels. Lead HG-AAS has demonstrated to be a valid alternative to ET-AAS, a technique that so far has been more extensively used for Pb determination. However, HG for Pb has traditionally offered poor sensitivity due to difficulties associated with Pb hydride formation, namely, low yield and low stability of volatile compounds. In spite of this it has been possible to generate PbH_4 in aqueous ethanolic media with an LoD of 24 $\mu g\ l^{-1}$ [64] by optimizing the influence of typical HG parameters and considering the influence of wine components (ethanol, tartaric acid, SO_2). HG efficiency decreased markedly in a generation medium containing up to

5 percent ethanol, above which the efficiency varied very little due to changes in pH and in the oxidation–reduction potential of the concomitants. Tartaric acid affects Pb HG efficiency, though to a lesser extent than ethanol. The presence of SO_2 does not interfere with the Pb signal at concentrations up to 1 g l^{-1} of the interferent. The analysis of 14 wines with different characteristics and origins showed good agreement between the proposed and the official analytical method.

Lead has been also determined in wine, other beverages, and fruit slurries by a simple and rapid FIA-HG-AAS method based on the generation of PbH_4 in an HNO_3–H_2O_2 medium [65]. Lead was analyzed directly in wine samples without mineralization or preconcentration, and results were compared with those obtained by ET-AAS. The employed coupling enabled these authors to reach a characteristic mass of 0.8 ng Pb from 100 μL injections. Lead concentrations of 59 ± 1 μg l^{-1} (FIA-HG-AAS) and 58 ± 1 μg l^{-1} (ET-AAS) were reported in red wine.

To improve the sensitivity and selectivity of PbH_4 generation by FAAS, a lactic acid–K_2CrO_4 medium was investigated and applied to the determination of Pb in food and beverages [66]. The increase in sensitivity obtained compared with that reached with ammonium peroxidisulfate–HNO_3 and H_2O_2–HNO_3 was attributed to the presence of lactic acid, which stabilized the Pb(IV) metastable compounds more effectively. These compounds act as a intermediate species in PbH_4 formation. When Pb was determined in wine without mineralization or preconcentration of the sample, no satisfactory results were obtained mainly owing to ethanol interference, which substantially reduced the Pb signal. The addition of 0.5 percent HNO_3 to the wine samples prior to analysis was necessary. The standard addition method was used for calibration and a Pb content of 0.10 ± 0.03 μg ml^{-1} was reported.

A comprehensive review on the application of HG to the AAS analysis of wines and beverages was authored by Baluja-Santos and González-Portal [67]. The paper aimed to describe the fundamentals of HG, evaluate factors causing interferences and their possible elimination, decide which was the spectrometric method best fit to be used in combination with HG systems, and compare the obtained results with those of other techniques for the analysis of wine and beverages. Attention was focused on key elements such as As, Sb, Bi, Pb, Sn, Se, and Te.

The combination of HG with ICP-AES has been also investigated. Marrero et al. [44] compared the effect of four acid oxidant mixtures and two gas–liquid separator designs on the determination of Pb in wine, beer, juice, and food samples. Ammonium persulfate was used as an oxidant agent and the effect of acetic, citric, nitric, and tartaric acids on hydride formation was assessed. The high efficiency of PbH_4 generation (85 percent) obtained with tartaric acid confirmed that the presence of a chelating agent is beneficial to stabilize the intermediate Pb(IV) metastable compounds, thus improving the kinetics of the reaction. In the analysis of wines and soft drinks it was necessary to account for the binding of Pb with organic groups. Consequently, samples were subjected to a preliminary oxidation process with HNO_3 and H_2O_2. The Pb content in red and white wines varied from 44.1 ± 1.9 to 76.1 ± 3.6 μg l^{-1}, while that in white and dark beer was 115 ± 5 and 101 ± 5 μg l^{-1}, respectively.

SIA combined with FAAS was applied to the determination of Cu, Fe, Mn, and Zn in Portuguese Verde, Maduro, and Porto wines [68]. By selecting an appropriate sample volume, the basic configuration of the system allowed concentrations up to milligrams per liter to be determined. The determination of Cu between 0.05 and 0.50 mg l^{-1} was carried out using a heterogeneous phase preconcentration unit composed of a commercial cartridge packed with silica C_{18} coupled to the system. Copper was retained as diethylammonium-*N,N*-diethyldithiocarbamate chelate. The results obtained were found to be in good agreement with those of a reference procedure with a relative standard deviation of less than 5 percent.

Only few papers have reported the determination of Sb in wines [69] because this element is usually present at concentrations close to the LoD of the majority of the spectroscopic techniques. The direct determination of As and Sb in wine by Zeeman ET-AAS was improved when the sample was decomposed with a mixture of HNO_3 and H_2O_2 prior to analysis [70]. Two decomposition systems were tested without losses of analyte. Under the experimental conditions used the concentrations for both elements were below 10 μg l^{-1}. To circumvent the problem of low sensitivity Wifladt et al. [71] generated SbH_3 from diluted wine and collected the hydride directly in a graphite furnace. Thiourea was added to reduce Sb(V) to Sb(III) and Pd was used as modifier in the collection step. The fact that the samples required no decomposition would indicate that Sb was not strongly bound as part of an organometallic compound. By processing 0.3 mL of sample, the combination of HG and ET-AAS allowed an LoD of 39 pg of Sb to be attained, which corresponds to 0.13 μg l^{-1} in wine. Very low concentrations, ranging from 0.6 to 5.7 μg l^{-1}, were found in five different red wines.

Arsenic is an element that raises much concern from the standpoint of human health. It is present in wines and beers at very low concentrations. Uncontaminated beverage samples contain concentrations of several micrograms per liter As. Only a few papers have reported [72] its determination in wine matrices. ET-AAS should be an adequate technique for As detection in alcoholic beverages, but sample decomposition is required in most cases to improve sensitivity and to achieve a quantitative recovery [70]. As already mentioned, AFS is a promising approach because it combines high sensitivity, a wide dynamic range of 4–6 orders of magnitude for concentration, and relatively simple equipment. In this context, samples of seven brands of wines and four brands of beers were investigated by Segura et al. [62] for the total As content. The authors employed a simple and rapid FIA-HG-AFS method, sufficiently sensitive and accurate for the direct determination of the element in untreated samples. LoDs were 0.3 and 0.5 μg l^{-1} for beer and wine, respectively. Arsenic concentrations in the samples analyzed varied from 2.0 ± 0.2 (Rioja Berberana wine) to 5.3 ± 0.1 μg l^{-1} (Damn beer without alcohol).

ET-AAS was applied to the determination of Cd, Cr, and Pb in samples taken during different stages of the wine-making process including grapes, must deposit, deposit of lees, must, and the final product [73]. With the exception of must and wine, which did not require a special treatment, high-pressure

MW-assisted digestion with HNO_3 and H_2O_2 was used to dissolve the metals. Several matrix modifiers were tested during the optimization of the temperature program for the graphite furnace atomization cell: $Mg(NO_3)_2$ and $NaVO_3$ for Cr and $NH_4H_2PO_4$ and $Pd(NO_3)_2$ for Pb and Cd. A nonspecific background absorption was reported which was 1.5–2 orders of magnitude higher for the ammonium-containing modifier than for the Pd compound.

Cvetković and coworkers [74] evaluated a mixed Ni and Sr nitrates matrix modifier for the determination of Se in wines by Zeeman ET-AAS. Samples were heated on a boiling water bath with small amounts of HNO_3 and H_2O_2. To eliminate interferences, especially sulfates and phosphates, Se was complexed with APDTC and extracted into methylisobutyl ketone (MIBK) and the graphite furnace temperature program was optimized for both aqueous and organic solutions. Selenium concentrations up to 0.93 $\mu g\ l^{-1}$ were detected in wines from the Republic of Macedonia.

The interest in assessing the presence of Al in wines is often motivated by its ability to produce bad taste and clouding effects. In an attempt to improve the sensitivity of ET-AAS instrumentation with end-capped transverse heated graphite tubes (ECTHGTs) for Al determination in Port wine, Almedia et al. [63] investigated the performance of several chemical modifiers. Compared with other commonly used and recommended chemical compounds, $K_2Cr_2O_7$ turned out to be better in terms of sensitivity and reduction of blank levels. The concentrations of Al in real samples were relatively high in comparison with other metals and varied from 273 to 803 $\mu g\ l^{-1}$.

Thallium is a nonessential, toxic element that has no known beneficial role and that may be present in wines from regions with naturally high Tl occurrence. In addition, knowledge of the Tl content in wines could be also valuable to identify their provenance. By means of a selective extraction procedure of Tl(I) and Tl(III) from digested samples, followed by ET-AAS measurement in the organic phase, the Tl species were quantified [75]. The authors concluded that the behavior of volatile species during the evolution of the furnace temperature program depended not only on the chemical and thermal properties of the compound, but also on tube geometry, internal gas flow and platform type, among other variables. The developed analytical procedure permitted a 50-fold preconcentration and the determination of 0.05 μg Tl l^{-1} in wine samples. The content of Tl in analyzed wines from Macedonia and Bulgaria varied between 0.1 and 0.8 $\mu g\ l^{-1}$.

Other authors have reported the use of a simultaneous multielement (SM) AAS instrument to identify the content of Pb and other metals in wines. In a 2000 review the principal characteristics and applications of the technique were discussed and the authors made reference to the determination of Cd and Pb in beverages and other matrices [76]. Freschi et al. [77] determined Pb and Cd in 10 white and red table wines. The characteristic mass was approximately 0.6 pg for Cd and 33 pg for Pb and the LoD was found to be 0.03 and 0.8 $\mu g\ l^{-1}$ for Cd and Pb, respectively. The comparative results of diluted and digested wines and recovery values indicated that the simple dilution of samples was sufficient to determine those elements in Brazilian wines. Fernandes and coworkers [78] evaluated Bi as internal standard and a mixture of $Pd(NO_3)_2$ plus $Mg(NO_3)_2$ was employed as chemical modifier to stabilize both the analyte and the internal standard. Wine samples were diluted 1 + 1 (v/v) in HNO_3 before analysis and the electrothermal responses of Bi and Pb from red, white, and rosé wines were compared. The high sensitivity of the technique allowed the authors to reach an LoD of 0.9 μg Pb l^{-1} with a characteristic mass of 45 pg Pb.

Other elements were identified in wines from different origins by applying multi-element techniques. The direct analysis of Al, B, Ba, Ca, Cu, Fe, K, Mg, Na, P, Rb, Sr, V, and Zn by ICP-AES was reported by Thiel and Danzer [79]. The method was applied to 17 white wines from six German wine-making regions. By using alcohol-containing calibration solutions, the results from the direct measurement were comparable with those from the control methods. In addition, typical patterns of elements obtained by the multicomponent analysis could be evaluated by multivariate data analysis to recognize the origin of the products.

The presence of yeast in beverages, as happens with certain wines, may contribute to the potential incorporation of Hg, a toxic element for humans. Capelo and coworkers [80] developed and investigated the performance of two different reactors for sample ozonization as a previous step to Hg determination in white wines by FIA combined with cold vapor (CV) AAS. The pretreatment with O_3 allowed the drastic reduction of the amount of chemical reagents used in the FIA-CV-AAS system. An LoD of 0.5 μg l^{-1} was obtained and the Hg content in four analyzed wines varied from 94 ± 9 to 101 ± 5 μg l^{-1}.

An Ar stabilized direct current (DC) arc plasma at atmospheric pressure was employed to determine Cu, Fe, and Mn in wine [81]. The experimental conditions were optimized by studying the lateral distribution of the spectral line intensities of the elements assessed in aqueous and ethanolic–aqueous solutions. The method was applied to quantify Cu, Fe, and Mn in six wines from three Serbian grape-growing regions and the accuracy was checked by FAAS.

ICP-MS is a multi-element technique suitable for the analysis of liquid samples, which provides higher selectivity and detection power and LoDs lower than other multi-element techniques, such as ICP-AES. These characteristics make ICP-MS an excellent tool for the detailed characterization of the elemental composition of beverages. In this context, ICP-MS was the technique selected for the determination of 31 trace elements in wine at concentrations ranging from 0.1 to 0.5 ng ml^{-1} [82]. Samples were diluted 1 + 1 and it was investigated whether a matrix effect derived from the presence of ethanol could be overcome by using a microconcentric nebulizer with membrane desolvation. The authors compared their results with those obtained using a conventional Meinhard nebulizer and concluded that the matrix effect observed could only be minimized by using an internal standard with the Meinhard nebulizer, but not with the other one.

ICP-MS allows different quantification procedures to be used depending on the accuracy and precision required. The suitability of the semiquantitative mode of analysis of an ICP-MS instrument for multi-element characterization of fortified and table wines was evaluated by Almeida and Vasconcelos [83]. As compared to table wines, fortified wines are much richer in alcohol content (19–22 percent volume), particles in suspension, and polymeric organic compounds (particularly sugars). Samples were first subjected to a UV-irradiation pretreatment because simple dilution of the wine was considered unsuitable for ICP-MS analysis. This more drastic sample treatment had been already successfully used by the authors [84]. The same samples were analyzed in parallel using the quantitative mode of the same equipment. The comparison of the results showed that, for most of the 30 elements analyzed, the semiquantitative mode provided two to six times higher LoDs than those obtained by the quantitative mode, while the precision of results was similar. The authors remarked that semiquantitative analysis was faster, required fewer reagents and could be considered an important tool for comparison of elemental composition of wines, including fortified wines. Almeida and coworkers [85] also analyzed 26 elements in three wine samples and compared the results using UV-irradiation pretreatment and direct injection of samples by means of a micro-concentric nebulizer (MCN). Both methodologies provided similar LoDs and precisions for most of the elements considered. Comparable results were obtained for Ba, Co, Cu, Li, Ni, Pb, Rb, Sr, V, and Zn in the three selected wines with relative differences in concentrations lower than 10 percent in most cases. For rare earth elements (REEs) the differences observed were slightly higher, while for As and Ga no comparable results were achieved. In a very interesting and thorough paper, Augaugneur et al. [86] optimized the performance of an MCN regarding the nebulizer gas flow rate and sample uptake rate in terms of signal intensity and S/N ratio for the direct determination of REEs in undiluted wine samples. The use of a MCN made it possible to suppress matrix interferences, thus increasing the detection power of the direct analysis. In addition to the considerably reduced sample consumption (30–50 μL), a gain in sensitivity allowed the authors to determine REEs directly without the need for either sample preconcentration or expensive ultrasonic nebulization. Jakubowski and coworkers [87] reported that the concentration pattern of REEs may be strongly affected by the wine-producing process and therefore it is generally unsuitable as a fingerprint for the provenance of wines. They concluded that the contamination from bentonites of different origins is mostly responsible for the presence of REEs in wines. Bentonites are used in the purification of wines from tarnishing components such as proteins.

Wine is a very complex matrix and the accurate, selective determination of species constitutes a challenge for analytical chemists. Furthermore, the speciation analysis of metals bound to biological ligands is a subject of increasing interest since complexation may reduce their toxicity and bioavailability. There is a limited number of studies concerning the speciation analysis of metals or metalloids in wines. Arsenite, arsenate, MMA, and DMA were separated in less than 10 min by means of an anion-exchange column [88]. Arsenic species detection was accomplished by the direct coupling of the column effluent to an HG system and AFS was used for detection. LoDs in white wine were 0.16, 0.33, 0.32, and 0.57 ng ml^{-1} for As(III), DMA, MMA, and As(V), respectively. In real samples of white and red wines, only inorganic species of As were detected with concentrations ranging from 1.41 $\pm$ 0.20 to 12.18 $\pm$ 0.26 ng ml^{-1}.

As regards fractionation methodologies for the selective determination of Pb and other metals compounds in wines, Lemos et al. [57] have developed a FIA method for the direct determination of available free Pb(II) and total Pb content in wine samples. Lead (II) was chemically retained on a packed polyurethane foam microcolumn, modified by immobilization of 2-(2-benzothiazolyazo)*p*-cresol (BTAC) and eluted with 0.1 M HCl. The eluate was determined on-line by FAAS. Total Pb was quantified after sample digestion with HNO_3 and H_2O_2, whereas free Pb was determined by direct sample on-line preconcentration and elution. No effect of ethanol on the enrichment procedures was observed. With a preconcentration factor of 26 an LoD of 1 $\mu g\ l^{-1}$ was reached. Four red and three white wines of different origins were analyzed and the total Pb content varied from 8 to 42 $\mu g\ l^{-1}$.

The presence of Fe in wines produces changes in stability and has a direct effect on its oxidation and ageing. Elevated concentrations of Fe produce undesirable turbidities. Iron is present in different forms depending on the conservation of the wine. When wine is protected from air, Fe(II) is the predominant species, while Fe(III), at much lower concentrations, is usually present as part of complexes. Investigations on the fractionation of Fe in wines are also worthy of note, as follows. SIA in conjunction with FAAS was proposed for the determination of Fe(III) and total Fe in Portuguese table wines [89]. The determination of Fe(III) was based on the extraction with MIBK of the complex formed between the metal and thiocyanate using an on-line liquid–liquid extraction module. A small sample aliquot was driven directly to the instrument for total Fe quantification. The results from the analysis of 10 red and white wines for Fe(III) and total Fe were compared with those given by a reference method [colorimetric thiocyanate method for Fe(III) and FAAS for total Fe]. Relative deviations of less than 10 and 5 percent were found for Fe(III) and total Fe, respectively. A micelle-mediated methodology was used for the determination of free and bound Fe in wines by FAAS [60]. The method was based on the use of the cloud point phenomenon for the determination of free and tannin-bound Fe. Tannins and other phenolic and insoluble compounds were precipitated with the aid of a nonionic surfactant mixture (TX-100 and TX-45) upon increase of temperature and subsequently separated by centrifugation. The surfactant-rich phase containing the tannins and the insoluble Fe fraction was directly aspirated into the nebulizer of the instrument after its uptake with a methanolic solution of HNO_3. The same cloud point extraction procedure was followed with the supernatant to determine free Fe species in the presence of a chelating agent with the aim of forming water-insoluble complexes with free Fe. The complexes extracted in the micelles were directly analyzed after dilution in a methanolic solution. The study revealed that a substantial amount of Fe in wine exists as insoluble complexes with tannins and other related compounds. The speciation analysis of suspended and dissolved Fe in 10 commercial and homemade wines was compared with a standard measurement method.

Karadjova and coworkers [90] in a detailed and comprehensive investigation established a scheme for fractionation of wine components and Cu, Fe, and Zn determination in the different fractions. Like Fe, the other two metals may analogously exist in wines as free ions, as complexes with organic acids and as complexes with proteins, polyphenols and polysaccharides. The resin XAD-8 was used for the separation of wine polyphenols. Dowex ion exchange resins were used for the separation of cationic and anionic species of metals that were subsequently quantified off-line in Bulgarian and Macedonian wines by FAAS or ET-AAS (depending on their concentration levels).

Ultrafiltration was applied to examine the size fractionation of Al, Ca, Cu, Fe, K, Na, and Pb in white and red wines [91]. Metal determinations were performed on the unfiltered wine, the 0.45 μm membrane-filtered wine and each ultrafiltrate fraction. Aluminum was determined by ET-AAS, while FAAS was employed for Cu and Fe. An electroanalytical technique, stripping potentiometry, was selected for Pb measurement, whereas flame photometry was chosen for K and Na quantification. Fractionation patterns were evaluated and discussed. Castiñeira et al. [92] combined on-line tangential-flow multistage ultrafiltration with a home-built carbon analyzer and ICP-MS for size fractionation of nonvolatile dissolved organic compounds and metal species in three German white wines. The study showed that the major part of the elements investigated (up to 25) were dissolved in the size fraction of <1 kDa, with the exception of Ba, Pb, and Sr, which also appeared in other fractions.

The determination of isotope ratios in wine by ICP-MS was carried out by several researchers. As regards Pb, the respective proportions of the four stable isotopes – originated from the genesis of the rock and ore deposits – vary with geological ages and consequently with geological locations. The precision offered by ICP-MS in the measurement of Pb isotope ratios was enough to differentiate natural variations of Pb isotope abundance in wines [93]. Lead isotope ratios in 24 samples of Port wine of different ages and characteristics were determined by Almeida and Vasconcelos [94]. Samples were pretreated by UV irradiation, filtered, and diluted 10 times for the subsequent ICP-MS determination. No significant differences were found between the results reported with two types of corrections (external correction with a Pb isotopic standard solution and internal correction with Tl as internal standard) and both methods were suitable for mass bias correction. Lead concentration varied from 47 to 804 $\mu g\ l^{-1}$ and a significant decrease was observed in Pb concentration as the age of the wine increased.

Different analyzers [quadrupole (Q), multicollector magnetic sector field (SF), and time-of-flight (TOF)] of ions produced in an plasma source were employed by Barbaste et al. [95] to evaluate the accuracy of the determination of Pb isotope ratios in 20 wines of different origin, including two reference wines, with concentrations ranging from 7 to 140 μg Pb l^{-1}. When TOF was evaluated, no acquisition of data was observed for wines containing less than 50 $\mu g\ l^{-1}$ owing to the poor sensitivity of the analyzer and matrix signal suppression. Good agreement on the level of accuracy was attained with the three instrumental approaches tested.

A method based on ICP-MS using an SF mass analyzer with multicollector unit was employed for the precise determination of $^{87}Sr-^{86}Sr$ in 11 wines [96]. To avoid the $^{87}Sr-^{87}Rb$ isobaric interference, Sr was separated from Rb and the Sr-rich fraction was introduced into the plasma. Precision values of 0.002–0.003 percent were routinely obtained. To check the accuracy, a CRM namely, the NIST SRM 987, was analyzed, measurements being within +0.02 percent of the target value.

Automatic systems based on FIA were also implemented for wine analysis. In a comprehensive review, Ferreira and coworkers [97] remarked that some systems had limitations as many of them were only tested with a particular wine type or demanded a previous treatment of the sample before injection. The lack of robustness of some manifold components (tubing of peristaltic pumps, some types of injection devices) was also thought to prevent the extensive use of FIA in industrial laboratories [98]. SIA has been proposed as a mechanically simple alternative to FIA [99]. As previously stated, SIA is based on the sequential aspiration of well-defined sample and reagent zones into a holding coil by means of a multiposition valve. The flow is then reversed and the stacked zones are mixed and propelled to the detector, where the reaction product is monitored. As already described for other beverage matrices (water, juices), the SIA of wine has been developed in recent years for the determination of more than 20 species and several aspects of these systems were reviewed by Segundo et al. [100] in a recent paper. The authors focused on the implementation of in-line treatment and the adaptation of system operation through software control to enable determination in different kind of wines.

The elements and compounds present in wines are related to grape varieties, soil, climate, culture, yeasts, and wine elaboration. In this context, different physical–chemical parameters, such as volatile compounds, organic acids, amino acids, isotopic parameters, and major and minor metal contents, in conjunction with chemometric procedures, have been used for the classification and monitoring of the quality, authenticity, and geographic origin of wines. However, given the direct relationship between soil composition and the capacity of grapes to absorb mineral substances, metals and metalloids are the best choice to differentiate wines according to the geographical origin. From this point of view, it appears clear that MS and AES techniques play a relevant role in supporting those activities. Since the pioneering studies of Kwan, Kowalski, and coworkers [101–103], numerous multivariate techniques were used to resolve wine identifcation problems based on diverse chemical descriptors. Rebolo et al. [96] applied a classification system to confirm the authenticity of Galician Certified Brand of Origin (CBO) wines from the Ribeira Sacra area. Thirty-four chemical variables, including metals and volatile and phenolic compounds, were determined in 39 red wines. The key metals selected on the basis of previous experiences were Li, Co, Fe, Mn, Na, Ni, Rb, and Zn [104, 105]. Potassium, Li, and Na were measured by AES, while AAS was employed for the quantification of the remaining elements. Through multivariate chemometric procedures wines were classified on the basis of selected chemical data. The procedure performed well in terms of classification, prediction, and differentiation among similar products originated in nearby geographical areas. In the same context Gonzáles-Larraina et al. [106] reported that K, Li, Fe, and Mn were the most important elements to classify the three areas of Rioja where wines are produced.

Trace elements namely, Al, B, Ba, Ca, Cu, Fe, K, Mg, Mn, Na, P, Rb, Sr, and Zn were measured by ICP-AES to classify wine samples from six regions using a three-layer artificial neural network (ANN) model with back-propagation (BP) of error [107]. The results obtained with ANN were compared with those obtained by cluster analysis (CA), principal component analysis (PCA), the Bayes discrimination method, and the Fisher discrimination method. A satisfactory prediction result (100 percent) was obtained with ANN for the classification of wine samples comprising six categories. As described by Frías et al. [108], CA and PCA showed differences in commercial wines from the Canary Islands (Spain) according to the island of origin and the ripening state of the grapes. Calcium, Cu, Fe, K, Mg, Mn, Na, Sr, and Zn were determined in dry and sweet wines by FAAS and Li and Rb by FAAS. The linear discriminant analysis gave 100 percent recognition ability and 95 percent prediction ability. The sensitivity and specificity reached using soft independent modeling of class analogy (SIMCA) as a multivariate technique were both 100 percent.

In a comprehensive study, Marengo and Aceto [109] analyzed 56 elements by ICP-MS in 68 samples of red Nebbiolo-based wines produced in different zones of Italy. Samples were analyzed without any preliminary treatment. A 1 + 4 dilution with 1 percent HNO_3 was performed according to procedures recommended by several authors. On the basis of the element distribution, chemometric techniques of discriminant analysis (DA) and PCA were applied to distinguish the five classes of wines from the province of Cuneo (Piedmont). The variables with the most discriminating power were found to be Mg, Mn, Mo, Si, and Ti concentrations. According to the data obtained, the authors considered that these variables are mainly influenced by the mineral properties of soil and/or the uptake capacity of grapes. Again, their results proved that the determination of metals and metalloids could be a fast way to characterize and classify wine samples. Two kinds of sparkling wines, namely, cava and champagne, were analyzed for their metal content and differentiated when applying pattern supervised learning recognition techniques such as linear discriminant analysis (LDA) and SIMCA [110]. Before analysis, samples were degassed and subjected to a digestion procedure with H_2O_2 and H_2SO_4. Aluminum, Ba, Ca, Cu, Fe, K, Mg, Mn, Na, P, Sr, and Zn were determined by ICP-AES. Measurements of Cd, Ni, and Pb were performed by ET-AAS, while As was determined by HG-AAS. The study showed that the differences between both classes of wines arouse mainly from the descriptors As, Cd, Cu, Na, Ni, P, Pb, Sr, and Zn and that the chemometric procedures used gave excellent classifications with a percentage of hits in both recalling and prediction of 100 percent.

15.5 BEERS

15.5.1 General Considerations

For centuries, beer has been one of the most widely consumed beverages throughout the world. In 1999, the estimated average beer consumption *per capita* around the world was 79.5 L, reaching 150 L in some countries [111]. Pilsen, ale, lager, stout, and porter are the most consumed types. Low alcoholic beers represent <1 percent of the total beer consumption, but this percentage is growing fast. The presence of trace metals in beer may be traced back to natural sources (soil, water, cereal, hops, and yeast), possible environmental contamination as a consequence of fertilizers or pesticides, industrial processing, and containers. The importance of determining trace elements in beer arises not only from their possible effect on humans, but also because they can influence the brewing process and affect the stability, quality, and organoleptic characteristics. The compounds responsible for the flavor of beer are most likely oxidation products which have been linked to the presence of oxygen and metal ions such as Cu and Fe [112]. A review on nutritive and energetic values of domestic beers [113] reports that beer contains small percentages of vitamins and significant proportions of recommended daily allowances of trace metals and minerals. The potential presence of toxic elements is also accounted for. In the case of As, several countries have specific legislation concerning tolerance limits in beer. As an example, this level was fixed at $100\ \mu g\ l^{-1}$ in Spain [26]. Ireland and the UK regulations prescribe $500\ \mu g\ l^{-1}$ and Hungary and Slovenia $200\ \mu g\ l^{-1}$ [114].

15.5.2 Applications of Individual Techniques

Conversion into hydrides was required in several procedures to increase detectability of most trace metalloids because of their low expected levels in beer samples. Even though HG techniques combined with atomic detectors have been widely used for the determination of elements in food, relatively few publications are devoted to the analysis of beer samples.

AFS and AAS were evaluated for the determination of As in beer and wine by HG sample introduction [115]. The direct generation of AsH_3 from inorganic As in the presence of KI and the subsequent determination by AFS made possible the fast determination of total inorganic As with a simple pretreatment of samples. It consisted of degassing followed by 1 : 10 dilution with HCl in the presence of an antifoaming agent and KI [116]. The LoD was $39\ ng\ ml^{-1}$. Data found for total As in commercial samples varied from 1.01 to $12.89\ ng\ ml^{-1}$, in relatively good agreement with those obtained after previous sample ashing.

In an attempt to determine low levels of As in beers by ET-AAS, Cervera et al. [117] introduced a preconcentration step that enhanced the sensitivity by a factor of 2.5, reaching an LoD of 0.7 ng of As per gram of beer. Samples were degassed and subjected to a preliminary dry ashing using $Mg(NO_3)_2$ and then evaporated until total dryness. A slurry was prepared by adding $100\ \mu L$ of HNO_3

and a chemical modifier solution of ascorbic acid and Ni to the ashes. Treated samples were injected directly into the Lvov platform. Zeeman correction was necessary to compensate for the high background produced by the matrix. Eleven beers were analyzed by HG-AAS and ET-AAS and results were comparable. The concentrations of As found ranged from 3 $\pm$ 1 to 65 $\pm$ 3 ng g^{-1}. An approach was described for the separation and determination of four As species using HPLC combined on-line with an HG system and AFS for the detection [118]. Inorganic (arsenite and arsenate) and organic (MMA and DMA) species were separated on a strong anion-exchange column using KH_2PO_4–K_2HPO_4 as the mobile phase. Prior to chromatographic analysis, beer samples were degassed by filtration and cleaned by passing through a C_{18} cartridge. Under the best experimental conditions, LoDs were found to be 0.12, 0.20, 0.27, and 0.39 μg l^{-1} for As(III), DMA, MMA, and As(V), respectively. The analysis of three samples evidenced that As(V) was the main species, with a maximum concentration of 6.0 $\pm$ 0.5 μg l^{-1}, followed by As(III), MMA, and DMA in that order.

Arsenic, Hg, Sb, Se, and Sn were determined in untreated beer samples by direct HG-ET-AAS using a batch system [119]. Elements were preconcentrated *in situ* onto the Pd- (for As, Sb, Se, and Sn) or Au-pretreated (for Hg) interior wall surface of a graphite furnace. The authors considered beneficial to degas beer samples by filtration and to generate the hydrides in the presence of an antifoam agent. An extra gas–liquid separator was necessary to minimize the amount of moisture from the reaction vessel reaching the graphite tube. With the combination of HG and ET-AAS LoDs of 28, 90, 21, 10, and 50 ng l^{-1} were reached for As, Hg, Sb, Se, and Sn, respectively.

A simple and rapid FI-HG-AAS method was developed and optimized for the determination of Pb in beer and other beverages by Cabrera et al. [65]. Beer samples were degassed in an ultrasonic bath prior to analysis in order to improve the precision of the determinations. Then they were acidified, introduced into a microwave oven and cooled in an ice bath to reduce the amount of water vapour. The Pb content in beer was found to be 41 $\pm$ 1 μg l^{-1}. Lead was also determined in beer by HG-ICP-AES and the effects of four acid oxidant mixtures on PbH_4 generation were evaluated [44]. Beer samples were treated with HNO_3 and H_2O_2 and evaporated to dryness before addition of tartaric acid. Lead HG was performed with 85 percent efficiency in a tartaric acid medium. Concentrations of 115 $\pm$ 5 and 101 $\pm$ 5 μg l^{-1} Pb in white and dark beers, respectively, were reported.

Pohl and Prusiz [120] used and compared the performance of the chelating resin Metalfix Chelamine and the strong cation exchanger resin Dowex 50W-X4 for the off-line preconcentration of Cd, Co, Cu, Ni, and Zn. The cation exchanger was selected as the best alternative for the analysis of commercial beer samples by ICP-AES. The preconcentration procedure allowed the selected metals to be determined at concentrations in the order of 1 ng ml^{-1}.

Twenty-three elements, including almost all the essential and toxic metals or metalloids such as As, Ag, Cd, Hg, Pd, and Tl, were quantified in 35 types of bottled and canned Polish beers by double-focusing SF-ICP-MS with ultrasonic nebulization after MW-assisted digestion of the samples [121]. Negligible levels of the more toxic elements were found in the samples tested. The interdependences among different trace elements were examined using the PCA method.

The characterization of beers according to their mineral content using various chemometric procedures has been investigated. Bellido-Milla and coworkers [122] used the concentrations of nine trace elements to discriminate and classify 25 beer samples. Aluminum, Ca, Cu, Fe, Mg, Mn, and Zn were determined by FAAS, and K and Na measured by flame AES after MW-assisted acid treatment. The concentrations of trace elements together with molecular absorption spectra provided information on the inorganic and organic components of the products. Alcázar and collaborators [123] studied the characterization of beers according to their concentrations of Al, B, Ba, Ca, Fe, K, Mg, Mn, Na, P, Sr, and Zn. ICP-AES was the technique chosen to analyze 32 mineralized samples corresponding to three types of beers (lager, dark, and nonalcoholic). Major concentrations of K and P appeared in all the samples, with average values of 406–216 mg l^{-1}, respectively. Calcium, Mg, and Na were present at lower concentrations, with respective average values of around 57, 87, and 39 mg l^{-1}. The remaining elements were present at concentrations lower than 1.0 mg l^{-1}, Ba and Zn being the metals with the lowest concentrations. The determined elements were considered as chemical descriptors in order to apply pattern recognition procedures, including display and supervised learning methods, such as LDA and ANN.

A large amount of beer produced in different countries is packed in cans. In spite of the importance of Al from the toxicological point of view, very few studies have been conducted to assess the can corrosion by beer. Vela and co-workers [124] demonstrated that beer cans made from Al are corroded over time by the canned product and that corrosion may be reduced through refrigeration. The researchers compared Al levels in fresh and stored canned beer representative of US quality draft. ICP-AES was employed for the analysis. A 2 × 2 × 4 statistical experimental design was used for two brands, stored at two different temperatures, 23°C (room temperature) and 5°C (refrigerated) for five months. The analytical results showed that beers stored at room temperature contained more Al than refrigerated beers. One of the brands tested contained significantly more Al (546 μg l^{-1}) than the other (414 μg l^{-1}) at the end of the storage period.

15.6 SPIRITS

15.6.1 General Considerations

Spirits are beverages of high alcoholic content, generally produced by fermentation of carbohydrate-containing cereals, followed by distillation. The studies concerning the determination of metals and metalloids in spirits are mainly related to a typical Brazilian sugarcane-derived spirit called cachaça. The inorganic fraction of this popular beverage is constituted by metallic ions, such as, among others, Al, Cd, Ca, Co, Cu, Cr, Fe, Hg, K, Mg, Mn, Na, Pb, Sb, Sn, and Zn. The presence of these elements in cachaça, as well as in other similar beverages, is attributed to the production process or to incorporation from soil to the raw material. The complete and accurate chemical composition of the spirit remains unknown, but the type and concentration of the different components vary depending upon the geographical region where the sugarcane is grown.

15.6.2 Applications of Individual Techniques

Silva et al. [125] developed a strategy for implementing the standard addition method (SAM) in a simple two-channel flow system based on merging zones and gradient exploitation. An advantage of this approach is worth noting: It requires only one standard solution and provides several addition levels using a single injection. The determination of Cu in spirits by FAAS was selected as application. The novel SAM system handled about 60 samples per hour and yielded precise results with a relative standard deviation <5 percent regardless of the ethanol content. The method was tested with synthetic samples. Other investigations relying on a similar methodology have been conducted for metal determinations in cachaça. By using a flow-batch system that presents the intrinsic favorable characteristics of both modalities, Almeida et al. [126] performed fast SAM with sufficient flexibility and versatility and employed only one standard solution per analysis. Using the proposed method, Cu was determined in cachaça by FAAS. This method is recommended for this kind of analysis because this sugar-derived alcoholic beverage presents a highly variable composition that may cause matrix effects, and even determinations that use matrix-matching standards can yield inaccurate results. By using five standard addition levels, the sample throughput reported was 70 h^{-1} and the consumptions of sample and standard solution were 1.5 and 0.5 mL per analysis, respectively.

Araujo Honorato et al. [127] have attempted to find out which methodology is more appropriate for calibration in the determination of Cu in sugarcane spirits when using AAS for detection. To select the best experimental design for the application of SAM, the influence of some factors was demonstrated, such as the number of additions and concentration levels, the location of the levels and the average concentration of the standard additions. The design with six standard additions, four concentration levels located near the inferior and superior levels, and the average concentration of the standard additions closer to zero led to a calibration method with an adequate compromise between precision, cost, and time of analysis. The proposed strategy can be applied to other instrumental techniques and samples.

Chromium and Sb have been determined in cachaça by ET-AAS using matrix matching calibration and Ru as a permanent modifier [128]. Calibrants containing both elements were prepared in a 40 percent v/v ethanol solution and spiked samples were used to check accuracy. LoDs of 2 and 43 pg were achieved for Cr and Sb, respectively. The results obtained from the analysis of 52 samples varied from nondetectable to 39.1 $\mu g\ l^{-1}$ for Sb and between 0.64 and 1.53 $\mu g\ l^{-1}$ for Cr.

15.7 INFUSIONS

Infusions included in the normal human diet are the product of the aqueous, hot extraction (brewing) of materials of vegetal origin (teas, coffee, herbs). The commercial scene is certainly dominated by tea and coffee of different varieties and origins and to a lesser extent by cocoa derivates. Perhaps as a result of the (apparently) wide acceptance of the ancient Chinese proverb Better to be deprived of food for three days, than of tea for one, tea may possibly be considered the most consumed beverage after water. There is a large number of countries that are tea producers, the major ones being in Asia and Africa. As one of the most important commodities in international trade, coffee also plays a relevant role in the economy of many developing countries. It is known that both coffee and tea infusions contain essential mineral components as well as numerous complexing organic compounds able to produce metal–ligand interactions that, in turn, affect the bioavailability of the former. The health benefits of tea were already documented in ancient Japanese and Chinese texts. Mineral components of tea include, among others, Al, B, Cu, K, Mg, Mn, Na, P, and Se. Tea infusions can be an important source of Mn, a metal related to bone growth, and the large amount of K contained in comparison with Na could be beneficial for hypertensive individuals. Boron is considered a stimulant of the brain function and together with other mineral components, such as Ca and Mg, in conjunction with vitamin D, may help to prevent osteoporosis. Organic compounds, such as metal binding polyphenols, are important to determine the properties of the infusions. Tea infusions are the principal source of Al dietary intake in some populations, a fact that has caused concern due to the association of that metal with neurodegenerative diseases, such as Alzheimer's. However, Al bioavailability is reduced in infusions because the metal is fully or partially bound in organic moieties, probably polyphenols. As for the leaching of metals during the brewing process, the different composition of fermented and non-fermented teas, as well as that of coffee of different varieties (green, roasted, soluble) and origins, has prompted metal speciation studies based on hyphenated methods with atomic spectrometric detectors.

15.7.1 Applications of Individual Techniques

Aluminum is one of the metals known to be present in tea and coffee infusions. The determination of this element at trace levels by FAAS methods usually requires preconcentration. Salacinski et al. [129] describe the application of a FIA method coupled to FAAS. An on-line cation exchange column is used for analyte enrichment. Tea and coffee infusion samples and tea leaf digests were employed for the evaluation. Organic components were prevented from interfering with the determination by precipitation in acid medium and separation of the solid phase by centrifugation. Flow-injection variables and N_2O–acetylene flame conditions were optimized. The Al concentrations found varied according to the type and origin of the tea analyzed. Values ranged from 299 ± 8 to 552 ± 38 μg ml^{-1} (for 5 percent m/v tea digests). Two tea infusions of 4 g leaf in 100 mL gave Al concentrations of 4.4 ± 0.6 and 12.6 ± 0.8 μg ml^{-1}. A ground coffee bean infusion gave 7 ± 1 μg Al ml^{-1}. The method is claimed to be subject to few potential interferences when preconcentration is carried out at pH lower than 4.0.

The limits of quantifications (LoQs) achieved by flame methods are often sufficient to characterize the composition of tea raw materials. By employing FAAS and also ICP-AES, Costa et al. [130] compared heating extraction procedures for Al, Ca, Mg, and Mn from tea leaves. By applying a focused-MW procedure, the extractants tested included diluted acids and solutions of water-soluble alkaline tertiary amines. Calcium, Mg, and Mn showed recoveries between quantitative and semiquantitative. The recovery of Al was generally poorer, a fact that is ascribed to the binding of this element to organic compounds that constitute the sample matrix. Variations were also observed in trace concentrations of elements extracted with different media from various types of teas. Total concentrations of Al, Cd, Ca, Cu, Fe, K, Mg, P, and Zn in tea leaves and hot distilled water extracts, measured by atomic spectrometric methods, have been also reported by Csikkel-Szolnoki and coworkers [131]. Small differences in metal concentrations in teas from different sources were observed. It was also suggested that potentially toxic effects of Al are greatly reduced by its presence in complexed forms rather than as free ionic metal. A procedure aimed at enhancing the sensitivity and freedom from interferences of FAAS is described in a paper by Narim and Soylak [132]. The methodology was based on the formation of metal chelates with ammonium pyrrolidine carbodithioate (APDC), subsequent collection of the complexes on a cellulose nitrate membrane filter, dissolution of the filter in a small volume of a suitable acid medium, and analysis of the solution by FAAS (air-acetylene flame). Microvolume (100 μL) sample injection was applied for sample introduction. Black tea samples were previously digested in HNO_3 with an MW-assisted digestor. The analytes studied were Cd, Co, Cu, Ni, and Pb. The highest concentration in a black tea sample was found for Pb ($1.0 \pm 0.1\ \mu g\ g^{-1}$). Cadmium concentration was found to be below the LoD ($0.03\ \mu g\ g^{-1}$). The influence of alkaline and alkaline earth metals, of some transition elements and of two anions (Cl^-, SO_4^{2-}) was very small or negligible.

Filgueiras and coworkers [133] developed a method of sample preparation aimed at the solubilization of the metals extracted from plant tissue, and applied it to the CRM GBW07605 Tea Leaves. The plant material was ground to particle size <50 μm and suspended in a dissolution medium consisting of 0.1 M EDTA at pH 10. A probe ultrasonic processor operating at 20 kHz and 100 W subjected the samples to high intensity ultrasonication during 3 min. After centrifugation, the supernatant was assayed for Ca, Cd, Mg, Mn, Pb, and Zn using FAAS and ET-AAS.

The multielement advantage ICP-AES was further demonstrated for the characterization of tea beverages by Fernández et al. [134]. The aim of the study was to establish the differences in metal content of 55 classes of teas, including some instant teas and several related soft drinks. Eleven elements (Al, Ba, Ca, Cu, Fe, K, Mg, Mn, Na, Sr, and Zn) were determined using several analytical lines per element and checking for matrix effects by the standard addition method. Differences among the beverages were discriminated by applying chemometric techniques.

As it has already been discussed, one important aspect regarding the bioavailability of metals is its dependence on the molecular moieties involving the binding of the elements. In tea infusions metals are generally bound to large organic molecules of natural origin, forming coordination complexes that participate in processes of absorption by the digestive tract. It is also known that the separation and isolation of those original species affect their distribution. These facts have been considered by Ødergård and Lund [135] in a paper describing the speciation analysis of metals in tea infusions by ICP-MS. Information on the extraction efficiency of 24 elements from tea leaves and tea infusions was obtained by two different procedures. With one procedure that employs solid-phase extraction with cation-exchange cartridges, the charge of the dissolved species was studied. The other method involved the use of size exclusion chromatography (SEC) to obtain complementary information on the molecular size of the metal-organic complexes. Decomposition of tea leaves was achieved by oxidation of the samples with HNO_3 and H_2O_2 in a polytrifluoroethylene (PTFE) bomb placed in a microwave oven. Infusions from a commercial brand of tea bags were obtained with boiling distilled water. The operating parameters for ICP-MS were optimized for calibration and quantitative determination of Ag, Al, As, Ba, Be, Bi, Ca, Cd, Co, Cr, Cu, Fe, Li, Mg, Mn, Pb, Rb, Se, Sn, Sr, Tl, V, and Zn. Rhodium was employed as the internal standard. Elemental concentrations in tea leaves and infusions are reported, together with percent ratios of soluble/total metals and noncationic soluble metals. The estimated molecular weights of the molecular compounds of several metals obtained from SEC are also reported. Interestingly, it was found that Al presents the more pronounced noncationic behavior, appearing associated to large molecular structures in the 400–6000 and 6500–8500 Da size ranges.

The extraction efficiency of about 40 elements from tea leaves was thoroughly investigated through an integral approach by Matsuura and coworkers [136] with the aid of ICP-AES and ICP-MS. The extraction under usual conditions (i.e., brewing black tea leaves in boiling water for 5 min) was assessed for each element as the ratio between its concentration in infusion and in leaves. Additionally, speciation analysis of major to trace metals was carried out by coupling the above mentioned spectrometric detectors with SEC. Elements were classified as highly extractable (>55 percent), moderately extractable (20–55 percent) and poorly extractable (<20 percent). For example, Al appears in the second group, while lanthanides are included in the third group. It was also found that, in tea infusions, numerous elements are associated to large organic molecules. The essential element Mn was subjected to a speciation study by Özdemir and Gücer [137]. FAAS was used as the determination technique. A series of experiments were conducted to quantify the different forms in which the metal is present in black tea leaves. Total Mn was determined after digestion with HNO_3 and $HClO_4$. Successive Soxhlet extractions with organic solvents and acid aqueous solutions were performed on the digests. Determinations of total Mn were also carried out on water infusions of Turkish teas. The same samples were employed to determine Mn(II) by coprecipitation on Al^{3+} with ammonia, to evaluate the total cationic and anionic Mn concentration by ion-exchange, and perform batch extractions with organic solvents. Five different groups of tea leaves showed Mn concentrations between 1107 and 2205 $\mu g\ g^{-1}$. The study of metal distribution indicated that 30 percent of the Mn dissolved by infusion was present as Mn(II), and that the Mn distributed among the different organic forms is about 2.5 percent.

Concern for the possible toxic effects of Al derived from metabolic processes after ingestion of tea has promoted toxicological studies where atomic spectrometric techniques played a crucial role. In a study by Drewitt et al. [138], 12 healthy volunteers were subjected to a controlled Al diet that included consumption of a tea infusion – with the addition of either milk or lemon juice – or mineral water, with the experiments following a three-way crossover design. ET-AAS and ICP-AES were employed to analyze the samples for Al. Metal concentrations were also determined in plasma samples collected before and up to 24 h after intake of tea or water. The authors conclude that, in the short term, the total body burden of Al is not significantly affected by drinking tea. Further interest on the speciation of Al in tea infusions derives primarily from the knowledge that polyphenolic compounds such as flavonoids are important in binding metals such as Al, Fe, and Cu. In this regard three categories of Al species were proposed by Zhou et al. [139], that is, large organic compounds, small stable organic compounds, and Al in free form. ET-AAS complemented by reverse phase-HPLC was employed by these authors to study the fate of leached Al in tea infusions. FAAS, complemented by molecular absorption spectrometry for the determination of phenolic compounds, was employed in a speciation study by Erdemoglu et al. [140]. A nonionic sorbent, Amberlite XAD-7, was used to separate species mostly consisting of Al bound to hydrolizable polyphenols, whilst a chelating resin (Chelex-100) allowed cationic Al species to be separated. The procedure consisted of pumping the effluent from a column made with the nonionic resin through a second column filled with the chelating resin. The infusion from black teas contained around 11–12 mg l^{-1} Al, representing 30 percent of the total metal content in the tea leaves. Of the Al in solution, 10–19 percent was found to be present as cations, and 28–33 percent as Al bound to hydrolizable polyphenols.

ET-AAS with graphite furnace tubes constituted the analytical support for the speciation analysis of Al and the concomitant determination of Al (III) in tea infusions, as described by Alberti and coworkers [141]. Lvov platforms incorporated into the graphite furnace atomizers enabled the authors to attain an LoD of 2 μg l^{-1} Al. The determination was not impaired by the relatively high solid contents. Adsorption of Al on ion Chelex-100 resin was employed to estimate the free metal content and the concentration and stability constants of complexed Al species. The metal is reported to be present in total concentrations from 0.09 to 0.26 mM, but mainly linked to strong complexes. Strong complexation is demonstrated by the inability of Chelex-100 to dissociate the complexes. These results are considered by the authors as an explanation of the low toxicity of the Al associated with tea infusions. The concentration of free Al was found to be very low (at the nM level). No structural or chemical information about the ligands was obtained from the method.

Boron is considered an essential element for the human organism. On the other hand, both its deficiency and excess affect the normal growing of plant organisms.

A paper on the analysis for B in black, green, and fruit tea classes, ground roasted coffee beans, and instant coffees available in the Czech Republic has been authored by Krejčová and Černohorský [142]. Before analysis, sample preparation was performed by two methods: total decomposition by means of an MW-assisted digestor and hot water extraction. The authors also assessed the effects of concomitant elements (Ca, Mg, K, and Na) and the presence of acids (HCl, HNO_3) on B measurements. Mean concentration values for B ranged from 3.21 ± 0.17 mg kg^{-1} in black teas to 21.3 ± 0.7 mg kg^{-1} in instant coffee.

The existence of a large number of tea-producing countries along with the considerable number of varieties that are distributed in commercial markets have generated a great deal of research on chemometric methods aimed at differentiating and classifying varieties and/or origins by using the metal contents as chemical descriptors. These approaches greatly benefit from the multielement characteristics of techniques based on plasma emission or ion generation. A preliminary attempt to establish the geographical origin of teas has been advanced by Marcos et al. [143]. Elements that may be present at relatively high concentration levels (Al, Ba, Ca, Cu, Fe, La, Mg, Mn, Sr, Ti, and Zr) were measured by ICP-AES, whilst ICP-MS was used to quantify metals present at trace and ultratrace levels (Cd, Co, Cr, Cs, Hg, La, Li, Nd, Ni, Pb, Pr, Rb, Se, Sn, Ti, V, and Zn). The methods were applied to 15 teas from 10 different countries. Multielemental analyses were carried out on HNO_3 digests, In being added as the internal standard. A Chinese tea CRM (GBW 08505) was employed to check the results. By applying the PCA to a large set of results, tea samples of African and Asian origin could be grouped and distinguished. The chemometric techniques might be refined by including additional factors, such as the different preparation procedures associated to each tea type.

In a related paper Herrador and González [144] described the application of PCA and CA and of two supervised techniques, LDA and back-propagated ANN on Al, Ba, Ca, Cu, K, Mg, Mn, and Zn data obtained from commercial Spanish tea samples. A minitorch ICP-AES instrument was used for the determinations. The characterization of three classes of tea was achieved. In a paper that expands previous research described in reference [47], trace metal concentrations measured by ICP-AES and ICP-MS were employed by Moreda–Piñeiro et al. [145] for a more elaborated chemometric treatment on 85 samples of tea of Asian, African, commercial, and unknown origin. Seventeen elements (Al, Ba, Ca, Cd, Co, Cr, Cu, Cs, Mg, Mn, Ni, Pb, Rb, Sr, Ti, V, and Zn) were determined. In addition to the techniques employed in the already mentioned papers (PCA, CA, LDA), soft independent modeling (SIM) of class analogy was also applied. The latter method resulted in the totally correct (100 percent) classification of Chinese teas.

Metal distribution studies are in no way limited to most known tea infusions. Other herbal products, some very popular in South American and Asian countries, have also attracted the attention of analytical chemists in connection with their possible contributions to human metal intake in large populations. Leaves and commercial products of Paraguayan *yerba mate* (*Ilex paraguaiensis S.H.*) were analyzed for their content of Ca, Cu, Fe, K, Mg, Mn, Na, and Zn by AAS. Significant differences in Fe content are reported between the herb leaves and the commercial products. Calcium and Fe contents also differ according to the commercial reputation of the products. Seasonal and soil-related variations are presented and discussed in the context of nutritional properties and product quality controls. AAS was applied by Wrobel et al. [146] to the analysis of black tea, green tea, *Hibiscus sabdariffa*, and mate (*Ilex paraguaiensis*). Results of metal extraction efficiencies indicated that mate, drunk as infusion, may be a good dietary source of the micronutrient Mn and that hibiscus infusions could be a source of Cu and Fe. It was found that a decrease in the leaching of the elements, with the exception of Ni, strongly correlates with the content of tannins in the beverage. Several macro- and microelements were determined in leaves and infusions of two herbs, peppermint (*Mentha piperitae folium*) and nettle (*Urticae folium*) by Łozak et al. [147]. Concentrations of As, Ba, Ca, Cd, Co, Cr, Cu, Fe, I, Li, Mg, Mn, Ni, Pb, Se, Sn, Sr, Ti, V, and Zn were measured by ICP-MS and AAS. Metal contents ranged from 244 and 107 mg kg^{-1} for the leaf of peppermint and nettle, respectively, to 0.10–0.08 mg kg^{-1} for Co for the same products. Differences in leaching between the elements were found, Sr, Se, and I being the most readily extracted ones.

Coffee infusion properties are of great dietary, social, and economic interest. Analytical spectrometric studies on mineral nutrients and toxic elements, as well as on the interactions of component metals and metalloids with the organic compounds naturally occurring in the different varieties (green, roasted, soluble) are being continuously reported. As an example, the presence of toxic metals such as Pb in instant coffee and tea powders was investigated during the early developments of ET-AAS techniques. One of the first applications of nonflame methods, that is, the carbon filament atomizer, was described by Kapur and West [148]. Both direct determination and determination after solvent extraction of the I^- complex of the metal with MIB K were employed for the determination of Pb at microgram and submicrogram per milliliter ml^{-1} concentrations. An evaluation of mineral nutrients in Brazilian coffee has been performed by dos Santos and de Oliveira [149]. Twenty-one samples of soluble coffee were digested with HNO_3 and H_2O_2 in a focused MW system, and determinations of Al, Ca, Cd, Cr, Cu, Fe, K, Mg, Mn, Na, Ni, P, Pb, S, Sb, Sn, and Zn were carried out by ICP-AES. Matrix interferences were detected (and corrected) only in the determination of Sn. One sample was found to contain Cr at a concentration of 0.52 ± 0.02 mg kg^{-1}, which exceeds the value specified by the Brazilian Food Legislation. To investigate the variability and the possible classification of the results, PCA and Agglomerative Hierarchical Clustering were applied. Three clusters were observed with similarities in mineral compositions.

Various other chemometric techniques, such as those applied to tea samples, turned out to be useful for the classification and characterization of coffee types and origins. In a study reported by Martín et al. [150] ICP-AES was used to analyze 41 samples of green coffee of the varieties *arabica* and *robusta*. PCA and CA were performed on concentration data for Ba, Ca, Cu,

Fe, K, Mg, Mn, Na, P, Sr, and Zn. Dried samples were dissolved by means of a wet ashing procedure employing H_2SO_4 and HNO_3, and conveniently diluted for introduction into the nebulizer. Significant differences between the two coffee classes were found, the contents of Cu, Mn, and P being different enough to permit discrimination. A similar strategy and set of measured elements were applied by the same authors [151] to roasted *arabica* and *robusta* varieties and coffee blends. The types were characterized by PCA and CA. In the case of coffee blends, the relative proportions of each variety in the blend were determined by partial least squares (PLS). Apart from the discrimination of the two coffee classes based on Cu, Mn, and P concentrations, it was possible to resolve mixtures with a low prediction error, employing those elements as chemical descriptors.

Not only the assessment of the metal content in food beverages, but also the accessibility of metals to biological organisms have greatly benefited from the application of multielemental spectrochemical techniques. In an interesting interdisciplinary paper Mounicou et al. [152] assessed the potential bioaccessibility of Cd and Pb present in cocoa powder and liquor. ICP-MS enabled the estimation of the solubility of Cd and Pb in enzymatic extracts produced *in vitro*, aiming at developing a model that simulates the activity of enzymes in the gastrointestinal tract. With that purpose, sequential extractions with solutions that simulated gastric and intestinal fluids were carried out. The residues from the extractions were then treated in parallel with phytase and cellulase enzymatic solutions. Bioaccessibility of Cd and Pb were found to differ. Cadmium was the metal potentially absorbed at a higher degree (10–50 percent), with an additional 20 percent recovered by phytase and 30 percent by cellulase. Not more than 5–10 percent of Pb was bioavailable under similar conditions. An evaluation of Al in chocolate and other beverages presented by Sepe and coworkers [153] made use of ET-AAS to establish the correlation between the percentages of cocoa and Al by processing 248 independent samples. Only chocolate and fruit beverages required a preliminary sample preparation. The mean concentration of Al found was 9.2 ± 7.5 mg kg^{-1}. As regards sample preparation, Maichin et al. [154] have recently described the adaptation of a commercial high-pressure MW device for wet digestion in open vessels using a Teflon liner. Cocoa, as well as other sample materials rich in organic components were treated with HNO_3–H_2O_2 mixtures and the resulting solutions analyzed by ET-AAS. Data obtained for nine CRMs showed that no losses of analytes occurred for volatile elements such as As, Hg, and Se. The residual carbon was found to vary between 0.2 and 10 percent and depended upon the amount of sample processed.

15.8 CONCLUSIONS AND FUTURE TRENDS

Admittedly, the challenges posed to the analytical chemist by the measurement of very low concentrations of both essential and potentially toxic elements in such complex types of matrices as the beverages considered in this chapter are being appropriately addressed by modern atomic spectrometric techniques.

FAAS and ET-AAS methods continue to share a significant fraction of the applications, in particular when complemented by off-line or on-line separation/preconcentration procedures of remarkable efficiency. In this regard, the degree of automation, concentration efficiency, and sample throughput attained with ingenious configuration manifolds will contribute to the acceptance of the method for the analytical control determinations required by legislations and product quality controls. Similar devices can be part of systems that involve derivatization (e.g., chemical generation of volatile compounds). The same liquid transport and chemical manipulation strategies can, of course, also be employed with ICP-AES or ICP-MS detection. Increased applications are envisaged for recently introduced ET-AAS instrumentation. In this case, the feasibility to compensate for nonspecific absorption by getting information from signals originated by structured background underlying the analytical lines can improve selectivity and analysis accuracy.

Due to its versatility, robustness, and relativel low cost, ICP-AES has started to be considered as routine instrumentation for elemental composition measurements in many laboratories involved in food analysis. Unquestionably, the acceptance of the technique has been further advanced by the commercial availability of rapid, simultaneous, and flexible, sequential instruments based on echelle grating crossed dispersion and solid state detectors. It is expected that novel nebulizer devices for liquid sample introduction will find increased applications in beverage analysis.

Similar concepts regarding (direct) sample introduction may constitute a field of great activity in ICP-MS analysis of liquid matrices. Such techniques are becoming increasingly attractive for total metal concentration measurements at extremely low levels, with isotopic discrimination. Few applications of double focusing magnetic/electrostatic SF instruments are reported, however, probably because of the current high purchase costs. TOF-ICP-MS will be a possible alternative, with commercial instruments now starting to appear on the market.

Although conventional wet-ashing procedures of sample preparation are still being extensively employed, digestion of solid samples with the assistance of MW in open focused or closed systems is becoming widely recognized as a clean and efficient method. Sample preparation may be carried out on-line by adding an MW treatment step to FIA manifolds, although this methodology has not been yet properly exploited in beverage analysis.

Speciation and fractionation analysis of metals, mostly covering the quantification of free ions and organometallic compounds present in liquid foods and infusions, is possibly the field in which atomic spectrometry has had the greatest impact. Virtually all the atomic detectors (absorption, emission, fluorescence, and ion generation) can be coupled on-line to high-resolution techniques, from GC through HPLC in its different modes and CE, expanding the dimensionality of analytical results. A great deal of complementation between these and other nonatomic techniques (e.g., molecular absorption spectrometry) is expected for the characterization of organic compound–metal interactions that may reduce the potential toxicity of soluble elements, as it has been frequently exemplified by studies on Al speciation.

The intrinsic information content of atomic spectrometric methods originates a considerable amount of data, for which proper evaluation cannot be conducted by conventional, univariate data treatment procedures. This is currently prompting the utilization of trace metal concentrations as chemical descriptors in chemometric methods employed as a mathematical tool, in some cases very sophisticated, aimed at the optimization of chemical and instrumental steps as well as the differentiation and grouping of samples of various origins. In addition, a great deal of activity is envisaged in connection to the routine application of calibration methods of higher dimensionalities, intended for accuracy improvement. This is expected to facilitate the validation of methodologies, somewhat impaired by the scarcity of CRMs of specific application in the field of beverage analysis.

ACKNOWLEDGMENTS

This chapter was prepared as a part of the CNEA Program P5, Project PID-36-2 (CAC).

REFERENCES

1. C. Ash, R. Stone, Metals: impacts on health and the environment. A question of dose, *Science*, **300** (2003), 925.
2. J. Kaiser, Metals: impacts on health and the environment. Manganese: a high octane dispute, *Science*, **300** (2003), 926–928.
3. P. Zatta, R. Lucchini, S. J. van Rensburg, A. Taylor, The role of metals in neurodegenerative processes: aluminum, manganese and zinc, *Brain Res. Bull.*, **62** (2003), 15–28.
4. B. Welz, M. Sperling, *Atomic Absorption Spectrometry*, Wiley VCH, New York, 1999.
5. A. Montaser, D. W. Golightly, *Inductively Coupled Plasma in Analytical Atomic Spectrometry*, 2nd edn, Wiley VCH, New York, 1992.
6. S. J. Hill (ed.), *Inductively Coupled Plasma Spectrometry and its Applications*, Sheffield Academic Press, 1999.
7. C. Minoia, S. Caroli (eds), *Applications of Zeeman Graphite Furnace Atomic Absorption Spectrometry in the Chemical Laboratory and in Toxicology*, Pergamon, New York, 1992.
8. A. Montaser (ed.), *Inductively Coupled Plasma Mass Spectrometry*, Wiley VCH, New York, 1998.
9. F. L. Fricke, Trace element analysis of food and beverages by atomic absorption spectrometry. *Prog. Anal. Atom. Spectrosc.*, **2** (1979), 185–286.

10. A. Taylor, S. Branch, D. Halls, M. Patriarca, M. White, Atomic spectrometry update. Clinical and biological materials, foods and beverages, *J. Anal. Atom. Spectrom.*, **17** (2002), 414–455.

11. N. A. Nkono, O. I. Asubiojo, Trace elements in bottled and soft drinks in Nigeria – a preliminary study, *Sci. Total Environ.*, **208** (1976), 161–163.

12. H. E. Allen, M. A. Halley-Henderson, C. N. Hass, Chemical composition of bottled mineral water, *Arch. Environ. Hlth*, **44** (1989), 102–116.

13. M. Claes, M. Bagar-Povse, J. Drinovec, B. Hołynska, P. Kump, B. Ostachowicz, M. Necemer, M. Olko, R. Van Grieken, D. Wegrzynek, Inventory of regulations and analysis techniques for trace elements in some water supply companies in eastern and western Europe, *Sci. Total Environ.*, **207** (1997), 141–148.

14. J. R. Coughlin, Sources of human exposure: overview of water supplies as sources of boron, *Biol. Trace Elem. Res.*, **66** (1998), 87–100.

15. P. Garzon, M. J. Eisenberg, Variation in the mineral content of commercially available bottled waters: implications for health and disease, *Am. J. Med.*, **105** (1998), 125–130.

16. E. M. García, C. Cabrera, J. Sánchez, M. L. Lorenzo, M. C. López, Chromium levels in potable water, fruit juices and soft drinks: influence on dietary intake, *Sci. Total Environ.*, **216** (1999), 181–192.

17. E. Rojas, L. A. Herrera, L. A. Poirier, P. Ostrosky-Wegman, Are metals dietary carcinogens? *Mutat. Res.*, **443** (1999), 157–181.

18. A. Misund, B. Frengstad, U. Siewers, C. Reimann, Variation of 66 elements in European bottled mineral waters, *Sci. Total Environ.*, **243** (1999), 21–41.

19. J. Korkisch, H. Krivanec, Atomic-absorption determination of vanadium and molybdenum in tap water and mineral waters after anion-exchange separation, *Anal. Chim. Acta*, **83** (1976), 111–118.

20. M. Soylak, U. Divrikli, M. Elci, M. Doğan, Preconcentration of Cr(III), Co(II), Cu(II), Fe(III), and Pb(II) as calmagite chelates on cellulose nitrate membrane filter prior to their flame atomic absorption spectrometric determination, *Talanta*, **56** (2002), 565–570.

21. R. M. Cespón Romero, M. C. Yebra-Biurrun, M. P. Bermejo-Barrera, Preconcentration and speciation of chromium by the determination of total chromium(III) in natural waters by flame atomic absorption spectrometry with a chelating ion-exchange flow injection system, *Anal. Chim. Acta*, **327** (1996), 37–45.

22. A. N. Araújo, R. C. C. Costa, J. L. F. C. Lima, B. F. Reis, Sequential injection system in flame atomic absorption spectrometry for the determination of calcium and magnesium in mineral waters, *Anal. Chim. Acta*, **358** (1998), 111–119.

23. V. Hudnik, S. Gomiscek, The atomic absorption spectrometric determination of arsenic and selenium in mineral waters by electrothermal atomization, *Anal. Chim. Acta*, **157** (1984), 135–142.

24. S. L. C. Ferreira, H. C. dos Santos, R. C. Campos, The determination of molybdenum in water and biological samples by graphite furnace atomic spectrometry after polyurethane foam column extraction, *Talanta*, **61** (2003), 789–795.

25. T. P. Flaten, Aluminium as a risk factor in Alzheimeŕs disease, with emphasis in drinking water, *Brain Res. Bull.*, **55** (2001), 187–196.

26. M. Plessi, A. Monzani, Aluminum determination in bottled mineral waters by electrothermal atomic absorption spectrometry, *J. Food Comp. Anal.*, **8** (1995), 21–26.
27. E. M. García, C. Cabrera, J. Sánchez, M. L. Lorenzo, M. C. López, Chromium levels in potable water, fruit juices and soft drinks: influence on dietary intake, *Sci. Total Environ.*, **216** (1999), 181–192.
28. M. El Himri, A. Pastor, M. de la Guardia M, Determination of uranium in tap water by ICP-MS, *Anal. Bioanal. Chem.*, **367** (2000), 151–156.
29. I. Al-Saleh, I. Al-Doush, Survey of trace elements in household and bottled drinking water samples collected in Riyadh, Saudi Arabia, *Sci. Total Environ.*, **216** (1998), 181–192.
30. A. Sanz-Medel (Ed.), *Flow Analysis with Atomic Spectrometric Detectors*, Elsevier, Amsterdam, 1999.
31. K. G. Heumann, S. M. Gallus, G. Rädlinger, J. Vogl, Accurate determination of element species by on-line coupling of chromatographic systems with ICP-MS using isotope dilution technique, *Spectrochim. Acta*, **53B** (1998), 273–287.
32. A. K. Das, M. de la Guardia, M. L. Cervera, Literature survey of on-line elemental speciation in aqueous solutions, *Talanta*, **55** (2001), 1–28.
33. J. A. Caruso, M. Montes-Bayon, Elemental speciation studies – new directions for trace metal analysis, *Ecotox. Environ. Saf.*, **56** (2003), 148–163.
34. M. A. López, M. M. Gómez, C. Cámara, Determination of six arsenic species by high-performance liquid chromatography – hydride generation – atomic absorption spectrometry with on-line thermo-oxidation, *Fresenius J. Anal. Chem*, **346** (1993), 643–647.
35. M. Careri, F. Bianchi, C. Corradini, Recent advances in the application of mass spectrometry in food-related analysis, *J. Chromatogr. A*, **970** (2002), 3–64.
36. M. Van Holderbeke, Y. Zhao, F. Vanhaecke, L. Moens, R. Dams, P. Sandra, Speciation of six arsenic compounds using capillary electrophoresis-inductively coupled plasma mass spectrometry, *J. Anal. Atom. Spectrom.*, **14** (1999), 229–234.
37. J. T. Van Elteren, V. Stibilj, Z. Šlejkovek, Speciation of inorganic arsenic in some bottled Slovene mineral waters using HPLC-HG-AFS and selective coprecipitation combined with FI-HG-AFS, *Water Res.*, **36** (2002), 2967–2974.
38. Y. He, M. L. Cervera, A. Pastor, M. de la Guardia, Determination of chromium (III) and chromium (VI) in mineral water by bidirectional electrostacking and electrothermal atomic absorption spectrometry, *Anal. Chim. Acta.*, **447** (2001), 135–142.
39. National Research Council, National Academy of Sciences, *Food Chem. News*, **1989**, 21–23.
40. Subcommittee on Recommended Dietary Allowances, Food and Nutrition Board, National Research Council, 10th edn, National Academy of Sciences, Washington, DC 1989.
41. S. G. Schäfer, W. Forth, The influence of tin, nickel, and cadmium on the intestinal absorption of iron, *Ecotox. Environ. Saf.*, **7** (1983), 87–95.
42. M. C. Valdés Hevia y Temprano, B. Aizpún Fernández, M. R. Fernández de la Campa, A. Sanz-Medel, Study of the influence of ordered media on the determination of lead by hydride generation inductively coupled plasma atomic emission spectrometry, *Anal. Chim. Acta*, **283** (1993), 175–182.

43. M. C. Valdés-Hevia y Temprano, M. R. Fernández de la Campa, A. Sánz-Medel, Sensitive method for determination of lead by potassium dichromate-lactic acid hydride generation inductively coupled plasma atomic emission spectrometry, *J. Anal. Atom Spectrom.*, **8** (1993), 821–825.
44. J. Marrero, S. Pérez-Arisnabarreta, P. Smichowski, Comparison of effects of four acid oxidant mixtures in the determination of lead in food and beverages by hydride generation – ICP – OES, *Atom Spectrosc.*, **24** (2002), 133–141.
45. R. C. Massey, D. Taylor, *Aluminium in food and the Environment*, Royal Society of Chemistry, London, 1991.
46. Ministerio de Agricultura, Pesca y Alimentación, *La alimentación en España*, Secretaría General de Alimentación, Madrid, 1997.
47. M. A. Zezzi Arruda, M. Gallego, M. Valcárcel, Determination of aluminium in slurry and liquid phase by flow injection analysis graphite furnace atomic absorption spectrometry, *Anal. Chem.*, **85** (1997), 3331–3335.
48. J. J. Beattied, T. N. Quoc, Manganese in pineapple juices, *Food Chem.*, **68** (2000), 37–39.
49. M. Cámara, C. Díez, E. Torija, Chemical characterization of pineapple juices and nectars. Principal components analysis, *Food Chem.*, **54** (1995), 98–100.
50. Y. Liu, B. Gong, Z. Li, Y. Xu, T. Lin, Direct determination of selenium in a wild fruit juice by electrothermal atomic absorption spectrometry, *Talanta*, **43** (1996), 985–989.
51. J. L. Burguera, P. Carrero, M. Burguera, C. Rondon, M. R. Brunetto, P. Gallignani, Flow injection for the determination of Se(IV) and Se(VI) by hydride generation atomic absorption spectrometry with microwave oven on-line pre-reduction of Se(VI) to Se(IV), *Spectrochim. Acta*, **51B** (1996), 1837–1847.
52. M. Gallignani, M. Valero, M. R. Brunetto, J. L. Burguera, M. Burguera, Y. Petit de Peña, Sequential determination of Se(IV) and Se(VI) by flow injection-hydride generation-atomic absorption spectrometry with HCl/HBr microwave aided pre-reduction of Se(VI) to Se(IV), *Talanta*, **52** (2000), 1015–1024.
53. N. Ulrich, Determination of antimony species with fluoride as modifier and flow injection hydride generation inductively coupled plasma emission spectrometry, *Anal. Chim. Acta*, **417** (2000), 201–209.
54. G. Weber, Speciation of tin in lemon juice: an example of trace metal speciation in food, *Anal. Chim. Acta*, **200** (1987), 79–88.
55. J. A. McHard, S. J. Foulk, J. L. Jorgensen, S. Bayer, J. D. Winefordner, Analysis of trace metals in orange juice, in S. Nagy, J. A. Attaway (eds), *Citrus Nutrition and Quality*, American Chemical Society Symposium Series no. 143, American Chemical Society, Washington, DC, 1980, 363–394.
56. W. A. Simpkins, H. Louie, M. Wu, M. Harrison, D. Goldberg, Trace elements in Australian orange juice and other products, *Food Chem.*, **71** (2000), 423–433.
57. V. A. Lemos, M. de la Guardia, S. C. L. Ferreira, An on-line system for preconcentration and determination of lead in wine samples by FAAS, *Talanta*, **58** (2002), 475–480.
58. C. S. Ough, E. A. Cromwell, J. Benz, Metal content of California wines, *J. Food Sci.*, **47** (1982), 825–828.
59. E. S. Lazos, A. Alexakis, Metal ion content in some Greek wines, *Int. J. Food Sci. Technol.*, **24** (1989), 39–46.

60. E. K. Paleologos, D. L. Giokas, S. M. Tzouwara-Karayanni, M. I. Karayannis, Micelle mediated methodology for the determination of free and bound iron in wines by flame atomic absorption spectrometry, *Anal. Chim. Acta*, **458** (2002), 241–248.

61. S. Pehkonen, Y. Erel, M. R. Hoffmann, Simultaneous spectrophotometric measurement of iron(II) and iron(III) in atmospheric water, *Environ. Sci. Technol.*, **26** (1992), 1731–1736.

62. M. Segura, Y. Madrid, C. Cámara, Evaluation of atomic fluorescence and atomic absorption spectrometric techniques for the determination of arsenic in wine and beer by direct hydride generation sample introduction, *J. Anal. Atom. Spectrom.*, **14** (1999), 131–135.

63. A. A. Almeida, M. I. Cardoso, L. F. C. Lima, Improved determination of aluminium in Port wine by electrothermal atomic absorption spectrometry using potassium dichromate chemical modification and end-capped graphite tubes, *J. Anal. Atom. Spectrom.*, **12** (1997), 837–840.

64. J. C. Cacho, V. Ferreira, C. Nerín, Determination of lead in wines by hydride generation atomic absorption spectrometry, *Analyst*, **117** (1992), 31–33.

65. C. Cabrera, Y. Madrid, C. Cámara, Determination of lead in wine, other beverages and fruit slurries by flow injection hydride generation atomic absorption spectrometry with on-line microwave digestion, *J. Anal. Atom. Spectrom.*, **9** (1994), 1423–1426.

66. Y. Madrid, J. Mesenguer, M. Bonilla, C. Cámara, Lead hydride generation in lactic acid-potassium dichromate medium and its application to the determination of lead in fish, vegetables and drink samples, *Anal. Chim. Acta*, **237** (1990), 181–187.

67. C. Baluja-Santos, A. González-Portal, Application of hydride generation to atomic-absorption spectrometric analysis of wines and beverages: a review, *Talanta*, **39** (1992), 329–339.

68. R. C. C. Costa, M. I. Cardoso, A. N. Araújo, Metals determination in wines by sequential injection analysis with flame atomic absorption spectrometry, *Am. J. Enol. Vitic.*, **51** (2000), 131–136.

69. H. Eschnauer, Trace elements in must and wine, *Am. J. Enol. Vitic.*, **33** (1982), 226–230.

70. B. I. Kildahl, W. Lund, Determination of arsenic and antimony in wine by electrothermal atomic absorption spectrometry, *Anal. Bioanal. Chem.*, **354** (1996), 93–96.

71. A. M. Wifladt, G. Wibetoe, W. Lund, Determination of antimony in wine by hydride generation graphite furnace atomic absorption spectrometry, *Fresenius J. Anal. Chem.*, **357** (1997), 92–96.

72. M. L. Cervera, A. Navarro, R. Montoro, Platform in furnace-Zeeman-effect atomic absorption spectrometric determination of arsenic in beer by atomization of slurries of sample ash, *J. Anal. Atom Spectrom.*, **6** (1991), 477–481.

73. J. Kristl, M. Veber, M. Slekovec, The application of ET-AAS to the determination of Cd, Cr, and Pb in samples taken during different stages of the winemaking, *Anal. Bioanal. Chem.*, **373** (2002), 200–204.

74. J. Cvetković, T. D. M. Stafilov, Nickel and strontium nitrates as modifiers for the determination of selenium in wine by Zeeman electrothermal atomic absorption spectrometry, *Anal. Bioanal. Chem.*, **370** (2001), 1077–1081.

75. J. Cvetković, S. Arpadjan, I. Karadjova, T. Stafilov, Determination of thallium in wine by electrothermal atomic absorption spectrometry after extraction preconcentration, *Spectrochim. Acta*, **57B** (2002), 1101–1106.

76. G. P. G. Freschi, C. S. Dakuzaku, J. A. G. Neto, M. Moraes, Simultaneous multielement graphite furnace atomic absorption spectrometry – a review of the techniques and applications, *Ecl. Quim.*, **25** (2000), 213–226.
77. G. P. G. Freschi, C. S. Dakuzaku, M. de Moraes, J. A. Nóbrega, Simultaneous determination of cadmium and lead in wine by electrothermal atomic absorption spectrometry, *Spectrochim. Acta*, **56B** (2001), 1987–1993.
78. K. G. Fernandes, M. de Moraes, J. A. Gomes Neto, J. A. Nóbrega, Evaluation and application of bismuth as an internal standard for the determination of lead in wines by simultaneous electrothermal atomic absorption spectrometry, *Analyst*, **127** (2002), 157–162.
79. G. Thiel, K. Danzer, Direct analysis of mineral components in wine by inductively coupled plasma optical emission spectrometry (ICP-OES), *Anal. Bioanal. Chem.*, **357** (1997), 553–557.
80. J. L. Capelo, H. A. Pedro, A. M. Mota, Ozone treatment for mercury determination in white wines, *Talanta*, **61** (2003), 485–491.
81. S. Ražic, M. Todorovic, I. Holclajtner-Antunovic, M. Stoiljkovic, Determination of metal traces in wine by argon stabilized d.c. arc, *Anal. Bioanal. Chem.*, **365** (1999), 367–370.
82. M. M. Castiñeira, R. Brandt, A. von Bohlen, Development of a procedure for the multi-element determination of trace elements in wine by ICP-MS, *Anal. Bioanal. Chem.*, **370** (2001), 553–558.
83. C. M. Almeida, M. T. S. D. Vasconcelos, Advantages and limitations of the semi-quantitative operation mode of an inductively coupled plasma-mass spectrometer for multi-element analysis of wines, *Anal. Chim. Acta*, **463** (2002), 165–175.
84. C. M. Almeida, M. T. S. D. Vasconcelos, UV-irradiation and MW-digestion pretreatment of Port wine suitable for the determination of lead isotope ratios by inductively coupled plasma mass spectrometry, *J. Anal. Atom Spectrom.*, **14** (1999), 1815–1821.
85. C. M. Almeida, M. T. S. D. Vasconcelos, M. Barbaste, B. Medina, ICP-MS multi-element analysis of wine samples – a comparative study of the methodologies used in two laboratories, *Anal. Bioanal. Chem.*, **374** (2002), 314–322.
86. S. Augagneur, B. Médina, J. Szpunar, R. Lobinski, Determination of rare earth elements in wine by inductively coupled plasma mass spectrometry using a microconcentric nebulizer, *J. Anal. Atom Spectrom.*, **11** (1996), 713–721.
87. N. Jakubowski, R. Brandt, D. Stuewer, H. R. Eschnauer, S. Gortges, Analysis of wines by ICP-MS: is the pattern of the rare elements a reliable fingerprint for the provenance?, *Anal. Bioanal. Chem.*, **364** (1999), 424–428.
88. E. Moreno, C. Cámara, W. T. Coros, D. W. Bryce, P. Stockwell, Arsenic speciation in beverages by direct injection-ion chromatography–hydride generation atomic fluorescence spectrometry, *J. Autom. Met. Manag. Chem.*, **22** (2000), 33–39.
89. R. Cerdeira de Campos Costa, A. Nova Araújo, Determination of Fe(III) and total Fe in wines by sequential injection analysis and flame atomic absorption spectrometry, *Anal. Chim. Acta*, **438** (2000), 227–233.
90. I. Karadjova, B. Izgi, S. Guccer, Fractionation and speciation of Cu, Zn, and Fe in wine samples by atomic absorption spectrometry, *Spectrochim. Acta*, **57B** (2002), 581–590.

91. A. J. McKinnon, G. R. Scollary, Size fractionation of metals in wine using ultrafiltration, *Talanta*, **44** (1997), 1649–1658.

92. M. M. Castiñeira, P. Burba, N. Jakubowski, J. Andersson, Size fractionation of non-volatile dissolved organic compounds and metal species in German white wines by combining on-line tangential-flow multistage ultrafiltration, a home-built carbon analyzer, and inductively coupled plasma mass spectrometry, *Anal. Bioanal. Chem.*, **376** (2003), 174–181.

93. J. R. Dean, L. Ebdon, R. C. Massey, Isotope ratio and isotope dilution analysis of lead in wine by inductively coupled plasma-mass spectrometry, *Food Addit. Contam.*, **7** (1990), 109–116.

94. C. M. R. Almeida, M. T. S. D. Vasconcelos, Determination of lead isotope ratios in port wine by inductively coupled plasma mass spectrometry after pre-treatment by UV-irradiation, *Anal. Chim. Acta*, **396** (1999), 45–53.

95. M. Barbaste, L. Halicz, A. Galy, B. Medina, H. Emteborg, F. C. Adams, R. Łobiński, Evaluation of the accuracy of the determination of lead isotope ratios in wine by ICP MS using quadrupole, multicollector magnetic sector and time-of-flight analyzers, *Talanta*, **307** (2001), 307–317.

96. S. Rebolo, R. M. Peña, M. J. Latorre, A. M. Botana, C. Herrero, Characterisation of Galician (NW Spain) Ribera Sacra wines using pattern recognition analysis, *Anal. Chim. Acta*, **417** (2000), 211–220.

97. A. M. R. Ferreira, J. L. F. C. Lima, T. I. M. S. Lopes, A. O. S. S. Rangel, Flow injection wine analysis: a critical review and future trends, *J. Int. Sci. Vigne Vin.*, **27** (1993), 67–97.

98. V. Cerdá, J. M. Estela, R. Forteza, A. Cladera, E. Becerra, P. Altimira, P. Stijar, Flow techniques in water analysis, *Talanta*, **50** (1999), 695–705.

99. J. Ruzicka, G. D. Marshall, Sequential injection: a new concept for chemical sensors, process analysis and laboratory assays, *Anal. Chim. Acta*, **237** (1990), 329–343.

100. M. A. Segundo, J. L. F. C. Lima, A. O. S. S. Rangel, Automatic flow system based on sequential injection analysis for routine determinations in wines, *Anal. Chim. Acta*, **513** (2004), 3–9.

101. W. O. Kwan, B. R. Kowalski, Classification of wines by applying pattern recognition to chemical composition data, *J. Food Sci.*, **43** (1978), 1320–1323.

102. W. O. Kwan, B. R. Kowalski, Pattern recognition analysis of gas chromatographic data. Geographic classification of wines of *Vitis vinifera* cv. Pinot Noir from France and the United States, *J. Agric. Food Chem.*, **28** (1980), 356–359.

103. W. O. Kwan, B. R. Kowalski, R. K. Schogerboe, Pattern recognition analysis of elemental data. Wines of *Vitis vinifera* cv. Pinot Noir from France and the United States, *J. Agric. Food Chem.*, **27** (1979), 1321–1326.

104. M. J. Latorre, C. Herrero, B. Medina, The use of mineral elements in the geographical classification of Galician wines, *J. Int. Sci. Vigne Vin.*, **26** (1992), 185–193.

105. M. J. Latorre, M. C. García, B. Medina, C. Herrero, Pattern recognition analysis applied to classification of wines from Galicia (Northwestern Spain) with certified brand of origin, *J. Agric. Food Chem.*, **42** (1994), 1451–1455.

106. M. Gonzáles-Larraina, A. Gonzáles, B. Medina, Les ions metalliques dans la différenciation des vins rouges des trois régions d́apellation d́origine Rioja, *Connaiss. Vigne Vin.*, **21** (1987), 127–140.

107. L.-X. Sun, K. Danzer, G. Thiel, Classification of wine samples by means of artificial neural networks and discrimination analytical methods, *Anal. Bioanal. Chem.*, **359** (1997), 143–149.

108. S. Frías, J. E. Conde, J. J. Rodríguez-Becomo, F. P. García-Montelongo, Classification of commercial wines from the Canary Islands (Spain) by chemometric techniques using metallic contents, *Talanta*, **59** (2003), 335–344.

109. E. Marengo, M. Aceto, Statistical investigation of the differences in the distribution of metals in Nebbiolo-based wines, *Food Chem.*, **81** (2003), 621–630.

110. A. Jos, I. Moreno, A. G. González, G. Repetto, A. M. Cameán, Differentiation of sparkling wines (cava and champagne) according to their mineral content, *Talanta*, **63** (2004), 377–382.

111. A. Alcázar, F. Pablos, M. J. Martín, A. G. González, Multivariate characterisation of beers according to their mineral content, *Talanta*, **57** (2002), 45–52.

112. M. L. Andersen, L. H. Skibsted, Electron spin resonance spin trapping identification of radicals formed during aerobic forced ageing of beer, *J. Agric. Food Chem.*, **46** (1998), 1272–1275.

113. R. Buliñski, J. Bloniarz, N. Koktysz, Z. Marzec, E. Szydlowska, Nutritive and energetic value of domestic beer, *Bromat. Chem.*, **19** (1986), 73–75.

114. Food Legislative Survey no. 6. *Contaminants in Food. A Survey of International Prescribed Limits*, 3rd edn, The British Food Manufacturing Industries Research Association, Leatherhead, 1993.

115. M. Segura, Y. Madrid C. Cámara, Evaluation of atomic fluorescence and atomic absorption spectrometric techniques for the determination of arsenic in wine and beer by direct hydride generation sample introduction, *J. Anal. Atom Spectrom.*, **14** (1999), 131–135.

116. A. Martínez, A. Morales-Rubio, M. L. Cervera, M. de la Guardia, Atomic fluorescence determination of total and inorganic arsenic species in beer, *J. Anal. Atom Spectrom.*, **16** (2001), 762–766.

117. M. L. Cervera, A. Navarro, R. Montoro, Platform in furnace-Zeeman-effect atomic absorption spectrometric determination of arsenic in beer by atomization of slurries of sample ash, *J. Anal. Atom Spectrom.*, **6** (1991), 477–481.

118. N. M. Melo Coelho, C. Parrilla, M. L. Cervera, A. Pastor, M. de la Guardia, High performance liquid chromatography-atomic absorption spectrometric determination of arsenic species in beer samples, *Anal. Chim. Acta*, **482** (2003), 73–80.

119. H. Matusiewicz, M. Mikolajczank, Determination of As, Sb, Sn, and Hg in beer and wort by direct hydride generation sample introduction – electrothermal AAS, *J. Anal. Atom Spectrom.*, **482** (2001), 652–657.

120. P. Pohl, B. Prusisz B, Pre-concentration of Cd, Co, Cu, Ni, and Zn using different off-line ion exchange procedures followed by the inductively coupled plasma atomic emission spectrometric detection, *Anal. Chim. Acta*, **502** (2004), 83–90.

121. B. Wyrzykowska, K. Szymczyk, H. J. Falandysz, B. Skwarzec, S. Yamasaki, Application of ICP sector field MS and principal component analysis for studying interdependences among 23 trace elements in polish beers, *J. Agric. Food Chem.*, **49** (2001), 3425–3421.

122. D. Bellido-Milla, J. M. Moreno-Perez, M. P. Hernández-Artiga, Differentiation and classification of beers with flame atomic spectrometry and molecular absorption

spectrometry and sample preparation assisted by microwaves, *Spectrochim. Acta*, **55B** (2000), 855–864.

123. A. Alcázar, F. Pablos, M. J. Martín, A. G. González, Multivariate characterisation of beers according to their mineral content, *Talanta*, **57** (2002), 45–52.

124. M. M. Vela, R. B. Toma, W. Reiboldt, A. Pierri, Detection of aluminium residue in fresh and stored canned beer, *Food Chem.*, **63** (1998), 235–239.

125. E. C. Silva, M. C. U. Araújo, R. S. Honorato, J. L. F. Costa Lima, E. A. Guidetti Zagatto, S. M. Boscolo Brienza, Standard additions in flow injection analysis based on merging zones and gradient exploitation: application to copper determination in spirits, *Anal. Chim. Acta*, **319** (1996), 153–158.

126. L. F. Almeida, V. Lacerda Martins, E. C. Silva, P. N. Teles Moreira, M. C. Ugolino Araujo, Implementation of an automatic standard addition method in a flow-batch system: application to copper determination in an alcoholic beverage by atomic absorption spectrometry, *Anal. Chim. Acta*, **486** (2003), 143–148.

127. R. S. Araujo Honorato, R. Saldanha Honorato, M. F. Pimentel, M. C. Ugulino Araujo, Analytical curve or standard addition method: how to elect and design a strategy applied to copper determination in sugarcane spirits using AAS, *Analyst*, **127** (2002), 1520–1525.

128. M. H. Canuto, H. G. Luna Siebald, G. Magela de Lima, J. B. Borda Silva, Antimony and chromium determination in Brazilian sugar cane spirit, cachaca, by electrothermal atomic absorption spectrometry using matrix matching calibration and ruthenium as permanent modifier, *J. Anal. Atom Spectrom.*, **18** (2003), 1404–1406.

129. H. J. Salacinski, P. G. Riby, S. J. Haswell, Coupled flow-injection analysis–flame atomic absorption spectrometry for the quantitative determination of aluminum in beverages and waters incorporating on-line cation exchange, *Anal. Chim. Acta*, **269** (1992), 1–7.

130. L. M. Costa, S. T. Gouveia, J. A. Nóbrega, Comparison of heating extraction procedures for Al, Ca, Mg, and Mn in tea samples, *Anal. Sci.*, **18** (2002), 313–318.

131. A. Csikkel-Szolnoki, S. A. Kiss, S. Veres, Elemental analysis of tea leaves by atomic spectroscopic methods, *Magnes. Res.*, **7** (1994), 73–77.

132. I. Narin, M. Soylak, Enrichment and determinations of nickel (II), cadmium(II), copper(II), cobalt(II), and lead(II) ions in natural waters, table salts, tea and urine samples as pyrrolydine dithiocarbamate chelates by membrane filtration–flame atomic absorption spectrometry combination, *Anal. Chim. Acta*, **493** (2003), 205–212.

133. A. V. Filgueiras, I. Lavilla, C. Bendicho, Ultrasound-assisted solubilization of trace and minor metals from plant tissue using ethylenediaminetetraacetic acid in alkaline medium, *Anal. Bioanal. Chem.*, **369** (2001), 451–456.

134. P. L. Fernández, F. Pablos, M. J. Martín, Multi-element analysis of tea beverages by inductively coupled plasma atomic emission spectrometry, *Food Chem.*, **76** (2002), 483–489.

135. K. E. Ødegård, W. Lund, Multi-element speciation of tea infusion using cation-exchange separation and size-exclusion chromatography in combination with inductively coupled plasma mass spectrometry, *J. Anal. Atom Spectrom.*, **12** (1997), 403–408.

136. H. Matsuura, A. Hokura, F. Katsuki, A. Itoh, Multielement determination and speciation of major-to-trace elements in black tea leaves by ICP-AES and ICP-MS with the aid of size exclusion chromatography, *Anal. Sci.*, **17** (2001), 391–398.
137. Y. Özdemir, S. Gücer, Speciation of manganese in tea leaves and tea infusions, *Food Chem.*, **61** (1998), 313–317.
138. P. N. Drewitt, K. R. Butterworth, C. D. Springall, S. R. Moorhouse, Plasma levels of aluminium after tea ingestion of healthy volunteers, *Food Chem. Toxicol.*, **31** (1993), 19–23.
139. C. Y. Zhou, J. Wu, H. Chi, M. K. Wong, L. L. Koh, Y. C. Wee, The behavior of leached aluminium in tea infusions, *Sci. Total Environ.*, **177** (1996), 9–16.
140. S. Erdemoglu, K. Pyrzyńska, S. Gucer, Speciation of aluminum in tea infusion by ion-exchange resins and flame AAS detection, *Anal. Chim. Acta*, **411** (2000), 81–89.
141. G. Alberti, R. Biesuz, A. Profumo, M. Pasavento, Determination of the total concentration and speciation of Al(III) in tea infusions, *J. Inorg. Biochem.*, **97** (2003), 79–88.
142. A. Krejcová, T. Černohorský, The determination of boron in tea and coffeee by ICP-AES method, *Food Chem.*, **82** (2003), 303–308.
143. A. Marcos, A. Fisher, G. Rea, S. J. Hill, Preliminary study using trace element concentrations and a chemometric approach to determine the geographical origin of tea, *J. Anal. Atom. Spectrom.*, **13** (1998), 521–525.
144. M. A. Herrador, A. G. A. González, Pattern recognition procedures for differentiation of Green, Black and Oolong teas according to their metal content from inductively coupled plasma atomic emission spectrometry, *Talanta*, **53** (2001), 1249–1257.
145. A. Moreda Piñeiro, A. Fisher, S. J. Hill, The classification of tea according to region of origin, *J. Food Compos. Analysis*, **16** (2003), 195–211.
146. K. Wrobel, E. M. Urbina, Determination of total aluminum, chromium, copper, iron, manganese, and nickel and their fractions leached to the infusions of black tea, green tea, *Hibiscus sabdariffa*, and *Ilex paraguariensis* (mate) by ETA-AAS, *Biol. Trace Elem. Res.*, **78** (2000), 271–280.
147. A. Łozak, K. Sołtyk, P. Ostapczuk, Z. Fijałek, Determination of selected trace elements in herbs and their infusions, *Sci. Total Environ.*, **289** (2002), 33–40.
148. J. K. Kapur, T. S. West, Determination of lead in 'instant' coffee and tea powders by carbon filament atomic absorption spectrometry, *Anal. Chim. Acta*, **73** (1974), 180–184.
149. E. J. Dos Santos, E. de Oliveira, Determination of mineral nutrients and toxic elements in Brazilian soluble coffee by ICXP-AES, *J. Food Comp. Anal.*, **14** (2001), 523–531.
150. M. J. Martín, F. Pablos, A. G. González, Characterization of green coffeee varieties according to their metal content, *Anal. Chim. Acta*, **358** (1998), 177–183.
151. M. J. Martín, F. Pablos, A. G. González, Characterization of *arabica* and *robusta* roasted coffee varieties and mixture resolution according to their metal content, *Food Chem.*, **66** (1999), 365–370.
152. S. Mounicou, J. Szpunar, D. Andrey, C. Blake, R. Lobinski, Development of a sequential enzymolisis approach for the evaluation of the bioaccessibility of Cd and Pb from cocoa, *Analyst*, **127** (2002), 1638–1641.

153. A. Sepe, S. Constantini, L. Ciaralli, M. Ciprotti, R. Giordano, Evaluation of aluminium concentrations in samples of chocolate and beverages by electrothermal atomic absorption spectrometry, *Food Addit. Contam.*, **18** (2001), 788–796.

154. B. Maichin, M. Zischka, G. Knapp, Pressurized wet digestion in open vessels, *Anal. Bioanal. Chem.*, **376** (2003), 715–720.

SECTION 3

SPECIATION ANALYSIS

16

SPECIES-SPECIFIC DETERMINATION OF METAL(LOID)-CONTAINING FOOD ADDITIVES AND CONTAMINANTS BY CHROMATOGRAPHY WITH ICP-MS DETECTION

A. Polatajko, B. Bouyssiere, and J. Szpunar

UMR 5034 CNRS, Hélioparc, 2 Avenue President Angot, 64053 Pau, France, E-mail: joanna.szpunar@univ-pau.fr

The Determination of Chemical Elements in Food: Applications for Atomic and Mass Spectrometry.
Edited by Sergio Caroli

16.1 INTRODUCTION

The nutritional, chemical, biological, and toxicological properties of a chemical element are known to be critically dependent on the form in which it occurs in food. The recognition of this fact has spurred the development of species-selective (speciation) analytical methods for food additives and contaminants. According to the IUPAC's definition, speciation analysis deals with the analytical activities of identification and/or measurement of the quantities of one or more individual chemical species in a given sample [1]. The analytical approach is usually based on the combination of a chromatographic separation technique with an element-specific detection technique. The former ensures that the analyte compound leaves the column unaccompanied by other species of the analyte element, whereas the latter enables a sensitive and specific detection of the target element. Coupled (also called hyphenated) techniques have become a fundamental tool for speciation analysis and have been discussed in many published reviews [2–6].

Originally, speciation analysis targeted well-defined analytes, usually anthropogenic organometallic compounds, such as methyl-Hg, alkyl-Pb, butyl- and phenyl-Sn compounds, simple organo-As, and organo-Se species, and products of their environmental degradation [2]. Calibration standards were either available or could be readily synthesized. The presence of a metal(loid)–C covalent bond assured a reasonable stability of the analyte(s) during sample preparation. The volatility of the species allowed for the use of gas chromatography (GC) with its inherent advantages, such as the high separation efficiency and the absence of the condensed mobile phase, thus enabling a sensitive (down to the femtogram level), element-specific detection by atomic spectrometry, basically atomic absorption spectrometry (AAS) and atomic emission spectrometry (AES) [7, 8].

Metalloids, such as As and Se, are known to be metabolized by living organisms in a way that leads to the formation of a covalent bond between the heteroatom and the C atom incorporated in a larger structure (e.g., arsenosugars, Se-containing amino acids, peptides, and proteins). The complexation of metals by components of food leads to a number of relatively poorly characterized metal complexes. The resulting compounds are difficult to convert into volatile species, thus preventing the use of GC for their separation. Moreover, standards for most of these species are often unavailable since many of the naturally synthesized compounds have not yet been identified and characterized. These analytical challenges favor the use of separation techniques in the liquid phase, such as high-performance liquid chromatography (HPLC) and capillary zone electrophoresis (CZE) prior to element-specific detection. Inductively coupled plasma mass spectrometry (ICP-MS) has definitely emerged as a sensitive and selective detector of choice in hyphenated methods for the speciation analysis of elements in foods. Molecule-specific techniques, especially electrospray mass spectrometry (ES-MS), need to be employed for the identification of unknown species for which analytical standards are not available.

This chapter reviews the state-of-the-art of the hyphenated techniques employing ICP-MS detection for species-selective determination of additives and contaminants in food.

16.2 ELEMENTS AND SPECIES OF INTEREST IN FOODSTUFFS

The presence of elemental species in foodstuffs results from an endogenous synthesis by a plant or an animal (methyl-Hg, organo-As, or organo-Se species) or is due to anthropogenic external contamination (of environmental origin, or caused by processing or leaching from packaging materials).

From the chemical point of view, the metal and metalloid species found in foodstuffs can be subdivided into:

(1) Those which contain a covalent bond between a C atom and a metal or metalloid atom such as, for example:
   (a) Tetraalkylated Pb compounds used as antiknock additives to gasoline, which are degraded to trialkyl or dialkyl species;
   (b) Ingredients of antifouling paints, such as butyl-, octyl-, and phenyl-Sn species, released into the aquatic environment and found in seafood products;
   (c) Methyl-Hg as the product of biomethylation of Hg by marine organisms;
   (d) Products of the metabolism of As by marine biota leading to the formation of the C–As bond, as, for example, in arsenobetaine or arsenosugars;
   (e) Selenoamino acids, selenopeptides, and selenoproteins, biosynthesized by plants and animals;
(2) Coordination complexes in which the metal is coordinated by a bio-ligand present in a biological matrix (e.g., metal-binding peptides which are enzymatically synthesized in living organisms exposed to heavy metal stress).

16.3 SAMPLE PREPARATION METHODS

16.3.1 General Aspects

Foodstuffs represent a chemically complex matrix. Therefore, sample preparation for species-selective analysis is often troublesome, and developments in this area are attracting considerable attention.

16.3.2 Sample Preparation for GC-Based Hyphenated Techniques

16.3.2.1 Preliminary Comments The most popular protocol for organometallic contaminant species is based on the solubilization of a tissue in strongly alkaline media, for example, tetramethylammonium hydroxide. An aliquot of the solution is diluted with water and analyzed after derivatization. The tissue solubilization procedure can be considerably accelerated by carrying it out in a microwave (MW) field [9].

Speciation analysis for Hg in seafood was reviewed by Carro et al. [10]. The increased use of MW-assisted extraction techniques in speciation analysis has also been discussed with regard to GC-ICP-MS [11, 12].

16.3.2.2 Derivatization Techniques Although some authors still use the classical DDTC extraction in the presence of EDTA followed by butylation for organo-Pb speciation analysis [6, 7], the role of tetraalkylborates allowing the derivatization in the aqueous phase, such as $NaBEt_4$ for organo-Hg and organo-Sn speciation analysis, and the later introduced $NaBPr_4$ [12] for organo-Pb, is well established.

Two careful comparison studies are worth noting. In one of them three derivatization approaches, that is, anhydrous butylation using a Grignard reagent, aqueous butylation by means of $NaBEt_4$, and aqueous propylation with $NaBPr_4$, were compared for Hg speciation [13]. The absence of transmethylation during the preparation of fish tissue samples was checked using a 97 percent enriched ^{202}Hg inorganic standard [13].

16.3.2.3 Purge and Trap Using Capillary Cryofocussing A semiautomated, compact interface for time-resolved introduction of gaseous analytes from aqueous solutions into an ICP MS without the need for a full-size GC-oven was described [14]. Using this accessory, a method for the analysis of fish samples was developed and validated by the analysis of the Certified Reference Materials (CRMs) DORM-1 and TORT-1 produced by the National Research Council of Canada (NRCC) [11].

16.3.2.4 Solid-Phase Microextraction Solid-phase microextraction (SPME) is a preconcentration technique based on the sorption of analytes present in a liquid phase or, more often, in a headspace gaseous phase, on a microfiber coated with a chromatographic sorbent and incorporated in a microsyringe [15]. The analytes sorbed in the coating are transferred to a GC injector for thermal desorption.

SPME is an emerging analytical tool for elemental speciation in biological samples [16]. This solvent-free technique offers numerous advantages such as simplicity, use of a small amount of liquid phase, low cost and the compatibility with on-line analytical procedures.

SPME is based on the equilibrium between the analyte concentrations in the headspace and in the solid phase fiber coating. Low extraction efficiencies are hence sufficient for quantification, but the amount of available analyte may be very small. Hence, there is an interest in the coupling of highly sensitive GC-ICP-MS with SPME.

Headspace SPME, using a polydimethylsiloxane fiber at no equilibrium conditions, was optimized as an extraction/preconcentration method for triphenyl-Sn residues in tetramethylammonium hydroxide (TMAH) and KOH–EtOH extracts of potato and mussel samples [17]. Derivatization was carried out with $NaBEt_4$ for 10–20 min [17]. Direct SPME led to a 10-fold lower sensitivity. Limits of detection (LoDs) in the low nanogram per gram range (dry weight) were reported [17]. Slightly better LoDs (0.6–20 pg l^{-1}) were obtained for fish tissue in another work [18].

The direct coupling of SPME with ICP-MS was described for the species-specific determination of methyl-Hg [19]. A fiber was inserted into a splitless-type GC injector, which was placed directly at the base of the torch. Immersion SPME was severely influenced by the matrix and led to a 70-fold decrease in sensitivity [19]. Analytical results showed good agreement between certified and measured values for the analysis of the NRCC seafood CRMs DORM-2 Dogfish muscle and DOLT-2 Dogfish liver [19].

16.3.3 Sample Preparation for Speciation Analysis by Liquid-Phase Separation Techniques Coupled with ICP-MS

16.3.3.1 Leaching with Neutral pH Solutions Leaching with water, water–methanol mixtures, and neutral buffers is an established technique for recovery of low molecular species of As and Se and metal complexes from solid foodstuffs.

Since selenoamino acids are water-soluble, leaching with hot water has been judged sufficient to recover Se species not incorporated into larger molecules. The sample is homogenized with water, sonicated or heated, and ultracentrifuged. The typical recovery of Se extracted in this way from selected yeast samples is usually ca. 10–20 percent [20, 21]. Free selenoamino acids can also be separated by ultrafiltration (breast milk) [22].

Twenty-seven different procedures employing nine buffering systems at a pH between 1 and 9 were compared for the extraction of Se compounds from high-Se-containing broccoli [23]. In extractions using nonbuffered solvents, more than 40 percent of the spiked Se-methylselenocysteine was not recovered in the filtrate. When buffer solutions were used for extraction, losses for Se-methylselenocysteine ranged from 10 to 20 percent. Mass balance indicated that approximately 30 percent of the naturally occurring Se in broccoli samples was volatilized and lost to the atmosphere in the course of extractions with buffer solutions [23].

Sonication with a water–methanol solution is the most popular method for the extraction of As species from rice powder [24, 25], algae, chicken meat [26], oyster tissue [27, 28], and baby foods [29]. Sample treatment with trifluoroacetic acid at 100°C was reported to be an efficient method for the extraction of As species present in different food matrices when compared with alternative methods that included sonication and accelerated solvent extraction. Extraction recoveries from 94 to 128 percent were obtained [29]. Low-molecular-weight Se compounds were extracted from nuts with $HClO_4$ to produce a fraction containing 3 to 15 percent of the total Se in different types of nuts [30].

16.3.3.2 Extraction of Se-Containing Proteins with Denaturating Conditions The use of sodium dodecyl sulfate (SDS) or 3-[(3-cholamidopropyl)dimethylammonio]-1-propanesulfonate (CHAPS) is particularly recommended when the element is incorporated or strongly complexed by the moiety of interest. The most widely investigated example is the recovery of the selenoprotein fraction from selenized yeast [31, 32]. A novel integrated approach was proposed for the analysis of intact yeast Se-containing proteins purified by a gel electrophoresis (GE) technique. The strategy consisted of: (1) high-resolution (2-DE) two-dimensional GE for proteins; (2) laser ablation (LA) – dynamic reaction cell (DRC) ICP-MS for Se detection; and (3) protein characterization by ES-MS. SDS–polyacrylamide–GE – a denaturating technique – is applicable in this case as Se is covalently bound to the proteins. Up to 10 Se-containing proteins were characterized in terms of molecular masses in the range from 9 to 20 kDa. This strategy can be successfully applied to the construction of a 2D reference map (Se content and molecular masses) for Se-containing proteins in yeast [33]. The proteins were isolated from nut samples by dissolution in 0.1 M NaOH, precipitated with acetone, and then dissolved in phosphate buffer at pH 7.5 prior to HPLC-ICP-MS analysis [30].

16.3.3.3 Enzymatic Extraction of Organometalloid Compounds The recovery of selenoamino acids from selenized-yeast-based food supplements can be increased to above 95 percent by degrading the species originally present with a mixture of proteolytic enzymes [31, 32]. The enzymes used in most digestion protocols were pure proteinase K [34] or a mixture of proteolytic enzymes [31, 32]. The use of protease XIV and proteinase K provided extraction yields of about 70 percent in plant tissues. The addition of cellulase to proteinase K was found not to offer any improvement in recovery, probably because the initial sample treatment with liquid nitrogen had already destroyed the cellulose walls of the tissue [34]. Proteinase K extraction was considered to be adequate only when the species of interest were not very labile. A dramatic enhancement of the activity of two proteolytic enzymes (protease XIV and subtilisin) was reported in a high-energy ultrasonic field [35].

Different enzymatic protocols were compared to elucidate the possible role of the cell-wall-digesting enzymes for the improvement of the efficiency of Se extraction from edible mushrooms [36]. A three-step enzymatic procedure gave the highest extraction efficiency (89 percent) [36]. A sequential sample preparation process was proposed for the same Se-enriched edible mushroom (*Agaricus bisporus*) after comparing five different sample extraction methods. The most efficient turned out to be a three-step protocol using water extraction and two proteolytic enzymes (pepsin and trypsin) with an extraction efficiency of 75 percent [37]. Hydrolysis carried out by boiling the sample with 4 M methanesulfonic acid under reflux was claimed to offer higher recovery of selenomethionine in comparison with the proteinase K–protease XIV procedure [38]. During chromatography of enzymatic digests, the recoveries from the column were not complete (40–90 percent). This was attributed to the presence of strongly hydrophobic peptides resulting from the incomplete decomposition of proteins to amino acids [34, 39–41].

Four different extraction procedures have been evaluated for a quantitative recovery of Se species during extraction from cod muscle. The highest Se recoveries were obtained in the presence of enzymes, whereas only 5 percent of total Se in cod was extracted when a "soft" extraction procedure (MeOH–HCl) was used [42]. The use of proteolytic enzymes allowed to extract more than 90 percent of Se present in the wheat flour samples and more than 85 percent of Se from yeast samples [43].

Enzymatic procedures were also reported for the recovery of organo-As species. A two-step procedure for the extraction of As from freeze-dried apples using a treatment with α-amylase enzyme followed by sonication with 40 percent acetonitrile was found to provide good extraction efficiency [44].

Cellulose and complex pectic polysaccharides are the main matrix of the water-insoluble residue after centrifugation of fruit and vegetable homogenates. The use of pectinolytic enzymes is therefore necessary to solubilize the solid sample. Pectinolysis is known to degrade efficiently large pectic polysaccharides, but some of them, for example, rhamnogalacturonan-II, are considered to be resistant to pectinolytic enzymes [45]. A mixture of commercial products: Rapidase LIQ™ and Pectinex Ultra-SPL™, was reported for the release of metal-complexes from the solid parts of edible plants, fruits, and vegetables [45].

Water-soluble polysaccharide species with higher molecular weights can be readily degraded by enzymic hydrolysis with a mixture of pectinase and hemicellulase to release the dRG-II complex [45]. The same mixture was found to be efficient to extract the dRG-II–metal complexes from water-insoluble residues of vegetables, owing to the destruction of the pectic structure [45].

16.3.3.4 Sequential Enzymatic Extractions for the Evaluation of the Bioaccessibility of Metals in Foodstuffs Some attention has been paid to the analysis of enzymatic digests of foodstuffs in the quest for molecular information to contribute to the knowledge of the bioavailability of some elements. Sequential enzymolyses in simulated gastric and gastrointestinal juice were proposed for an estimation of bioavailability of heavy metal species from meat [46] and cocoa samples [47, 48]. The soluble fractions of the stomach and upper intestinal contents of guinea pigs on different diets were investigated for the species of Al, Cu, Mn, Rb, Sr, and Zn [49].

An in vitro model simulating enzymatic activity in the gastrointestinal tract of monogastric species was developed for the assessment of the potential bioavailability of Cd and Pb in cocoa powder and liquor of different geographical origins. The model was based on the sequential extraction with simulated gastric and intestinal juices. The residue after the extraction was further investigated by using, in parallel, solutions of phytase and cellulase [47].

The chemical stability of four arsenosugars (molecular weight within parentheses) {3-[5′-deoxy-5′-(dimethylarsinoyl)-β-ribofuranosyloxy]-2-hydroxypropylene glycol (328); 3-[5′-deoxy-5′-(dimethylarsinoyl)-β-ribofuranosyloxy]-2-hydroxy propanesulfate (392); 3-[5′-deoxy-5′-(dimethyl-arsinoyl)-β-ribofuranosyloxyl-2-hydroxy propylhydrogen sulfate (408), and 3-[5′-deoxy-5′-(dimethylarsinoyl)-β-ribofuranosyloxy]-2-hydroxypropyl-2,3-hydroxypropyl phosphate (482)}, isolated from seaweed extracts was investigated [50]. The arsenosugars were subjected to simulated gastric juice and acidic artificial stomach conditions. They were found to degrade at a rate of 1.4 percent per h at 38°C and 12.2 percent per h at 60°C. A single common degradation product was observed in all the stability studies. It was tentatively identified by ES-MS/MS as a fragment with molecular weight 254. An acid hydrolysis mechanism was proposed for its formation from each of the native arsenosugars [50].

The in vitro gastric and intestinal digests of selenized yeast food supplements were investigated by successive HPLC-ICP-MS and HPLC-ES-MS/MS [51]. The main compound extracted by both gastric and intestinal juice was Se-methionine, which was also the main Se compound extracted by proteolytic digestion from the yeast supplements. Two other minor compounds could be identified as Se-cystine and Se(O)-methionine, a degradation product of Se-methionine [51].

16.3.3.5 Fractionation of Metal-Species According to the Molecular Weight Prior to Analytical Chromatography or Capillary Electrophoresis The complexity of food matrices often makes it necessary to separate the molecular mass fraction containing the species of interest from high-molecular-weight polysaccharides or inorganic salts that would otherwise foul the analytical column used in the subsequent analysis. The simplification of the matrix often allows the analytes in the isolated fraction to be preconcentrated by means of lyophilization [20]. Care should be taken to avoid possible losses of volatile species. The two most common techniques used for the fractionation of extracts of biological tissues include ultrafiltration and gel filtration. The most frequent is filtration using 10 kDa cut-off filters as described for selenoamino acids [52] and metal complexes [53] in milk. The recent advances in the size-exclusion (SE) gel manufacturing technology allow for a relatively high peak capacity of the columns. The fractionation of Se species using a Superdex Peptide gel allowed several organo-Se fractions to be separated in a selenized-yeast water extract [31, 32]. It should be noted that separation mechanisms other than SE often play a role in the fractionation process and the calibration of the column with the molecular weight markers may be problematic.

16.4 HYPHENATED TECHNIQUES IN SPECIATION ANALYSIS

A fit-for-purpose analytical technique for speciation analysis should address three major issues:

(1) Selectivity of the separation technique – this should allow the target analyte species to arrive at the detector well separated from potential matrix interferents and from other species of the same element.
(2) Sensitivity of the element or molecule-specific detection technique – the already low total concentrations of trace elements in environmental samples are obviously even lower in the case of the their various chemical forms.
(3) Species identification – the retention time matching that is usually employed requires the availability of standards. When standards are not available, the use of a molecule-specific detection technique is mandatory.

The choice of hyphenated techniques available for speciation analysis in foods is schematically shown in Figure 16.1. In the most frequent cases a separation technique, for example, GC, HPLC, electrochromatography or GE, is combined with ICP-MS.

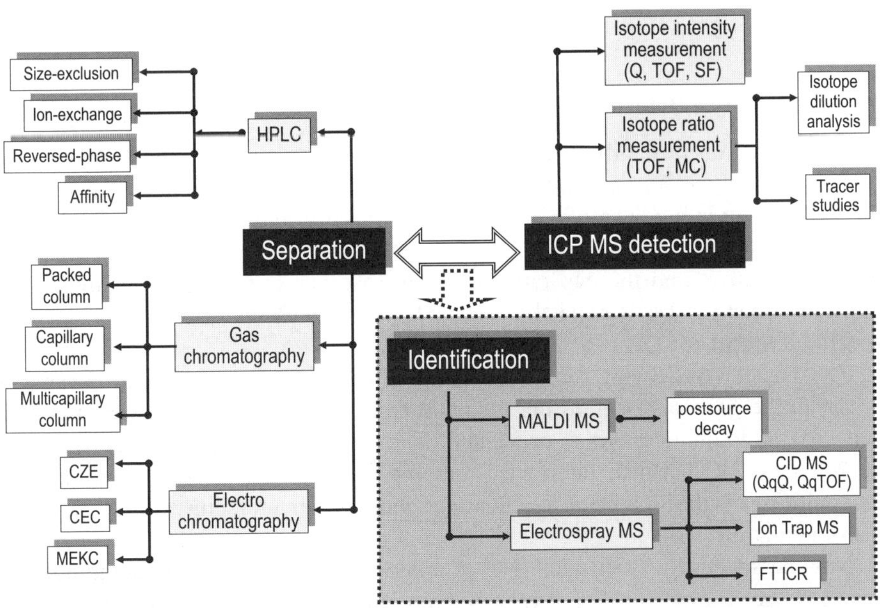

Figure 16.1. Coupled techniques used in speciation analysis of foods.

The separation component of the coupled system becomes of particular concern when the target species have similar physical–chemical properties. GC should be chosen wherever possible because it offers a high separation efficiency and very low LoDs can be achieved because of the absence of the condensed mobile phase. For nonvolatile species liquid phase column separation techniques, such as HPLC and CE, are the usual choice. They can be easily coupled on-line with ICP-MS. A variety of separation mechanisms and mobile phases can be used to allow the species identity to be preserved.

For element-specific detection in GC, a number of dedicated spectrometric detection techniques can be used, for example, quartz furnace AAS or atomic fluorescence spectrometry (AFS) for Hg, or microwave-induced plasma atomic emission spectrometry (MIP-AES) for Pb or Sn. However, ICP-MS is virtually the only technique capable of coping, in the on-line mode, with the trace element concentrations in liquid chromatography (LC) and capillary electrophoresis (CE) effluents. The femtogram level absolute LoDs may still turn out to be insufficient if an element present at the nanogram per milliliter level splits into a number of species, or when the actual amount of sample analyzed is limited to some nanoliters as in the case of CE or nanoflow HPLC. The isotope specificity of ICP-MS offers a still underexploited potential for tracer studies and for improved accuracy via isotope dilution analysis.

The third, often indispensable, component of an analytical strategy is the identification and characterization of metallospecies, especially those for which standards are unavailable. This can be achieved by ES-MS or MALDI. Structural information can be acquired by collision-induced dissociation (CID) of an ion selected by a quadrupole (Q) mass filter followed by a product ion scan using a Q or a time-of-flight (TOF) mass analyzer.

16.4.1 ICP-MS Detection in GC

The combination of capillary GC with ICP-MS has become an ideal approach to speciation analysis for organometallic compounds in complex matrices because of the high resolving power of GC and the sensitivity and specificity of ICP-MS. Indeed, features of ICP-MS such as low LoDs reaching the femtogram level, high matrix tolerance, and the capacity to measure isotope ratios, thus enabling accurate quantification by isotope dilution (ID), make ICP-MS the leading element-specific detector for GC.

The basic requirement for an interface is that the analytes should be maintained in the gaseous form during transport from a GC column to the plasma, in a way that any condensation is prevented. The interface designs were reviewed in detail [8]. A GC-ICP-MS interface is commercially available which indicates the recognition of the maturity of this coupling by manufacturers of analytical instrumentation.

New perspectives for GC-ICP-MS in a variety of research areas including food sciences are in sight on the levels of both sample introduction and mass spectrometry [54], namely, sample preparation methods including MW-assisted, solid-phase microextraction or purge and capillary trap automated sample introduction systems, miniaturization of GC hardware allowing for the time-resolved introduction of gaseous analytes into the plasma, for example, based on microcolumn multicapillary GC, wider availability of collision cell (CC) technology and TOF detection, and last, but not least, the emergence of GC-sector field (SF) ICP-MS with ion multicollection. The most popular application of GC-ICP-MS in food speciation analysis is the determination of methyl-Hg in seafood samples [10, 11, 13, 19].

16.4.2 Coupling of Liquid Separation Techniques with ICP-MS

16.4.2.1 HPLC-ICP-MS Coupling The principal HPLC separation mechanisms used in speciation analysis of foodstuffs include SE chromatography (SEC), ion-exchange chromatography (IEC), and reversed-phase chromatography (RPC). Capillary electrophoresis (CE) is less mature, but offers exciting possibilities for speciation analysis owing to the high separation efficiency, the nanoliter sample requirement, and the absence of packing susceptible to interacting with metals and affecting the complexation equilibria. With the combination of electrophoretic and electroosmotic flows, a wide variety of positive, neutral, and negative ions and compounds can be separated in one run.

The key to a successful HPLC-ICP-MS coupling is the interface. In the simplest case the outlet of an HPLC column (4.6–10 mm) is connected to a conventional pneumatic or crossflow nebulizer. The use of capillary or megabore (0.32–1.0 mm) HPLC systems that are becoming popular, especially for RPC, requires the use of micronebulizers, either direct injection nebulizers (DIN) or micronebulizers (e.g., Micromist, DS-5) fitted with a small-volume nebulization chamber.

SEC coupled to ICP-MS allows the presence of stable metal complexes and metalloid species in liquid samples to be monitored, for example, fruit juices [55], milk [56, 57], and tea infusions. The peak width is sufficiently large to allow a Q mass analyzer to be used for the simultaneous monitoring of up to 12 isotopes. The molecular weight of the analyte can be determined through the correlation of the elution volume with the molecular weight of the eluted molecule. On this basis a hypothesis regarding the identity of the eluted species can be put forward.

A finer characterization can be achieved by ion-exchange (IE) or reversed-phase (RP) HPLC-ICP-MS. Both cation and anion exchangers were widely used for the separation of metal species, especially organo-As [24, 58, 59] and organo-Se compounds [60]. The high content of organic modifier makes RP-HPLC poorly compatible with ICP-MS, resulting in plasma instability and carbon build-up. The use of a cooled spray chamber, low-flow nebulizers, and the addition of oxygen usually circumvents these problems.

16.4.2.2 CZE-ICP-MS Coupling The coupling of CZE to ICP-MS is less straightforward than that of HPLC. The problems due to the laminar flow generated by the nebulizer suction – loss of sensitivity because of electroosmotic flow dilution by the make-up liquid and peak broadening in the spray chamber – were resolved in the commercially available interface based on a total-consumption, self-aspirating micronebulizer fitted with a small-volume spray chamber [61].

16.4.3 ICP-MS Detection in Hyphenated Systems

The use of a Q mass analyzer in ICP-MS detection is the most widespread. The instrumentation undergoes constant improvement, which leads to a wider availability of more sensitive, less interference-prone, smaller, and cheaper instruments. All these features contribute to a wider use of ICP-MS as the chromatographic detector. The latest generation of instruments offers sub-femtogram absolute LoDs for many metals. The isobaric overlaps are generally not a problem because of the on-line separation from the potential interferents, for example, ^{35}Cl ($^{40}Ar^{35}Cl$) in the case of the ^{75}As determination, although ghost peaks may appear. The application of a double-focusing SF instrument offers the higher resolution that may be required for the interference-free determination of S or of the isotope ratios of some elements, for example, Cr, Fe, and V. An increase in resolution inevitably leads to a dramatic decrease in sensitivity. It should also be noted that the sensitivity of the latest generation Q instruments is only a 2- to 3-fold lower than that of SF-ICP-MS operated in the low-resolution mode. A good trade-off between sensitivity, freedom from isobaric interferences, and price is offered by ICP-MS instruments equipped with a CC, which have recently proliferated on the market.

Both Q and SF mass spectrometers are scanning (sequential) analyzers and multiisotope analysis can be achieved at the expense of the measurement sensitivity and precision. The sequential measurement of m/z at different points within a time-dependent concentration profile of a transient signal can result in peak distortions and quantification errors commonly referred to as spectral skew. The alternative is TOF-MS which features the ability to produce a complete atomic mass spectrum in less than 50 μs and thus allows very brief transient signals to be recorded with high fidelity. This is especially useful in the on-line isotope ratio determination. However, a 10-fold loss in sensitivity of a TOF-ICP-MS instrument in comparison with the latest Q instruments often creates an obstacle for the wider application of TOF-ICP-MS as a detector in the CE of metallobiomolecules in biological samples.

16.5 SELECTED APPLICATIONS OF CHROMATOGRAPHY WITH ICP-MS DETECTION FOR SPECIES-SPECIFIC DETERMINATION OF METAL(LOID)-CONTAINING FOOD ADDITIVES AND CONTAMINANTS

16.5.1 Alkylmetal Species

16.5.1.1 Organo-Pb Compounds The widespread use of organo-Pb compounds [$Me_nEt_{4-n}Pb$ ($n = 0–4$)] as antiknock additives to gasoline has made Pb one of the most ubiquitous metal contaminants in the environment. However, owing to severe restrictions regarding the use of leaded petrol, the concentration levels of organo-Pb compounds are decreasing and, although GC-ICP-MS is potentially a highly suitable method for alkyl-Pb speciation, there are no literature reports on its application to food analysis.

16.5.1.2 Methyl-Hg Determination in Seafood The interest in the speciation of Hg is due to the enhanced toxicity of some of its specific chemical forms. Methyl-Hg (MeHg), the most common organo-Hg compound, is of special concern because of the ease of penetrating biological membranes, high stability and long residence time in tissues. The ability of MeHg to accumulate in fish tissues causes Hg poisoning in marine life and is responsible for a risk associated with seafood consumption by man.

In terms of methodology, the speciation analysis of Hg has reached its maturity. Because of a convenient conversion of Hg species into volatile compounds, GC has been the dominant separation technique prior to Hg specific ICP-MS detection [10, 11, 13, 19]. HPLC methods cannot compete in terms of figures of merit with GC. HPLC-ICP-MS is, however, a valuable independent analytical technique that allows the formation of artifacts during derivatization in GC to be controlled. Owing to the availability of a number of CRMs, analytical procedures for the analysis of seafood samples were extensively validated and can be applied in routine analysis.

Homogeneity, storage conditions, and stability of analytical species and sample materials have great influence on analytical results. Sub-sampling of half-frozen, partly thawed fish and analysis of refrozen fish samples were reported to give significantly different concentrations, which on average differed by a factor of 2. The concentration range of total Hg measured in the shark fillets was between 0.9 and 3.6 $\mu g\ g^{-1}$ in thawed shark fillet. Speciation analysis demonstrated that about 94 percent Hg was present as MeHg [62]. Figure 16.2 shows a multicapillary GC-ICP- MS chromatogram of a CRM analyzed for Hg speciation [11]. An attractive feature of the system used is of the possibility to separate Hg species isothermally and at ambient temperature, thus eliminating the need for a chromatographic oven [11].

Speciation analysis for Hg in seafood was reviewed [63]. Method performance evaluation for MeHg determination in fish has been discussed in detail [64].

16.5.1.3 Speciation of Organotins in Seafood Organo-Sn compounds have the general formula R_nSnX_{4-n}, where n denotes an alkyl or aryl group, usually butyl-, octyl-, cyclohexyl-, or phenyl-, and X stands for the counterion, usually a halide, hydroxide, or acetate. They have a number of industrial applications, but particular attention has been paid to the fate of the most toxic trialkyl and triaryl derivatives released directly into the marine environment from the leaching of the tributyl-Sn-based antifouling paints used on boats and ships. Thus, the species selective determination of organo-Sn compounds in seafood is of

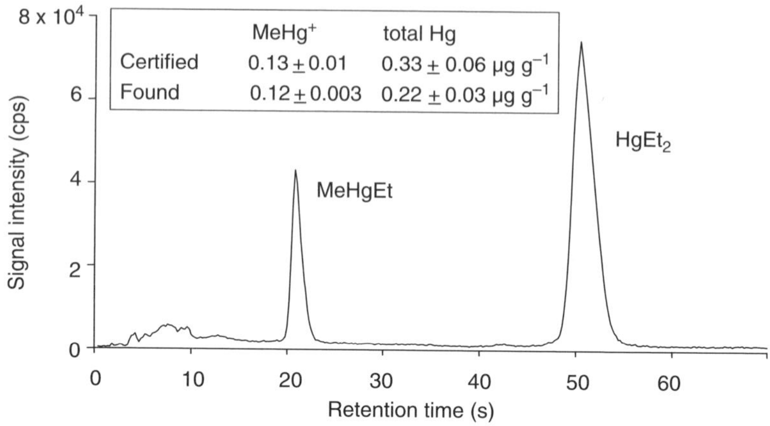

Figure 16.2. Multicapillary GC-ICP-MS chromatogram of a seafood CRM (crab tissue, TORT-1, NRCC).

paramount importance. A typical set of organo-Sn species to be determined in marine food samples includes mono-, di-, and tributyl-Sn and mono-, di-, and triphenyl-Sn.

The use of HPLC for the separation of organo-Sn compounds has been scarce. The technique allows for the baseline separation of the butyl- and phenyl-Sn compounds although the LoDs obtained by HPLC-ICP-MS are much worse than with GC-ICP-MS. Therefore, despite the need for a prior derivatization of organo-Sn compounds, GC has been accepted as the separation method of choice because of its robustness and high sample throughput. HPLC-ICP-MS, however, should be considered as a valuable independent technique to be used for method validation and production of CRMs.

ICP-MS is only one of several detection techniques used for species-selective analysis of Sn in food matrices; other detectors in use include electron impact (EI) MS, flame photometric detection (FPD), pulsed FPD and MIP-AES.

An attractive feature of ICP-MS is its isotope specificity, which allows isotope dilution (ID) analysis to be used. In vitro gastrointestinal digestion was employed in combination with species-specific ID analysis to study the solubilization and transformation of butyl-Sn species throughout a simulated human digestion tract. A biological matrix (mussel tissue) was spiked with isotopically labelled butyl-Sn compounds prior to enzymatic digestion and the degradation of organo-Sn species was monitored. The analysis of the soluble and insoluble fractions resulting from the simulated gastric and gastointestinal digestion of the mussel tissue provided information on the solubilization and the degradation of the butyl-Sn species during the digestion procedures [65].

Speciation analysis for organo-Sn compounds in seafood was the topic of many interlaboratory studies [66, 67] and resulted in the production of CRMs (Table 16.1).

TABLE 16.1. Standard Reference Materials for Species Selective Analysis in Food Matrices

Element	Standards Reference Material	Species Certified (Concentration)
Sn	BCR 477 Mussel Tissue	MBT (1.50 ± 0.28 mg kg^{-1}) DBT (1.54 ± 0.12 mg kg^{-1}) TBT (2.20 ± 0.19 mg kg^{-1})
Sn	NIES 11 Sea Bass	TBT(1.3 ± 0.1 mg kg^{-1})
Hg	BCR 463 Tuna Fish	$CH_3H_g^+$ (2.85 ± 0.16 mg kg^{-1})
Hg, As	NRCC DORM-2 Dogfish muscle	$CH_3H_g^+$ (4.47 ± 0.32 mg kg^{-1} as Hg) Arsenobetaine ($16.4 + 1.1$ mg kg^{-1} as As) Tetramethylarsonium ($0.248 + 0.054$ mg kg^{-1} as As)
Hg	NIST 1566b Oyster Tissue	0.0132 ± 0.00007 mg kg^{-1} as Hg
Hg	NIST 2976 Mussel Tissue	$CH_3H_g^+$ (61.0 mg kg^{-1})
Hg	NRCC TORT-2 Lobster Hepatopancreas	$CH_3H_g^+$ (0.152 ± 0.013 mg kg^{-1} as Hg)
As	BCR 627 Tuna Fish	Arsenobetaine (52 ± 3 μmol kg^{-1}) Dimethylarsenic acid (2 ± 0.3 μmol kg^{-1})
Hg	NRCC LUTS-1 Nondefatted Lobster Hepatopancreas	CH_3Hg^+ (0.0094 ± 0.0006 mg kg^{-1} as Hg)

16.5.2 Species Selective Analysis of As in Food

Speciation of As in food is of particular interest due to the potential accumulation of this element in the food chain and the risk to man. The metabolism of inorganic As by marine and terrestrial plants and animals leads to the formation of a range of organic As species that may be considered as naturally occurring compounds. The most widely referred species of this group are a quaternary arsenocompound (arsenobetaine), which is the major organo-As compound in marine animals, and arsinoyl-ribosides (arsenosugars), which are products of the As metabolism in marine plants and some bivalves. Health risk associated with dietary As intake may be different for infants and adults. Seafood is the main contributor to As intake for adults, while terrestrial-based food is the primary source for infants. Another field of interest includes studies of the metabolism of As following its administration to humans and experimental animals. The most important As species of interest for food analysis are listed in Figure 16.3.

The literature on speciation of As in food is extremely rich and concerns mostly seafood samples. A survey of the analytical techniques used can be found in several extensive reviews [59, 68, 69].

Anion-exchange (AE) HPLC-ICP-MS seems to be the most attractive technique for a comprehensive speciation analysis of organo-As species in biological materials. An example chromatogram for such an analysis is shown in Figure 16.4. Cation-exchange HPLC has often been used in parallel to AE-HPLC to confirm

Arsenous acid (arsenite)	$OH\text{-}As(OH)_2$
Arsenic acid (arsenate)	$O{=}As(OH)_3$
MMA	$CH_3AsO(OH)_2$
DMA	$(CH_3)_2AsO(OH)$
Tetramethylarsonium oxide	$(CH_3)_3AsO$
TMAs+	$(CH_3)_4As^+$
Arsenobetaine	$(CH_3)_3As\text{-}CH_2\text{-}COOH$
Arsenocholine	$(CH_3)_3As\text{-}CH_2\text{-}CH_2\text{-}OH$
DMAsEt	$(CH_3)_2AsO(C_2H_5)$
DMAsAc	$(CH_3)_2AsO(CH_3CO)$

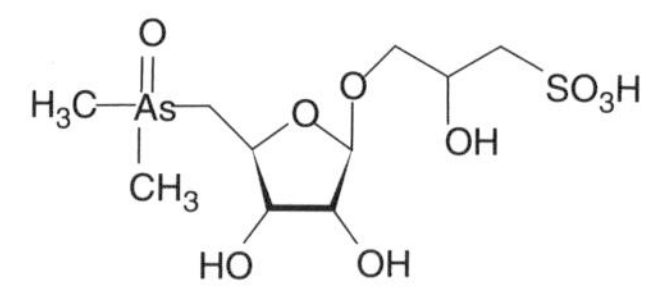

Arsenosugar 392

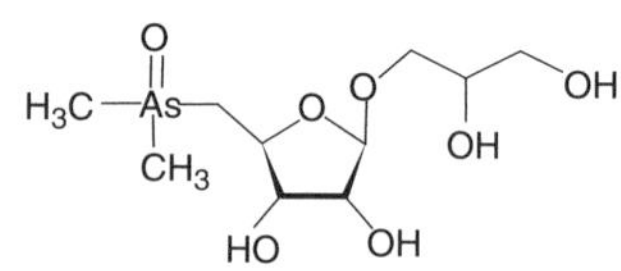

Arsenosugar 328

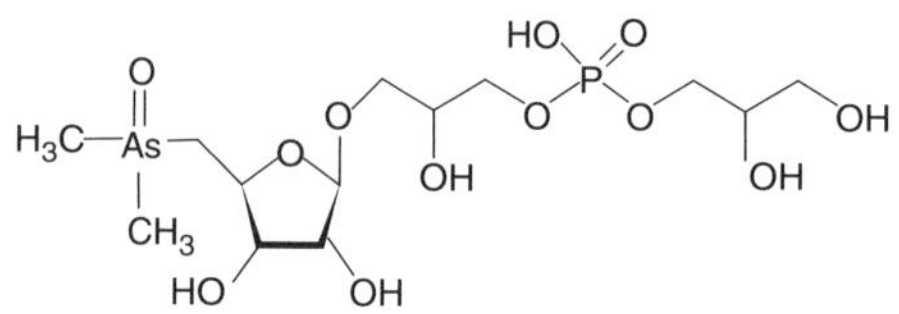

Arsenosugar 482

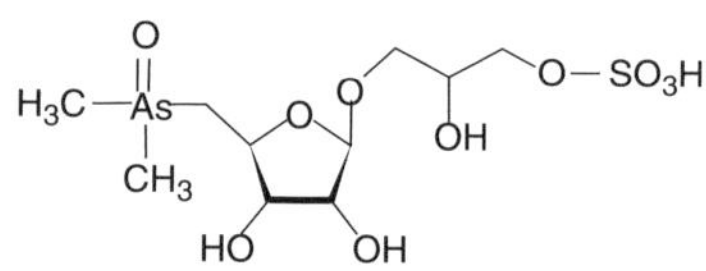

Arsenosugar 408

Figure 16.3. Arsenic species of interest in food speciation analysis.

the identity of the analyte species, especially for samples that might contain arsinoylribosides. Most organo-As species are not retained on an RP column; the addition of an ion-pairing reagent is required to achieve their separation.

Relatively new areas of species-selective analysis of As in food concern terrestrial plants [44, 70], especially rice [24, 29, 71], and chicken meat [26].

Four forms of As, namely, inorganic As (III), dimethylarsinic acid (DMA), monomethylarsonic acid (MMA), and inorganic As (V), were detected in rice [25], but in another study only As(V) and MMA were found [24]. As the concentration of As in the rice grain was low, As speciation was performed only on rice straw to predict the risk associated with feeding contaminated straw to the cattle. Speciation of As in tissues by HPLC-ICP-MS revealed that the predominant species present in straw was arsenate followed by arsenite and dimethylarsinic acid (DMA). As DMA was only present at low concentrations, it was considered unlikely this would greatly alter the toxicity of As present in rice [71].

Processed infant food products such as rice-based cereals, mixed rice/formula cereals, milk-based infant formula, apple sauce, and puree of peaches, pears, carrots, sweet potatoes, green beans, and squash were analyzed [29]. Inorganic

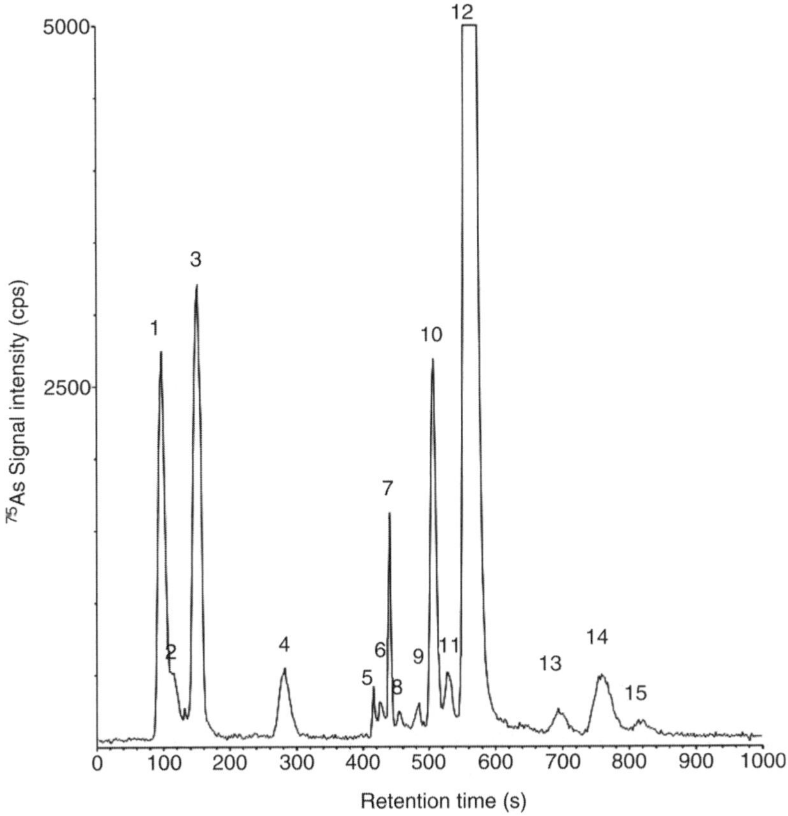

Figure 16.4. HPLC-ICPMS chromatogram of oyster tissue: (1) arsenite; (2) unknown; (3) MMA–PO_4-arsenosugar; (4) DMA; (5) SO_4-arsenosugar; (6) arsenate; (7), (8) and (9) unknown; (10) OH-arsenosugar; (11) unknown; (12) arsenobetaine; (13) TMAO; (14) arsenocholine; (15) TMAs (courtesy of Dr Ute Kohlmeyer, GALAB Laboratories, Geesthacht).

As and DMA were found to be the main species in rice-based and mixed rice/formula cereals, although traces of MMA were also detected. Inorganic As was present in freeze-dried sweet potatoes, carrots, green beans, and peaches. MMA and DMA were not detected in these samples [29].

Concentrations and speciation of As in poultry are of particular interest since this is the most widely eaten meat in many countries. Chickens are extensively fed fish meal and water that may contain a relatively high amount of As. A method was developed for the speciation analysis of As in chicken meat. Different procedures were optimized for the recovery of As compounds without destroying the original compounds. Two AE-HPLC columns were compared for the separation of As species prior to on-line detection by ICP-MS. The two species found were dimethylarsinic acid and arsenobetaine [26].

16.5.3 Selenium in Plants and Selenized Yeast Food Supplements

16.5.3.1 General Aspects Selenium has been shown to be both essential for life and toxic at levels little above those required for health. Indeed, dietary levels of the desired amount of Se are in a very narrow range: consumption of food containing less than 0.1 mg kg^{-1} of this element will result in its deficiency whereas dietary levels above 1 mg kg^{-1} will lead to toxic manifestations [72]. Selenium exists in food samples in the form of inorganic species such as SeO_3^{2-} or SeO_4^{2-} or in the form of organic species having a range of molecular masses and charges, starting from the simplest MeSeH and ending at complex selenoproteins. The development of analytical methodologies for Se speciation has been gaining momentum since the study of Clark et al. indicated the putative role of selenized yeast in cancer prevention [73]. Chemical forms of Se reported in the context of speciation studies in food are summarized in Figure 16.5. The most widely used technique for Se speciation in food is ion-pairing-ICP-MS [39, 41, 74–77], cation-exchange-ICP-MS [60], or AE-HPLC-ICP-MS [43]. The method based on reaction with CNBr followed by GC-ICP-MS was recommended [78]. ID GC-MS was used for quantification of selenomethionine in wheat-based CRMs [78], cod muscle [42], yeast, and wheat flour [43].

Fish, shellfish, red meat, wheat, liver, and some vegetables (e.g., garlic and broccoli) are all good dietary sources of Se, but speciation studies published so far have mainly focused on vegetables and selenized yeast-based food supplements.

16.5.3.2 Se Speciation in Plants The presence of a number of volatile Se species was reported in edible allium plants such as garlic by GC-AES [79]. Selenomethionine, Se-methylselenocysteine, and γ-glutamyl-Se-methyl-L-seleno cysteine were identified in garlic and onion by HPLC-ICP-MS and ES-MS/MS [80]. Selenomethionine is the primary species found in all types of nuts (19–25 percent of the total Se) [30], sunflower [81], and mushrooms [36, 37]. The distribution of Se among different fractions (lipid extract, low molecular weight, and protein fractions) of nuts and speciation analysis was studied [30]. Selenium was not detected in any of the lipid extracts obtained from the different types of nuts [30]. Results obtained for Brazil nuts by SEC with on-line ICP-MS detection showed that approximately 12 percent of total Se was weakly bound to proteins [30].

Values of the total Se and selenomethionine content of four wheat-based CRMs were obtained by GC-stable ID MS methods [78]. The data obtained indicate that the four wheat samples (wheat gluten, durum wheat, hard red spring wheat, and soft winter wheat), though having a 30-fold range in total Se content, all have about 45 percent of their total Se values in the form of selenomethionine. Additional experiments were required to verify that all selenomethionine in the wheat samples was accounted for [78].

16.5.3.3 Selenium Speciation in Selenized Yeast-Based Food Supplements One of the most economic sources of organic forms of Se is yeast grown in Se-enriched media [83]. The popularity of selenized yeast-based food supplements

Species	Formula
Selenite	$\mathbf{Se}O_3^-$
Selenate	$\mathbf{Se}O_4^-$
Trimethylselenonium ion	$Me_3\mathbf{Se}^+$
Selenocysteine	H_3N^+-CH(COO$^-$)-CH_2-**Se**H
Selenocystine	H_3N^+-CH(COO$^-$)-CH_2-**Se-Se**-CH_2-CH(COO$^-$)-NH_3^+
Selenomethionine	H_3N^+-CH(COO$^-$)-CH_2-CH_2-**Se**H
Semethylselenocysteine	H_3N^+-CH(COO$^-$)-CH_2-**Se**-CH_3
γ-Glutamyl-Se-methylselenocysteine	H_3N^+-CH_2-CH_2-CO-NH-CH(COO$^-$)-CH_2-**Se**-CH_3
Selenocystathionine	H_3N^+-CH(COO$^-$)-CH_2-CH_2-**Se**-CH_2-CH(COO$^-$)-NH_3^+
Selenohomocysteine	H_3N^+-CH(COO$^-$)-CH_2-CH_2-**Se**H
Selenocystamine	H_3N^+-CH_2-CH_2-**Se-Se**-CH_2-CH_2-NH_2

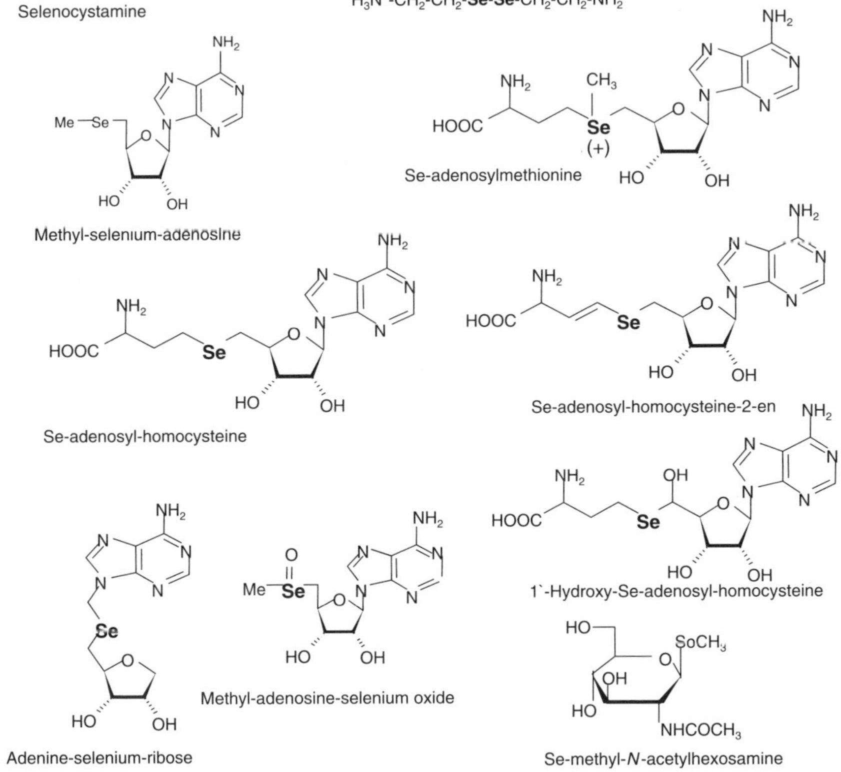

Figure 16.5. Selenium species of interest in food speciation analysis.

makes yeast one of the most often investigated samples in food speciation analysis. Analytical approaches to speciation of Se species in yeast, usually by HPLC-ICP-MS, have been discussed in detail in review and research papers [31, 32, 82]. The most abundant species found in enzymatic extracts of selenized yeast is selenomethionine [34, 39, 43, 60, 76, 84]. The water soluble fraction of selenized yeast has been the subject of many studies with parallel ICP-MS and ES-MS/MS detection in HPLC, which allowed a number of seleno amino acid adenosyl derivatives [20, 21] and Se-containing proteins to be detected and identified [85].

16.5.3.4 Selenium Speciation in Edible Animal Tissues Reports on Se speciation analysis in edible animal tissues have been scarce. Speciation analysis of Se in cod muscle tissue was performed by separating the species using both RP- and SE-HPLC prior to ICP-MS detection. The main Se compound found in enzymatic hydrolysates was selenomethionine [42]. This selenocompound was absent in MeOH–HCl extracts, indicating that Se was mainly incorporated into proteins. A number of unidentified Se species were also detected in cod muscle tissue, the separated Se compounds being quantified on-line by post-column isotope dilution [42]. Soluble Se compounds extracted from muscles of chicken, turkey, duck, ostrich, lamb, cattle, and pig were separated by SEC with ICP-MS detection. Four peaks were observed, but distribution of Se among these peaks varied considerably in tissues from different animal species [86].

16.5.4 Protein Complexes with Heavy Metals

Quantitative determination of the Fe-containing protein myoglobin in different foods, including meat, processed meat, fish, and shellfish, by HPLC coupled to a double-focusing SF-ICP-MS instrument was reported [87, 88]. By using a double-focusing ICP-MS instrument, direct on-line detection of the most abundant iron isotope ^{56}Fe was possible without interference from the major polyatomic interference ($^{40}Ar^{16}O$). Separation of myoglobin from other Fe-containing compounds was facilitated by the use of a gel filtration column and Tris buffer (pH 7.2). In order to ensure sufficient quality control (QC) throughout the study, a raw beefsteak sample was developed as an in-house reference material. The LoD of the reported method for myoglobin was 0.85 ng l^{-1} (as Fe) [87, 88].

16.5.5 Polysaccharide Complexes with Heavy Metals

Plants contain significant concentrations of polysaccharides of which the potentially negatively charged oxygen functions can bind cations electrostatically or chelate them via polyhydroxy groups [89]. Particular attention was attracted by a structurally complex pectic polysaccharide rhamnogalacturonan-II (RG-II) [90]. This ubiquitous component of primary plant cell walls forms dimers cross-linked by 1 : 2 borate diol esters (dRG-II) that were found to complex in vitro specific divalent cations and the majority of Ba, Pb, Sr, and rare earth elements (REEs) in fruit and vegetables [45, 91].

SEC with parallel ICP-MS and refractometric detection was applied to investigate the metal complexes with polysaccharides in plants and related samples [41, 91]. Aqueous extracts of plants contained the dRG-II complex, but the presence of water-soluble polysaccharide species with higher molecular weights was also demonstrated [45]. The latter were probably of pectic origin since they could be readily decomposed by enzymic hydrolysis with a mixture of pectinase and hemi-cellulase to release the dRG-II complex [45]. The same mixture of enzymes was reported to be efficient to extract the dRG-II–metal complexes from water-insoluble residues of vegetables owing to the destruction of the original pectic structure [45].

In plant tissue, B exists as borate monoester and diester, together with H_3BO_3. A method for the speciation of B in radish juice with separation of the three forms by HPLC followed by ICP-MS detection was developed [92]. Most B compounds were of low molecular weight, but some B macromolecule complexes could be detected.

16.5.6 Miscellaneous

Speciation of trace elements, especially Al, in tea infusions attracted some attention. The metal was found to be bound to the same range of organic molecules in the infusions, regardless of the origin of the tea [93]. Applications of plasma spectrometric techniques for speciation analysis in beverages was reviewed [94]. The predominant Fe species in apple juice were Fe(II)-malate and Fe(III)-citrate, whereas Fe(II) and Fe(III) tartrates were the most abundant in white wine [95]. Human milk was investigated as for Mn speciation; inorganic Mn species and an Mn–citrate complex were identified [96].

16.6 QUALITY ASSURANCE AND CONTROL

16.6.1 Method Validation

Validation of the methods for speciation of anthropogenic organometallic contaminants was comprehensively discussed by Quevauviller [66]. A number of interlaboratory round-robin exercises were carried out which resulted in the awareness of the problem and in the availability of CRMs for organo-Hg, organo-Sn, and organo-As species in food matrices (Table 16.1). In contrast to anthropogenic contaminants, the problem of validation of speciation analyses concerning endogenous metal species in biological materials is still open. The chromatographic purity of peaks and the identity of species of which produce signals in chromatography often remain unknown. Also, it is usually unknown whether the species observed at the detector existed in the original sample or was simply created during the analytical procedure, for example, by oxidation and ligand exchange.

In order to cope with the lack of CRMs for endogenous species, the practice of laboratory internal QC materials is developing. A Laboratory Reference Material (LRM) was prepared from Brazil nuts (*Bertholletia excelsa*) for QC of the selenomethionine determinations. The concentration of selenomethionine was 79.9 $\mu g\ g^{-1}$. The homogeneity and stability of this candidate CRM passed the relevant tests recommended by the Measurement and Testing Programme [97].

16.6.2 Stability Studies

In order to obtain reliable information on speciation analysis it is necessary to evaluate the stability of the species of interest in a given matrix or sample. Also, when sample treatment is time-consuming, the knowledge of how to maintain the species integrity in the extracts to be stored is of paramount importance.

The stability of arsenobetaine in baby foods under different experimental conditions was evaluated. The highest As levels in the analyzed baby foods corresponded to those containing plaice (2–3 $\mu g\ g^{-1}$). Arsenobetaine was the only As compound present. Its levels remained unchanged in baby foods stored for different times or when the samples were freeze-dried, thus confirming the stability of arsenobetaine and the absence of the formation of other As species by interconversion during storage [98].

The stability of As species in the water–methanol rice extract stored at +4°C for at least one month was demonstrated. Once the rice grains were ground, MMA and As(V) were unstable under any storage conditions probably due to microbiological activity. When ground rice was γ-irradiated, As species remained stable [24]. The stability of As species in a chicken meat candidate CRM for at least 12 months was demonstrated [26].

The stability of SeMet and $TMSe^+$ in freeze-dried oyster and in the enzymatic extracts stored at different temperatures (−18, 4, and 20°C) was studied. The results obtained for the freeze-dried sample showed that SeMet and $TMSe^+$ were stable for at least 12 months under all the conditions tested. However, Se species in the enzymatic extracts were only stable for 10 days if stored at 4°C in Pyrex containers [99].

16.7 CONCLUSIONS

The use of speciation information is becoming of paramount importance for studies aiming at the elucidation of the role trace elements in nutrition at the molecular level. They are also essential for the characterization of supplementation with essential elements. Hence, accurate, reliable, precise, and, if possible, straightforward and inexpensive analytical methodologies for the identification and quantification of metal(loid)-containing food additives are required. The variety of analytical techniques resorted to for element speciation is impressive, but the use of GC with ICP-MS for monitoring of organometallic contaminants and of HPLC-ICP-MS (where GC fails) seems nowadays to be an established trend. As most of the endogenous element species in natural products still remain unknown, the development of generic procedures based on the tandem use of ICP-MS and electrospray MS/MS is required. It should not be forgotten at any stage of method development that the use of hyphenated techniques represents the final stage of an analytical procedure. Equal attention should be given to the validation of sampling, storage, and sample preparation protocols.

REFERENCES

1. D. Templeton, F. Ariese, R. Cornelis, L.-G. Danielsson, H. Muntau, H. P. van Leeuwen, R. Lobinski, Guidelines for terms related to chemical speciation and fractionation of trace elements. Definitions, structural aspects, and methodological approaches (IUPAC Recommendations 2000), *Pure Appl. Chem.*, **72** (2000), 1453–1470.

2. R. Lobinski, Elemental speciation and coupled techniques, *Appl. Spectrosc.*, **51** (1997), 260A–278A.
3. G. K. Zoorob, J. W. McKiernan, J. A. Caruso, ICP-MS for elemental speciation studies, *Mikrochim. Acta*, **128** (1998), 145–168.
4. J. A. Caruso, K. L. Sutton, K. L. Ackley (eds), *Elemental Speciation: New Approaches for Trace Element Analysis*, Comprehensive Analytical Chemistry, Elsevier, Amsterdam, 2000.
5. J. Szpunar, Bioinorganic speciation analysis by hyphenated techniques, *Analyst*, **125** (2000), 963–988.
6. R. Cornelis, J. Caruso, H. Crews, K. G. Heumann, *Handbook of Element Speciation – Techniques and Methodology*, Wiley, Chichester, 2003.
7. R. Lobinski, F. C. Adams, Speciation analysis by gas chromatography with plasma source spectrometric detection, *Spectrochim. Acta*, **52B** (1997), 1865–1903.
8. B. Bouyssiere, J. Szpunar, R. Lobinski, Gas chromatography with inductively coupled plasma mass spectrometric detection in speciation analysis, *Spectrochim. Acta*, **57B** (2002), 805–828.
9. J. Szpunar, V. O. Schmitt, J. L. Monod, Rapid speciation of butyltin compounds in sediments and biomaterials by capillary gas chromatography-microwave induced plasma atomic-emission spectrometry after microwave-assisted leaching-digestion, *J. Anal. Atom. Spectrom.*, **11** (1996), 193–199.
10. A. M. Carro, R. A. Lorenzo, M. J. Vazquez, M. Abuin, R. Cela, Different extraction techniques in the preparation of methylmercury biological samples: classic extraction, supercritical-fluid and microwave extraction, *Int. Lab.*, **28** (1998), 23–28.
11. S. Slaets, F. Adams, I. R. Pereiro, R. Lobinski, Optimization of the coupling of multicapillary GC with ICP-MS for mercury speciation analysis in biological materials, *J. Anal. Atom. Spectrom.*, **14** (1999), 851–857.
12. T. De Smaele, L. Moens, R. Dams, P. Sandra, Van der Eycken, J. Vandyck, Sodium tetra(*n*-propyl)borate: a novel aqueous *in situ* derivatization reagent for the simultaneous determination of organomercury, -lead and -tin compounds with capillary gas chromatography-inductively coupled plasma mass spectrometry, *J. Chromatogr. A*, **793** (1998), 99–106.
13. R. G. García Fernandez, M. M. Bayon, I. Garcia Alonso, A. Sanz-Mendel, Comparison of different derivatization approaches for mercury speciation in biological tissues by gas chromatography/inductively coupled plasma mass spectrometry, *J. Mass Spectrom.*, **35** (2000), 639–646.
14. A. Wasik, I. Rodriguez Pereiro, C. Dietz, J. Szpunar, R. Lobinski, Speciation of mercury by ICP-MS after on-line capillary cryofocussing and ambient temperature multicapillary gas chromatography, *Anal. Commun.*, **35** (1998), 331–335.
15. J. B. Pawliszyn, *Solid-Phase Microextraction. Theory and Practice*, Wiley VCH, Weinheim, 1997.
16. Z. Mester, R. Sturgeon, J. Pawliszyn, Review on speciation by SPME, *Spectrochim. Acta*, B56 (2001), 233–260.
17. J. Vercauteren, A. De Meester, T. De Smaele, F. Vanhaecke, L. Moens., R. Dams, P. Sandra, Headspace solid-phase microextraction-capillary gas chromatography-ICP mass spectrometry for the determination of the organotin pesticide fentin in environmental samples, *J. Anal. Atom. Spectrom.*, **15** (2000), 651–656.

18. S. Aguerre, G. Lespes, V. Desauziers, M. Potin-Gautier, Speciation of organotins in environmental samples by SPME-GC: comparison of four specific detectors: FPD, PFPD, MIP-AES and ICP-MS, *J. Anal. Atom. Spectrom.*, **16** (2001), 263–269.

19. Z. Mester, J. Lam, R. Sturgeon, J. Pawliszyn, Determination of methylmercury by solid-phase microextraction inductively coupled plasma mass spectrometry: a new sample introduction method for volatile metal species, *J. Anal. Atom. Spectrom.*, **15** (2000), 837–842.

20. S. McSheehy, J. Szpunar, V. Haldys, J. Tortajada, Identification of selenocompounds in yeast by electrospray quadrupole–time of flight mass spectrometry, *J. Anal. Atom. Spectrom.*, **17** (2002), 507–514.

21. S. McSheehy, F. Pannier, J. Szpunar, M. Potin-Gautier, R. Lobinski, Speciation of selenocompounds in yeast aqueous extracts by three dimensional liquid chromatography with ICP MS and electrospray MS detection, *Analyst*, **127** (2002), 223–229.

22. B. Michalke, Capillary electrophoresis methods for a clear identification of selenoamino-acids in complex matrices like human milk, *Fresenius J. Anal. Chem.*, **351** (1995), 670–677.

23. M. T. Roberge, A. J. Borgerding, J. W. Finley, Speciation of selenium compounds from high selenium broccoli is affected by the extracting solution, *J. Agric. Food Chem.*, **51** (2003), 4191–4197.

24. I. Pizarro, M. Gomez, M. A. Palacios, C. Camara, Evaluation of stability of arsenic species in rice, *Anal. Bioanal. Chem.*, **376** (2003), 102–109.

25. M. D'Amato, G. Forte, S. Caroli, Identification and quantification of major species of arsenic in rice, *J. AOAC Int.*, **87** (2004), 238–243.

26. A. Polatajko, J. Szpunar, Speciation of arsenic in chicken meat by anion-exchange liquid chromatography with inductively coupled plasma–mass spectrometry, *J. AOAC Int.*, **87** (2004), 233–237.

27. D. Velez, N. Ybanez, R. Montoro, Determination of arsenobetaine in manufactured seafood products by liquid chromatography, microwave-assisted oxidation and hydride-generation atomic-absorption spectrometry, *J. Anal. Atom. Spectrom.*, **12** (1997), 91–96.

28. S. McSheehy, P. Pohl, R. Lobinski, J. Szpunar, Investigation of arsenic speciation in oyster test reference material by multidimensional HPLC-ICP MS and electrospray tandem mass spectrometry (ES MS/MS), *Analyst*, **126** (2001), 1055–1062.

29. N. P. Vela, D. T. Heitkemper, Total arsenic determination and speciation in infant food products by ion chromatography-inductively coupled plasma-mass spectrometry, *J. AOAC Int.*, **87** (2004), 244–252.

30. S. S. Kannamkumarath, K. Wrobel, K. Wrobel, J. A. Caruso, HPLC-ICP MS determination of selenium distribution and speciation in different types of nuts, *Anal. Bioanal. Chem.*, **373** (2002), 454–460.

31. J. Ruiz Encinar, M. Sliwka-Kaszynska, A. Polatajko, V. Vacchina, J. Szpunar, Methodological advances for selenium speciation analysis in yeast, *Anal. Chim. Acta*, **500** (2003), 171–183.

32. A. Polatajko, M. Sliwka-Kaszynska, M. Dernovics, R. Ruzik, J. Ruiz-Encinar, J. Szpunar, A systematic approach to selenium speciation in selenized yeast, *J. Anal. Atom. Spectrom.*, **19** (2004), 114–120.

33. H. Chassaigne, C. C. Chery, G. Bordin, F. Vanhaecke, A. R. Rodriguez, Two-dimensional gel electrophoresis technique for yeast selenium-containing proteins.

Sample preparation and MS approaches for processing 2-D gel protein spots, *J. Anal. Atom. Spectrom.*, **19** (2004), 85–95.

34. C. B'Hymer, J. A. Caruso, Evaluation of yeast-based selenium food supplements using HPLC-ICP MS, *J. Anal. Atom. Spectrom.*, **15** (2000), 1531–1539.
35. J. L. Capelo, P. Ximenez-Embun, Y. Madrid-Albarran, C. Camara, Enzymatic probe sonication: enhancement of protease-catalyzed hydrolysis of selenium bound to proteins in yeast, *Anal. Chem.*, **76** (2004), 233–237.
36. M. Dernovics, Z. Stefanka, P. Fodor, Improving selenium extraction by sequential enzymatic processes for Se-speciation of selenium-enriched *Agaricus bisporus*, *Anal. Bioanal. Chem.*, **372** (2002), 473–480.
37. Z. Stefánka, I. Ipolyi, M. Dernovics, P. Fodor, Comparison of sample preparation methods based on proteolytic enzymatic processes for Se-speciation of edible mushroom (*Agaricus bisporus*) samples, *Talanta*, **55** (2001), 437–447.
38. K. Wrobel, S. S. Kannamkumarath, K. Wrobel, J. A. Caruso, Hydrolysis of proteins with methanesulfonic acid for improved HPLC-ICP MS determination of selenomethionine in yeast and nuts, *Anal. Bioanal. Chem.*, **375** (2003), 133–138.
39. S. M. Bird, P. C. Uden, J. F. Tyson, E. Block, E. Denoyer, Speciation of selenoamino-acids and organoselenium compounds in selenium-enriched yeast using high-performance liquid chromatography–inductively coupled plasma mass spectrometry, *J. Anal. Atom. Spectrom.*, **12** (1997), 785–788.
40. M. Kotrebai, M. Birringer, J. F. Tyson, E. Block, P. C. Uden, Selenium speciation in enriched and natural samples by HPLC-ICP-MS and HPLC-ESI-MS with perfluorinated carboxylic acid ion-pairing agents, *Analyst*, **125** (2000), 71–78.
41. M. Kotrebai, J. F. Tyson, E. Block, P. C. Uden, HPLC of selenium compounds utilizing perfluorinated carboxylic acid ion-pairing agents and inductively coupled plasma and electrospray ionization mass spectrometric detection, *J. Chromatogr. A*, **866** (2000), 51–63.
42. V. Diaz Huerta, M. L. Fernandez Sanchez, A. Sanz-Medel, Quantitative selenium speciation in cod muscle by isotope dilution ICP MS with a reaction cell: comparison of different reported extraction procedures, *J. Anal. Atom. Spectrom.*, **19** (2004), 644–648.
43. V. Diaz Huerta, L. Hinojosa Reyes, J. M. Marchante-Gayón, M. L. Fernández Sánchez, A. Sanz-Medel, Total determination and quantitative speciation analysis of selenium in yeast and wheat flour by isotope dilution analysis ICP MS, *J. Anal. Atom. Spectrom.*, **10** (2003), 1243–1247.
44. J. A. Caruso, D. T. Heitkemper, C. B'Hymer, An evaluation of extraction techniques for arsenic species from freeze-dried apple samples, *Analyst*, **126** (2001), 136–140.
45. J. Szpunar, P. Pellerin, A. Makarov, T. Doco, P. Williams, R. Lobinski, Speciation of metal-carbohydrate complexes in fruit and vegetable samples by size-exclusion HPLC-ICP MS, *J. Anal. Atom. Spectrom.*, **14** (1999), 639–644.
46. H. M. Crews, J. R. Dean, L. Ebdon, R. C. Massey, Application of HPLC-ICP MS to the investigation of cadmium speciation in pig kidney following cooking and *in vitro* gastro-intestinal digestion, *Analyst*, **114** (1989), 895–899.
47. S. Mounicou, J. Szpunar, D. Andrey, C. J. Blake, R. Lobinski, Development of a sequential enzymolysis approach for the evaluation of the bioaccessibility of Cd and Pb from cocoa, *Analyst*, **127** (2002), 1638–1641.

48. S. Mounicou, J. Szpunar, R. Lobinski, D. Andrey, C. J. Blake, Bioavailability of cadmium and lead in cocoa: comparison of extraction procedures prior to size-exclusion fast-flow liquid chromatography with inductively coupled plasma mass spectrometric detection (SEC-ICP-MS), *J. Anal. Atom. Spectrom.*, **17** (2002), 880–886.
49. L. M. W. Owen, H. M. Crews, R. C. Massey, N. J. Bishop, Determination of copper, zinc and aluminium from dietary sources in the femur, brain and kidney of guinea pigs and a study of some elements in *in vivo* intestinal digesta by size-exclusion chromatography–inductively coupled plasma mass spectrometry, *Analyst*, **120** (1995), 705–712.
50. B. M. Gamble, P. A. Gallagher, J. A. Shoemaker, X. Wei, C.A. Schwegel, J. T. Creed, An investigation of the chemical stability of arsenosugars in simulated gastric juice and acidic environments using IC-ICP-MS and IC-ESI-MS/MS, *Analyst*, **127** (2002), 781–785.
51. E. Dumont, F. Vanhaecke, R. Cornelis, Hyphenated techniques for speciation of Se in *in vitro* gastrointestinal digests of *Saccharomyces cerevisiae*, *Anal. Bioanal. Chem.*, **379** (2004), 504–511.
52. B. Michalke, P. Schramel, Application of capillary zone electrophoresis–inductively coupled plasma mass spectrometry and capillary isoelectric focusing-inductively coupled plasma mass spectrometry for selenium speciation, *J. Chromatogr. A*, **807** (1998), 71–80.
53. A. Theobald, L. Dunemann, Capillary zone electrophoresis with UV-scanning detection for the analysis of metal species and their organic binding partners in biological systems, *J. High Resolut. Chromatogr.*, **19** (1996), 608–612.
54. R. Lobinski, I. R. Pereiro, H. Chassaigne, A. Wasik, J. Szpunar, Elemental speciation and coupled techniques – towards faster and reliable analyses, *J. Anal. Atom. Spectrom.*, **13** (1998), 859–867.
55. J. Szpunar, Trace element speciation analysis of biomaterials by high performance liquid chromatography with inductively coupled plasma mass spectrometric detection (HPLC-ICP MS), *Trends Anal. Chem.*, **19** (2000), 127–137.
56. L. Fernandez Sanchez, J. Szpunar, Speciation analysis for iodine in milk by size-exclusion chromatography with inductively coupled plasma mass spectrometric detection (SEC-ICP MS), *J. Anal. Atom. Spectrom.*, **14** (1999), 1697–1703.
57. R. R. de la Flor St. Remy, M. L. Fernández Sánchez, J. B. López Sastre, A. Sanz-Medel, Multielemental distribution patterns in premature human milk whey and pre-term formula milk whey by size exclusion chromatography coupled to inductively coupled plasma mass spectrometry with octopole reaction cell, *J. Anal. Atom. Spectrom.*, **19** (2004), 1104–1110.
58. U. Kohlmeyer, J. Kuballa, E. Jantzen, Simultaneous separation of 17 inorganic and organic arsenic compounds in marine biota by means of high-performance liquid chromatography/inductively coupled plasma mass spectrometry, *Rapid Commun. Mass Spectrom.*, **16** (2002), 965–974.
59. K. A. Francesconi, D. Kuehnelt, Determination of arsenic species: a critical review of methods and applications, 2000–2003, *Analyst*, **129** (2004), 373–395.
60. E. H. Larsen, M. Hansen, H. Paulin, S. Moesgaard, M. Reid, M. Rayman, Speciation and bioavailability of selenium in yeast-based intervention agents used in cancer chemoprevention studies, *J. AOAC Int.*, **87** (2004), 225–232.

61. D. Schaumlöffel, A. Prange, A new interface for combing capillary electrophoresis with inductively coupled plasma–mass spectrometry, *Fresenius J. Anal. Chem.*, **364** (1999), 452–456.

62. P. Krystek, R. Ritsema, Mercury speciation in thawed out and refrozen fish samples by gas chromatography coupled to inductively coupled plasma mass spectrometry and atomic fluorescence spectroscopy, *Anal. Bioanal. Chem.*, **381** (2005), 354–359.

63. A. M. Carro, M. C. Mejuto, Review: application of chromatographic and electrophoretic methodology to the speciation of organomercury compounds in food analysis, *J. Chromatogr. A*, **882** (2000), 283–307.

64. P. Quevauviller, M. Filippelli, M. Horvat, Method performance evaluation for methylmercury determination in fish and sediment, *Trends Anal. Chem.*, **19** (2000), 157–166.

65. P. Rodriguez-Gonzalez, J. R. Encinar, J. I. Alonso, A. Sanz-Medel, Monitoring the degradation and solubilisation of butyltin compounds during in vitro gastrointestinal digestion using "triple spike" isotope dilution GC-ICP-MS, *Anal. Bioanal. Chem.*, **381** (2005), 380–387.

66. P. Quevauviller (ed.), *Method Performance Studies for Speciation Analysis*, Royal Society of Chemist, Cambridge, **1998**.

67. P. Quevauviller, M. Astruc, R. Morabito, F. Ariese, L. Ebdon, Collaborative evaluation of methods for tributyltin determinations in sediments and mussel tissue, *Trends Anal. Chem.*, **19** (2000), 180–188.

68. T. Guerin, A. Astruc, M. Astruc, Speciation of arsenic and selenium compounds by HPLC hyphenated to specific detectors: a review of the main separation techniques, *Talanta*, **50** (1999), 1–24.

69. Z. L. Gong, X. F. Lu, M. Ma, C. Watt, X. C. Le, Arsenic speciation analysis, *Talanta*, **58** (2002), 77–96.

70. H. Helgesen, E. H. Larsen, Bioavailability and speciation of arsenic in carrots grown in contaminated soil, *Analyst*, **123** (1998), 791–796.

71. M. J. Abedin, M. S. Cresser, A. A. Meharg, J. Feldmann, J. Cotter-Howells, Arsenic accumulation and metabolism in rice (*Oryza sativa L.*), *Environ. Sci. Technol.*, **36** (2002), 962–968.

72. L. H. Foster, S. Sumar, Selenium in health and disease: a review, *Crit. Rev. Food Sci. Nutr.*, **37** (1997), 211–228.

73. L. C. Clark, J. G. Combos, B. W. Turnbill, E. H. Slate, D. K. Chalker, J. Chow, K. S. Davis, R. A. Glover, G. F. Graham, E. G. Gross, A. Krongrad, J. L. Lesher, K. Park, B. B. Sanders, C. L. Smith, J. R. Taylor, Effects of selenium supplementation for cancer prevention in patients with carcinoma of the skin: a randomized controlled trial, *J. Am. Med. Assoc.*, **276** (1996), 1957–1963.

74. S. M. Bird, H. Ge, P. C. Uden, J. F. Tyson, E. Block, E. Denoyer, High-performance liquid chromatography of selenoamino acids and organo-selenium compounds: speciation by inductively coupled plasma mass spectrometry, *J. Chromatogr. A.*, **789** (1997), 349–359.

75. M. Kotrebai, M. Biringer, J. F. Tyson, E. Block, P. C. Uden, Identification of the principal selenium compounds in selenium-enriched natural sample extracts by ion-pair liquid chromatography with inductively coupled plasma- and electrospray ionization–mass spectrometric detection, *Anal. Commun.*, **36** (1999), 249–252.

76. M. Kotrebai S. M. Bird, J. F. Tyson, E. Block, P. C. Uden, Characterization of selenium species in biological extracts by enhanced ion-pair liquid chromatography with inductively coupled plasma mass spectrometry and by referenced electrospray ionization mass spectrometry, *Spectrochim. Acta*, **54B** (1999), 1573–1591.

77. P. Uden, H. T. Boakye, C. Kahakachchi, R. Hafezi, P. Nolibos, E. Block, S. Johnson, J. F. Tyson, Element selective characterization of stability and reactivity of selenium species in selenized yeast, *J. Anal. Atom. Spectrom.*, **19** (2004), 65–73.

78. W. R. Wolf, R. J. Goldschmidt, Selenomethionine contents of NIST wheat reference materials, *Anal. Bioanal. Chem.*, **378** (2004), 1175–1181.

79. P. C. Uden, Modern trends in the speciation of selenium by hyphenated techniques, *Anal. Bioanal. Chem.*, **373** (2002), 422–431.

80. J. Auger, W. Yang, I. Arnault, F. Pannier, M. Potin-Gautier, High-performance liquid chromatographic-inductively coupled plasma mass spectrometric evidence for Se-"alliins" in garlic and onion grown in Se-rich soil, *J. Chromatogr. A*, **1032** (2004), 103–107.

81. P. Ximenez-Embun, I. Alonso, Y. Madrid-Albarran, C. Cámara, Establishment of selenium uptake and species distribution in lupine, Indian mustard, and sunflower plants, *J. Agric. Food Chem.*, **52** (2004), 832–838.

82. Y. Ogra, K. Ishiwata, J. Ruiz Encinar, R. Lobinski, K. Suzuki, Speciation of selenium in selenium-enriched shiitake mushroom, *Lentinula edodes*, *Anal. Bioanal. Chem.*, **379** (2004), 861–866.

83. M. Rayman, The use of high-selenium yeast to raise selenium status: how does it measure up? *Br. J. Nutr.*, **92** (2004), 557–573.

84. J. Zheng, W. Goessler, W. Kosmus, The chemical forms of selenium in selenium nutritional supplements: an investigation by using HPLC/ICP MS and GF/AAS, *Trace Elem. Electrol.*, **15** (1998), 70–75.

85. J. Ruiz Encinar, L. Ouerdane, W. Buchmann, J. Tortajada, R. Łobiński, J. Szpunar, Identification of water-soluble selenium-containing proteins in selenized yeast by size-exclusion-reversed-phase HPLC/ICPMS followed by MALDI-TOF and electrospray Q-TOF Mass Spectrometry, *Anal. Chem.*, **75** (2003), 3765–3774.

86. C. Daun, T. Lundh, G. Önning, B. Åkesson, Separation of soluble selenium compounds in muscle from seven animal species using size exclusion chromatography and inductively coupled plasma mass spectrometry, *J. Anal. Atom. Spectrom.*, **19** (2004), 129–134.

87. C. F. Harrington, S. Elahi, S. A. Merson, P. Ponnampalavanar, A method for the quantitative analysis of iron speciation in meat by using a combination of spectrophotometric methods and high-performance liquid chromatography coupled to sector field inductively coupled plasma mass spectrometry, *Anal. Chem.*, **73** (2001), 4422–4427.

88. C. F. Harrington, S. Elahi, S. A. Merson, P. Ponnampalavanar, Quantitative analysis of iron-containing protein myoglobin in different foodstuffs by liquid chromatography coupled to high-resolution inductively coupled plasma mass spectrometry, *J. AOAC Int.*, **87** (2004), 253–258.

89. D. M. Whitfield, S. Stoijkovski, B. Sarkar, Metal coordination to carbohydrates. Structures and function, *Coord. Chem. Rev.*, **122** (1993), 171–225.

90. P. Pellerin, T. Doco, S. Vidal, P. Williams, J. M. Brillouet, M. A. O'Neill, Structural characterization of red wine rhamnogalacturonan II, *Carbohydr. Res.*, **290** (1996), 183–197.

91. J. Szpunar, P. Pellerin, A. Makarov, T. Doco, P. Williams, B. Medina, R. Lobinski, Speciation analysis for biomolecular complexes of lead in wine by size-exclusion high-performance liquid chromatography–inductively coupled plasma mass spectrometry, *J. Anal. Atom. Spectrom.*, **13** (1998), 749–754.

92. T. Matsunaga, T. Ishii, H. Oda, Speciation of water-soluble boron compounds in radish roots by size-exclusion HPLC-ICP MS, *Anal. Sci.*, **12** (1996), 673–675.

93. A. K. Flaten, W. Lund, Speciation of aluminium in tea infusions studied by size exclusion chromatography with detection by post-column reaction, *Sci. Total Environ.*, **207** (1997), 21–28.

94. J. Meija, S. Mounicou, J. A. Caruso, Plasma spectrometry for elemental speciation and characterization in beverages, *J. AOAC Int.*, **87** (2004), 205–224.

95. G. Weber, Speciation of iron using HPLC with electrochemical and flame-AAS detection, *Fresenius J. Anal. Chem.*, **340** (1991), 161–165.

96. B. Michalke, P. Schramel, Manganese speciation in human milk using size exclusion chromatography combined with strong anion exchange chromatography and inductively coupled plasma mass spectrometry detection, *J. Anal. Atom. Spectrom.*, **19** (2004), 121–128.

97. E. T. Bodo, Z. Stefanka I. Ipolyi, C. Soros, M. Dernovics, P. Fodor, Preparation, homogeneity and stability studies of a candidate LRM for Se speciation, *Anal. Bioanal. Chem.*, **377** (2003), 32–38.

98. P. Vinas, I. Lopez-Garcia, B. Merino-Merono, N. Campillo, M. Hernandez-Cordoba, Stability of arsenobetaine levels in manufactured baby foods, *J. Food Prot.*, **66** (2003), 2321–2324.

99. P. Moreno, M. A. Quijano, A. M. Gutierrez, M. C. Perez-Conde, C. Camara, Stability of total selenium and selenium species in lyophilised oysters and in their enzymatic extracts, *Anal. Bioanal. Chem.*, **374** (2002), 466–476.

17

ELEMENTAL SPECIATION IN HUMAN MILK AND SUBSTITUTE FOOD FOR NEWBORNS

BERNHARD MICHALKE

GSF National Research Center for Environment and Health, Institute for Ecological Chemistry, Neuherberg/München, Ingolstädter Landstrasse 1 D-85764, Germany

MARÍA LUISA FERNÁNDEZ SÁNCHEZ AND ALFREDO SANZ-MEDEL

Department of Physical and Analytical Chemistry. Faculty of Chemistry, University of Oviedo, Oviedo, C/ Julián Clavería 8, E-33006, Spain, E-mail: bernhard.michalke@gsf.de, marisafs@uniovi.es, asm@uniovi.es

The Determination of Chemical Elements in Food: Applications for Atomic and Mass Spectrometry.
Edited by Sergio Caroli

17.1 INTRODUCTION

As discussed in Chapter 13 "Total Analysis and Distribution of Trace Elements in Human, Cow, and Formula Milk," the determination of trace elements has gained outstanding importance in life and nutritional sciences. As a point of fact, vital functions can be drastically affected by the smallest changes in the concentration of trace elements in biological tissues. Studies of human physiological and pathophysiological processes require the determination of elements levels from microgram per liter (or nanogram per gram) down to nanogram per liter (picogram per gram). The presence of substantially higher concomitant amounts of organic and inorganic components makes the determination of trace elements rather difficult in both living organisms and nutritional samples. Moreover, the complexity of the process that leads from the first step of trace element analysis to the final statement of bioavailability and related biological implications is urgently demanding close collaboration between analytical chemists, nutrition experts, and life scientists.

Over the last two decades analytical chemists have increasingly realized that total concentrations of chemical elements cannot give, as a rule, enough information about mobility, bioavailability, cellular uptakes, metabolism, and so on and the eventual impact of elements on biological organisms. Only the knowledge of the chemical species of the elements can provide an understanding of chemical and biochemical reactions, which in turn leads to more information about toxicity or essentiality. In this context, it is also worth stressing that new trace elements and some of their species are being added to the list of those that are known or suspected to be essential.

The usual method of elemental speciation – also in human milk samples – is the use of already well-established separation and detection methods, which have to be combined in novel ways and modified according to the relevant speciation problems. Off-line combination and on-line hyphenation of separation techniques and element-selective or molecule-selective detection systems are generally the basis for speciation analysis. Many methodological developments (mostly for adequate sampling to preserve speciation) are still necessary, but also hyphenation and quality strategies are increasingly demanding. Investigations on quality control and assurance have shown that, during sampling, sample preparation, and storage, separation and detection changes in the original species information can easily occur. Such errors are obviously relevant also to speciation investigations in human milk as well. Historically, the speciation of elements in human milk was already performed at the very onset of speciation efforts in other fields. Although such investigations could not always provide complete quality-based speciation strategies, they considerably contributed to establishing speciation concepts for this biological fluid. The necessity for element speciation information in nutrition and specifically in human milk was soon recognized and reported early in speciation literature.

Human milk is usually the first diet for the infant [1]. It supposedly meets all the needs of the newborn, mostly because it provides essential trace elements as well as proteins and enzymes along with all the other organic components of milk reported in the literature, as discussed in the Chapter 13 "Total Analysis and Distribution of Trace Elements in Human, Cow, and Formula Milk." Trace elements and organic compounds may quickly interact chemically or be already bound together when secreted by the mammary gland. Thus, elements may occur as the building blocks of macromolecules (proteins, enzymes, hormones, etc.), being associated with bioligands or present in metabolites (small molecules) with different possible oxidation states. Therefore, the nature and amount of an element species in a food matrix, rather than the total element concentration, are responsible for its bioavailability (e.g., in human milk) and eventually for its biochemical, essential, or toxicological role [2]. Therefore, only information about the particular species of an element would allow a sound assessment of how this element is metabolized, whether its effects are toxic or rather it has no adverse impact at a given specific concentration, or even whether the element may be taken as an essential, nutrient under that particular form. Problem-related speciation analysis may become a most useful tool to effectively assess the role of chemical elements in food.

In brief, today the scientific community worldwide recognizes the need to carry out speciation analysis, not just total determinations, of both essential and potentially toxic elements in milk.

17.2 DEFINITIONS OF TERMS RELATED TO SPECIATION

Terms related to speciation have been developed and defined on the basis of the increasing understanding of needs in speciation investigations resulting from the growing experience in this field. The usage of concepts and terms related to chemical speciation in recent years still mirrors a certain degree of inconsistency in the scientific community. In recognition of the importance of harmonized terminology from the viewpoint of both interdisciplinary communication and constructive interaction with decision-makers, three divisions of the IUPAC have worked together and reached consensus on some basic definitions to which specialists in this discipline can now resort with confidence [2]. One of the major conclusions of the above working group was that the term "speciation" should be restricted to the distribution of an element among its well-defined chemical forms. A clear distinction was also made between speciation and fractionation. As it is deemed crucial for those working in this field to fully abide by such definitions, they are reported below verbatim (see Table 17.1).

Thus, speciation analysis, the analytical activity of identifying and measuring species, includes the clear identification of the species present, their exact quantification in representative samples, as well as adequate quality of quantitative results [3, 4]. If the identification and quantification of a given chemical species cannot be performed, there are still available analytical procedures leading to useful "operationally" defined species distribution studies. Operationally and functionally defined characterization of species has to be distinguished from chemical speciation analysis [5] as it provides only the characterization of molecule groups (not single species) showing similar behavior during an analytical procedure (operation), such as liquid–solid extraction. The characterization of molecule groups is strongly dependent on the selected analytical procedure, and usually original species information (identity) is lost. In this sense, fractionation can be regarded as an operationally defined characterization that still provides useful information about the function of some related species group [3, 6].

TABLE 17.1. Definitions (Short Forms)

Chemical Species	Specific Form of an Element (Specific Molecule, Complex, Oxidation State)
Speciation analysis	Analytical activity of identifying and measuring species in a real and representative sample or matrix Identifying and measuring strictly includes quality control
Operationally defined characterization	Not chemical speciation analysis. No species identification provided. Characterization of molecule groups according to a given analytical procedure
Functionally defined characterization	No chemical speciation analysis. No species identification provided. Characterization of molecule groups according to their impact on organisms (essential, toxic)
Speciation	Distribution of species in a particular sample or matrix

17.3 NUTRITION AND SPECIATION

Nowadays speciation information is essential in several fields (e.g., environment, biology, medicine, nutrition). Food chemistry and nutrition, in particular, can greatly benefit from the speciation approach, which acts between these fields as an "interface." Bioavailability is directly linked to biochemical mechanisms within the organism. The mode of absorption of an element in the intestine and its subsequent transport within the body depend on its chemical form, and information on speciation can be used to predict and interpret how the element will be metabolized. Thus, the fate of the chemical species (e.g., adsorption to membranes, transport and incorporation into larger molecules as enzymes, enrichment, excretion, etc.) may be modified and result in an unbroken path from environmentally changed species to toxicity, deficiency, or growth in biological systems. However, the importance of trace element speciation in nutritional sciences is still less established than in toxicology, where the link between chemical form and toxic effect is better known [7].

The achievement of reference concentration values and intervals for elements in nutritional samples, especially in human milk, is of paramount importance to allow for the detection of imbalances which can adversely affect human and, particularly, the newborn's health. However, no ranges of element species in human milk have been summarized so far. Moreover, publications investigating species of the same elements in human milk very often report widely differing results. One reason for this unsatisfactory situation is given by the fact that results were gained by different methods, while strategies for quality control and assurance are either not reported in detail or their role in validating the speciation results is missing in the report.

Thus, this chapter will not only summarize speciation results in human milk, but also address the important topic of the current need to report upon methods and quality management in terms of quality control (QC) and quality assurance (QA).

17.4 SPECIATION ANALYSIS IN HUMAN MILK: SAMPLING, STORAGE, PREPARATION, SEPARATION, AND DETECTION

As detailed in the earlier chapter on milk, from an analytical viewpoint the total determination of trace elements in human milk samples as such is still an open issue because decomposition, preconcentration, and storage of trace analytes in solutions are often a source of concern. The accuracy of the analytical results can, in fact, be endangered by these pretreatments. There are relatively few methods for such total direct analytical determinations, and in many cases they are still inadequate to meet the needs of the analyst.

These analytical problems regarding total element determination in human milk are magnified in milk speciation analysis as this requires a series of carefully planned steps of which chemical and/or physical pretreatments of the sample are almost always mandatory. In this context, both sampling and sample preparation are of prime importance in milk speciation analysis. Without proper sampling and sample treatment procedures, there is little chance that any later speciation analysis can provide reliable data. Thiers and Vallee stated several decades ago that "unless the complete history of samples is known with certainty, the analyst is well advised not to spend his or her time analysing them" [8]. First, the container itself where the human milk sample is stored is a potential source of contamination. Secondly, if unexpected changes in the form of the element occur (e.g., in oxidation state, extent of chelation, decomposition of the organometallic compound, etc.), then clearly the species of interest has been changed, and therefore its original identity and amount cannot be ascertained. On the other hand, highly reliable data can be obtained with careful evaluation of all potential chemical species changes in the sampling and sample preparation processes.

17.4.1 Sampling

The sampling step can be the most critical one in speciation analysis [9]. Sampling should be designed so as to preserve the original information about the existing native species. In human milk sampling, however, no specific technical problems arise. Milk is naturally excreted and may be collected by an experienced nurse or, after an adequate training to avoid possible contamination phenomena, by the mother herself. The precleaned sampling containers and doubly distilled water for breast cleaning before sample collection will be typically provided by the laboratory. The nurse, or the mother after leaving the hospital, will store the samples immediately after collection. A host of sampling problems are known that range from container wall adsorption effects to contamination from the sampler resulting in alteration of the chemical equilibrium [3]. In biological samples metal contamination is sometimes observed from metal tools, as such changes can subsequently alter also the species pattern of other elements. Species alterations are also known from bacterial activity [9]. On the other hand, representativity of the sample in terms of chemical species is often not guaranteed. Owing to density differences of milk compounds, fat and other biocompounds may rapidly separate when human milk is allowed to stand, even for only a few minutes, without vortexing for homogenization.

Major precautions to preserve the integrity of milk sample can be summarized as follows:

- Make sure representative aliquot is sampled.
- Keep sampling time short.
- Avoid contamination.
- Keep the volume-to-surface ratio as high as possible to minimize wall effects at the container inner surface.
- Do not use stainless steel instruments for biological samples; keep metal tools away from human milk.

17.4.2 Storage and Processing of Human Milk

Typically, human milk samples cannot be analyzed on site. Therefore, storage becomes necessary (although it should be as short as possible). Reliable storage precautions generally used for total trace element analysis, such as acidification, are inappropriate for element speciation studies as pH changes will affect species composition, thus altering information on native species. Many trace element content data are unreliable due to contamination by gloves and rubber stoppers. In fact, there are many potential sources of contamination with them: dust, dirt, cosmetics, disinfectants, talc, and dust on the gloves, metallic corrosion products, and so on. Laboratory dust can contain up to 0.3 percent Al, 0.3 percent Ca, 0.3 percent Fe, 0.8 percent K, 0.2 percent Mg, 0.3 percent Na, 0.2 percent Pb, 2 percent S, and 0.2 percent Zn. These contaminants are prone to disturbing trace element analysis, and, of course, they also endanger element speciation. The drying of samples can, in turn, result in the loss of element species (predominantly volatile species). Subsampling problems must be also considered. In most regulations and directives, analytical methods and sampling are only briefly specified, although such methods should definitely be more accurately detailed in protocols.

Of course, samples should not be contaminated or destabilized during storage, and appropriate sample handling, clean room conditions, and pre-cleaned vials should be secured throughout. As a general rule, sample preparation should be kept at a minimum to reduce the risk of alterations. Storage time should be very short (preferably at 4°C). In any case, when long-term storage is unavoidable, aliquots of the homogenous sample should be deep-frozen (freezing at −80°C is recommended). If freeze-drying is planned, the freeze-drying system should allow for sample temperature control and volatile species loss should be taken into account. It also goes without saying that for milk speciation analysis no acid addition or other pH-changing agents, nor repetitive slow thawing/freezing cycles are allowed. Glassware can show ion-exchange properties; hence, it is less suitable than polymer materials [10]. Errors due to such processes are, however, in most cases independent of the analytical method used and should be assessed on a case-by-case basis. Table 17.2 summarizes some important quality control precautions during storage, sample preparation and derivatization for trace element speciation in milk.

17.4.3 Speciation Approaches: Direct and Combined (Hyphenated) Techniques

After sampling, storage, and sample preparation, species are to be identified and quantified. Direct speciation approaches can provide full information about the species in a sample without any additional (separation) method, and quantify the species directly. Such methods, for example, chemical sensors, biosensors, and nuclear magnetic resonance (NMR), however, have many limitations in sensitivity and/or selectivity when applied to real-world samples as human milk.

TABLE 17.2. Summary of Quality Control Measures for Storage, Sample Preparation and Derivatization

Storage of Human Milk Samples

- Stroage at 4°C should be as short as possible. Longer storage after shock freezing at −80°C
- For replicate measurements provide aliquots before freezing (no replicate freeze/thaw cycles)

Sample Preparation of Human Milk Samples

- Mass balances and recovery rates (species spikes) should be determined. However, species spikes may show a different behavior compared to native species
- Species spikes may change species equilibriums and lead to altered species pattern in the sample
- Low temperature helps for species preservation

The most frequent, reliable approach for element speciation in real samples today uses combined – or hyphenated – techniques, where species are selectively separated (e.g., by a chromatographic technique) and then the elements in the various chemical forms are selectively detected. To enhance the quality of measurements, molecule-selective detection is also coupled to separation devices. These combinations provide extended flexibility and broad applicability. Disadvantages are that, as complexity increases, the risk increases of system failure. Besides, species equilibrium can be drastically altered during separation due to dilution, some components can be removed, and so on. Under such conditions, species transformation and destruction is a likely consequence. Therefore, the total separation time should be shorter than the transformation rate of species. Figure 17.1 gives an overview of the strategy and various speciation methods mostly used for element speciation in human milk.

17.5 ELEMENTS SPECIATED IN HUMAN MILK

The subsequent sections will focus on the various elements speciated so far in human milk. Descriptions will be given of the different speciation approaches, reasons for choosing a given element for speciation in human milk, type of samples used (sample pooled or individual, lactation state), separation and detection techniques, identification of species, QC and QA. Lastly, the results available will be briefly overviewed.

17.5.1 Calcium and Magnesium

Calcium is important for several functions in the body, including bone and teeth growth, function of nerves and muscles, proper heart function and blood clotting [11]. Likewise, Mg is needed for protein and fatty acid formation, cell renewal,

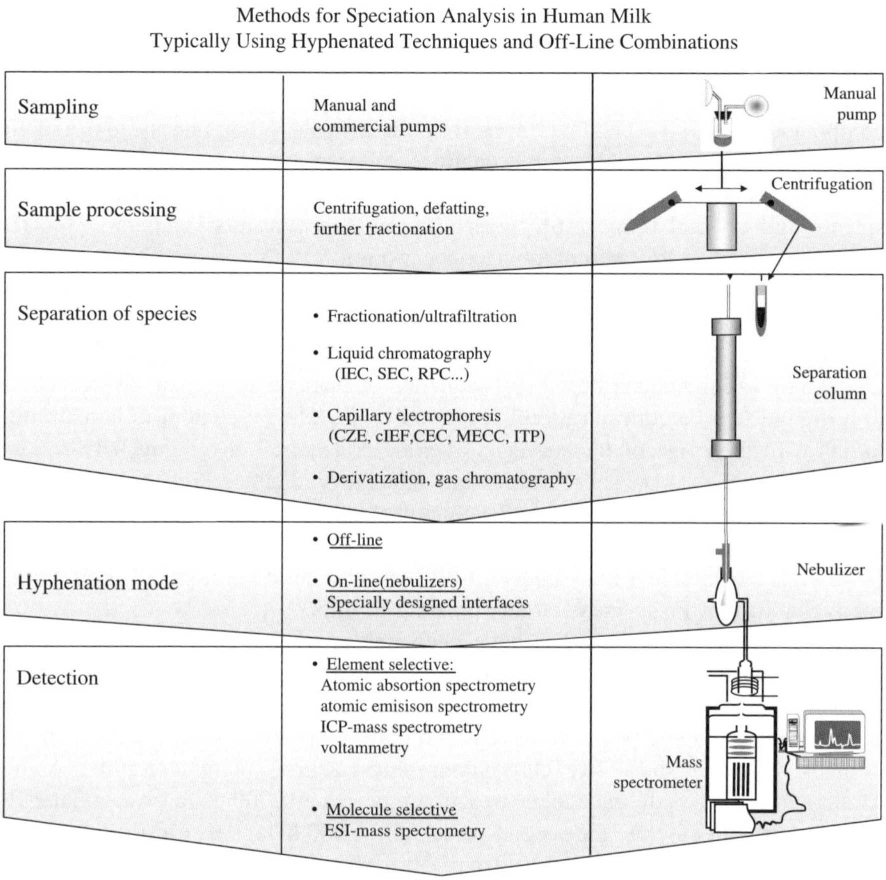

Figure 17.1. Speciation scheme with typical analytical steps, as employed in usual speciation methods.

B vitamin activation, muscle relaxation, and adenosin triphosphate (ATP) formation – the energy the body runs on. Concentrations of Ca and Mg in human milk range between 250–300 and 25–35 mg l^{-1}, respectively, but they remain constant along the entire lactating period [11].

It has been shown that the major part of Ca and Mg are found in the soluble fraction (whey) of human milk, whereas only small amounts are present in insoluble caseins or in fat [11–13]. Calcium and Mg speciation studies in milk whey using size exclusion chromatography (SEC) combined with inductively coupled plasma atomic emission spectrometry (ICP-AES) [14] indicated that they were preferably associated with the milk nonprotein fraction. A further work [15] showed similar chromatographic profiles by SEC hyphenated with inductively coupled plasma mass spectrometry (ICP-MS), with Ca and Mg eluting in the lowmolecular-weight region (LMW) (<1.4 kDa). Thus, there seem to be weak associations, if any, of Ca^{2+} or Mg^{2+} with higher molecular mass biocompounds of milk.

17.5.2 Iron

Iron is a well-known essential element forming part of the oxygen-carrying proteins (hemoglobin in red blood cells and myoglobin in muscles). It is also a component of various enzymes necessary for a healthy immune system and for energy production. Iron is stored mainly as ferritin, but also as hemosiderin. These proteins are present primarily in the liver, but also in the bone marrow, spleen, and skeletal tissue [16]. Severe Fe deficiency results in anemia, and red blood cells with a low hemoglobin concentration. On the other hand, anemia in pregnancy increases the risk of premature or low-weight birth. Owing to its rapid growth rate, the demand of the body for Fe is the greatest during the first year of life, and the Recommended Daily Allowance (RDA) is 18 mg per day [16]. Deficiency in infants can be the consequence of inadequate amount of storage Fe at birth and low Fe concentration in maternal milk. The Fe content in human milk varies with the stage of lactation, its average concentration ranging between 0.2 and 0.6 $\mu g\ ml^{-1}$ [11]. These values are three to five times higher in colostrum than in mature milk. The concentration of total Fe in breast milk has always been considered low. Interestingly, however, breast-fed children rarely show Fe-deficiency anemia (most likely due to a compensating high absorption and bioavailability of Fe species actually existing in milk).

In this sense, several studies have been published that address Fe speciation. Caroli and co-workers studied the analytical capabilities of SEC coupled to ICP-AES for the speciation of Ca, Cu, Fe, Mg, Mn, and Zn in defatted human milk. Five chromatographic peaks were reported and Fe appeared to be rather homogeneously spread in those five chromatographic fractions in mature milk. A first fraction (species with molecular weight higher than 2000 kDa) was related to association with caseins; the second peak (500–2000 kDa) was ascribed to immunoglobulins; the peak between 100 and 500 kDa corresponded to human serum albumin and lactoferrin, and the fourth fraction (2–100 kDa) would include lactoalbumin. The fifth peak (at less than 2 kDa) would represent Fe bound to LMW compounds (e.g., orotic acid, citrate, etc.). According to Coni et al. [17], however, in colostrum Fe is mainly bound to fraction numbers 1 and 4.

In mature breast milk most Fe was found in the whey fraction [11, 13]. While some authors suggested that Fe was mainly bound to compounds with molecular weight between 3 and 76 kDa, Rivero Martino et al. [15] found this element widely distributed in compounds that ranged from high molecular (higher than 160 kDa) to LMW. Studies on multielemental distribution patterns of Al, Br, Cd, Co, Cr, Cu, Fe, I, Mn, P, Pb, S, Se, and Zn in human milk whey along the lactating period were carried out by SEC coupled on-line to ICP-MS with a collision cell (CC) [18]. The use of H_2 as collision gas reduced the background due to polyatomic interferences at the most abundant isotope ^{56}Fe [18]. These authors showed (Fig. 17.2) that Fe in human milk whey appears mainly associated with species with molecular weight from >600 to 160 kDa (e.g., immunoglobulin) and to low-molecular-weight species (<10 kDa). Moreover, the Fe pattern notably changes in the immunoglobulin region along the lactation period. A small Fe

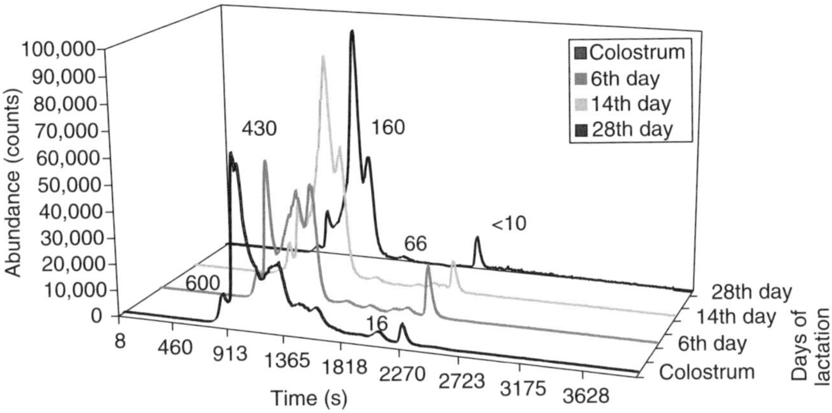

Figure 17.2. Iron distribution profile in colostrum and human milk with lactating period whey. From de la Flor St Reny et al. [18], with permission of RSC.

peak at >600 kDa appears in colostrum, and decreases in mature milk (this could be ascribed to Immunoglobulin-M, which is known to be present in colostrum in higher amounts than in mature milk). Another two small Fe peaks at 16 and <10 kDa are also present in all human samples. In addition, a peak at 67 kDa, possibly associated with lactoferrin, appears in colostrum. Iron saturation of lactoferrin protein is very low (1–4 percent) and this may be important in relation to the role of lactoferrin as a bacteriostatic agent. Dramatic daily changes in concentration and species of Cu, Fe, P, and S in colostrum have also been reported [19].

17.5.3 Copper

Copper is an essential trace element absorbed in the gut and transported to the liver bound to albumin. It is found in a variety of enzymes, including superoxide dismutase. In the bloodstream Cu is present mostly in ceruloplasmin. Tissues with a relatively high content of Cu are liver, heart, and brain. The RDA for Cu in normal, healthy adults is 0.9 mg day^{-1}, but newborns usually have liver levels higher than those of adults. The concentration of Cu in mature milk ranges between 0.2 and 0.3 mg l^{-1}; in colostrum it is higher (0.4–0.6 mg l^{-1}), but decreases along the lactation period (see the Chapter 13 by de la Flor St Remy et al.).

Distribution of Cu in milk has been studied by SEC coupled to ICP-AES [12, 14], and ICP-MS [15, 17–19], or even using electrothermal atomization atomic emission spectrometry (ET-AAS) for the detection [20]. Breast milk Cu seems to be distributed all over the biocompounds from high (caseins, immunoglobulins, lactoferrin, and serum albumin) to lower-molecular-weight ligands (lactoalbumin, peptones, free aminoacids, citrates, etc.). The distribution patterns of Cu have been shown to be very different in mature milk and colostrum [18] (Fig. 17.3).

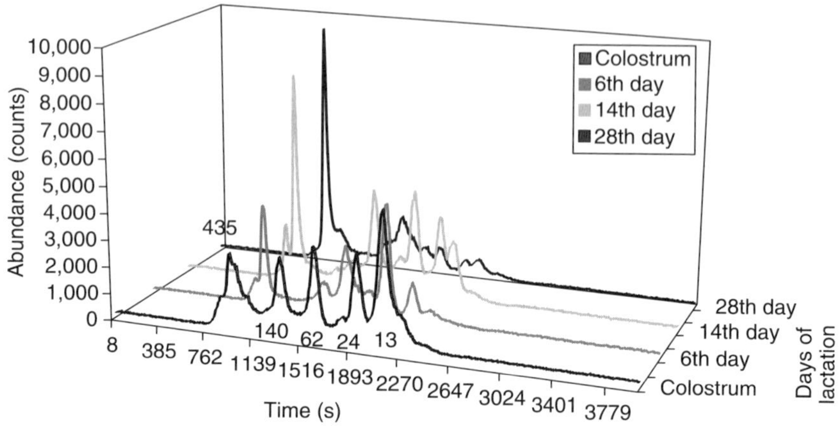

Figure 17.3. Copper distribution profile in colostrum and human milk lactating period whey. From de la Flor St Reny et al. [18], with permission of RSC.

Figure 17.3 shows that the peak for colostrum and human milk whey at 13 kDa decreases when passing to mature milk, whereas the peak at 435 kDa (probably associated with Immunoglobulin A) increases dramatically in mature milk. Conversely, the amount of Cu associated with the 62 kDa bioligand (probably serum albumin) and the 24 kDa fraction seems to remain unchanged throughout the lactating period. An off-line method to determine Cu bound to relatively LMW compounds in human and formula milk using SEC separation and ET-AAS detection has also been reported [21].

A tentative identification of Cu-bound species by comparison with serum proteins was carried out by Suzuki et al. [19], where a compound obtained at 140 kDa was ascribed to ceruloplasmin. However, it was only detected in colostrum, not in transitional or mature milk.

17.5.4 Zinc

Zinc is a very important essential trace element, especially for newborns. Babies need Zn mostly for cell proliferation, DNA, RNA, and protein synthesis [22, 23], and in general for all fast-growing tissues. Many essential enzymes are regulated or activated by Zn. Breast milk contains lower concentrations of Zn than formula milks. This clearly indicates differences in Zn speciation between the two types of milk. Early investigations thus focused on specific Zn ligands in human milk. Much work was done in the 1980s by Lönnerdal and co-workers, who set milestones in Zn speciation. These authors reported a LMW Zn binding ligand in human milk [24]. They used SEC for separation with an eluent containing NH_4AcO to reduce charged groups. A second separation dimension was introduced by the use of ion exchange chromatography (IEC) to confirm the Zn ligand identity. SEC separation resulted in two Zn peaks at >70,000 Da and another major signal around 600–650 Da, which was identified as citrate (24). Concentrations determined were 1830 μg l^{-1} for Zn (total) and 0.99 mM for citrate. These results were discussed in a manner that LMW compounds might be of greater importance in binding essential metals and in their transport or function in the animal/human body.

On the other hand, Zn concentration in cow's milk seems to be 10 times higher than in humans, but as it is bound to casein it becomes inaccessible for infants. The presence of Zn citrate would enhance the availability of this element to newborns. These results were further verified [25]. Again the authors found Zn in human milk predominantly bound to a low-molecular-weight compound and critically studied the "modified SEC separation" mode. The paper by Johnson and Evans [26] was mentioned to show the generation of nonnative Zn ligands when using this approach.

Martin et al. [27] and Blakeborough et al. [28] identified Zn citrate in milk. The speciation changes during SEC separations was systematically and critically investigated by Arnaud et al. [29] using ultrafiltration as a quality control alternative tool. In this study, the samples were human transitional milk (14–30 day), having 65 μmol (4200 ng ml^{-1}) Zn, 19 g l^{-1} protein and 2.5 μmo l^{-1} (480 mg/l^{-1}) citrate. Possible Zn ligands were found to be albumin and citrate, and most of the Zn was associated to albumin when using acetate buffers. Other Zn ligands proposed were albumin and lactoferrin in agreement with some findings [24, 30], but in contrast to others [28]. In a subsequent paper, Lönnerdal et al. [25] assigned the main three peaks found to casein, albumin, and LMW ligands, respectively.

At the beginning of the 1990s two papers from Michalke and Schramel focused on Zn speciation in human milk. The first one [31] followed the finding of metallothioneins in human milk, but other Zn species were also studied [31–33]. Zinc ligands were identified according to peak retention times matching retention times of Zn species standards and additionally by analyzing citrate in Zn containing fractions using a citrate specific determination kit for citrate identification. Further, Zn species were analyzed with IEF (multi-dimensional approach). For QC mass balances, re-injection experiments and Zn-transfer experiments were also carried out. Summarizing, the results of Zn speciation studies showed 95 percent of Zn in human milk (3181 ± 2.5 μg Zn l^{-1}) to be bound to citrate, and 5 percent associated with proteins [31, 33]. These proteins were identified as lactoferrin (25 ± 2.1 μg Zn l^{-1}), casein (69 ± 4.4 μg Zn l^{-1}), albumin (33 ± 3.3 μg Zn l^{-1}), and metallothionein (18 ± 2.5 μg Zn l^{-1}).

In 1995 a paper by Negretti de Brätter et al. [51] investigated Zn speciation in human milk whey samples from Venezuela. Special attention was given to QC-implemented sampling by thoroughly cleaning breast nipples and avoiding contamination. Two detecting modes, namely ICP-AES and instrumental neutron activation analysis (INAA), were used for verifying the reliability of Zn determination during the speciation process. Total Zn was determined at 2.4–5.2 mg l^{-1} levels. The columns were a combination of Asashipak GS 620 (7.6 × 500 mm) and GS 520 (7.6 × 500 mm), and were mass calibrated between 17.8 and 669 kDa, employing standard compounds isolated from human milk. The experimental

Zn profile showed a predominant elution of Zn in the region around 67 kDa, then a considerable Zn amount between 10 and 1 kDa, and nearly no Zn below 1 kDa.

The quantification in the column eluate was checked by INAA, which confirmed these findings. Possible SEC gel adsorption effects [29] were not considered and SEC eluted Zn species identification was not further pursued. Later on, the same group [34] investigated the elution patterns of various trace elements from human milk using an Asashipak GS 620 HQ column (300×7.6 mm) and Tris–HCl, pH 7.1 buffer, reporting that most Zn eluted at less than 10 kDa. Further identification work showed the main Zn species to be citrate. Other Zn species were seen in not totally resolved peaks, which were assigned to casein and albumin. Interestingly, an opposite effect was observed between Zn citrate and Se concentration in human milk. A high Se supplementation to the mother resulted in high Se concentration in milk, which was linked to lower total Zn and Zn citrate levels in this milk [35].

Caroli and his team [36] determined the total Zn content and its binding patterns in different polluted regions of Italy. Using SEC high performance liquid chromatography (HPLC)–ICP-AES they determined the elution pattern of Ca, Cu, Fe, Mg, Mn, and Zn in human milk from 60 mothers in different areas. They found Zn at 2720 ± 70 μg l^{-1}, with a higher trend in mothers under 30 years of age. These authors reported an elution pattern with five peaks, with peak 1 assigned as casein aggregates (>2000 kDa), peak 2 assigned as immunoglobulins (500–2000 kDa), peak 3 assigned as albumin (100–500 kDa), peak 4 assigned as (probably) α-lactalbumin (2–100 kDa), and peak 5 all Zn compounds with a molecular weight lower than 2 kDa. Finally, de la Flor St Reny et al. [18] investigated human milk whey at different lactation stages (Fig. 17.4). In agreement with the literature, the predominant Zn species eluted in the LMW range.

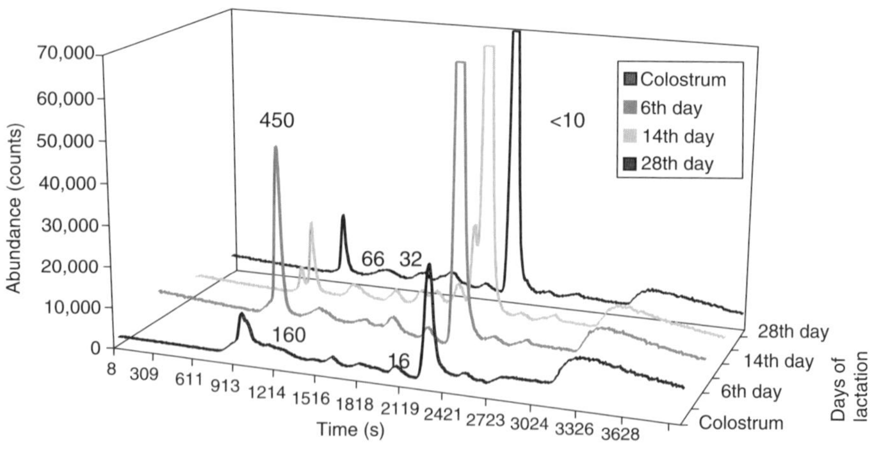

Figure 17.4. Zinc distribution profile in colostrum and human milk lactating period whey. From de la Flor St Reny et al. [18], with permission of RSC.

Nearly all investigations found Zn citrate to be the predominant Zn species in human milk. Several other species were found to be of minor importance, but their presence was confirmed by at least two groups. These species were Zn lactoferrin, Zn casein, Zn albumin, and Zn α lactalbumin. The high bioavailability of Zn in human milk is believed to be linked to Zn citrate.

17.5.5 Selenium

Selenium is an essential trace element. Sufficient Se supplementation can protect against several heart diseases and its role in cancer prevention has been highlighted [37, 38]. The detoxification effects of Se have been proven and described widely [38, 39]. Selenium deficiency is associated with several diseases like haemolysis, muscular dystrophy [38–40], cystic fibriosis, congestive cardiomyopathy, Kashin–Beck disease, Legionnaires' disease, multiple sclerosis, myotonic dystrophy, neural ceroid lipofuscinosis, and rheumatic arthritis [39]. Furthermore, thyroid metabolism seems to be impaired, because many de-iodinases are Se-proteins [41]. Selenium is of particular importance for pregnant women, as the fetus receives much Se during gestation [42]. Selenium deficiency is most critical for the brain and growth of infants [41]. Newborns initially show nearly normal serum Se levels [3] or slightly decreased ones (50–60 $\mu g\ l^{-1}$) [39, 40], which then decrease during the first half year to reach the initial values again after 12 months [38, 43]. Furthermore, the total Se concentration in breast milk decreases from 20 $\mu g\ l^{-1}$, [44] or 40–80 $\mu g\ l^{-1}$ [45] in colostrum to 5–18 $\mu g\ l^{-1}$, [43–48] in mature milk, with slight variations among lactating women and depending on nutrition [34]. The bioavailability of Se depends on its binding form (species) [49, 50]. Therefore, an absolutely clear identification and quantification of Se species in human milk is felt to be necessary.

In a pioneer work, Brätter et al. [30] coupled a SEC column in order to digest on-line the eluate followed by hydride generation (HG) ICP-AES detection. Limits of detection (LoDs) of 1–2 $\mu g\ l^{-1}$ were achieved. Selenium was found associated with LMW compounds. In 1995 and 1998 the group of Brätter repeated the experiments with human milk samples from the seleniferous region of Venezuela using ICP-MS detection and INAA for QC and found that Se was predominantly associated with high molecular weight compounds [34, 51]. They extended their work to two additional countries (Portugal and Germany). Figure 17.5 shows the elution pattern they obtained for Se. Furthermore, these authors found variations in the elution pattern between samples from different countries and this result was related to the different Se supplementation of mothers living in different regions.

Selenium shows a similar pattern along the lactating period (four peaks can be clearly detected at 350, 64, 19 and less than 10 kDa), and most Se is bound in the 350 kDa fraction. The Se content in this fraction is higher in colostrum than in mature milk whey [18]. Michalke and co-workers published a series of papers on Se speciation in human milk, with special attention to QC. Generally, they used pooled human milk samples between days 7 and 14 after delivery. A first

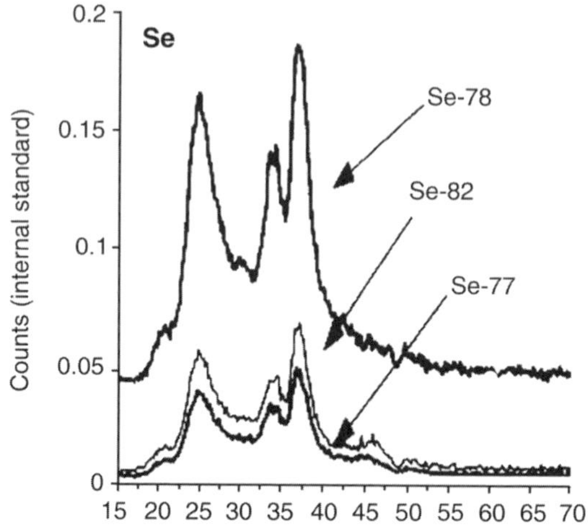

Figure 17.5. Selenium speciation in a human milk sample. The Se elution pattern is that of an SEC column which is directly coupled to an ICP-MS detector. From Brätter et al. [34] with permission from RSC.

methodical paper provided techniques for the clear identification of Se species using different capillary electrophoresis (CE) methods as a secondary separation approach for identification. First, defatted human milk samples were SEC fractionated and the size-characterized Se-containing fractions were analyzed by two different capillary zone electrophoresis (CZE) methods or by a CZE and a capillary isoelectric focusing (CIEF) method. Selenium was determined using electrothermal volatilization (ETV) ICP-MS in SEC fractions and in CE fractions, gathered after consecutive SEC and CE separations. Using this two-dimensional purification, identification and direct assignment of Se to the (two-dimensionally) purified Se compound were possible (52, 53). The Se concentrations per species were determined as 2.5 ($\pm$0.2) (Se-carrying glutathione, GSeH) $\mu g\ l^{-1}$, 3.1 ($\pm$ 0.3) $\mu g\ l^{-1}$ (Se-cystamine, SeCM), 5.2 ($\pm$0.4) $\mu g\ l^{-1}$ (Se-cystine, SeC), and 1.1 ($\pm$0.1) $\mu g\ l^{-1}$ (Se-methionine, SeM). The results were confirmed by a set of QC measurements. All these compounds were in the LMW range. It should be noted that both groups, Michalke and Schramel [53] and Brätter et al. [30], used samples from the slightly Se-depleted Germany, whilst Brätter's group in 1995 and 1998 [34, 51] predominantly analyzed samples from the seleniferous areas of Venezuela.

Finally, Michalke and Schramel analyzed Se speciation in human milk using CE coupled to ICP-MS to obtain orthogonal, independent results from a SEC fractionation step. CZE was employed as well as CIEF [54, 55]. These experiments confirmed the findings with the SEC-CE-ETV-ICP-MS approach of Michaulke and Schramel [53]. Electropherograms from both CZE-ICP-MS and CIEF-ICP-MS are illustrated in Figure 17.6.

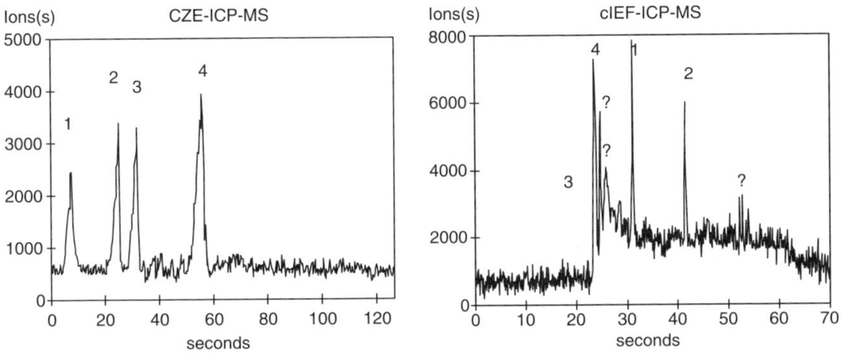

Figure 17.6. Analysis of a human milk sample in parallel by CZE-ICP-MS and CIEF-ICP-MS. From Michalke and Schramel [54] and Michalke [55], with permission of Elsevier and Advanstar.

Based on the coincidence of CZE- and CIEF-ICP MS results with the formerly two-dimensionally gained results these fast techniques were used to elucidate the trend of Se species in human milk at different lactation stages [33]. These trends are shown in Figure 17.7. It is obvious that the observed behavior (protein-associated Se decreases as lactation proceeds) is similar to glutathione-bound Se, whilst SeC increases. This may be of specific importance due to the central role of SeC in the human metabolism of Se.

On the basis of the data available, for samples from low-Se geographical regions Se is mostly associated with LMW compounds, while for those from very high Se-content regions Se is predominantly bound to proteins.

17.5.6 Iodine

Iodine is known to be an essential micronutrient that the thyroid gland uses for the biosynthesis of the thyroid hormones thyroxin (T4) and tri-iodothyronine (T3) [56–58]. These hormones have a strong influence on a wide range of biochemical reactions. Immune defense and antibody production depend on reliable thyroid function and availability of T4 and T3 hormones [59, 60]. The knowledge about the status of the different I species (such as thyroxin, tri-iodothyronine or I^- in body fluids can give information about malfunction of the thyroid gland and may also explain other T4/T3-related metabolic abnormalities [58]. This was the rationale behind several investigations on I speciation in human body fluids in general, predominantly using chromatographic methods [61]. Furthermore, a lack of I supplementation in newborns is thought to result in slower brain development and even in severe damages to central nervous system [62]. An I transporter and a peroxidase enzyme are supposed to be involved in I accumulation in mammary glands [63]. These facts render iodine speciation a subject matter for human milk investigations.

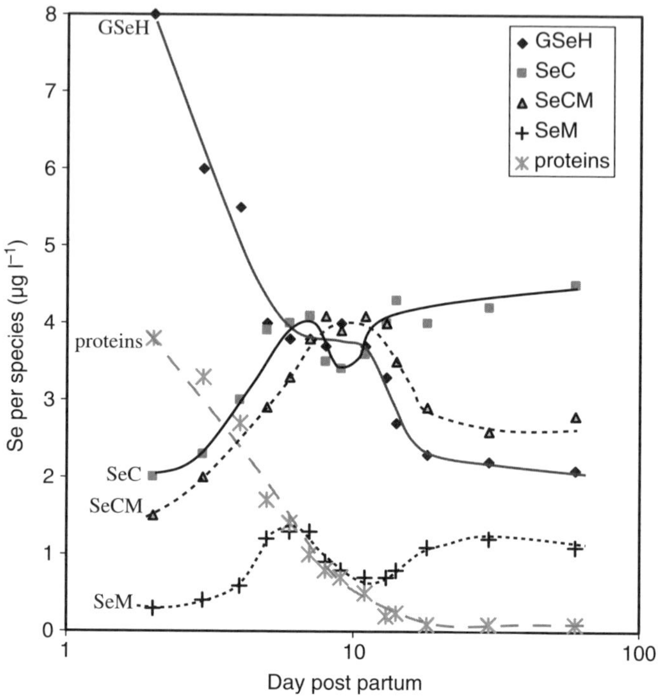

Figure 17.7. Trends of various Se species in human milk depending on lactation state. From Michalke [33].

Brätter et al. [34] used the same technique for I speciation described above for Se and Zn. Iodine was monitored at ^{127}I on-line by ICP-MS after SEC separation. It turned out that 80 percent of I in human milk was present as I^-. In a further work by Brätter et al. [62] human milk whey showed a predominant I^- peak as well as six further peaks in the high molecular weight (HMW) range (5–300 kDa). A partial co-elution of I-peaks with Se peaks was obvious. Total I in European breast milk samples was 95 ± 60 μg l^{-1} references, according to data in [18, 64]. Furthermore, a negative correlation was found between Se concentration in human milk and I species. The authors hypothesized that a high Se intake may impair the synthesis of I binding organic compounds. In parallel to the Se speciation work [53], I speciation in breast milk was investigated as well [54] using the same samples and sample preparation steps. Total I varied according to lactation state, beginning at 60 μg l^{-1} on day 2 (post partum), reaching 100 μg l^{-1} on day 3 and decreasing to 80 μg l^{-1} (day 6) or 60 μg l^{-1} from days 9 to 60. The prefractionation by centrifugation showed that I was associated to fat at approximately 30 percent throughout (70 percent to the LMW fraction). Characterization of I species from milk whey was done with SEC-ICP-MS and strong anion exchange chromatography (AEC) ICP-MS. The SEC fractionation showed I^- to predominate with ca. 37 μg l^{-1}, as well as two more I species (with 1.5 and 1 μg l^{-1} I) which could be T4 and T3 hormones. AEC-ICP-MS experiments confirmed I^- to be the major I species in human milk.

In conclusion, I speciation in breast milk points out that I^- is the predominant I species. A nutritional direct and indirect dependency is also seen, that influences the concentration of organic I species, by simultaneous Se supplementation to the mother.

17.5.7 Manganese

Manganese is a trace element known to activate many enzymes involved in metabolic processes. It is the key element at the active sites of several enzymes. Manganese is needed for protein and fat metabolism, healthy nerves, and immune system as well as sugar regulation. Manganese is one of the key elements for enzymes in energy production and increases the level of antioxidative protection, especially via mitochondrial Mn-superoxide dismutase [65–67]. This element is involved in the utilization of Vitamin B1 and vitamin E, and it is required for normal bone growth and for the prevention of clotting defects. On the other hand, Mn is used as anti-knock agent in gasoline, resulting in increased Mn-blood levels as monitored in Canadian children [68]. Increased Mn levels are known to damage the central nervous system, resulting in motor abnormalities and psychic disorders [66–68].

In human milk Mn is an ultratrace element. Depending on the post partum stage, Mn concentrations between 2 and 5 μg l^{-1} for transitory and mature milk or up to 10 μg l^{-1} for colostrum have been published for the European region [12, 14, 69–71]. The environment and nutrition have also been reported to have some influence on such levels [69].

Only few groups have so far investigated Mn speciation in human milk. Owing to the weak binding of Mn to some organic ligands, most Mn species in human milk are considered to be labile. Therefore, typically the first (or sole) analytical step was again a SEC fractionation in most investigations. In their multi-element speciation approach Brätter et al. [34] also discussed Mn SEC fractionation in breast milk compared with cow's milk. Two peaks were reported: one attributed to lactoferrin and then a sharp main peak which might be citrate.

Caroli and his co-workers [14] published multielement speciation results including data about Mn speciation. These data regarded samples from 60 mothers (approximately day 30 after delivery) living in different areas of Italy. The investigation was based on SEC fractionation using a mass calibrated column and total concentration of Mn was found to be around 3 μg l^{-1}. They found 28 percent of Mn in the void fraction (>2000 kDa), 30 percent bound to the nonprotein fraction, (relatively LMW compounds), and the remaining Mn was found at low levels in the other fractions.

Recently, Michalke and Schramel [72] published a multidimensional investigation on Mn speciation in human milk. Total Mn was determined as 2.8 μg l^{-1}.

These authors used prefractionation via centrifugation, gaining a fat-supernatant fraction, a protein fraction and a LMW-supernatant fraction, followed by two different SEC fractions, again followed by AEC-ICP-MS measurements. In this work it turned out that Mn was predominantly LMW bound (ca. 90 percent of Mn), while the remaining Mn was widely spread associated to HMW molecules (peaks between 75 and 107 kDa). Standard addition experiments at the LMW-SEC step as well as in the SEC-AEC-ICP-MS step pointed to Mn citrate as the most important Mn species in breast milk. Free inorganic Mn was also seen in low amounts. Figure 17.8 shows the consecutive chromatograms of SEC-ICP-MS and AEC-ICP-MS. These findings are in good agreement with those of Brätter et al. [34] who reported Mn lactoferrin (78 kDa) and the predominant species at the citrate retention time. On the basis of these results, it can be said that citrate plays a major role for Mn in milk, but other Mn compounds seem also to be present.

17.5.8 Cadmium

A number of potentially toxic metals have been reported in breast milk, including Cd, Hg, and Pb. Unlike the persistent organic pollutants (POPs), metals do not bind to the fat, and so they do not usually accumulate at concentrations higher in breast milk than in blood. As a result, infants are likely to be exposed to higher levels of toxic metals before birth than during breastfeeding. Nonetheless, knowledge of concentrations of potentially toxic metals in breast milk is important as an indication of both likely prenatal and postnatal exposure.

Cadmium levels in breast milk have been shown to be significantly associated with cigarette smoking. A German study showed a direct relationship between the number of cigarettes a mother smokes per day and the level of Cd in her milk. The investigations on Cd speciation in human milk were aimed at clarifying whether Cd in human milk is solely a matter of environmental load or whether there might exist a homeostatic control by the mammary gland. In general, total element concentrations have been determined in the body fluids and placenta of mothers living in areas with different environmental metal loads (rural, urban, industrial). The Cd concentration in human milk was found to be independent of the environmental load [73]. Vuori et al. [74] found only a slight decline of total Cd in human milk going from 2 $\mu g\ l^{-1}$ in the first lactation month to approximately 1 $\mu g\ l^{-1}$ in month 6 after delivery. This constancy of Cd concentration in human milk, being independent of the environmental load of the metal, ignited a discussion about a Cd binding protein, most probably linked to some homeostatic control by the mammary gland. Therefore, speciation investigations were started on Cd in human milk.

Along this line of thought, Michalke and Schramel [75] studied pooled human milk samples from days 7–14 after delivery. They developed a method for the preparative separation of the proteins from skimmed human milk, which was proven to preserve the metal protein complexes as well as keeping the samples and fractions free from Cd contamination. For this purpose an SEC separation was applied and fractions were collected for Cd determination by voltammetry

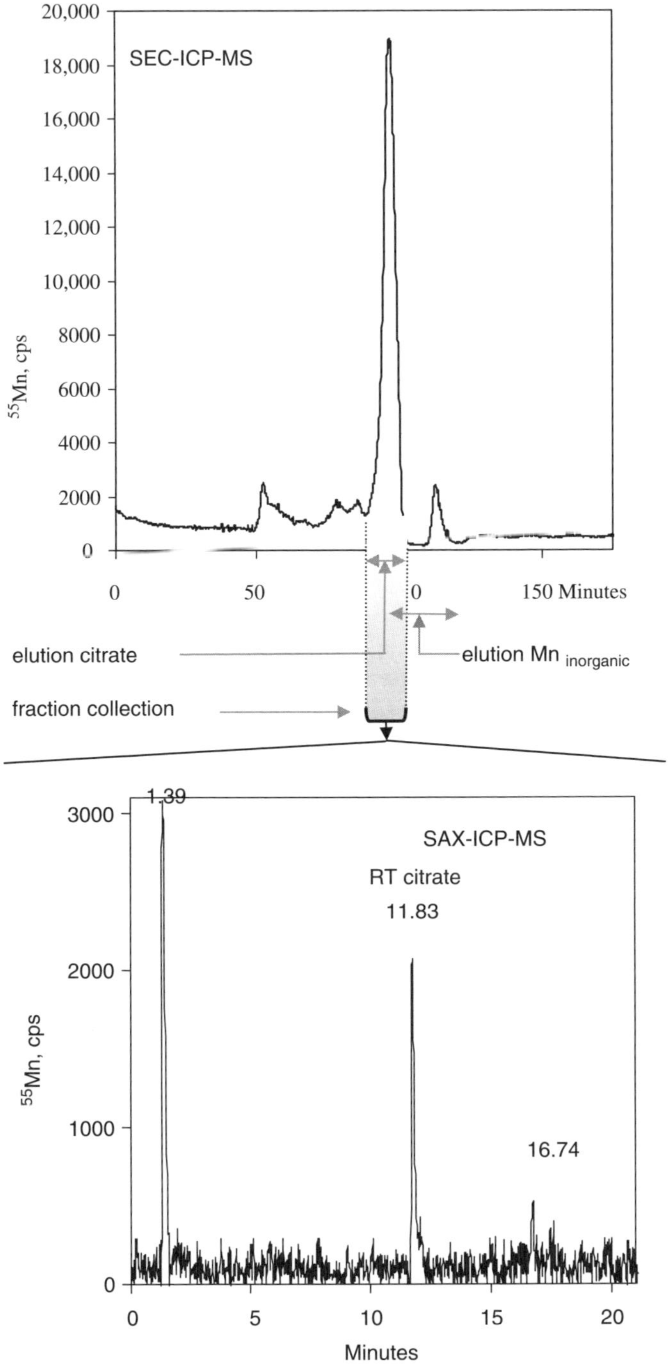

Figure 17.8. Two-dimensional speciation of Mn in human milk. First a separation according to molecular size was performed, followed by AEC-ICP-MS analysis. Citrate and inorganic Mn appeared when using both methods in series (unpublished data).

and for protein quantification. Ten fractions were collected, although Cd was found only in the metallothionein (MT) fraction. Both Cd and MT were constant (1 $\mu g\ l^{-1}$ of Cd and 10 $mg\ l^{-1}$ of MT) in various pooled samples. Citrate, possibly a coeluting ligand of Cd, was excluded to be the Cd carrying species when using isoelectric focusing. The whole speciation procedure as well as the result was quality checked by using strategies like re-injection experiments, mass balances and two-dimensional identification by SEC and CIEF. Woitiez concluded that the analytical procedure used was performed in an "extremely careful way" where the "accuracy of data was convincingly demonstrated" [76].

It is likely that only one more paper investigating Cd species pattern in human milk is available [34]. These latter authors used SEC-separation, whilst Cd detection was carried out on-line using ICP-MS. The elution pattern of Cd showed only one small peak in the final volume of the column. The separation range of this SEC column was given as 10–300 kDa, and proved no differentiation of species below 10 kDa size (Cd MTs having 7 kDa as molecular mass).

17.6 ELEMENTS SPECIATED IN FORMULA MILKS

Human milk can be considered the optimal food for infants. It contains all the macronutrients and micronutrients necessary for the correct development of the newborn and at the adequate levels. When infants are not breast-fed, or breast-feeding is discontinued very early, formula milks are used instead. Therefore, it is important to evaluate the capacity of formulas to deliver satisfactory quantities of minerals and trace elements that can be bioavailable to the children in order to cover their biological needs. This fact explains the need to perform trace elements speciation analysis in formula milk. Thus, the speciation results obtained in formulas and published so far are critically evaluated here and compared with those obtained in human milk.

17.6.1 Calcium and Magnesium

Calcium turned out to be around 500 $mg\ ml^{-1}$ and Mg 40–50 $mg\ l^{-1}$ in fortified formulas. Comparing with human breast milk, higher concentrations of these elements are required in formulas because of their lower bioavailability in cow's milk-based formulas. Some studies have demonstrated that the absorption of Ca and Mg is similar from both fortified formulas and human milk [11]. On the other hand, both elements showed the same chromatographic profile in formulas and in breast milk as they eluted in the LMW fraction [12, 15].

17.6.2 Iron

It seems that infants fed with nonfortified Fe formulas frequently develop anemia. Therefore, most infant formulas are fortified with Fe at different levels. Infants fed with fortified formulas may have an intake of 4.8–10 mg per day. Thus, more

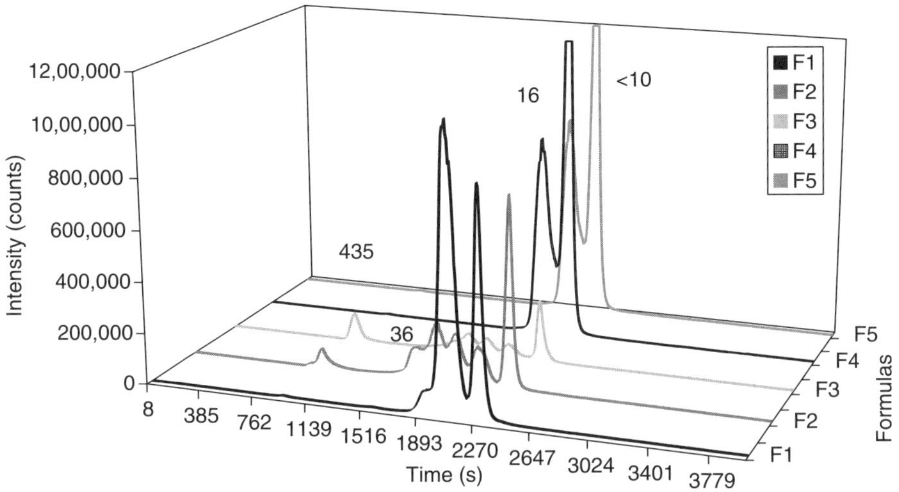

Figure 17.9. Iron distribution profile in different commercial formula whey. From de la Flor St Reny et al. [18] with permission of RSC.

than 10–100 fold the breast-fed infant's Fe intake per day seems necessary to provide the adequate amount of Fe in a bioavailable form. Thus, Fe is much better absorbed from breast milk than from formulas [11]. Usually the composition of formula milk whey seems with be completely different compared with human milk whey [12, 15, 18, 21]. As can be seen in Figure 7.9 (to be compared with Fig. 17.2), in five formulas from different manufacturers, only a small fraction of Fe was bound to HMW biomolecules. In formulas, most Fe appeared associated to be with the LMW region, probably as a result of the corresponding Fe artificial fortification of the cow's milk base used for formula milks.

17.6.3 Copper

Most infant formulas are fortified with Cu to a level of 0.4–0.6 mg l^{-1}. The bioavailability of Cu was evaluated [11] for full-term infants and it was demonstrated that Cu retention from human milk is higher than from fortified/nonfortified formulas [11]. However, the amount of Cu absorbed from milk formulas seems to be adequate, since Cu levels in infants fed this way had indexes of Cu status very similar to those of breast-fed infants [77]. It has to be pointed out, however, that premature infants receiving only formula milk with no Cu fortification frequently develop Cu deficiency [11].

These observed differences in absorption of the element can be explained by the different distribution (speciation) of Cu in human and formula milks (see Fig. 17.10 and compare with Fig. 17.3). In premature formula milk whey, most Cu was distributed as follows: a small amount in the HMW region (450 kDa) and a large amount in the LMW region (13 and <10 kDa), the latter corresponding to

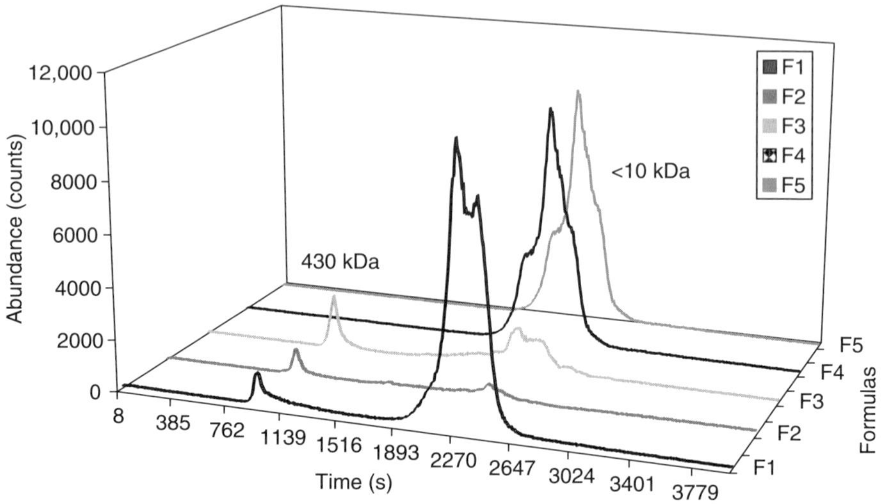

Figure 17.10. Copper distribution profile in different commercial formula whey. From de la Flor St Reny et al. [18] with permission of RSC.

citrates or ionic forms of the element [18]. These results agree well with most reports available in the literature on formula milks [12, 15, 21, 36]. Interestingly, the total Cu present in the analyzed formulas was not so dependent on the individual manufacturer (except for F2 and F3), but its level was usually higher than in breast milk as a result of fortification.

17.6.4 Zinc

Studies in human adults using radioisotopes or stable isotopes of Zn showed that Zn absorption is higher from human milk (41 percent) than from cow (28 percent) or soy formulas (14 percent) [77]. In any case, it seems that there is no need to fortify cow-based formulas with Zn, since cow milk contains much more Zn than breast milk [77]. However, Zn bioavailavility from breast milk is clearly higher than from formulas, which indicates a different composition in Zn species. Experiments in speciation of Zn in formula milk whey carried out by SEC showed that Zn was mostly found in the LMW (<10 kDa) fraction, as a complex of citrate and only a small fraction of the metal was bound to whey proteins (435 kDa) [15, 18, 21]. Although such species are detected in human milk, important differences have been found between the distribution patterns of Zn in breast milk and in cow-based formula milks (see Fig. 17.11 and compare with Fig. 17.4). The levels of Zn in formulas are higher than in human milk and in colostrums [18].

17.6.5 Selenium

Selenium levels vary in infant formulas, depending mainly on the Se content of their cow's milk source. In any case, Se concentrations in formulas are usually

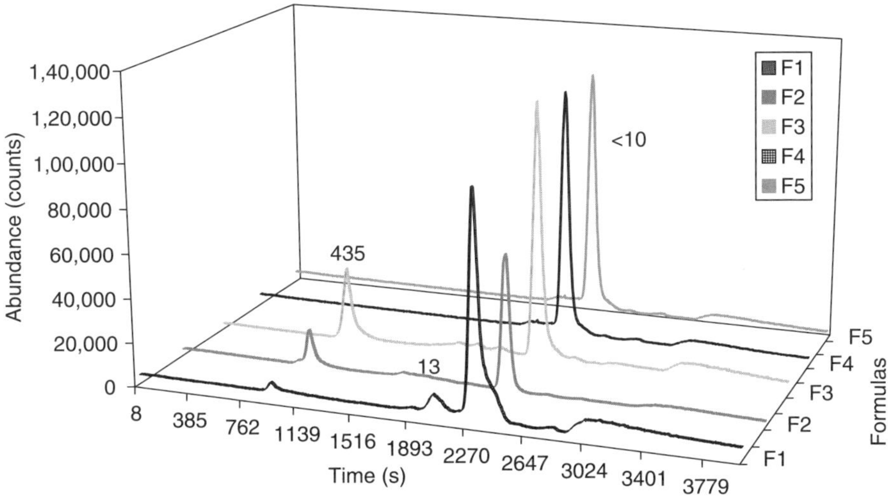

Figure 17.11. Zinc distribution profile in different commercial formula whey. From de la Flor St Reny et al. [18] with permission of RSC.

lower than in breast milk. In Europe clinical studies showed that serum levels are higher in breast-fed than in formula-fed infants. For that reason formulas are currently supplemented with the element. In Brätter's 1998 study [34] on binding patterns of trace elements in human and formula milk whey, concentration and binding patters of Se in the raw material were given. In all soy-based formulas the observed patterns were similar and nearly independent of the soy source. However, in cow-based formulas Se concentration depended on the origin of the raw material. Moreover, the profiles of soy- and cow's milk-based formula were completely different; although the Se peaks observed in those types of formula whey occurred in the LMW region, retention times were very different from those observed in human milk. In preterm formula whey two main peaks were reported, at 25 and $<$10 kDa, respectively [18]. Curiously enough, among all analyzed formulas, only one of them contained Se associated with the HMW biocompounds. A method to determine Se distribution in formula milk whey by SEC-HG-ET-AAS was developed by Bermejo et al. [20], requiring a previous sample preconcentration. Selenium was distributed in three fractions ($>$160, 40, and mostly in the $<$10 kDa fraction).

17.6.6 Iodine

In spite of the fact that I is an essential micronutrient to man, no studies on the absorption of I in infants are available [11]. In studies on the distribution of I in milk whey by SEC-ICP-MS, I^- seems to be the only I species in formulas, while human milk whey showed more I species [18]. However, Fernandez et al. [78] carried out a study on I speciation in milk from animals (cow and goat), humans of different geographic origin and infant formulas from different manufactures. They found that whey obtained from fresh milk samples (animal or humans) contained more than 95 percent of the iodine initially present in milk, whereas the formula whey contained only 15–50 percent of the total I. More than half of the iodine present in formula whey was bound to HMW species (>1000 kDa).

17.6.7 Manganese

Manganese is found in much higher concentrations in infant formulas than in breast milk. In fact, soy-based formulas were shown to contain about 80 times more Mn than breast milk [11]. Also, animal protein-based formulas contained also about 30 times more Mn than human milk. Infants are not able to absorb and excrete excess Mn during their first year of life, a period of rapid development. As Mn is neurotoxic and impairs intestinal Fe absorption, it would be desirable to reduce Mn levels in infant formulas down to those found in human milk (4–8 $\mu g\ l^{-1}$). Coni et al. [12] reported that in formulas more than 25 percent of total Mn is bound to caseins. On the other hand, Mn present in soy- and in cow's milk-based formula whey is mainly associated with LMW components (<10 kDa), as is the case with human milk whey [15, 18, 34, 72].

17.6.8 Toxic Metals in Formulas

Potentially toxic metals such as Al, Cd, and Pb have been reported in breast milk. As they may also contaminate drinking water it is not surprising that they are found in infant formulas at levels even higher than in breast milk. Cow's milk-based formulas exhibit lower toxic metals levels [13]. For example, the exposure to Cd for an infant fed a soy formula turned out to be about 20 times higher than for a breast-fed infant. Interestingly, gastrointestinal Cd uptake is believed to be higher in newborns than in adults. So far studies on Cd speciation in formula milk are very scarce, but it has been reported that Cd in formulas is distributed between caseins and LMW compounds [12].

Considering that most infant formulas are based on cow's milk, it is understandable that the observed elemental distributions of the protein fraction of such formulas follow a pattern similar to that of UHT-cow's milk. Some discrepancies observed for several elements are most likely due to additional mineral fortification during the manufacturing process of formulas.

17.7 CONCLUSIONS

Human milk (or the substitute food, formula milk) is the only source of essential trace elements for newborns. To ensure the optimum development of the baby, mother's milk contains essential elements at the adequate levels. However, today it is well known that bioavailability, biological activity, nutritional value, as well as toxicity and metabolism of trace metals and metalloids strongly depend of their chemical form. This fact explains the present need to carry out studies on trace element speciation in milk samples. Most methodologies developed so far for elemental speciation in milk are based on coupling on-line a separation technique as HPLC (SEC) with an ICP-MS specific detector, although CE-ICP-MS could be a useful alternative tool.

Elemental speciation studies show that human milk, especially colostrum and transitional milk, is very rich in HMW species associated with metals. Of course, many more studies are needed for a reliable speciation of such HMW compounds. In this sense, validation approaches, both for the reliable identification and exact determination of such species, should be urgently developed. In the same way as the elemental composition and/or distribution of human milk can be considered ideal for feeding the newborn, the composition of formula milks for newborns should ideally be as similar as possible to maternal milk at every lactating stage. However, essential element speciation in formula milks is far from that of human milk (the ideal composition). This could explain why the bioavailability of essential elements (including Cu, Fe, I, Mn, Se, and Zn) from formula milk is much lower than from human milk. Hence, artificial formulas are usually supplemented with such essential elements. More scientific knowledge is definitely necessary on the composition (speciation) of such elements and more attention must be paid to the chemical form in which they are added to formula milks.

REFERENCES

1. G. Harzer, J. G. Bindels, Changes in human milk immunoglobulin A and lactoferrin during early lactation, in J. Schaub (ed.), *Composition and Physiological Properties of Human Milk*, Elsevier Science, Amsterdam, 1985, pp. 285–295.
2. D. M. Templeton, F. Ariese, R. Cornelis, L. G. Danielsson, H. Muntau, H. P. van Leeuwen, R. Łobinski, Guidelines for terms related to chemical speciation and fractionation of elements. Definitions, structural aspects, and methodological approaches, *Pure Appl. Chem.*, **72** (2000), 1453–1470.
3. S. Caroli, Chemical speciation: a decade of progress, in S. Caroli (ed.), *Element Speciation in Bioinorganic Chemistry*, Wiley, New York, 1996, pp. 1–18.
4. B. Michalke, Quality control and reference materials in speciation, *Fresenius J. Anal. Chem.*, **363** (1999), 439–445.
5. A. M. Ure, Ph. Quevauviller, H. Muntau, B. Griepink, Speciation of heavy metals in soils and sediments. An account of the improvement and harmonization of extraction techniques undertaken under the auspices of the BCR of the Commission of the European Communities, *Int. J. Environ. Anal. Chem.*, **51** (1993), 135–151.
6. A. M. Mota, L. Simaes Gonçalves, Direct method of speciation of heavy metals in natural waters, in S. Caroli (ed.), *Element Speciation in Bioinorganic Chemistry*, Wiley, New York, 1996, pp. 21–96.
7. S. Fairweather-Tait, The importance of trace element speciation in nutritional sciences. *Fresenius J. Anal. Chem.*, **363** (1999), 536–540.
8. R. E. Thiers, B. L. Vallee, Distribution of metals in subcellular fractions of rat liver, *J. Biol. Chem.*, **226** (1957), 911–920.

9. L. Dunemann, J. Begerow, *Kopplungstechniken zur Elementspeziesanalytik*, VCH, Weinheim, 1995, pp. 1–213.
10. I. T. Urasa, Development of new methods of speciation analysis, in S. Caroli (ed.), *Element Speciation in Bioinorganic Chemistry*, Wiley, New York, 1996, pp. 121–154.
11. B. Lönnerdal, Effects of milk and milk components on calcium, magnesium, and trace element absorption during infancy, *Physiol. Rev.*, **77** (1997), 643–669.
12. E. Coni, A. Alimonti, A. Bocca, F. La Torre, D. Pizzuti, S. Caroli, Speciation of trace elements in milk by high-performance liquid chromatography combined with inductively coupled plasma atomic emission spectrometry, in S. Caroli (ed.), *Element Speciation in Bioinorganic Chemistry*, Wiley, New York, 1996, pp. 255–283.
13. F. A. Rivero Martino, M. L. Fernández Sánchez, A. Sanz-Medel, The potential of double focusing ICP-MS for studying elemental distribution patterns in whole milk, skimmed milk and milk whey of different milks, *Anal. Chim. Acta*, **442** (2001), 191–200.
14. B. Bocca, A. Alimonti, E. Coni, M. Di Pasquale, L. Giglio, A. Bocca, S. Caroli, Determination of the total content and binding pattern of elements in human milk by high performance liquid chromatography–inductively coupled plasma atomic emission spectrometry, *Talanta*, **53** (2000), 295–303.
15. F. A. Rivero Martino, M. L. Fernández Sánchez, A. Sanz-Medel, Multi-elemental fractionation in milk whey by size exclusion chromatography coupled on line to ICP-MS, *J. Anal. Atom. Spectrom.*, **17** (2002), 1271–1277.
16. E. R. Morris, Iron, in W. Mertz (ed.), *Trace Elements in Human and Animal Nutrition*, Vol. 1, Academic Press, San Diego, CA, 1987, p. 87.
17. E. Coni, A. Bocca, B. Galoppi, A. Alimonti, S. Caroli, Identification of chemical species of some trace and minor elements in mature breast milk, *Microchem. J.*, **67** (2000), 187–194.
18. R. R. de la Flor St Remy, M. L. Fernández Sánchez, J. B. López Sastre, A. Sanz-Medel, Multielemental distribution patterns in premature human milk whey and preterm formula milk whey by size exclusion chromatography coupled to inductively coupled plasma mass spectrometry with octapole reaction cell, *J. Anal. Atom. Spectrom.*, **19** (2004), 1104–1110.
19. K. T. Suzuki, H. Tamagawa, S. Hirano, E. Kobayashi, K. Takahashi, N. Shimojo, Changes in element concentration and distribution in breast-milk fractions of a healthy lactating mother, *Biol. Trace Elem. Res.*, **28** (1991), 109–121.
20. P. Bermejo, E. M. Peña, D. Fonpedriña, R. Dominguez, A. Bermejo, J. A. Cocho, J. R. Fernández, J. M. Fraga, Speciation of zinc in low molecular weight proteins of breast milk and infant formulas by size exclusion chromatography/flame atomic absorption spectroscopy, *Res. Trace Elem.*, **4** (2001), 847–852.
21. P. Bermejo, J, Barciela, R. M. Peña, A. Bermejo, J. M. Fraga, J. A. Cocho, Determination of selenium in infant formulas whey fractions by SEC-HPLC-HG-ETAAS, *J. Anal. Atom. Spectrom*, **16** (2001), 189–193.
22. B. J. Mills, R. D. Lindeman, Zink, in H. Zumkley (ed.), *Spurenelemente*, Georg Thieme, Stuttgart, 1983, pp. 197–227.
23. M. P. Waalkes, P. L. Goering, Metallothionein and other cadmium-binding proteins: recent developments, *Chem. Res. Toxicol.*, **3** (1990), 281–288.
24. B. Lönnerdal, A. G. Stanislowski, L. S. Hurley, Isolation of a low molecular weight zinc binding ligand from human milk, *Inorg. Biochem.*, **12** (1980), 71–78.

25. B. Lönnerdal, B. Hoffmann, L. Hurley, Zinc and copper binding proteins in human milk, *Am. J. Clin. Nutr.*, **36** (1982), 1170–1176.

26. P. E. Johnson, G. W. Evans, Relative zinc availability in human breast milk, infant formulas, and cow's milk, *Am. J. Clin. Nutr.*, **31** (1978), 416–421.

27. M. T. Martin, K. F. Licklider, J. G. Brushmiller, F. A. Jacobs, Detection of low molecular weight copper (II) and zinc (II) binding ligands in ultrafiltered milks, the citrate connection, *J. Inorg. Biochem.*, **15** (1981), 55–65.

28. P. Blakeborough, D. N. Salter, M. J. Gurr, Zinc binding in cow's milk and human milk, *Biochem. Soc.*, **209** (1983), 505–512.

29. J. Arnaud, D. Andre, M. C. Bouillet, D. Kia, A. Favier, Problems associated with the use of exclusion-diffusion chromatography for identification of zinc ligands in human milk, *J. Trace Elem. Electrol. Health Dis.*, **6** (1992), 81–90.

30. P. Brätter, B. Gercken, U. Rösick, A. Tomiak, Speciation of trace elements in human milk and milk formulas, in P. Brätter, P. Schramel (eds), *Trace Element Analytical Chemistry in Biology and Medicine*, Vol. 5, Walter de Gruyter, Berlin, 1988, pp. 145–156.

31. B. Michalke, D. Münch, P. Schramel, Contribution to Zn-speciation in human milk: fractionation of organic compounds by HPLC and subsequent Zn-determination by DCP AES, *J. Trace Elem. Electrol. Health Dis.*, **5** (1991), 251–258.

32. B. Michalke, D. Münch, P. Schramel, Advanced quality control for zinc speciation investigations in human breast milk, *Fresenius J. Anal. Chem.*, **344** (1992), 306–310.

33. B. Michalke, *Elementspeziesanalytik in bio-medizinischen und Umweltproben*, Habilitationsschrift an der Technischen Universität Graz, 1988.

34. P. Brätter, I. N. Blasco, V. E. Negretti de Brätter, A. Raab, Speciation as an analytical aid in trace element research in infant nutrition, *Analyst*, **123** (1998), 821–826.

35. P. Brätter, V. E. Negretti de Brätter, S. Recknagel, R. Brunetto, Maternal selenium status influences the concentration and binding pattern of zinc in human milk, *J. Trace Elem. Med. Biol.*, **11** (1997), 203–209.

36. B. Bocca, A. Alimonti, E. Coni, M. Di Pasquale, L. Giglio, A. Bocca, S. Caroli, Determination of the total content and binding pattern of elements in human milk by high performance liquid chromatography–inductively coupled plasma atomic emission spectrometry, *Talanta*, **53** (2000), 295–303.

37. M. P. Rayman, The importance of selenium to human health. *Lancet*, **356** (2000), 233–241.

38. P. D. Whanger, Selenium in the treatment of heavy metal poisoning and chemical carcinogenesis, *J. Trace Elem. Electrol. Health Dis.*, **6** (1992), 209–221.

39. I. Lombeck, K. Kasperek, H. D. Harbisch, L. E. Feinendegen, H. Bremer, The selenium state of healthy children. *J. Eur. Pediatr.*, **125** (1977), 81–88.

40. S. Bro, H. Berendtsen, J. Norgaard, P. J. Jorgensen, Serum selenium concentration in maternal and umbilical cord blood. Relation to course and outcome of pregnancy, *J. Trace Elem. Electrol. Health Dis.*, **2** (1988), 165–169.

41. G. Örndahl, U. Sellden, S. Hallin, H. Wetterqvist, Selenium treatment of myotonic dystrophy, in P. Brätter and P. Schramel (eds), *Trace Element Analytical Chemistry in Medicine and Biology*, Vol. 4, Walter de Gruyter, Berlin, 1988, pp. 597–604.

42. I. Lombeck, K. Kasperek, H. D. Harbisch, K. Becker, E. Schumann, W. Schröter, L. E. Feinendegen, H. J. Bremer, The selenium state of children, *Eur. J. Pediatr.*, **128** (1978), 213–223.
43. T. W. Westermarck, Consequences from low selenium intake for man, in P. Brätter, P. Schramel (eds), *Trace Element Analytical Chemistry in Medicine and Biology*, Vol. 3, Walter de Gruyter, Berlin, 1984, pp. 49–70.
44. B. Akesson, R. Wälivaara, L. Jansson, Selenium content of human milk and its relation to other nutrients, in J. Schaub (ed.), *Composition and Physiological Properties of Human Milk*, Elsevier Science, Amsterdam, 1985, pp. 87–91.
45. J. R. Arthur, F. Nicol, G. J. Beckett, The role of selenium in thyroid hormone metabolism and effects of selenium deficiency on thyroid hormone and iodine metabolism, *Biol. Trace Elem. Res.*, **33** (1992), 37–42.
46. K. Dörner, K. Schneider, E. Sievers, G. Schulz-Lell, H.-D. Oldings, Selenium balances in young infants fed on breast milk and adapted cow's milk formula, *J. Trace Elem. Electrol. Health Dis.*, **4** (1990), 37–40.
47. M. F. Picciano, J. A. Milner, Selenium in human milk: factors exerting an influence, form and distribution, in J. Schaub (ed.), *Composition and Physiological Properties of Human Milk*, Elsevier Science, Amsterdam, 1985, pp. 77–86.
48. P. Van Dael, H. Deelstra, Distribution of Selenium in cow's and human milk, in P. Brätter, P. Schramel (eds), *Trace Element Analytical Chemistry in Medicine and Biology*, Vol. 6, Walter de Gruyter, Berlin, 1988, pp. 136–144.
49. P. Van Dael, H. Deelstra, Selenium distribution over protein fractions of human, cow, goat and sheep milk, in P. Brätter, B. Ribas, P. Schramel (eds), *Trace Element Analytical Chemistry in Medicine and Biology*, Vol. 6, Consejo Superior de Investigationes Cientificas, Madrid, 1994, pp. 321–330.
50. J. B. Luten,. W. Bouquet, M. M Burgraaf, J. Rus, The *in-vitro* availability of copper, zinc, and selenium and the speciation of selenium in fishery products, in P. Brätter, P. Schramel (eds), *Trace Element Analytical Chemistry in Medicine and Biology*, Vol. 4, Walter de Gruyter, Berlin, pp. 509–519.
51. V. E. Negretti de Brätter, S. Recknagel, D. Gawlik, Speciation of Se, Fe and Zn in human milk whey: the use of instrumental neutron activation analysis (INAA) to corroborate element profiles measured with inductively coupled plasma atomic emission spectrometry (ICP-AES), *Fresenius J. Anal. Chem.*, **353** (1995), 137–142.
52. B. Michalke, P. Schramel, CE-methods for a clear identification of selenoamino acids in complex matrices such as human milks, *J. Chromatogr. A*, **716** (1995), 323–329.
53. B. Michalke, P. Schramel, Selenium speciation in human milk with special respect to quality control. *Biol. Trace Element Res.*, **59** (1997), 45–56.
54. B. Michalke, P. Schramel, Application of CZE-ICP-MS and cIEF-ICP-MS for selenium speciation. *J. Chromatogr. A*, **807** (1998), 71–80.
55. B. Michalke, CE-ICP-MS: improvements in selenium speciation, *Spectroscopy*, **15** (2000), 30–34.
56. E. J. Underwood, *Trace Elements in Human and Animal Nutrition*, 4th edn, Academic Press, New York, 1977.
57. A. S. Prasad, Trace elements and iron in human metabolism, Plenum, New York, 1978.
58. H. Keller, *Klinische-chemische Labordiagnostik für die Praxis. Analyse, Befund, Interpretation*, Georg Thieme, Stuttgart, 1991, pp. 362–379.

59. W. Reinhardt, M. Luster, K. H. Rudorff, C. Heckmann, S. Petrasch, S. Lederbogen, R. Haase, B. Saller, C. Reiners, D. Reinwein, K. Mann, Effect of small doses of iodine on thyroid function in patients with Hashimotós thyroiditis residing in an area of mild iodine deficiency, *Eur. J. Endocrin.*, **139** (1998), 23–28.

60. L. M. Robison, P. W. Sylvester, P. Birkenfeld, J. P Lang, R. J. Bull, Comparison of the effects of iodine and iodide on thyroid functions in humans. *J. Toxicol. Environ. Health, Part A*, **55** (1998), 93–106.

61. G. Knapp, B. Maichin, P. Fecher, S. Hasse, P. Schramel, Iodine determination in biological materials. *Fresenius J. Anal. Chem.*, **362** (1998), 508–513.

62. P. Brätter, I. N. Blasco, V. E. Negretti de Brätter, A. Raab, Maternal selenium status influences the concentration and binding form of iodine in human milk, *Metal Ions Biol. Med.*, **6** (2000), 244–247.

63. J. A. Rillema, D. L. Rowady, Characteristics of the prolactin stimulation of iodide uptake into mouse mammary gland explants, *Proc. Soc. Expl. Biol. Med.*, **215** (1997), 366–369.

64. N. Etling, F. Gehin-Fouque, Iodinated compounds and thyroxine binding to albumin in human breast milk, *Pediatr. Res.*, **18**(9) (1984), 901–903.

65. W. Zidek, Mangan, in H. Zumkley (ed.), *Spurenelemente, Grundlagen, Ätiologie, Diagnosen, Therapie*, Georg Thieme, Stuttgart, 1983, pp. 140–151.

66. F. C. Wedler, Biological significance of manganese in mammalian systems, *Prog. Med. Chem.*, **30** (1993), 89–133.

67. M. Aschner, K. E. Vrana, W. Zheng, Manganese uptake and distribution in the central nervous system (CNS), *Neurotoxicology*, **20** (1999), 173–180.

68. R. T. Ingersoll, E. B. Montgomery, H. V. Aposhian, Central nervous system toxicity of manganese. II. Cocaine or reserpine inhibit manganese concentration in the rat brain. *Neurotoxicology*, **20** (1999), 467–476.

69. G. V. Iyengar, Reference values for the concentrations of As, Cd, Cr, Cu, Fe, I, Hg, Mn, Mo, Ni, Pb, Se and Zn in selected human tissues and body fluids, *Biol. Trace Elem. Res.*, **12** (1987), 263–295.

70. B. Lönnerdal, Regulation of mineral and trace elements in human milk: exogenous and endogenous factors, *Nutr. Rev.*, **58**(8) (2000), 223–229.

71. R. M. Parr, E. M. DeMaeyer, V. G. Iyengar, A. Byrne, G. F. Kirkbright, G. Schöch, L. Niinisto, O. Pineda, H. L. Vis, Y. Hofvander, A. Omoloulu, Minor and trace elements in human milk from Guatemala, Hungary, Nigeria, Philippines, Sweden, and Zaire, *Biol. Trace Elem. Res.*, **29** (1991), 51–75.

72. B. Michalke, P. Schramel, Manganese speciation in human milk using size exclusion chromatography combined with strong anion exchange chromatography and inductively coupled plasma mass spectrometry detection, *J. Anal. Atom. Spectrom.*, **19** (2004), 121–128.

73. P. Schramel, G. Lill, S. Hasse, B. J. Klose, Mineral- and trace element concentrations in human breast milk, placenta, maternal blood, and the blood of the newborn, *Biol. Trace Elem. Res.*, **16** (1988), 67–75.

74. E. Vuori, M. Vetter, P. Kuitunen, S. Salmela, Cadmium in finnish breast milk, a longitudinal study, *Arch. Toxicol.*, **53** (1983), 207–211.

75. B. Michalke, P. Schramel, Protein fraction and Cd-speciation in human breast milk by HPLC and voltammetry, *J. Trace Elem. Electrol. Health Dis.*, **4** (1990), 163–167.

76. J. R. W. Woittiez, On the use of separation techniques for elemental analysis of protein fractions, in P. Brätter, B. Ribas, P. Schramel (eds), *Trace Element Analytical Chemistry in Medicine and Biology*, Vol. 6, Consejo Superior De Investigationes Cientificas, Madrid, 1994, pp. 1–32.
77. C. Ekmekcioglu, Intestinal bioavailability of minerals and trace elements from milk and beverages in humans, *Nahrung*, **44** (2000), 390–397.
78. M. L. Fernández, J. Szpunar, Speciation analysis for iodine in milk by size-exclusion chromatography with inductively coupled plasma mass spectrometric detection (SEC-ICP MS), *J. Anal. Atom. Spectrom.*, **14** (1999), 1697–1702.

18

MEASUREMENT OF TOTAL ARSENIC AND ARSENIC SPECIES IN SEAFOOD BY Q ICP-MS

WILLIAM A. MAHER, JASON KIRBY, AND FRANK KRIKOWA

Ecochemistry Laboratory, Institute of Applied Ecology, University of Canberra, ACT 2601, Australia, E-mail: bill.maher@canberra.edu.au

The Determination of Chemical Elements in Food: Applications for Atomic and Mass Spectrometry.
Edited by Sergio Caroli

18.1 INTRODUCTION

Arsenic is present in seafood at high concentrations with typical ranges of 1–100 $\mu g\ g^{-1}$ dry mass [1, 2]. Fish and shellfish contribute the majority of ingested arsenic [3, 4] as most foods for human consumption contain less than 0.5 $\mu g\ g^{-1}$ of As. It has been known for a long time that the toxicity and carcinogenicity of As is species and valency dependent [5–7] and that seafood contains organic As compounds that do not appear to be harmful [8, 9]. Most As in animal tissues is in nontoxic forms with little As (less than 1 percent) present as toxic inorganic forms [10]. A comprehensive list of As species that have been isolated from marine algae and animals has been published by McSheedy et al. [11], Edmonds and Francesconi [2] and Francesconi and Kuehnelt [12]. Table 18.1 and Figure 18.1 give the structures of the common As species discussed in this chapter by name and accepted acronyms. Arsenobetaine (AB), the major As compound in most marine fish, molluscs, and crustaceans muscle tissues, is metabolically inert, nontoxic [13–17], and rapidly excreted by humans. However, in some marine tissues such as dugong livers and silver drummer tissues, AB is a minor constituent [18, 19]. Other forms of As, such as monomethylarsonic acid (MA), dimethylarsinic acid (DMA), tetramethylarsonium (TETRA), trimethylarsine oxide (TMAO), and arsenosugars can also be present [20] and risks must be assessed based on the As species present, the element concentration and dietary consumption. Seaweeds contain high concentrations of As and in certain Asian countries contribute substantially to human diets. Edible seaweeds contain mostly arsenosugars [21] that are considered to be nontoxic based on cytotoxicity studies [22] and the absence of incidents of poisoning within communities that consume large quantities of seaweed. However, some seaweeds can contain large amounts of inorganic As [23, 24]. In addition, studies of the metabolism of seaweeds by humans have shown that human urine often contains MA, DMA, TMAO, and dimethylarsinoyl ethanol [25–28]. DMA has been shown to be a potential human carcinogen [29]. Arsenolipids are also found in marine biota [30–34] and appear to be the products of the esterification of arsenosugars. At present the daily intake

TABLE 18.1. Common As Species Found in Seafood

Name	Acronym	Chemical Formula
Arsenous acid	As (III)	$As(OH)_3$
Arsenic acid	As (V)	$AsO(OH)_3$
Methylarsonic acid	MA	$CH_3AsO(OH)_2$
Dimethylarsinic acid	DMA	$(CH_3)_2AsO(OH)$
Arsenobetaine	AB	$(CH_3)_3As^+CH_2COO^-$
Arsenocholine	AC	$(CH_3)_3As^+CH_2CH_2OH$
Trimethylarsine oxide	TMAO	$(CH_3)_3AsO$
Tetramethylarsonium ion	TETRA	$(CH_3)_3As^+$
Trimethylarsonioproprionate	TMAP	$(CH_3)_3As^+CH_2CH_2COO^-$

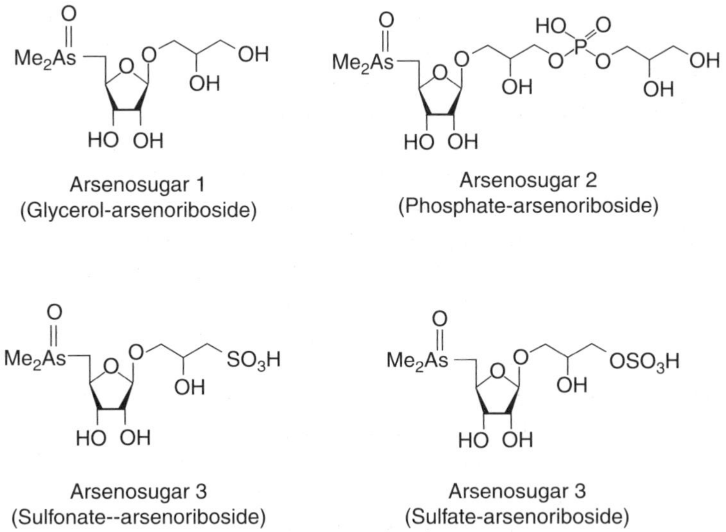

Figure 18.1. Structure of arsenosugars.

of As is set at 1–2 μg g^{-1} wet mass of inorganic As [35] with no regulations governing other As species in seafood.

Inductively coupled plasma mass spectrometry (ICP-MS) is one of the most significant analytical advances to occur in the last 20 years as it allows multielement analysis of solutions and solids to be performed at subnanogram concentrations. Instrumental advances have occurred such that Quadrupole (Q) ICP-MS units are now in routine use in many laboratories. In this chapter, the use of Q-ICP-MS and high performance liquid chromatography (HPLC)–ICP-MS is discussed as regards the quantification of total As and As species in seafood. To highlight the strengths and weaknesses of the use of ICP-MS, data are used that were mainly produced in the laboratories of the authors of this chapter.

18.2 GENERAL CONSIDERATIONS

18.2.1 Instrumentation

There are three basic types of scanning Q-ICP-MS systems in use, namely, single Q as well as dynamic reaction cell (DRC) and collision cell (CC) ICP-MS. The most widespread system in use employs a single Q mass spectrometer. These can typically resolve ~0.5 mass units. DRC instruments that have dual Q mass spectrometers and CC hexapole/octapole-Q instruments are becoming more and more common [36–39]. A reactive gas (NH_3, H_2, or CH_4) is passed into the DRC and used to remove interfering ions via charge transfer reactions [40]. For example, $^{40}Ar^{35}Cl^+$ has a mass equivalent to that of $^{75}As^+$ and the Q mass detector cannot discriminate between these ions. Addition of H_2 will react with this ion, that is, multiple reactions including $^{40}Ar^{35}Cl^+ + H_2 \rightarrow HCl + {}^{40}Ar^1H^+$ occur. The Q mass spectrometer in the reaction cell is used to remove any charged reaction products. Uncharged reaction products are swept out of the ion path while $^{75}As^+$ continues on to the second Q mass spectrometer for analysis [37].

The difference between a DRC and a CC ICP-MS is that a reaction cell uses a Q mass spectrometer with a defined stability region, allowing the rejection of a defined mass region from the ion beam, whereas a CC uses a hexapole or octapole mass spectrometer as an ion guide. DRC have no limit on the gas type used, whereas CCs can only use inert gases or a simple reactive gas such as He or H_2. Limits of detection (LoDs) for As are $\sim$0.01 ng l^{-1}. However, LoDs can only be achievable with careful control of contamination and minimization of polyatomic interferences. Arsenic LoDs may be improved by the introduction of the element into the ICP-MS in the gaseous form, for example, by formation of AsH_3, which markedly improves the LoD as nearly 100 percent of the analyte in solution reaches the mass detector.

To get the best productivity out of a Q-ICP-MS instrument, this should have a stabilized power supply and be coupled to a flow injection (FI) system. The stabilized power supply ensures that the instrument is stable and protects the turbo-molecular vacuum pumps from power surges. The FI system allows samples to be diluted on-line, internal standards to be added, and AsH_3 to be generated, this improving accuracy, precision, and laboratory productivity [41].

18.2.2 Interferences

As with all atomic spectroscopy techniques, ICP-MS also suffers from a number of interferences.

18.2.2.1 Matrix Interferences Four processes (as well as detector response) control the mass counts measured: the generation and transport of the aerosol, plasma dynamics, the fraction of ions produced, and the efficiency of ion transportation from the plasma torch to the detector. The sample matrix affects all of these processes and enhancement or suppression of signals may occur [42, 43]. Acid matrices may reduce the total aerosol and vapor load to the plasma [43] and result in a decrease in plasma temperature [44, 45]. Addition of large quantities of matrix elements will also reduce the plasma energy available for ionization of analyte elements, resulting in loss of sensitivity [43]. Space charge effects bias transmission to ion optics in favor of higher mass ions and cause suppression of analyte signal by matrix elements [46]. The presence of high concentrations of an organic solvent can have a cooling effect on the plasma, increase reflected power and result in plasma extinction. In addition, the ionization characteristics of the Ar plasma are changed, resulting in changes in sensitivity of As species [47]. High C content will also cause carbonization of torches and cones. These problems can be often alleviated by the use an appropriate sample introduction system [48] and internal reference standard(s) such as Li, Rh, and Y, provided the chosen internal standard is not present in the sample at any appreciable concentration. Similar to the analyte of interest, internal reference standards may also be prone to molecular interferences. Internal reference standards can correct for instrument instability, signal drift, and nonspectral interferences, such as changes in viscosity, surface tension, and aerosol characteristics, but not polyatomic interferences [49, 50].

18.2.2.2 Molecular Interferences High abundance of ions is likely to lead to the formation of polyatomic or molecular ions with the same mass as the elements of interest [51, 52]. Polyatomic interferences arise mainly from various combinations of Ar, C, Ca, Na, Cl, N, O, and S matrix, elements, which are present in sample matrices, plasma gas, and reagents used in the digestion of samples [53]. Arsenic has serious problems from polyatomic interferences arising from Ar plasma gas and matrix constituents (C and Cl). The use of correction equations, the addition of N_2 or the use of a DRC or CC can minimize and sometimes eliminate these interferences [36–40, 54]. The use of correction equations should be avoided, especially if the As mass counts are small compared with the interference present. Correction equations suffer from inherent problems in that the masses used to correct for interferences may themselves suffer from interferences.

Examples of potential interferences for As are $^{43}Ca^{16}O^{16}O^+$, $^{40}Ar^{35}Cl^+$, and $^{38}Ar^{37}Cl^+$. These interferences cannot be separated from analytes using a Q detector because of the insufficient resolution power. Many papers in the literature report different interferences when measuring As in similar matrices. This is because the extent and magnitude of interferences depend on the ICP-MS used, the type of nebulizer, the plasma torch conditions, the mass to final dilution volume, and the analyte concentration being measured [55, 56].

18.2.2.3 Living with Interferences To measure As accurately in sample digests or extracts it is necessary to:

- Understand the samples being analyzed, that is, have an appreciation of the As concentrations and potential interferences.
- Decide on the final mass-to-volume ratios as this allows the expected interference concentrations to be determined.
- Know the apparent analyte concentrations for potential interferences if chemical resolving power is not available.
- Determine potential errors and what is acceptable. This depends on the As concentration and apparent analyte concentration. From these, potential errors can be estimated.
- Remove the effects of interferences by use of chemical resolution associated with a DRC or CC, or correct for interferences using correction equations after analysis, for example, ^{75}As must be corrected for interference from $^{40}Ar^{35}Cl^+$. The use of correction equations are fraught with difficulties and

should only be used when extensive knowledge of interferences has been gained [50, 57].

18.3 TOTAL As MEASUREMENTS

18.3.1 Sample Storage and Preservation

If samples are to be stored for long periods, the presence of microorganisms may cause a deterioration of samples and change the species composition. It has been reported that the storage of fish samples at $-10^{\circ}C$ resulted in the formation of trimethylarsine [58]. On the other hand, it has been found that, if samples are stored frozen at $-10^{\circ}C$ or freeze-dried, As is not lost from samples and species are preserved [59].

18.3.2 Sample Digestion

Dry ashing or wet digestion procedures are commonly used for the mineralization of organic matter to release As [60, 61]. Dry ashing at temperatures above 500°C must employ an ashing aid such as $MgO-MgNO_3$ to prevent loss of the element as volatile chloride or oxychloride compounds [60, 62, 63]. However, large losses have been reported even with the use of an ashing aid [64]. Traditional wet digestion usually uses a combination of acids such as HNO_3, H_2SO_4, and $HClO_4$, and oxidants such as $K_2S_2O_8$, $KMnO_4$, and V_2O_5 to quantitatively convert As species to inorganic As [61, 65, 66]. Microwave (MW)-assisted digestion with HNO_3 or $HNO_3-H_2O_2$ using closed Teflon vessels has replaced traditional digestion procedures. The advantages of resorting to MW heating include less acid being used, less laboratory contamination, and lower analytical blanks. Digestion temperatures are limited to about 250°C because the vessels melt at this temperature. The use of HNO_3 alone is advantageous for ICP-MS, as, compared with other acids, it produces the least background interference in ICP-MS [51]. It has been found that, for a Q-ICP-MS instrument, variations in HNO_3 concentration (0–6 percent, v/v) do not influence As mass counts [56].

Generally digestion with HNO_3 produces digests with a light yellow colour, indicating that samples are not completely digested. The problem with incomplete digestion is that residual C present in digests [67] may assist in the ionization of As in the ICP-MS plasma, causing the enhancement of As mass counts [47].

MW-assisted digestion at 200–250°C does not convert all As species to inorganic As [68–71]. At these temperatures AB is converted to TMAO. Only the use of an MW-assisted autoclave with temperatures $>500^{\circ}C$ quantitatively converts all As species to inorganic As [71].

Hydride generation (HG) techniques that rely on As to be in inorganic forms will underestimate total As in extracts. MA and DMA also form hydrides, but at different rates and extents to arsenite [72], while TMAO, TETRA, AB, AC, and arsenosugars do not form hydrides to an appreciable extent under normal working conditions [73].

18.3.3 ICP-MS Analysis

Unlike some published studies [47], the authors of this chapter have found that the different As species give similar mass counts when analyzed by ICP-MS using a cross-flow nebulizer.

Extensive analyses have been undertaken using 14 marine Certified Reference Materials (CRMs) of possible polyatomic interferences that may arise during the analysis of marine biological materials when using a Q-ICP-MS instrument [56]. Some matrix elements concentrations, at sample:solution ratios of 1 : 100 routinely used in the authors' laboratory, are relatively high: Al (0.02–2 mg l^{-1}), Ca (0.1–65 mg l^{-1}), Cl (6.8–60 mg l^{-1}), Fe (0.3–1.2 mg l^{-1}), K (0.01–61 mg l^{-1}), Mg (1–9 mg l^{-1}), Mn (0.001–0.3 mg l^{-1}), Na (1–37 mg l^{-1}), P (2.6–13 mg l^{-1}), and S (6.8–12 mg l^{-1}). High K and Na concentrations may depress As mass counts because of space charge effects or reduced ionization of As because of the large K or Na ion populations [74]. For As, a major interference is Cl^-. The worse case scenario using a sample/solution ratio 1 : 100 and an extreme Cl^- concentration (60 mg l^{-1}) gives an apparent As concentration of 0.16 μg l^{-1}. Therefore, for <5 percent error from Cl^-, digests need to contain ~3.2 μg l^{-1} of As corresponding to ~3.2 μg g^{-1} in the original sample. For <10 percent error from Cl^-, digests need to contain ~1.6 μg g^{-1} in the original sample. Marine organisms typically have 1–100 μg g^{-1} As. Thus, Cl^- is unlikely to cause an interference unless the sample has a very low As concentration or contains the worse-case high concentration of Cl^-. In practice, it has been found that one can accurately measure As concentrations in samples and CRMs [56] down to ~1 μg l^{-1} (1 μg g^{-1} dry mass), but at this level one needs to carefully check for Cl^- interferences. However, it should be stressed that the interference caused by Cl^- (as is the case with most interferences) will depend to some extent on the instrument used, instrumental conditions such as nebulizer flow rate, and the specific sample being analyzed. It should be noted that these interferences are avoidable when using Q-ICP-MS instruments with chemical or spectral resolving capability, that is, a DRC or a CC [36–40, 54].

As previously mentioned, the presence of C may enhance As mass counts [47]. It has been found that MW-assisted digestion of tissues with HNO_3 (0.1 g sample per 1 mL of HNO_3) yields less than 50–100 mg l^{-1} of C in extracts and causes no enhancement of As mass counts [41].

HG procedures are often advocated as a means of elimination of interferences due to Cl^-. However, careful consideration of the gas–liquid separator design is required to minimize or eliminate transport of Cl^- to the plasma [75]. The efficiency of AsH_3 production is dependent on the acidity, $NaBH_4$ concentration, and prereductant used [76]. The use of cysteine as a reductant has all but replaced traditional reductants such as KI and KI–ascorbic acid as the hydride yield is greater, less acid is required (thus blanks are reduced), less H is evolved and interferences from transition elements are minimized [72]. Again, it is stressed that HG can only be used for total As measurements if all As species in digests have been degraded to inorganic As. Otherwise, total As in extracts will be underestimated.

18.4 ARSENIC SPECIATION MEASUREMENTS USING HPLC-ICP-MS

18.4.1 Introduction

The use of ICP-MS provides some important benefits when coupled to an HPLC that separates As species prior to measurement. ICP-MS has fit-for-purpose limits of detection (LoDs); thus, analytical, not preparative, separation systems can be used with a high sampling rate ($\sim$1 ms) in order to measure transient signals and to separately quantify closely eluting compounds from HPLC. HPLC-ICP-MS coupling can be achieved simply via a 50 cm piece of PEEK tubing without loss of peak resolution, whereas chromatographic flow rates are compatible with conventional ICP-MS spray chambers.

HPLC separation and ICP-MS need to be compatible in terms of the solvents selected. The optimum solvent for extraction of species may not be compatible with the other steps; for example, chlorinated solvents may result in ICP-MS polyatomic interferences, that is, $^{40}Ar^{35}Cl^{+}$. High concentrations of organic solvents used for C_{18} reverse-phase separations may extinguish the ICP-MS plasma discharge [77]. Thus, the percentage of organic phase needs to be reduced or O_2 added to the plasma to ensure removal of the additional C entering the ICP torch. Alternatively, for some organic solvents, a chilled spray chamber (5–10°C) or a desolvation device can be used to reduce or eliminate the organic C load [48, 77, 78].

Figure 18.2 presents an overall flow chart of extraction and separation procedures that can be used to isolate, separate, and measure lipid, water-soluble, and residue-associated As species in seafoods.

18.4.2 Sample Extraction

A variety of extraction conditions employing methanol–water mixtures [11], phosphoric acid [79], and enzymes [80, 81] in combination with heating [82], sonification [83, 84], shaking [85–87], accelerated solvent extraction [88, 89], and MW-assisted extraction [90-93] have been used to isolate As species from marine algae and animal tissues.

Two important considerations are the necessity to lyse cells and whether samples need to be defatted. Cells and membranes need to be ruptured to ensure quantitative extraction. The use of freeze-drying and grinding will normally break up biological materials, and it has been shown that As species in extracts of fresh and freeze-dried material are identical [91]. The presence of fat may lower extraction efficiencies and nondefatted samples can change HPLC column performance [94, 95]. Fats can be easily removed with acetone or ether with little coextraction of As [82, 92, 96, 97].

Most methods only determine As species that are soluble in water or methanol–water solutions [11]. This is not a problem when analyzing animal muscle tissues, as As is normally present as AB and readily extractable [92]. For seaweeds, no single extraction method will remove all As species. It has been shown for seaweeds that the optimal methanol–water composition varied for different genera and species of seaweeds [93]. The glycerol–arsenoriboside lacks an acidic group in the aglycone and is less polar than other arsenosugars and better extracted into methanol [79, 98]. Inorganic As is not readily extracted by methanol [12, 79]. In other non-muscle tissues, such as lobster hepatopancreas, As can be present in lipids [31] and only isolated by solvent extraction with chloroform–methanol [32, 33, 99]. It is essential that both total As and extractable As be measured to allow the percentage of unextracted As species to be determined.

It has been found that As can be extracted from most marine animal tissues after freeze-drying, grinding, and defatting with acetone, using 50 percent v/v methanol–water with MW-assisted heating at 70°C for 5 min [92]. The success of

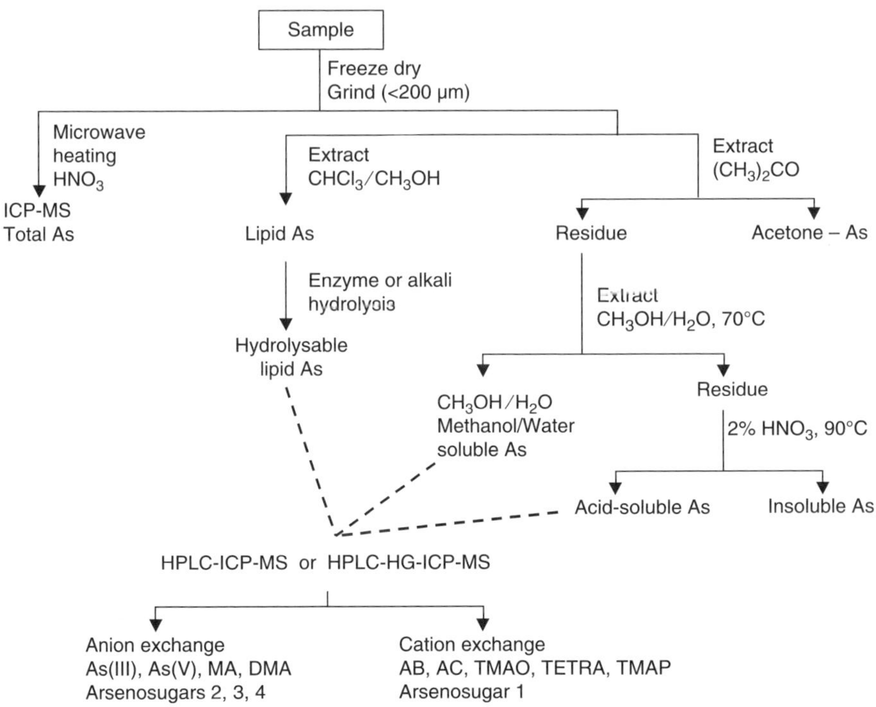

Figure 18.2. Extraction and separation procedures that can be used to isolate, separate, and measure lipid, water-soluble, and residue associated As species in seafoods.

MW-assisted extraction procedures rely on heating the extraction solvent to near its boiling point to obtain convection currents to ensure thorough mixing. Arsenosugars can be extracted from seaweeds using water or dilute HNO_3, but methanol–water mixtures can also be equally effective (93) with the advantages of minimizing the coextraction of water-soluble material and easier concentration. Generally, aqueous extracts of organic As species stored at 4–20°C are stable [12, 91, 98, 100]. Phosphate and sulfonated arsenosugars can hydrolyze to glycerol sugars, thus arsenosugar extracts are best stored as freeze-dried extracts [98]. Arsenosugars also degrade slowly at low pH [101]. The stability of arsenite and arsenate is unclear, but standards have been reported to be unstable [12] and it is likely that arsenite is oxidized to arsenate in extracts [100].

The As species not readily extracted by methanol–water can be extracted using 20 percent v/v HNO_3 with MW-assisted heating at 90°C for 8 min. Using these conditions, the common As species are stable except arsenosugars that are hydrolyzed to a single form, but allow total arsenosugars to be measured.

At present hydrolysis with alkali or phospholipase D enzyme is used to cleave the As moiety in lipid extracts and the water-soluble As species liberated are analyzed by HPLC-ICP-MS [32, 33, 99]. Lipid-soluble As species such as phosphatidyl AC and phosphatidyl arsenosugar on treatment release DMA, MA, AC, and glycerol arsenosugar.

18.4.3 Separation and Quantification of As Species

Arsenic species have a wide variety of ionic characteristics that are pH-dependent, thus finding one separation scheme that will separate all the common As species is difficult. Consequently, different approaches using anion exchange chromatography (AEC), cation exchange chromatography (CEC), reverse phase chromatography (RPC), and size-exclusion chromatography (SEC) have been used to separate As species [102–104]. Normally, a combination of these techniques in sequence is used [24, 105–108]. The use of AEC and CEC allows 12 of the common As species to be separated and quantified by HPLC-ICP- MS (Figs. 18.3 and 18.4). Examples of As species found in seaweed [93] and marine animals [108] are given in Figure 18.5. Note the presence of unknown anionic and cationic As species. This illustrates the limitation of undertaking speciation measurements using ICP-MS. As many as 17 As species have been separated in one run using an anion exchange column, an ion-pairing reagent, and gradient elution with HNO_3 [109]. However, longer elution times are better as there is less chance of coeluting unknown species, and the use of complementary separation reduces problems from matrix components. For example, when using AEC and CEC, cations are voided from the anion column and anions voided from the cation column, virtually eliminating matrix interferences.

Sample matrix components can affect chromatography [12] and spiking is essential to account for matrices influencing retention times. For example, Sloth et al. [95] have shown that a four-fold increase in sample loading of the CRM TORT-2 extract decreased the retention of AB from 10 to 7 min. Peak

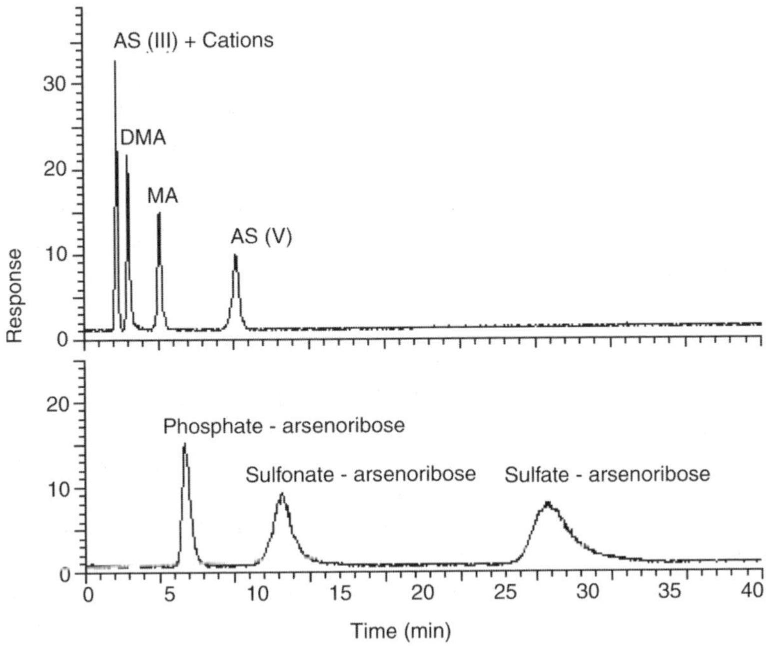

Figure 18.3. Separation and quantification of anionic As species by HPLC-ICP-MS using a Hamilton PRP X-100 anion exchange column, 20 mM $NH_4H_2PO_4$, pH 5.6, 1.5 ml min^{-1}, 40°C.

suppression and splitting of AB by K and Na has been observed during CEC [83, 85], but can be eliminated by dilution or adjustment of eluent pH to separate AB from these cations. Potassium, Na and other unretained material migrate with the solvent front in AEC and may interfere with arsenite and glycerol arsenoriboside measurements. Glycerol-arsenoriboside is better quantified using a cation exchange column (see Fig. 18.3). It has been found that As (III) [and low concentrations of MA, DMA, and As (V)] can be reliably measured using AEC and HG-ICP-MS (Fig. 18.6) [24]. Many published methods do not use column separation, but discriminate between nontoxic and toxic As based on hydride active versus hydride nonactive species, based on the assumption that only arsenite, arsenate, MA, and DMA are hydride active. Arsenosugars can also give responses when analyzed by HG without a decomposition step [72, 110]. Thus, the use of HG coupled to ICP-MS alone may overestimate the presence of potentially toxic As species.

18.5 DIRECT STRUCTURAL INFORMATION

Many samples contain As species that do not match As standards (see Fig. 18.5). If the As species has not been identified or the standard compound is not available,

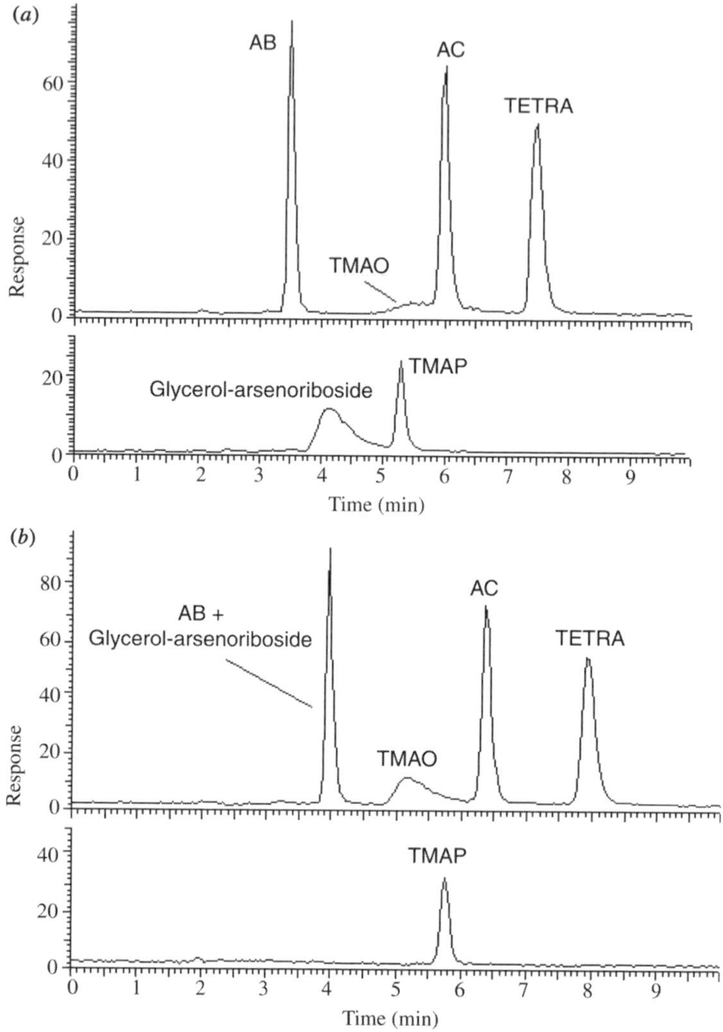

Figure 18.4. Separation and quantification of cationic As species by HPLC-ICP-MS using a Supelcosil LC-SCX cation exchange column, 20 mM pyridine–formic acid, 1.5 mL min^{-1}, 40°C. (*a*) pH 2.6 and (*b*) pH 2.2.

the As species will not be able to be identified or quantified by HPLC-ICPMS. An HPLC instrument coupled to a tandem mass spectrometry (MS) system (HPLC-MS-MS) is required for this purpose [111, 112]. For example, it is difficult to synthesize natural occurring arsenosugars. HPLC-MS-MS and HPLC-ion trap multistage MS can be used to characterize natural arsenosugars in seaweeds [41, 113] for use as HPLC-ICP-MS standards (see Fig. 18.7). Tandem mass spectrometers with electrospray ionization (ESI) using collision-induced dissociation

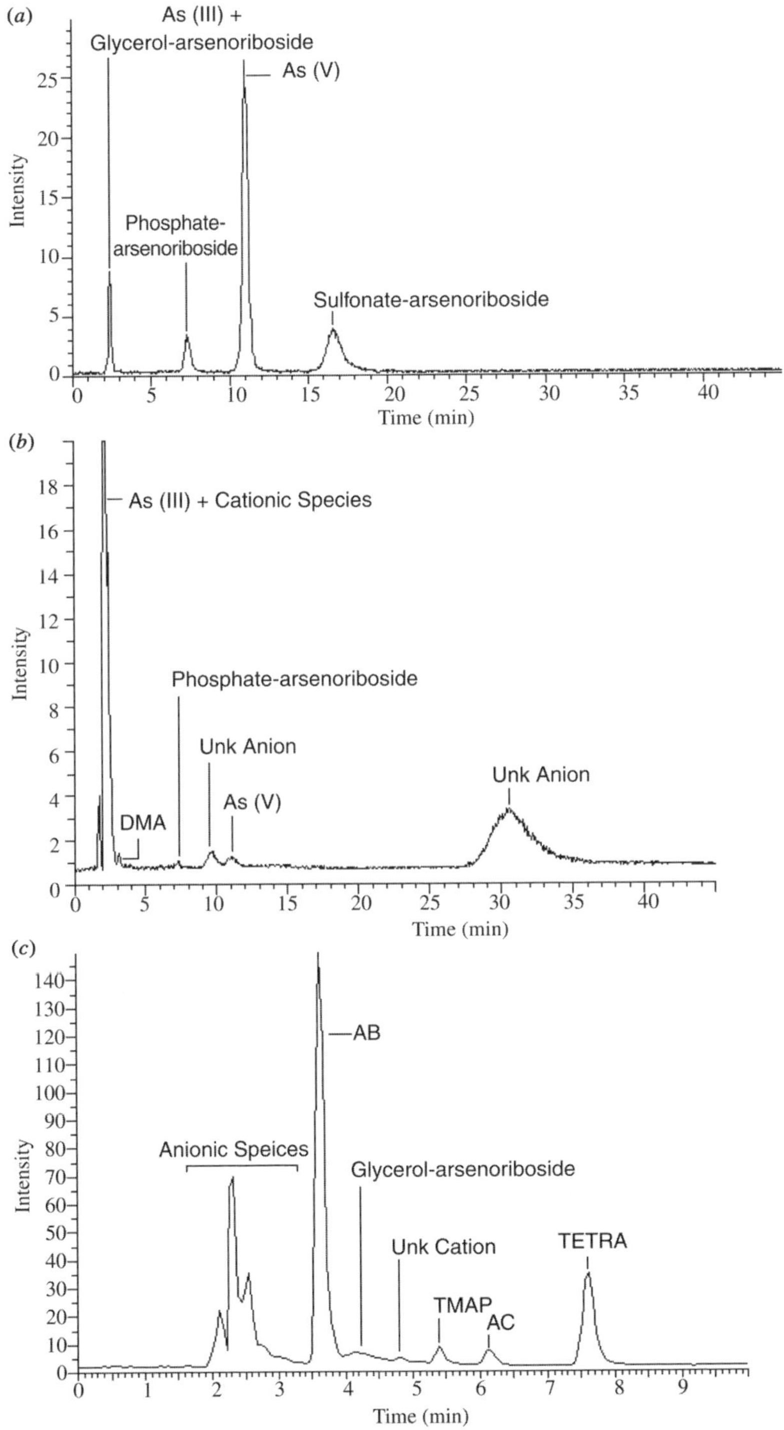

Figure 18.5. Separation and quantification of As species in marine seaweeds and animals. (*a*) *Phyllospora comosa*; (*b*) *Mictyris longicarpus* (whole tissues); (*c*) *Bembicium auratum* (whole tissues).

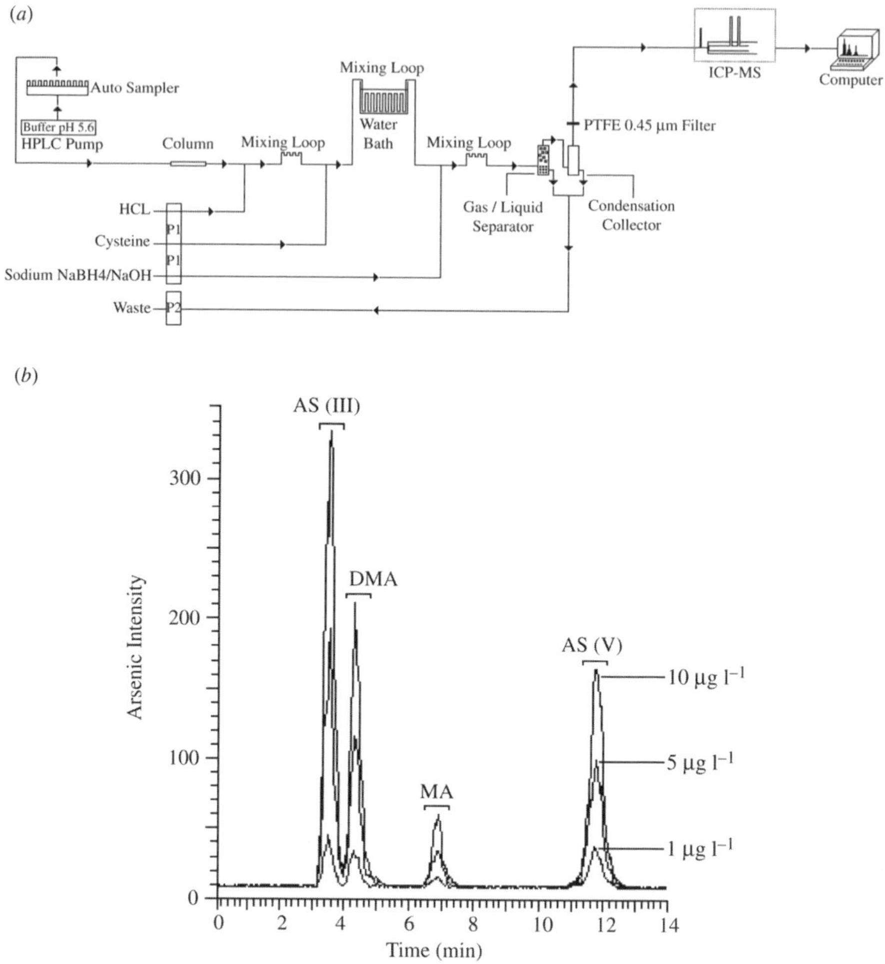

Figure 18.6. Schematic of an HPLC flow injection-ICP-MS hydride generation system (*a*) and real-time measurement of As species via hydride formation (*b*).

gives characteristic fragments of methylated As species [114]. The use of ESI with a signal Q mass spectrometer is limited, as the signal cannot always be unambiguously assigned. However, the use of a single Q mass spectrometer with variable fragmentor voltages has been used successfully to identify TMAP in dogfish muscle [115] and arsenosugars in seaweed extracts [98, 116]. Multidimensional liquid chromatography where different separation methods are used in series, normally involving a primary separation from which a fraction is collected for a secondary separation [117], with ICP-MS and ESI-MS has been used to identify a number of As species in marine CRMs [118] and in other tissues such as the kidney of the giant clam *Tridacna derasa* [119].

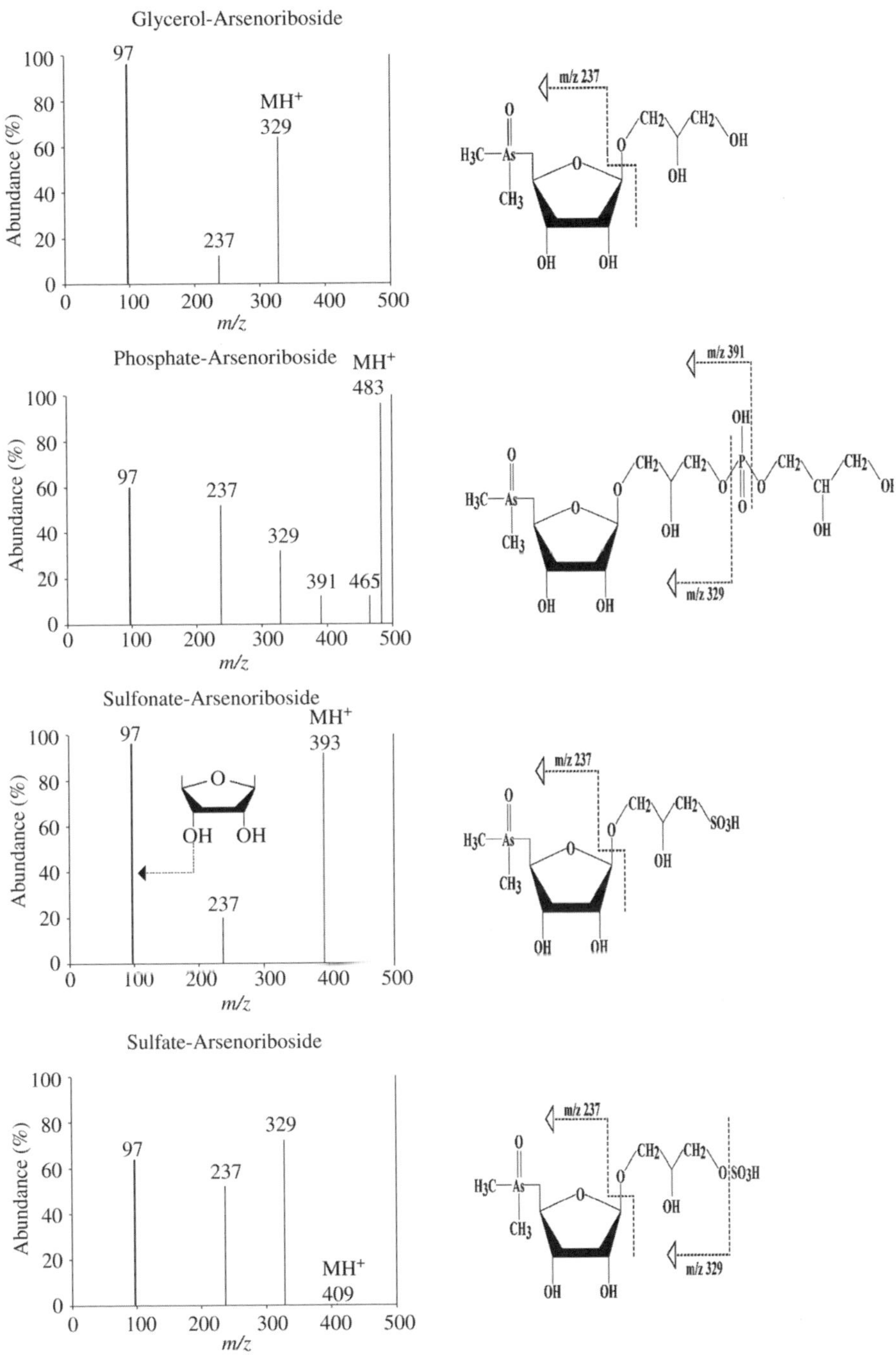

Figure 18.7. HPLC-MS-MS spectra of arsenosugars.

18.6 AVAILABILITY AND USE OF CRMS

Digestion, extraction, separation, and measurement processes are all subject to errors including contamination, degradation, matrix effects, and calibration errors. Mass balances and recovery rates must be determined for all steps of the analytical procedure. CRMs are available to assist in quantifying errors.

There are at least 23 marine CRMs certified for total As concentrations (Table 18.2). As well, many laboratories have published values for As species in CRMs (Tables 18.3–18.6).

TABLE 18.2. CRMs Certified for Total As

Reference Material	Certified Concentration ($\mu g\ g^{-1}$ dry mass)
BCR CRM 414 Plankton	6.82 ± 0.28
BCR 279 Sea Lettuce (*Ulva lactuca*)	3.09 ± 0.20
IAEA 149/TM Fucus	44.3 ± 2.1
NIES No 9 Sargasso Seaweed	115 ± 9
NIH MURST-ISS-A2 Antarctic Krill	5.02 ± 0.44
IAEA-MA-A 1^{TM} Copepod homogenate	6.7 ± 0.6
NIST SRM 1566 Oyster Tissue	13.4 ± 1.9
NIST SRM 1566a Oyster Tissue	14.0 ± 1.2
AGAL-3 Prawn	3.85 ± 0.28
BCR CRM 278R Mussel Tissue (*Mytilus edulis*)	6.07 ± 0.13
NIES Mussel No. 6 (*Mytilus edulis*)	9.4 ± 0.4
NRCC TORT-1 Lobster Hepatopancreas	24.6 ± 2.2
NRCC TORT-2 Lobster Hepatopancreas	21.6 ± 1.8
AGAL-2 Shark	23.27 ± 1.91
NRCC LUTS-1 Lobster Hepatopancreas	19.0 ± 0.9
BCR CRM 422 Cod muscle	21.1 ± 0.5
BCR CRM 627 Tuna Fish Tissue	4.8 ± 0.3
IAEA MA-A-2 Fish Flesh	2.6 ± 0.1
NRCC DOLT-1 Dogfish Liver	10.1 ± 1.4
NRCC DOLT-2 Dogfish Liver	16.6 ± 1.1
NRCC DORM-1 Dogfish Muscle	17.7 ± 2.1
NRCC DORM-2 Dogfish Muscle	18.0 ± 1.1
NIST CRM 50 Albacore Tuna	3.3 ± 0.4

BCR, Commission of the European Communities, Community Bureau of Reference, Belgium. IAEA, International Atomic Energy Agency, Austria.
NIES, National Institute of Environmental Science, Japan.
NIH, National Institute of Health, Italy.
NIST, National Institute of Standards and Technology, USA.
AGAL, Australian Government Analytical Laboratory, Australia.
NRCC, National Research Council of Canada.

TABLE 18.3. Arsenic Species Reported in the CRM DORM 1

Arsenic Species (μg As^{-1}g dry mass)								
As (III)	As (V)	MA	DMA	AB	AC	TMAO	TETRA	References
	0.18 ± .04		0.47 ± 0.02	15.7 ± 0.8	0.11 ± 0.05			[82]
				15.7			0.46	[128]
<0.030	<0.027	<0.024		14.2	0.019	0.019	0.34	[83]
				15.1 ± 0.6				[80]
				16.5 ± 0.9				[121]
	<0.15	<0.08	<0.13	14.5 ± 1.5				[122]
	<0.04	0.11 ± 0.03	0.61 ± 0.12	15.9 ± 0.5		0.43 ± 0.13		[96]
	0.129 ± 0.003	0.036 ± 0.010	0.25 ± 0.01	17.0 ± 0.7		0.121 ± 0.006		[96]
				16.5 ± 0.6				[123]
<0.03	<0.03	<0.03	0.49 ± 0.04	15.6 ± 0.07	0.043 ± 0.001	<0.03	0.38 ± 0.04	[85]

TABLE 18.4. Arsenic Species Reported in CRM DORM 2

Arsenic Species (μg As g^{-1} dry mass)									
As(III)	As(V)	MA	DMA	AB	AC	TMAO	TETRA	TMAP	References
				16.6 ± 0.6					[120]
<0.03	<0.03	<0.03	0.28 ± 0.01	16.0 ± 0.7	0.024 ± 0.01	<0.03	0.25 ± 0.01	0.24 ± 0.01	[85]
				16.5	0.08				[124]
				16.4 ± 1.1			0.248 ± 0.054		[125]
			0.42 ± 0.02	17.2 ± 0.5			0.26 ± 0.01		[78]
				16.3 ± 0.1	0.098 ± 0.002	<0.001	0.127 ± 0.001		[87]
				16.0 ± 0.6					[126]
					16.4 ± 0.8	<0.05	0.25 ± 0.01	0.24 ± 0.01	[79]
								0.15 ± 0.04	[127]
0.05 ± 0.01	0.05 ± 0.02	0.14 ± 0.02	0.49 ± 0.03	16.1 ± 0.7		0.30 ± 0.03	0.30 ± 0.02		[109]
			0.280 ± 0.004	16.8 ± 0.14	0.023 ± 0.002		0.24 ± 0.02	0.17 ± 0.01	[20]
			0.25 ± 0.03	16.0 ± 0.6	<0.04		0.28 ± 0.04		[19]
			0.30 ± 0.01	16.9 ± 1.0	0.010 ± 0.002	0.021 ± 0.002	0.26 ± 0.02	0.16 ± 0.01	[95]

TABLE 18.5. Arsenic Species Reported in Other CRMs

Arsenic Species (μg As g^{-1} dry mass)

Reference Material	As (III)	As (V)	MA	DMA	AB	AC	TMAO	TETRA	TMAP	References
NRCCDOLT-1				0.80	5.69					[128]
NIES No. 6 Mussel				0.71	1.38					[129]
NIES No. 6 Mussel				0.85	1.24					[106]
NIST 1566a Oyster Tissue				0.86	1.24					[106]
NRCC TORT-1				1.01	16.1					[106]
NIST 1566a Oyster Tissue				1.54 ± 0.22	3.18 ± 0.33			0.16 ± 0.03		[96]
NRCC DOLT-1			0.092 ± 0.022	0.18 ± 0.02	6.07 ± 0.22			0.017 ± 0.003		[96]
BCR CRM 627 Tuna Fish Tissue				0.15 = 0.02	3.9 = 0.2					[130]
NRCC TORT-1									0.086 ± 0.003	[127]
NRCC TORT-2									0.920 ± 0.030	[127]
NRCC DOLT-1									0.470 ± 0.020	[127]
NRCC DOLT-2									0.110 ± 0.020	[127]
BCR CRM 627 Tuna Fish Tissue									0.072 ± 0.005	[127]
BCR CRM 422 Cod Muscle									0.032 ± 0.004	[127]
BCR CRM 627 Tuna Fish Tissue	0.35 ± 0.03	0.10 ± 0.02	0.84 ± 0.06	0.94 ± 0.04	3.49 ± 0.17	0.17 ± 0.02	0.14 ± 0.02			[109]
NRCC TORT-2		0.41 ± 0.03	0.20 ± 0.01	1.03 ± 0.10	13.10 ± 0.08		0.055 ± 0.005	1.20±0.03		[20]
BCR CRM 627 Tuna Fish Tissue				0.14 = 0.01	3.7 = 0.2	0.012 = 0.002	0.010 = 0.002	0.037 = 0.002	0.033 = 0.002	[95]

TABLE 18.6. Arsenosugars Reported in CRMs

Arsenic Species (μg As g^{-1} dry mass)							
Reference Material	Arsenosugar 1	Arsenosugar 2	Arsenosugar 3	Arsenosugar 4	As(V)	DMA	References
NIES No. 6 Mussel	0.61	0.99				0.71	[129]
NIES No. 6 Mussel	0.69	0.95				0.85	[106]
NIST 1566a Oyster	1.13	1.22				0.86	[106]
NRCC TORT-1	0.71	0.18				1.01	[106]
BCR CRM 525 Mussel	1.34	3.34					[106]
Fucus serratus extract[a]	0.01 ± 0.04	0.086 ± 0.003	0.62 ± 0.02	0.40 ± 0.01	0.001	0.005 ± 0.001	[98]
BCR 477 Mussel				0.49 ± 0.04			[109]
NRCC TORT-2		0.13 ± 0.03					[20]
Fucus serratus extract[a]	0.098 ± 0.004	0.082 ± 0.005	0.58 ± 0.03	0.39 ± 0.02			[73]

[a]Concentration is μg As.

18.7 CONCLUDING REMARKS

ICP-MS can detect nanogram per liter concentrations of As, but suffers from polyatomic interferences generated from the sample matrix. Efficient digestion of samples is required to eliminate C enhancement of As signals within the plasma. HG coupled to ICP-MS should not be used for total As measurements unless As species in digests have been degraded to inorganic As, as total As concentrations will be underestimated.

The coupling of anion and cation HPLC and ICP-MS allows As species in seaweed and marine animal tissues to be measured. Most As species are water-soluble and can be easily extracted. Many As species found in environmental samples are not known and analytical standards are not available, thus the concurrent use of HPLC-MS-MS is needed to confirm the identity of As species. It is essential that CRMs are used to confirm the validity of total As and As species results.

REFERENCES

1. W. A. Maher, G. E. Batley, Organometallic in the nearshore marine environment of Australia, *Appl. Organomet. Chem.*, **4** (1990), 419–437.
2. J. S. Edmonds, K. A. Francesconi, Organoarsenic compounds in the marine environment, in P. J. Craig (ed.), *Organometallic Compounds in the Environment*, Wiley., New York, 2003, pp. 195–223.
3. E. L. Gunderson, FDA total diet study July 1986 to April 1991. Dietary intake of pesticides, selected elements and other chemicals, *J. Off. Anal. Chem. Int.*, **78** (1995), 1353–1363.
4. M. A. Adams, P. M. Bolger, E. L. Gunderson, Dietary intake and hazards of arsenic, in W. R. Chappell, C. O. Abernathy, R. C. Cothern (eds), *Arsenic Exposure and Health*, Science and Technology. Letters, Northwood, London, 1994, pp. 41–49.
5. B. L. Vallee, D. D. Ulmer, W. E. C. Wacker, Arsenic toxicology and chemistry, *Arch. Ind. Hlth*, **21** (1960), 132–151.
6. M. Vahter, Mechanisms of arsenic biotransformation, *Toxicology*, **181** (2002), 211–217.
7. X. C. Le, X.-F. Li, Arsenic speciation, *Anal. Chem.*, **76** (2004), A27–A33.
8. A. C. Chapman, On the presence of compounds of arsenic in marine crustaceans and shell fish, *Analyst*, **51** (1926), 548–563.
9. E. A. Coulson, R. E. Remington, K. M. Lynch, Metabolism in the rat of the naturally occurring arsenic in shrimp as compared with arsenic trioxide, *J. Nutr.*, **10** (1935), 255–270.
10. W. A. Maher, Inorganic arsenic in marine organisms, *Mar. Pollut. Bull.*, **14** (1983), 308–310.
11. S. McSheehy, J. Szpunar, R. Morabito, Ph. Quevauviller, The speciation of arsenic in biological tissues and the certification of reference materials for quality control, *Trends Anal. Chem.*, **22** (2003), 191–209.

12. K. A. Francesconi, D. Kuehnelt, Determination of arsenic species: a critical review of methods and applications, 2000–2003, *Analyst*, **129** (2004), 373–395.

13. W. M. F. Jongen, J. M. Cardinaals, P. M. Bos, P. Hagel, Genotoxicity testing of arsenobetaine, the predominant form of arsenic in marine fisheries products, *Food Chem. Toxicol.*, **23** (1985), 669–674.

14. T. Kaise, S. Watanabe, K. Itoh, The acute toxicity of arsenobetaine, *Chemosphere*, **14** (1985), 1327–1332.

15. E. Sabbioni, M. Fischbach, G. Pozzi, R. Pietra, M. Gallorini, J. L. Piette, Cellular retention, toxicity and carcinogenic potential of seafood arsenic. 1. Lack of cytotoxicity and transforming activity of arsenobetaine in the BALB/3T3 cell line, *Carcinogenesis*, **12** (1991), 1287–1291.

16. T. Kaise, S. Fukui, The chemical form and acute toxicity of arsenic compounds in marine organisms, *Appl. Organomet. Chem.*, **6** (1992), 155–160.

17. C. Kojima, T. Sakurai, M. Ochiai H. Kumata, W. Qu, M. P. Waalkes, K. Fujiwara, Cytotoxicological aspects of the organic arsenic compound arsenobetaine in marine animals, *Appl. Organomet. Chem.*, **16** (2002), 421–426.

18. J. Edmonds, Y. Shibata, K. Francesconi, R. Rippinggale, M. Morita, Arsenic transformations in short marine food chains studied by HPLC-ICPMS, *Appl. Organomet. Chem.*, **11** (1997), 281–287.

19. R. Kubota, T. Kunito, S. Tanabe, Chemical speciation of arsenic in the livers of higher trophic marine animals, *Mar. Pollut. Bull.*, **45** (2002), 218–223.

20. J. Kirby, W. A. Maher, Tissue accumulation and distribution of arsenic compounds in three marine fish species: relationship to trophic position, *Appl. Organomet. Chem.*, **16** (2002), 108–115.

21. R. Tukai, W. A. Maher, I. J. McNaught, M. J. Ellwood, M. Coleman, Occurrence and chemical form of arsenic in marine macroalgae from the east coast of Australia, *Austr. Mar. Freshwat. Res.*, **53** (2002), 971–980.

22. T. Kaise, Y. Oya-Ohta, T. Ochi, T. Okubo, K. Hanaoka, K. J. Irgolić, T. Sakurai, C. Matsubara, Toxicological study of organic arsenic compound in marine algae using mammalian cell culture techniques, *J. Food Hyg. Soc. Jap.*, **37** (1996), 135–141.

23. J. S. Edmonds, M. Morita, Y. Shibata, Isolation and identification of arsenic containing ribofuranosides and inorganic arsenic from Japanese edible seaweed *Hizikia fusiforme*, *J. Chem. Soc. Perkin Trans.*, **1** (1987), 577–580.

24. J. Kirby, W. Maher, M. Ellwood, F. Krikowa, Arsenic species in marine biological tissues by HPLC-ICP-MS and HPLC-HG-ICP-MS, *Austral. J. Chem.*, **57** (2004), 957–966 .

25. M. Van Hulle, C. Zhang, B. Schotte, L. Mees, F. Vanhaecke, R. Vanholder, X. R. Zhang, R. Cornelis, Identification of some arsenic species in human urine and blood after ingestion of Chinese seaweed *Laminaria*, *J. Anal. Atom. Spectrom.*, **19** (2004), 58–64.

26. K. A. Francesconi, R. Tanggaard, C. J. McKenzie, W. Goessler, Arsenic metabolites in human urine after ingestion of an arsenosugar, *Clin. Chem.*, **48** (2002), 92–101.

27. M. Ma, X. C. Le, Effect of arsenosugar ingestion on urinary arsenic speciation, *Clin. Chem.*, **44** (1998), 539–550.

28. X. C. Le, W. R. Cullen, K. J. Reimer, Human urinary arsenic excretion after one-time ingestion of seaweed, crab and shrimp, *Clin. Chem.*, **40** (1994), 617–624.

29. S. Yamamoto, Y. Konishi, T. Murai, M. A. Shibata, T. Matsuda, K. Kuroda, G. Endo, S. Fukushima, Enhancing effects of an organic arsenic compound dimethylarsenic (cacodylic acid) in a multi-organism carcinogenesis bioassay, *Appl. Organomet. Chem.*, **8** (1994), 197–201.

30. M. Morita, Y. Shibata, Isolation and identification of arsenolipids from a brown alga, *Undaria pinnatifida* (Wakame), *Chemosphere*, **17** (1988), 1147–1152.

31. J. S. Edmonds, Y. Shibata, K. A. Francesconi, J. Yoshinaga, M. Morita, Arsenic lipids in the digestive gland of the western rock lobster *Panulirus cygnus*: an investigation by HPLC-ICP-MS, *Sci. Total Environ.*, **122** (1992), 321–335.

32. K. Hanaoka, W. Goessler, K. Yoshida, Y. Fujittaka, T. Kaise, K. Irgolić, Arsenocholine and dimehtylated arsenic containing lipids in starspotted shark *Mustelus manazo*, *Appl. Organomet. Chem.*, **13** (1999), 765–770.

33. K. Hanaoka, Y. Tanaka, Y. Nagata, K. Yoshida, T. Kaise, Water-soluble arsenic residues from several arsenolipids occurring in the tissues of the starspotted shark *Mustelus manazo*, *Appl. Organomet. Chem.*, **15** (2001), 299–305.

34. K. Ebisuda, T. Kunito, J. Fujihara, R. Kobuta, Y. Shibata, S. Tanabe, Lipid-soluble and water-soluble arsenic compounds in blubber of ring seal (*Pusa hispida*), *Talanta*, **61** (2002), 779–787.

35. J. S. Edmonds, K. A. Francesconi, Arsenic in seafoods: human health aspects and regulations, *Mar. Pollut. Bull.*, **12** (1993), 665–674.

36. I. Feldmann, N. Jakubowski, D. Stueuer, Application of a hexapole collision and reaction cell in ICP-MS. Part 1: instrumental aspects and operational optimisation, *Fresenius J. Anal. Chem.*, **365** (1999), 415–421.

37. S. D. Tanner, V. I. Baranov, U. Volkopf, A dynamic reaction cell for inductively coupled plasma mass spectrometry (ICP-DRC-MS). Part III. Optimisation and analytical performance, *J. Anal. Atom. Spectrom.*, **15** (2000), 1261–1269.

38. S. D. Tanner, V. I. Baranov, D. R. Bandura, Reaction cells and collision cells for ICP-MS: a tutorial review, *Spectrochim. Acta*, **57 B** (2002), 1361–1452.

39. D. W. Koppenall, G. C. Eiden, C. J. Barinaga, Collision and reaction cells in atomic mass spectrometry: development status and applications, *J. Anal. Atom. Spectrom.*, **19** (2004), 561–570.

40. S. D. Tanner, V. I. Baranov, Theory, design and operation of a dynamic reaction cell for ICP-MS, Atom. *Spectrosc.*, **20** (1999), 45–52.

41. W. Maher, F. Krikowa, J. Kirby, A. T. Townsend, P. Snitch, Measurement of trace elements in marine environmental samples using solution ICPMS. Current and future applications, *Austral. J. Chem.*, **56** (2003), 103–116.

42. G. R. Gillson, D. J. Douglas, J. E. Fulford, K. W. Halligan, S. D. Tanner, Non-spectroscopic interelement interferences in inductively coupled plasma mass spectrometry, *Anal. Chem.*, **60** (1988), 1472–1474.

43. I. I. Stewart, J. W. Olesik, Steady state acid effects in ICP-MS, *J. Anal. Atom. Spectrom.*, **13** (1998), 1313–1320.

44. M. J. Campbell, C. Demesmay, M. Olle, Determination of total arsenic concentrations in biological matrices by inductively coupled plasma mass spectrometry, *J. Anal. Atom. Spectrom.*, **9** (1994), 1379–1385.

45. M. J. Campbell, A. Torvényi, Non-spectroscopic suppression of zinc in ICP-MS in a candidate biological reference material (IAEA 392, Algae), *J. Anal. Atom. Spectrom.*, **14** (1999), 1313–1317.

46. A. Montaser, *Inductively Coupled Plasma Mass Spectrometry*, Wiley-VCH, New York, 1998, pp. 1–964.

47. E. H. Larsen, S. Stürup, Carbon-enhanced inductively coupled plasma mass spectrometric detection of arsenic and selenium and its application to arsenic speciation, *J. Anal. Atom. Spectrom.*, **9** (1994), 1099–1105.

48. J. Mora, S. Maestre, V. Hernandis, J. L. Todoli, Liquid-sample introduction in plasma spectrometry, *Trends Anal. Chem.*, **22** (2003), 123–132.

49. S. Munro, L. Ebdon, Application of inductively coupled mass spectrometry (ICP-MS) for trace metal determination in foods, *J. Anal. Atom. Spectrom.*, **1** (1986), 211–221.

50. J. W. Tromp, R. T. Tremblay, J.–M. Mermet, E. D. Salin, Matrix interference diagnostics for the automation of inductively coupled plasma mass spectrometry (ICP-MS), *J. Anal. Atom. Spectrom.*, **15** (2000), 617–627.

51. S. H. Tan, G. Horlick, Background spectral features in inductively coupled plasma mass spectrometry, *Appl. Spectrosc.*, **40** (1986), 446–460.

52. T. W. May, R. H. Wiedmeyer, A table of polyatomic interferences in ICP-MS. *Atom. Spectrosc.*, **9** (1997), 150–155.

53. N. M. Reed, R. O. Cairns, R. C. Hutton, Y. Takaku, Characterisation of polyatomic ion interferences in inductively coupled plasma mass spectrometry using a high resolution mass spectrometer, *J. Anal. Atom. Spectrom.*, **9** (1994) 881–897.

54. S. J. Hill, M. J. Ford, L. Ebdon, Simplex optimisation of nitrogen-argon plasmas in inductively coupled plasma mass spectrometry for the removal of chloride based interferences, *J. Anal. Atom. Spectrom.*, **7** (1992), 719–727.

55. S. Wu, X. Feng, A. Wittmeier, Microwave digestion of plant and grain reference materials in nitric acid or a mixture of nitric acid and hydrogen peroxide for the determination of multielements by inductively coupled mass spectrometry, *J. Anal. Atom. Spectrom.*, **12** (1997), 797–807.

56. W. Maher, S. Forster, F. Krikowa, P. Snitch, G. Chapple, P. Craig, Measurement of trace elements and phosphorus in marine animal and plant tissues by low-volume microwave digestion and ICP-MS, *Atom. Spectrosc.*, **22** (2001), 361–369.

57. J. L. M. De Boer, Real time adjustment of ICP-MS elemental equations, *J. Anal. Atom. Spectrom.*, **15** (2000), 1157–1161.

58. F. B. Whitfield, D. J. Freeman, K. J. Shaw, Trimethylarsine: an important off-flavour in some prawn species, *Chem. Ind.*, **20** (1983), 786–787.

59. W. A. Maher, An investigation of trace element losses during lyophilization of marine biological samples, *Sci. Total Environ.*, **26** (1983), 173–181.

60. K. H. Tam, H. B. S. Conacher, The suitability of the dry ashing procedure for the determination of arsenic in marine samples, *J. Environ. Health*, **B12** (1977), 213–217.

61. W. A. Maher, A decomposition procedure for the determination of arsenic in marine samples, *Talanta*, **30** (1983), 534–536.

62. T. T. Gorsuch, Radiochemical investigations on the recovery for analysis of trace elements in organic and biological materials, *Analyst*, **84** (1959), 135–173.

63. R. G. Lewis, *Determination of Arsenic and Arsenicals in Food and Biological* Materials, F. A. Gunther, J. D. Gunther (eds), *Research. Reviews*, Vol.68, Springer, Berlin), **68** 1977, 123–149.

64. E. Vassileva, H. Docekalova, H. Baelen, S. Vanhentenrijk, M. Hoenios, Revisitation of mineralisation modes for arsenic and selenium determination in environmental samples, *Talanta*, **54** (2001), 187–196.

65. K. Jin, H. Ogawa, M. Taga, Wet digestion method for determination of total arsenic in marine organisms. 1. Study on wet digestion method for determination of total arsenic in marine organisms by continuous flow arsine generation and atomic absorption spectrometry using model compounds, *Bunseki Kagaku*, **31** (1983), E171–E176.
66. W. A. Maher, Some observations on the determination of total arsenic in biological tissues, *Microchem. J.*, **40** (1989), 132–135.
67. G. C. L. Araujo, M. H. Gonzalez, A. G. Ferreira, A. R. A. Noggueira, J. A. Nobrega, Effect of acid concentration on closed-vessel microwave-assisted digestion of plant materials, *Spectrochim. Acta*, **57B** (2002), 2121–2132.
68. A. Chatterjee, Behaviour of anionic arsenic compounds in microwave system with nitric acid and hydrogen peroxide – preliminary laboratory study, *Sci. Total Environ.*, **228** (1999), 25–34.
69. A. Chatterjee, Determination of total cationic and total anionic arsenic species in oyster tissue using microwave-assisted extraction followed by HPLC–ICP-MS, *Talanta*, **51** (2000), 303–314.
70. Z. Slejkovec, J. T. van Elteren, U. D. Woroniecka, Underestimation of the total arsenic concentration by hydride generation techniques as a consequence of the incomplete mineralization of arsenobetaine in acid digestion procedures, *Anal. Chim. Acta*, **443** (2001), 277–282.
71. W. Goessler, M. Pavkov, Accurate quantification and transformation of arsenic compounds during wet ashing with nitric acid and microwave assisted heating, *Analyst*, **128** (2003), 796–802.
72. A. G. Howard, C. Salou, Cysteine enhancement of the cryogenic trap hydride AAS determination of dissolved arsenic species, *Anal. Chim. Acta*, **333** (1996), 89–96.
73. E. Schmeisser, W. Goessler, N. Kienzl, K. A. Francesconi, Volatile analytes formed from arsenosugars: determination by HPLC-HG-ICPMS and implications for arsenic speciation analyses, *Anal. Chem.*, **76** (2004), 418–423.
74. I. Rodushkin, T. Ruth, D. Klockare, Non-spectral interferences caused by a saline water matrix in Q and high resolution inductively coupled plasma mass spectrometry, *J. Anal. Atom. Spectrom.*, **13** (1998), 159–166.
75. W. C. Story, J. A. Caruso, T. Heitkemper, L. Perkins, Elimination of the chloride interference on the determination of arsenic using hydride generation inductively coupled plasma mass spectrometry, *J. Chromatogr. Sci.*, **30** (1992), 427–432.
76. M. Morita, J. S. Edmonds, Determination of arsenic species in environmental and biological samples, *Pure Appl. Chem.*, **64** (1992), 575–590.
77. B. Michalke, Element speciation definitions, analytical methodology, and some examples, *Ecotoxicol. Environ. Safety*, **56** (2003), 122–139.
78. K. Falk, H. Emons, Speciation of arsenic compounds by ion-exchange HPLC-ICP-MS with different nebulizers, *J. Anal. Atom. Spectrom.*, **15** (2000), 643–649.
79. D. Kuehnelt, K. Irgolić, W. Goessler, Comparison of three methods for the extraction of arsenic compounds from the NRCC standard reference material DORM-2 and the brown alga *Hijiki fuziforme*, *Appl. Organomet. Chem.*, **15** (2001), 445–456.
80. S. Branch, L. Ebdon, P. O'Neil, Determination of arsenic species in fish by directly coupled high performance liquid chromatography-inductively coupled plasma mass spectrometry, *J. Anal. Atom. Spectrom.*, **9** (1994), 33–37.
81. M. Pardo-Martinez, P. Vinas, A. Fisher, S. J. Hill, Comparison of enzymatic extraction procedures for use with directly coupled high performance liquid

chromatography-inductively coupled plasma mass spectrometry for the speciation of arsenic in baby foods, *Anal. Chim. Acta*, **441** (2001), 29–36.

82. D. Beauchemin, M. E. Bednas, S. S. Berman, J. W. McLaren, K. W. Siu, R. E. Sturgeon, Identification and quantification of arsenic species in dogfish muscle reference material for trace elements, *Anal. Chem.*, **60** (1988), 2209–2212.
83. E. H. Larsen, G. Pritzl, S. H. Hansen, Carbon enhanced inductively coupled plasma mass spectrometric detection of arsenic and selenium and its application to arsenic speciation, *J. Anal. Atom. Spectrom.*, **8** (1993), 1075–1085.
84. V. M. M. Lai, W. R. Cullen, S. Ray, Arsenic speciation in sea scallop gonads. *Appl. Organomet.Chem.*, **15** (2001), 533–538.
85. W. Goessler, D. Kuehnelt, C. Schlagenhaufen, Z. Slejkovec, K. Irgolić, Arsenobetaine and other arsenic compounds in the National Research Council of Canada certified reference materials DORM 1 and DORM 2, *J. Anal. Atom. Spectrom.*, **13** (1998), 183–187.
86. W. Maher, W. Goessler, J. Kirby, G. Raber, Arsenic concentrations and speciation in the tissues and blood of sea mullet (*Mugil cephalus*) from Lake Macquarie, NSW, Australia, *Mar. Chem.*, **68** (1999), 169–182.
87. M. A. Suner, V. Devesa, I. Rivas, D. Velez, R. Montoro, Speciation of cationic arsenic species in seafood by coupling liquid chromatography with hydride generation atomic fluorescence detection, *J. Anal. Atom. Spectrom.*, **15** (2000), 1501–1508.
88. J. W. McKiernan, J. T. Creed, C. A. Brockhoff, J. A. Caruso, R. M. Lorenzana, A comparison of automated and traditional methods for extraction of arsenicals in fish. *J. Anal. Atom. Spectrom.*, **14** (1999), 607–613.
89. P. A. Callagher, S. Murray, X. Wei, C. A. Schwegel, J. T. Creed, An evaluation of sample dispersion media used with accelerated solvent extraction for the extraction and recovery arsenicals from LFB and DORM-2, *J. Anal. Atom. Spectrom.*, **17** (2002), 581–586.
90. K. L. Ackley, C. B'Hymer, K. L. Sutton, J. A. Caruso, Speciation of arsenic in fish tissue using microwave-assisted extraction followed by HPLC-ICP-MS, *J. Anal. Atom. Spectrom.*, **14** (1999), 845–850.
91. M. Vilano, R. Rubio, Determination of arsenic species in oyster tissue by microwave-assisted extraction and liquid chromatography-atomic fluorescence detection, *Appl. Organomet.Chem.*, **15** (2001), 658–666.
92. J. Kirby, W. A. Maher, Measurement of water-soluble arsenic species in freeze-dried marine animal tissues by microwave-assisted extraction and HPLC-ICP-MS, *J. Anal. Atom. Spectrom.*, **17** (2002), 838–843.
93. R. Tukai, W. A. Maher, I. J. McNaught, M. Ellwood, Measurement of arsenic species in marine macroalgae by microwave-assisted extraction and high performance liquid chromatography-inductively coupled plasma mass spectrometry, *Anal. Chim. Acta*, **457** (2002), 173–185.
94. T. Dagnac, A. Padro, R. Rubio, G. Rauret, Speciation of arsenic in mussels by the coupled system liquid chromatography–UV irradiation–hydride generation–inductively coupled plasma mass spectrometry, *Talanta*, **48** (1999), 763–772.
95. J. J. Sloth, E. H. Larsen, K. Julshamn, Determination of organoarsenic species in marine samples using gradient elution cation exchange HPLC-ICP-MS, *J. Anal. Atom. Spectrom.*, **18** (2003), 452–459.

96. Z. Slejkovec, A. R. Byrne, B. Smodis, M. Rossbach, Preliminary studies on arsenic species in some environmental samples, *Fresenius J. Anal. Chem.*, **354** (1996), 592–595.
97. L. S. Milstein, A. Essader, C. Murrell, E. D. Pellizzari, R. A. Fernando, J. H. Raymer, O. Akinbo, Sample preparation, extraction efficiency and determination of six arsenic species present in food composites, *J. Agricul. Food Chem.*, **51** (2003), 4180–4184.
98. A. D. Madsen, W. Goessler, S. N. Pedersen, K. A. Francesconi, Characterisation of an algal extract by HPLC-ICP-MS and LC-electrospray MS for use in arsenic speciation studies, *J. Anal. Atom. Spectrom.*, **15** (2000), 657–662.
99. S. Dęvalla, J. Feldmann, Determination of lipid-soluble arsenic species in seaweed-eating sheep from Orkney, *Appl. Organomet. Chem.*, **17** (2003), 906–912.
100. I. Pizarro, M. Gomez, M. A. Palacios, C. Camara, Evaluation of stability of arsenic species in rice, *Anal. Bioanal. Chem.*, **376** (2003), 102–109.
101. B. M. Gamble, P. A. Gallagher, J. A. Shoemaker, X. Wei, C. A. Schwegel, J. T. Creed, An investigation of the chemical stability of arsenosugars in simulated gastric juice and acidic environments using IC-ICP-MS and IC-ESI-MS/MS, *Analyst*, **127** (2002), 781–785.
102. M. Burguera, J. L. Burguera, Analytical methodology for speciation of arsenic in environmental and biological samples, *Talanta*, **44** (1997), 1581–1604.
103. T. Guerin, A. Astruc, M. Astruc, Speciation of arsenic and selenium compounds by HOLC hyphenated to specific detectors: a review of the main separation techniques, *Talanta*, **50** (1999), 1–24.
104. Z. Gong, X. Lu, M. Ma, C. Watt, C. Le, Arsenic speciation analysis, *Talanta*, **58** (2002), 77–96.
105. S. H. Hansen, E. H. Larsen, G. Pritzl, C. Cornett, Separation of seven arsenic compounds by high-performance liquid chromatography with on-line detection by hydrogen–argon flame atomic absorption spectrometry and inductively coupled plasma mass spectrometry, *J. Anal. Atom. Spectrom.*, **7** (1992), 629–634.
106. E. H. Larsen, Speciation of dimethylarsinyl-riboside derivatives (arsenosugars) in marine reference materials by HPLC-ICP-MS, *Fresenius J. Chem.*, **352** (1995), 582–588.
107. G. Raber, K. A. Francesconi, K. J. Irgolić, W. Goessler, Determination of 'arsenosugars' in algae with anion-exchange chromatography and inductively coupled plasma mass spectrometry as element specific detector, *Fresenius J. Anal. Chem.*, **367** (2000), 181–188.
108. J. Kirby, W. Maher, A. Chariton, F. Krikowa, Arsenic concentrations in a temperate mangrove ecosystem, NSW, Australia, *Appl. Organomet. Chem.*, **16** (2002), 192–201.
109. U. Kohlmeyer, J. Kuballa, E. Jantzen, Simultaneous separation of 17 inorganic and organic arsenic compounds in marine biota by means of high-performance liquid chromatography/inductively coupled plasma mass spectrometry, *Rapid Commun. Mass Spectrosc.*, **16** (2002), 965–974.
110. D. Sanchez-Rodas, A. Geiszinger, J. L. Gomez-Ariza, K. A. Francesconi, Determination of an arsenosugar in oyster extracts by liquid chromatography electrospray mass spectrometry and liquid chromatography-ultraviolet photooxidation–hydride generation atomic fluorescence spectroscopy, *Analyst*, **127** (2002), 60–65.

111. J. J. Corr, E. H. Larsen, Arsenic speciation by liquid chromatography coupled with ion-spray tandem mass spectrometry, *J. Anal. Atom. Spectrom.*, **11** (1996), 1215–1224.

112. K. A. Francesconi, S. A. Pergantis, Application of selected reaction monitoring tandem mass spectrometry to the quantitative determination of an arsenic-containing nucleoside in a crude biological extract, *Analyst*, **129** (2004), 398–399.

113. M. Miguens-Rodriguez, R. Pickford, J. E. Thomas-Oates, S. A. Pergantis, Arsenosugar identification in seaweed extracts using high-performance liquid chromatography/electrospray ion trap mass spectroscopy, *Rapid Commun. Mass Spectrom.*, **16** (2002), 323–331.

114. B. R. Larsen, C. Astorga-Llorens, M. H. Florencio, A. M. Nettencourt, Fragmentation pathways of organoarsenical compounds by electrospray ion trap multiple mass spectrometry (MS^6), *J. Chromatog. A*, **926** (2001), 167–174.

115. K. A. Francesconi, S. Khokiattiwong, W. Goessler, S. N. Pedersen, M. Pavkov, A new arsenobetaine from marine organisms identified by liquid chromatography-mass spectrometry, *Chem. Commun.*, 2000, 1083–1084.

116. S. N. Pedersen, K. A. Francesconi, Liquid chromatography electrospray mass spectrometry with variable fragmentor voltages gives simultaneous elemental and molecular detection of arsenic compounds, *Rapid Commun. Mass Spectrom.*, **14** (2000), 641–645.

117. S. McSheehy, J. Szpunar, Speciation of arsenic compounds in edible algae by bidimensional size-exclusion anion-exchange HPLC with dual ICP-Ms and electrospray MS/MS detection, *J. Anal. Atom. Spectrom.*, **15** (2000), 79–87.

118. S. McSheehy, Z. Mester, Arsenic speciation in marine certified reference materials. Part 1. Identification of water-soluble arsenic species using multidimensional liquid chromatography combined with inductively coupled plasma, electrospray and electrospray high-field asymmetric waveform ion mobility spectrometry with mass spectrometric detection, *J. Anal. Atom. Spectrom.*, **19** (2004), 373–380.

119. S. McSheehy, J. Szpunar, R. Lobinski, V. Haldys, J. Tortajada, J. S. Edmonds, Characterisation of arsenic species in kidney of the clam *Tridacna derasa* by multidimensional liquid chromatography-ICP-MS and electrospray time of flight tandem mass spectrometry, *Anal. Chem.*, **74** (2002), 2370–2378.

120. J. J. Corr, Measurement of molecular species of arsenic and tin using elemental and molecular dual mode analysis by ion spray mass spectrometry, *J. Anal. Atom. Spectrom.*, **12** (1997), 537–547.

121. N. Ybanez, D. Velez, W. Tejedor, R. Montoro, Optimisation of the extraction, clean up and determination of arsenobetaine in manufactured seafood products by coupling liquid chromatography with inductively coupled plasma atomic emission spectrometry, *J. Anal. Atom. Spectrom.*, **10** (1995), 459–467.

122. K. J. Lamble, S. J. Hill, Arsenic speciation in biological samples by on-line high performance liquid chromatograph microwave digestion hydride generation atomic absorption spectrometry, *Anal. Chim. Acta*, **334** (1996), 261–270.

123. D. Velez, N. Ybanez, R. Montoro, Analysis of geological materials for bismuth, antimony, selenium and tellurium by continuous flow hydride generation inductively coupled plasma mass spectrometry, *J. Anal. Atom. Spectrom.*, **12** (1997), 91–97.

124. J. Mattusch, R. Wennrich, Determination of anionic and cationic species of arsenic by ion chromatography with ICP-MS detection in environmental samples, *Anal. Chem.*, **70** (1998), 3649–3655.

125. *NRCC, Certification Sheet: Dogfish Muscle and Liver Certified Reference Materials for Trace Metals*, National Research Council of Canada, Ottawa, 1999.

126. J. C. Wu, Z. Mester, J. Pawliszyn, Speciation of organoarsenic compounds by polypyrrole-coated capillary in tube solid phase microextraction coupled with liquid chromatography/electrospray ionisation mass spectrometry, *Anal. Chim. Acta*, **424** (2000), 211–222.

127. M. Pavkov, K. A. Francesconi, W. Goessler, A new arsenobetaine in eight reference materials. *5th International Conference on Environmental and Biological Aspects of Main-group Organometals*, Schielleiten, Austria, 5–9 June, 2001.

128. Y. Shibata, M. Morita, Exchange of comments on the identification and quantitation of arsenic species in dogfish muscle reference material for trace elements, *Anal. Chem.*, **61** (1989), 2116–2118.

129. Y. Shibata, M. Morita, Characterisation of organic arsenic in bivalves, *Appl. Organomet. Chem.*, **6** (1992), 343–349.

130. F. Lagarde, M. B. Amran, M. J. F. Leroy, C. Demesmay, M. Ollc, A. Lamotte, H. Muntau, P. Michel, P. Thomas, S. Caroli, E. Larsen, P. Bonner, G. Rauret, M. Foulkes, A. Howard, B. Griepink, E. A. Maier, Certification of total arsenic, dimethylarsinic acid and arsenobetaine contents in a tuna fish powder (BCR-CRM 627), *Fresenius J. Chem.*, **363** (1999), 18–22.

19

SAMPLE PREPARATION PRIOR TO As AND Se SPECIATION

MIHÁLY DERNOVICS AND PÉTER FODOR

Department of Applied Chemistry, Faculty of Food Science, Corvinus University of Budapest, Hungary, E-mail: mihaly.dernovics@uni-corvinus.hu, fodor.peter@uni-corvinus.hu

The Determination of Chemical Elements in Food: Applications for Atomic and Mass Spectrometry. Edited by Sergio Caroli

19.1 INTRODUCTION

Sample preparation procedures prior to As and Se speciation analysis of food and near-food (e.g., over-the-counter drugs and supplements) samples should not be considered more sophisticated or more time-consuming than those usually resorted to for the determination of the total concentrations of the two elements. On the other hand, there are two aspects that are basically different. First, in the case of chemical speciation, the available analytical instrumentation and some characteristics of the analytes of interest, such as stability, sensitivity to oxidation, and the type of chemical and physical binding of the analytes to the major components of the food matrix require stricter pre-analytical conditions in terms of reagents, pH, temperature, and so on. Second, the analyst should choose, depending on the purpose of the analysis, among different sample preparation methods even in the case of the same food matrix and the same analyte species. In other words, 100 percent extraction and identification are not always essential. As an example, on the one hand, the nutritional point of view may require the release (and the identification) only of the presumably bioavailable species from the food matrix, namely, the more or less toxic As species and the Se species with or without anticarcinogenic properties. On the other hand, the analyst should arrive at a compromise between efficiency and ease of sample preparation, especially as regards qualitative speciation analysis. In any case, the analysis ends up in numerical values, namely the extraction efficiency and the rate of identification, sometimes also called chromatographic recovery [1]. The former, in particular, is useful for the characterization of the sample preparation chosen, in order to assess how much the method is fit for purpose. It is noteworthy that total extraction and total recovery are usually difficult or impractical to achieve [2].

Sample preparation methods for the speciation analysis of As and Se have been improving almost at the same pace as the development of speciation methods for other elements. In the case of As, the study by Braman and Foreback [3] was one of the first key experiments, while path-finding Se speciation analyses were conducted during the 1960s, among others by Martin and Cummins [4]. Afterwards, each new sample preparation method that was proved to be suitable for speciation purposes (i.e., that did not alter the chemical species of a given sample) was taken as a reference for the next step ahead. This feedback process can be easily seen in the development of new As and Se speciation approaches. This is the reason why it would be difficult to group the sample preparation methods for As and Se speciation on the basis of "typical" samples, as the same sample can often be prepared by different methods characterized by similar extraction efficiency.

This chapter aims at presenting the various approaches to sample preparation prior to As and Se speciation from the standpoint of method types rather than of sample nature. As already noted above, there is in fact usually no ideal method for a given sample type, since the purpose of the analysis may require different degrees of extraction efficiency. In the following, the most common sample types have been associated with the most frequently recommended preparation methods as a function of specific requirements. Examples are given to illustrate such sample–method couples.

19.2 SAMPLE PREPARATION FOR TOTAL As AND Se ANALYSIS PRIOR TO SPECIATION

19.2.1 General Aspects

The term "sample preparation prior to speciation" usually covers the methods capable of releasing or extracting the species from the samples for speciation analysis without any alteration of their speciation state. At the same time, speciation analysis should always be accompanied by total As and Se determination following sample digestion. The latter is considered a well-controlled phase of the analysis for which several matrix Certified Reference Materials (CRMs) are available. There are two main reasons why total As and Se analysis (and the relevant sample preparation steps) should be taken into account in speciation analysis. First, information on the total element concentration of a given sample is useful to decide whether it is necessary or desirable to carry out a complete speciation analysis, for example, in agreement with the relevant food safety limits. Second, the total concentrations of As and Se – nowadays sometimes only optional – may play a key role in the quality control (QC) of speciation analysis by allowing mass balances to be calculated as regards: (1) the original untreated sample; (2) the fractions and residues associated with each step of the sample preparation; and (3) the resulting sample solution ready for the final analysis (e.g., chromatography-coupled spectrometric detection). The total element concentration versus the sum of those relevant to the various chemical species provides information on the extraction efficiency and the chromatographic recovery as a whole [1, 5].

The sample digestion procedure should result in a homogeneous solution containing all species in the same chemical – usually inorganic – form at high oxidation state. This task is not easy to achieve in practice because both As and Se may possess organic species in the same sample which are rather stable even under extreme oxidation and pH conditions. A mineralization process carried out inadequately in terms of digestion parameters can cause a high C content and an inhomogeneous distribution of species. This may lead to erroneous analytical results with the instrumentation normally used for quantifying elements, for example, atomic absorption spectrometry (AAS), atomic emission spectrometry (AES), and mass spectrometry (MS). This problem is of particular concern when hydride generation (HG) is applied to total As or Se determination if the markedly different HG-forming capabilities of the various chemical species are overlooked [6].

Nowadays total As and Se determinations are carried out by means of wet digestion, mostly through microwave (MW) assisted techniques with the addition of oxo- and halogenic acids using polytetrafluoroethylene (PTFE) vessels. Although the required digestion parameters (temperature, pressure, and irradiation time) can be achieved in household as well as in laboratory pressure devices, it is understandable that microprocessor-controlled instrumentation affording shorter digestion time, traceable and tunable digestion parameters, and QC-oriented data recording offers a more reliable, though more expensive, option. On the other hand, whatever the analyst's choice may be, the wet digestion process should be done in closed vessel systems because of the possible presence of volatile As and Se species [7].

Last, but not least, it should be borne clearly in mind that even the most careful sample preparation cannot compensate for a wrong sampling procedure resulting, for example, in inhomogeneous and/or not representative samples. This is especially true in the field of food analysis, since food matrices can be extremely diverse. Therefore, prior to analysis sample homogenization is mandatory.

19.2.2 Examples of Sample Preparation Methods for Total Se Determination

The one-step closed-vessel wet digestion carried out with either HNO_3 or the HNO_3–$HClO_4$ mixture (with or without H_2O_2) usually yields ready-to-analyze sample solutions from organic samples. The MW-assisted or high-pressure ashing procedures based only on HNO_3 can adequately digest relatively complex samples such as mullet [8], yeast-based food supplement [1], Se-enriched lactic acid bacteria [9], and edible spirulina [10]. Occasionally, the addition of $HClO_4$ to HNO_3 may become necessary to achieve complete destruction of the organic matrix if there is a high lipid content, as in the cases, for example, of Brazil nuts [11], soybean [12], and shellfish [13]. However, the use of $HClO_4$ is not recommended owing to explosion hazards.

The addition of H_2O_2 acting as an additional oxidizing agent is also intended for more aggressive and thorough digestion. Its application is not sample-dependent and the existing literature does not state unambiguously when it is really necessary to resort to it. There are studies describing the simultaneous use of HNO_3 and H_2O_2 for the digestion of, for example, Se-enriched yeast [14] and yeast-based food supplements [15]. In this context, it is noteworthy that there are samples like higher edible fungi that are always digested with both HNO_3 and H_2O_2 by several independent research groups, since the chitin-based cell wall may well require the stronger oxidizing power of H_2O_2 [16–19]. Uden et al. [15] used

MW-assisted digestion with a mixture of HNO_3 and H_2O_2 for several types of samples (*Allium* ssp., alfalfa, CRMs based on vegetable and animal matrices), thus achieving adequate recovery figures. As a rule of thumb, if the digestion mixture after the addition of H_2O_2 does not exceedingly dilute the sample solution, its use is more than expedient. It should not be overlooked, however, that high purity HNO_3 is available in a more cost-effective way, for example, after sub-boiling distillation of the AR-graded reagent, than high purity H_2O_2.

The use of the highly basic reagent tetramethylammoniumhydroxide (TMAH) is sometimes reported in the literature for the total digestion of organic samples. Its applicability for speciation purposes has been debated over time, for example, as regards the treatment of wheat samples for Se speciation at a reported concentration of 25 percent m/m [20]. This caused significant variations in the concentrations of the Se species. Such changes were unambiguous even at a concentration level as low as 2.5 percent m/m [1, 21, 22]. Hence, TMAH is accepted as a digestion means only for the determination of the total Se concentration [10]. Also moderate basic conditions such as those provided by the use of a 4 percent m/m NH_3 solution (definitely milder than those characteristic of TMAH) were found to alter the concentrations of Se species [23]. Taking all this into account, it can be stated that a pH range higher than the biological range (pH = 6–8) cannot be recommended for Se speciation analysis.

19.2.3 Examples of Sample Preparation Methods for Total As Determination

The sample preparation methods applied prior to total As determination can be subdivided into two major groups depending on the analytical instrumentation. When the digested sample solution is analyzed by a non-HG technique, for example, by inductively coupled plasma (ICP)-AES or ICP-MS with conventional nebulization, the MW-assisted digestion either with HNO_3 alone [24] or with HNO_3–H_2O_2 mixture [25–27] is considered adequate and reliable. There is widely documented evidence that this approach does not result in measurement bias, even when the As standard used for calibration [usually As(V)] and the As species formed in the sample solution during the digestion process are not checked for their identity. The analysis of green algae [28], shellfish [29], fish [30], baby foods [31], and different rice species [32] along with the relevant matrix CRMs are representative of this assumption.

If HG instrumentation is available for total As determination, the sample matrix should be carefully considered. For instance, seafood and other sea-related samples usually contain high amounts of arsenobetaine (AB), an As species that cannot be directly subjected to HG and is one of most resistant species to oxidative acidic digestion. According to Šlejkovec et al. [33, 34], the temperature necessary to decompose AB into As(V) is around 300°C. This temperature can be achieved either with the application of a digestion mixture of H_2SO_4, HNO_3, and H_2O_2, where H_2SO_4 is required due to its higher boiling point compared to that of HNO_3, or with focused MW digestion devices.

The sample preparation to be followed prior to the use of an electrothermal atomization (ET) based detection system can be described as a borderline case between the two groups described above. This is due to the different vaporization temperatures of the As species, which are able to escape from the furnace during the heating stage of the measurement cycle prior to detection, thus paving the way to continuous As loss and erroneous mass balance of As. There are two ways of preventing this phenomenon; the analyst should either apply the digestion method described for the hydride-based instrumentation so as to achieve the total conversion of all As species into a homogenous sample solution containing only As(V), or use matrix modifier compounds (e.g., salts of Pd and Mg) to minimize the untimely As vaporization [35].

19.3 SAMPLE PREPARATION METHODS PRIOR TO Se SPECIATION ANALYSIS

19.3.1 Background Information

The sample preparation methods the analyst can choose for Se speciation purposes are generally limited by three factors. First, inorganic Se species may be affected by temperatures higher than 70°C, depending on the sample matrix [36], resulting in the oxidation of Se(IV) into Se(VI), whereas a considerable change in pH may alter the redox balance between the two inorganic forms in the sample [37]. Second, proteolysis – the most frequently used method at the moment for Se speciation – is not suitable to study intact proteins. Third, since some characteristics of the newly identified Se species and derivatives are unknown as yet, it is definitely reasonable to emphasize the use of mild, though poorly extracting, methods able to keep the original speciation information. Despite all these limitations, there is still a wide range of methods available.

19.3.2 Extraction with Nonbuffered and Buffered Solutions

Extraction is considered to be the mildest sample preparation method for Se speciation. The pH of the extracting solutions is usually set in the biological buffer range (pH = 6–8). Extraction is accompanied by shaking and stirring for 1–2 h, while the temperature is set between room temperature and 80–90°C. This process results mostly in the extraction of inorganic Se, of free selenoamino acids and their derivatives, and – in the case of neutral pH – of 20–30 percent of the proteins from the sample. Since the extraction recovery achieved with these methods depends more on the type of samples than on the extracting solution chosen, the relevant practical examples are given by sample groups.

When analyzing samples of vegetable origin, water extraction may end up in Se species that differ not only in quantitative, but also in qualitative terms even in the case of the same plant species. The reason for this is that some plant species (e.g., sunflower) accumulate high amounts (up to 900 $\mu g\ g^{-1}$) of Se from Se-rich soil or

Se-enriched hydroponic media. The element is not metabolized and can be extracted even up to 100 percent by common water extraction [38]. Nevertheless, the lower the Se concentration in the soil is, the higher the amounts of Se metabolized (e.g., by sunflower plants).

Garlic shows a different Se accumulation pattern. Up to a certain extent of Se supply, garlic synthesizes the highly anticarcinogenic γ-glutamyl-Se-methyl-selenocysteine that can be almost 100 percent extracted with water at 80–90°C [39, 40]. However, when higher amounts of Se are available from the soil or from the hydroponic medium, methylselenocysteine (SeCys) and selenomethionine (SeMet) – a selenoamino acid mostly built in proteins – are synthesized, this reducing the efficiency of water extraction [41]. Certain garlic species [41, 42] synthesize the protein-incorporated SeMet at higher extents than water-soluble Se species. Thus, under such circumstances, the efficiency of water extraction is usually less than 90 percent.

Selenium-enriched yeast, regarded as one of the most interesting materials for Se speciation, has also been studied through extraction procedures with water and buffered solutions as well, typically with 30 mmol l^{-1} tris(hydroxymethyl)amino-methane (Tris)–HCl at pH = 7.0. The investigations performed by several independent research groups almost always led to the same results: both water extraction (at 50–90°C for 1–2 h) and Tris-buffered extraction (at 37°C for 1 h) achieved about 10 percent extraction of Se, the main Se species identified being Se-adenosyl-homocysteine, a small amount of free SeMet and some nonmetabolized, inorganic Se(IV) in the fermentation-media [21, 40, 41, 43, 44].

A useful, yet almost never mentioned aspect of water or buffered extraction is its ability to remove water-soluble metal ions that can hamper a scheduled sample preparation step, for example, enzymatic proteolysis, through enzyme inhibition [45].

Water and buffered extraction solutions can also be used for the release of water-soluble Se-containing peptides and polypeptides, often followed by size exclusion chromatography (SEC) of the extracted molecules to map their size and distribution. This approach can be well illustrated with the examples of soybean [12], edible mushroom *Agaricus bisporus* [46], fish species [47], muscles of animal species intended for food [48], and Se-enriched yeast [49, 50]. On the other hand, Se speciation nowadays cannot be fully grasped without mentioning the Se-proteomics-oriented experiments that create a common link between bio-analytical chemistry, food chemistry, and human physiology [51–53]. The basic sample preparation methods addressed in the field of Se proteomics are also based on the mild extraction techniques with water and buffered solutions that are often completed with sodium dodecylsulfate (SDS) to denature and bring into solution proteins for gel techniques polyacrylamide gel electrophoresis (PAGE), 2-dimensional gel electrophoresis (2D-GE), isoelectric focusing (IEF), and capillary zone electrophoresis (CZE). Although these studies often go beyond the purpose of plain speciation analysis and are sometimes focused basically on obtaining structural information, there are recent relevant investigations where both aspects are dealt with, for example, those on Se-enriched yeast [54–56], liver [57], and mushroom [58].

19.3.3 Extraction with Organic Solvents Prior to Se Speciation

Up to now, no hydrophobic Se species has been identified. At the same time, some experiments carried out on samples that had never or seldom been analyzed before included an extraction step with 1 : 2 or 2 : 1 (v/v) mixture of chloroform–methanol, for example, soybean [12], mullet, and cockles [8]. From these samples, only some parts (especially the gill) of the cockle contained considerable amounts of lipid-soluble Se (up to 30 percent of the total Se content). As no speciation analysis was carried out on this cockle tissue, it is uncertain whether the presence of real lipophilic Se species or the incorporation of formerly identified Se species into lipoproteins could explain the occurrence of this particularly large lipophilic Se-containing fraction. Incidentally, it is the weak extracting power of organic solvents that prompted other types of sample preparation methods for Se speciation, for example, enzymatic hydrolysis [59]. Organic solvents are usually used to eliminate the lipoid fractions of the samples, ameliorate analytical precision, and remove compounds that worsen chromatographic conditions. Kannamkumarath et al. [60] and Bodó et al. [61] used a mixture of chloroform–methanol or cyclohexane, respectively, to defat ground Brazil nuts, while Gómez-Ariza et al. [13] used dichloromethane for the cleaning up process of shellfish samples.

The use of methanol and ethanol, two solvents that can be well mixed with water, should be discussed separately, as they are used for several types of sample preparation on the basis of similar expected effects, but for different purposes. The feature common to both solvents is the observation that most of the sample proteins precipitate when the concentration of these compounds exceeds ca. 40 percent (v/v), thus enabling the analyst to separate the protein fraction by centrifugation or microfiltration. This way, either the proteins not intended for further analysis (e.g., enzymes or proteins that were inadequately hydrolyzed) can be removed [62], or, the purification of Se-containing proteins by successive solvent extractions can be achieved [12]. At concentrations of less than 40 percent (v/v), both methanol and ethanol are usually mixed with 0.1 mol l^{-1} HCl; however, the use of these mixtures entails a relatively low extraction efficiency of Se (10–14 percent). Therefore, they are intended only for the extraction of water-soluble Se species, generally free selenoamino acids [15, 21, 63, 64].

19.3.4 Sample Preparation at pH Values Lower than the Biological Range

19.3.4.1 The Use of HCl at Concentrations Higher than 1.0 mol l^{-1} This appears to be one of the most antinomic parts of Se speciation. The basic question is rather simple: Can concentrated HCl be used for the purpose of Se speciation without altering the Se species during sample preparation? The traditional protein hydrolysis carried out with 6 mol l^{-1} HCl under anaerobic conditions at 110–120°C for 24 h has been considered a consolidated, automated technique since the 1960s, and has been used for the hydrolysis of Se-containing proteins.

The redox balance of the two inorganic Se species, Se(IV) and Se(VI), in the sample can be significantly affected. Therefore, acid hydrolysis is unfit for the analysis of Se inorganic species. Accordingly, if no independent analysis of the protein fraction of the sample under study is carried out, a substantial part of Se can only be identified as total inorganic Se fraction. Apart from this drawback, one of the most important advantages of acid hydrolysis is the short-term analysis provided by the introduction of MW-assisted techniques [65].

In recent years several applications of the HCl proteolysis have been published in the field of Se speciation, for example, as regards Se-enriched lactic acid bacteria [66], mullet and cockles [8], and algae [67], where the technique provided extraction efficiencies of greater than 90 percent and preserved the integrity of the selenoamino acids. The general usefulness of this method of Se speciation is, however, questionable. Sometimes the authors do not state clearly whether phenol – an essential compound for the prevention of oxidation of SeCys – was used or not. In practice, neither phenol nor the short-duration MW-assisted irradiation can prevent the alteration of selenoamino acids [68–71]. At the moment, no final conclusion on the applicability of HCl proteolysis can be drawn, as CRMs certified for SeCys are still unavailable. On the other hand, an Se extraction efficiency of 80–90 percent can be achieved with this method only if either proteins are at least partly separated from the other components of the matrix, for example, separate analysis of fish muscles is carried out [8], or a considerable portion of Se is originally contained in inorganic forms in the sample, as observed by B'Hymer and Caruso [1] in the case of Se-enriched food supplements.

19.3.4.2 The Use of Acids at Concentrations Lower than 1.0 mol l^{-1} One of the most typical compounds from this standpoint is trichloroacetic acid (TCA), which is employed as a rule in ice-cold solutions at a concentration of 1–12 percent m/m. It is used to precipitate of high molecular weight proteins through the reduction of their hydration potential and pH to allow low molecular weight, water-soluble, Se-containing molecules to be studied, for example, inorganic Se species and free or methylated selenoamino acids and their derivatives. The use of TCA, in practice, allows Se-containing analytes to be characterized on the basis of their molecular weight where there is no possibility for either ultrafiltration or SEC. However, the border line between the precipitated or still soluble, or in other terms, big or small molecules cannot be unequivocally defined, since both the size of the proteins and their hydrophobic nature determine their solubility in the TCA solution. It can be stated as a rule that peptides containing less than seven amino acids are not precipitated by TCA [72]. The practical field of the application of TCA is quite extended: green algae (*Chlorella* ssp.) [67, 73], codfish and flounder [72], Brazil nuts [62], Se-enriched lactic acid bacteria [74], and soybean [12] can be listed as relevant sample types. It is noteworthy that volatility of TCA opens the door to its elimination by speed-vacuum instruments, this being by far a beneficial feature when analyzing sample fractions with a low content of Se.

Besides TCA, $HClO_4$, and HCl (usually at concentrations of 0.4 and 0.1–1.0 mol l^{-1}, respectively) are also used for the extraction of low-molecular-weight Se species. Since the application of HCl, especially when its use is associated with MW-assisted irradiation, alters the original concentration ratio of Se(IV) to Se(VI) of the sample, the speciation data related to the inorganic Se species should be given as total inorganic Se [75]. Taking this drawback into account, the question of why diluted acids are used instead of water or buffers is unavoidable. The answer is provided by the observation that diluted acids extract significantly higher amounts of free or weakly bound amino acids and related compounds from the sample than water and buffers do. This fact has been confirmed by comparative analyses on several samples, for example, soybean [76], Se-enriched yeast [21], Se-enriched lactic acid bacteria [77], and Brazil nuts [62].

Only one study has reported Se extraction efficiency for diluted HCl lower than those afforded by other extraction media. Lintschinger et al. [38] carried out an experiment on wheat and alfalfa sprouts cultivated in inorganic Se-enriched hydroponic media. They found that the highest Se extraction efficiency was achieved with the use of 0.1 mol l^{-1} NaOH solution. However, a considerable amount of the extracted Se could not be identified. The authors ascribed this fact to the likely degradation of the original species or to the presence of unknown compounds. On the other hand, extraction with a 0.1 mol l^{-1} HCl solution resulted in an extraction efficiency even lower than that of water extraction. Both phenomena are considered unusual and show that there is no general-purpose sample preparation for Se speciation; that is, an ad hoc sample preparation approach might become necessary when dealing with samples that have never been analyzed before.

19.3.5 The Role of Enzymatic Hydrolysis with Proteolytic and Nonproteolytic Enzymes in Sample Preparation Methods Prior to Se Speciation

19.3.5.1 Preliminary Considerations At the beginning of the 1980s a new sample preparation approach was developed that almost completely replaced the formerly dominant 6 mol l^{-1} HCl proteolysis: the enzymatic method. In addition to the doubts on the applicability of the acidic method discussed in Section 19.3.4.1, the main reason for the success of the enzymatic approach was the concern about the bioavailability of Se from different foodstuffs. Namely, there was no well-established approach to link data on the total Se content of food to the generally acknowledged narrow Recommended Daily Allowance (RDA) range of Se, with the exception of relatively expensive feeding studies. This new approach filled a substantial gap since human metabolism and digestion could not be simulated merely via the HCl-based techniques and the interpretation of the results of relevant feeding studies was difficult without any information on how the digestive tract acts on the Se-containing molecules.

From the publication of the first study on the enzymatic sample preparation for Se speciation of soybean [78], the more than 20-year long period of progress in enzymatic sample preparation can be divided into three phases on the basis of current analytical tasks and enzymes used. As regards the first phase, roughly until the mid 1990s the most important goal of the studies was to assess the bioavailability of Se from foods. Therefore, the most frequent approach was based on the simulation of human digestion, usually by applying multiple stage enzymatic treatments, either focusing on proteolytic enzymes only or taking into account all digestive enzymes available. Achievement of the maximum extraction efficiency of Se was out of question, as the typical goal was to decide whether Se could be assimilated from the given food sample and to what extent. This kind of investigation can still be found, though less frequently, in today's scientific literature. The second phase was initiated by the well-known study by Gilon et al. [79] devoted to a relatively simple, one-step enzymatic sample preparation in the field of Se speciation with an extraction efficiency as high as possible. Partly due to the novelty of the approach and partly to the new chromatographic and MS instrumentation available at the time, dozens of newly identified Se species were reported between 1995 and 2001. The third phase is definitely associated with proteomics and has paved the way to a more sophisticated speciation approach which focuses much more on the intact selenoproteome and other Se-containing macromolecules rather than on selenoamino acids and their derivatives. This latter strategy requires special enzymes that differ from those of the former two phases, that is, for sequencing purposes, which transforms traditional sample preparation into an instrumentation-oriented and instrumentation-dictated approach.

All of the three phases of enzymatic methods were dominated by proteolytic enzymes. However, nonproteolytic ones were also addressed for special purposes to indirectly assist proteolysis, for example, by providing access to proteins to be hydrolyzed through the enzymatic decomposition of the matrix constituents. At the same time, the use of nonproteolytic enzymes might be promoted in the future to study the possibly nonprotein-associated Se macromolecules which have already been observed in the case of numerous real samples, for example, edible mushroom [58, 80], but so far have never been investigated in detail.

19.3.5.2 General Characterization of Enzymatic Sample Preparation Prior to Se Speciation Besides providing the possibility to assess bioavailability and simulate human digestion, enzymatic methods possess all the required characteristics of Se speciation-oriented sample preparation. As they show adequate specific activity between 30 and 50°C and in the range of biological buffers (pH = 6–8), their application requires neither high temperature nor extreme acidic/basic conditions, except for the enzymatic systems mimicking stomach activity. Compared with the MW-assisted HCl proteolysis, their disadvantages are the relatively time-consuming operation (up to 2–3 days), the costliness of some of the enzymes, and the sensitivity to sample inhomogeneities that may result in low extraction efficiency and poor reproducibility, especially in the case of big particle size (>125–250 μm).

19.3.5.3 The Use of Proteolytic Enzymes in the Simulation of Human Digestion Proteolytic enzymes can particularly contribute to the extraction of Se species when the samples addressed: (1) are of biological origin; (2) are presumed to contain a considerable amount of protein; and (3) feature a protein content accessible to the enzymes used during sample preparation. If the sample meets all these requirements, an optimized set of working parameters (pH, T, and t) are set up and no inhibiting agents are present, then the enzyme(s) hydrolyze(s) the protein of the sample into amino acids, or peptides. The most important goal here is the choice of the right enzyme for the given analytical purpose.

In the human body food proteins are hydrolyzed into amino acids with 95–98 percent efficiency. This fact served as a basis for Yasumoto et al. [76, 78] to set up for the first time an enzymatic sample preparation scheme for the analysis of Se-containing soybean proteins. The multiple-stage sample preparation was based on a preliminary treatment with a 1 percent m/m pepsin solution at pH and temperature values simulating those of the human stomach, that is, pH = 1 and 37°C. The next step focused on the use of a mixture of pancreas-derived enzymes, pancreatin (1 percent m/m, pH = 7.5), whereas the final step was centred on pronase E, a mixture of nonsubstrate-specific proteolytic enzymes of bacterial origin (1 percent m/m, pH = 7.5). Pepsin and pancreatin hydrolyzed the protein content mostly into peptides, while pronase E hydrolyzed soybean proteins and peptides into amino acids. The total duration of the process was 63 h. Dialysis tubes were used so as to monitor the amount of low-molecular-weight, Se-containing by- and end-products appearing during proteolysis. The conclusion of the study was the identification and quantification of SeMet, the main Se species of soybean protein.

Crews et al. [81] studied cooked cod by means of a two-step *in vitro* gastrointestinal enzymolysis. For the first step of sample preparation they employed gastric juice (1 percent m/v pepsin, pH = 2.0, in 0.15 mol l^{-1} NaCl) at 37°C for 4 h. Afterwards a pancreatin-based mixture was added to the sample solution containing 1.5 percent m/v pancreatin, 0.5 percent m/v amylase, and 0.15 percent m/v bile salts in 0.15 mol l^{-1} NaCl at pH = 6.9 for a further 4 h at 37°C. The relatively short (8 h) enzymatic activity and the lack of enzymes capable of hydrolyzing proteins directly into amino acids resulted in the identification of inorganic Se (IV) only, as no selenoamino acids could be detected.

Thomas et al. [82] carried out an Se speciation experiment on pork meat, American toast (presumed to contain a high, or at least moderate, quantity of Se), and Se food supplements by combining the two methods described above, that is, dialysis tubes were used for a two-step in vitro gastrointestinal enzymolysis. The amount of Se diffused across the molecular dialysis membrane from the digestion mixture was used as an indicator of bioavailability and, following the general approach of bioavailability assessment, the total hydrolysis of food proteins was not targeted. According to their results about 40–50 percent of Se proved to be bioavailable from the toast and pork meat samples in the form of SeMet, while Se food supplements provided 70–100 percent bioavailable Se in the forms of SeMet and Se(VI).

Martín et al. [22] treated Se-enriched yeast with simulated gastric juice containing 1 percent m/m pepsin (pH = 1.8, in 0.9 percent m/m NaCl) in a one-step procedure for 4 h. As expected, pepsin itself could provide partial proteolysis. Of the total Se content 11 percent could be identified as SeMet and Se(IV).

As a conclusion one can say that enzymes with the ability of hydrolyzing proteins into amino acids are required to carry out more detailed studies both on the assessment of Se-bioavailability from foods and on the extraction and identification of more Se species. At the same time, the mechanism of human digestion is still partly unknown, as stated by Dumont et al. [83], who adopted in vitro gastrointestinal conditions in one of the latest studies published in the area of simulated human digestion. Interestingly, the authors treated three kinds of Se-enriched yeast supplements (two of which were Se-speciated as well) with simulated gastric fluid (SGF), that is, 0.32 percent m/v pepsin and 0.2 percent m/v NaCl in 84 mmol l^{-1} HCl at pH 1.2, and with simulated intestinal fluid (SIF), that is, 1 percent m/v pancreatin in 50 mmol l^{-1} KH_2PO_4 at pH 6.8. According to their results, extraction efficiencies between 5 and 80 percent were achieved and SeMet, selenomethionine Se-oxide (SeOMet), and selenocystine ($SeCys_2$) were liberated from the supplements, depending on the sample type and the extraction fluid. The identification of SeOMet, an Se species with unknown bioavailability, from one of the digested samples highlights the fact that human digestion may turn the original Se species into less characterized ones, thus querying the Se bioavailability assessments based on the original speciation information.

19.3.5.4 The Use of Proteolytic Enzymes in One-Step Sample Preparation Methods The time-consuming sample preparation schemes (usually 1–2 days are necessary), the intermediate activity (e.g., changes in the dialyzing solutions and buffers, the addition of new enzymes, the addition of freeze-drying steps, etc.), and the risk of species deterioration because of microbial activity in the absence of sterile conditions prove that the approach to simulate human digestion should be attempted only in particular cases. On the other hand, there is no reason against the assumption that food proteins are almost totally decomposed by the human digestive tract, as only 2–5 percent of proteins escape proteolysis [84] and special protein folding inhibiting the access of proteolytic enzymes is unusual in foods, except for prions, for example. The exploitation of a one-step enzymatic proteolysis to hydrolyze sample proteins into amino acids would then aim at three different targets at the same time, namely: (1) a more realistic description can be achieved of the bioavailability of Se from the sample under testing; (2) the length of sample preparation can be shortened; and (3) the almost complete hydrolysis would result in outstanding extraction and identification efficiency.

All this is valid from a theoretical point of view. In practice, foods contain several minor and major constituents; they form a special rheological matrix of water, carbohydrates, lipids, proteins, fibers, mineral compounds, and so on, and human metabolism resorts to more than 100 types of enzymes to decompose it.

In one word, a possibly low in vitro extraction efficiency does not necessarily go together with low Se bioavailability. The powdered basic samples of Se speciation (Se-enriched yeast, *Allium* ssp., wheat flour, etc.) – albeit they are food materials or food supplements – are far less complex than real foodstuffs. At the same time, they facilitate sample preparation and the assessment of Se supply from prepared food materials through the analysis of some raw food materials.

In light of these considerations, it is understandable why the one-step enzymatic method published by Gilon et al. [79] could become a basic technique in Se speciation analysis. The principle behind their method was the addition, at high excess, of a mixture of proteolytic enzymes, pronase E (EC 3.4.24.4 and EC 3.4.24.31 from *Streptomyces griseus*, sometimes named actinase E or protease XIV, depending on the crude or partly purified product and the supplier), to the well-homogenized sample at the almost optimal working parameters of the enzymes (pH = 7.4–8.8 at 37–40°C). The term "high excess" usually means 10 mg enzyme per 100 mg sample. This is a huge ratio as compared with the usual application of the enzyme to make DNA/RNA extracts free from proteins [85] and the average natural enzyme: substrate molar ratio of 1 : 10,000. Incidentally, this excess may be the reason why the actual enzymatic activity of the products used is usually not provided by the authors. The background of such extreme ratios is obvious: the highest hydrolysis and extraction efficiency must be achieved in vitro, that is, under suboptimal conditions, on proteins that can be irreversibly precipitated/denaturated during the pretreatment process of the sample material. Nevertheless, pronase E may still reach an almost 100 percent proteolysis after 20–24 h, depending on the sample material (see Table 19.1), thanks to its composition of at least 10 proteolytic enzymes showing different activities (endo- and exo-, -protease and -peptidase, acting on both C- and N-terminals) [86]. The resulting mixture is a nonsubstrate-specific, crude preparation that is able to hydrolyze almost all kinds of proteins, except for some special structures like keratin [29]. Practically, two products have been in use in speciation analysis: a low activity protease XIV (4–5 U mg^{-1}, supplied by Sigma-Aldrich) containing a partly purified serine protease constituent of the basic crude extract, and a high-activity pronase E (ca. 4000 U mg^{-1}, supplied by Merck) that is less purified. The former has a leading role in the relevant scientific literature, as can be seen in Table 19.1, although the rationale behind its choice has not been completely elucidated. Gómez-Ariza et al. [13] observed considerable recovery problems of inorganic Se species during the digestion of prawn tissues with protease XIV, while Yang et al. [87] recorded this phenomenon for pronase E when organic Se species were to be extracted from spiked Se-enriched yeast. Clearly, further investigations are needed to better grasp these aspects.

The data presented in Table 19.1 possess some special features. First, the dominant sample is Se-enriched yeast, which is a clear consequence of at least three factors, that is: (1) Se-enriched yeast is relatively easy to produce with the instrumentation available in a conventional microbiological or biotechnological laboratory; (2) numerous Se-enriched yeast-based products are available on the market as food supplements; and (3) the often quoted Clark report [88] proved

TABLE 19.1. Examples of Se Speciation Studies Based on the Application of One-step Enzymatic Proteolytic Sample Preparation Methods

Se Species Identified [a]	Chromatographic Recovery (as % of Se Extracted)	Extraction Efficiency (as % of Total Se)	Reaction Medium	Enzyme(s) Applied	Total Se Content, $\mu g\ g^{-1}$	Samples Addressed	References
2, 6, 7	80–84	90–92	Citric acid–phosphate buffer, pH = 7.5	Protease XIV	1038	Se-enriched yeast	[79]
6, 7, 9	>90	90	Deionized water	Protease XIV	1922	Se-enriched yeast	[142]
7, 9, 19	~52	~84	Deionized water	Protease XIV	138	Indian mustard (*Brassica juncea*)	[15]
7, 14 7, 10	82 86	80–90 ~ 100	Deionized water	Protease XIV	1922 296	Se-enriched yeast Se-enriched garlic	[40, 43]
1, 2, 7, 9 1, 2, 7	71 99	15 95	Deionized water	Protease XIV	2071 592	Phytoplankton Se-enriched yeast	[143]
1, 2, 7	94	88	Phosphate buffer, pH = 7.5	Protease XIV (+lipase VII)	1180	Se-enriched yeast	[21]
2, 6, 7, 9, 10, 11, 14, 15, 17, 9, 10, 11, 1, 6, 7, 9, 10, 11, 1, 6, 7, 9, 10, 11, 1, 7, 9	67–93[b] 60–85[b] 87–96[b] 92–96[b] 90[b]		Deionized water	Protease XIV	1200–2100 48–524 0.2–1355 0.2–140 138	Se-enriched yeast Se-enriched and normal ramps (*Allium tricoccum*) Se-enriched and normal garlic Se-enriched and normal onion Indian mustard (*Brassica juncea*)	[41, 42]

(*Continued*)

TABLE 19.1. *Continued*

Se Species Identified [a]	Chromatographic Recovery (as % of Se Extracted)	Extraction Efficiency (as % of Total Se)	Reaction Medium	Enzyme(s) Applied	Total Se Content, $\mu g\ g^{-1}$	Samples Addressed	References
1, 2, 7	42–103	12–99	$NaCO_3$-buffer, pH = 7–7.2	Proteinase K	107–222	Se-enriched yeast based food supplements	[1]
7, 9, 14, 15, 9, 10, 1, 2, 9, 10	NR	NR	HCl + NaCl, pH = 1	Pepsin	1922 296 140	Se-enriched yeast Se-enriched garlic Se-enriched onion	[144]
1, 2, 1, 2, 1	~68, ~62 ~100	62 74 100	Citric acid–phosphate buffer, pH = 7.5	Protease XIV	112 164 900	Se-enriched wheat sprouts Se-enriched alfalfa sprouts Se-enriched sunflower sprouts	[38]
6, 7	81.4	74.1	Ammonium-phosphate buffer, pH = 7.5	Protease XIV	426	Se-enriched yeast based food supplement	[14]
1, 2, 6, 7, 20	35.3	25	TRIS–HCl, pH = 7.3	Protease XIV	148	Se-enriched lactic acid bacteria	[9]
3, 7	~100	29.5–35	TRIS–HCl, pH = 7.5	Subtilisin	1.22	Oyster RM	[89, 145]
7	~43–47 (Brazil nut)	~60 (Brazil nut)	phosphate-buffer, pH = 7.5	Proteinase K	0.1–35.1	Different kinds of nut (walnut, Brazil nut, pecan, cashew)	[60]
7	38	83	TRIS–HCl, pH = 7.5	Proteinase K	35	Brazil nuts	[62]

7	NR	73 70	Deionized water	Proteinase K protease XIV	202	Indian mustard (*Brassica juncea*)	[75]
2 NR 2, 6, 7	~91 NR ~100	60 29 44	Deionized water	Subtilisin (+lipase VII)	0.79 2.93 6.88	Prawn clam CRM (TORT-1)	[13]
6, 7, 20, 1, 2, 6, 7, 20	~26.6 ~37.0	~8.3 ~ 12.1	TRIS–HCl, pH = 7.3	Protease XIV protease XIV +lisozyme	140	Se-enriched lactic acid bacteria	[77]
2, 6, 7 6, 7	93 94.6	77 15.7	Phosphate-buffer, pH = 7.5	Protease XIV	512 6.06	Se-enriched yeast based food supplement CRM (DOLT-2)	[146]
7	~59	90	Deionized water	Protease XIV	1390	Se-enriched yeast	[92]
2, 7 7	~71[b] 68[b]	87 92	Deionized water	Protease XIV+ lipase VII	1279 0.58	Se-enriched yeast wheat flour	[147]
7	98	98	Phosphate-buffer, pH = 7.4	Pronase E	82.9	Brazil nut LRM	[61]
7	53	70	Deionized water	Protease XIV	1.63	Cod muscle CRM	[64]
7	63[b]	NR	TRIS–HCl, pH = 7.5	Protease XIV	1374	Se-enriched yeast	[148]
7	80	81	Phosphate-buffer, pH = 7.4	Pronase E	1.0	Hulled sesame seeds	[149]

[a]See Table 19.A1 in the Appendix for the identified Se-species.
[b]Related to the total Se content.
NR, not reported.

that the consumption of Se-enriched yeast significantly reduced overall cancer mortality and the occurrence of some specific types of cancer (prostate, colorectal, and lung tumors). That is why studies on Se-enriched yeast play a key role in the development of speciation analysis as well as in the elucidation of the anticarcinogenic effects assumed by Clark et al. [88].

Another important feature is summarized in the first column of Table 19.1. About two-thirds of the Se species identified and listed there have been available only to the laboratories that can synthesize them, as several Se species are still commercially unavailable. This general lack of Se species standards can be one of the most important reasons behind the sometimes low identification efficiency referred to in Table 19.1. On the other hand, if the speciation analysis and species identification were carried out with a single chromatographic separation technique, the risk of misidentification because of co-elution or inadequate chromatographic resolution could not be excluded.

The optimal pH range for pronase E is 6–9, which is generally set with inorganic or organic buffers (see Table 19.1). At the same time, it is not rare to find experimental studies where deionized water is used as the medium only to avoid the use of buffers, mostly to render the sample preparation fit for the requirements of the instrumentation in terms of matrix complexity and load of inorganic ions. However, the acidity of an organic sample solution during proteolysis may drop by 2–3 pH values over some hours because of either the introduction of acidic by- and end-products of the hydrolysis or microbial activity – let alone the case when the original sample has a naturally acid pH (e.g., liver, some vegetables and fruits) – which is far from the optimal working pH range of the enzyme. This might be one of the causes of the sometimes low extraction efficiency of sample preparations with deionized water shown in Table 19.1.

After the introduction of pronase E, other more or less nonsubstrate-specific proteolytic enzymes have been applied to assist Se speciation. Most of them were derived from DNA/RNA clean-up protocols. The "new" enzymes (subtilisin from *Bacillus licheniformis*, also named protease VIII, EC 3.4.21.14; proteinase K from *Tritirachium album*, EC 3.4.21.64; the crude Novo Nordisk product of Flavourzyme from *Aspergillus oryzae*) proved to be capable of extracting Se with varying yields and chromatographic recovery of Se species. It is important to highlight that the latter parameter also depends on the instrumentation available. In this regard, different recovery values for the same samples reported by independent research groups do not necessarily indicate successful or unsuccessful sample preparation. Similarly, extraction efficiency (defined as the ratio of extracted Se to total Se in the sample) cannot be used as such for comparison purposes because sample preparation may include some extra steps, for example, TCA precipitation or ultrafiltration, which may reduce this value even by 10–20 percent.

The enzymes mentioned above have been compared several times in an attempt to elaborate a generally applicable sample preparation protocol. Moreno et al. [89] compared pronase E and subtilisin in the case of oyster and mussel samples. They resorted to phosphate buffer during proteolysis, which resulted in a precipitation problem with the Ca salt in the Sigma-Aldrich pronase. The study was therefore continued only with subtilisin. Without the possibility of switching to another buffer or another enzyme producer, no definitive conclusions could be reached. Gómez-Ariza et al. [13] treated spiked shellfish samples with pronase E and protease VIII. According to their results, neither of the two enzymes was applicable to recover the four spiked Se species [$SeCys_2$, SeMet, Se(IV), and Se(VI)] from the samples because of the applied clean-up procedure and the incompatibility of the reagent additives. Montes-Bayón et al. [75] compared proteinase K and pronase E on Indian mustard samples, achieving higher Se extraction efficiency and chromatographic recovery with proteinase K. However, the investigation was carried out without buffering the sample solution, thus probably working outside the optimal pH range of the enzymes.

With no claims of potential future importance, a new approach called enzymatic probe sonication should be mentioned here. Introduced by Capelo et al. [90], it practically seems to bypass all relevant drawbacks or difficulties of the one-step proteolysis, for example, the relatively long extraction time required for the complete hydrolysis of proteins and the possible need for optimal pH setting during proteolysis. This goal is achieved through ultrasonic assisted proteolysis carried out with protease XIV, which requires not more than 30 s for the complete extraction of SeMet from Se-enriched yeast samples. Clearly, this outstanding enhancement of efficiency and significant drop in the length of time usually demanded for hydrolysis are due to the synergic combination of special circumstances, that is (1) small particle-sized samples (similar to CRMs) with good homogeneity were considered, thus allowing small amounts (in the order of magnitude of 10–50 mg) to be subsampled for the analysis; (2) the enhancement of enzymatic activity induced by sonication during the very short sample treatment was not negatively affected by the possible decrease in enzymatic activity due to enzyme unfolding or denaturing also caused by sonication; (3) the formerly mentioned high excess (ratio of 1 : 10) of the enzyme over the amount of the sample generally used in Se speciation was not altered. Nevertheless, if successfully adopted for the treatment of less ideal samples, this method may become the dominant sample preparation technique in Se speciation.

19.3.5.5 The Use of Nonproteolytic Enzymes Prior to Se Speciation It goes without saying that food samples contain components other than proteins. This fact can be related to the observation that some sample preparation attempts based on the sole use of proteolytic enzymes have failed, for example, protease XIV could not extract more than 8 percent of Se from Se-enriched lactic acid bacteria with a chromatographic recovery of 27 percent [77]. There are two possible explanations for this phenomenon: (1) formerly unidentified and nonprotein-bound Se species might be present in the samples under test, which cannot be released by proteolytic enzymes from the matrix constituents under the experimental conditions adopted; (2) the unextracted Se species are protein-bound, but the proteolytic attack is hampered by a matrix constituent, hindering any enzymatic access.

Usually there are two main components deemed responsible for the decrease in extraction efficiency, that is the lipid content and the cell wall in the case of samples of plant or fungal origin that may remain partly intact during sample pretreatment. This fact prompted sample preparation including also cell wall degrading and/or lipolytic enzymes, as reported soon after the initiation of the second phase of enzymatic sample preparation techniques in the publication by Gilon et al. [79].

One of the first remarkable attempts in this field was made in the Se speciation study of Se-enriched yeast with a yeast-derived lipolytic enzyme (lipase VII from *Candida rugosa*, EC 3.1.1.3) as a simultaneously assisting reagent to proteolysis by pronase E [21]. Degradation of lipids might become necessary to decrease the high-performance liquid chromatography (HPLC) column load due to the fat matrix. As no data were provided on the possible independent use of the two active enzymes, pronase E and lipase, it cannot be concluded whether sample preparation really benefited from this approach. On the other hand, the simultaneous one-step application of the enzymes in question may result in the decomposition of the lipolytic one, as it can be a substrate for the proteolytic enzyme and is found in intact, that is, easily accessible, form, unlike the sample proteins that could be preliminarly coagulated/precipitated during sample pretreatment. However, this simultaneous technique was not set aside, as it was also applied in other investigations, for example, by Cases et al. [10] for the Se speciation of Se-enriched spirulina and by Gómez-Ariza et al. [13] for oyster and shellfish samples.

The study by Casiot et al. [21] introduced the use of a cell-wall-degrading enzyme as well, which led to more meaningful results as compared with lipase. Driselase, the higher fungi (*Basidiomycetes* ssp.) derived crude enzyme mixture containing, laminarinase, xylanase, pectinase, amylase, protease, and cellulose activity, was chosen to release the potentially cell-wall-bound Se species or molecules. The novel aspects of this work is due to a protease inhibiting agent, phenylmethanesulfonyl fluoride (PMSF), which blocks proteolysis caused probably by either the yeast enzymes or by Driselase, the product originally applied for protoplast formation from fungi and plant cells. The use of Driselase significantly increased Se-extraction efficiency as compared with the water or buffered extraction options, thus indirectly confirming the presence of Se in non-protein sample constituents that, unfortunately, could not be identified.

The approach of Casiot et al. [21] was soon accepted and followed in the field of Se speciation. Wróbel et al. [91] applied a bacterium (*Arthrobacter luteus*) derived lysing enzyme mixture added with PMSF to study the intermediary molecules of Se metabolism of Se-enriched yeast without proteolysis. In order to tailor the cell wall degrading mechanism to the samples under test, Michalke et al. [77] used bacterial lisozyme and pronase E, either alone or in combination, for the Se speciation of Se-enriched lactic acid bacteria. Independent and simultaneous experiments were carried out with the two enzymes, thus achieving outstanding total Se-extraction efficiency (85–105 percent) with the sole application of pronase E and relatively low chromatographic recovery (8–12 percent) (still the highest among all the investigated changes in sample preparation methods) only with the use of lisozyme. The simultaneous use of the two enzymes was not accompanied by unambiguous advantages. These results can be traced back to the inadequate number of commercially available Se species, the possibly incomplete hydrolysis, and presumably no use of invasive sample pretreatment steps (drying, grinding).

The self-digestion problem arising during the simultaneous use of a proteolytic and a nonproteolytic enzyme can be eliminated in two different ways. Larsen et al. [73, 92] set up a sequential enzymatic sample preparation technique for the Se speciation of Se-enriched algae and yeast samples to achieve the highest extraction efficiency. The enzymatic treatment in their study was performed by adding first the cell-wall-degrading enzyme, a Novo-type β-glucosidase based Novozym 234, to the sample solution. Three hours later the addition of a wide-spectrum proteolytic crude product by Novo, Flavourzyme 1000L, followed. One of the most important results of their studies was the detection of SeOMet, which was proved to be a conversion product of SeMet and to arise from the use of the cell-wall-degrading enzyme. This fact, along with the results of Dumont et al. (see Section 19.3.5.3) [83], calls for a more detailed investigation of this selenoamino acid derivative, as it may possess different bioavailability compared with SeMet, similarly to the methionine–methionine sulfoxide couple [93, 94].

Dernovics et al. [95] succeeded in achieving high Se recovery from the selenized edible mushroom *Agaricus bisporus*, and aimed at the separation of Se released through cell wall degradation. To this end, each aliquot of the sequential enzymatic sample preparation was subjected to centrifugation to recover the Se extracted at the given step. Two cell wall degrading crude enzymes were used, i.e., Driselase and the lysing enzyme derived from *Trichoderma harzianum*, a wood decaying mould that is also often used for the preparation of protoplasts due to its outstanding lysing activity. However, there was no significant difference in the extraction efficiency as achieved by the two cell wall-degrading enzymes. The lysing enzyme-assisted proteolysis liberated more Se species than the Driselase-based sample preparation did.

As discussed in Section 19.3.5.3, mimicking human digestion for the assessment of Se-bioavailability could be practically regarded as early examples of sequential enzymatic sample preparation applying, for example, proteolytic (pepsin), mixed (pancreatin), and non-proteolytic (amylase) enzymes, and digestive compounds like bile salts for lipid emulsification. On the other hand, the simultaneous use of enzymes was not considered a source of self-digestion problems as this phenomenon is common in the digestive tract.

19.3.6 Doubts on the Applicability of Enzymatic Sample Preparation: Seeking Alternative Methods

In most cases, as detailed in Table 19.1, the reported extraction efficiency and chromatographic recovery values for Se speciation analysis carried out on the usual samples (*Allium* ssp., Se-enriched yeast, and food supplements) have been around 80–100 percent. However, there has been more and more evidence that for certain samples the unsuccessful speciation results cannot be ascribed to the lack of fit-for-purpose CRMs. Rather, incomplete enzymatic hydrolysis was responsible for the inadequate chromatographic recovery. Montes-Bajón et al. [96] needed an extra enzymatic step with aminopeptidase-M to complete the one-step method based on proteinase K in order to achieve an acceptable hydrolysis rate of an Se-enriched yeast food supplement. Moreno et al. [89] had to apply a double enzymatic treatment with subtilisin on oyster samples in order to arrive at good extraction efficiency and chromatographic recovery, although a relatively high buffer capacity was provided (0.1 mol l^{-1} TRIS–HCl, pH = 7.5) in the analysis of cell wall free, albeit fatty, samples. Vonderheide et al. [62] could only achieve 11 percent chromatographic recovery at 83 percent extraction efficiency when analyzing defatted Brazil nuts after proteinase K treatment.

The authors of the studies cited above found that the enzymes used during sample preparation could not completely hydrolyze sample proteins into amino acids. This is partly understandable as both proteinase K and subtilisin are known to show hydrolyzing preference for specific residues of proteins. Accordingly, they do not necessarily arrive at total hydrolysis. At the same time, neither pronase E nor protease XIV could always provide 100 percent extraction of Se from the samples (see Table 19.1). Therefore, alternative methods should be developed in the field of sample preparation to replace enzymatic methods.

Finally, Wróbel et al. [97] adapted the application of 4 mol l^{-1} methanesulfonic acid to hydrolyze sample proteins for the Se speciation of defatted Brazil nuts and Se-enriched yeast samples. Their results show that an 8-hour long, methanesulfonic acid reflux digestion significantly increases the amount of extracted SeMet as compared with the parallel sequential enzymatic methods based on proteinase K and protease XIV. They concluded that the method could not assure the integrity of the original speciation information because of two factors, that is, the highly reductive reaction medium containing β-mercaptoethanol to prevent oxidation and the increase in temperature, and about 2 percent of SeMet was found to be damaged during hydrolysis. It seems that the reductive medium may be mainly responsible for the damage reported. Yang et al. [98] concluded that there was no analyte degradation during a 16 h methanesulfonic acid reflux digestion of Se-enriched yeast without β-mercaptoethanol. Accordingly, this new and promising method may provide a real alternative to the widely used enzymatic approach.

19.4 SAMPLE PREPARATION METHODS PRIOR TO As SPECIATION ANALYSIS

19.4.1 Background Information

Basically, As speciation regards nowadays four main groups of samples, namely: water samples (drinking, irrigation, and wastewater); food samples (mostly seafood and vegetables); soils, sediments, and sludge; and biological fluids to monitor human As metabolism (urine, blood). This section focuses only on food-related samples. Sample preparation methods for the analysis of the first two groups of material is examined in detail hereafter.

Water samples in As speciation have a place of their own in terms of sample preparation, as they usually feature the two inorganic As species, As(III) and As(V), the major chemical properties of which (e.g., redox stability) have been thoroughly studied in the recent past. The analyst is usually well aware of the importance of preserving the original ratio of inorganic As species. Hence, problems with low chromatographic recovery are rarely faced.

Food samples, especially seafood (fish, shellfish, and seaweed), are particularly interesting from the viewpoint of As speciation as they contain most of the newly identified As species. On the other hand, the apparent contradiction between the relatively high total As content of seafood and the absence of the expected adverse effects of As intoxication was brilliantly solved in the 1970s [99]. Nowadays two general approaches of sample preparation should be mentioned in this field: (i) extraction and quantification of the toxic As species only, for example, by an indirect determination of inorganic As species through the hydride-forming capacity, (*ii*) achievement of the highest extraction to fully map the As species distribution in a given material. Needless to say, the two approaches substantially overlap in the case of some samples and methods, as testified also by recent publications.

19.4.2 Samples Requiring no Sample Preparation: Liquid (Water) Samples in As Speciation

Unlike solid samples, liquid samples require no sample preparation whenever there is no need to bring the analyte into a measurable form, for example, by solubilization. At the same time there are a few inescapable pretreatment steps that should be carried out in the case of liquid samples, such as filtration through 0.45 μm membranes to clean (and theoretically to sterilize) the sample in order to prevent blockage of the chromatographic system by solid particles, cells, or precipitates. Although filtration greatly contributes to the stability of the sample, storage might be also a key step when the final analysis does not take place immediately.

The stability and the redox balance of As(III) and As(V), which account for almost all of the As content of water, are basically affected by temperature, concentration of each species, pH, and the type of liquid matrix. As a rule, calibrants for these two As species dissolved in deionized water at a concentration higher than 20 $\mu g\ ml^{-1}$ as well as filtered ground, potable, and waste water samples can be stored for several weeks without significant alteration in their redox state if kept in a refrigerator, that is, protected from sunlight and at a temperature of 0–4°C (no freezing) [100, 101]. The use of preservatives should be avoided as no conclusive evidence has been provided of their positive effect on the stability of As species; moreover, antioxidants like ascorbic acid may cause either the reduction of As(V) (thus modifying the original redox balance) or chromatographic difficulties [102–104].

If water samples containing a high concentration of Fe are to be speciated for As, the samples should be preliminarily acidified to prevent coprecipitation of Fe oxo-hydroxides and As. The pH should be set at 1–2; however, further acidification should be avoided because of the sensitivity of the As(III)–As(V) redox balance to lower pH values [105, 106].

19.4.3 The Role of Organic Solvents and the Mixture of Methanol–Water in the Preparation of Solid Samples Prior to As Speciation

Arsenic speciation strongly depends on the ratio of methanol to water used in the extraction phase. Since the introduction of one of the first methanolic extraction protocols for As speciation [107], almost all organic samples for As speciation were prepared using this extraction medium with the aid of sonication or shaking. This method is easy to use because methanol can be quickly removed after extraction, for example, by evaporation or speed-vacuum instruments. On the other hand, As metabolism is dependent on proteins at each step of the As detoxification process. Arsenic is transported, methylated, and excreted via combination with proteins [108]. Any kind of sample preparation irreversibly changing the protein structure may result in the release of As at each detoxification stage.

On the basis of the experience accrued so far, the extraction efficiency can be affected by both the nature of the sample under test and the As species actually present in the sample. The lipid content of the sample and the presence of As species with hydrophobic residues play a key role in this context. As a consequence, it may be useful to remove or partition lipids with an organic solvent prior to the methanol–water extraction in order to increase the extraction efficiency. This approach is also viable to prevent emulsification of methanol and lipids, which could otherwise significantly reduce the extraction efficiency [26]. Thus, sample preparation prior to As speciation usually starts with the chloroform–methanol extraction, followed by centrifugation of the sample solution to remove solid particles. Subsequently, after the addition of methanol and water, the supernatant is separated in a funnel. The amount of As remaining in the organic phase is usually quantified as total As by wet digestion after solvent evaporation. Obviously, this entails the loss of all speciation information related to the lipid-soluble As species [2]. Speciation analysis is finally carried out on the methanol–water phase, usually after solvent evaporation (see Table 19.2).

When the sample does not contain any considerable amount of lipids, no organic solvents are required and sample preparation can simply consist of repeated extractions with the methanol–water mixture with a centrifugation step after each run. The supernatants are then pooled and treated as described above. The number of repeated extractions usually does not exceed five, although, under circumstances, one single extraction may suffice [2].

TABLE 19.2. Examples of As Speciation Studies Based on the Application of Methanolic Extraction as the Main Sample Preparation Method

Identified As Species[a]	Hydrophobic As Species (% of Total As Content)	Chromatographic Recovery (% of As Extracted)	Extraction Efficiency (% of Total As)	Applied (Occasionally Optimal) Ratio of Methanol–Water Mixture	Extraction Media[c]	Total As Content, $\mu g\ g^{-1}$	Samples Addressed	References
AsBet, AsC, As(III), As(V), MMA(V), DMA(V), TMAO, TETRA	1–18	NR	46–99	Step 1, 2 : 5 : 0 Step 2, 5 : 0 : 7	Chloroform–methanol–water	3.2–118	Shrimp, crab, fish, fish liver, shellfish, lobster digestive gland	[151]
AsBet, DMA(V), As(V)	0.08–2.2	NR	37–87	Step 1, 1 : 2 : 0 Step 2, 1 : 0 : 1	Chloroform–methanol–water	3.5–196.1	7 kinds of fish	[24]
AsBet, DMA(V)	NR	66–150	59.6–144	1 : 1	Methanol–water	3.8–96	7 kinds of fish snail, shrimp, clam	[27]
AsBet, AsC, DMA(V), As(III), As(V), TETRA	NR	?–147	85–100	9 : 1	Methanol–water	10.9–30.0	Edible mushroom	[18]
As(III), As(V), MMA(V), DMA(V)	NR	~100[d]	~100[d]	Exceeding the ratio of 2 : 3	Methanol–water	2480–14,900	Green algae LRM	[28]
AsBet, TETRA	NR	93.2–99.0	99.4	Step 1, 2 : 5 : 0 Step 2, 5 : 0 : 7	Chloroform–methanol–water	42.2–43.2	Shrimp and plaice CRM	[141]
DMA(V), MMA(V), As(V), AsBet, TMAO	NR	~100[d]	68–85	1 : 9	Methanol–water	23–1420	Edible mushroom	[121]
As(III), As(V), MMA(V)	NR	~100[d]	83–98	Step 1, acetone (for defatting) Step 2, 1 : 9	Acetone; methanol–water	22.4–206	As-enriched yeast	[100]
As(III), As(V)	NR	~100[d]	46–69	1 : 9	Methanol–water	0.11–1.85	Carrots	[115]

(*Continued*)

TABLE 19.2. *Continued*

Identified As Species[a]	Hydrophobic As Species (% of Total As Content)	Chromatographic Recovery (% of As Extracted)	Extraction Efficiency (% of Total As)	Applied (Occasionally Optimal) Ratio of Methanol–Water Mixture	Extraction Media[c]	Total As Content, $\mu g\ g^{-1}$	Samples Addressed	References
AsBet, DMA(V)	NR	~100[d]	88–99	1 : 1	Methanol–water	4.8	Tuna CRM	[113]
AsBet + AsC, As(V), MMA(V), DMA(V)	0.4–5.1	~98	84.9–89.1	Step 1, acetone (for defatting) Step 2, 1 : 1 (m/m)	Acetone; methanol–water	2.99–60.6	4 kinds of fish and CRM DORM-2	[118]
AsBet	NR	~100 for AsBet	46–100	Methanol–water	4 : 1	2.6–19.4	4 kinds of fish and CRM DORM-2	[122]
As(III), As(V), DMA(V)	NR	~100[d]	30–100	Methanol–water	1 : 9	0.093–9.90	3 kinds of plant material and CRM	[120]
AsBet, AsBet-2, AsC, TETRA, DMA(V), Arseno-sugars B, D	NR	71	53	Methanol–water	1 : 1	15	Oyster CRM	[114]
As(III), As(V), DMA(V), Arseno-sugars A, B, C, D	NR	~100[d]	25.6–77.4	Methanol–water	3 : 7	0.045–109.4	3 kinds of seaweed	[119]
DMA(V), As(III) + As(V)	NR	~100[d]	31–76	Methanol–water	1 : 1	0.11–0.34	7 kinds of rice and grain CRM	[126]

As(III), As(V), MMA(V), DMA(V)	NR	~100[d]	60.5–97.1	Methanol–water	1 : 1	0.0082–0.0809	3 kinds of apple	[127]
As(III), As(V)	NR	~100[d]	60–100	Methanol–water	1 : 1	168–3894	As accumu- lating fern	[152]
AsBet, AsBet-2, AsC, As(V), MMA(V), DMA(V), TETRA, TMAO	0.21–10.5	81–100	58–103	Acetone; methanol–water	Step 1, acetone (for de-fatting) Step 2, 1 : 1	3.3–23.2	19 kinds of fish, shrimp, oyster and clam CRM	[123]
DMA(V), DMAA, AsBet, TMAO, TMAP, AsC, TMAs	NR	99[b] 85[b]	NR	Methanol–water	1 : 1	18.0 4.8	CRM DORM-2 CRM 627	[116]
As(III), DMA(V), TMAO, MMA(V), As(V), Arsenosugars A, B As(III), DMA(V), MMA(V), As(V), AsBet, TMAP, TETRA	NR	2.3–29[b] 97–120[b]	50–70 NR	Methanol	100 % methanol	0.062–0.363 2.5–4.72	Freshwater carp freshwater catfish	[112]
AsBet AsBet, DMA(V) AsBet, DMA(V), TMAO AsBet, As(III), DMA(V)	NR	~115 ~85 ~42–88 ~98–133	0.4 77.2 2.4–36.2 53.5–100	Methanol–water	9 : 1	0.806 0.097 0.088–0.247 0.80–1.235	Freshwater catfish freshwater turbot freshwater carp fresh-water trout	[34]

[a]See Table 19.A2 in Appendix for the identified As- species.
[b]Related to the total As content.
[c]Without clean-up procedures.
[d]Presumed value.
NR, not reported.

The methanol-to-water ratio in the extraction mixture is usually determined on the basis of the water content of the sample. Methanol alone is generally used if fresh, untreated samples are dealt with, because the natural water content (reaching even 90 percent m/m) in practice adds to methanol [109, 110]; however, in some studies, nonaqueous methanol has used been also for freeze-dried samples [26, 111, 112]. The methanol-to-water ratio mostly chosen is around 1 : 1 v/v (40 : 60, 50 : 50, 60 : 40). However, also the 10 : 90 ratio is also used [100], not to mention cases where the full mixture scale from 0 to 100 percent methanol has been exploited [113, 114]. The study by Helgesen and Larsen [115], where a 10 percent (v/v) methanol mixture was used for the extraction of As species from carrots because of its antibacterial effect, is noteworthy.

Table 19.2 epitomizes the extraction efficiencies achieved in the As speciation analysis of some food and food-related samples along with the ratio used for methanol–water mixtures. Only the studies focusing either on the highest possible extraction efficiency or the extraction of the major As-compound, for example, AB, are mentioned. Interestingly, not all authors report the mass balance of As. In some cases the evaluation of both extraction efficiency and chromatographic recovery was a complex task. Even data on total As content were sometimes ambiguously presented. When interpreting the data shown in the column headed Applied (occasionally optimal) ratio of methanol–water mixture, it should not be overlooked that the data relate to the extraction mixture added to the sample, whereas the original water content of the sample was either ignored or not explicitly reported. Therefore, these data are intended only for general information. The studies cited therein often addressed also other sample preparation techniques than the extraction with the methanol–water mixture. Nevertheless, the data shown in Table 19.2 are always related to the methanolic method.

The column headed chromatographic recovery (% of As extracted) calls for explanation. Speciation analytical publications do not always report chromatographic recovery values that would show the ratio of identified species to the total As injected onto the chromatographic column. Theoretically, these values might be useful for the characterization of the entire speciation analysis, from the sample preparation method applied to the final identification and quantification of the species. In practice, the assessment of these values requires a complex approach. The quantification of As species is affected by several factors such as the purity of the available As standards, their actual concentration in the sample, the reproducibility of the sample preparation, the accuracy and the precision of measurements, the peak shape observed during the chromatographic separation, and so on. In the case of water and higher plant samples containing mostly inorganic As species, only a chromatographic recovery of around 100 percent may provide correct results. In the case of other organic samples, for example, seaweed, seafood, or freshwater fish, a substantial part of the As species is still neither assigned a defined chemical structure nor commercially available, thus demanding in-house synthesis or direct isolation from well characterized samples. An example that well describes the situation is the group of presumably nontoxic arsenoylribosides (also named arsenosugars), consisting of about 10 species from which none is commercially available at the time of writing. Therefore, high (>80–90 percent) chromatographic recovery can be expected only for some well-characterized samples, such as white fish and some well-studied CRMs [116, 117].

As shown in Table 19.2, extraction with the mixture of methanol–water provides adequate extraction efficiency for most of the samples. However, this technique should be automated and the duration of sample preparation should be reduced. The usual methanolic extraction is carried out with the help of a shaking device and the total time of sample preparation including centrifugation and manual separation of the phases may be as long as 10–12 h. Recently, two new techniques have been introduced that can replace the traditional approach, that is, the accelerated solvent extraction (ASE), sometimes referred to as pressurized liquid extraction (PLE), and the MW assisted extraction (MAE).

ASE was introduced in 1999 [118]. All basic operational parameters (time, temperature, pressure, solvent removal) are computer controlled to provide high reproducibility. However, ASE has not displaced the traditional method, as the recommended high pressure (3–21 MPa) and temperature (150–200°C) easily achievable by ASE for As-extraction [119], did not significantly contribute to higher extraction efficiency [118]. Moreover, ASE achieved adequate extraction efficiency mostly when the temperature was over 100°C, this definitely leading to the variation of the As(III) : As(V) ratio in the sample [120]. It is noteworthy that, for some CRMs, ASE provided results comparable with the traditional extraction techniques as regards the major As species, AB and dimethylarsinic acid (DMA), although with some substantial method constraints [117].

In its turn, MAE, the other promising technique, is still waiting for the optimal use in As speciation. In the beginning typical teething problems, such as overheating, were reported [115, 121], which required rather time-consuming schemes comparable with the traditional technique. The development of fit-for-purpose instrumentation solved some of the early problems [122]; however, it turned out that the optimal working conditions of the equipment were strongly sample dependent and even response surface method (RSM) optimization could not arrive at a well-defined set of parameters to be used for samples slightly differing from those addressed during the optimization process [123]. In other words, even when the sample preparation method is optimized, there can still be a difference of 20–30 percent in the extraction efficiency that is related to the nature of the sample.

In conclusion, it can be stated that extraction with the methanol–water mixture is still considered the leading sample preparation approach in As speciation, except for lipid-soluble As species. These last were out of the scientific focus for a long time, but nowadays are being investigated again as highlighted by Francesconi and Kuehnelt [124].

19.4.4 The Role of Enzymatic Methods in Sample Preparation Prior to As Speciation

When comparing the enzymatic methods used prior to Se or As speciation, an unambiguous and essential difference comes out. In Se speciation, enzymatic methods are meant to act directly on the extraction of the Se species, for example, by hydrolyzing proteins that contain selenoamino acid residues. In As speciation, enzymes are used in an indirect manner to act on a main matrix constituent of the sample, thus making As species accessible, for example, for methanolic extraction. According to this criterion, there is no need for either non-specific or high-purity, high-activity enzymes and relatively inexpensive products can be selected. Since it is assumed that As species do not form polymers requiring digestion with enzymes, a possibly incomplete enzymatic hydrolysis does not pose the problems that are faced in Se speciation, that is, a false identification due to Se-containing peptides resulting from incomplete digestion.

The first enzymatic attempts for As speciation were reported in the mid 1990s. Branch et al. [24] studied seven fish species through several sample preparation methods, trypsin enzymolysis among others. Their results on the applicability of the enzyme did not show unequivocal advantages, as the enzymatic technique proved to be sample-dependent and the time required for the enzymatic process was definitely longer than for methanolic extraction. Lamble and Hill [125] adopted the tryptic method for the analysis of two CRMs and achieved extraction efficiency better than 87 percent for both samples. The difference between the applicability of trypsin in the two cases may have been caused by the characteristics of the samples, that is, CRMs are definitely easier to analyze than real-world samples. Pardo-Martínez et al. [31] carried out the As speciation analysis of baby foods containing hake, sole, and angler using enzymolysis (trypsin and pancreatin) for sample preparation. According to their findings, trypsin proved to be a useful tool for the complete extraction of the As species from the samples, while the use of pancreatin altered the As species investigated and its application was consequently rejected.

Besides proteins, starch is also considered a matrix-forming constituent that can be present at considerable amounts in the samples. Its degradation or elimination can be achieved by enzymolysis. Heitkemper et al. [126] compared four sample preparation methods, including one based on α-amylase, to extract As from different rice species. The enzymatic method applied as a pretreatment step prior to methanolic extraction turned out to be the second most effective way after trifluoroacetic acid (TFA) extraction, freeing significantly higher amounts of As than methanolic extraction itself. However, the time required by the α-amylase method was more than 12 h. Therefore, no further studies employed it. Conversely, the study by Caruso et al. [127] on lyophilized apple samples recorded the clear advantage of the use of α-amylase by reaching a very high extraction efficiency. The 6 h-long enzymatic method extracted 10–20 percent more As from the apple samples than the methanolic methods did. It is important to note that in both types of sample – rice and apple – starch was either originally or indirectly (i.e., after lyophilization) the major matrix component. Therefore, the increase in As extraction due to the degradation of starch was expectedly substantial. On the other hand, neither of the two studies presented the amylolytic method as an individual sample preparation step, since the role of enzymatic starch hydrolysis was only to assist the subsequent methanolic step by providing access to As species.

19.4.5 The Role of Concentrated Acid and Basic Solutions in Sample Preparation Prior to As Speciation

An As extraction efficiency lower than 100 percent in the analysis of seafood or other samples is not a rare finding (see Table 19.2) and several attempts have been made to clarify the reason for this fact. The first explanation is connected with the available analytical instrumentation. Typical sample preparation methods for water-soluble As species aim at generating analytes compatible with the widely used HPLC techniques, for example, with ion-exchange columns [2]. Therefore, the hardly known lipid soluble As species that may predominate are usually not extracted and analyzed. This is obviously rather serious as the relevant samples are sometimes foods consumed in large quantities, for example, freshwater fish and mussels in inland countries [34, 112, 128]. This scientific gap in As speciation has just been identified and is being worked out.

The second reason is species-specific. Arsenic (III) was discovered in 1997 to be selectively bound to free thiol groups of proteins [129], thus creating a covalent As–S bond. When the sample preparation is not aggressive enough to denature proteins, the As–S bonds may remain unaltered, thus blocking this As species in the extraction residue. As the amount of inorganic As species is usually up to only a few percent of total As in the sample [130], the lack of As(III) may go unnoticed in the mass balance of As because of the uncertainty caused by other steps of the analysis. To overcome this problem, Muñoz et al. [131, 132] presented a selective extraction technique for inorganic As species. Their method consists of the HCl-assisted hydrolysis of sample proteins, followed by either a separation step with the chloroform–methanol mixture [131] or the MW-assisted distillation of inorganic [As(III) + As(V)] As species [132]. The complete distribution of As species can be determined by completing one of these inorganic As-specific techniques with the conventional methanolic extraction that frees water-soluble As species, such as AB, from the sample.

Another acid reagent used for As speciation is TFA, a compound often applied for ion-pairing purposes in liquid chromatography. Heitkemper et al. [126] achieved 84–97 percent of As extraction from rice species with the help of a 2 mol l^{-1} TFA solution. Since the redox balance of As(III) and As(V) is affected by TFA through the reduction of As(V), the amount of these two species must be pooled and presented as total inorganic As in the mass balance of As, similarly to the methods presented by Muñoz et al. [131, 132].

From a practical point of view, the latter loss in speciation is less of a problem. In as much as the toxicity of As(III) and As(V) cannot be differentiated in terms of food safety reasons, their pooled concentrations can be used to assess the overall risk and the analyst can focus on the much less known As species to determine their structure and toxicity. This approach was followed by Parks et al. [133], who addressed the availability of As species from matrix components in difficult-to-extract seafood samples. TMAH was used, a highly basic compound formerly rejected for As speciation because of its incompatibility with chromatography [122], to create a three-step sequential extraction technique. TMAH was used in the first step (i.e., basic hydrolysis), whereas acetic acid was used in the second step to neutralize TMAH. Heating up to 80°C ended the treatment. This method proved to be especially useful for the extraction of arsenosugars, which were actually found at higher concentration than AB, the As species often thought or proved to be the dominant As compound in seafood. Since arsenosugars are often difficult to free from seafood samples [119], the report by Parks et al. [133] can considerably contribute to current knowledge in the field of As speciation.

19.5 CONCLUSIONS

Although decades of continuing development in the field of instrumentation and application of As and Se speciation have led to landmarks as regards sample preparation methods, no universal approach can fulfil all the requirements of today's speciation analytical challenges. On the other hand, a few key steps have been consolidated in the optimization of sample preparation for materials that have never been analyzed before, namely, methanolic extraction for As speciation and enzymatic proteolysis for Se speciation.

At the same time, new challenges have arisen in the two speciation fields. In the case of As speciation, the qualitative and quantitative analysis of water-soluble As species or, in other terms, of As species that can be extracted by methanol, poses almost no problems: it is worth stressing that the (incidentally increasing) availability of calibrants for the determination of As species is no longer a major difficulty, if one considers the analytical potential of current ICP-MS instrumentation [116]. The new challenges call for novelty in sample preparation and analytical approach as well. On the one hand, there has been an increasing need for the analysis of As species that cannot be made accessible or may not be studied by methanolic extraction, for example, lipid-soluble As species in fish and mussels [34, 112, 124, 128]. On the other hand, further investigations should be promoted as regards the human digestion of As-containing food products with the help of simulated *in vitro* digestion [134]. The sample preparation technique addressed in the latter case is well known in the field of Se speciation, whereas it is still in its infancy in the case of As speciation.

The current analytical targets of Se speciation have been slightly changed as well, arriving at a narrower or more specific area for enzymolysis. The clear-cut, trend-setting proteomics that adopts well-established methods for the analysis of whole and intact proteins either excludes the application of nonspecific proteolytic enzymes or utilizes them for special purposes, for example, for additional characterization of separated proteins [50, 58]. The investigation of water-soluble and free (i.e., not protein-bound) Se species with antitumoral properties may also lead proteolysis being abandoned, as evidenced by the production of Se-enriched onions as a dietary source of Se-methylselenocysteine [135], in an attempt to focus on Se species that are more valuable from a physiological point of view than protein-bound selenoamino acids.

Hopefully, in the near future both areas will be endowed with new CRMs for speciation analysis, the lack of which has been a substantial drawback over the past years. A ground-breaking CRM in the field of Se speciation, a CRM certified for SeMet, has been recently introduced [136], while the first CRM intended for As speciation was released in 1999 [113]. Moreover, an EU-funded project named SEAS has further prompted the development of speciation analytical techniques to produce and use CRMs for As and Se speciation [137]. It is understandable that the bottleneck for any further progress is financial as the scientific community involved in speciation has already confirmed that there is a keen need for such QC materials [5, 138–140]. For the time being, either the production of working reference materials for speciation analysis [28, 61, 141] or the speciation-oriented recertification of already available CRMs [117] may provide viable alternatives.

In conclusion, whatever the origin of the sample for As or Se speciation and the target of the analysis are, one thing is for sure: the whole process must be designed so as to preserve the integrity of the speciation information. To achieve this goal, the analyst should first select a fit-for-purpose sample preparation method, simple or complex as it may be. Hopefully, what has been presented in this chapter can be of assistance to this end.

REFERENCES

1. C. B'Hymer J. A. Caruso, Evaluation of yeast-based selenium food supplements using high-performance liquid chromatography and inductively coupled plasma mass spectrometry, *J. Anal. Atom. Spectrom.*, **15** (2000), 1531–1539.
2. K. A. Francesconi, Working methods. Complete extraction of arsenic species: a worthwhile goal?, *Appl. Organomet. Chem.*, **17** (2003), 682–683.
3. R. S. Braman, C. C. Foreback, Methylated forms of arsenic in the environment, *Science*, **182** (1973), 1247–1249.
4. J. L. Martin, L. M. Cummins, Separate elution of selenocystine and selenomethionine by ion-exchange technique, *Anal. Biochem.*, **15** (1966), 530–532.
5. E. H. Larsen, Method optimization and quality assurance in speciation analysis using high performance liquid chromatography with detection by inductively coupled plasma mass spectrometry, *Spectrochim. Acta*, **53B** (1998), 253–265.
6. L. Abrankó, Z. Stefánka, P. Fodor, Possibilities and limits of the simultaneous determination of As, Bi, Ge, Sb, Se, and Sn by flow injection–hydride generation–inductively coupled plasma–time-of-flight mass spectrometry (FI–HG–ICP–TOFMS), *Anal. Chim. Acta*, **493** (2003), 13–21.
7. A. D'Ulivo, Critical review. Determination of selenium and tellurium in environmental samples, *Analyst*, **122** (1997), 117R–144R.
8. W. Maher, M. Deaker, D. Jolley, F. Krikowa, B. Roberts, Selenium occurrence, distribution and speciation in the cockle *Anadara trapezia* and the mullet *Mugil cephalus*, *Appl. Organomet. Chem.*, **11** (1997), 313–326.
9. B. Michalke, H. Witte, P. Schramel, Developments of a rugged method for selenium speciation, *J. Anal. Atom. Spectrom.*, **16** (2001), 593–597.

10. J. Cases, I. A. Wysocka, B. Caporiccio, N. Jouy, P. Besançon, J. Szpunar, J. Rouanet, Assessment of selenium bioavailability from high-selenium spirulina subfractions in selenium-deficient rats, *J. Agric. Food Chem.*, **50** (2002), 3867–3873.

11. J. C. Chang, W. H. Gutenmann, C. M. Reid, D. J. Lisk, Selenium content of Brazil nuts from two geographic locations in Brazil, *Chemosphere*, **30** (1995), 801–802.

12. Z. Wang, S. Xie, A. Peng, Distribution of Se in soybean samples with different Se concentration, *J. Agric. Food Chem.*, **44** (1996), 2754–2759.

13. J. L. Gómez-Ariza, M. A. Caro de la Torre, I. Giráldez, D. Sánchez-Rodas, A. Velasco, E. Morales, Pretreatment procedure for selenium speciation in shellfish using high-performance liquid chromatography–microwave-assisted digestion–hydride generation–atomic fluorescence spectrometry, *Appl. Organomet. Chem.*, **16** (2002), 265–270.

14. J. Zheng, M. Ohata, N. Furuta, W. Kosmus, Speciation of selenium compounds with ion-pair reversed-phase liquid chromatography using inductively coupled plasma mass spectrometry as element-specific detection. *J. Chromatogr. A*, **874** (2000), 55–64.

15. P. C. Uden, S. M. Bird, M. Kotrebai, P. Nolibos, J. F. Tyson, E. Block, E. Denoyer, Analytical selenoamino acid studies by chromatography with interfaced atomic mass spectrometry and atomic emission spectral detection, *Fresenius J. Anal. Chem.*, **362** (1998), 447–456.

16. J. Vetter, Mineral element content of edible and poisonous macrofungi, *Acta Aliment. Hung.*, **19** (1990), 27–40.

17. J. Vetter, Selenium content of some higher fungi. *Acta Aliment. Hung.*, **22** (1993), 383–387.

18. D. Kuehnelt, W. Goessler, K. J. Irgolić, Arsenic compounds in terrestrial organisms I: *Collybia maculata*, *Collybia butyracea* and *Amanita muscaria* from arsenic smelter sites in Austria, *Appl. Organomet. Chem.*, **11** (1997), 289–296.

19. L. Rácz, A. Bumbálová, M. Harangozó, J. Tölgyessy, O. Tomeček, Determination of cesium and selenium in cultivated mushrooms using radionuclide X-ray fluorescence technique, *J. Radioanal. Nucl Chem.*, **245** (2000), 611–614.

20. M. B. De La Calle-Guntiñas, C. Brunori, R. Scerbo, S. Chiavarini, P. Quevauviller, F. Adams, R. Morabito, Determination of selenomethionine in wheat samples: comparison of gas chromatography–microwave-induced plasma atomic emission spectrometry, gas chromatography–flame photometric detection and gas chromatography–mass spectrometry, *J. Anal. Atom. Spectrom.*, **12** (1997), 1041–1046.

21. C. Casiot, J. Szpunar, R. Łobiński, M. Potin-Gautier, Sample preparation and HPLC separation approaches to speciation analysis of selenium in yeast by ICP-MS, *J. Anal. Atom. Spectrom.*, **14** (1999), 645–650.

22. C. Martín, M. Gómez, M. A. Palacios, C. Cámara, Evaluation of different types of hydrolysis of proteins for selenium speciation in yeast. Poster presentation at the *II. International Conference on Trace Element Speciation in Biomedical, Nutritional and Environmental Sciences*, Munich, 7–10 May 2001.

23. H. Emteborg, G. Bordin, A. R. Rodriguez, Speciation of organic and inorganic selenium in a biological certified reference material based on microbore ion-exchange chromatography coupled to inductively coupled plasma atomic emission spectrometry via a direct injection nebulizer or coupled to electrothermal atomic absorption spectrometry, *Analyst*, **123** (1998), 245–253.

24. S. Branch, L. Ebdon, P. O'Neill, Determination of arsenic species in fish by directly coupled high-performance liquid chromatography–inductively coupled plasma mass spectrometry, *J. Anal. Atom. Spectrom.*, **9** (1994), 33–37.
25. H. Matusiewicz, R. E. Sturgeon, S. S. Berman Trace element analysis of biological material following pressure digestion with nitric acid–hydrogen peroxide and microwave heating, *J. Anal. Atom. Spectrom.*, **4** (1989), 323–327.
26. N. Ybáñez, D. Velez, W. Tejedor, R. Montoro, Optimization of the extraction, clean-up and determination of arsenobetaine in manufactured seafood products by coupling liquid chromatography with inductively coupled plasma atomic emission spectrometry, *J. Anal. Atom. Spectrom.*, **10**, (1995), 459–465.
27. A. El Moll, R. Heimburger, F. Lagarde, M. J. F. Leroy, E. Maier, Arsenic speciation in marine organisms: from the analytical methodology to the constitution of reference materials, *Fresenius J. Anal. Chem.*, **354** (1996), 550–556.
28. W. Goessler, J. Lintschinger, J. Száková, P. Mader, J. Kopecký, J. Doucha, K. J. Irgolić, *Chlorella* sp. and arsenic compounds: an attempt to prepare an algal reference material for arsenic compounds, *Appl. Organomet. Chem.*, **11** (1997), 57–66.
29. P. Bermejo-Barrera, S. Fernández-Nocelo, A. Moreda-Piñeiro, A. Bermejo-Barrera, Usefulness of enzymatic hydrolysis procedures based on the use of pronase E as sample pre-treatment for multi-element determination in biological materials, *J. Anal. Atom. Spectrom.*, **14** (1999), 1893–1900.
30. B. Maichin, P. Kettisch, G. Knapp, Investigation of microwave assisted drying of samples and evaporation of aqueous solutions in trace element analysis, *Fresenius J. Anal. Chem.*, **366** (2000) 26–29.
31. M. Pardo-Martínez, P. Viñas, A. Fisher, S. J. Hill, Comparison of enzymatic extraction procedures for use with directly coupled high performance liquid chromatography-inductively coupled plasma mass spectrometry for the speciation of arsenic in baby foods, *Anal. Chim. Acta*, **441** (2001), 29–36.
32. S. D'ilio, M. Alessandrelli, R. Cresti, G. Forte, S. Caroli, Arsenic content of various types of rice as determined by plasma-based techniques, *Microchem. J.*, **73** (2002), 195–201.
33. Z. Šlejkovec, J. T. van Elteren, U. D. Woroniecka, Underestimation of the total arsenic concentration by hydride generation techniques as a consequence of the incomplete mineralization of arsenobetaine in acid digestion procedures, *Anal. Chim. Acta*, **443** (2001), 277–282.
34. Z. Šlejkovec, Z. Bajc, D. Z. Doganoc, Arsenic speciation patterns in freshwater fish, *Talanta*, **62** (2004), 931–936.
35. M. Deaker, W. Maher, Determination of arsenic in arsenic compounds and marine biological tissues using low volume microwave digestion and electrothermal atomic absorption spectrometry, *J. Anal. Atom. Spectrom.*, **14** (1999), 1193–1207.
36. Y. Q. Zhang, J. N. Moore, Changes in selenium speciation in wetland sediments induced by laboratory testing, *Commun. Soil Sci. Plant.*, **28** (1997), 341–350.
37. S. Sharmasarkar, G. F. Vance, F. Cassel-Sharmasarkar, Analysis and speciation of selenium ions in mine environments, *Environ. Geol.*, **34** (1998), 31–38.
38. J. Lintschinger, N. Fuchs, J. Moser, D. Kuehnelt, W. Goessler, Selenium-enriched sprouts. A raw material for fortified cereal-based diets, *J. Agric. Food Chem.*, **48** (2000), 5362–5368.

39. S. McSheehy, W. Yang, F. Pannier, J. Szpunar, R. Łobiński, J. Auger, M. Potin-Gautier, Speciation analysis of selenium in garlic by two-dimensional high-performance liquid chromatography with parallel inductively coupled plasma mass spectrometric and electrospray tandem mass spectrometric detection, *Anal. Chim. Acta*, **421** (2000), 147–153.
40. C. Ip, M. Birringer, E. Block, M. Kotrebai, J. F. Tyson, P. C. Uden, D. J. Lisk, Chemical speciation influences comparative activity of selenium-enriched garlic and yeast in mammary cancer prevention, *J. Agric. Food Chem.*, **48** (2000) 2062–2070.
41. M. Kotrebai, M. Birringer, J. F. Tyson, E. Block, P. C. Uden, Selenium speciation in enriched and natural samples by HPLC-ICP-MS and HPLC-ESI-MS with perfluorinated carboxylic acid ion-pairing agents, *Analyst*, **125** (2000), 71–78.
42. P. D. Whanger, C. Ip, C. E. Polan, P. C. Uden, G. Welbaum, Tumorigenesis, metabolism, speciation, bioavailability, and tissue deposition of selenium in selenium-enriched ramps (*Allium tricoccum*), *J. Agric. Food Chem.*, **48** (2000), 5723–5730.
43. M. Kotrebai, M. Birringer, J. F. Tyson, E. Block, P. C. Uden. Identification of the principal selenium compounds in selenium-enriched natural sample extracts by ion-pair liquid chromatography with inductively coupled plasma- and electrospray ionization-mass spectrometric detection, *Anal. Commun.*, **36** (1999), 249–252.
44. S. McSheehy, P. Pohl, J. Szpunar, M. Potin-Gautier, R. Łobiński, Analysis for selenium speciation in selenized yeast extracts by two-dimensional liquid chromatography with ICP-MS and electrospray MS-MS detection. *J. Anal. Atom. Spectrom.*, **16** (2001), 68–73.
45. F. R. Abou-Shakra, M. Rayman, N. I. Ward, V. Hotton, G. Bastian, Enzymatic digestion for the determination of trace elements in blood serum by inductively coupled plasma mass spectrometry, *J. Anal. Atom. Spectrom.*, **12** (1997), 429–433.
46. J. T. van Elteren, U. D. Woroniecka, K. J. Kroon, Accumulation and distribution of selenium and cesium in the cultivated mushroom *Agaricus bisporus* – a radiotracer-aided study, *Chemosphere*, **36** (1998), 1787–1798.
47. G. Önning, Separation of soluble selenium compounds in different fish species, *Food Chem.*, **68** (2000), 133–139.
48. C. Daun, T. Lundh, G. Önning, B. Åkesson, Separation of soluble selenium compounds in muscle from seven animal species using size exclusion chromatography and inductively coupled plasma mass spectrometry, *J. Anal. Atom. Spectrom.*, **19** (2004), 129–134.
49. J. Ruiz Encinar, L. Ouerdane, W. Buchmann, J. Tortajada, R. Łobiński, J. Szpunar, Identification of water-soluble selenium-containing proteins in selenized yeast by size-exclusion-reversed-phase HPLC/ICPMS followed by MALDI-TOF and electrospray Q-TOF mass spectrometry, *Anal. Chem.*, **75** (2003), 3765–3774.
50. A. Połatajko, M. Śliwka-Kaszyńska, M. Dernovics, R. Ruzik, J. Ruiz Encinar, J. Szpunar, A systematic approach to selenium speciation in selenized yeast, *J. Anal. Atom. Spectrom.*, **19** (2004), 114–120.
51. D. Behne, C. Hammel, H. Pfeifer, D. Röthlein, H. Gessner, A. Kyriakopoulos, Speciation of selenium in the mammalian organism, *Analyst*, **123** (1998), 871–873.
52. H. Zeng, Genomic and proteomic techniques and their application in selenium research, *Curr. Pharmacogenom.*, **1** (2003), 59–65.

53. R. Łobiński, J. S. Edmonds, K. T. Suzuki, P. C. Uden, Species-selective determination of selenium compounds in biological materials (technical report), *Pure Appl. Chem.*, **72** (2000), 447–461.

54. C. C. Chéry, E. Dumont, R. Cornelis, L. Moens, Two-dimensional gel electrophoresis of selenized yeast and autoradiography of 75Se-containing proteins, *Fresenius. J. Anal. Chem.*, **371** (2001), 775–781.

55. H. Chassaigne, C. C. Chéry, G. Bordin, F. Vanhaecke, A. R. Rodriguez, 2-Dimensional gel electrophoresis technique for yeast selenium-containing proteins – sample preparation and MS approaches for processing 2-D gel protein spots, *J. Anal. Atom. Spectrom.*, **19** (2004), 85–95.

56. S. Mounicou, S. McSheehy, J. Szpunar, M. Potin-Gautier, R. Łobiński, Analysis of selenized yeast for selenium speciation by size-exclusion chromatography and capillary zone electrophoresis with inductively coupled plasma mass spectrometric detection (SEC-CZE-ICP-MS), *J. Anal. Atom. Spectrom.*, **17** (2002), 15–20.

57. C. Chen, J. Zhao, P. Zhang, Z. Chai, Speciation and subcellular location of Se–containing proteins in human liver studied by sodium dodecyl sulfate–polyacrylamide gel electrophoresis and hydride generation–atomic fluorescence spectrometric detection, *Anal. Bioanal. Chem.*, **372** (2002), 426–430.

58. V. Gergely, E. Kápolna, A. Süle, G. Hajós, M. Dernovics, P. Fodor, Preparative liquid isoelectric focusing (Rotofor IEF) based Se-speciation of Se-enriched *Agaricus bisporus*, *J. Anal. Atom. Spectrom.*, **19** (2004), 1485–1488.

59. G. Alsing Pedersen, E. H. Larsen, Speciation of four selenium compounds using high performance liquid chromatography with on-line detection by inductively coupled plasma mass spectrometry or flame atomic absorption spectrometry, *Fresenius J. Anal. Chem.*, **358** (1997), 591–598.

60. S. S. Kannamkumarath, K. Wróbel, K. Wróbel, A. Vonderheide, J. A. Caruso, HPLC-ICP-MS determination of selenium distribution and speciation in different types of nut, *Anal. Bioanal. Chem.*, **373** (2002), 454–460.

61. E. T. Bodó, Z. Stefánka, I. Ipolyi, C. Sörös, M. Dernovics, P. Fodor, Preparation, homogeneity and stability studies of a candidate LRM for Se speciation, *Anal. Bioanal. Chem.*, **377** (2003), 32–38.

62. A. Vonderheide, K. Wróbel, S. S. Kannamkumarath, C. B'Hymer, M. Montes-Bayón, C. Ponce de León, J. A. Caruso, Characterization of selenium species in Brazil nuts by HPLC-ICP-MS and ES-MS, *J. Agric. Food Chem.*, **50** (2002), 5722–5728.

63. X. C. Le, X. F. Li, V. Lai, M. Ma, S. Yalcin, J. Feldmann, Simultaneous speciation of selenium and arsenic using elevated temperature liquid chromatography separation with inductively coupled plasma mass spectrometry detection, *Spectrochim. Acta*, **53B** (1998), 899–909.

64. V. Díaz Huerta, M. L. Fernández Sánchez, A. Sanz-Medel, Quantitative selenium speciation in cod muscle by isotope dilution ICP-MS with a reaction cell: comparison of different reported extraction procedures, *J. Anal. Atom. Spectrom.*, **19** (2004), 644–648.

65. S. Cavalli, N. Cardellicchio, Direct determination of seleno-amino acids in biological tissues by anion-exchange separation and electrochemical detection, *J. Chromatogr. A*, **706** (1995), 429–436.

66. M. Calomme, J. Hu, K. Van den Branden, D. A. Vanden Berghe, Seleno-lactobacillus – an organic selenium source, *Biol. Trace Elem. Res.*, **47** (1995), 379–383.

67. T. W. M. Fan, A. N. Lane, D. Martens, R. M. Higashi, Synthesis and structure characterization of selenium metabolites, *Analyst*, **123** (1998), 875–884.

68. R. E. Huber, R. S. Criddle, Comparison of the chemical properties of selenocysteine and selenocystine with their sulfur analogs, *Arch. Biochem. Biophys.*, **122** (1967), 164–173.

69. L. N. Mackey, T. A. Beck, Quantitative high-performance liquid chromatographic determination of sulfur amino acids in protein hydrolysates, *J. Chromatogr.*, **240** (1982), 455–461.

70. J. Kroll, H. Rawel, R. Krock, Microwave digestion of proteins, *Z. Lebensm. Forsch.*, **207** (1998), 202–206.

71. M. A. Sochaski, A. J. Jenkins, T. J. Lyons, S. R. Thorpe, J. W. Baynes, Isotope dilution gas chromatography/mass spectrometry method for the determination of methionine sulfoxide in protein, *Anal. Chem.*, **73** (2001), 4662–4667.

72. G. Önning, I. A. Bergdahl, Fractionation of soluble selenium compounds from fish using size-exclusion chromatography with on-line detection by inductively coupled plasma mass spectrometry, *Analyst*, **124** (1999), 1435–1438.

73. E. H. Larsen, M. Hansen, T. Fan, M. Vahl, Speciation of selenoamino acids, selenonium ions and inorganic selenium by ion exchange HPLC with mass spectrometric detection and its application to yeast and algae, *J. Anal. Atom. Spectrom.*, **16** (2001), 1403–1408.

74. M. Calomme, K. Van den Branden, D. A. Vanden Berghe, Selenium and Lactobacillus species, *J. Appl. Bacteriol.*, **79** (1995), 331–340.

75. M. Montes-Bayón, E. G. Yanes, C. Ponce de León, K. Jayasimhulu, A. Stalcup, J. Shann, J. A. Caruso, Initial studies of selenium speciation in *Brassica juncea* by LC with ICP-MS and ES-MS detection: an approach for phytoremediation studies, *Anal. Chem.*, **74** (2002), 107–113.

76. K. Yasumoto, T. Suzuki, M. Yoshida, Identification of selenomethionine in soybean protein, *J. Agric. Food Chem.*, **36** (1988), 463–467.

77. B. Michalke, H. Witte, P. Schramel, Effect of different extraction procedures on the yield and pattern of Se-species in bacterial samples, *Anal. Bioanal. Chem.*, **372** (2002), 444–447.

78. K. Yasumoto, K. Iwami, M. Yoshida, Nutritional efficiency and chemical form of selenium, an essential trace element, contained in soybean protein, *Diazu Tanpakushitu Eiyo Kenkyuukai Kaishi (Nutr. Sci. Soy Protein Jpn.)*, **4** (1983), 35–40.

79. N. Gilon, A. Astruc, M. Astruc, M. Potin-Gautier, Selenoamino acid speciation using HPLC-ET-AAS following an enzymic hydrolysis of selenoprotein. *Appl. Organomet. Chem.*, **9** (1995), 623–628.

80. S. Piepponen, M. J. Pellinen, T. Hattula, The Selenium of Mushrooms, in B. Peter, S. Peter (eds), *Trace Element Analytical Chemistry in Medicine and Biology*, Vol. 3, Walter de Gruyter, Berlin **3** (1984), pp. 159–166.

81. H. M. Crews, P. A. Clarke, D. J. Lewis, L. M. Owen, P. R. Strutt, A. Izquierdo, Investigation of selenium speciation in in vitro gastrointestinal extracts of cooked cod by high-performance liquid chromatography – inductively coupled plasma mass spectrometry and electrospray mass spectrometry, *J. Anal. Atom. Spectrom.*, **11** (1996), 1177–1182.

82. C. Thomas, I. Buth, H. Waldner, Bioavailability investigations of selenium from foods and supplements after enzymatic digestion using HPLC-ICP-MS with hexapole collision cell. Poster presentation at the *II. International Conference on Trace*

Element Speciation in Biomedical, Nutritional and Environmental Sciences, München, Germany, 7-10 May 2001.

83. E. Dumont, F. Vanhaecke, R. Cornelis, Hyphenated techniques for speciation of Se in *in vitro* gastrointestinal digests of *Saccharomyces cerevisiae*, *Anal. Bioanal. Chem.*, **379** (2004), 504–511.
84. W. F. Ganong, *Review of Medical Physiology*, Appleton and Lange, München, 1987 (Hungarian edition: *Medicina*, Budapest, 1990, p. 488).
85. G. Davey, B. C. Richardson, Purification and some properties of diplococcin from *Streptococcus cremoris* 346, *Appl. Environ. Microb.*, **41** (1981), 84–89.
86. I. A. Yamskov, T. V. Tichonova, V. A. Davankov, Pronase-catalyzed hydrolysis of amino acid amides, *Enzyme Microb. Tech.*, **8** (1986), 241–244.
87. L. Yang, R. E. Sturgeon, S. McSheehy, Z. Mester, Comparison of extraction methods for quantitation of methionine and selenomethionine in yeast by species specific isotope dilution gas chromatography-mass spectrometry, *J. Chromatogr. A*, **1055** (2004), 177–184.
88. L. C. Clark, G. F. Combs, B. W. Turnbull, E. H. Slate, D. K. Chalker, J. Chow, L. S. Davis, R. A. Glover, G. F. Graham, E. G. Gross, A. Krongrad, J. L. Lesher, H. K. Park, B. B. Sanders, C. L. Smith, J. R. Taylor, Effects of selenium supplementation for cancer prevention in patients with carcinoma of the skin. A randomized controlled trial, *J. Am. Med. Assoc.*, **276** (1996), 1957–1963.
89. P. Moreno, M. A. Quijano, A. M. Gutiérrez, M. C. Pérez-Conde, C. Cámara, Fractionation studies of selenium compounds from oysters, and their determination by high-performance liquid chromatography coupled to inductively coupled plasma mass spectrometry, *J. Anal. Atom. Spectrom.*, **16** (2001), 1044–1050.
90. I. L. Capelo, P. Ximénez-Embún, Y. Madrid-Albarrán, C. Cámara, Enzymatic probe sonication: enhancement of protease-catalyzed hydrolysis of selenium bound to proteins in yeast, *Anal. Chem.*, **76** (2004), 233–237.
91. K. Wróbel, K. Wróbel, J. A. Caruso, Selenium speciation in low molecular weight fraction of Se-enriched yeasts by HPLC-ICP-MS: detection of selenoadenosylmethionine, *J. Anal. Atom. Spectrom.*, **17** (2002), 1048–1054.
92. E. H. Larsen, J. Sloth, M. Hansen, S. Moesgaard, Selenium speciation and isotope composition in ^{77}Se-enriched yeast using gradient elution HPLC separation and ICP-dynamic reaction cell-MS, *J. Anal. Atom. Spectrom.*, **18** (2003), 310–316.
93. I. Iwami, Y. Higashiyama, F. Ibuki, Reevaluation of L-methionine sulfoxide bioavailability and a difference between its epimers in normal young-rats, *Nutr. Res.*, **12** (1992), 1155–1166.
94. R. Puchała, H. Piór, M. Von Keyserlingk, J. A. Shelford, W. Barej, Determination of methionine sulfoxide in biological materials using HPLC and its degradability in the rumen of cattle, *Anim. Feed Sci. Tech.*, **48** (1994), 121–130.
95. M. Dernovics, Z. Stefánka, P. Fodor, Improving selenium extraction by sequential enzymatic processes for Se-speciation of selenium-enriched *Agaricus bisporus*, *Anal. Bioanal. Chem.*, **372** (2002), 473–480.
96. I. Montes-Bayón, C. B'Hymer, C. Ponce de León, J. A. Caruso, Resolution of seleno-amino acid optical isomers using chiral derivatization and inductively coupled plasma mass spectrometric (ICP-MS) detection, *J. Anal. Atom. Spectrom.*, **16** (2001), 945–950.

97. K. Wróbel, S. S. Kannamkumarath, K. Wróbel, J. A. Caruso, Hydrolysis of proteins with methanesulfonic acid for improved HPLC-ICP-MS determination of selenomethionine in yeast and nuts, *Anal. Bioanal. Chem.*, **375** (2003), 133–138.
98. L. Yang, Z. Mester, R. E. Sturgeon, Determination of methionine and selenomethionine in yeast by species-specific isotope dilution GC/MS, *Anal. Chem.*, **76** (2004), 5149–5156.
99. J. S. Edmonds, K. A. Francesconi, J. R. Cannon, C. L. Raston, B. W. Skelton, A. H. White, Isolation, crystal structure and synthesis of arsenobetaine, the arsenical constituent of the western rock lobster *Panulirus longipes cygnus george*, *Tetrahedron Lett.*, **18** (1977), 1543–1546.
100. Z. Jókai, J. Hegóczki, P. Fodor, Stability and optimization of extraction of four arsenic species, *Microchem. J.*, **59** (1998), 117–124.
101. T. Lindemann, A. Prange, W. Dannecker, B. Neidhart, Stability studies of arsenic, selenium, antimony and tellurium species in water, urine, fish and soil extracts using HPLC/ICP-MS, *Fresenius J. Anal. Chem.*, **368** (2000), 214–220.
102. E. H. Larsen, G. Pritzl, S. H. Hansen, Speciation of eight arsenic compounds in human urine by high-performance liquid chromatography with inductively coupled plasma mass spectrometric detection using antimonate for internal chromatographic standardization, *J. Anal. Atom. Spectrom.*, **8** (1993), 557–563.
103. M. Edwards, S. Patel, L. McNeill, H. W. Chen, M. Frey, A. D. Eaton, R. C. Antweiler, H. E. Taylor, Considerations in As analysis and speciation, *J. Am. Water Works Assoc.*, **90** (1998), 103–113.
104. G. E. M. Hall, J. C. Pelchat, G. Gauthier, Stability of inorganic arsenic (III) and arsenic (V) in water samples, *J. Anal. Atom. Spectrom.*, **14** (1999), 205–213.
105. J. Feldmann, V. W. M. Lai, W. R. Cullen, M. S. Ma, X. F. Lu, X. C. Le, Sample preparation and storage can change arsenic speciation in human urine, *Clin. Chem.*, **45** (1999), 1988–1997.
106. M. Raessler, B. Michalke, S. Schulte-Hostede, A. Kettrup, Long-term monitoring of arsenic and selenium species in contaminated groundwaters by HPLC and HG-AAS, *Sci. Total Environ.*, **258** (2000), 171–181.
107. Y. Shibata, M. Morita, Speciation of arsenic by reversed-phase high performance liquid chromatography – inductively coupled plasma mass spectrometry, *Anal. Sci.*, **5** (1989), 107–109.
108. M. Stýblo, Z. Drobná, I. Jaspers, S. Lin, D. J. Thomas, The role of biomethylation in toxicity and carcinogenicity of arsenic: a research update, *Environ. Health Persp.*, **110** (2002), Suppl. 5, 767–771.
109. S. McSheehy, M. Marcinek, H. Chassaigne, J. Szpunar, Identification of dimethylarsinoyl-riboside derivatives in seaweed by pneumatically assisted electrospray tandem mass spectrometry, *Anal. Chim. Acta*, **410** (2000), 71–84.
110. J. Szpunar, Bio-inorganic speciation analysis by hyphenated techniques, *Analyst*, **125** (2000), 963–988.
111. K. Shiomi, Y. Sugiyama, K. Shimakura, Y. Nagashima, Arsenobetaine as the major arsenic compound in the muscle of two species of freshwater fish, *Appl. Organomet. Chem.*, **9** (1995), 105–109.

112. C. Soeroes, W. Goessler, K. A. Francesconi, N. Kienzl, R. Schaeffer, P. Fodor, D. Kuehnelt, Arsenic speciation in farmed Hungarian freshwater fish, *J. Agric. Food Chem.*, **23** (2005), 9238–9243.

113. F. Lagarde, M. B. Amran, M. J. F. Leroy, C. Demesmay, M. Ollé, A. Lamotte, H. Muntau, P. Michel, P. Thomas, S. Caroli, E. H. Larsen, P. Bonner, G. Rauret, M. Foulkes, A. Howard, B. Griepink, E. A. Maier, Certification of total arsenic, dimethylarsinic acid and arsenobetaine contents in a tuna fish powder (BCR-CRM 627), *Fresenius J. Anal. Chem.*, **363** (1999), 18–22.

114. S. McSheehy, P. Pohl, R. Łobiński, J. Szpunar, Investigation of arsenic speciation in oyster test reference material by multidimensional HPLC-ICP-MS and electrospray tandem mass spectrometry (ES-MS-MS), *Analyst*, **126** (2001), 1055–1062.

115. H. Helgesen, E. H. Larsen, Bioavailability and speciation of arsenic in carrots grown in contaminated soil, *Analyst*, **123** (1998), 791–796.

116. J. J. Sloth, E. H. Larsen, K. Julshamn, Determination of organoarsenic species in marine samples using gradient elution cation exchange HPLC-ICP-MS, *J. Anal. Atom. Spectrom.*, **18** (2003), 452–459.

117. R. Wahlen, S. McSheehy, C. Scriver, Z. Mester, Arsenic speciation in marine certified reference materials. Part 2. The quantification of water-soluble arsenic species by high-performance liquid chromatography – inductively coupled plasma mass spectrometry, *J. Anal. Atom. Spectrom.*, **19** (2004), 876–882.

118. J. W. McKiernan, J. T. Creed, C. A. Brockhoff, J. A. Caruso, R. M. Lorenzana, A comparison of automated and traditional methods for the extraction of arsenicals from fish, *J. Anal. Atom. Spectrom.*, **14** (1999), 607–613.

119. A. Gallagher, J. A. Shoemaker, X. Wei, C. A. Brockhoff-Schweigel, J. T. Creed, Extraction and detection of arsenicals in seaweed via accelerated solvent extraction with ion chromatographic separation and ICP-MS detection, *Fresenius J. Anal. Chem.*, **369** (2001), 71–80.

120. C. Schmidt, W. Reisser, J. Mattusch, P. Popp, R. Wennrich, Evaluation of extraction procedures for the ion chromatographic determination of arsenic species in plant materials, *J. Chromatogr. A*, **889** (2000), 83–91.

121. E. H. Larsen, M. Hansen, W. Gössler, Speciation and health risk considerations of arsenic in the edible mushroom *Laccaria amethystina* collected from contaminated and uncontaminated locations, *Appl. Organomet. Chem.*, **12** (1998), 285–291.

122. K. L. Ackley, C. B'Hymer, K. L. Sutton, J. A. Caruso, Speciation of arsenic in fish tissue using microwave-assisted extraction followed by HPLC-ICP-MS, *J. Anal. Atom. Spectrom.*, **14** (1999), 845–850.

123. J. Kirby, W. Maher, Measurement of water-soluble arsenic species in freeze-dried marine animal tissues by microwave-assisted extraction and HPLC-ICP-MS, *J. Anal. Atom. Spectrom.*, **17** (2002), 838–843.

124. K. A. Francesconi, D. Kuehnelt, Determination of arsenic species: a critical review of methods and applications, 2000–2003, *Analyst*, **129** (2004), 373–395.

125. K. J. Lamble, S. J. Hill, Arsenic speciation in biological samples by on-line high performance liquid chromatography – microwave digestion–hydride generation–atomic absorption spectrometry, *Anal. Chim. Acta*, **334** (1996), 261–270.

126. D. T. Heitkemper, N. Vela, K. R. Stewart, C. S. Westphal, Determination of total and speciated arsenic in rice by ion chromatography and inductively coupled plasma mass spectrometry, *J. Anal. Atom. Spectrom.*, **16** (2001), 299–306.

127. J. A. Caruso, D. T. Heitkemper, C. B'Hymer, An evaluation of extraction techniques for arsenic species from freeze-dried apple samples, *Analyst*, **126** (2001), 136–140.

128. C. Soeroes, W. Goessler, K. A. Francesconi, E. Schmeisser, R. Raml, N. Kienzl, M. Kahn, P. Fodor, D. Kuehnelt, Thio-arsenosugars in freshwater mussels from the Danube in Hungary, *J. Environ. Monitor.*, **7** (2005), 688–692.

129. M. Stýblo, D. J. Thomas, Binding of arsenicals to proteins in an in vitro methylation system, *Toxicol. Appl. Pharm.*, **147** (1997), 1–8.

130. J. Brooke, W. H. Evans, Determination of total inorganic arsenic in fish, shellfish and fish products, *Analyst*, **106** (1981), 514–520.

131. Muñoz, D. Vélez, R. Montoro, Optimization of the solubilization, extraction and determination of inorganic arsenic (As(III) + As(V)) in seafood products by acid digestion, solvent extraction and hydride generation atomic absorption spectrometry, *Analyst*, **124** (1999), 601–607.

132. Muñoz, D. Vélez, M. L. Cervera, R. Montoro, Rapid and quantitative release, separation and determination of inorganic arsenic (As(III) + As(V)) in seafood products by microwave-assisted distillation and hydride generation atomic absorption spectrometry, *J. Anal. Atom. Spectrom.*, **14** (1999), 1607–1613.

133. N. Parks, P. A. Gallagher, C. A. Schwegel, A. H. Ackerman, J. T. Creed, The liberation of arsenosugars from matrix components in difficult to extract seafood samples utilizing TMAOH/acetic acid sequentially in a two-stage extraction process, Poster presentation at the *2003 European Winter Conference on Plasma Spectrochemistry*, Garmisch-Partenkirchen, 12–17 January 2003.

134. M. Gamble, P. A. Gallagher, J. A. Shoemaker, X. Wei, C. A. Schwegel, J. T. Creed, An investigation of the chemical stability of arsenosugars in simulated gastric juice and acidic environments using IC-ICP-MS and IC-ESI-MS/MS, *Analyst*, **127** (2002), 781–785.

135. Katarzyna Wróbel, Kazimierz Wróbel, S. S. Kannamkumarath, J. A. Caruso, I. A. Wysocka, E. Bułska, J. Świątek, M. Wierzbicka, HPLC–ICP-MS speciation of selenium in enriched onion leaves – a potential dietary source of Se-methylselenocysteine, *Food Chem.*, **86** (2004), 617–623.

136. Z. Mester, S. Willie, L. Yang, R. Sturgeon, J. A. Caruso, M. L. Fernández, P. Fodor, R. J. Goldschmidt, H. Goenaga-Infante, R. Łobiński, P. Maxwell, S. McSheehy, A. Polatajko, B. B. M. Sadi, A. Sanz-Medel, C. Scriver, J. Szpunar, R. Wahlen, W. Wolf, L. Yang, Certification of a new selenized yeast reference material (SELM-1) for methionine, selenomethinone and total selenium content and its use in an intercomparison exercise for quantifying these analytes, *Anal. Bioanal. Chem.*, **385** (2006), 168–180.

137. SEAS, *Feasibility Studies for Speciated CRMs for Arsenic in Chicken, Rice, Fish and Soil, and Selenium in Yeast and Cereal*, EC Contract GROWTH GRD1-2000-25019.

138. S. McSheehy, J. Szpunar, R. Morabito, P. Quevauviller, The speciation of arsenic in biological tissues and the certification of reference materials for quality control, *Trends Anal. Chem.*, **22** (2003), 191–209.

139. B. Michalke, Quality control and reference materials in speciation analysis, *Fresenius J. Anal. Chem.*, **363** (1999), 439–445.

140. Ph. Quevauviller, Requirements for production and use of Certified Reference Materials for speciation analysis: a European Commission perspective, *Spectrochim. Acta*, **53B** (1998), 1261–1279.

141. E. H. Larsen, G. A. Pedersen, J. W. McLaren, Characterization of National Food Agency shrimp and plaice reference materials for trace elements and arsenic species by atomic and mass spectrometric techniques, *J. Anal. Atom. Spectrom.*, **12** (1997), 963–968.

142. M. Bird, P. C. Uden, J. F. Tyson, E. Block, E. Denoyer, Speciation of selenoamino acids and organoselenium compounds in selenium-enriched yeast using high-performance liquid chromatography–inductively coupled plasma mass spectrometry, *J. Anal. Atom. Spectrom.*, **12** (1997), 785–788.

143. M. Kotrebai, S. M. Bird, J. F. Tyson, E. Block, P. C. Uden, Characterization of selenium species in biological extracts by enhanced ion-pair liquid chromatography with inductively coupled plasma–mass spectrometry and by referenced electrospray ionization–mass spectrometry, *Spectrochim. Acta*, **54B** (1999), 1573–1591.

144. C. A. Ponce de Leon, K. L. Sutton, J. A. Caruso, P. C. Uden, Chiral speciation of selenoamino acids and selenium enriched samples using HPLC coupled to ICP-MS, *J. Anal. Atom. Spectrom*, **15** (2000), 1103–1107.

145. P. Moreno, M. A. Quijano, A. M. Gutiérrez, M. C. Pérez-Conde, C. Cámara, Stability of total selenium and selenium species in lyophilised oysters and in their enzymatic extracts, *Anal. Bioanal. Chem.*, **374** (2002), 466–476.

146. J. Zheng, Y. Shibata, N. Furuta, Determination of selenoamino acids using two-dimensional ion-pair reversed phase chromatography with on-line detection by inductively coupled plasma mass spectrometry, *Talanta*, **59** (2003), 27–36.

147. V. Díaz Huerta, L. Hinojosa Reyes, J. M. Marchante-Gayón, M. L. Fernández Sánchez, A. Sanz-Medel, Total determination and quantitative speciation analysis of selenium in yeast and wheat flour by isotope dilution analysis ICP-MS, *J. Anal. Atom. Spectrom.*, **18** (2003), 1243–1247.

148. L. Hinojosa Reyes, F. Moreno Sanz, P. Herrero Espílez, J. M. Marchante-Gayón, J. I. García Alonso, A. Sanz-Medel, Biosynthesis of isotopically enriched selenomethionine: application to its accurate determination in selenium-enriched yeast by isotope dilution analysis-HPLC-ICP-MS, *J. Anal. Atom. Spectrom.*, **19** (2004), 1230–1235.

149. E. Kápolna, V. Gergely, M. Dernovics, P. Fodor, Selenium speciation of some raw food materials. Poster presentation at the *8th International Symposium on Metal Ions in Biology and Medicine*, Budapest, 19–22 May 2004.

150. Y. Kobayashi, Y. Ogra, K. Ishiwata, H. Takayama, N. Aimi, K. T. Suzuki, Selenosugars are key and urinary metabolites for selenium excretion within the required to low-toxic range, *Proc. Natl Acad. Sci. USA*, **99** (2002), 15932–15936.

151. E. H. Larsen, G. Pritzl, S. H. Hansen, Arsenic speciation in seafood samples with emphasis on minor constituents: an investigation using high-performance liquid chromatography with detection by inductively coupled plasma mass spectrometry, *J. Anal. Atom. Spectrom.*, **8** (1993), 1075–1084.

152. W. Zhang, Y. Cai, C. Tu, L. Q. Ma, Arsenic speciation and distribution in an arsenic hyperaccumulating plant, *Sci. Total Environ.*, **300** (2002), 167–177.

153. J. J Corr, E. H. Larsen, Arsenic speciation by liquid chromatography coupled with ionspray tandem mass spectrometry, *J. Anal. Atom. Spectrom.*, **11** (1996), 1215–1224.

154. Z Gong, X Lu, M Ma, C Watt, XC. Le, Arsenic speciation analysis, *Talanta*, **58** (2002), 77–96.

APPENDIX

TABLE 19.A1. Some of the Most Often Studied Se Species [41, 53, 73, 91, 137, 150]

	Names and Some Common Abbreviations	Chemical Formula
1	Selenate, Se(VI)	SeO_4^{2-}
2	Selenite, Se(IV)	SeO_3^{2-}
3	Trimethylselenonium ion, $TMSe^+$	$(CH_3)_3Se^+$
4	Dimethylselenide, DMSe	$(CH_3)_2Se$
5	Selenocysteine, SeCys	$H_2N—CH(COOH)—CH_2—SeH$
6	Selenocystine, $SeCys_2$	$H_2N—CH(COOH)—CH_2—Se—Se—CH_2—CH(COOH)—H_2N$
7	Selenomethionine, SeMet	$H_2N—CH(COOH)—CH_2—CH_2—Se—CH_3$
8	Selenomethionine-Se-oxide, SeOMet	$H_2N—CH(COOH)—CH_2—CH_2—Se(O)—CH_3$
9	(Se)-Methylselenocysteine	$H_2N—CH(COOH)—CH_2—Se—CH_3$
10	γ-Glutamyl-Se-methylselenocysteine	$H_2N—CH(COOH)—CH_2—CH_2—CO—NH—CH(COOH)—CH_2—Se—CH_3$
11	Selenocystathionine	$H_2N—CH(COOH)—CH_2—CH_2—Se—CH_2—CH(COOH)—H_2N$
12	Selenohomocysteine	$H_2N—CH(COOH)—CH_2—CH_2—SeH$
13	Selenocystamine	$H_2N—CH_2—CH_2—Se—Se—CH_2—CH_2—NH_2$
14	Se-adenosyl-selenohomocysteine	$H_2N—CH(COOH)—CH_2—CH_2—Se—CH_2—C_4H_5O_3C_5N_4NH_2$
15	Se-lanthionine	$H_2N—CH(COOH)—CH_2—Se—CH_2—CH(COOH)—H_2N$
16	Selenoethionine, SeEt	$H_2N—CH(COOH)—CH_2—CH_2—Se—CH_2—CH_3$
17	Dimethylselenonium propionate, DMSeP	$(CH_3)_2Se^+CH_2—CH_2—COO^-$
18	1β-Methylseleno-*N*-acetyl-D-galactosamine	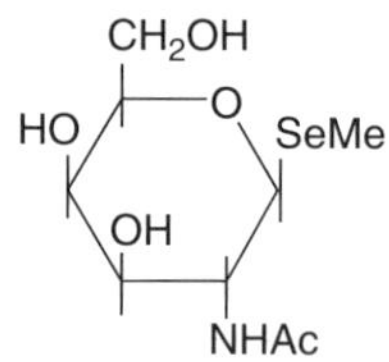
19	Se-adenosyl-methionine	$H_2N—CH(COOH)—CH_2—CH_2—Se^+(CH_3)—CH_2—C_4H_5O_3C_5N_4NH_2$
20	Se-allylselenocysteine	$H_2N—CH(COOH)—CH_2—Se—CH_2—CH=CH_2$
21	Selenourea, SeU	$H_2N—C(Se)—NH_2$

TABLE 19.A2. Some of the Most Often Studied As Species [63, 119, 128, 131, 153, 154]

Name	Chemical Structure	Common Abbreviations	LD_{50} Value (mg kg^{-1})
Arsenite (arsenous acid)	$OH\text{-}As(OH)_2$	As(III)	14–34.5
Arsenate (arsenic acid)	$O{=}As(OH)_3$	As(V)	20–41
Monomethylarsonous acid	$CH_3As(OH)_2$	MMA(III)	? (presumably highly toxic)
Monomethylarsonic acid	$CH_3AsO(OH)_2$	MMA(V)	700–1800
Dimethylarsinous acid	$(CH_3)_2AsOH$	DMA(III)	? (presumably highly toxic)
Dimethylarsinic acid	$(CH_3)_2AsO(OH)$	DMA(V)	700–2600
Dimethylarsinoyl ethanol	$(CH_3)_2AsOCH_2CH_2OH$	DMAE (DMAsEt)	? (presumably nontoxic)
Dimethylarsinoyl-acetic acid	$(CH_3)_2AsOCH_2CH_2COOH$	DMAA (DMAsAc)	
Trimethylarsine oxide	$(CH_3)_3AsO$	TMAO	>10,000
Tetramethylarsonium ion	$(CH_3)_4As^+$	$TMAs^+$ (TETRA)	890
Arsenobetaine	$(CH_3)_3As^+CH_2COO^-$	AB	>10,000
Arsenobetaine 2, trimethylarsonio proprionate	$(CH_3)_3As^+CH_2CH_2COO^-$	AB-2 (TMAP)	? (presumably nontoxic)
Arsenocholine	$(CH_3)_3As^+CH_2CH_2OH$	AsC	6500–6900
Arsines	$(CH_3)_xAsH_{3-x}$, where $x = 0–3$	Me_xAsH_{3-x}, where $x = 0–3$	? (extremely toxic)
Ethylmethylarsines	$(CH_3CH_2)_xAs(CH_3)_{3-x}$, where $x = 0–3$	Et_xAsMe_{3-x}, where $x = 0–3$	

(Continued)

TABLE 19.A2. *Continued*

Name	Chemical Structure	Common Abbreviations	LD_{50} Value (mg kg^{-1})
	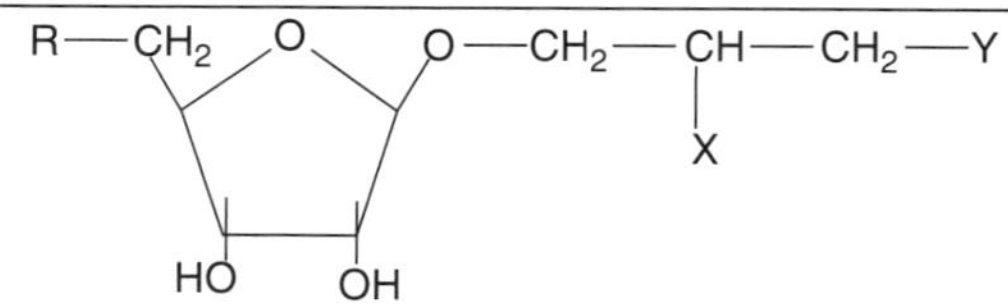		
	R = $(CH_3)_2AsO$—, X = —OH, Y = —OH	Arsenosugar A, X, 392, 11, glycerol	
	R = $(CH_3)_2AsS$—, X = —OH, Y = —OH	Thioglycerol arsenosugar	
	R = $(CH_3)_2AsO$—, X = —OH, Y = —$OPO_3HCH_2CH(OH)CH_2OH$	Arsenosugar B, XI, 328, 13, phosphate	
	R = $(CH_3)_2AsS$—, X = —OH, Y = —$OPO_3HCH_2CH(OH)CH_2OH$	Thio phosphate arsenosugar	
Arsenoyl-ribosides	R = $(CH_3)_2AsO$—, X = —OH, Y = —SO_3H	Arsenosugar C, XII, 408, 10, sulfonate	? (presumably nontoxic)
	R = $(CH_3)_2AsO$—, X = —OH, Y = —OSO_3H	Arsenosugar D, XIII, 482, 12, sulfate	
	R = $(CH_3)_2AsO$—, X = —NH_2, Y = —SO_3H	Arsenosugar E, XIV, 390	
	R = $(CH_3)_3As^+$—, X = —OH, Y = —OSO_3H	Arsenosugar F, XV, 407	
	R = $(CH_3)_3As^+$—, X = —NH_2, Y = —OSO_3H	Arsenosugar G, XVI, 406	

20

MEASUREMENT OF TOTAL Se AND Se SPECIES IN SEAFOOD BY QUADRUPOLE INDUCTIVELY COUPLED PLASMA MASS SPECTROMETRY, ELECTROTHERMAL ATOMIZATION ATOMIC ABSORPTION SPECTROMETRY, AND HIGH-PERFORMANCE LIQUID CHROMATOGRAPHY INDUCTIVELY COUPLED PLASMA MASS SPECTROMETRY

WILLIAM A. MAHER AND FRANK KRIKOWA

Ecochemistry Laboratory, Institute of Applied Ecology, University of Canberra, ACT 2601, Australia, E-mail: bill.maher@canberra.edu.au

The Determination of Chemical Elements in Food: Applications for Atomic and Mass Spectrometry. Edited by Sergio Caroli

20.1 INTRODUCTION

Selenium has been classified as an essential element [1]. It is incorporated in selenoproteins including glutathione peroxidase and selenoprotein P [2–6] as the amino acids selenocysteine and selenomethionine, through specific tRNA molecules [7, 8]. Selenoproteins play an important role in preventing cellular oxidation [3, 9]. Selenium can also occur as Se nucleosides [10, 11]. Selenium deficiency in humans manifests itself as Keshan disease, or cardiomyopathy, a heart disease in the form of tissue degeneration and scar tissue formation, and Kaschin–Beck disease, a degenerative bone disease [12]. However, Se is toxic to organisms at elevated concentrations [13, 14]. Diseases that caused malformations in children as well as loss of hair and fingernails were described by Marco Polo in 1295 and by Father Pedro Simon in 1560 [15]. These are now attributed to toxic intakes of Se. Diets containing more than 5 $\mu g\ g^{-1}$ are considered poisonous to man. At present the regulatory level of Se in seafood is set at 1 $\mu g\ g^{-1}$ wet mass of Se based on the maximum daily allowable intake of Se [16]. Acute toxicity in humans is rare; it leads to convulsions, diarrhea, hypochromic anemia, and respiratory failure. Selenium is of nutritional concern because there is a very small difference between a dose of Se that is beneficial and one that is toxic; toxicity may be induced at concentrations as low as twice those nutritionally required [17].

A significant portion of the world's diet consists of seafood. Selenium is present in seafood at low concentrations with typical values of 0.02–4 $\mu g\ g^{-1}$ dry mass in molluscs, crustaceans, and fish [18–24]. In a comprehensive survey of Se concentrations in Australian marine organisms, Maher et al. [21] reported that 42 percent of mollusc tissues, 50 percent of crustacean tissues, and 86 percent of fish tissues had Se concentrations less than 0.5 $\mu g\ g^{-1}$ wet mass. Selenium concentrations in herbivores, omnivores, and carnivores are not significantly different [21, 25]. However, although dietary sources do not appear to play an important role in Se accumulation and retention by marine organisms, variations may be obscured by differences in animal ages, sex, habitat, or prevailing physical–chemical conditions. Nonpreferential accumulation and dietary independence of Se suggests that this element is assimilated by marine animals for specific roles such as in the glutathione peroxidase pathway [26], with excess Se being excreted [27]. Seaweeds contain 0.01–0.6 $\mu g\ g^{-1}$ dry mass with rhodophyta containing significantly higher Se concentrations than chlorophyta and phaeophyta [28–30].

In contaminated environments, such as areas receiving inputs from coal-fired power stations, Se concentrations can be considerably higher. Selenium is biomagnified through food webs [31]. Selenium concentrations up to 70 μg g^{-1} dry mass have been measured in polychaete tissues, 12 μg g^{-1} dry mass in mollusc tissues, and 60 μg g^{-1} dry mass in fish tissues from contaminated areas [32–36]. Selenium is predominantly associated with amino acids and proteins and not present as characterizable inorganic Se species [25, 28, 30, 32, 37, 38]. In some cases Se appears to be present as selenocysteine and selenomethionine [32, 34]. No conclusive unequivocal evidence exists for the production and storage of selenoamino acids in marine algae and animals. However, virtually all naturally occurring Se compounds are S analogues [39]. The structures, names, and acronyms of Se species that may occur in marine algae and animals, based on the assumption that Se metabolism follows the S pathway [32], are given in Table 20.1. Many of these species have been identified in terrestrial bacteria, plants, and animals [40]. Recently, an Se sugar (see Table 20.1) was identified as a human excretion product to low-level exposure of Se [41, 42], and it possibly occurs as an excretion product of marine organisms as well.

Direct evidence for the presence of Se in combined and free analogs of S-containing amino acids in microalgae comes from the inorganic Se uptake studies performed by Wrench [43] and Bottino et al. [44]. They reported the presence of selenocysteic acid, selenomethionine selenoxide, selenomethylcysteine, selenocystine, selenomethionine, Se-methylselenomethionine, and Se-methyl selenocysteine selenone. These studies were performed at elevated Se concentrations, and microalgae may have different biochemical mechanisms for metabolizing Se at lower, naturally occurring Se levels. In addition, inorganic Se may not be incorporated, but covalently bound to lipids [45], that is, an artifact of the isolation

TABLE 20.1. Common Se Species Found in Seafood

Name	Acronym	Chemical Formula
Selenious acid	Se(IV)	$Se(OH)_3$
Selenic acid	Se(VI)	$SeO(OH)_3$
Selenocysteine	SeCyst	$HSeCH_2CH(NH_2)COOH$
Senocystine	$SeCyst_2$	$SeCH_2CH(NH_2)COOH$
Se-methylselenocysteine	MSeCyst	$CH_3SeCH_2CH(NH_2)COOH$
Selenohomocysteine	MSeHCyst	$HSeCH_2\ CH_2CH(NH_2)COOH$
Selenomethionine	SeMeth	$CH_3SeCH_2\ CH_2CH(NH_2)COOH$
Se-methylselenomethionine	MSeMeth	$(CH_3)_2SeCH_2\ CH_2CH(NH_2)COOH$
Trimethylselenonium ion[a]	TMSe	$(CH_3)_3Se^+$
Methylselenic acid[a]	MSe(V)	$CH_3SeO(OH)$
1β-methylseleno-*N*-acetyl-D-galatosamine	Selenosugar	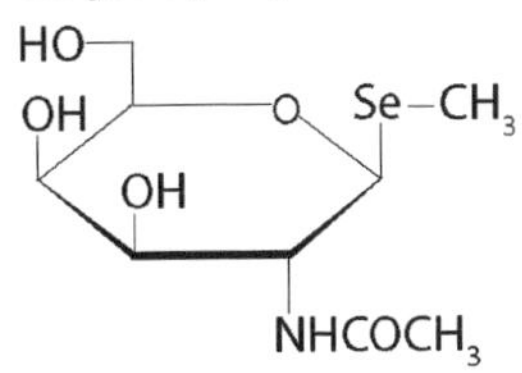

[a]Possible marine animal excretion compounds.

experiment. The production of seleno amino acids and proteins is, however, a general phenomenon of terrestrial plants with chloroplasts [46, 47]. Wrench [48] also isolated selenocystic acid, selenomethionine, selenomethionine selenoxide, and selenohomocysteine from oysters fed high concentrations of inorganic Se. Again, these may be artifacts of the high dosages of inorganic Se used. Selenium-dependent glutathione peroxidase has been isolated from black sea bass *Centropristis striata* [26], indicating that some Se may be present as selenocysteine. Selenocysteine is a normal component of glutathione peroxidase [11]. Alternatively, Se may be in a nonprotein moiety tightly held to proteins, but not covalently bound. For example, Se can readily form selenotrisulphides with thiols such as cysteine and glutathione [49, 50]. These can be incorporated and stabilized within protein structures [51]. Selenium has also been found in dolphin livers as Se–Hg granules [52, 53]. It is clear that the role that Se plays is dependent on its chemical form and how it is metabolized [13, 54]. Thus, measuring only total concentrations of Se without quantifying Se species is inadequate to understand its effects. In this chapter, the use of quadrupole (Q) inductively coupled plasma mass spectrometry (ICP-MS) and electrothermal atomization atomic absorption spectrometry (ET-AAS) to measure total Se concentrations and high-performance liquid chromatography (HPLC) combined with Q-ICP-MS to measure Se-containing proteins and amino acid concentrations in seafood is discussed. To highlight the strengths and weaknesses of the use of Q-ICP-MS and ET-AAS data mainly produced in the authors' laboratory have been used.

20.2 TOTAL Se MEASUREMENTS

20.2.1 Sample Storage and Preservation

If samples are to be stored for long periods, the presence of microorganisms may cause deterioration of samples and change Se species. It has been found that, if samples are stored frozen at $-10^{\circ}C$ or freeze-dried, Se is not lost from samples [55–57] and species are preserved [57].

20.2.2 Sample Digestion

Dry ashing and wet digestion techniques are used to digest biological samples for Se analysis. Procedures for the determination of total Se using dry ashing usually use an HNO_3–$Mg(NO_3)_2$ [58, 59] or a MgO–$Mg(NO_3)_2$ mixture [60] as an ashing aid to prevent loss of Se during decomposition. Wet digestion procedures using HNO_3 [61], HNO_3–H_2SO_4 [62, 63], HNO_3–$HClO_4$ [64–67], HNO_3–H_2SO_4–$HClO_4$ [25, 59, 61, 68, 69], and HNO_3–H_2O_2 [70] have been reported. The use of HCl is avoided because of the loss of Se as volatile chlorides and oxychlorides [71].

Many authors have reported difficulties in digesting biological samples using traditional techniques and in obtaining complete recovery of Se from Certified Reference Materials (CRMs) [61, 71–76] and normally use mixtures of concentrated acids [61, 70, 77–81]. Convection heating using HNO_3 alone leads to incomplete digestion and erroneous results [67, 72, 80]. For example, when comparing digestion techniques, Maher [67] showed that the best recoveries of Se from fish muscle (*Sillaginodes punctatulus*) were obtained with HNO_3–$HClO_4$ and HNO_3–$HClO_4$–H_2SO_4 mixtures with complete breakdown of organo-Se compounds containing C–Se covalent bonds. The use of HNO_3 alone only recovered 53 percent of Se as inorganic Se from fish tissue.

Microwave (MW) assisted digestion with HNO_3 or HNO_3–H_2O_2 using closed Teflon vessels has replaced traditional digestion procedures. The advantages of using MW heating include less acid used, less laboratory contamination, lower analytical blanks, and better reproducibility [82, 83]. Digestion temperatures are limited to about 250°C because the vessels melt at this temperature. Hocquellet and Candiller [80], Ducros et al. [84], Saraswati et al. [85], and Bryce et al. [86] have demonstrated that there is no significant difference between traditional wet digestion or reflux methods and MW digestion using open vessels when similar acid (oxidizing agents) mixtures are used. LamLeung et al. [76] showed that the efficiency of digestion using HNO_3 was greatly improved with the use of closed vessel MW heating over conventional open vessel digestion procedures. Maher and coworkers have shown that complete recovery of Se from 14 marine CRMs can be obtained using a low-volume MW digestion technique [87, 88]. These results are in good agreement with published values: $Se_{found} = 0.9877$ $Se_{certified} \pm 0.093\ \mu g\ g^{-1}$ dry mass, $r^2 = 0.9878$, $n = 55$. Complete recovery of added Se compounds (selenocystine and selenomethionine, but not trimethylselenonium iodide) was obtained within experimental error when added to three marine tissues [87]. However, it should be noted that Se was measured using Q-ICP-MS or ET-AAS and Se species only had to be released from the matrix since Se measured by these techniques is not dependent on the Se species present in the final digest solution.

MW-assisted digestion at 160°C using closed vessels does not quantitatively convert all Se species into inorganic Se (Fig. 20.1). Selenocysteine is completely degraded to H_2SeO_3 and selenomethionine is degraded to a number of Se species, while trimethylselenonium ion is not degraded to any extent. Only the use of an MW-assisted autoclave with temperatures higher than 500°C is likely to quantitatively convert all Se species to inorganic Se, as is found for As species [89]. Hydride generation (HG) techniques that rely on Se to be in an inorganic form will underestimate total Se concentrations in extracts.

The use of HNO_3 alone is advantageous for Q-ICP-MS, as compared with other acids. HNO_3 produces the least background interference in Q-ICP-MS [90]. Variation in HNO_3 concentration has been shown to suppress trace element mass counts [91]. Maher et al. [88] have found that for a Q-ICP-MS instrument, variations in HNO_3 concentration (0–6 percent v/v) significantly influence Se mass counts. Thus, care needs to be taken to match the HNO_3 concentration of standards and samples. Generally, digestion with HNO_3 produces digests with a light yellow color, indicating that samples are not completely digested. The problem with incomplete digestion is that residual C present in digests [92] may assist in the ionization of Se in the plasma, causing the enhancement of Se mass counts [93–98].

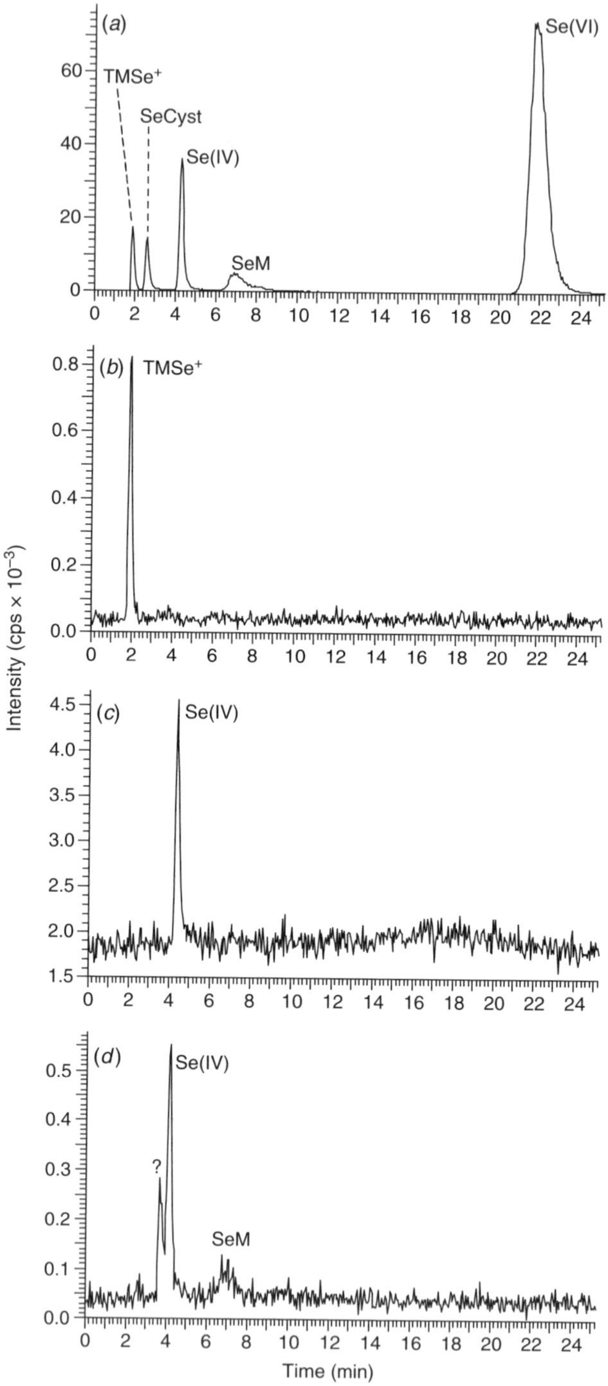

Figure 20.1. Se species degradation by MW digestion (160°C) with HNO_3. A Hamilton PRP X-100 anion exchange column, 20 mM $NH_4H_2PO_4$, pH 6, 1.5 ml min^{-1}, 40°C (*a*) standards; (*b*) $TMSe^+$; (*c*) selenocystine; (*d*) selenomethionine.

20.2.3 Q-ICP-MS Analysis

Selenium measurement by Q-ICP-MS suffers from polyatomic interferences [99, 100] arising from Ar plasma gas (e.g., $^{40}Ar^{36}Ar^{+}$, $^{40}Ar^{38}Ar^{+}$, and $^{40}Ar^{40}Ar^{+}$) and matrix constituents (e.g., $^{40}Ar^{37}Cl^{+}$, and $^{81}Br^{1}H^{+}$). The use of correction equations, addition of nitrogen or a C source such as methanol or butanol, or the use of a dynamic reaction cell (DRC) or collision cell (CC) can minimize and sometimes eliminate these interferences [101–115]. Utilization of cooled spray chambers (at less than 15°C) have also been shown to reduce $^{36}Ar^{40}Ar^{+}$ and $^{40}Ar^{37}Cl^{+}$ interferences [102, 116].

The use of correction equations should be avoided, especially if the Se mass counts are small compared with the interference present. Correction equations suffer from inherent problems in that the masses used to correct for interferences may themselves suffer from interferences [113, 114]. Examples of potential polyatomic interferences for Se are given in Table 20.2. These interferences cannot be separated from analytes using a Q detector. Many papers in the literature report different interferences when measuring Se in similar matrices. This is because the extent and magnitude of interferences depend on the Q-ICP-MS instrument used, the type of nebulizer, the plasma torch conditions, the final dilution volume of the sample, and the isotope concentration being measured [88].

Maher et al. [88] have undertaken extensive analyses of possible polyatomic interferences that may arise during the analysis of marine biological materials when using a Q-ICP-MS instrument. Some concentrations of matrix elements typically found in digests, at the sample : solution ratio of 1 : 100 routinely used in the authors' laboratory, are relatively high: Al (0.02–2 mg l^{-1}), Ca (0.1–65 mg l^{-1}), Cl (6.8–60 mg l^{-1}), Fe (0.3–1.2 mg l^{-1}), K (0.01–61 mg l^{-1}), Mg (1–9 mg l^{-1}), Mn (0.001–0.3 mg l^{-1}), Na (1–37 mg l^{-1}), P (2.6–13 mg l^{-1}), and S (6.8–12 mg l^{-1}). High K and Na concentrations may depress Se mass counts because of space charge effects, or reduce ionization of Se because of the large K or Na ion populations [117, 118]. For Se, a major interference is Cl^{-}, as this can occur in sample matrices at concentrations up to 60 mg l^{-1}. In practice it was found that Se concentrations can be accurately measured in samples and reference materials [88] down to ~0.5–1 μg l^{-1} (0.5–1 μg g^{-1} dry mass), but at these levels one needs to carefully check for Br^{-} and Cl^{-} interferences and C enhancement. In practice, low Se concentrations in new sample matrices are cross-checked using ET-AAS (see Section 20.2.4). However, it should be stressed that the interference caused by Br^{-}, Cl^{-}, and C (as is the case with most interferences) will depend to some extent on the Q-ICP-MS instrument used, instrumental conditions such as nebulizer flow rate, and the specific sample being analyzed. It should be noted that that these interferences are avoidable when using Q-ICP-MS instruments with chemical resolving capability, that is, a DRC or a CC [102–114].

As previously mentioned, the presence of C may enhance Se mass counts [93, 94, 96, 98, 105, 107]. Maher et al. [88] found that MW-assisted digestion of tissues with HNO_3 (0.1 g sample per 1 mL of HNO_3) yield less than

TABLE 20.2. Polyatomic Ions and Interfering Isotopes

Isotope	Natural Abundance (Mass %)	Polyatomic Ion							
		Ar	Cl	Br	Ca	S	P	N	Si
$^{74}Se^+$	0.83		$^{37}Cl^{37}Cl^+$			$^{40}Ar^{34}S^+$			
$^{76}Se^+$	9.00	$^{40}Ar^{36}Ar^+$	$^{38}Ar^{38}Ar^+$ $^{39}K^{37}Cl^+$		$^{44}Ca^{32}S^+$ $^{42}Ca^{34}S^+$ $^{44}Ca^{32}S^+$ $^{40}Ca^{36}S^+$	$^{40}Ar^{36}S^+$		^{31}P $^{31}P^{14}N^+$	
$^{77}Se^+$	7.43	$^{38}Ar^{39}K^+$ $^{36}Ar^{41}K^+$ $^{40}Ar^{36}Ar^1H^+$	$^{40}Ar^{37}Cl^+$ $^{42}Ca^{35}Cl^+$		$^{40}Ca^{37}Cl^+$				
$^{78}Se^+$	23.47	$^{40}Ar^{38}Ar^+$	$^{41}K^{37}Cl^+$		$^{38}Ar^{40}Ca^+$ $^{44}Ca^{34}S^+$	$^{42}Ar^{36}S^+$	^{31}P $^{31}P^{16}O^+$		
$^{80}Se^+$	50.21	$^{40}Ar^{40}Ar^+$			$^{36}Ar^{44}Ca^+$ $^{40}Ar^{40}Ca^+$	$^{32}S^{16}O$ $^{16}O^{16}O^+$			
$^{82}Se^+$	9.06	$^{40}Ar^{40}Ar^1H^1H^+$	$^{12}C^{35}Cl^{35}Cl^+$	$^{81}Br^1H^+$	$^{40}Ar^{42}Ca^+$ $^{38}Ar^{44}Ca^+$				^{34}Si ^{16}O $^{16}O^{16}O^+$

50–100 mg l^{-1} of C in extracts and causes no enhancement of Se mass counts. Similiar to other authors (e.g., [119]), Maher et al. [88] found that, when a cross-flow nebulizer is used, Se species (H_2SeO_3, H_2SeO_4, selenocystine, selenomethionine, and trimethylselenonium iodide) give a similar Q-ICP-MS response. The limits of detection (LoDs) can be markedly lowered using ultrasonic [119, 120] or hydraulic high-pressure nebulizers [106, 121, 122]. However, desolvation is needed to reduce aerosol water and acid loading.

HG procedures are often advocated as a means of eliminating interferences due to Cl^- and a means of increasing sensitivity [123]. However, careful consideration of the gas–liquid separator design is required to minimize or eliminate transport of Cl^- to the plasma [124]. Traditionally, boiling with HCl is used to reduce H_2SeO_4 to H_2SeO_3 prior to the formation of SeH_4 [78]. HBr is now preferred as it reduces Se(VI) faster and HCl often contains significant quantities of Cl_2, which reoxidizes Se(IV) [125]. A drawback is that the carry-over of Br^- may form $^{81}Br^{1}H^{+}$, a major interferent on $^{82}Se^{+}$. Again, it is stressed that HG can only be used for total Se measurements if all Se species in digests have been quantitatively degraded to inorganic Se, as otherwise total Se concentrations in extracts will be underestimated.

20.2.4 ET-AAS Analysis

The effectiveness of ET-AAS critically depends on the chemical modifier used to prevent the loss of Se during the thermal pretreatment to remove matrix elements prior to atomization. Deaker and Maher [126] have compared the stabilizing effects of the commonly used modifiers Cu, Ni, and Pd in conjunction with the use of $Mg(NO_3)_2$ and shown that, when compared on an equimolar basis, Pd is superior. Digested biological tissues require higher amounts of chemical modifier than pure reference Se standards (H_2SeO_3, H_2SeO_4, selenocystine, selenomethionine, trimethyselenonium iodide, and selenocystamine dihydrochloride). The use of surface response curves using varying amounts of $Pd(NO_3)_2$ and $Mg(NO_3)_2$ allows compromise conditions to be selected for reference Se standards and marine CRMs and comparisons to be made against external inorganic Se standards without the need to use standard additions. ET-AAS does not suffer from Br^-, Cl^-, or C interferences, but is not as sensitive as Q-ICP-MS, with a measurement level of ~5 μg l^{-1} (1 μg g^{-1} dry mass) for a normal 20–50 μL injection. To obtain the necessary LoDs, multilayering of sample extracts, that is, multiple injections, drying, and ashing before atomization, is required.

20.3 SELENIUM SPECIES MEASUREMENT USING HPLC-Q-ICP-MS

20.3.1 General Aspects

As defined by IUPAC, chemical species are the specific forms of an element with respect to its isotopic composition (e.g., ^{77}Se, ^{80}Se, ^{82}Se), electron or oxidation state (e.g., H_2SeO_3, H_2SeO_4), and/or complex (e.g., selenomethionine, trimethylselenonium ion) or molecular structure (Se-proteins) [127]. Here the focus is on the isolation of Se-containing proteins and the presence of selenoamino acids as these are the likely forms of Se in seafood [25, 30, 32, 37].

20.3.2 Sample Extraction

A variety of extraction conditions employing methanol–water mixtures [43], ethanol [25, 30, 37, 43, 44], water [38, 128, 130, 134], trichloroacetic acid [44, 130], enzymes [43, 80, 128, 133, 135, 136], and buffers [25, 26, 30, 32, 37] in combination with heating, sonification, shaking, and MW-assisted extraction have been used to isolate Se proteins and selenoamino acids from marine tissues. Efficient extraction requires the rupture of cell membranes by freeze-drying or mechanical homogenization in water or a buffer solution. Isolation of compounds such as Se trisulfides will be difficult as during extraction they will be destabilized and this will result in precipitation of elemental Se [137]. Maher [30, 37] found that methanol–water mixtures and water only extract small amounts of Se, although some authors have reported extraction of large amounts of Se (60–80 percent) with water [129, 130–134, 138, 139]. To extract Se from marine animal tissues, after freeze-drying and grinding, a detergent such as sodium dodecylsulfate (SDS) is normally required [30, 32, 37]. SDS forms ion-pairs with proteins and renders them water soluble. Therefore, most Se in seafood is either membrane-bound or associated with lipoprotein material.

Enzymes such as protease in conjunction with pancreatin and amylase have been extensively used to liberate Se species from proteins for analysis [43, 57, 128, 133–136]. Relatively long times ($\sim$24 h) are required to fully hydrolyze proteins using enzymes. However, not all Se is released as simple amino acids. Some peptides, and small molecular weight proteins remain. Thus, ultrafiltration ($<$1 kDa) before analysis will be needed to separate amino acids from other material with higher molecular weight. In the presence of cysteine, selenomethionine and selenocysteine are stable to enzyme attack (Fig. 20.2). However, although large amounts of Se are released from marine tissues (30–60 percent), little (less than 10–20 percent) is characterizable by HPLC-ICP-MS.

Alternatively, MW-assisted acid hydrolysis using 6 M HCl at 100–110°C under anaerobic conditions can be used to hydrolyze proteins [32]. Conflicting results as to the stability of selenomethionine have been reported in the literature [133, 140–142]. Stability is probably dependent on sample matrix and procedural variations. Selenocysteine is widely reported to be unstable [143], and normally is reacted with either 1-fluoro-2,4-dinitrobenzene, iodoacetic acid, or ethylenimine to form a stable 2,4-dinitrophenyl, carbomethyl or aminoethyl derivative before hydrolysis [137, 144–146].

Maher and coworkers [25, 30, 32, 37] found that selenomethionine and selenocysteine are stable during acid hydrolysis as long as a reducing agent such as cysteine is added and oxygen is rigorously excluded. A selenocysteine–cysteine (Cyst–SeCyst) complex is formed and some selenomethionine can be oxidized, probably to selenomethionine–selenoxide (Fig. 20.3).

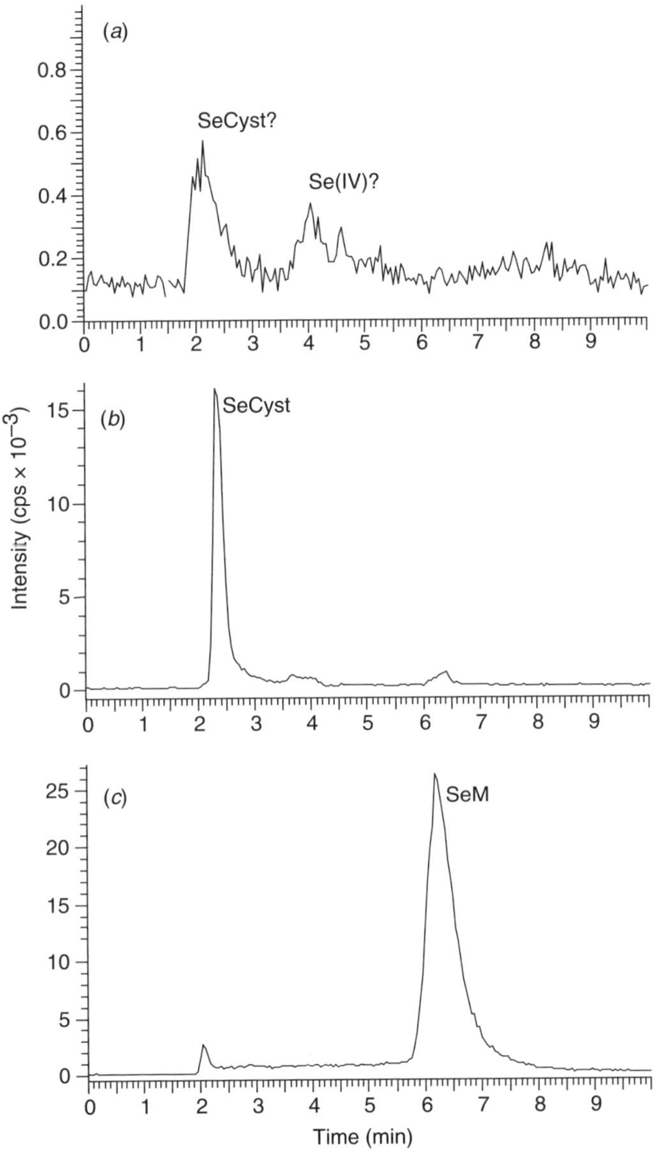

Figure 20.2. HPLC-ICP-MS separation and quantification of selenoamino species in lobster hepatopancreas after digestion with protease type XIV (10 mg per 0.1 g sample), 0.05% m/v cysteine, pH 7.5, 37°C for 24 h. Chromatography conditions as in Figure 20.1. (*a*) Extract; (*b*) extract spiked with selenocystine; (*c*) extract spiked with selenomethionine.

Sequential extractions with various solvents have been used to isolate discrete biochemical fractions and to investigate the biochemical associations of Se in marine animals and seaweeds [30, 32, 37]. Figure 20.4 presents an overall flow chart of extraction and separation procedures that can be used to isolate, separate,

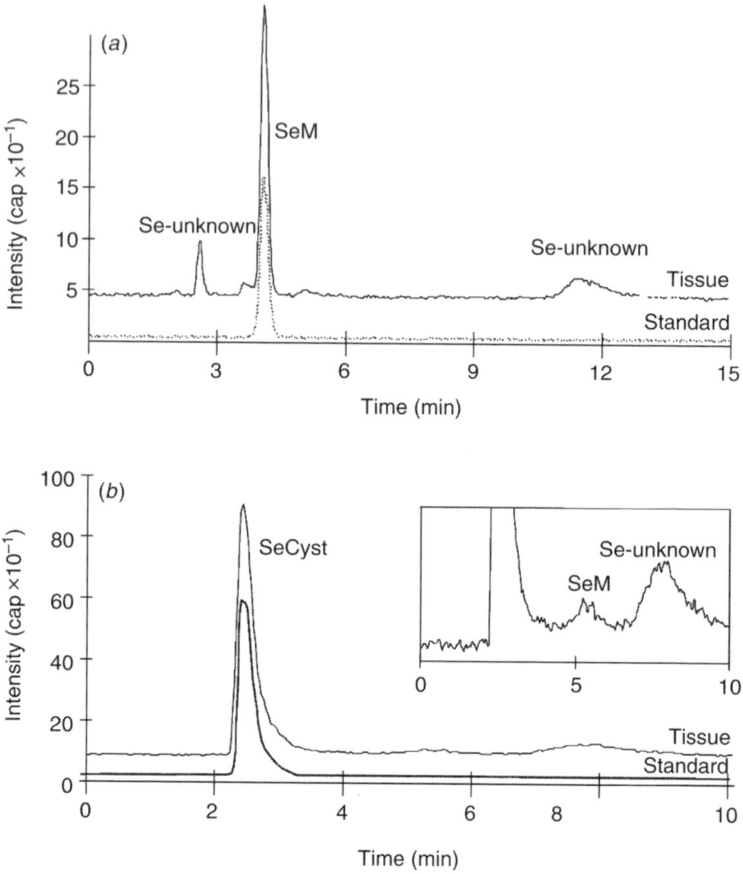

Figure 20.3. HPLC-ICP-MS separation of 6 M HCl–0.05% m/v cysteine hydrolyzed selenoamino acid standards and *Mugil cephalus* muscle tissue. (*a*) Cation exchange of hydrolyzed selenomethionine and muscle tissue, Supelcosil LC-SCX cation exchange column, 20 mM pyridine-formic acid, 1.5 ml min^{-1}, 40°C, pH 2.5. (*b*) Anion exchange of hydrolyzed selenocystine and muscle tissue. Chromatography conditions as in Figure 20.1.

and measure lipid, cytosol-soluble, membrane-bound, and residue and associated Se species in seafoods.

Chloroform–methanol is used to extract lipid and lipoprotein material according to Folch et al. [147], with the addition of aqueous KCl to isolate lipid and lipoproteins. Alternatively, if lipid material is not to be investigated, fats can be removed with acetone with little co-extraction of Se. The presence of fats is not desirable as lower aqueous extraction efficiencies are obtained and nondefatted samples can degrade HPLC performance. Extraction with phosphate buffer will remove cytosol-soluble Se, while extraction with phosphate buffer containing SDS will release and solubilize membrane-bound Se. Phenylmethane sulfonyl fluoride or a similar reagent needs to be added to buffers to retard proteolytic

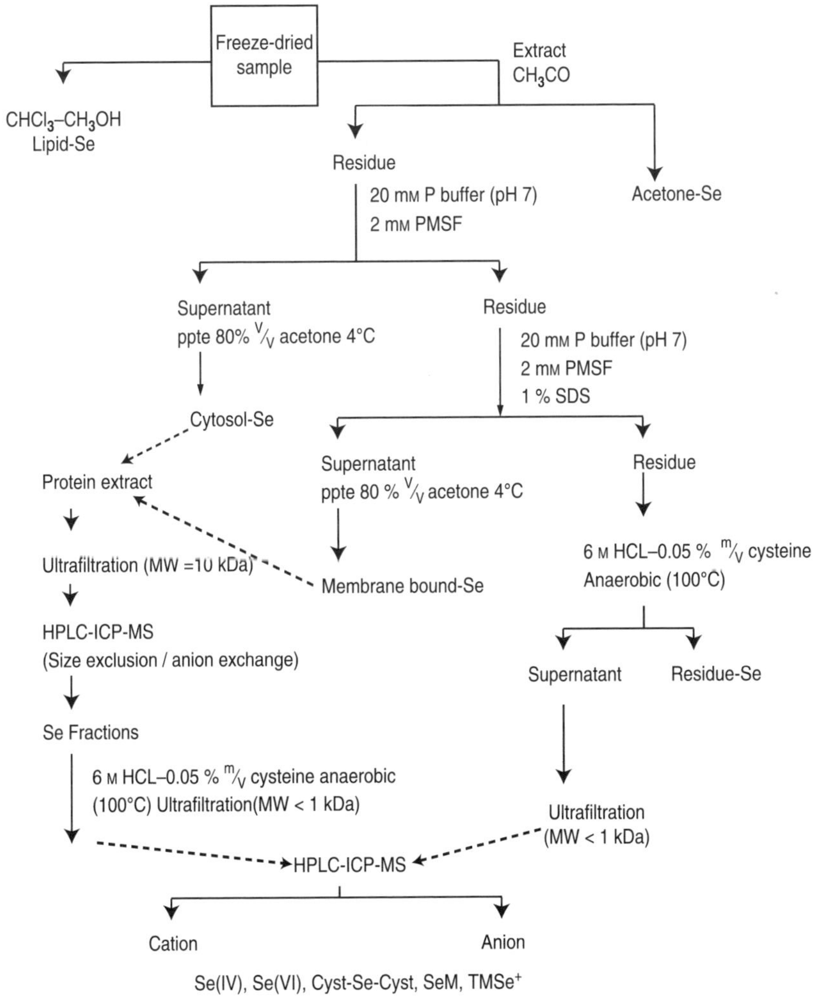

Figure 20.4. Extraction and separation procedures that can be used to isolate, separate, and measure lipid, water-soluble, and residue associated Se species in seafood.

activity. Insoluble Se can be released from pellets with protease or hydrolyzed with 6 M HCl. Free Se species are unlikely to be present in buffer extracts; these can be thus stored frozen or freeze-dried to prevent the bacterial degradation of proteins. Moreno et al. [57] have shown that Se species in enzymatic extracts are stable for at least 10 days if stored in Pyrex containers at 4°C.

20.3.3 Separation and Quantification of Se Species

20.3.3.1 HPLC Separation of Se Proteins Selenium proteins and amino acids have a wide variety of molecular weights and ionic characteristics. Initially, a

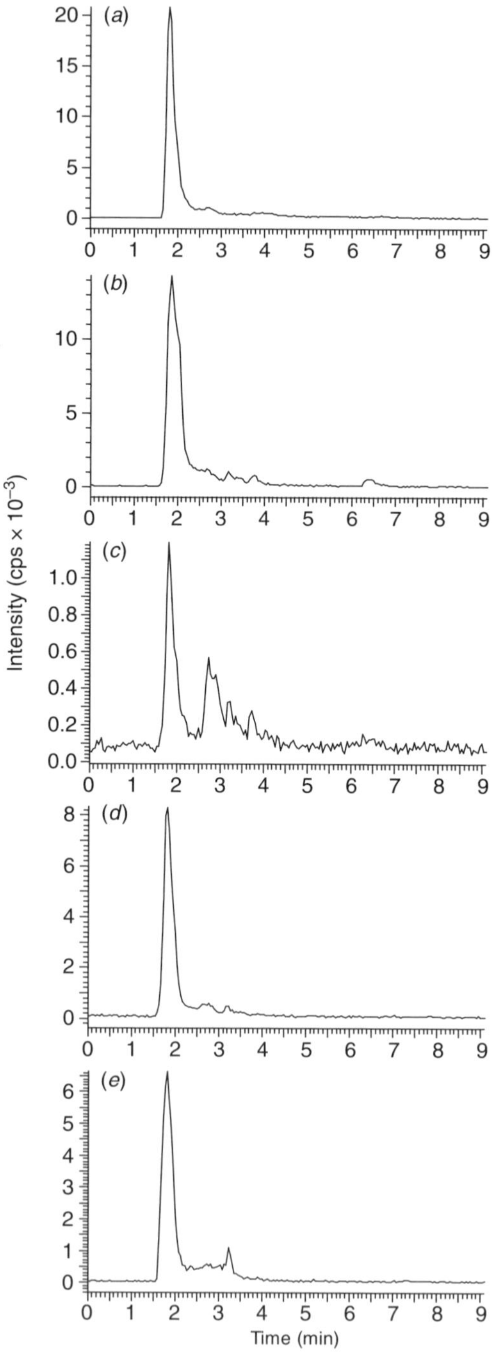

Figure 20.5. Separation of Se-containing proteins in marine animal tissues by SEC using a Macrosphere GCP (100A, 250 × 4.6 mm) column eluted with 20 mM K_2HPO_4, pH 7, 25°C. (*a*) *Mugil cephalus* muscle; (*b*) *Mugil cephalus* stomach; (*c*) prawn muscle; (*d*') oyster homogenate; (*e*) lobster hepatopancreas.

preliminary separation of proteins based on molecular weight using size exclusion chromatography (SEC) can be carried out [26, 38, 53, 134, 148]. Given that proteins are denatured and that interaction (adsorption, ion-exchange) occurs with the stationary phase, the molecular weight, as determined by comparison with cochromatographed molecular weight standards, will only be an "apparent" or "indicative" molecular weight. Two size exclusion columns (1–10 kDa, 1–300 kDa) can be used to separate high- and low-molecular-weight proteins. Extracts should be therefore ultrafiltered before chromatography to separate proteins into 0–10 and >10 kDa fractions. Further purification and separation of coeluting Se proteins can be achieved by using anion or affinity chromatography [38, 132, 145, 146, 148, 149]. Using preparative HPLC rather than analytical HPLC has the advantage that higher loadings of proteins can be applied to columns [150], thus allowing individual Se fractions to be collected for further chromatography and hydrolysis to constituent amino acids to take place. Examples of the separation of membrane-bound Se proteins isolated from marine animal tissues by SEC are given in Figure 20.5.

20.3.3.2 HPLC Separation of Selenoamino Acids Selenium proteins and amino acids have a wide variety of ionic characteristics that are pH dependent [149, 151, 152] and can be separated by anion, cation, and reversed-phase chromatography [94, 97, 98, 133, 135, 152–157]. Maher and his coworkers [30, 32, 37] used both anion and cation HPLC to confirm the identity of Se species (Figs 20.1 and 20.3). It should be noted that selenocysteine is released from proteins complexes with cysteine and chromatographs differently than the commercially available selenocystine [158]. Although the major selenoamino acid in mullet has been tentatively identified as selenocysteine (Fig. 20.3), further confirmation is required.

Selenomethionine forms optical enantiomers and a chiral column is needed to separate the two forms [160–162]. As the presence of salts and organic matter can cause retention times to vary, it is essential that spiking is used to verify them. Also, it should be noted that retention time alone does not allow peaks to be unequivocally assigned. It has been found that the removal of lipid material prior to protein extraction dramatically improves the reproducibility of the chromatographic system and the lifetime of the columns.

20.4 DIRECT STRUCTURAL INFORMATION

At present identification and quantification of selenoamino acids in seafood relies solely on matching retention time of peaks with available standards. To unambiguously assign peaks, the use of HPLC coupled to a mass spectrometer is required, as for the identification of As species [158, 159]. However, at present the low concentrations of Se species, the low detection power, and the lack of knowledge about Se species does not make this possible [131, 135, 136, 163]. Preparative scale isolation and purification of Se proteins as done for bacterial proteins is required to allow Se proteins and selenoamino acids to be characterized by mass spectrometry [164]. Preparative gel and two-dimensional electrophoresis offers promise to obtain pure Se protein fractions. Selenium-containing proteins from *Anadaria trapezia* have been isolated and separated using SDS-polyacrylamide gel electrophoresis (PAGE) [165]. However, it was not possible to determine which specific proteins contained Se, as at that time only ET-AAS was available to measure Se. Moreover, SDS interferes with the determination of Se by ET-AAS [165]. Elution of proteins from the gel and determination by Q-ICP-MS or direct in situ determination by laser ablation Q-ICP-MS offers considerable promise [166, 167].

20.5 AVAILABILITY AND USE OF CRMs

Digestion, extraction, separation, and measurement processes are all subject to errors including contamination, degradation, matrix effects, and calibration errors. Mass balances and recovery rates must be determined for all steps of the

TABLE 20.3. CRMs for Total Se

CRM	Certified Concentration ($\mu g\ g^{-1}$ Dry Mass)
BCR CRM 414 Plankton	1.75 ± 0.10
BCR 279 Sea Lettuce (*Ulva lactuca*)	0.593 ± 0.032
IAEA 149/TM Fucus[a]	0.079 ± 0.032
NIES No. 9 Sargasso Seaweed[a]	0.05
MURST-ISS-A2 Antarctic Krill	7.37 ± 0.91
IAEA-MA-A 1™ Copepod Homogenate	3.0 ± 0.2
NIST SRM 1566 Oyster Tissue	2.1 ± 0.5
NIST SRM 1566a Oyster Tissue	2.21 ± 0.24
AGAL-3 Prawn	2.74 ± 0.24
BCR CRM 278R Mussel Tissue (*Mytilus edulis*)	1.84 ± 0.10
NIES Mussel No. 6 (*Mytilus edulis*)[a]	1.65 ± 0.24
NRCC TORT-1 Lobster Heptapancreas	6.88 ± 0.47
NRCC TORT-2 Lobster Heptapancreas	5.63 ± 0.67
AGAL-2 Shark	1.67 ± 0.19
NRCC LUTS-1 Lobster Heptapancreas	4.3 ± 0.36
BCR CRM 422 Cod Muscle	1.63 ± 0.07
IAEA MA-A-2 Fish Flesh	1.7 ± 0.3
NRCC DOLT-1 Dogfish Liver	7.34 ± 0.42
NRCC DOLT-2 Dogfish Liver	6.06 ± 0.49
NRCC DORM-1 Dogfish Muscle	1.62 ± 0.12
NRCC DORM-2 Dogfish Muscle	1.4 ± 0.09
NIST CRM 50 Albacore Tuna[a]	3.6 ± 0.4

[a]Reference value only.

BCR = Commission of the European Communities, Community Bureau of Reference, Belgium. IAEA = International Atomic Energy Agency, Austria. NIES = National Institute of Environmental Science, Japan. ISS = Istituto Superiore di Sanità (National Institute of Health), Italy. NIST = National Institute of Standards and Technology, USA. AGAL = Australian Government Analytical Laboratory, Australia. NRCC = National Research Council of Canada, Canada.

analytical procedure. Certified Reference Materials (CRMs) are available to assist in quantifying errors. There are at least 18 marine CRMs for total Se concentrations (Table 20.3). In addition, many laboratories have published values for Se species in CRMs, but to date no unequivocal evidence has been presented that the correct identification of Se species has occurred, and published information on Se species should be used with caution.

20.6 CONCLUDING REMARKS

Q-ICP-MS can detect microgram per liter concentrations of Se, but suffers from polyatomic interferences generated from the sample matrix. Efficient digestion of samples is required to eliminate C enhancement of Se signals within the plasma. ET-AAS offers a good alternative to cross-check Se concentrations in digests. HG coupled to Q-ICP-MS should not be used for total Se measurements, unless Se species in digests have been degraded to inorganic Se, as total Se concentrations will be underestimated.

SEC and anionic HPLC-Q-ICP-MS can be used to determine the approximate molecular weight of Se-containing proteins and identify fractions for selenoamino acid analysis. The coupling of anion and cation HPLC and Q-ICP-MS also allows selenoamino acids in hydrolyzed protein extracts to be measured. However, the lack of knowledge about species, low Se concentrations, and lack of detection power make quantification of Se species difficult. The concurrent use of HPLC-MS-MS is needed to confirm the identity of Se species, but at present the low concentrations of Se in extracts preclude its use. It is essential that CRMs are used to confirm the validity of results regarding total Se and Se species.

REFERENCES

1. K. Schwartz, C. M. Faltz, Selenium as an integral part of Factor 3 against dietary necrotic liver degeneration, *J. Am. Chem. Soc.*, **79** (1957), 3292–3296.
2. H. E. Ganther, Selenoproteins, *Chem. Scripta*, **8A** (1975), 79–84.
3. R. F. Burk, Molecular biology of selenium with implications for metabolism, *FASEB J.*, **5** (1991), 2274–2279.
4. T. Mizutani, H. Kurata, K. Yamada, Study of mammalian selenocysteyl-tRNA synthesis with $[^{75}Se]HSe^-$, *FEBS J.*, **289** (1991), 59–63.
5. R. Read, T. Bellew, J.-G. Yang, K. E. Hill, I. S. Palmer, R. F. Burk, Selenium and amino acid composition of selenoprotein P, the major selenoprotein in rat serum, *J. Biol. Chem.*, **265** (1990), 17899–17905.
6. B. Akesson, T. Bellew, R. Burk, Purification of selenoprotein P from human plasma, *Biochim. Biophys. Acta*, **1204** (1994), 243–249.
7. B. J. Lee, M. Rajagopalan, Y. S. Kim, K. H. You, K. B. Jacobson, D. Hatfield, Selenocysteine $tRNA^{[Ser]Sec}$ gene is ubiquitous within the animal kingdom, *Mol. Cell Biol.*, **10** (1990), 1940–1949.

8. D. Hatfield, B. J. Lee, L. Hampton, A. Diamond, Selenium induces changes in the selenocysteine $tRNA^{[Ser]Sec}$ population in mammaliam cells, *Nucl Acid Res.*, **19** (1991), 939–943.
9. S. G. Patching, P. H. E. Gardiner, Recent developments in selenium metabolism and chemical speciation: a review, *J. Trace Elem. Med. Biol.*, **13** (1999), 193–214.
10. A. J. Wittwer, L. Tsai, W.-M. Ching, T. C. Stadtman, Identification and synthesis of a naturally occurring selenonucleoside in bacterial tRNAs 5-[(methylamino)methyl]-2-selenouridine, *Biochemistry*, **23** (1984), 4650–4655.
11. T. C. Stadtman, Specific occurrence of selenium in enzymes and amino acid tRNAs, *FASEB*, **1** (1987), 375–379.
12. Y. Thomassen, Y. J. Aaseth, Human tissues, in M. Inhat (ed.), *Occurrence and Distribution of Selenium*, CRC Press, Boca Raton, FL, 1989, Chapter 5, pp. 169–213.
13. H. E. Ganther, Biochemistry of Selenium, in R. A. Zingaro, W. C. Cooper (eds), *Selenium*, Van Nostrand and Reinhold, New York, 1974, Chapter 9, pp. 546–614.
14. A. L. Moxon, O. E. Olson, Selenium in Agriculture, in R. A. Zingaro, W. C. Cooper (eds), *Selenium*, Van Nostrand and Reinhold, New York, 1974, Chapter 12, pp. 675–707.
15. R. J. Turner, J. M. Finch, Selenium and the immune response. *Proc. Nutr. Soc.*, **50** (1991), 275–285.
16. ANFA, *Australian Food Standards Code*, Australian and New Zealand Food Authority, ACT, Australia, 1992.
17. A. D. Lemly, Environmental implications of excess selenium: a review, *Biomed. Environ. Sci.*, **10** (1997), 415–435.
18. W. A. Maher, Selenium in marine organisms from St. Vincent's Gulf, South Australia, *Mar. Pollut. Bull.*, **14** (1983), 35–36.
19. K. W. M. Sui, S. S. Berman, The marine environment, in M. Ihnat (ed.), *Occurrence and Distribution of Selenium*, CRC Press, Boca Raton, Fl, 1989, Chapter 11, pp. 264–293.
20. W. A. Maher, G. E. Batley, Organometallics in the nearshore marine environment of Australia, *Appl. Organomet. Chem.*, **4** (1990), 419–437.
21. W. A. Maher, S. Baldwin, M. Deaker, M. Irving, Review: characteristics of selenium in Australian marine biota, *Appl. Organomet. Chem.*, **6** (1992), 103–112.
22. S. Baldwin, W. A. Maher, E. Kleber, F. Krikowa, Selenium in marine organisms of seagrass habitats (*Posidonia australis*) of Jervis Bay Australia, *Mar. Pollut. Bull.*, **32** (1996), 310–316.
23. S. A. Baldwin, W. A. Maher, Spatial and temporal variation of selenium concentration in five species of intertidal molluscs from Jervis bay, Australia, *Mar. Environ. Res.*, **44** (1997), 243–262.
24. D. F. Jolley, W. A. Maher, J. Kyd, Selenium accumulation in the cockle *Anadara trapezia*, *Environ. Pollut.*, **132** (2004), 203–212.
25. W. A. Maher, Distribution of selenium in marine animals: relationship to diet, *Comp. Biochem. Phys.*, **86C** (1987), 131–133.
26. S. Braddon-Galloway, J. E. Balthrop, Se-dependent GSH-peroxidase isolated from Black Sea Bass (*Centropristis striata*), *Comp. Biochem. Physiol.*, **82C** (1985), 297–300.
27. P. Zatta, P. Buso, G. Moschini, Selenium distribution in the tissues of *Carcinus maenas*, *Comp. Biochem. Physiol.*, **81C** (1985), 469–470.

28. Y. Horiguchi, H. Noda, M. Naka, Biochemical studies on marine algae. VI. Concentration of selenium in marine algae and its importance as a trace metal for the growth of lavers. *Bull. Jap. Soc. Sci. Fish.*, **37** (1971), 996–1001.

29. J. D. Burton, W. A. Maher, P. J. Statham, Aspects of the distribution and chemical form of arsenic and selenium in ocean waters and marine organisms, *Thalassiosira Jugoslavia*, **16** (1980), 155–164.

30. W. A. Maher, Selenium in marine macroalgae, *Bot. Mar.*, **2** (1985), 269–274.

31. M. Barwick, W. A. Maher, Biotransference and biomagnification of selenium, copper, cadmium, zinc, arsenic and lead in a temperate seagrass community from lake Macquarie, NSW, Australia, *Mar. Environ. Res.*, **56** (2003), 471–502.

32. W. A. Maher, M. Deaker, D. Jolley, F. Krikowa, B. Roberts, Selenium occurrence, distribution and speciation in the cockle *Anadara trapezia* and the mullet *Mugil cephalus*, *Appl. Organomet. Chem.*, **11** (1997), 313–326.

33. G. M. Peters, W. A. Maher, F. Krikowa, A. C. Roach, H. K. Jeswani, J. P. Barford, V. G. Gomes, D. D. Reible, Selenium in sediments, pore waters and benthic infauna of Lake Macquarie, New South Wales, Australia, *Mar. Environ. Res.*, **47** (1999), 491–508.

34. G. M. Peters, W. A. Maher, D. Jolley, B. I. Carroll, V. G. Gomes, A. V. Jenkinson, G. D. McOrist, Selenium contamination, redistribution and remobilisation in sediments of Lake Macquarie, NSW, *Org. Geochem.*, **30** (1999), 1287–1300.

35. J. Kirby, W. A. Maher, F. Krikowa, Selenium, cadmium, copper, and zinc concentrations in sediment and mullet (*Mugil cephalus*) from the southern basin of Lake Macquarie, NSW, Australia, *Arch. Environ. Contam. Toxicol.*, **40** (2001), 246–256.

36. J. Kirby, W. A. Maher, D. Harasti, Changes in selenium, copper, cadmium and zinc concentrations in mullet (*Mugil cephalus*) from the southern basis of Lake Macquarie, Australia in response to alteration of coal fire power station fly ash handing procedures, *Arch. Environ. Contam. Toxicol.*, **41** (2001), 171–181.

37. W. A. Maher, Characteristics of selenium in marine animals, *Mar. Pollut. Bull.*, **16** (1985), 33–34.

38. S. Braddon-Galloway, C. R. Sumpter, A unique selenoprotein isolated from yellowfin tuna (*Thunnus albacares*) liver, *Comp. Biochem. Physiol.*, **83C** (1986), 13–17.

39. J. S. Edmonds, M. Morita, The identification of selenium species in biological samples, *Appl. Organomet. Chem.*, **14** (2000), 133–145.

40. P. Craig, W. A. Maher, Organoselenium compounds in the environment, in P. Craig (ed.), *Organometallic Compounds in the Environment*, Longman, Harlow, 2003, Chapter 10, pp. 391–399.

41. Y. Kobayashi, Y. Ogra, K. Ishiwata, H. Takayama, N. Aimi, K. T. Suzuki, Selenosugars are key and urinary metabolites for selenium excretion with the required to low toxic range, *Proc. Natl. Acad. Sci. USA*, **99** (2002), 15933–15936.

42. B. Gammelgaard, K. E. Madsen, J. Bjerrum, L. Bendahl, O. Jons, J. Olsen, U. Sidenius, Separation, purification and identification of the major selenium metabolite from human urine by multidimensional HPLC-ICP-MS and APCI-MS, *J. Anal. Atom. Spectrom.*, **18** (2003), 65–70.

43. J. J. Wrench, Selenium metabolism in marine phytoplankters *Tetraselmis tetrathele* and *Dunaliiella minuta*, *Mar. Biol.*, **49** (1978), 231–236.

44. N. R. Bottino, C. H. Banks, K. J. Irgolić, P. Micks, A. E. Wheeler, R. A. Zingaro, Selenium containing amino acids and proteins in marine algae, *Phytochemistry*, **23** (1984), 2445–2452.
45. J. M. Gennity, N. R. Bottino, R. A. Zingaro, A. E. Wheeler, K. J. Irgolić, The binding of selenium to the lipids of two unicellular marine algae, *Biochem. Biophys. Res. Commun.*, **118** (1984), 176–182.
46. B. H. Ng, J. W. Anderson, Light-dependent incorporation of selenite and sulphite into selenocysteine and cysteine by isolated pea chloroplasts, *Phytochemistry*, **18** (1979), 573–580.
47. T. A. Brown, A. Shrift, Selenium: toxicity and tolerance in higher plants, *Biol. Rev.*, **57** (1982), 59–83.
48. J. J. Wrench, Uptake and metabolism of selenium by oysters, *Mar. Sci. Commun.*, **5** (1979), 47–49.
49. E. P. Painter, The chemistry and toxicity of selenium compounds with special reference to the selenium problem, *Chem. Rev.*, **28** (1941), 179–213.
50. H. E. Ganther, Selenotrisulfides. Formation by reaction of thiols with selenious acid, *Biochemistry*, **7** (1968), 2898.
51. J. R. Holker, J. B. Speakman, Action of selenium dioxide on wool. *J. Appl. Chem.*, **8** (1958), 1–3.
52. R. Martoja, D. Viale, Accumulation of mercuric selenide granules in the liver of Odontocetes (mammifer cetaced): a possible method of detoxification of methymercury by selenium, *Comp. Rend. Hebdom. Sean., Paris, Series D: Sci. Natur.*, **285** (1977), 109–112.
53. K. Matsumoto, Speciation and determination of selenium and mercury accumulated in a Dolphin liver, in K. S. Subramanian, G. V. Iyengar, K. Okamoto, (eds), *Biological Trace Element Research: Multidisciplinary Perspectives, ACS Symposium Series*, no. 445, American Chemical Society, Washington, DC, 1991, pp. 278–289.
54. H. M. Crews, Speciation of trace elements in foods, with special reference to cadmium and selenium: is it necessary? *Spectrochim. Acta Part B*, **53** (1998), 213–219.
55. W. A. Maher, An investigation of trace element losses during lyophilization of marine biological samples, *Sci. Total Environ.*, **26** (1983), 173–181.
56. M. A. Z. Arruda, M. Gallego, M. Valcarcel, Semi-on-line microwave assisted digestion of shellfish tissue for the determination of selenium by electrothermal atomic absorption spectrometry, *J. Anal. Atom. Spectrom.*, **11** (1996), 169–173.
57. P. Moreno, M. A. Quijano, A. M. Gutierrez, M. C. Perez-Conde, C. Camara, Stability of total selenium and selenium species in lyophilised oysters and their enzymatic extracts, *Anal. Bioanal. Chem.*, **374** (2002), 466–476.
58. D. C. Reamer, C. Veillon, Preparation of biological materials for determination of selenium by hydride generation–atomic absorption spectrometry, *Anal. Chem.*, **53** (1981), 1192–1195.
59. K. W. M. Siu, S. S. Berman, Comparison of two digestion methods used in the determination of selenium in marine biological tissues by gas chromatography with electron capture detection, *Talanta*, **31** (1984), 1010–1012.
60. G. K. H. Tam, G. Lacroix, Recovery studies of selenium using dry ashing procedure with ashing aid, *J. Environ. Sci. Health B*, 1**4** (1979), 515–524.

61. B. Welz, M. Melcher, Decomposition of marine biological tissues for determination of arsenic, selenium and mercury using hydride-generation and cold vapour atomic absorption spectrometeries, *Anal. Chem.*, **57** (1985), 427–431.

62. B. Llyod, P. Holt, P. H. T. Delves, Determination of selenium in biological samples by hydride generation and atomic absorption spectroscopy, *Analyst*, **107** (1982), 927–933.

63. S. B. Adeloju, A. M. Bond, M. H. Briggs, Critical evaluation of some wet digestion methods for the stripping voltametric determination of selenium in biological materials, *Anal. Chem.*, **56** (1984), 2397–2401.

64. J. H. Watkinson, Fluorimetric determination of selenium in biological material with 2, 3-diaminonaphthalene, *Anal. Chem.*, **38** (1966), 92–97.

65. H. Uchida, Y. Shimoishi, K. Toei, Rapid determination of trace amounts of selenium in biological materials by gas chromatography with electron capture detection, *Analyst*, **106** (1981), 757–762.

66. J. Piwonka, G. Kaiser, G. Tölg, A survey of selenium in the environment and a critical review of its determination, *Fresenius Anal. Chem.*, **321** (1985), 225–234.

67. W. A. Maher, Decomposition of marine biological materials for the determination of selenium by fluorescence spectroscopy, *Microchem. J.*, **35** (1987) 125–129.

68. M. Ihnat, Atomic absorption spectrometric determination of selenium with carbon furnace atomisation, *Anal. Chim. Acta*, **82** (1976), 293–309.

69. W. A. Maher, Fluorimetric determination of selenium in marine geological materials, *Anal. Lett.*, **A16** (1983), 491–500.

70. K. Julshamn, O. Ringdal, K. E. Slinning, O. R. Braekkan, Optimisation of determination of selenium in marine samples by atomic absorption spectrometry: comparison of a flameless graphite furnace atomic absorption system with a hydride generation atomic absorption system, *Spectrochim. Acta*, **37B** (1982), 473–482.

71. M. Verlinden, On the acid decomposition of human blood and plasma for the determination of selenium. *Talanta*, **29** (1982), 875–882.

72. J. Neve, M. Hanocq, L. Molle, B. Le Febvre, Study of some systematic errors during the determination of total selenium and some of its ionic species in biological materials, *Analyst*, **107** (1982), 934–941.

73. M. Hoenig, P. Van Hoeyweghen, Determination of selenium and arsenic in animal tissues with platform furnace atomic absorption spectrometry and deuterium background correction, *Int. J. Environ. Anal. Chem.*, **24** (1986), 193–202.

74. B. Welz, M. Schubert-Jacobs, G. Schlemmer, M. Sperling, Trace elements in analytical chemistry in medicine and biology, in P. Brätter, P. Schramel (eds), *Proceedings of the Fifth International Workshop on Trace Elements in Medicine and Biology, Neuherberg*, Vol. 5, Walter de Gruyter, Berlin (1988), pp. 25–42.

75. H. Matusiewicz, R. E. Sturgeon, S. S. Berman, Trace element analysis of biological materials following pressure digestion with nitric acid–hydrogen peroxide and microwave heating, *J. Anal. Atom. Spectrom.*, **4** (1989), 323–327.

76. S. Y. LamLeung, V. K. M. Cheng, Y. W. Lam, Application of a microwave oven for drying and nitric acid extraction of mercury and selenium from fish tissue, *Analyst*, **116** (1991), 957–959.

77. O. E. Olson, I. S. Palmer, E. E. Cary, Modification of the official fluorimetric method for selenium in plants, *J. AOAC*, **58** (1975), 117– 121.

78. W. A. Maher, Determination of selenium in marine organisms using hydride generation and electrothermal atomic absorption spectroscopy, *Anal. Lett.*, **A16** (1983), 801–809.
79. A. J. Krynitsky, Preparation of biological tissue for determination of arsenic and selenium by graphite furnace atomic absorption spectrometry, *Anal. Chem.*, **59** (1987), 1884–1886.
80. P. Hocquellet, M. P. Candiller, Evaluation of microwave digestion and solvent extraction for the determination of trace amounts of selenium in feeds and plant and animal tissues by electrothermal atomic absorption spectrometry, *Analyst*, **116** (1991), 505–509.
81. M. A. Quijano, P. Moreno, M. Gutierrez, M. C. Perez-Conde, C. Camara, Selenium speciation in animal tissues after enzymatic digestion by high-performance liquid chromatography coupled to inductively coupled plasma mass spectrometry. *J. Mass Spectrom.*, **35** (2000), 878–884.
82. H. M. Kingston, L. B. Jassie, Microwave energy for acid decomposition at elevated temperatures and pressures using biological and botanical samples, *Anal. Chem.*, **58** (1986), 2534–2541.
83. J. K. Friel, C. S. Skinner, S. E. Jackson, H. P. Longerich, Analysis of biological reference materials prepared by microwave dissolution using inductively coupled plasma mass spetrometry, *Analyst*, **115** (1990), 269–273.
84. V. Ducros, Ruffieux, N. Belin, A. Favier, A comparison of two digestion methods for the determination of selenium in biological samples, *Analyst*, **119** (1994), 1715–1717.
85. R. Saraswati, T. W. Vetter, R. L. Watters, Jr, Comparison of reflux and microwave oven digestion for the determination of arsenic and selenium in sludge reference materials using flow-injection and atomic absorption spectrometry, *Analyst*, **120** (1995), 95–99.
86. D. W. Bryce, A. Izquierdo, M. D. Luque de Castro, Use of focussed microwaves for expeditive shortening of sample pre-treatment. Digestion and reduction procedures prior to selenium speciation as selenium (IV) or selenium (VI) *Analyst*, **120** (1995), 2171–2174.
87. M. Deaker, W. A. Maher, Low volume microwave digestion for the determination of selenium in marine biological tissues by graphite furnace atomic absorption spectrometry, *Anal. Chim. Acta*, **350** (1997), 287–294.
88. W. A. Maher, S. Forster, F. Krikowa, P. Snitch, G. Chapple, P. Craig, Measurement of trace elements and phosphorus in marine animal and plant tissues by low-volume microwave digestion and ICP-MS, *At. Spectrosc.*, **22** (2001), 361–369.
89. W. Goessler, M. Pavkov, Accurate quantification and transformation of arsenic compounds during wet ashing with nitric acid and microwave assisted heating, *Analyst*, **128** (2003), 796–802.
90. S. H. Tan, G. Horlick, Background spectral features in inductively coupled plasm mass spectrometry, *Appl. Spectrosc.*, **40** (1986), 446–460.
91. M. J. Campbell, A. Torvényi, Non-spectroscopic suppression of zinc in ICP-MS in a candidate biological reference material (IAEA 392, Algae), *J. Anal. Atom. Spectrom.*, **14** (1999), 1313–1317.
92. G. C. L. Araujo, M. H. Gonzalez, A. G. Ferreira, A. R. A. Noggueira, J. A. Nobrega, Effect of acid concentration on closed-vessel microwave-assisted digestion of plant materials, *Spectrochim. Acta*, **57B** (2002), 2121–2132.

93. G. A. Pedersen, E. H. Larsen, Speciation of four selenium compounds using high performance liquid chromatography with on-line detection by inductively coupled plasma mass spectrometry or flame atomic absorption spectroscopy, *Fresenius J. Anal. Chem.*, **358** (1997), 591–598.

94. E. H. Larsen, G. Pritzl, S. H. Hansen, Carbon enhanced inductively coupled pasma mass spectrometric detection of arsenic and selenium and its application to arsenic speciation, *J. Anal. Atom. Spectrom.*, **8** (1993), 1075–1085.

95. E. H. Larsen, S. Stürup, Carbon-enhanced inductively coupled plasma mass spectrometric detection of arsenic and selenium and its application to arsenic speciation, *J. Anal. Atom. Spectrom.*, **9** (1994), 1099–1105.

96. E. H. Larsen, J. Sloth, M. Hansen, S. Moesgaard, Selenium speciation and isotope composition in ^{77}Se-enriched yeast using gradient elution HPLC separation and ICP-dynamic reaction cell-MS, *J. Anal. Atom. Spectrom.*, **18** (2003), 310–316.

97. B. Gammelgaard, L. Bendahl, U. Sidenius, O. Jons, Selenium speciation in urine by ion-pairing chromatography with perfluorinated carboxylic acids and ICP-MS detection, *J. Anal. Atom. Spectrom.*, **17** (2002), 570–575.

98. K. Tirez, W. Brusten, S. Van Roy, N. De Brucker, L. Diels, Characterization of inorganic selenium species by ion chromatography with ICP-MS detection in microbial-treated industrial waste water, *J. Anal. Atom. Spectrom.*, **15** (2000), 1087–1092.

99. T. W. May, R. H. Wiedmeyer, A table of polyatomic interferences in ICP-MS, *Atom. Spectrosc.*, **9** (1997), 150–155.

100. A. Montaser, *Inductively Coupled Plasma Mass Spectrometry*, Wiley-VCH, New York, 1998, pp. 1–964.

101. S. J. Hill, M. J. Ford, L. Ebdon, Simplex optimisation of nitrogen-argon plasmas in inductively coupled plasma mass spectrometry for the removal of chloride based interferences, *J. Anal. Atom. Spectrom.*, **7** (1992), 719–727.

102. F. Laborda, M. J. Baxter, H. M. Crew, J. Dennis, Reduction of polyatomic interferences in inductively coupled plasma mass spectrometry by selection of instrumental parameters and using an argon-nitrogen plasma: effect on multi-element analyses, *J. Anal. Atom. Spectrom.*, **9** (1994), 727–736.

103. H. T. Delves, C. E. Sieniawska, Simple method for the accurate determination of selenium in serum by using inductively coupled plasma mass spectrometry, *Anal. Atom. Spectrom.*, **12** (1997), 387–389.

104. I. Feldmann, N. Jakubowski, D. Stueuer, Application of a hexapole collision and reaction cell in ICP-MS. Part 1. Instrumental aspects and operational optimisation, *Fresenius J. Anal. Chem.*, **365** (1999), 415–421.

105. S. D. Tanner, V. I. Baranov, Theory, design and operation of a dynamic reaction cell for ICP-MS, *Atom. Spectrom.*, **20** (1999), 45–52.

106. J. Marchante-Gayon, C. Thomas, I. Feldmann, N. Jakubowski, Comparison of different nebulisers and chromatographic techniques for the speciation of selenium in nutritional commercial supplements by hexapole collision and reaction cell ICP-MS. *J. Anal. Atom. Spectrom.*, **15** (2000), 1093–1102.

107. J. J. Sloth, E. H. Larsen, The application of inductively coupled plasma dynamic reaction cell mass spectrometry for measurement of selenium isotopes, isotope ratios and chromatographic detection of selenoamino acids, *J. Anal. Atom. Spectrom.*, **15** (2000), 669–672.

108. S. D. Tanner, V. I. Baranov, U. Volkopf, A dynamic reaction cell for inductively coupled plasma mass spectrometry (ICP-DRC-MS). Part III. Optimisation and analytical performance, *J. Anal. Atom. Spectrom.*, **15** (2000), 1261–1269.

109. S. Mazan, N. Gilon, G. Cretier, J. L. Rocca, J.-M. Mermet, Inorganic selenium speciation using HPLC-ICP-hexapole collision/reaction cell-MS, *J. Anal. Atom. Spectrom.*, **17** (2002), 366–370.

110. S. D. Tanner, V. I. Baranov, D. R. Bandura, Reaction cells and collision cells for ICP-MS: a tutorial review, *Spectrochim. Acta*, **57B** (2002), 1361–1452.

111. G. P. Jackson, F. I. King, D. C. Duckworth, Efficient polyatomic interference reduction in plasma source mass spectrometry *via* collision induced dissociation, *J. Anal. Atom. Spectrom.*, **18** (2003), 1026–1032.

112. L. H. Reyes, J. M. M. Gayon, J. I. G. Alonso, A. Sanz-Medel, Determination of selenium in biological materials by isotope dilution analysis with an octapole reaction system ICP-MS, *J. Anal. Atom. Spectrom.*, **18** (2003), 1210–1216.

113. J. J. Sloth, E. H. Larsen, S. H. Bugel, S. Moesgaard, Determination of total selenium and ^{77}Se in isotopically enriched human samples by ICP-dynamic reaction cell-MS, *J. Anal. Atom. Spectrom.*, **18** (2003), 317–322.

114. D. W. Koppenall, G. C. Eiden, C. J. Barinaga, Collision and reaction cells in atomic mass spectrometry: development status and applications, *J. Anal. Atom. Spectrom.*, **19** (2004), 561–570.

115. R. C. Hutton, A. N. Easton, Role of aerosol water vapour loading in inductively coupled plasma mass spectrometry, *J. Anal. Atom. Spectrom.*, **2** (1987), 595–598.

116. J. L. M. DeBoer, Real time adjustment of ICP-MS elemental equations, *J. Anal. Atom. Spectrom.*, **15** (2000), 1157–1161.

117. J. W. Tromp, R. T. Tremblay, J.-M. Mermet, E. D. Salin, Matrix interference diagnostics for the automation of inductively coupled plasma mass spectrometry (ICP-MS), *J. Anal. Atom. Spectrom.*, **15** (2000), 617–627.

118. I. Rodushkin, T. Ruth, D. Klockare, Non-spectral interferences caused by a saline water matrix in quadrupole and high resolution inductively coupled plasma mass spectrometry, *J. Anal. Atom. Spectrom.*, **13** (1998), 159–166.

119. B. Gammelgaard, O. Jons, Comparison of an ultrasonic nebulizer with cross-flow nebulizer for selenium speciation by ion-chromatography and inductively coupled plasma mass spectrometry, *J. Anal. Atom. Spectrom.*, **15** (2000), 499–505.

120. K. L. Yang, S. J. Jiang, Determination of selenium-compounds in urine samples by liquid chromatography–inductively coupled plasma–mass spectrometry with an ultrasonic nebuliser, *Anal. Chim. Acta*, **307** (1995), 109–115.

121. E. Puskel, Z. Mester, P. Fodor, Determination of selenoamino acids by high-performance liquid chromatography–hydraulic high pressure nebulization–atomic fluorescence spectrometry, *J. Anal. Atom. Spectrom.*, **14** (1999), 973–976.

122. J. Marchante-Gayon, I. Feldmann, C. Thomas, N. Jakubowski, Speciation of selenium in human urine by HPLC-ICP-MS with a collision and reaction cell, *J. Anal. Atom. Spectrom.*, **16** (2001), 457–463.

123. J. M. G. Lafuente, M. L. F. Sanchez, A. Sanz-Medel, Speciation of inorganic selenium and selenoaminoacids by on-line reversed phase high performance liquid chromatography–focused microwave digestion–hydride generation–atomic detection, *J. Anal. Atom. Spectrom.*, **11** (1996), 1163–1169.

124. W. C. Story, J. A. Caruso, T. Heitkemper, L. Perkins, Elimination of the chloride interference on the determination of arsenic using hydride generation inductively coupled plasma mass spectrometry, *J. Chromatogr. Sci.*, **30** (1992), 427–432.

125. D. Wallschlager, N. S. Bloom, Determination of selenite, selenate and selenocyanate in waters by ion chromatography–hydride generation–atomic fluorescence spectroscopy (IC-HG-AFS), *J. Anal. Atom. Spectrom.*, **16** (2001), 1322–1328.

126. M. Deaker, W. Maher, Determination of selenium in seleno-compounds and marine biological tissues using stabilised temperature platform furnace atomic absorption spectroscopy, *J. Anal. Atom. Spectrom.*, **10** (1995), 423–434.

127. D. M. Templeton, F. Ariese, R. Cornelis, L.-G. Danielsson, H. Muntau, H. P. Van Leeuwen, R. Lobinski, Guidelines and terms related to chemical speciation and fractionation of elements. Definitions, structural aspects and methodological approaches, *Pure Appl. Chem.*, **72** (2000), 1453–1470.

128. G. Lunde, Location of lipid-soluble selenium in marine fish to the lipoproteins, *J. Sci. Food Agric.*, **23** (1972), 987–994.

129. C. J. Cappon, J. C. Smith, Mercury and selenium content and chemical form in fish muscle, *Arch. Environ. Contam. Toxicol.*, **10** (1981), 305–319.

130. C. J. Cappon, J. C. Smith, Chemical form and distribution of mercury and selenium in edible seafood, *J. Anal. Toxicol.*, **6** (1982), 10–21.

131. C. J. Cappon, Speciation of selected trace elements in edible seafood, in J. O. Nriagu, M. S. Simmons (eds), *Food Contamination from Environmental Sources, Advances in Environmental Science and Technology*, J. O. Nriagu (ed.), Wiley-Interscience, New York, 1990, pp. 145–194.

132. S. A. Braddon, Selenoproteins in marine animals, in J. E. Spallholz, J. L. Martin, H. E. Ganther (eds), *Selenium in Biology and Medicine*, AVI Westport, CT, 1981, pp. 499–504.

133. P. Moreno, M. A. Quijano, A. M. Gutierrez, M. C. Perez-Conde, C. Camara, Fractionation studies of selenium compounds from oysters and their determination by high-performance liquid chromatography coupled to inductively coupled plasma mass spectrometry, *J. Anal. Atom. Spectrom.*, **16** (2001), 1044–1050.

134. P. Moreno, M. A. Quijano, A. M. Gutierrez, M. C. Perez-Conde, C. Cámara, Study of selenium species distribution in biological tissues by size exclusion and ion exchange chromatography inductively coupled plasma–mass spectrometry, *Anal. Chim. Acta*, **524** (2004), 315–327.

135. M. Kotrebai, S. M. Bird, J. F. Tyson, E. Block, P. C. Uden, Characterization of selenium species in biological extracts by enhanced ion-pair liquid chromatography with inductively coupled plasma–mass spectrometry and by referenced electrospray ionisation-mass spectrometry, *Spectrochim. Acta*, **54B** (1999), 1573–1591.

136. H. M. Crews, P. A. Clarke, D. J. Lewis, L. M. Owen, P. R. Strutt, A. Izquierdo, Investigation of selenium speciation in *in vitro* gastrointestinal extracts of cooked cod by high performance liquid chromatography–inductively coupled plasma mass spectrometry and electrospray mass spectrometry, *J. Anal. Atom. Spectrom.*, **11** (1996), 1177–1182.

137. H. E. Ganther, R. J. Kraus, S. J. Foster, Identification of selenocysteine by high performance liquid chromatography and mass spectrometry, *Meth. Enzymol.*, **107** (1984), 587–592.

138. K. Itano, Selenium in the extracts of skipjack liver, *Nippon Suisan Gakkaishi*, **49** (1983), 1933.

139. N. Sochi, Mercury and selenium in tuna muscle, *Chudokuhen Ho*, **12** (1979), 13–14.

140. L. Shepherd, R. E. Huber, Some chemical and biochemical properties of selenomethionine, *Can. J. Biochem.*, **47** (1969), 877–881.

141. M. B. De La Calle Guntinas, C. Brunori, R. Scerbo, S. Chiavarini, Ph. Quevauviller, F. C. Adams, R. Morabito, Determination of selenomethionine in wheat samples: comparison of gas chromatography–microwave induced plasma atomic emission spectrometry, gas chromatography–flame photometric detection and gas chromatography–mass spectrometry, *J. Anal. Atom. Spectrom.*, **12** (1997), 1041–1047.

142. M. A. Palacios, A. Varga, M. Gomez, C. Camara, F. Gavilanes, Evaluation of acid hydrolysis of proteins on Se-aminoacids and trimethylselenonium species by liquid chromatography–microwave digestion–hydride generation–atomic absorption spectrometry, *Quìm. Anal.*, **18** (1999), 163–168.

143. R. E. Huber, R. S. Criddle, Comparison of the chemical properties of selenocysteine and selenocystine with their sulfur analogues, *Arch. Biochem. Biophys.*, **122** (1967), 164–173.

144. J. W. Fostrom, J. J. Zakowski, A. L. Tappel, Identification of the catalytic site of rat liver glutathione peroxidase as selenocysteine, *Biochemistry*, **17** (1978), 2639–2644.

145. R. J. Kraus, S. J. Foster, H. E. Ganther, Indentification of selenocysteine in glutathione peroxidase by mass spectroscopy, *Biochemistry*, **22** (1983), 5853–5858.

146. A. L. Tappel, W. C. Hawkes, E. I. Wihelmsen, M. A. Motsenbocker, Selenocysteine containing proteins and glutathione peroxidase, *Meth. Enzymol.*, **107** (1984), 602–619.

147. J. Folch, M. Lee, G. H. Sloane-Stanley, A simple method for the isolation and purification of total lipids from animal tissues, *J. Biol. Chem.*, **226** (1957), 497–509.

148. L. H. Reyes, J. M. M. Gayon, J. I. G. Alonso, A. Sanz-Medel, Quantitative speciation of selenium in human serum by affinity chromatography coupled to post-column isotope dilution analysis ICP-MS, *J. Anal. Atom. Spectrom.*, **18** (2003), 11–16.

149. L-Ping, H. Nagasawa, K. Matsumoto, A. Suzuki, K. Fuwa, Extraction and purification of a new compound containing selenium and mercury accumulated in dolphin liver, *Biol. Trace Elem. Res.*, **11** (1986), 185–199.

150. R. Cornelis, J. DeKimpe, X. Zhang, Trace elements in clinical samples revisited – speciation is knocking at the door. Sample preparation, separation of species and measurement methods, *Spectrochim. Acta*, **53B** (1998), 187–196.

151. M. R. Da Silva, R. M. Olivas, O. F. X. Donard, M. Lamotte, Determination of the deprotonation constants of seleno-DL-cystine and seleno-DL-methionine and implication to their separation by HPLC, *Appl. Organomet. Chem.*, **11** (1997), 21–30.

152. F. Li, W. Goessler, K. J. Irgolić, Determination of trimethylselenonium iodide, selenomethionine, selenious acid and selenic acid using high performance liquid chromatography with on-line detection by inductively coupled plasma mass spectrometry or flame atomic absorption spectrometry, *J. Chromatogr. A*, **830** (1999), 337–344.

153. R. M. Muňoz Olivas, O. F. X. Donard, N. Gilon, M. Potin-Gautier, Speciation of organic selenium compounds by high-performance liquid chromatography–inductively coupled plasma mass spectrometry in natural samples, *J. Anal. Atom. Spectrom.*, **11** (1996), 1171–1176.

154. M. A. Quijano, M. Gutierrez, M. C. Perez-Conde, C. Camara, Determination of selenocystine, selenomethionine, selenite and selenate by high-performance liquid chromatography coupled to inductively coupled plasma mass spectrometry, *J. Anal. Atom. Spectrom.*, **11** (1996), 407–411.

155. S. M. Bird, P. C. Uden, J. F. Tyson, E. Block, E. Denoyer, Speciation of selenoamino acids and organoselenium compounds in selenium–enriched yeast using high-performance liquid chromatography-inductively coupled plasma mass spectrometry, *J. Anal. Atom. Spectrom.*, **12** (1997), 785–788.

156. J. Zheng, Y. Shibata, A. Tanaka, Study of the stability of selenium compounds in human urine and determination by mixed ion-pair reversed phase chromatography with ICP-MS detection, *Anal. Bioanal. Chem.*, **374** (2002), 348–353.

157. W. Goessler, D. Kuehnelt, C. Schlagenhaufen, K. Kalcher, M. Abegaz, K. J. Irgolić, Retention behaviour of inorganic and organic selenium compounds on a silica-based strong-cation exchange column with an inductively coupled plasma mass spectrometer as selenium specific detector. *J. Chromatogr. A*, **789** (1997), 233–245.

158. J. T. Deagen, M. A. Beilstein, P. D. Whanger, Chemical forms of selenium in selenium containing proteins from human plasma, *J. Inorg. Biochem.*, **41** (1991), 261–268.

159. K. L. Sutton, C. A. Ponce de Leon, K. L. Ackley, R. M. C. Sutton, A. M. Stalcup, J. A. Caruso, Development of chiral HPLC for selenoamino acids with ICP-MS detection: application to selenium nutritional supplements, *Analyst*, **125** (2000), 281–286.

160. A. Sanz-Medel, E. Blanco-Gonzalez, Chiral speciation of trace elements: approaches to the speciation of selenoaminoacid enantiomers in biological samples, *J. Anal. Atom. Spectrom.*, **16** (2001), 957–963.

161. J. J. Corr, E. H. Larsen, Arsenic speciation by liquid chromatography coupled with ion-spray tandem mass spectrometry, *J. Anal. Atom. Spectrom.*, **11** (1996), 1215–1224.

162. K. A. Francesconi, S. A. Pergantis, Application of selected reaction monitoring tandem mass spectrometry to the quantitative determination of an arsenic-containing nucleoside in a crude biological extract, *Analyst*, **129** (2004), 398–399.

163. M. Kotrebai, J. F. Tyson, E. Block, P. C. Uden, High performance liquid chromatography of selenium compounds utilising perfluorinated carboxylic acid ion pairing agents and inductively coupled plasma and electrospray ionisation mass spectrometric detection, *J. Chromatogr. A*, **866** (2000), 51–63.

164. J. R. Encinar, M. Sliwka-Kaszynska, A. Polatajko, V. Vacchina, J. Szpunar, Methodological advances for selenium speciation analysis in yeast, *Anal. Chim. Acta*, **500** (2003), 171–183.

165. D. F. Jolley, The accumulation and storage of selenium in *Anadara trapezia*, PhD thesis, University of Canberra, 1999, pp. 1–200.

166. H. Chassaigne, C. C. Chery, G. Bordin, F. Vanhaecke, A. R. Rodriguez, 2-Dimensional gel electrophoresis technique for yeast selenium-containing proteins – sample preparation and MS approaches for processing 2-D gel protein spots, *J. Anal. Atom. Spectrom.*, **19** (2004), 85–95.

167. T. W. M. Fan, E. Pruszkowski, S. Shuttleworth, Speciation of selenoproteins in Se-contaminated wildlife by gel electrophoresis and laser ablation-ICPMS, *J. Anal. Atom. Spectrom.*, **17** (2002), 1621–1623.

21

APPLICATION OF ICP-MS FOR THE EVALUATION OF Se SPECIES IN FOOD RELATED PRODUCTS AND IN DIETARY SUPPLEMENTS

KATARZYNA WROBEL AND KAZIMIERZ WROBEL

Instituto de Investigaciones Científicas, Universidad de Guanajuato, L. De Retana 5, 36000 Guanajuato, Mexico

JOSEPH A. CARUSO

Department of Chemistry, University of Cincinnati, OH 45221-0172, USA, E-mail: joseph.caruso@uc.edu, katarzyn@quijote.ugto.mx

The Determination of Chemical Elements in Food: Applications for Atomic and Mass Spectrometry. Edited by Sergio Caroli

21.1 INTRODUCTION

The scientific community and society in general are aware of the importance of Se for health. Depending on its chemical form and total concentration, the element may exhibit essential or toxic effects with a very narrow tolerance band [1]. The proteinogenic amino acid Se-cysteine forms an active center of a number of proteins, including selenoenzymes of glutathione peroxidase, iodothyronine deionidase, and thioredoxine reductase [2–4]. Experimental evidence exists of Se bioactivity in cancer prevention [5]. The ability of the element to reduce toxic effects of heavy metals has also been observed [6]. On the other hand, the toxicity of Se originates in its high affinity to S, which results in nonspecific competition of the two elements for S-binding sites in different biomolecules.

The main sources of Se for humans are food products and drinking water. The recommended Se Reference Nutrient Intake (RNI) is 75 and 60 $\mu g\ day^{-1}$ for adult males and females, respectively [7]. According to the UK Department of Health, an intake of 750–900 $\mu g\ day^{-1}$ is toxic and the maximum safe intake is 450 $\mu g\ day^{-1}$ for adult males [8]. However, the main concern is that the supply of Se in many geographical regions is below the recommended values, as low as 10–20 $\mu g\ day^{-1}$ in Keshan (Iran), China, the UK, Finland, and others [8, 9]. To deal with this problem, the use of Se supplements has become common practice. As the bioavailability, retention, and fate of the element in the human body is species-dependent, the chemical form of Se in such supplements is important. To assure efficient uptake of Se and to avoid possible adverse effects, the food products naturally rich in this element are preferred as well as vegetables or yeast enriched during growth. The identification and quantification of Se species in these materials is vital because it may help to link the observed beneficial/adverse effects of the element to its specific chemical form [10].

In this chapter the recent literature on speciation of Se in food related products and in dietary supplements is reviewed, and some recent experiences from the authors' laboratory are discussed [11–21].

21.2 NUTRITIONAL ASPECTS OF Se SPECIES

The essentiality of Se for the health of higher animals is well documented. The results of several epidemiological trials and studies carried out in different experimental systems have proved that Se deficiency involves higher disease susceptibility [8]. Low Se status has been related to the etiology of serious health disorders including immunity impairment, viral infections, male fertility failure, thyroid dysfunction, cardiovascular disease, and cancer [4]. Furthermore, the protective role of Se supplementation in element deficient animals and humans has been demonstrated [5, 22, 23]. It is also known that the biological functions of Se depend on about 20 selenoproteins, containing selenocysteine in their active site [24]. These proteins may arise from 25 genes identified by computer searching of the human genome, with alternate splicing resulting in the selenoproteins synthesis [25, 26].

Recent progress in Se biochemistry and, in particular, better understanding of its specific biological functions justify the necessity for defining suitable strategies for the supplementation of Se-deficient individuals. Full characterization of Se forms in supplements to be used, their stability, as well as species-dependent bioavailability and pharmacokinetics in human body are of primary importance. The food products naturally rich in Se should also been considered. On the other hand, the risk of intoxication due to excessive Se intake needs to be minimized [27].

The majority of Se in human diets is present as selenomethionine (plant and animal products) and selenocysteine (animal products) [28]. Inorganic forms of Se are also present in dietary products, but their contribution in nonfortified food is considerably lower. When entering the mammal organisms, all Se compounds are reduced to inorganic form, chemically similar to selenide, which is considered a common pool of Se. This inorganic, reduced Se is then used in combination with serine for the synthesis of selenocysteine, which is incorporated specifically in selenoproteins [24, 29]. Such a metabolic route could suggest that the form of Se administered would not affect element incorporation to the target proteins. However, different bioavailability, higher chemical reactivity of inorganic versus organic Se species as well as the production of biologically active, but nonprotein-bound, Se metabolites have to be considered. The excessive intake of selenate may cause oxidative stress, and selenite reacts rapidly and directly with glutathione, forming reduced Se of high affinity to thiol groups [30]. At high concentration rates, selenite was shown to be mutagenic [8, 31]. At the present time, the consensus has been reached that Se should be handled mainly as relatively unreactive, organic forms that also present higher bioavailability with respect to the inorganic species [24, 32, 33]. However, adverse effects of selenomethionine related to its non-specific incorporation into proteins have been reported [34, 35].

The biological activity of different Se species for tumor prevention has been studied by Ip et al. in several experimental models [36–38]. On the basis of their results, the highest activity was found for methylselenocysteine followed by selenite, selenocystine, and dimethylselenoxide [36]. In a more recent study that included different alk(en)ylselenocysteines and their γ-glutamyl derivatives, high cancer preventive activity was also observed for methylselenocysteine [39]. In addition to inorganic forms and possible Se metabolites, the activity of the synthetic compound [1,4-phenylenebis(methylene)selenocyanate] has been tested [40, 41].

An important aspect of products amenable to Se supplementation is the stability of the active form of the element. That is, neither total element content nor species identity may change over time or depending on conditions of storage or food processing. Since some Se species are highly volatile, they can be easily lost at elevated temperature, especially if present in the form of methylselenoaminoacids [8]. On the other hand, Se compounds can affect the sensory qualities of foods, especially odor and color. With natural foods, the highest Se levels have been found in Brazil nuts (8–35 $\mu g\ g^{-1}$) [12] and the protective effect of these nuts against tumor formation was observed in rats [42]. Different plants, mainly of the *Allium* family, have been enriched during their growth, yielding Se-rich products, which were then tested for possible cancer preventive activity [22, 23, 38, 43, 44]. However, the main source of additional Se is through selenized yeast-based commercially available supplements [11, 17].

21.3 PRESENT TRENDS IN ANALYTICAL METHODOLOGY FOR Se SPECIATION IN FOOD-RELATED PRODUCTS

21.3.1 Introductory Aspects

The design of any speciation scheme requires a careful selection of both the pretreatment procedure and the analytical methodology to be applied for separation and detection or species confirmation/identification. In the early stage of analytical speciation, research was focused on method development, while using simple model solutions of the elemental species of interest. With progress in the analytical technology, the interest of research has moved toward the characterization of complex biological samples, and the choice of suitable pretreatment procedures has become an important task. The biological matrices of interest in studies of Se speciation in food-related products encompass different plant tissues (including sprouts, crops, and nuts) and yeast.

21.3.2 Sample Pretreatment

This topic is exhaustively dealt with in Chapter 19. Here, only some basic concepts are recalled and a few major findings in this field are reported in order to present a thorough and self-consistent survey on the current problems in Se speciation analysis. The aim of pretreatment is to convert the original sample into a form amenable for introduction in the separation/detection system, yet avoiding possible losses and/or species interconversion. The preparation of food-related products for Se speciation involves the homogenization of the initial material, followed by species-targeted leaching or other fractionation protocol to separate large molecules from low-molecular-weight compounds, extraction of lipids and/or protein hydrolysis [13–15, 20, 21, 45–49]. Thus, a common approach for Se in yeast has been extraction with hot water to leach water-soluble proteins and low-molecular-weight, nonprotein-bound Se [50, 51]. To release the selenoamino acids covalently bound to proteins (about 90 percent of total Se), acid or enzymatic digestions were carried out [11, 14, 17, 52, 53]. In the analysis of *Allium* vegetables, no significant differences were observed between speciation results obtained in hot water and enzymatic digests, indicating that the primary Se species are not bound into the protein structure [37]. Advanced techniques, such as accelerated solvent extraction (ASE), supercritical fluid extraction (SFE), and microwave (MW)-assisted extraction (AE), are generally used [54–60]. In the cited studies both, the efficiency of extraction and preservation of native speciation have been addressed. Thus, the enhanced activity of two proteolytic enzymes (protease XIV and subtilisin) was observed by utilizing an ultrasonic probe, which enabled the rapid and efficient extraction of Se from enriched yeast at room temperature and without the need for pH adjustment [61]. Roberge et al. systematically studied the effect of extraction conditions on the results of Se speciation in enriched broccoli (876 $\mu g\ g^{-1}$) by high-performance liquid chromatography (HPLC) inductively coupled plasma mass spectrometry (ICP-MS) [58]. Twenty-seven extraction conditions were explored involving nine different buffering systems between pH 1 and 9. Mass balance indicated that approximately 30 percent of naturally occurring Se in broccoli samples was volatilized and lost to the atmosphere when buffered extractions were made. The recovery obtained for methylselenocysteine spike was about 80–90 percent in buffered extractions and significantly poorer (about 60 percent) in water extraction. SPE indicated that the polarity of Se compounds in solution was also dependent on the extracting solution. Changes in speciation results by HPLC-ICP-MS were analyte-, pH-, and buffer-dependent, but generally, a higher pH resulted in more highly oxidized Se compounds [58].

Using *Brassica juncea* as a model to study Se speciation in plants, Mounicou et al. evaluated different sample preparation protocols for the analysis of Se-containing proteins [21]. The authors considered the typical procedures for the extraction of plant proteins too aggressive for Se species and examined: (1) the feasibility of using Tris–HCl buffer (pH 7.5) to recover water-soluble proteins; (2) the capacity of 1 percent sodium dodecylsulfate (SDS) in Tris–HCl (pH 7.5) to solubilize proteins by forming ion pairs; and (3) the effect of sonication on protein solubility with SDS containing Tris–HCl. In each case, the obtained protein fraction was purified by precipitation with acetone (80 percent), re-solubilized with SDS-containing Tris buffer, and analyzed by two-dimensional liquid chromatography (LC), namely, size exclusion chromatography (SEC) and anion exchange (AE) fast protein liquid chromatography (FPLC) followed by determination with ICP-MS. Several Se-containing proteins of different solubility characteristics were observed. A similar approach had been earlier applied to Se speciation in enriched yeast [62]. After sequential extraction with Tris–HCl buffer (1) and SDS in Tris–HCl (2), the nonsolubilized fraction was acid digested for total residue Se (3). The recovery evaluated as the sum of total Se in each fraction was 91.5 percent [15.2 percent in (1), 36.8 percent in (2), and 39.5 percent in (3)].

To assess Se distribution and speciation in Brazil and other types of nuts, the defatted and powdered samples were fractionated into proteins and cytosol as shown by a flow chart in Figure 21.1 [13, 14]. For the evaluation of Se binding to proteins by SEC, protein precipitate was solubilized with SDS-containing acetate buffer. For the speciation of low-molecular-weight (LMW) Se compounds released from proteins, two procedures of protein hydrolysis were applied and the results obtained were compared. It was shown that acid hydrolysis with methanesulfonic acid allowed for more efficient release of selenomethionine from proteins as compared with enzymatic proteolysis [14].

When chemically labile Se compounds are to be characterized, there is a need for species-targeted sample preparation. As a representative example, the study of

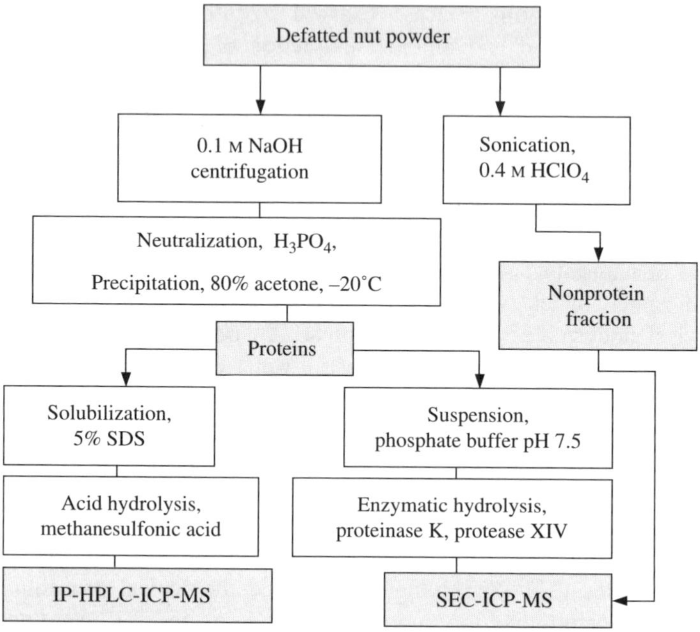

Figure 21.1. Flow chart of the pretreatment procedure applied to study Se distribution and speciation in defatted nut samples [13, 14].

LMW yeast Se compounds is given [46]. It was assumed that adenosylselenohomocysteine (AdoSe-Hcy), which had been identified in yeast by different authors [51, 63], could be a demethylation product of adenosylselenomethionine (AdoSe-Met). The latter compound is formed when selenomethionine enters the pathway of its S analog in enzymatic transmethylation. For the safe detection of AdoSe-Met, an LMW fraction of Se enriched yeast was obtained by homogenization of fresh cells and precipitation of proteins with $HClO_4$ at 0°C. The identification and/or confirmation of species observed by ion-pairing (IP) HPLC-ICP-MS was accomplished by spiking the samples with both commercial and laboratory-synthesized standards, and by electrospray (ES) MS.

21.3.3 Separation Techniques

A strong preference in speciation analysis is to use a separation step that can be combined with a detection step in an on-line system [45]. In such coupling, analytical selectivity relies on the application of different chromatographic or electrophoretic methods, while the use of atomic spectrometric techniques assures high sensitivity and fit-for-purpose limits of detection (LoDs). However, hyphenated techniques with element-specific detection do not provide structural information on the species. If appropriate standards are available, the assignation of chromatographic peaks can be accomplished by spiking experiments. On the other hand, the identification of unknown forms and/or ultimate confirmation of unexpected compounds observed in the sample require the use of complementary techniques, for example, molecular mass spectrometry (MMS) or nuclear magnetic resonance (NMR).

In the analysis of food-related products, LC coupled to ICP-MS has been preferentially used for nonvolatile Se species. SEC provides elution profiles of protein-bound Se [15, 20, 21]. AE columns have been applied for the separation of two inorganic forms [Se(IV) and Se(VI)] as well as selenomethionine, selenocysteine, methylselenocysteine, and trimethylselonium ion in vegetable and yeast extracts [45, 64, 65]. A number of organo-Se compounds (selenoamino acids, peptides, trimethylselonium ion, etc.) are cations, or can be easily converted to cationic species by lowering the pH of the solution, thus enabling separation by cation exchange mechanisms. Up to 10 Se compounds (methylselenocysteine, allylselenocysteine, propylselenocysteine, selenomethionine, selenoethionine, selenocystine, dimethyl-seloniumpropionic acid, selenohomocystine, trimethylselonium ion, and methylselenomethionine) have been resolved with different mobile phases at pH 2–5.7 (pyridine, ammonium formate, etc.) [66–68]. The representative ICP-MS spectrum is presented in Figure 21.2 [69]. The conjunctive use of AE- and cation exchange-HPLC was shown to provide information on the ionic

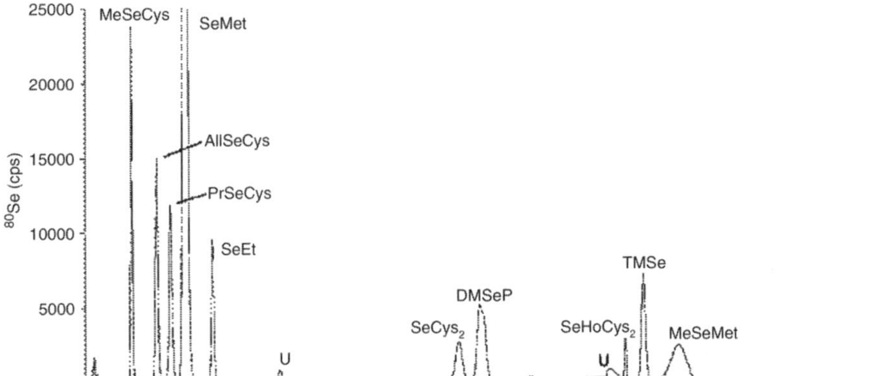

Figure 21.2. Cation exchange HPLC-ICP-MS analysis of a mixture of 10 organic Se compounds (natural isotope abundance) using ^{80}Se for the detection. Gradient elution was carried out with three mobile phases: (*a*) 0.75 mM pyridinium formate, pH 3.0; (*b*) 2.0 mM pyridinium formate, pH 3.0; (*c*) 8.0 mM pyridinium formate, pH 3.2), in water–methanol 97 : 3. The concentration of selenomethionine (SeMet) was 6000 ng ml^{-1} as Se and the concentration of methylselenocysteine (MeSeCys), allylselenocysteine (AllSeCys), propylselenocysteine (PrSeCys), selenoethionine (SeEt), selenocystine (SeCys$_2$), dimethylseloniumpropionic acid (DMSeP), selenohomocystine (SeHoCys$_2$), trimethylselonium ion (TMSe), and methylselenomethionine (MeSeMet) was 200 ng ml^{-1}. Reproduced from reference [69] with permission of The Royal Society of Chemistry.

nature of the retained and separated Se species with a high degree of selectivity [53]. The majority of Se speciation analyses in food-related products have been done with IP HPLC [12, 70]. Different perfluorinated carboxylic acids have been examined as the IP agents (pH 2.5–4.5), and the feasibility for separation of a large number of species was demonstrated in the enriched yeast and vegetable extracts [65, 71]. Alkylsulfonic acids have been successfully used in speciation analyses carried out on nuts, onion leaves, and yeast [13–15, 46]. A multidimensional chromatographic approach to Se speciation in yeast and plant-related materials has been discussed [21, 72–76]. A typical scheme involves the use of SEC, the collection of Se-containing fractions of interest, their preconcentration and separation by reversed-phase or ion-exchange HPLC. The enhanced purity of the Se species collected is of special interest for their further characterization by mass spectrometry [63, 74–76].

Owing to the different biological activity of D- and L-enantiomers of selenoamino acids, the chiral separation of optical isomers has been undertaken in selenized yeast and in yeast-based commercial supplements. Both, chiral stationary phase (crown ether) and chiral derivatization prior to reversed-phase HPLC were used [16, 77, 78].

Gas chromatography (GC) is suitable for the determination of volatile Se species, and the direct coupling to ICP-MS detection has been used in few studies. Dimethylselenide and dimethyldiselenide were found as the primary Se species in the headspace of genetically modified *Brassica juncea* seedlings [79]. GC separation of volatile, chiral derivatives of Se-Met enantiomers has also been studied [19, 80, 81].

Only a few applications of capillary electrophoresis (CE) have been reported for Se speciation in biological materials [82]. This technique allows two inorganic forms of the element and several organic compounds to be separated in a short time. Michalke and Schramel separated six Se species [selenocystamine, selenate, selenocystine, selenite, selenomethionine, and Se-carrying glutathione (GS-Se-SG)] by CE with alkaline background electrolyte (Na_2CO_3–KOH, pH 11.5) [83]. The procedure was applied to the analysis of human milk [84]. Chiral speciation of selenomethionine in the enzymatic digest of selenized yeast was reported by Day et al. [85]. Marfey's reagent (1-fluoro-2,4-dinitrophenyl-5-l-alanine amide) was used for derivatization of the free selenoamino acids and the separation was achieved using a phosphate buffer (pH 3.3).

Gel electrophoresis (GE) is a common separation technique in protein analysis and it has also been used for the speciation of metals bound to proteins [86]. In most applications, metals have been detected by autoradiography, limiting the studies to those elements for which a relatively stable radionuclide exists [87]. As an example, ^{75}Se radiotracer allowed Se to be detected after two-dimensional GE (2-DE) separation [88]. Owing to the high sensitivity and isotopic capability of ICP-MS, this technique has been proposed as the detection tool of choice for elements in gel. The efficient transport of the sample from the protein spot on gel to plasma has been achieved by laser ablation (LA) [89, 90] and electrothermal (ET) atomization [62, 91] techniques. The particular advantage of ICP-MS in such analyses lies in the simultaneous acquisition of multielemental data for each gel fraction, thus enhancing any biologically significant information obtained [92]. Furthermore, the list of metals/metalloids determined by this technique could be extended to those presenting nonradioactive isotopes. The separation of Se-containing proteins in aqueous yeast extracts by one-dimensional GE (1-DE) followed by ET-ICP-MS [62] and by 2-DE with LA-ICP-MS [93] has been reported. In the latter study, the yeast extract was first denatured with a nonreducing SDS treatment and the proteins were derivatized with iodoacetic acid to prevent the formation of S–S, Se–S, or Se–Se bridges. The separation was performed on polyacrylamide gel, first by isoelectric focusing with immobilized pH gradient (pH 3–10 linear) and then in the presence of SDS to obtain separation of proteins, according to their molecular weights. After Ag staining, the spots were laser ablated. The movement of the ablation cell was software-controlled and the displacement of the sample to another spot of interest took approximately 30 s. Dynamic reaction cell (DRC) technology was applied for more sensitive and selective Se determinations. The results obtained revealed the presence of two Se-containing proteins (Fig. 21.3) [62].

21.3.4 Mass Spectrometric Detection

Among the element-specific detectors available for chromatography-based speciation, ICP-MS has proved to be an excellent choice [94]. The compatibility of ICP-MS with different separation techniques has been discussed in a series of comprehensive reviews [67, 82, 95–100]. In addition to relatively easy coupling, the particular suitability of ICP-MS relies on its high sensitivity, multielement and isotopic capabilities, and practically interference-free response over a wide

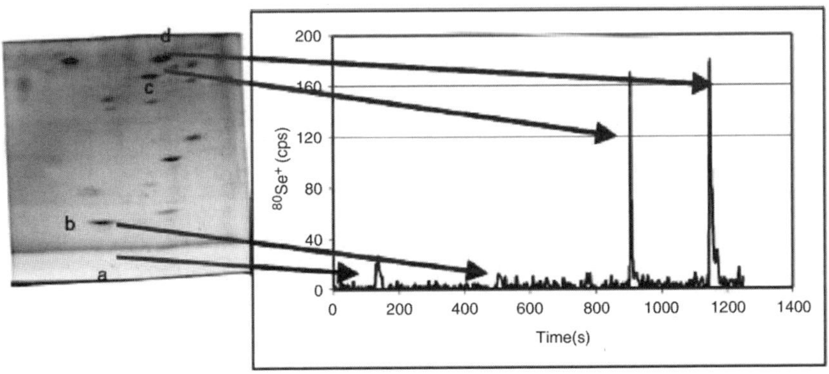

Figure 21.3. 2-DE of selenized yeast extract: the part of gel that was analyzed is shown (molecular weight <30 kDa) together with the LA-ICP-MS results for Se: (*a*, *b*) two proteins not containing Se; (*c*, *d*) two Se-containing proteins. Reproduced from Reference [93] with permission of Wiley-VCH.

range of concentrations [45, 80]. These attributes also make ICP-MS a technique of choice for low-flow and low-volume chromatographic methods (microbore LC columns or cation exchange) with the advantage of outstanding resolution, low sample loading, minute amount of column effluent entering plasma, hence better transport efficiency, and enhanced analytical performance of the plasma. However, the application of conventional ICP-MS to Se involves some difficulties. The ionization potential of Se (9.8 eV) lies relatively close to the first ionization potential of Ar (15.8 eV). Calculations based on the assumption of local thermodynamic equilibrium within a typical ICP predict that only 33 percent of Se atoms would be ionized in Ar-sustained plasma [101]. Furthermore, the most abundant Se isotope (^{80}Se, 49.61 percent) suffers from spectral interferences caused by Ar dimer ion. The detailed list of possible isobaric interferences in the determination of Se in biological samples using Ar ICP-MS is given in Table 21.1 [102, 103]. The resolutions of most popular quadrupole (Q) mass analyzers (0.7–1.0 amu) are not adequate to discriminate the analyte signal from those of interfering species [104]. Owing to these limitations, the sensitivity in Se speciation is often compromised by incomplete ionization and the necessity of monitoring less abundant isotopes (^{77}Se or ^{78}Se). It is noted that newer speciation studies utilize ICP-MS instrumentation equipped with DRC or collision cell (CC) for polyatomic ion removal with subsequent enhanced detection capabilities. Furthermore, the effect of the chromatographic mobile phase composition on Se response in ICP-MS should be mentioned. In the applications of IP chromatography, the addition of organic modifier influences both the stability of plasma and the magnitude of the Se signal. In the presence of 20 percent of methanol, about 45 percent lower Se signal is observed as compared to that obtained with 5 percent methanol in the mobile phase [46]. Consequently, compromise chromatographic conditions, including isocratic elution, had to be used. The reported LoD for Se species by ICP-MS coupled to chromatographic separation range from 0.5 to 50 μg l^{-1} (as Se concentration in the solution introduced onto the column) [11, 19, 52, 53].

The progress in the ICP-MS instrumentation and in particular commercial availability of high-resolution sector field (SF) ICP-MS instruments and DRC and

TABLE 21.1. Potential Spectral Interferences in the Determination of Se in Biological Materials by Ar Sustained ICP-MS [102, 103]

Isotope	Abundance (%)	Interfering Species
74	0.89	$^{38}Ar^{36}Ar^+$, $^{37}Cl_2^+$, $^{40}Ar^{34}S^+$
76	9.37	$^{40}Ar^{36}Ar^+$, $^{40}Ar^{36}S^+$, $^{31}P_2^{14}N^+$
77	7.63	$^{40}Ar^{36}ArH^+$, $^{38}Ar_2H^+$, $^{40}Ar^{37}Cl^+$, $^{59}Co^{18}O^+$, $^{65}Cu^{12}C^+$, $^{63}Cu^{14}N^+$
78	23.8	$^{40}Ar^{38}Ar^+$, $^{31}P_2^{16}O^+$
80	49.6	$^{40}Ar_2^+$, $^{79}BrH^+$
82	8.73	$^{40}Ar_2H_2^+$, $^{12}C^{35}Cl_2^+$, $^{34}S^{16}O_3^+$, $^{82}Kr^+$, $^{81}BrH^+$, $^{65}Cu^{17}O^+$, $^{64}Zn^{18}O^+$, $^{68}Zn^{14}N^+$

CC technology shows improved conditions for Se determination as suggested above. In a recent study, Featherstone et al. demonstrated that the use of high resolution ($m/\Delta m = 7500$) SF mass analyzer enabled suppression of a number of possible polyatomic interferences while monitoring $^{82}Se^{+}$ in human plasma samples [103]. Even though the lower limit of quantification (LoQ) (0.2 versus 0.3 $\mu g\ l^{-1}$) and better precision (0–3 versus 1–6 percent) were observed in the low – with respect to the high – resolution mode, the latter assured enhanced accuracy of Se determination in reference serum (Seronorm^{TM}). Concerning food-related products, Jakubowski et al. examined low and high mass resolution modes of SF instrument to study Se speciation in herring gull eggs [105]. In high resolution mode, the interference from $^{81}BrH^{+}$ was not observed while monitoring $^{82}Se^{+}$. In earlier work by this group, it was shown that, at higher mass resolution ($m/\Delta m = 1400$), the solvent-induced interferences can be eliminated with better detection capabilities [106].

Alternatively to a high-mass-resolution system, the elimination of polyatomic interferences can be achieved by the means of DRC and CC technology [107]. Collisions, reactions, and energy discrimination are modes to eliminate problematic interferences. While the collisions of ions with buffer gas (He, Xe) lead to a thermalization process that favors ions focusing in the Q-, hexapole- or octapole-based cell, the addition of reactive gases (H_2, CH_4, NH_3, N_2O, O_2) may help neutralize and/or chemically convert the ionic species of interest to that at a higher mass [108–110]. In particular, the use of N_2O or O_2 as the reaction gases, allowed the m/z value of $^{80}Se^{+}$ to be shifted to that of $^{80}SeO^{+}$, which is free from the interference by $^{40}Ar_2^{+}$ [109]. With the use of He, Mounicou et al. observed slightly lower background signals while monitoring $^{80}Se^{+}$ and $^{78}Se^{+}$ after chromatographic separation of different plant protein extracts [21]. When H_2 was added to the cell, the background signal for $^{80}Se^{+}$ and $^{78}Se^{+}$ significantly decreased (with 1 ml min^{-1} H_2, by two orders of magnitude at $m/z = 78$). These same two gases (4 ml min^{-1} He and 2 ml min^{-1} H_2) have been successfully used for eliminating $^{40}Ar_2^{+}$ interference on $^{80}Se^{+}$ and improving the LoD from 25 ng l^{-1} in the absence of DRC or CC down to 2 ng l^{-1} [110]. In another study, the hexapole collision and reaction cell with He and H_2 was used as a detection tool with the aim of comparing different chromatographic techniques and nebulization systems as applied for Se speciation in nutritional commercial supplements [70]. The lowest relative LoD for inorganic Se [Se(IV), Se(VI)], selenocystine, selenomethionine, and selenoethionine were in the range 35–90 ng l^{-1} while using IP HPLC, hydraulic high pressure nebulizer, and monitoring isotope ^{80}Se.

The isotope dilution (ID) technique provides highly accurate and precise analytical results for elements that have at least two isotopes that can be monitored by ICP-MS under interference-free conditions. The strength of ID is in its ability to compensate for many sources of errors, including plasma fluctuations, instrumental drift, and signal suppressions by sample matrix. Furthermore, the species-specific ID mode enables correction for analyte losses and/or conversion during the entire analytical procedure and reliable quantification results to be obtained [111]. The application of species-unspecific spiking for ID provides conditions for accurate quantification of both known and unknown element species (with species concentrations referred to Se) [64, 112]. Measuring the isotope ratio instead of one isotope intensity helps to eliminate the effect of mobile phase composition on the element response, thus enabling the use of gradient elution without compromising the sensitivity [69]. For Se, the most abundant isotope of which suffers from Ar dimer interferences, the combination of ID with DRC or CC ICP-MS is of special interest. Such instrumental design has been used for both total element determination and speciation analysis in yeast, algae, and wheat flour matrices [64, 68, 69, 102].

The identification of Se-containing compounds observed in the application of hyphenated techniques with ICP-MS detection is not always straightforward. The lack of commercially available standards and sometimes unclear Se pathways in the studied system are the main problems. The characterization of a number of species, including adenosylselenohomocysteine, adenosylselenomethionine, and Se-containing glutathione *S*-conjugates in yeast extracts, γ-glutammylmethylselenocysteine in garlic, methylselenocysteine in different plants of the *allium* family, and Se-containing peptide in Brazil nuts, has been obtained by complementary ES ionization mass spectrometry (IMS) experiments [12, 46, 52, 63, 76, 113–115].

21.4 APPLICATIONS FOCUSING THE Se ENRICHMENT PROCESS IN PLANTS AND YEAST

Since soil is a primary source of Se in food, considerable variation in the element levels worldwide has a direct impact on its dietary intake in the local populations [8]. As already noted, both Se deficiency and excessively high Se levels in soils have been detected in different geographical regions and, to improve Se content in plant-derived food, remediation strategies are needed. A common approach is to use Se compounds as additives in fertilizers. On the other hand, the ability of different plants to accumulate Se has been examined for phytoremediation purposes.

Enhancing the Se levels in crops can be achieved by adding organic amendments (manure of Se-supplemented farmed animals) or inorganic Se to mineral fertilizers [116, 117]. The use of sodium selenate-enriched fertilizers in Finland resulted in increased Se levels in different foods and, consequently, the average serum Se in the population improved over the period 1984–1988. ICP-MS was used to study the feasibility of wheat enrichment by selenate addition to soil fertilizers [118]. AE-HPLC-ICP-MS was optimized for the separation of selenite, selenate, selenocysteine, and selenomethionine. Total Se determination and speciation analysis were performed in water extracts and in enzymatic digests of wheat samples. It was shown that a major part of the selenate taken up by cereals was converted to selenomethionine.

Phytoremediation has been proposed as a strategy against different toxic contaminants in soils [119]. In the case of excessive soil Se, the use of *Brassica juncea*, *Arabidopsis thaliana*, and some species of *Astragalus* has been considered

[120]. It was shown that the accumulation rate depends on the form of Se in the growing medium. The highest uptake was observed for selenocysteine and selenomethionine, followed by selenite and selenate [121]. To enhance Se assimilation by the plant, a transgenic *Brassica juncea* was obtained, which overexpresses the enzyme catalyzing methylation of selenocysteine to methylselenocysteine (SMT, Se-cysteine methyltransferase) [122]. Using HPLC–ICP-MS, the species-dependent uptake and speciation of Se in these plants were studied [49, 123]. The highest total Se uptake was observed for plants grown in the presence of selenate, followed in decreasing order by selenomethionine and selenite [49]. The speciation results were also dependent on the Se form present in the growth medium. It was demonstrated that the genetic modification did not affect the ability of the plant grown in the presence of selenite to accumulate more Se, but it caused more efficient conversion of the element to methylselenocysteine. For the identification of unknown species, observed mainly in plants enriched with selenomethionine and selenate, the chromatographic fractions were collected, concentrated and analyzed by ES IMS. The results seemed to support the proposed Se cycle in plants [120].

When selenized natural products are considered as the source of supplementation, the efficiency of the fortification process becomes an issue. In particular, the tolerance of biological species to different concentrations of Se compounds present in the growth medium, the element uptake rates, and its distribution and speciation have been addressed. In an early study, enriched garlic was obtained by fertilizing the crop with water-soluble selenate and selenite salts [22]. Using similar conditions for growth (peat moss–vermiculite mixture), the uptakes of Se to garlic and onion bulbs were compared, showing considerably lower total Se in onion plants (28 μg Se g^{-1} in dry onion versus 110–150 μg g^{-1} in garlic) [43]. Whanger et al. compared the Se uptake to ramp bulbs under hydroponic conditions and while using a peat moss–vermiculite mixture as the growth medium [38]. Sodium selenate solutions were used at different Se levels, up to 300 mg Se kg^{-1} in solid medium and up to 120 mg l^{-1} in liquid medium. No significant inhibition of plant growth was observed at any concentrations. However, the assimilation of Se from solution was more efficient as referred to the solid medium and shown by total acid-digest Se determination. The relationship between selenate and sulfate uptake and assimilation in onion plants was studied by Barak et al. [124]. Plants were grown in a hydroponic system and a partial factorial design was applied to study the effect of Se : S molar ratio in the growth medium on the element content in bulbs. Total acid-digested Se and S were determined by ICP-MS (Se) and -atomic emission spectrometry (ICP-AES). If the concentration of one element (Se or S) in the growth medium was kept at a low and constant level, the increasing content of the second element in plant tissue was observed, proportionally to its concentration in the solution. However, at a relatively high rate of selenate, Se uptake was suppressed by increasing sulfate concentration in the culture solution. The antagonistic effect observed suggested that careful management of both S and Se levels would be necessary for the production of selenized onion [124]. Lintschinger et al. used a selenate solution to enrich wheat, alfalfa (*Medicago sativa*), and sunflower (*Helianthus annuus*) seeds during the sprouting stage [125]. Uptake rates were assessed by the determination of total Se. To attain Se metabolism information, Se speciation was carried out in extracts of the sprouts using AE-HPLC-ICP-MS [125].

The fortification of onion leaves (*Allium cepa* L.) grown hydroponically in selenate- or selenite-containing medium was also studied [15]. Both element forms (5 mg l^{-1} Se in medium) significantly inhibited leaf growth with respect to the control plants and a more pronounced effect was observed in the presence of selenite. ICP-MS determination of total Se showed better uptake to leaves as compared to bulbs. Furthermore, higher levels of the element were found in selenate-enriched leaves (about 600 versus 155 mg kg^{-1} in leaves grown in selenite). SEC-UV-ICP-MS analysis of NaOH leaf extracts indicated more efficient incorporation of selenite to high molecular weight compounds (HMW, >10 kDa). About 33 percent of total acid digested Se was found in the HMW fraction of the extract from selenite-enriched leaves and 3 percent in selenate-enriched leaves. In a further development, the enrichment process was studied on green onions (*Allium fistulosum*), which are characterized by a relatively large biomass of leaves [20]. In contrast to most other onion species, bulbs and leaves are usually eaten. Green onions were grown in cells containing commercial solid substrate (a mixture of sphagnum, peat moss, perlite, vermiculite, dolomite, and calcitic lime stone), kept in a greenhouse (16–21°C), and watered/fertilized daily. Two months after seeding, Na_2SeO_3 was added to the cells at different concentration levels (1–15 mg Se kg^{-1} of solid substrate) and, after four months the plants were harvested. Using this protocol, no significant inhibition of plant growth was observed at any Se concentration used. The total Se (MW-assisted acid digestion of the freeze-dried samples) accumulated by the green onion plants was proportional to the amount of selenite added to the growth media, but lower with respect to the results obtained in hydroponically grown plants (30.3 mg kg^{-1} found in plants enriched with 15 mg kg^{-1} Se in the growth medium). In the application of SEC-ICP-MS to NaOH leaf extracts, the incorporation of Se to HMW compounds (>10 kDa) was observed, similarly as in *Allium cepa* extracts (Fig. 21.4*a* and *b*). The elution of Se in the region of LMW (<2 kDa) was also observed in the two studies.

Selenized yeast is a common dietary source of Se. Concerning the enrichment process, Ponce de Leon et al. evaluated four enrichment protocols at various points in the growth cycle in terms of total Se accumulation and forms of Se incorporated [17]. Na_2SeO_3 was used as the Se source and the enrichment was carried out at different levels during the growth phase and the nongrowth phase. In addition, enrichment by seeding in a fermentable carbon source (glucose) and with a nonfermentable carbon source (glycerol) were examined. The speciation results obtained in enzymatic yeast digests revealed that the most efficient incorporation of Se to L-selenomethionine is obtained during the growth phase using small concentrations of Na_2SeO_3. To study the absorption rate of Se given as selenized yeast in humans, ^{77}Se-enriched yeast was prepared [69]. The isotope, in the form of Na_2SeO_3, was gradually added to the standard growth medium over 18 h. The harvested ^{77}Se enriched yeast contained 1380 ± 10 μg g^{-1} ^{77}Se in dry mass.

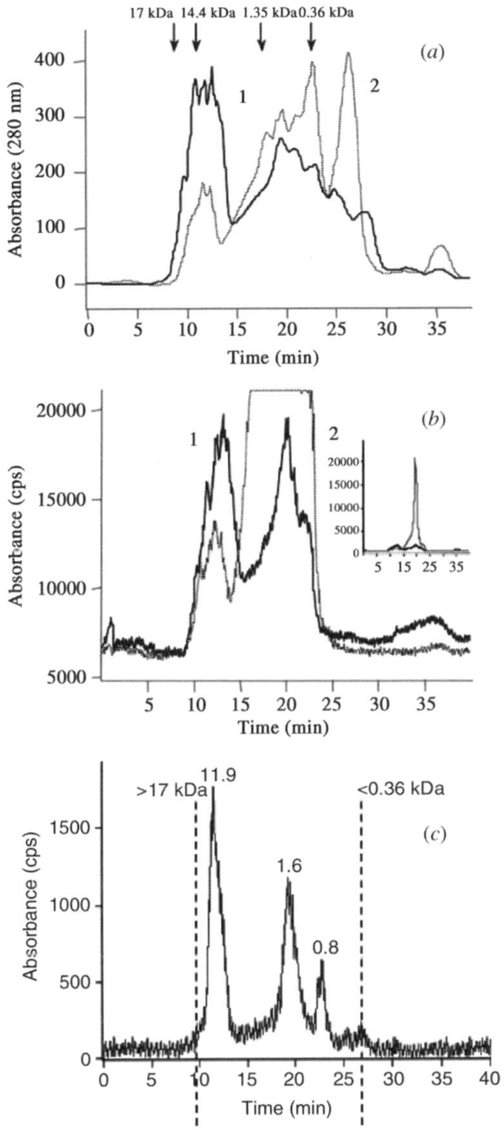

Figure 21.4. Typical SEC profiles of NaOH extracts of onion leaves grown in the presence of Se. (*a*) Extract of *Allium cepa* leaves grown hydroponically in the presence of (1) Se(IV) and (2) Se(VI). UV detection, 280 nm (the elution of molecular weight standards marked in the upper part). Reproduced from reference [15] with permission from Elsevier. (*b*) Extract of *Allium cepa* leaves grown hydroponically in the presence of (1) Se(IV) and (2) Se(VI). ICP-MS detection, ^{82}Se. Reproduced from reference [15] with permission from Elsevier. (*c*) Extract from *Allium fistulosum* grown in the presence of Se(IV) on the solid substrate, ICP-MS detection, ^{82}Se (molecular weights estimated according to column calibration with molecular weights standards). Reproduced from reference [20] with permission from The Royal Society of Chemistry.

Literature is available describing Se bioavailability, pharmacokinetics, and/or possible biological effects of selenized natural products in different clinical trials and in animal experiments. ICP-MS detection has been used in several studies [37, 38, 125–130].

21.5 EVALUATION OF Se SPECIES IN FOOD-RELATED PRODUCTS

The results of most representative Se speciation studies are shown in Table 21.2. As already noted, the food-related products of interest include enriched yeast, garlic, onion, ramps, and broccoli as well as naturally Se-rich Brazil nuts.

The great majority of commercially available supplements are based on Se-enriched yeast. Protein-bound selenomethionine is the primary element form, providing the long-term stability and biological activity of these products [4, 5, 35, 129, 131]. Different studies have focused on speciation analysis of Se in enriched yeast. The identification and/or confirmation have been undertaken of the Se compounds found in water soluble extracts (nonprotein-bound Se) [37, 46, 50, 52, 65, 132] and of those incorporated into the protein structure [37, 50–52, 61, 62, 91, 132, 133]. In the earlier studies, a relatively large number of species were observed, many of them unidentified, in part because of the lack of available standards [60, 65, 133]. In further developments, the application of ES-MS to the analysis of nonprotein extracts enabled the characterization and confirmation of adenosylselenohomocysteine, adenosylselenomethionine, as well as selenodiglutathione (GS-Se-SG) and the mixed selenotrisulfide of glutathione and cysteinylglycine (GS-Se-SCG) [46, 51, 52, 63, 76, 132]. The chromatographic properties of the Se species observed under different experimental conditions (pH, separation mechanism applied, the presence of reducing agents) has also been helpful for species separation and assignment [53, 74, 134]. There was concern that not all Se peaks observed on ICP-MS chromatograms would correspond to real species extracted from the original sample. The incomplete cleavage of selenoamino acids from proteins during enzymolysis has been observed with formation of several Se-containing peptides, which were eluted both in SEC and in IP-HPLC [14]. Species interconversion during pretreatment procedure has been studied [58] and the formation of selenomethionine oxide demonstrated [53, 69].

Another approach to selenized yeast is to characterize Se-containing proteins. The application of polyacrylamide gel electrophoresis (PAGE) has been proposed with the introduction of the protein spot to ICP-MS via LA or ET atomization [62, 93]. The feasibility of matrix-assisted laser desorption ionization (MALDI)–time–of-flight mass spectrometry (TOF-MS) prior to ES tandem MS as applied to tryptic protein digests has also been explored [135, 136].

Since selenized yeast is used in human nutrition, there is a need for full characterization of Se species. However, even more importantly, the quantitative analysis of biologically active selenomethionine is demanded. After a long-term intake of selenized yeast, the undesired accumulation in proteins occurs with increased

TABLE 21.2. Representative Results of Se Speciation in Food-related Products, Obtained Using Different Fractionation Schemes and Analytical Procedures with Mass Spectrometric Detection

Sample	Fractionation/ Pretreatment	Analytical Methodology	Analytical Results	Reference
1355 μg g^{-1} Se garlic	Extraction with hot water or MeOH–0.2 M HCl (1 : 9)	IP-HPLC-UV-ICP-MS (254 nm, ^{82}Se) Use of available Se compounds for peak assignation	Primary species: MeSe-DL-Cys Also found: inorganic Se, Se-Met Unidentified peaks observed	[60, 65]
235 μg g^{-1} Se garlic			Primary species: unidentified Also found: inorganic Se, MeSe-Cys, Se-Met Unidentified peaks observed	
68–1355 μg g^{-1} Se garlic	Enzymatic hydrolysis (protease XIV)	IP-HPLC-ICP-MS (^{82}Se) IP-HPLC-ESI-MS	Recovery of speciation procedure: 68–95% Primary species: γ-glutamyl MeSe-Cys for lower and MeSe-Met for higher Se garlic Also detected: inorganic Se, Se-Met Three organic species identified by ES IMS	[37, 52]
296 μg g^{-1} Se garlic	Hot water extraction	IP-HPLC-ICP-MS (^{82}Se) IP-HPLC-ESI-MS	Primary species: γ-glutamyl MeSe-Cys (90% of total Se), possibly γ-glutamyl MeSe-Met Species identity deduced from ESI-MS spectra	[51]
96, 140 μg g^{-1} Se onion	Enzymatic hydrolysis (protease XIV)	IP-HPLC-ICP-MS (^{82}Se) IP-HPLC-ESI-MS	Recovery of speciation procedure: 92–96% Primary species: γ-glutamyl MeSe-Cys Also found: inorganic Se, Se-Cis, Se-cystathionine, MeSe-Cys, Se-Met Identified by ESI-MS: γ-glutamyl MeSe-Cys	[52]

(Continued)

TABLE 21.2. *Continued*

Sample	Fractionation/ Pretreatment	Analytical Methodology	Analytical Results	Reference
154–601 μg g^{-1} Se onion leaves and 16–51 μg g^{-1} Se onion bulbs, enriched with Se(IV) or Se(VI) during hydroponic growth	Protein fraction extracted with 0.1 M NaOH Extraction of LMW fraction: (1) MeOH–$CHCl_3$–H_2O (12 : 5 : 3) (2) 0.4 M $HClO_4$: 80% EtOH	SEC-UV-ICP-MS (280 nm, ^{82}Se) IP-HPLC-ICP-MS Use of available Se compounds for peak assignation	Better assimilation of Se to onion leaves with respect to bulbs; higher total Se in plants enriched with Se(VI); more efficient incorporation of Se to proteins in Se(IV) enriched plants (33 and 26% of total Se in leaves and bulbs versus 3 and 5% observed in Se(VI) enriched plants) Primary species observed in two LMW extracts: MeSe-Cys, 4.0 and 1.9% of total Se respectively in Se(IV)- and Se(VI)-enriched leaves	[15]
1–30 μg g^{-1} Se green onion enriched with Se(IV) during growth in solid medium	Protein fraction extracted with 0.1 M NaOH Extraction on nonprotein-bound selenium with 0.1 M HCl Enzymatic hydrolysis (proteinase K, protease XIV)	SEC-UV-ICP-MS (280 nm, ^{82}Se) IP-HPLC-ICP-MS ESI-MS (ion trap) ESI-MS-MS	About 44% of total Se bound to HMW compounds (MW > 10 kDa) Primary species in HCl extract: γ-glutamyl MeSe-Cys, MeSe-Cys (Se-Met also detected) Species found in enzymatic extract: Se-Met, Se-Cys, γ-glutamyl MeSe-Cys, inorganic Se γ-glutamyl MeSe-Cys identity confirmed by ESI-MS and ESI-MS-MS	[20]
48–524 μg g^{-1} Se ramp	Enzymatic hydrolysis (protease XIV)	IP-HPLC-ICP-MS (^{82}Se)	Recovery of speciation procedure: 60–85% Primary species: MeSe-Cys Also found: inorganic Se, Se-Cystathionine, Se-Met, γ-glutamyl MeSe-Cys	[52]

48–524 μg g^{-1} Se ramps grown hydroponically or on vermiculite-peat moss	Hot water and enzymatic (protease XIV) extractions	IP HPLC-ICP-MS (^{82}Se) Use of available Se compounds for peak assignment	Recovery evaluated as the sum of identified species versus total Se in ramps: 67–83% Primary species: MeSe-cys (35–50%) and Se(VI) (15–42%) Also found: Se-cystathionine, γ-glutamyl MeSe-Cys In rats fed with Se ramps the reduction in chemically induced tumors was observed	[38]
876 μg g^{-1} Se in broccoli	27 extraction conditions tested (effect of pH, buffer composition)	HPLC-ICP-MS Use of Se available compounds for peak assignment	Species interconversion and/or volatilization found to be dependent on the extraction conditions	[58]
1922 μg g^{-1} Se yeast	Hot water extraction	IP HPLC-ICP-MS Use of available Se compounds for peak assignation	Primary species: Se-Met Also found: inorganic Se, Se-Cys, MeSe-DL-Cys Unidentified peaks observed	[65]
1.2–2.1 mg g^{-1} Se yeast	Hot water extraction	IP-HPLC-ICP-MS (^{82}Se) IP-HPLC-ESI-MS ESI-MS	About 10% of total Se extracted Primary species: Se-Met Identified by ESI-MS: AdoSe-Hcy and Se-Met	[37, 50, 52, 132]
	Enzymatic hydrolysis (protease XIV)		Recovery of speciation procedure: 67–93% Primary species: Se-met (60–85%) Also found: inorganic Se, Se-Cis, Se-lathionine, Se-cystathionine, MeSe-Cys	
1922 μg g^{-1} Se yeast	Enzymatic hydrolysis (protease XIV)	IP-HPLC-ICP-MS IP-HPLC-ESI-MS	Primary species: Se-Met, AdoSe-Hcy (the two account for 85% of total Se) Species identity deduced from ESI-MS spectra	[51]

(*Continued*)

TABLE 21.2. *Continued*

Sample	Fractionation/ Pretreatment	Analytical Methodology	Analytical Results	Reference
400 μg g^{-1} Se yeast-based supplement	Enzymatic hydrolysis (protease XIV)	IP HPLC-ICP-MS (^{77}Se, ^{78}Se, ^{82}Se) the use of available Se compounds for peak assignation	Low recovery of extraction (74% of total Se) ascribed to matrix effects Primary species: Se-Met (55% of total Se in supplement) Also found: Se-Cys 2 unknown peaks detected	[133]
Se-enriched yeast	Enzymatic digestion (trypsin)	SEC-ICP-MS and reversed phase HPLC-ICP-MS	Primary species: Se-Met (74.8%) Also found: Se-Cis (9.9%), selenite (5.1%) Unidentified species observed (10.2%)	[129]
Se-enriched yeast	Enzymatic hydrolysis (protease XIV and subtilisin) with an ultrasonic probe	HPLC-ICP-MS	Efficient extraction of Se-Met at room temperature without necessity of pH adjustment and in short time	[61]
^{77}Se-enriched yeast (1390 μg g^{-1} Se)	Enzymatic hydrolysis (β-glucosidase and the mixture of proteases)	Gradient elution cation exchange LC-ICP-MS (^{77}Se, ^{80}Se) with DRC	Extraction efficiency: 90–95% Separation of over 30 Se species feasible Se-Met accounted for 53% Se in yeast Oxidation of Se-Met occurred during sample preparation	[69]
Speciated candidate CRMs of wheat flour and yeast	Enzymatic hydrolysis (protease XIV, lipase VII)	AE-C-ICP-MS with octapole CC, species-unspecific ID analysis, measuring 78/77 Se isotopes ratio	Quantitative results for Se-Met (59 and 68% of total Se in wheat flour and in yeast) in good agreement with the mean values obtained in collaborative interlaboratory exercise	[64]
Commercial selenized yeast	Hot water extraction	HPLC-ICP-MS The collected fraction analyzed by ESI-MS and ESI-MS-MS	Identification/confirmation of AdoSe-Hcy Species structure deduced from ESI-MS spectra and respective fragmentation pattern	[63]

Laboratory Se-enriched yeast	Homogenization of fresh cells and extraction of LMW fraction with 0.4 M $HClO_4$ at 0°C	IP-HPLC-ICP-MS, ESI-MS	Primary species: Se-Cis, AdoSe-Hcy Identification of AdoSe-Met using laboratory synthesized standard and confirmation by ESI-MS and ESI-MS-MS analysis	[46]
Se-enriched yeast	Water extraction	2-DE (SEC followed by HPLC on porous graphite carbon column) ICP-MS, nanoESI-MS-MS	Identification of selenodiglutathione (GS-Se-SG) and the mixed selenotrisulfide of glutathione and cysteinylglycine (GS-Se-SCG)	[76]
Nutritional yeast supplements	Water extraction (50°C)	2-DE (SEC, followed by reversed-phase HPLC) – ICP-MS ESI-MS, ESI-MS-MS	Species identified/confirmed: AdoSe-Hcy, Se-Met Characterization of unidentified peaks by their behavior against reducing agents and by interpretation of ESI-MS and ESI-MS-MS spectra	[74]
Se-enriched yeast	Hot water extraction	3-Dimensional chromatography (SEC, followed by AE and cation exchange)-ICP-MS, ESI-MS	More than 30 SE compounds detected, commercially available standards and ESI-MS analysis used for species identification	[134]
Commercial yeast-based supplements	Hot water extraction	IP HPLC-ICP-MS with hexapole collision and reaction cell, hydraulic high pressure nebulizer	Extraction efficiency: 10–25% Quantitative determination of Se species with the detection limits for Se(IV), Se(VI), Se-Cys, Se-Met, Se-Et in the range 35–90 ng l^{-1}	[70]

(*Continued*)

TABLE 21.2. *Continued*

Sample	Fractionation/ Pretreatment	Analytical Methodology	Analytical Results	Reference
1.3 mg g^{-1} Se yeast	Sequential extraction yielding water-soluble and nonsoluble fractions	2-Dimensional chromatography (SEC and AE-LC)-UV-ICP-MS (280 nm, ^{78}Se, ^{82}Se), SDS-PAGE-ETV-ICP-MS	Recovery from sequential extraction: 93% LMW Se species eluted from preparative SEC were then separated on AE column, quantitative data for Se-Met, Se-Cys and inorganic Se HMW fraction from SEC (10–100 kDa) analyzed by GE Feasibility of ET-ICP-MS for specific Se detection in protein spots demonstrated	[62]
Commercial selenized yeast	Protein extraction with 5% SDS in Tris–HCl pH 8.0, denaturation with nonreducing SDS treatment, derivatization with iodoacetic acid	2-Dimensional PAGE-LA-ICP-MS (^{77}Se, ^{80}Se, ^{82}Se)	Feasibility of the technique to detection of Se-containing proteins and Se quantification demonstrated (limit of detection 0.07 μg Se g^{-1} gel) Two Se-containing proteins detected	[93]
Se-enriched yeast	Water extraction followed by tryptic digestion	SEC, reversed-phase HPLC, MALDI-TOF-MS, ESI-MS, ESI-MS-MS	Characterization of Se peptides and SE-containing proteins approached	[135, 136]
Commercial yeast	Solid phase microextraction	Multicapillary GC-ICP-MS (among other element-specific detectors)	Volatile alkylselenides analyzed [primary species $(CH_3)_2Se$] The pasteurized product was shown to be active in production of volatile Se species	[137]

6 Brands of yeast-based supplements (91–253 μg Se per tablet)	Different extraction conditions tested Most effective for Se-Met: MW heating with 2 M HCl or enzymolysis with proteinase K	IP HPLC-ICP-MS (^{82}Se)	Only 2 brands showed high levels of Se-Met (up to 130 μg Se per tablet)	[11]
Se-enriched yeast	Enzymatic hydrolysis	Different chiral separation modes HPLC-ICP-MS with or without precolumn derivatization	Separation on Chirobiotic T column was shown to be the best suited procedure	[77]
1.25 mg g^{-1} Se yeast and commercial supplements	Enzymatic hydrolysis (proteinase K or its mixture with aminopeptidase M)	Reversed phase HPLC-UV-ICP-MS (340 nm, ^{78}Se, ^{82}Se) Precolumn derivatization with Mafrey's reagent	Derivatization yield about 70% Resolution of Se-Met and Se-Et enantiomers in one run Quantification results given for Se-L-Met and Se-D-Met	[19]
1922 μg g^{-1} Se yeast 296 μg g^{-1} Se garlic 140 μg g^{-1} Se onion	Hot water and enzymatic (pepsin) extractions	Chiral (crown ether) HPLC-ICP-MS	Feasibility to separate enantiomers of 8 selenoamino acids demonstrated Primary species in enzymatic extracts: γ-glutamyl MeSe-Cys, MeSe-L-Cys in garlic and onion Trace amounts of 9 Se species detected in hot water yeast extract	[16]
Yeast-based Se supplements	Enzymatic extraction	Capillary GC-ICP-MS Precolumn derivatization with ethylchloroformate	Quantification of Se-Met enantiomers achieved with the limit of detection 4 pg	[81]
Nonenriched garlic, onion and yeast (<0.5 μg g^{-1} Se)	Enzymatic hydrolysis (protease XIV)	IP-HPLC-ICP-MS (^{82}Se)	Recovery of speciation procedure: 100% Primary species: inorganic Se in onion, Se-Met in yeast, Se-Met, γ-glutamyl MeSe-Cys, MeSe-Cys in garlic	[52]

(*Continued*)

TABLE 21.2. *Continued*

Sample	Fractionation/ Pretreatment	Analytical Methodology	Analytical Results	Reference
Garlic harvested from naturally seleniferous soil (205 μg g^{-1} Se)	Hot water extraction	2-Dimensional chromatography (SEC followed by reversed-phase HPLC)–ICP-MS ESI-MS, ESI-MS-MS	Identification/confirmation of γ-glutamyl MeSe-Cys as the primary species	[73]
Fish (plaice, cod)	Different fractionation of extracts: ultrafiltration, protein precipitation with trichloroacetic acid (TCA)	SEC-ICP-MS	Extraction efficiency: 23–47% SEC column recovery: 70% Ultrafiltration and TCA treatment: 20–38% of Se in plaice and 69.77% in cod fish bound to proteins (MW >10 kDa)	[48]
Nuts	Fractionation yielding lipids proteins and cytosol (see Fig. 21.1) Enzymatic hydrolysis of proteins (proteinase K)	SEC-ICP-MS (^{78}Se), IP-HPLC-ICP-MS	Protein-bound Se found (binding through Se-C and RS-Se-SR′ or RS-SeCH$_3$ demonstrated) Primary species: Se-Met (19–25% of total Se in proteins and 3–12% in cytosol) incomplete release of Se-Met from proteins observed	[13]
Brazil nuts	Different extraction conditions tested and the enzymatic hydrolysis (proteinase K) applied as the most effective	IP-HPLC-ICP-MS and ESI-MS	In addition to Se-Met a considerable number of unidentified peaks observed, the characterization of these compounds by ESI-MS undertaken and the structure of selenopeptide tentatively proposed	[12]
Brazil nuts 1261 μg g^{-1} Se yeast	Acid hydrolysis with methanesulfonic acid and enzymatic hydrolysis (proteinase K and protease XIV)	SEC-ICP-MS IP-HPLC-ICP-MS (^{82}Se)	More efficient Se-Met extraction observed in acid hydrolysis Quantitative results for Se-Met: 65 and 46% of total Se in acid and enzymatic hydrolysis of yeast; 75 and 25% in nuts	[14]

Commercial Se yeast 566 μg g^{-1} Se algae	Nonprotein bound Se in algae: 10% TCA extraction (0°C) Yeast: enzymolysis (commercial enzymes: Novozyme 234™ and Flavourzyme 1000L™)	Cation exchange- and AE-HPLC-ICP-MS ESI-MS	The use two ion exchange modes of separation enabled for enhanced species characterization Primary species in yeast: Se-Met; unknown peak tentatively assigned to SeO-Met Primary species in algae: DMeSeP (confirmed by ESI-MS), few unidentified Se species observed	[53]
Various grains obtained growing the plants in Se-supplemented soils	Hot water extraction and enzymolysis (pronase)	AE-LC-ICP-MS (^{82}Se)	Extraction efficiency: 3–9% in water and 80–95% in enzymatic extractions Se-Met, Se-Cys and Se(VI) accounted for 62–86, 15–22 and 12–16% of total Se	[118]
Herring gull eggs, 706 ng g^{-1} Se (primarily used for evaluation of heavy metal pollution of marine ecosystem, but also a food-related product)	MeOH–water (9 : 1) extraction	HPLC-SFMS (low and high resolution mode)^{82}Se identification of species based on their chromatographic properties and the results of standard addition	Extraction efficiency: about 8% Column recovery: 106 ± 5%. Primary species: Se-Cystamine (about 25% in extract) Also found: Se-Cis, Se-Met Unidentified peaks present	[105]

AdoSe-Hcy, adenosylselenohomocysteine; AdoSe-Met, adenosylselenomethionine; DmeSeP, dimethyl-Se propionate ion; MeSe-Cys, methylselenocysteine; MeSe-Met, methylselenomethionine; Se-Et, selenoethionine; Se-Cys, selenocysteine; Se-Cis, oxidized selenocysteine, Se-Met, selenomethionine; SeO-Met, selenomethionine oxide.

risk of intoxication [34, 35]. The improved sample preparation [11, 14, 61] and the application of high-resolution MS and DRC or CC technology [70, 105] and of ID [64] have been reported for enhanced determination of selenomethionine. Furthermore, selenoamino acid enantiomers have been analyzed [77, 78, 81].

The Se-enriched vegetables of the *Allium family* have been proved to be more effective in cancer prevention as compared with yeast [23, 37, 38]. Different applications of hyphenated techniques with ICP-MS detection and of ES-MS are presented in Table 21.2. In constrast to yeast analysis, similar speciation results have been observed in water extracts and after enzymolysis of bulbs, indicating that the plant material contained mainly nonprotein-bound selenium [50, 52]. However, in onion leaves, the considerable incorporation of Se to proteins (26–33 percent of total Se) has been reported [15, 20]. The LMW species found in bulbs and in leaves were methylselenocysteine, γ-glutamylmethylselenocysteine (two primary organic species), inorganic Se, selenomethionine, selenocystine, and selenocystathionine [15, 20, 37, 38, 52, 60, 65, 73]. It is worth mentioning that essentially these same Se species were found in nonselenized plants, in enriched plants grown in laboratory conditions (liquid or solid medium) and in plants harvested from seleniferous soils [52, 73]. However, depending on the biological species and total Se in the plant, different relative contributions of methylselenocysteine and γ-glutamylmethylselenocysteine have been observed [37, 38, 52, 58]. Even though Se-alk(en)ylcysteines and their γ-glutamyl derivatives present higher biological activity as compared to selenomethionine, the enriched plants have not been made commercially available. Since these compounds are precursors of volatile species, the stability problem is of concern. Furthermore, the cultivation of *Saccharomyces cerevisiae* is relatively easy and economically attractive with high rates of biomass in few days, while the plant growth takes months. Within this context, the enriched green onions or onion leaves seem to be interesting alternatives for Se supplementation. The considerable biomass of leaves can be obtained within a week. They assimilate Se from both liquid and solid media and convert selenate to bioactive organic species relatively efficiently [15, 20].

Another approach to improving Se intake in human diets is products naturally rich in Se, and in particular Brazil nuts. Speciation studies carried out in the authors' laboratory showed that Se in nuts incorporates in protein structure in the form of selenomethionine. This species accounts for about 75 percent of total nut Se [12, 13, 46].

21.6 CONCLUSIONS AND FUTURE TRENDS

It is clear that the nutritional properties of Se rely on the physical–chemical forms of the element actually present in the dietary sources. Both the bioavailability and the specific biological activity are species-dependent. Hence, analytical speciation has been a necessary tool to assess which species could be responsible for beneficial health effects observed after the administration of high Se yeast, garlic, onion, broccoli, or Brazil nuts. Suitability and versatility of hyphenated techniques with ICP-MS detection have been proved in a large number of applications. The progress in sample preparation has allowed for more efficient leaching of species from the original sample, while the enhanced selectivity has been achieved by refining the separation processes. On the other hand, the application of SF-MS instruments and DRC or CC technology allowed for more sensitive Se detection and the elimination of polyatomic interferences in ICP-MS.

Present challenges in Se speciation focus on the following topics: (1) highly precise and accurate determination of biologically active species; (2) ultimate characterization of elemental species with the emphasis on selenoproteins and Se-containing proteins; (3) studies on bioavailability, digestibility, and pharmacokinetics of Se in fortified products; and (4) search for the alternative dietary sources of biologically active Se. Concerning the first topic, full validation of speciation results would rely on the availability of the so-called speciated CRMs. Some recent reports have highlighted the problems related to incomplete protein cleaving, species oxidation, and/or interconversion, indicating the necessity for revisiting Se speciation in food-supplements. On the other hand, coupling GE with ICP-MS Se detection has proved to be a suitable tool for Se-containing proteins. The use of stable Se isotopes for labeling the fortified products seems very attractive for evaluation of Se uptake and pharmacokinetics in higher animals (third topic). In addition to the enhanced quality of the analytical results, the isotope capabilities of ICP-MS offer important bioethical aspects in animal/human models since the stable isotopes are preferred when compared to radioactive ^{75}Se. Finally, commercial supplements in use are generally based on selenized yeast. There is no doubt of their cancer-preventive properties, though the undesired Se accumulation in tissues is of concern. The use of more effective and safe enriched garlic and onion is compromised by the instability of methylated Se species and is not a well-established process of production as compared to yeast. Thus, the need for stable Se bioactive products amenable to large-scale production is emerging.

REFERENCES

1. S. R. Stapleton, Introduction: the selenium conundrum, *Cell. Mol. Life Sci.*, **57** (2000), 1823–1824.
2. T. C. Stadtman, Selenocysteine, *A. Rev. Biochem.*, **65** (1996), 83–100.
3. H. E. Ganther, Selenium metabolism, selenoproteins and mechanisms of cancer prevention: complexities with thioredoxin reductase, *Carcinogenesis*, **20** (1999), 1657–1666.
4. N. P. Rayman, The importance of selenium to human health, *Lancet*, **356** (2000), 233–241.
5. L. C. Clark, B. W. Turnball, E. H. Slate, D. K. Chalker, J. Chow, L. S. Davis, R. A. Glover, D. K. Graham, E. G. Gross, A. Krongrad, J. L. Lesher, H. K. Park, B. B. Sanders, C. L. Smith, J. R. Taylor, D. S. Alberts, R. J. Allison, J. C. Bradshaw, D. Curtus, D. R. Deal, M. Dellasega, J. D. Hendrix, J. H. Herlong, L. J. Hixon, J. Knight, J. Moore, J. S. Rice, A. I. Rogers, B. Schuman, E. H. Smith, J. C. Woodward, Effects of selenium

supplementation for cancer prevention in patients with carcinoma of the skin: A randomized controlled trial, *J. Am. Med. Assoc.*, **276** (1996), 1957–1963.

6. C. Sasakura, K. T. Suzuki, Biological interaction between transition metals (Ag, Cd and Hg), selenide/sulfide and selenoprotein P, *J. Inorg. Biochem.*, **71** (1998), 159–162.
7. M. N. I. Barelay, A. MacPherson, J. Dixon, Selenium content of a range of UK foods, *J. Food Comp. Anal.*, **8** (1995), 307–318.
8. C. Reilly, Selenium: A new entrant into the functional food arena, *Trends Food Sci. Tech.*, **9** (1998), 114–118.
9. L. H. Foster, S. Sumar, Selenium in the environment, food and health, *Nutr. Food Sci.*, **5** (1995), 17–23.
10. F. Cubadda, Inductively coupled plasma–mass spectrometry for the determination of elements and elemental species in food: a review, *Int. J. AOAC*, **87** (2004), 173–204.
11. C. B'Hymer, J. A. Caruso, Evaluation of yeast-based selenium food supplements using high-performance liquid chromatography and inductively coupled plasma mass spectrometry, *J. Anal. Atom. Spectrom.*, **15** (2000), 1531–1539.
12. A. P. Vonderheide, Kazimierz Wrobel, S. S. Kannamkumarath, C. B'Hymer, M. Montes-Bayon, C. Ponce De Leon, J. A. Caruso, Characterization of selenium species in Brazil nuts by HPLC-ICP-MS and ES-MS, *J. Agric. Food Chem.*, **50** (2002), 5722–5728.
13. S. S. Kannamkumarath, Kazimierz Wrobel, Katarzyna Wrobel, A. Vonderheide, J. A. Caruso, HPLC-ICP-MS determination of selenium distribution and speciation in different types of nut, *Anal. Bioanal. Chem.*, **373** (2002), 454–460.
14. Katarzyna Wrobel, S. S. Kannamkumarath, Kazimierz Wrobel, J. A. Caruso, Hydrolysis of proteins with methanesulfonic acid for improved HPLC-ICP-MS determination of seleno-methionine in yeast and nuts, *Anal. Bioanal. Chem.*, **375** (2003), 133–138.
15. Katarzyna Wrobel, Kazinierz Wrobel, S. S. Kannamkumarath, J. A. Caruso, I. A. Wysocka, E. Bułska, J. Swiatek, M. Wierzbicka, HPLC-ICP-MS speciation of selenium in enriched onion leaves – a potential dietary source of Se-methylselenocysteine, *Food Chem.*, **86** (2004), 617–623.
16. C. A. Ponce de Leon, K. L. Sutton, J. A. Caruso, P. C. Uden, Chiral speciation of selenoamino acids and selenium enriched samples using HPLC coupled to ICP-MS, *J. Anal. Atom. Spectrom.*, **15** (2000), 1103–1107.
17. C. A. Ponce de Leon, M. Montes-Bayon, C. Paquin, J. A. Caruso, Selenium incorporation into *Saccharomyces cerevisiae* cells: a study of different incorporation methods, *J. Appl. Microbiol.*, **92** (2002), 602–610.
18. J. Meija, A. Vonderheide, J. Bryson, M. Montes Bayon, J. A. Caruso, Studies on selenium containing volatiles in roasted coffee. *J. Agric. Food Chem.*, **51** (2003), 5116.
19. M. Montes Bayon, C. B́Hymer, C. Ponce De Leon, J. A. Caruso, Resolution of selenoamino acids optical isomers using chiral derivatization and ICP-MS detection, *J. Anal. Atom. Spectrom.*, **16** (2001), 945–950.
20. M. Shah, S. S. Kannamkumarath, J. C. A. Wuilloud, R. G. Wuilloud, J. A. Caruso, Identification and characterization of selenium species in enriched green onion (*Allium fistulosum*) by HPLC-ICP-MS and ESI-ITMS, *J. Anal. Atom. Spectrom.*, **19** (2004), 381–386.
21. S. Monicou, J. Meija, J. A. Caruso, Preliminary studies on selenium-containing proteins in *Brassica juncea* by size exclusion chromatography and fast protein liquid chromatography coupled to ICP-MS, *Analyst*, **129** (2004), 116–123.

22. C. Ip, D. J. Lisk, G. S. Stoewsand, Mammary cancer prevention by regular garlic and selenium enriched garlic, *Nutr. Cancer*, **17** (1992), 279–286.
23. P. D. Whanger, Selenium and its relationship to cancer: an update dagger, *Br. J. Nutr.*, **91** (2004), 11–28.
24. J. R. Arthur, Micronutrient Group Symposium on "Micronutrient supplementation: when and why?", *Proc. Nutr. Soc.*, **62** (2003), 393–397.
25. D. L. Hatfield, V. N. Gladyshev, How selenium has altered our understanding of the genetic code, *Mol. Cell Biol.*, **22** (2002), 3565–3576.
26. G. V. Kryukov, V. N. Gladyshev, Mammalin selenoprotein gene signature: Identification and functional analysis of selenoprotein genes using bioinformatic methods, *Meth. Enzymol.*, **347** (2002), 84–100.
27. Y. Y. Kim, D. C. Mahan, Prolonged feeding of high dietary levels of organic and inorganic selenium to gilts from 25 kg body weight through one parity, *J. Anim. Sci.*, **79** (2001), 956–966.
28. G. F. Combs, Selenium in global food system, *Br. J. Nutr.*, **85** (2001), 517–547.
29. K. T. Suzuki, K. Ishiwata, Y. Ogra, Incorporation of selenium into selenoprotein P and extracellular glutathione peroxidase: HPLC-ICPMS data with enriched selenite, *Analyst*, **124** (1999), 1749–1753.
30. R. C. McKenzie, J. R. Arthur, S. Miller, T. S. Rafferty, G. J. Beckett, Selenium and the immune system, in H. Hill (ed.), *Nutrition and Immune Function*, CABI/Nutrition Society, Wallingford (2002), pp. 229–250.
31. Anonymous, Possible enhancement of carcinogenesis by selenium in an animal tumor model, *Nutr. Rev.*, **47** (1989), 173–175.
32. M. Sanz Alaejos, F. J. Díaz Romero, C. Díaz Romero, Selenium and cancer: some nutritional aspects, *Nutrition*, **16** (2000), 376–383.
33. Y. Y. Kim, D. C. Mahan, Comparative effects of high dietary levels of organic and inorganic selenium on selenium toxicity of growing-finishing pigs, *J. Anim. Sci.*, **79** (2001), 942–948.
34. G. N. Schrauzer, Selenomethionine: a review of its nutritional significance, metabolism and toxicity, *J. Nutr.*, **130** (2000), 1653–1656.
35. G. N. Schrauzer, The nutritional significance, metabolism and toxicology of seleno methionine, *Adv. Food Nutr. Res.*, **47** (2003), 73–112.
36. C. Ip, C. Hayes, R. M. Budnick, H. E. Ganther, Chemical form of selenium, critical metabolites, and cancer prevention, *Cancer Res.*, **51** (1991), 595–603.
37. C. Ip, M. Birringer, E. Block, M. Kotrebai, J. F. Tyson, P. C. Uden, D. J. Lisk, Chemical speciation influences comparative activity of selenium-enriched garlic and yeast in mammary cancer prevention, *J. Agric. Food Chem.*, **48** (2000), 2062–2070.
38. P. D. Whanger, C. Ip, C. E. Polan, P. C. Uden, G. Welbaum, Tumorgenesis, metabolism, speciation, bioavailability, and tissue deposition of selenium in selenium-enrichedramps (*Allium tricoccum*), *J. Agric Food. Chem.*, **48** (2000), 5723–5730.
39. E. Block, M. Birringer, W. Jiang, T. Nakahodo, H. J. Thompson, P. J. Toscano, H. Uzar, X. Zhang, Z. Zhu, Allium chemistry: Synthesis, natural occurence, biological activity, and chemistry of Se-alk(en)ylselenocysteines and their g-glutamyl derivatives and oxidation products, *J. Agric. Food Chem.*, **49** (2001), 458–470.

40. A. Das, D. Desai, B. Pittman, S. Amin, K. El-Bayoumy, Comparison of the chemopreventive efficacies of 1,4-phenylenebis(methylene)selenocyanate and selenium-enriched yeast on 4-(methylnitrosamino)-1-(3-pyridyl)-1-butanone induced lung tumorigenesis in A/J mouse, *Nutr. Cancer*, **46** (2003), 179–185.

41. J. B. Guttenplan, T. E. Spratt, M. Khmelnitsky, W. Kosinska, D. Desai, K. El-Bayoumy, Effects of 3H-1,2-dithiole-3-thione, 1,4-phenylenebis(methylene)selenocyanate, and selenium-enriched yeast individually and in combination on benzo[a]-pyrene-induced mutagenesis in oral tissue and esophagus in lacZ mice, *Mutat. Res.*, **559** (2004), 199–210.

42. J. C. Chang, W. H. Gutenmann, C. M. Reid, D. J. Lisk, Selenium content of Brazil nuts from two geographic locations in Brazil, *Chemosphere*, **30** (1995), 801–902.

43. C. Ip, D. J. Lisk, Enrichment of selenium in allium vegetables for cancer prevention, *Carcinogenesis*, **15** (1994), 1881–1885.

44. P. D. Whanger, Selenocompounds in plants and animals and their biological significance, *J. Am. Coll. Nutr.*, **21** (2002), 223–232.

45. B. Michalke, Element speciation definitions, analytical methodology, and some examples, *Ecotox. Environ. Saf.*, **56** (2003), 122–139.

46. Kazimierz Wrobel, Katarzyna Wrobel, J. A. Caruso, Selenium speciation in low molecular weight fraction of Se-enriched yeats by HPLC-ICP-MS: detection of selenoadenosylmethionine, *J. Anal. Atom. Spectrom.*, **17** (2002), 1048–1054.

47. J. S. C. Casiot, R. Lobinski, M. Potin-Gautier, Sample preparation and HPLC separation approaches to speciation analysis of selenium in yeast by ICP-MS, *J. Anal. Atom. Spectrom.*, **14** (1999), 645–650.

48. G. Onning, I. A. Bergdahl, Fractionation of soluble selenium compounds from fish using size-exclusion chromatography with on-line detection by inductively coupled plasma mass spectrometry, *Analyst*, **124** (1999), 1435–1438.

49. M. Montes Bayon, T. D. Grant, J. Meija, J. A. Caruso, Selenium in plants by mass spectrometric techniques: Developments in bio-analytical methods, *J. Anal. Atom. Spectrom.*, **17** (2002), 1015–1023.

50. S. M. Bird, P. C. Uden, J. F. Tyson, E. Block, E. Denoyer, Speciation of selenoamino acids and organoselenium compounds in selenium-enriched yeast using high-performance liquid chromatography-inductively coupled plasma mass spectrometry, *J. Anal. Atom. Spectrom.*, **12** (1997), 785–788.

51. M. Kotrebai, M. Birringer, J. F. Tyson, E. Block, P. C. Uden, Identification of the principal selenium compounds in selenium-enriched natural sample extracts by ion-pair HPLC with ICP-MS and ESI-MS detection, *Anal. Commun.*, **36** (1999), 249–252.

52. M. Kotrebai, M. Birringer, J. F. Tyson, E. Block, P. C. Uden, Selenium speciation in enriched and natural samples by HPLC-ICP-MS and HPLC-ESI-MS with perfluorinated carboxylic acid ion-pairing agents, *Analyst*, **125** (2000), 71–78.

53. E. H. Larsen, M. Hansen, T. Fan, M. Vahl, Speciation of selenoamino acids, selonium ions and inorganic selenium by ion exchange HPLC with mass spectrometric detection and its application to yeast and algae, *J. Anal. Atom. Spectrom.*, **16** (2001), 1403–1408.

54. J. Szpunar, B. Bouyssiere, R. Lobinski, Sample preparation techniques for elemental speciation studies, in K. L. Ackley (ed.), *Elemental Speciation. New Approaches for Trace Element Analysis*, Elsevier, Amsterdam, 2000, pp. 7–40.

55. K. Wrobel, S. S. Kannamkumarath, K. Wrobel, J. A. Caruso, Environmentally friendly sample pretreatment for speciation analysis by hyphenated techniques, *Green Chem.*, **5** (2003), 250–259.

56. J. L. Gomez-Ariza, E. Morales, I. Giraldez, D. Sanchez-Rodas, A. Velasco, Sample treatment in chromatography-based speciation of organometallic pollutants, *J. Chromatogr. A*, **938** (2001), 211–224.

57. J. A. Brisbin, J. A. Caruso, Comparison of extraction procedures for the determination of arsenic and other elements in lobster tissue by inductively coupled plasma mass spectrometry, *Analyst*, **127** (2002), 921–929.

58. M. T. Roberge, A. J. Borgerding, J. W. Finley, Speciation of selenium compounds from high selenium broccoli is affected by the extracting solution, *J. Agric. Food Chem.*, **51** (2003), 4191–4197.

59. A. P. Vonderheide, M. Montes-Bayon, J. A. Caruso, Solid-phase microextraction as a sample preparation strategy for the analysis of seleno amino acids by gas chromatography–inductively coupled plasma mass spectrometry, *Analyst*, **127** (2002), 49–53.

60. P. C. Uden, S. M. Bird, M. Kotrebai, P. Nolibos, J. F. Tyson, E. Block, E. Denoyer, Analytical selenoamino acid studies by chromatography with interfaced atomic mass spectrometry and atomic emission spectral detection, *Fresenius J. Anal. Chem.*, **362** (1998), 447–456.

61. J. L. Capelo, P. Ximenez-Embun, Y. Madrid-Albarran, C. Camara, Enzymatic probe sonication: enhancement of protease-catalyzed hydrolysis of selenium bound to proteins in yeast, *Anal. Chem.*, **76** (2004), 233–237.

62. H. Chassaigne, C. C. Chery, G. Bordin, A. R. Rodriguez, Development of new analytical methods for selenium speciation in selenium-enriched yeast material, *J. Chromatogr. A*, **976** (2002), 409–422.

63. C. Casiot, V. Vacchina, H. Chassaigne, J. Szpunar, M. Potin-Gautier, R. Lobinski, An approach to the identification of selenium species in yeast extracts using pneumatically-assisted electrospray tandem mass spectrometry, *Anal. Commun.*, **36** (1999), 77–80.

64. V. Diaz-Huerta, L. Hinojosa Reyes, J. M. Marchante-Gayon, M. L. Fernandez-Sanchez, A. Sanz-Medel, Total determination and quantitative speciation analysis of selenium in yeast and wheat flour by isotope dilution analysis ICP-MS, *J. Anal. Atom. Spectrom.*, **18** (2003), 1243–1247.

65. S. M. Bird, H. Ge, P. C. Uden, J. F. Tyson, E. Block, E. Denoyer, High-performance liquid chromatography of selenoamino acids and organo selenium compounds. Speciation by inductively coupled plasma mass spectrometry, *J. Chromatogr. A*, **789** (1997), 349–359.

66. W. Goessler, D. Kuehnelt, C. Schlagenhaufen, Z. Slejkovec, K. J. Irgolić, Arsenobetaine and other arsenic compounds in the National Research Council of Canada Certified Reference Materials DORM 1 and DORM 2, *J. Anal. Atom. Spectrom.*, **13** (1998), 183–187.

67. M. Montes-Bayon, K. DeNicola, J. A. Caruso, Liquid chromatography-inductively coupled plasma mass spectrometry, *J. Chromatogr. A*, **1000** (2003), 457–476.

68. J. J. Sloth, E. H. Larsen, The application of inductively coupled plasma dynamic reaction cell mass spectrometry for measurement of selenium isotopes, isotope ratios and chromatographic detection of selenoamino acids, *J. Anal. Atom. Spectrom.*, **15** (2000), 669–672.
69. E. H. Larsen, J. Sloth, M. Hansen, S. Moesgaard, Selenium speciation and isotope composition in ^{77}Se-enriched yeast using gradient elution HPLC separation and ICP-dynamic reaction cell-MS, *J. Anal. Atom. Spectrom.*, **18** (2003), 310–316.
70. J. M. Marchante Gayón, C. Thomas, I. Feldmann, N. Jakubowski, Comparison of different nebulizers and chromatographic techniques for the speciation of selenium in nutritional commercial supplements by hexapole collision and reaction cell ICP-MS, *J. Anal. Atom. Spectrom.*, **15** (2000), 1093–1102.
71. M. Kotrebai, J. F. Tyson, E. Block, P. C. Uden, High-performance liquid chromatography of selenium compounds utilizing perfluorinated carboxylic acid ion-pairing agents and inductively coupled plasma and electrospray ionization mass spectrometric detection, *J. Chromatogr. A*, **866** (2000), 51–63.
72. J. Szpunar, R. Lobinski, Multidimensional approaches in biochemical speciation analysis, *Anal. Bioanal. Chem.*, **373** (2002), 404–411.
73. S. McSheehy, W. Yang, F. Pannier, J. Szpunar, R. Lobinski, J. Auger, M. Potin-Gautier, Speciation analysis of selenium in garlic by two-dimensional high performance LC with parallel ICP-MS and ESI-MS detection, *Anal. Chim. Acta*, **421** (2000), 147–153.
74. S. McSheehy, P. Pohl, J. Szpunar, M. Potin-Gautier, R. Lobinski, Analysis of selenium speciation in selenized yeast extracts by two-dimensional liquid chromatography with ICP-MS and electrospray MS-MS detection, *J. Anal. Atom. Spectrom.*, **16** (2001), 68–73.
75. B. Gammelgaard, A. D. Madsen, J. Bjerrum, L. Bendhal, O. Jons, J. Olsen, U. Sidenius, Separation, purification and identification of the major selenium metabolite from human urine by multi-dimensional HPLC-ICP-MS and APCI-MS, *J. Anal. Atom. Spectrom.*, **18** (2003), 65–70.
76. T. Lindemann, H. Hintelmann, Identification of selenium-containing glutathione S-conjugates in a yeast extract by two-dimensional liquid chromatography with inductively coupled plasma MS and nanoelectrospray MS/MS detection, *Anal. Chem.*, **74** (2002), 4602–4610.
77. S. P. Mendez, E. B. Gonzalez, A. Sanz-Medel, Hybridation of different chiral separation techniques with ICP-MS detection for the separation and determination of selenomethionine enantiomers: chiral speciation of selenized yeast, *Biomed. Chromatogr.*, **15** (2001), 181–188.
78. K. L. Sutton, C. A. Ponce de Leon, K. L. Ackley, R. M. Sutton, A. M. Stalcup, J. A. Caruso, Development of chiral HPLC for selenoamino acids with ICP-MS detection: application to selenium nutritional supplements, *Analyst*, **125** (2000), 281–286.
79. J. Meija, M. Montes-Bayon, D. L. Le Duc, N. Terry, J. A. Caruso, Simultaneous monitoring of volatile selenium and sulfur species from se accumulating plants (wild type and genetically modified) by GC/MS and GC/ICPMS using solid-phase microextraction for sample introduction, *Anal. Chem.*, **74** (2002), 5837–5844.
80. J. A. Caruso, M. Montes-Bayon, Elemental speciation studies–new directions for trace metal analysis, *Ecotox. Environ. Saf.*, **56** (2003), 148–163.

81. C. Devos, K. Sandra, P. Sandra, Capillary gas chromatography inductively coupled plasma mass spectrometry (CGC-ICPMS) for the enantiomeric analysis of D,L-selenomethionine in food supplements and urine, *J. Pharm. Biomed. Anal.*, **27** (2002), 507–514.

82. S. S. Kannamkumarath, Katarzyna Wrobel, Kazimierz Wrobel, C. B'Hymer, J. A. Caruso, Capillary electrophoresis–inductively coupled plasma–mass spectrometry: an attractive complementary technique for elemental speciation analysis, *J. Chromatogr. A*, **975** (2002), 245–266.

83. B. Michalke, P. Schramel, Selenium speciation by interfacing capillary electrophoresis with inductively coupled plasma–mass spectrometry, *Electrophoresis*, **19** (1998), 270–275.

84. B. Michalke, P. Schramel, Selenium speciation in human milk with special respect to quality control, *Biol. Trace Elem. Res.*, **59** (1997), 45–56.

85. J. A. Day, S. S. Kannamkumarath, E. G. Yanes, M. M. Bayon, J. A. Caruso, Chiral speciation of Marfey's derivatized DL-selenomethionine using capillary electrophoresis with UV and ICP-MS detection, *J. Anal. Atom. Spectrom.*, **17** (2002), 27–31.

86. S. Lustig, D. Lampaert, K. De Cremer, J. De Kimpe, R. Cornelis, P. Schramel, Capability of flatbed electrophoresis (IEF and native PAGE) combined with sector field ICP-MS and autoradiography for the speciation of Cr, Ga, In, Pt and V in incubated serum samples, *J. Anal. Atom. Spectrom.*, **14** (1999), 1357–1362.

87. C. C. Chéry, Gel electrophoresis for speciation purposes, in K. G. Heumann (ed.), *Handbook of Elemental Speciation*, Wiley, Chichester, 2003, pp. 224–239.

88. C. C. Chéry, E. Dumont, R. Cornelis, L. Moens, Two-dimensional gel electrophoresis of selenized yeast and autoradiography of ^{75}Se-containing proteins, *Fresenius J. Anal. Chem.*, **371** (2001), 775–781.

89. M. Wind, I. Feldmann, N. Jakubowski, W. D. Lehmann, Spotting and quantification of phosphoproteins purified by gel electrophoresis and laser ablation-element mass spectrometry with phosphorus-31 detection, *Electrophoresis*, **24** (2003), 1276–1280.

90. T. W. M. Fan, E. Pruszkowski, S. Shuttleworth, Speciation of selenoproteins in Se-contaminated wildlife by gel electrophoresis and laser ablation-ICP-MS, *J. Anal. Atom. Spectrom.*, **17** (2002), 1621–1623.

91. C. C. Chery, H. Chassaigne, L. Verbeeck, R. Cornelis, F. Vanhaecke, L. Moens, Detection and quantification of selenium in proteins by means of gel electrophoresis and electrothermal vaporization ICP-MS, *J. Anal. Atom. Spectrom.*, **17** (2002), 576–580.

92. D. R. Bandura, V. I. Baranov, S. D. Tanner, Detection of ultratrace phosphorus and sulfur by quadrupole ICP-MS with dynamic reaction cell, *Anal. Chem.*, **74** (2002), 1497–1502.

93. C. C. Chéry, D. Gunther, R. Cornelis, F. Vanhaecke, L. Moens, Detection of metals in proteins by means of polyacrylamide gel electrophoresis and laser ablation-ICP-MS: Application to selenium, *Electrophoresis*, **24** (2003), 3305–3313.

94. J. A. Caruso, B. Klaue, B. Michalke, D. M. Rocke, Group Assessment: Element Speciation, *Ecotox. Environ. Safe.*, **56** (2003), 32–44.

95. K. L. Sutton, J. A. Caruso, Liquid chromatography-inductively coupled plasma mass spectrometry, *J. Chromatogr. A*, **856** (1999), 243–258.

96. C. A. Ponce de Leon, M. Montes-Bayon, J. A. Caruso, Elemental speciation by chromatographic separation with inductively coupled plasma mass spectrometry detection, *J. Chromatogr. A*, **974** (2002), 1–21.

97. K. Sutton, R. M. Sutton, J. A. Caruso, Inductively coupled plasma mass spectrometric detection for chromatography and capillary electrophoresis, *J. Chromatogr. A*, **789** (1997), 85–126.

98. G. K. Zoorob, J. W. McKiernan, J. A. Caruso, ICP-MS for elemental speciation studies, *Mikrochim. Acta*, **128** (1998), 145–168.

99. N. P. Vela, J. A. Caruso, Element selective detection for supercritical-fluid chromatography, *J. Biochem. Biophys. Meth.*, **43** (2000), 45–58.

100. B. Bouyssiere, J. Szpunar, R. Lobinski, Gas chromatography with inductively coupled plasma mass spectrometric detection in speciation analysis, *Spectrochim. Acta*, **57B** (2002), 805–828.

101. R. S. Houk, Mass spectrometry of inductively coupled plasmas, *Anal. Chem.*, **58** (1986), 97A–105A.

102. L. Hinojosa Reyes, J. M. Marchante Gayón, J. I. Garcia Alonso, A. Sanz-Medel, Determination of selenium in biological materials by isotope dilution analysis with an octapole reaction system ICP-MS, *J. Anal. Atom. Spectrom.*, **18** (2003), 11–16.

103. A. M. Featherstone, A. T. Townsend, G. A. Jacobson, G. M. Peterson, Comparison of methods for the determination of total selenium in plasma by magnetic sector inductively coupled plasma mass spectrometry, *Anal. Chim. Acta*, **512** (2004), 319–327.

104. R. Thomas, A beginner's guide to ICP-MS: Part VII: Mass separation devices – double-focusing magnetic-sector technology, *Spectroscopy*, **16** (2001), 22–27.

105. N. Jakubowski, D. Stuewer, D. Klockow, C. Thomas, H. Emons, Speciation of organic selenium compounds using reversed-phase LC and ICP-MS. Part III. Application of a sector field instrument with low and high mass resolution for selenium speciation in herring gull eggs, *J. Anal. Atom. Spectrom.*, **16** (2001), 135–139.

106. I. Feldmann, N. Jakubowski, D. Stuewer, C. Thomas, Speciation of organic selenium compounds by reversed-phase liquid chromatography and inductively coupled plasma mass spectrometry, *J. Anal. Atom. Spectrom.*, **15** (2000), 371–376.

107. Katarzyna Wrobel, K. DeNicola, Kazimierz Wrobel, J. A. Caruso, ICP-MS: metals and much more, in J. J. Monaghan (ed.), *Advances in Mass Spectrometry*, Elsevier Science, Amsterdam, 2004, pp. 231–275.

108. S. Mazan, N. Gilon, G. Crétier, J. L. Rocca, J. M. Mermet, Inorganic selenium speciation using HPLC-ICP-hexapole collision/reaction cell-MS, *J. Anal. Atom. Spectrom.*, **17** (2002), 366–370.

109. D. R. Bandura, V. I. Baranov, S. D. Tanner, Reaction chemistry and collisional processes in multiple devices for resolving isobaric interferences in ICP-MS, *Fresenius J. Anal. Chem.*, **370** (2001), 454–470.

110. M. Iglesias, N. Gilon, E. Poussel, J. M. Mermet, Evaluation of an ICP-collision/reaction cell-MS system for the sensitive determination of spectrally interfered and non-interfered elements using the same gas conditions, *J. Anal. Atom. Spectrom.*, **17** (2002), 1240–1247.

111. K. G. Heumann, L. Rottman, J. Vogl, Elemental speciation with liquid chromatography–inductively coupled plasma isotope dilution mass spectrometry, *J. Anal. Atom. Spectrom.*, **9** (1994), 1351–1355.

112. W. R. Wolf, H. Zainal, Methylseleno-amino acid content of food materials by stable isotope dilution mass spectrometry, *Food Nutr. Bull.*, **23** (2002), 120–123.

113. J. F. T. M. Kotrebai, E. Block, P.C. Uden, High-performance liquid chromatography of selenium compounds utilizing perfluorinated carboxylic acid ion-pairing agents and inductively coupled plasma and electrospray ionization mass spectrometric detection, *J. Chromatogr. A*, **866** (2000), 51–63.

114. H. M. Crews, D. J. Lewis, L. M. Owen, P. R. Strutt, Investigation of selenium speciation in in vitro gastrointestinal extracts of cooked cod by high-performance liquid chromatography–inductively coupled plasma mass spectrometry and electrospray mass spectrometry, *J. Anal. Atom. Spectrom.*, **11** (1996), 1177–1182.

115. M. Montes-Bayon, E. G. Yanes, D. Ponce, K. Jayasimhulu, A. Stalcup, J. Shann, J. A. Caruso, Initial studies of selenium speciation in *Brassica juncea* by LC with ICPMS and ES-MS detection: an approach for phytoremediation studies, *Anal. Chem.*, **74** (2002), 107–113.

116. S. Blagojevic, M. Jakovljevic, B. Zarkovic, Influence of long-term fertilization on the selenium content of calcareous chernozem soil, *J. Environ. Pathol. Toxicol. Oncol.*, **17** (1998), 183–187.

117. D. Strauch, G. Ballarini, Hygienic aspects of the production and agricultural use of animal wastes, *Zentralbl. Veterinarmed. B*, **41** (1994), 176–228.

118. M. Stadlober, M. Sager, K. J. Irgolić, Effects of selenate supplemented fertilisation on the selenium levels in cereals – identification and quantification of selenium compounds by HPLC-ICP-MS, *Food Chem.*, **73** (2001), 357–366.

119. M. Hansen, P. Duda, A. M. Zayed, N. Terry, Selenium removal by constructed wetlands: role of biological volatilization, *Environ. Sci. Technol.*, **32** (1998), 591–597.

120. N. Terry, A. M. Zayed, M. P. De Souza, A. S. Tarun, Selenium in higher plants, *A. Rev. Plant. Physiol. Plant. Mol. Biol.*, **51** (2000), 401–432.

121. M. P. De Souza, C. M. Lytle, M. M. Mulholland, M. L. Otte, N. Terry, Selenium assimilation and volatilization from dimethylselenoniopropionate by Indian mustard, *Plant Physiol.*, **122** (2000), 1281–1288.

122. F. Bourgis, S. Roje, M. L. Nuccio, D. B. Fisher, M. C. Tarczynski, C. Li, C. Herschbach, H. Rennenberg, M. J. Pimenta, T. L. Shen, D. A. Gage, A. D. Hanson, *S*-methylmethionine plays a major role in phloem sulfur transport and is synthesized by a novel type of methyltransferase, *Plant Cell.*, **11** (1999), 1485–1498.

123. M. Montes Bayon, D. D. L. Duc, N. Terry, J. A. Caruso, Selenium speciation in wild-type and genetically modified selenium accumulating plants with HPLC separation and ICP-MS/SE-MS detection, *J. Anal. Atom. Spectrom.*, **17** (2002), 872–879.

124. P. Barak, I. L. Goldman, Antagonistic relationship between senenate and sulfate uptake in onion (*Allium cepa*): implications for the production of organosulfur and organoselenium compounds in plants, *J. Agric. Food Chem.*, **45** (1997), 1290–1294.

125. J. Lintschinger, N. Fuchs, J. Moser, D. Kuehnelt, W. Goessler, Selenium-enriched sprouts. A raw material for fortified cereal-based diets, *J. Agric. Food Chem.*, **48** (2000), 5362–5368.

126. J. W. Finley, C. D. Davis, Y. Feng, Selenium from high-selenium broccoli is protective against colon cancer, *J. Nutr.*, **130** (2000), 2384–2389.

127. E. Dumont, Emmie, F. Vanhaecke, R. Cornelis, Hyphenated techniques for speciation of Se in *in vitro* gastrointestinal digests of Saccharomyces cerevisiae, *Anal. Bioanal. Chem.*, **379** (2004), 504–511.

128. P. B. Moser-Veillon, A. R. Mangels, K. Y. Patterson, C. Veillon, Utilization of two different chemical forms of selenium during lactation using stable isotope tracers: an example of speciation in nutrition, *Analyst*, **117** (1992), 559–562.

129. M. Yoshida, S. Sugihara, T. Suenaga, C. Naito, K. Fukunaga, H. Tsuchita, Digestibility and chemical species of selenium contained in high-selenium yeast, *J. Nutr. Sci. Vitaminol. (Tokyo)*, **48** (2002), 401–404.

130. J. J. Sloth, E. Larsen, S. Bugel, S. Moesgaard, Determination of total selenium and ^{77}Se in isotopically enriched human samples by ICP-dynamic reaction cell-MS, *J. Anal. Atom. Spectrom.*, **18** (2003), 317–322.

131. K. El-Bayoumy, J. P. Richie, Jr., T. Boyiri, D. Komninou, B. Prokopczyk, N. Trushin, W. Kleinman, J. Cox, B. Pittman, S. Colosimo, Influence of selenium-enriched yeast supplementation on biomarkers of oxidative damage and hormone status in healthy adult males: a clinical pilot study, *Cancer Epidemiol. Biomarkers Prev.*, **11** (2002), 1459–1465.

132. M. Kotrebai, S. M. Bird, J. F. Tyson, E. Block, P. C. Uden, Characterization of selenium species in biological extracts by enhanced ion-pair liquid chromatography with ICP-MS and by referenced ESI-MS, *Spectrochim. Acta*, **54B** (1999), 1573–1591.

133. J. Zheng, M. Ohata, N. Furuta, W. Kosmus, Speciation of selenium compounds with ion-pair reversed-phase liquid chromatography using inductively coupled plasma mass spectrometry as element-specific detection, *J. Chromatogr. A*, **874** (2000), 55–64.

134. S. McSheehy, F. Pannier, J. Szpunar, M. Potin-Gautier, R. Lobinski, Speciation of seleno compounds in yeast aqueous extracts by three-dimensional liquid chromatography with ICM-MS and ESI-MS detection, *Analyst*, **127** (2002), 223–229.

135. J. R. Encinar, L. Ouerdane, W. Buchmann, J. Tortajada, R. Lobinski, J. Szpunar, Identification of water-soluble selenium-containing proteins in selenized yeast by size exclusion reversed phase HPLC-ICP-MS followed by MALDI-TOFMS and ESI-Q-TOF MS, *Anal. Chem.*, **75** (2003), 3765–3774.

136. J. R. Encinar, R. Ruzik, W. Buchman, J. Tortajada, R. Lobinski, J. Szpunar, Detection of selenocompounds in a tryptic digest of yeast selenoprotein by MALDI time-of-flight MS prior to their structural analysis by electrospray ionization triple quadrupole MS., *Analyst*, **128** (2003), 220–224.

137. C. Dietz, J. Sanz Landaluze, P. Ximénez-Embún, Y. Madrid-Albarran, C. Camara, SPME-multicapillary GC coupled to different detection systems and applied to volatile organo-selenium speciation in yeast, *J. Anal. Atom. Spectrom.*, **19** (2004), 260–266.

22

DETERMINATION OF Hg SPECIES IN SEAFOOD

PETRA KRYSTEK

Formerly at the National Institute for Public Health and the Environment (RIVM), The Netherlands, Present E-mail: petra.krystek@freenet.de

ROB RITSEMA

National Institute for Public Health and the Environment (RIVM), PO Box 1, 3720 BA Bilthoven, The Netherlands, E-mail: rob.ritsema@rivm.nl

22.1 INTRODUCTION

Mercury, as the only liquid metal, is always playing an extraordinary role. The element Hg as well as its compounds show physical–chemical properties quite different from those of other metals. The most important Hg compounds can be classified in the following three groups: (1) elementary Hg; (2) inorganic Hg(I)

The Determination of Chemical Elements in Food: Applications for Atomic and Mass Spectrometry.
Edited by Sergio Caroli

and Hg(II) salts; (3) organometallic Hg compounds (Hg compounds with a direct covalent bond between Hg and C) [1].

Mercury is extremely toxic and can cause neurological and kidney damage, and blindness. It is also associated with birth defects. Given the extremely high toxicity, especially of organic Hg compounds, particular safety precautions are absolutely necessary during handling. For instance, direct contact of methyl-Hg (Me-Hg) with skin can be lethal [2, 3]. Mercury is produced by mining and smelting of cinnabar ore, and has many different uses: in chloralkali plants (where chlorine and sodium hydroxide are produced), in paints as preservative or pigment, in electrical switching equipment and batteries, in measurement and control equipment (thermometers, medical equipment), in copper and silver amalgams in tooth-filling materials, and as a fungicide in agriculture (especially as seed dressings) [4].

Besides its low natural concentrations, Hg is mostly anthropogenically introduced into the environment. The anthropogenic sources of Hg are numerous and worldwide. The annual emissions are estimated to be between 2700 and 6000 tons. The global cycle of Hg involves the emission of the metal from land and water surfaces into the atmosphere, its transport in the atmosphere on a global scale, possible conversion to unidentified soluble species, and return to land and water by various depositional processes [4]. In the aquatic system numerous transformations take place, which are mainly microbiological processes. An overview of the most common Hg species in the environment is given in Figure 22.1.

Elements are abundant in different chemical states and binding forms, the so-called chemical species. Natural methylation in the environment is now well established for a number of elements. Mercury was one of the first cases studied owing to its methylation potential and the high toxicity associated with the Me-Hg form and also with dimethylmercury (Me_2Hg). In general, interest is growing in the correct determination of species as they exhibit quite different degrees of toxicity [5]. As is the case with many other elements, the organic species of Hg are more poisonous than the free inorganic Hg(II) species [6]. Me-Hg is considered to be one of the most dangerous pollutants with an extremely high toxicity [3, 7]. Mercury species are very noxious to aquatic life. In most cases of Hg poisoning from seafood the predominant form was found to be alkyl-Hg, Me-Hg in particular. If methylated species occur, their concentrations increase through bioaccumulation processes via seafood into the food chain [6, 8].

Between 1956 and 1960 severe accidents related to Me-Hg, the health consequences of which were afterwards grouped under the name of Minamata disease, occurred in Japan following long-term consumption of contaminated fish and fish products. A large amount of Hg-polluted effluents (from 200 to 600 tons) reached Minamata Bay, Japan, where they accumulated not only in the bottom sediment, but also in fish and shellfish. The consumption of such foods resulted in a mass outbreak of Me-Hg poisoning [9].

European legislation regulates the maximum permitted concentration of Hg in different kinds of food, for example, predatory fish and eels, 1.0 $\mu g\ g^{-1}$ wet weight, and mackerel, herring, sprat, and other fish, 0.5 $\mu g\ g^{-1}$ wet weight. No differentiation is made between Me-Hg and inorganic Hg [10]. The World Health

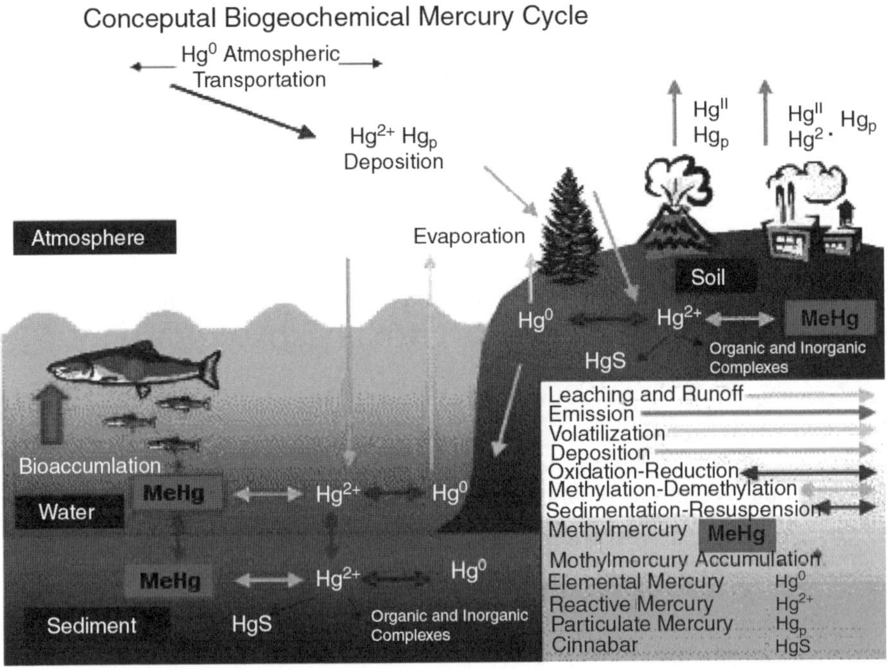

Figure 22.1. Biochemical circle of Hg [75].

Organization (WHO) recommends limits for the safe intake of chemical elements. The limit for Hg is 300 μg per week of which a maximum of 200 μg is Me-Hg [11]. In April 2004 the European Food Safety Authority (EFSA) published new limits for Me-Hg. On the one hand, a Provisional Tolerable Weekly Intake (PTWI) of 1.6 $\mu g\ kg^{-1}$ body weight is given; on the other hand, a significantly lower intake limit of 0.7 $\mu g\ kg^{-1}$ body weight per week from a previous evaluation by the United States National Research Council (USNRC) is established [12].

The speciation of Hg in all kinds of environmental matrices (air, water, sediment, and seafood/fish) has been studied for several years now and it is still a hot topic in ongoing studies. Different environmental compartments show different ratios between the abundance of species; for example, for natural waters the abundance of Me-Hg and Hg^{2+} depends on the geological origin and the pretreatment (filtered or not filtered). From less than 1 percent [13] to around 5 percent [14, 15], especially for unfiltered seawater, Hg as Me-Hg has been reported. While phytoplankton contains around 15 percent, in zooplankton around 20 percent of total Hg is present in the form of Me-Hg [15, 16]. In foods they can occur as different species, normally in trace amounts. The level of Hg in the environment is generally low, but there are some species with a greater tendency to accumulate. The accumulation of Hg species is higher in fish than in any other aquatic organisms. The contribution of Me-Hg increases in herbivorous fish to approximately 70 percent [17], while in predatory fish a maximum of nearly 100 percent as Me-Hg is possible [18]. Besides the differences for fish species, the concentration of Me-Hg varies for the different parts of the body and also with the age and size of the fish [15, 19, 20]. Investigations on binding forms of Me-Hg in fish, especially in swordfish skeletal muscle, have been recently published. Results obtained by X-ray absorption near-edge spectra revealed that Hg binds directly with a C atom and a S atom, forming most likely Me-Hg-cysteine (or structurally related species) with potential toxicological implications [21].

22.2 COMPENDIUM OF THE ANALYSIS OF Hg SPECIES SAMPLE PROCEDURES, PREPARATION, CHROMATOGRAPHY, MEASUREMENT, AND DETECTION

The traditional method for extracting and separating Hg species from organic matrices for electron capture detection (ECD) is well established and commonly known as the Westöö method since 1966–1967 [22, 23]. It is mostly used for the determination of Me-Hg. Most of the protocols used to date are derived from this method and the general scheme can be summarized as follows:

- Liberation of Me-Hg from its protein bond by displacing the mercapto group with halogen ion at low pH.
- Selective extraction of organic Hg species in an organic solvent (especially toluene).
- Purification of the organic extract from interfering impurities by extraction into aqueous solution in the presence of a thiol compound (cysteine).
- Redissociation of the organo-Hg–thiol complex.
- Reextraction into the organic solvent and possible preconcentration by evaporation.
- Measurement by gas chromatography (GC) analysis with packed columns and ECD.

The chromatographic step of this procedure needs to be carefully controlled since the Hg–halide bond exhibits a very polar character and interacts strongly with the column, leading to severe tailing of the chromatographic peaks. Several approaches were followed to overcome this problem [8]. In all cases, conditioning was necessary, but all approaches gave satisfactory results and the use of a low polarity column for routine analysis was recommended. With the introduction of the capillary columns a clear advantage over traditionally packed columns was not always found [24]. Nevertheless, in order to avoid chromatographic problems, several alternative approaches have been used. Ultimately, to eliminate the problem generated by the polar Hg–halide bond, Hg species can be butylated by a Grignard reagent to yield nonpolar dialkyl derivatives. As the electron-capturing halide moiety is absent from these derivatives, Hg specific detection is necessary and is achieved by GC coupled to microwave-induced plasma atomic emission spectrometry (MIP-AES) [25].

Further developments are expected due to the nonspecificity of the ECD detector. The possibility of coelution with other compounds has triggered the interest in interfacing chromatography systems to Hg-specific detectors. The volatility of zerovalent Hg and the ease of thermal decomposition of alkyl-Hg compounds have allowed for a wide variety of solutions that use cold vapor (CV) atomic absorption spectrometry (AAS) in a fused silica quartz cell. Significant improvements were observed when the atomization and detection were performed directly in the fused silica furnace at 780°C using an O_2 flow and after considerable reduction of the transfer lines [26].

The use of GC-MIP-AES is advantageous because it avoids the predecomposition step required in the AAS detection mode. The first applications of the MIP-AES detector for Hg speciation and detection were reported in the 1970s [27–29]. Despite the overall good detection ability of the detectors, however, most of the above methods require large sample volumes, tedious solvent extraction procedures, and usually lead to the final determination of only the Me-Hg species. The description of the feasibility of quantitative in situ aqueous ethylation of Hg^{2+} and Me-Hg followed by on-line preconcentration and detection by atomic fluorescences pectrometry (AFS) or AAS certainly produces a wealth of information since it allows all Hg species to be detected in the same chromatographic run. Also on-line speciation of Hg and Me-Hg by chromatography–AFS hydride generation (HG) was used [30].

A general overview about traditionally used methods for the analysis of Me-Hg is given in Table 22.1. It may be stated that nowadays the measurement of Hg species is carried out basically by AAS as well as by hyphenated techniques, namely high-performance liquid chromatography (HPLC) coupled with inductively coupled plasma mass spectrometry (ICP-MS) [31–34] and GC-ICP-MS [35–41]. In particular, the combination of GC-ICP-MS is a powerful tool thanks to the high resolving power of the capillary GC technology and the sensitive multielemental detection capabilities of ICP-MS. One of the main advantages of GC-ICP-MS over GC-MIP-AES is that solvent venting (to prevent unstable plasma conditions) is not necessary in GC-ICP-MS. Interfaces/transfer lines between GC and ICP-MS are nowadays also commercially available. An example is given in Figure 22.2. Various setups exist with the addition of plasma carrier gas at different points of the interface. One major requirement consists of the isothermal conditions over the entire transfer line up to 600°C, depending on applications. Nevertheless, it is less an issue for the Hg species than for less volatile substances like high-boiling-point compounds of other elements.

The transfer line consists of an electrically heated stainless steel tube, through which an uncoated, yet deactivated fused silica transfer capillary is passed until the end of the plasma injector. All parts of the stainless steel transfer tube are heated, including the part inside the torch box. The ICP-MS instrumentation is prone to signal suppressions and/or instrumental drift. These problems can be compensated by the use of internal standards. In the case of GC-ICP-MS the internal standard can be added to the carrier gas of the GC apparatus. A suitable internal standard is Xenon (Xe) [41]. The ^{126}Xe signal is monitored simultaneously with the other isotopes of interest. In this way instrumental drift and signal suppression can be corrected.

TABLE 22.1. Basic Methods for the Analysis of Me-Hg [72]

Separation Technique	Method
GC	GC-AAS
	GC-AED
	GC-AFS
	GC-ECD[a]
	GC-FTIR[b]
	GC-HC-GD[c]-AED
	GC-MIP-AED
	GC-MS
	GC-ICP-MS
LC	HPLC[d]-AAS
	HPLC[d]-AFD
	HPLC[d]-ICP-MS
	IC[e]-AAS
Others	CE[f]-ICP-MS
	SFC[g]-ICP-MS

[a]Electron capture detector.
[b]Fourier transformation infrared spectrometry.
[c]Hollow cathode glow discharge.
[d]High performance liquid chromatography.
[e]Ion chromatography.
[f]Capillary electrophoresis.
[g]Supercritical fluid chromatography.

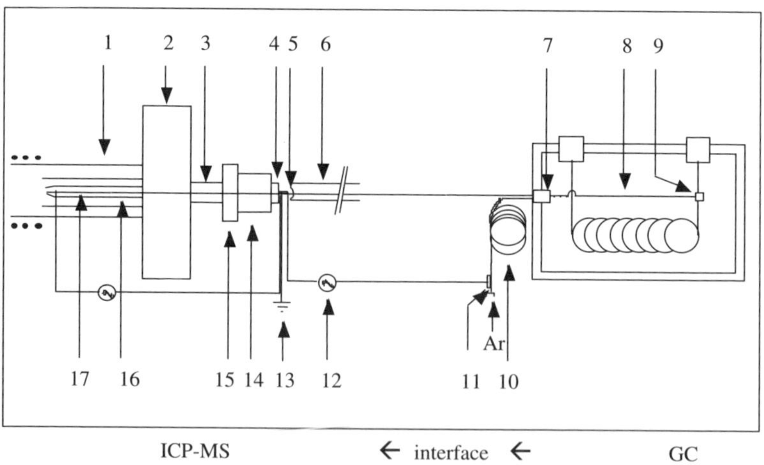

Figure 22.2. Example of a GC-ICP-MS instrumental set-up [76]. 1, torch; 2, torch box; 3, injector supply; 4, teflon adapter; 5, electrical contact point; 6, thermal isolation; 7, T-joint; 8, transfer capillary; 9, effluent splitter; 10, Ar gas heating coil; 11, female Swagelok adapter; 12, variable AC supply; 13, earthing; 14, Teflon coupling piece; 15, device attaching screw to torch; 16, stainless steel transfer tube; and 17, transfer capillary.

22.3 SELECTED PROCEDURES FOR THE DETERMINATION OF Hg SPECIES IN SEAFOOD

22.3.1 Introduction and Overview

Apart from the development of a general method for the analysis of Hg species (see the previous section) during recent years, many applications have been developed and refined for seafood matrices. For the determination of Hg species in seafood, a wide range of methods primarily focus on the determination of Me-Hg. Some publications focus on selected aspects of sample pretreatment, while others deal with the features of the analytical instruments used. When higher detection power for speciation analysis is requested, GC-ICP-MS is often the method of choice. The procedure is based on three steps, as illustrated in Figure 22.3.

Recently published aspects are highlighted and discussed in the following sections. A general overview of the limits of detection (LODs) achievable for different species by recent techniques is given in Table 22.2.

22.3.2 Key Aspects of Sampling and Quality Control

22.3.2.1 Sampling and Subsampling of Seafood Seafood can be analyzed directly as fresh or after deep-freezing as thawed out material. Handling deep-frozen material should follow exact rules of sub-sampling, otherwise the aliquots taken are not representative for the original sample. To avoid inhomogeneity effects, the fine-cut fish material must be totally thawed out and homogenized by stirring before taking subsamples, for example, by analyzing partially thawed fish material can lead to concentrations that are lower by a factor of two [42].

Nevertheless, sample material is often dried by any fit-for-purpose procedure to obtain the lyophilizate. Especially for fish material, three different drying methods were examined and it was noted that freeze-drying or microwave (MW) drying led to losses of Hg, whereas no losses were observed with oven drying [43].

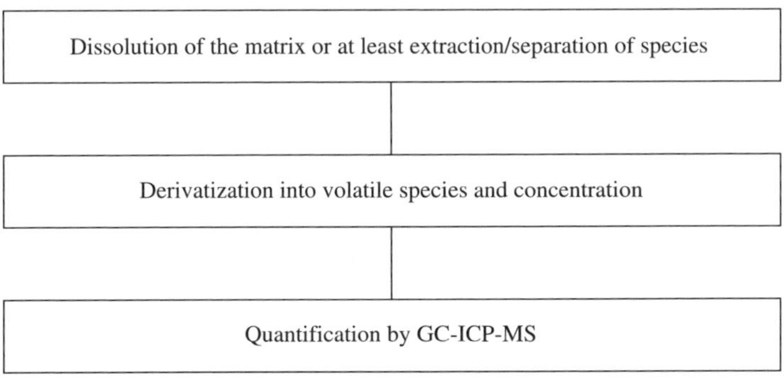

Figure 22.3. Basic procedure for the analysis of Hg species by GC-ICP-MS.

TABLE 22.2. Recent Applications for the Matrix Fish. Techniques and LoDs Obtained

Technique	LoD (Me-Hg)	LoD (Hg^{2+})	LoD (Hg_{tot})	Reference
CV-AAS with slurry technique	11 ng g^{-1}	0.977 ng g^{-1}	4.04 ng g^{-1}	[60]
CV-AAS with slurry technique	ND	0.12 μg g^{-1}	0.11 μg g^{-1}	[61]
GC-AFS	1.2 pg (absolute)	ND	ND	[43]
GC-MS	40 ng g^{-1}	ND	ND	[50]
GC-ICP-MS	2.8 ng g^{-1}	ND	ND	[64]
GC-MIP-AES	0.1 μg g^{-1}	0.6 μg g^{-1}	ND	[53]
GC-MIP-AES	4.7 ng g^{-1}	ND	ND	[64]
GC-MIP-AES	3 pg g^{-1}	ND	ND	[48]
GC-MIP-AES with PIT	12.5 pg g^{-1}	ND	ND	[48]
LC-ICP-MS	Range: 0.05–0.09 ng ml^{-1}		ND	[62]
MCGC-MIP-AES	20 ng g^{-1}	80 ng g^{-1}	ND	[49]
MCGC-ICP-TOFMS	16 fg g^{-1}	257 fg g^{-1}	ND	[65]
SPME-GC-FAPES	1.5 ng g^{-1}	0.7 ng g^{-1}	ND	[68]
SPME-GC-ID-ICP-MS	37 ng g^{-1}	ND	ND	[67]
SPME-GC-ICP-MS	2.1 ng g^{-1}	ND	ND	[66]

ND, not determined.

22.3.2.2 Accuracy and Analytical Quality Control Aspects The analysis of certified reference materials (CRMs) following the same analytical procedures was performed for assessment of the accuracy of the procedure and for quality control (QC). Yet, the available CRMs are mostly freeze-dried, not fresh or deep-frozen. In Table 22.3 an overview of CRMs in a seafood matrix with respect to organic Hg is given. It is certainly beneficial that more and more CRMs are becoming commercially available. Recently, a new CRM for trace elements in a matrix of lyophilized tuna fish (IMEP-20) has been produced [44]. Apart from total Hg (4.32 mg kg^{-1} dry mass) and Me-Hg (4.24 mg kg^{-1} dry mass), this material is also certified for other elements such as As, Pb, and Se.

22.3.2.3 Dissolution of the Sample Material, Extraction and Separation of Hg Species

Dissolution and Extraction in General For the dissolution of fish tissues different procedures are reported. Acid or alkaline digestion or solvent extraction is usually used for the separation of Hg species from biological matrices [45, 46]. For the quantitative extraction of Hg, four methods were examined: HCl leaching, alkaline–methanol extraction with tetramethylammonium hydroxide (TMAH) and

TABLE 22.3. Organic Hg CRMs in a Matrix of Seafood [73]

Name	Code	Value Type	Concentration ($\mu g\ g^{-1}$)	Remark
Tuna fish	BCR-CRM 464	Certified	5.5	As Hg, Me-Hg
Dogfish muscle	NRCC-DORM-2	Certified	4.47	
Tuna fish	BCR-CRM 463	Certified	3.04	As Hg, Me-Hg
Lobster hepatopancreas	NRCC-TORT-2	Certified	0.152	
Organics in mussel tissue	NIST-SRM-1974a	Certified	0.0773	As Hg
Nondefatted lobster hepatopancreas	NRCC-LUTS-1	Certified	0.063	
Mussel tissue	NIST-SRM-2977	Certified	0.0362	
Mussel tissue	NIST-SRM-2976	Certified	0.0278	As Me-Hg, trace elements
Oyster tissue	NIST-SRM-1566b	Certified	0.0132	As Hg

Remarks: In several of these CRMs other compounds and/or elements are also certified which are not explicitly reported in this table.
Suppliers: BCR, Institute for Reference Materials and Measurements (IRMM), Belgium; NIST, National Institute of Standards and Technology, United States; NRCC, National Research Council of Canada. Classifications: CRM, Certified Reference Material; DORM, Dogfish Muscle Reference Materials for Trace Metals; LUTS, Nondefatted Lobster Hepatopancreas Reference Material for Trace Metals; SRM, Standard Reference Material; TORT, Lobster Hepatopancreas Marine Reference Material for Trace Metals.

with KOH, and sodium dodecyl sulfate (SDS). Only HCl leaching led to >97 percent quantitative extraction of Hg [43].

The use of TMAH and shaking at room temperature is preferred and recommended in many applications. Depending on the sample material, total dissolution may take several hours [47]. However, different experiences about the use of TMAH have been reported. For instance, Me_2Hg was found to form when using TMAH to extract Hg from fish tissues [43]. Upon sample pretreatment, abiotic methylation of inorganic Hg occurred during pH adjustment of digested samples, but this was significantly reduced by prolonged treatment (over 24 h) of tissue samples with TMAH [31].

As the dissolution with TMAH became more practical and relevant, MW-assisted digestion was used since it enables mild, quick, and complete dissolution of the fish tissues [48]. It was optimized so that a simple and rapid procedure of only 2.5 min of MW-assisted solubilization of biomaterial with TMAH could be set up [49]. The combination of MW-assisted digestion with TMAH and the addition of promoters like Cu^{2+} is also possible [50].

Dissolution and Separation by Special Applications: Enzymes and Microextraction Several enzymes such as trypsin, protease type XIV, lipase and/or cellulase, are used for enzymatic hydrolysis. For the determination of Hg^{2+} and Me-Hg in fish tissue, enzymatic treatment with protease, extraction to an ODS-PEEK column, and elution of Hg species with 5 percent methanol containing 0.06 M CH_3COONH_4–cysteine were applied [45].

For the speciation of Me-Hg in seafood after volatilization of its hydride or chloride derivative, the direct coupling between headspace solid-phase microextraction (SPME) in 2 M HCl and quartz tube-AAS detection has been evaluated. Following variable optimization in both procedures, a critical influence of the fiber sorbent material was observed. Different types of liquid-phase coatings were tested whereby the best LoDs were provided by silica fibre for Me-HgH and by polydimethylsiloxane-divinylbenzene for Me-HgCl. In general, repeatability was about 6 percent for the preconcentration of both Hg derivatives in the different fiber coatings. The affinity for sorption of Me-HgH was slightly improved when using silica fibres loaded with L-cysteine or $Pd(NO_3)_2$. Generation of the chloride derivates showed an enhanced selectivity as compared with generation of the hydride. Nevertheless, experimental evidence of decomposition of the chloride derivative upon heating is also given [51].

22.3.2.4 Aspects of Derivatization and Measurement of Hg Species

Introduction to Derivatization For many years derivatization by alkylation – especially as ethylation, but also as propylation – has been applied to transform Hg species into volatile Hg species before measurement with hyphenated techniques [2, 50, 52]. Sodium tetraphenylborate ($NaBPh_4$) was also used for derivatization prior to measurement with GC-MIP-AED [53]. Studies of possible species transformation, for example, during the analytical procedure, have been carried out with isotope-specific determination methods. The results showed that a direct ethylation of Me-Hg in an atmospheric precipitation sample by $NaBEt_4$ produced no significant amount of artifactual Me-Hg [54]. Others investigated the species transformation processes in synthetic solutions to simulate environmental matrices. From the experiments it could be concluded that the species conversion, for example, of Me-Hg into zerovalent Hg, depends on the concentration levels of the halide [2]. Furthermore, the procedural order is of great importance, for example, ethylation should be done after addition of the organic phase to avoid species transformation [55].

Studies of Species Stability The analysis of Me-Hg from fish and other environmental samples by GC-MIP-AES after the ethylation and preconcentration of the volatile product in hexane has been critically investigated. In order to correct potential sources of random error along the analytical procedure affecting the overall repeatability of the analysis, the use of inorganic Hg naturally occurring in environmental samples as internal standard is proposed. A study to establish the best conditions to achieve a quantitative recovery of Me-Hg without altering its chemical structure has also been carried out. Magnetic stirring (without heating) of fish tissue with 2 M HNO_3 or 10 percent methanolic KOH, over 90 min, has been considered the most effective procedure to release Me-Hg while preserving its structure [56].

To study the abiotic formation of Me-Hg from Hg^{2+} as well as demethylation of Me-Hg to Hg^{2+} in biological tissues during treatment with TMAH, enriched isotope standards from $Me^{198}Hg$ and $^{201}Hg^{2+}$ were added to the samples to monitor species transformation and to apply species-specific isotope dilution (SSID) calibration. The HPLC instrumental set-up with a C_{18} column and an aqueous phase eluent was directly coupled to ICP-MS [31].

Derivatization and Measurement by Different AAS Techniques For the selective speciation of Me-Hg from biological matrices a novel, nonchromatographic method was developed. Me-Hg was separated from ethyl-Hg phenyl-Hg and Hg^{2+} through a displacement reaction of Me-Hg with a Cu(II) complex of diethyldithiocarboamate (DDTC) which was presorbed onto a microcolunm packed with cigarette filter sorbent. The bound Me-Hg was subsequently eluted with 50 μL of ethanol directly into the graphite furnace for determination by electrothermal (ET) AAS [57]. A selective reduction of Hg^{2+} and Me-Hg with $NaBH_4$, using two gas–liquid separators in series and CV-AAS, was resorted to for the analysis of biological materials; Hg^{2+} was reduced with 0.01 percent $NaBH_4$ in the first separator, and Me-Hg was reduced with 0.3 percent $NaBH_4$ and $FeCl_3$ in the second separator [58]. Using CV-AAS with a sequential reduction of alkaline digested fish samples with $SnCl_2$ and $NaBH_4$ led to a simple method for measuring Hg^{2+} and Me-Hg. Owing to matrix interference, calibration curves with matrix addition were needed for Hg determination [59].

Two procedures of slurry sampling were tested by CV-AAS. The fish tissues were first suspended in HCl containing Triton X-100 as dispersing agent and then the slurries were on-line injected in an acid carrier stream, which was either sequentially mixed with H_2SO_4–$K_2S_2O_8$ oxidizing agent and $SnCl_2$ reducing agent streams or merged with $NaBH_4$ reducing agent stream. Mercury speciation was carried out as a function of the oxidation coil temperature or $NaBH_4$ concentration. The flow injection system without on-line oxidation was simpler, faster and more sensitive for inorganic Hg and total Hg determination. Furthermore, the latter system allowed Me-Hg and Hg^{2+} to be separately determined by selecting an adequate HCl concentration in the suspension medium and $NaBH_4$ as reducing agent [60]. A special application of the matrix fish egg oil is presented using a three-phase surfactant–oil–water emulsion and a peristaltic pump to introduce the highly viscous sample into the CV-AAS system. The optimum proportion of the oil–water mixture was 2 : 3 v/v with a Tween 20 surfactant concentration in the emulsion of 0.008 percent v/v [61].

Derivatization and Measurement by Liquid Chromatography Systems Apart from classical applications with liquid chromatography (LC)-ICP-MS (see above) two new applications have been reported. A vapor generation system and LC combined with ICP-MS were used to determine Hg species in extracts of biological tissues. The efficiency of the mobile phase, a mixture of L-cysteine and 2-mercaptoethanol, was evaluated for the LC separation of three Hg-species (Hg^{2+}, Me-Hg, and ethyl-Hg). The LoDs (see Table 22.2) and repeatability of the

LC-ICP-MS system with a vapor generator were comparable to, or better than, that of an LC-ICP-MS system with conventional pneumatic nebulization [62].

For the multispecies determination of a maximum of four species like inorganic Hg^{2+}, Me-Hg, ethyl-Hg, and phenyl-Hg an automatic system was devised based on the on-line coupling of HPLC with CV-AFS via a postcolumn MW digestion. The digestion medium was $K_2S_2O_8$ in HCl. Parameters influencing on-line digestion efficiency and the separation effects were optimized, which allowed for the separation of the four Hg species within 13 min. The absolute LoDs were 0.3 ng for Hg^{2+}, 0.2 ng for Me-Hg, 0.17 ng for ethyl-Hg, and 0.14 ng for phenyl-Hg^+ [63].

Derivatization and Measurement by Different GC-ICP-MS Techniques A simple and rapid sample pretreatment procedure is based on acid leaching (5 min) of the biological material, followed by simultaneous in situ derivatization and extraction (40 min) in the presence of $NaBH_4$ and nonane, buffered at pH 7.0, in an ultrasonic bath. The separation is done with GC, while two independent on-line detecting techniques (MIP-AES and ICP-MS) are used. This approach affords LoDs better by a factor of 2 than those offered by ICP-MS [64].

Purge-and-trap-injection (PTI) for the determination of Hg species in fish samples was used for the first time in 1997. A cooled injection system and HG with $NaBH_4$ was used [48]. As for time-saving procedures, a rapid simultaneous quantitative ethylation-extraction of Hg species into hexane (15 min) and flash isothermal separation (15–30 s) using a multicapillary regular column (100 cm) or a minicolumn (22 cm) must be mentioned. The chromatographic hardware necessary for sample introduction into an MIP was downscaled to a split injection port and a $20 \times 20 \times 20$ cm compartment (housing the 1 m column) or a 22×1 cm thermally insulated tube that was kept at a constant temperature and connected directly to the AES detector [49]. Another simple, rapid, and accurate method by Time-Of-Flight (TOF) ICP-MS was combined with Multicapillary (MC)-GC. Furthermore two injection systems PTI and Split Injections (SI) were tested. The former was preferred [65]. The advantages of TOF-ICP-MS are based on the detection of fast transient signals with peak widths significantly shorter than 1 s and high time stability.

Use of Internal Standards The use of internal standards envisages different possibilities. The procedure described here is based on two internal standards. Once thawed, fish sample were dissolved in TMAH, ethylated with $NaBEt_4$, extracted into iso-octane and subjected to GC-ICP-MS for the identification and quantification of Me-Hg and inorganic Hg^{2+}. For the correction of procedural errors two internal standards were used. The sample pretreatment was corrected by the recovery factor of the spiked dibutyl-dipentyl-Sn (DBT-pe), while the GC-ICP-MS measurements were controlled by the signal stability of Xe added to the GC carrier gas [47]. In another application propyl-Hg was used as an internal standard to correct for matrix-induced ion signal variation and instrumental drift [65].

Applications with Solid-Phase Microextraction (SPME) in Combination with SSID Solid-phase microextraction (SPME) and SSID with GC-ICP-MS were also applied for the determination of Me-Hg in fish tissue. The samples were digested with methanolic KOH, derivatized with sodium tetrapropoyl bromate and adsorbed onto a polydimethylsiloxane-coated capillary fiber for determination by GC-ICP-MS. The isotope dilution (ID) technique with a laboratory synthesized Me-Hg enriched with ^{198}Hg gave a three- to four-fold improvement in precision compared with the standard addition calibration [66, 67].

For another evaluation with SPME, fish tissue was digested with methanolic KOH, derivatized with $NaBH_4$, and extracted by SPME. Species separation and measurement were carried out using tandem GC-furnace atomization plasma emission spectrometry (SPME-GC-FAPES) [68].

22.4 CONCENTRATION RANGES OF Hg IN VARIOUS KINDS OF SEAFOOD

Commonly, multielemental analysis of As, Cd, Hg, and Pb are reported as total concentrations in different kinds of fish. Interest is, however, growing in the

TABLE 22.4. Mercury Levels for Different Seafood Species [74]

Species	Mean ($\mu g\ g^{-1}$)	Range ($\mu g\ g^{-1}$)	Number of Samples
	Highest Hg Levels		
Tilefish	1.45	0.65–3.73	60
Swordfish	1.00	0.10–3.22	598
King mackerel	0.73	0.30–1.67	213
Shark	0.96	0.05–4.54	324
	Lower Hg Levels		
Grouper (*Mycteroperca*)	0.43	0.05–1.35	64
Tuna (fresh or frozen)	0.32	ND–1.30	191
Lobster, northern (American)	0.31	0.05–1.31	88
Grouper (*Epinephelus*)	0.27	0.19–0.33	48
Halibut	0.23	0.02–0.63	29
Sablefish	0.22	ND–0.70	102
Pollock	0.20	ND–0.78	107
Tuna (canned)	0.17	ND–0.75	248
Crab, blue	0.17	0.02–0.50	94
Crab, dungeness	0.18	0.02–0.48	50
Crab, tanner	0.15	ND–0.38	55
Crab, king	0.09	0.02–0.24	29
Scallop	0.05	ND–0.22	66
Catfish	0.07	ND–0.31	22
Salmon (fresh, frozen or canned)	ND	ND–0.18	52
Oysters	ND	ND–0.25	33
Shrimps	ND	ND	22

ND, not detectable.

accumulation of Hg in fish from contaminated waters, and national and international legislation imposes controls to ensure that the concentrations of Hg do not exceed the maximum permitted values [10–12]. It was found that even in uncontaminated regions Hg concentrations do exceed the limits set by WHO [69, 70].

Apart from total concentrations of Hg, the toxic species Me-Hg raises the highest concern. To obtain a first overview of the potential occurrence of Hg and Me-Hg in seafood species and the variation of their concentration levels, two tables (Tables 22.4 and 22.5) from the U.S. Food and Drug Administration (USFDA) reporting data obtained in May 2001 are presented. Table 22.4 shows that significant differences in Hg levels are found depending on the seafood species analyzed. Fish higher in the food chain, like large predatory fish, accumulate more Hg and contain significantly higher concentration levels. Data on Me-Hg in seafood species are presented in Table 22.5. For comparison purposes, it must be considered that the data in Table 22.5 are based on small samples.

The results of the analysis for Me-Hg and Hg^{2+} in shark fillets are given in Table 22.6. Slight variations are found for Me-Hg, which forms between 94 and 98.9 percent of the abundant Hg species in the shark fillets analyzed [47]. Other

TABLE 22.5. Seafood Species with Me-Hg Levels [74]

Species	Mean ($\mu g\ g^{-1}$)	Range ($\mu g\ g^{-1}$)	Number of Samples
Red snapper	0.60	0.07–1.46	10
Marlin	0.47	0.25–0.92	13
Moonfish	0.60	0.60	1
Orange roughy	0.58	0.42–0.76	9
Bass saltwater	0.49	0.10–0.91	9
Trout freshwater	0.42	1.22 (max)	U
Bluefish	0.30	0.20–0.40	2
Croaker	0.28	0.18–0.41	15
Trout seawater	0.27	ND–1.19	4
Cod (Atlantic)	0.19	ND–0.33	11
Mahi mahi	0.19	0.12–0.25	15
Ocean perch	0.18	ND–0.31	10
Haddock (Atlantic)	0.17	0.07–0.37	10
Whitefish	0.16	ND–0.31	2
Herring	0.15	0.016–0.28	8
Spiny lobster	0.13	ND–0.27	8
Perch freshwater	0.11	0.10–0.31	4
Perch saltwater	0.10	0.10–0.15	6
Flounder/sole	0.04	ND–0.18	17
Clams	ND	ND	6
Tilapia	ND	ND	8

ND, not detectable; U, unknown.

TABLE 22.6. Balances of $Me\text{-}Hg^{+}$ and Hg^{2+} as well as Total Hg in Three Shark Fillets Determined by Two Techniques

Sample	$c(Me\text{-}Hg^{+})$ ($\mu g\ g^{-1}$ Hg)[a]	$c(Hg^{2+})$ ($\mu g\ g^{-1}$ Hg)[a]	Percentage ($Me\text{-}Hg^{+}$)in ΣHg (%)[a]	c(ΣHg) ($\mu g\ g^{-1}$ Hg)[a]	c(Hg, total) ($\mu g\ g^{-1}$ Hg)[b]
Shark 1	1.43 ± 0.08[c]	0.08 ± 0.02	94.0	1.52	1.85 ± 0.16
Shark 2	0.94 ± 0.17	0.01 ± 0.002	98.9	0.95	1.05 ± 0.09
Shark 3	3.42 ± 0.28	0.11 ± 0.02	96.9	3.53	3.33 ± 0.19

[a]Determined by GC-ICP-MS.
[b]Determined by AFS.
[c]Standard deviation from replicate determinations ($n = 3$).

determinations of Hg species were done in different tissues of Arctic marine mammals (belugas, seals, and narwhals) as well as in liver, muscle, and skin (muktuk). A robust linear regression of Me-Hg on total Hg and Me-Hg on age of animals was performed and a strong correlation between the two variables was found in each case. The ratio of Me-Hg to total Hg as indicated by the regression coefficients was close to 1 for muscle and skin (muktuk), while for liver it was less than 1. The mean percentage of Me-Hg in the liver of marine mammals was 3–12 percent of the total Hg in this tissue depending on species and location [71].

22.5 CONCLUSIONS AND TRENDS

Trace metals such as Hg and in particular Hg organic compounds pose a serious risk to human health and the environment because they bioaccumulate through the food chain. For this reason, the routine determination of the highly toxic Me-Hg and of other organic and inorganic Hg compounds in marine and terrestrial animals, vegetables, fruit, and fresh water is of utmost importance in health and environmental control programmes throughout the world.

More attention should be paid to specific analytical aspects. For instance, the handling of frozen or lyophilized fish material entails other effects and problems than fresh fish material does. Important issues with fresh fish material are, for example, inhomogeneity and formation of fat and/or water layers which could lead to incorrect subsampling and potentially to wrong analytical results and misinterpretations.

Many procedures for the analysis of Hg species have shown good potential for the study of different kinds of fish material. In particular, instrumental optimization and new developments bring about more specific and selective determinations, and better LoDs often ensue. Automated analyses will probably become more and more frequent.

Last, but not least, analytical methods for the determination of Hg species in seafood should be developed and validated, whereas new legislation is necessary to bring speciation analysis into routine use.

REFERENCES

1. M. R. Greenwood, Quecksilber, in E. Merian (ed.), *Metalle in der Umwelt*, Verlag Chemie, 1984, pp. 511–539.
2. N. Demuth, K. G. Heumann, Validation of methylmercury determinations in aquatic systems by alkyl derivatization methods for GC analysis using ICP-IDMS, *Anal. Chem.*, **73** (2001), 4020–4027.
3. A. C. Gutleb, C. Schenck, E. Staib, Giant otter (*Pteronura brasiliensis*) at risk? Total mercury and methylmercury levels in fish and otter scats, *Peru. AMBIO*, **26** (1997), 511–514.
4. World Health Organization, *Air Quality Guidelines, Mercury*, Chapter 6.9, WHO, Copenhagen, 2000.
5. F. Adams, S. Slaet, Improving the reliability of speciation analysis of organometallic compounds, *Trends Anal. Chem.*, **19** (2000), 80–85.
6. P. J. Craig, *Organometallic Compounds in the Environment*, Harlow, Longman, 1986, pp. 65–101.
7. H. Strasdeit, Methylquecksilber – eine Warnung, Nachrichten aus Chemie, *Techn. Labor.*, **46** (1998), 846–848.
8. R. Ritsema, O. X. F. Donard, Organometallic compound determination in the environment by hyphenated techniques, in D. Barcelo (ed.), *Sample Handling and Trace Analysis of Pollutants: Technique, Applications and Quality Assurance*, Amsterdam, Elsevier, 1999, pp. 1003–1073.
9. P. Weigert, Accidents affecting food and food poisoning, in K. van der Heijden (ed.), *International Food Safety Handbook*, Marcel Dekker, New York, 1999, pp. 477–491.
10. Commission Regulation (EC) No. 466/2001 of 8 March 2001 setting maximum levels for certain contaminants in foodstuffs, *Off. J. Eur. Commun.*, **L77/1–13** (2001), 16.3.
11. World Health Organization, *Environmental Health Criteria 101, Methylmercury*, WHO, Geneva, 1990.
12. Request EFSA-Q-2003-030, Opinion of the Scientific Panel on Contaminants in the Food Chain on a request from the Commission related to mercury and methylmercury in food, *EFSA J.*, **34** (6 April 2004), 1–14.
13. R. P. Mason, A. L. Lawrence, Concentration, distribution, and bioavailability of mercury and methylmercury in sediments of *Baltimore Harbor* and *Chesapeake Bay, Maryland, USA*, *Environ. Toxicol. Chem.*, **18** (1999), 2438–2447.
14. M. Logar, M. Horvat, H. Akagi, B. Pihlar, Simultaneous determination of inorganic mercury and methylmercury compounds in natural waters, *Anal. Bioanal. Chem.*, **374** (2002), 1015–1021.
15. H. Potgeter, Entwicklung und Anwendung eines neuartigen Analysenverfahrenszur Bestimmung quecksilberorganischer Verbindungen in Sedimenten mit Hilfe eines gekoppelten SFC-AFS-Systems, Dissertation, University of Hamburg, 1998.
16. C. J. Watras, N. S. Bloom, Mercury and methylmercury in individual zooplankton: Implications for bioaccumulation, *Limnol. Oceanogr.*, **37** (1992), 1313–1318.
17. R. P. Mason, J. R. Reinfelder, F. M. M. Morel, Bioaccumulation of mercury and methylmercury, *Water Air Soil Pollut.*, **80** (1995), 915–921.

18. B. Lasorsa, S. Allen-Gil, The methylmercury to total mercury ratio in selected marine, freshwater, and terrestrial organisms, *Water Air Soil Pollut.*, **80** (1995), 905–913.

19. R. Wagemann, E. Trebacz, R. Hunt, G. Boila, Percent methylmercury and organic mercury in tissues of marine mammals and fish using different experimental and calculation methods, *Environ. Toxicol. Chem.*, **16** (1997), 1859–1866.

20. NST, Determination of mercury and organic mercury content in local seafood, *NST Quarterly*, January 1996, www.mint.gov.my/papers/NST/1996/Nst1_96.htm (accessed 27 April 2004).

21. H. H. Harris, I. J. Pickering, G. N. George, The chemical form of mercury in fish, *Science*, **301** (2003), 1203.

22. G. Westöö, Determination of methylmercury compounds in foodstuffs: I. Methylmercury compounds in fish, identification and determination, *Acta Chem. Scandin.*, **20** (1966), 2131–2137.

23. G. Westöö, Determination of methylmercury coumpounds in foodstuffs. II. Determination of methylmercury in fish, eff, meat, and liver, *Acta Chem. Scandin.*, **21** (1967), 1790–1800.

24. P. Lansens, C. C. Laino, C. Meuleman, W. Baeyens, Evaluation of gas chromatographic columns for the determination of methylmercury in aqueous head space extracts from biological samples, *J. Chromatogr.*, **586** (1991), 329–340.

25. E. Bułska, D. C. Baxter, W. Fresh, Capillary column gaschromatography for mercury speciation, *Anal. Chim. Acta*, **249** (1991), 545–554.

26. J. G. Gonzalez, R. T. Ross, Interfacing of an atomic absorption spectrophotometer with a gas–liquid chromatograph for the determination of trace quantities of alkylmercury compounds in fish tissue, *Anal. Lett.*, **5** (1972), 683–694.

27. W. E. L. Grossman, J. Eng, Y. C. Tong, Selectivity of a microwave discharge detector in the gas-chromatographic determination of dimethylmercury, *Anal. Chim. Acta*, **60** (1972), 447–453.

28. B. D. Quimby, P. C. Uden, R. Barnes, Atmospheric pressure helium microwave detection system for gas chromatography, *Anal. Chem.*, **50** (1978), 2112–2118.

29. Y. Talmi, The rapid sub-picogram determination of volatile organo-mercury compounds by gas chromatography with a microwave emission spectrometric detector system, *Anal. Chim. Acta*, **74** (1975), 107–117.

30. R. Ritsema, O. X. F. Donard, On-line speciation of mercury and methylmercury in aqueous samples by chromatography atomic fluorescence spectrometry after hydride generation, *Appl. Organomet. Chem.*, **8** (1994), 571–576.

31. J. Qvarnström, W. Frech, Mercury species transformations during sample pre-treatment of biological tissues studied by HPLC-ICP-MS, *J. Anal. Atom. Spectrom.*, **17** (2002), 1486–1491.

32. J. Morton, V. A. Carolan, P. H. E. Gardiner, The speciation of inorganic mercury and methylmercury in human hair by high-performance liquid chromatography with inductively coupled plasma mass spectrometry, *J. Anal. Atom. Spectrom.*, **17** (2002), 377–381.

33. C. F. Harrington, T. Catterick, Problems encountered during the development of a method for the speciation of mercury and methylmercury by high-performance liquid chromatography coupled to inductively coupled plasma mass spectrometry, *J. Anal. Atom. Spectrom.*, **12** (1997), 1053–1056.

34. M. J. Bloxham, A. Gachanja, S. J. Hill, P. J. Worsfold, Determination of mercury species in sea-water by liquid chromatography with inductively coupled plasma mass spectrometric detection, *J. Anal. Atom. Spectrom.*, **11** (1996), 145–148.

35. J. Leenaers, W. van Mol, H. G. Infante, F. C. Adams, Gas chromatography-inductively coupled plasma-time-of-flight mass spectrometry as a tool for speciation analysis of organomercury compounds in environmental and biological samples, *J. Anal. Atom. Spectrom.*, **17** (2002), 1492–1497.

36. G. R. Peters, D. Beauchemin, Versatile interface for gas chromatographic detection or solution nebulization analysis by inductively coupled plasma mass spectrometry: preliminary results, *J. Anal. Atom. Spectrom.*, **7** (1992), 965–970.

37. A. W. Kim, M. E. Foulkes, L. Ebdon, S. J. Hill, R. L. Patience, A. G. Barwise, S. J. Rowland, Communications. Construction of a capillary gas chromatography inductively coupled plasma mass spectrometry transfer line and application of the technique to the analysis of alkyllead species in fuel, *J. Anal. Atom. Spectrom.*, **7** (1992), 1147–1150.

38. W. G. Pretorius, L. Ebdon, S. J. Rowland, Development of a high-temperature gas chromatography – inductively coupled plasma mass spectrometry interface for the determination of metalloporphyrins, *J. Chromatog.*, **646** (1993), 369–375.

39. A. Prange, E. Jantzen, Determination of organometallic species by gas chromatography inductively coupled plasma mass spectrometry, *J. Anal. Atom. Spectrom.*, **10** (1995), 105–109.

40. T. de Smaele, P. Verrept, L. Moens, R. Dams, A flexible interface for the coupling of capillary gas chromatography with inductively coupled plasma mass spectrometry, *Spectrochim. Acta*, **50B** (1995), 1409–1416.

41. T. de Smaele, L. Moens, R. Dams, P. Sandra, ICP-MS – a sensitive detector for metal speciation with capillary GC, *LC-GC Int.*, **3** (1996), 138–142.

42. P. Krystek, R. Ritsema, Mercury speciation in thawed out and refrozen fish samples by gas chromatography coupled to inductively coupled plasma mass spectrometry and atomic fluorescence spectrometry, *Anal. Bioanal. Chem.*, **381** (2005), 354–359.

43. A. I. C. Ortiz, Y. M. Albarran, D. D. Rica, Evaluation of different sample pretreatment and extraction procedures for mercury speciation in fish sample, *J. Anal. Atom. Spectrom.*, **17** (2002), 1595–1601.

44. Institute for Reference Materials and Measurements, Belgium, 2003, www.irmm.jrc.be/imep/imep20/IMEP-20_certificate.pdf (accessed 29 October 2003).

45. R. Rai, W. Maher, F. Kirkowa, Measurement of inorganic and methylmercury in fish tissues by enzymatic hydrolysis and HPLC-ICP-MS, *J. Anal. Atom. Spectrom.*, **12** (2002), 1560–1563.

46. C. M. Tseng, A. de Diego, F. M. Martin, O. X. F. Donard, Rapid and quantitative microwave-assisted recovery of methylmercury from standard reference sediments, *J. Anal. Atom. Spectrom.*, **12** (1997), 629–635.

47. P. Krystek, R. Ritsema, Determination of methylmercury and inorganic mercury in shark filets, *Appl. Organomet. Chem.*, **18** (2004), 640–645.

48. C. Gerbersmann, M. Heisterkamp, F. C. Adams, J. A. C. Broekaert, Two methods for the speciation analysis of mercury in fish involving microwave-assisted digestion and gas chromatography atomic emission spectrometry, *Anal. Chim. Acta*, **350** (1997), 273–285.

49. I. R. Pereiro, A. Wasik, R. Lobinski, Determination of mercury species in fish reference materials by isothermal multicapillary gas chromatography with atomic emission detection after microwave-assisted solubilization and solvent extraction, *J. Anal. Atom. Spectrom.*, **13** (1998), 743–747.

50. S. S. Chen, S. S. Chou, D. F. Hwang, Determination of methylmercury in fish using focused microwave digestion following by Cu^{2+} addition, sodium tetrapropoylborate derivatization, *n*-heptane extraction, and gas chromatography–mass spectrometry, *J. Chromatogr.*, **1024** (2004), 209–215.

51. S. Fragueiro, I. Lavilla, C. Bendicho, Direct coupling of solid phase microextraction and quartz tube-atomic absorption spectrometry for selective and sensitive determination of methylmercury in seafood: an assessment of chloride and hydride generation, *J. Anal. Atom. Spectrom.*, **19** (2004), 250–254.

52. S. Rapsomanikis, O. X. F. Donard, J. H. Weber, Speciation of lead and methyllead ions in water by chromatography/atomic absorption spectrometry after ethylation with sodium tetraethylborate, *Anal. Chem.*, **58** (1986), 35–37.

53. H. E. L. Palmieri, L. V. Leonel, Determination of methylmercury in fish tissue by gas chromatography with microwave-induced plasma atomic emission spectrometry after derivatization with sodium tetraphenylborate, *Fresenius J. Anal. Chem.*, **366** (2000), 466–469.

54. J. Holz, J. Kreutzmann, R. D. Wilken, R. Falter, Methylmercury monitoring in rainwater samples using in situ ethylation in combination with GC-AFS and GC-ICP-MS techniques, *Appl. Organomet. Chem.*, **13** (1999), 789–794.

55. K. G. Heumann, University of Mainz (personal communication), 2003.

56. J. S. Landaluze, A. de Diego, J. C. Raposo, J. M. Madariaga, Methylmercury determination in sediments and fish tissues from the *Nerbioi-Ibaizabal* estuary (*Basque Country, Spain*), *Anal. Chim. Acta*, **508** (2004), 107–117.

57. X. P. Yan, Y. Li, Y. Jiang, Selective measurement of ultratrace methylmercury in fish by flow injection on-line microcolumn displacement sorption preconcentration and separation coupled with electrothermal atomic absorption spectrometry, *Anal. Chem.*, **75** (2003), 2251–2255.

58. A. D. P. Monteiro, L. S. N. de Andrade, A. S. Luna, R. C. de Campos, Sequential quantification of methyl mercury in biological materials by selective reduction in the presence of mercury(II), using two gas–liquid separators, *Spectrochim. Acta*, **57B** (2002), 2103–2112.

59. F. Ubillus, A. Algeria, R. Barbera, R. Farre, M. J. Lagarda, Methylmercury and inorganic mercury determination in fish by cold vapour generation atomic absorption spectrometry, *Food Chem.*, **71** (2000), 529–533.

60. S. R. Segade, J. F. Tyson, Evaluation of two flow injection systems for mercury speciation analysis in fish tissue samples by slurry sampling cold vapor atomic absorption spectrometry, *J. Anal. Atom. Spectrom.*, **18** (2003), 268–273.

61. J. L. Burguera, I. A. Quintana, J. L. Salager, M. Burguera, C. Rondon, P. Carrero, R. A. de Salager, Y. P. de Pena, The use of emulsions for the determination of methylmercury and inorganic mercury in fish-eggs oil by cold vapor generation in a flow injection system with atomic absorption spectrometric detection, *Analyst*, **124** (1999), 593–599.

62. C. S. Chiou, S. J. Jiang, K. S. K. Danadurai, Determination of mercury compounds in fish by microwave-assisted extraction and liquid chromatography–vapor generation–inductively coupled plasma mass spectrometry, *Spectrochim. Acta*, **56B** (2001), 1133–1142.

63. L. N. Liang, G. B. Jiang, J. G. Liu, J. T. Hu, Speciation analysis of mercury in seafood by using high-performance liquid chromatography on-line coupled with cold-vapor atomic fluorescence spectrometry via a post column microwave digestion, *Anal. Chim. Acta*, **477** (2003), 131–137.

64. Q. Tu, J. Qian, W. Frech, Rapid determinatison of methylmercury in biological materials by GC-MIP-AES or GC-ICP-MS following simultaneous ultrasonic-assisted in situ ethylation and solvent extraction, *J. Anal. Atom. Spectrom.*, **12** (2000), 1583–1588.

65. P. Jitaru, H. G. Infante, F. C. Adams, Multicapillary gas chromatography coupled to inductively coupled plasma-time-of-flight mass spectrometry for rapid mercury speciation analysis, *Anal. Chim. Acta*, **489** (2003), 45–57.

66. L. Yang, Z. Mester, R. E. Sturgeon, Determination of methylmercury in fish tissues by isotope dilution SPME-GC-ICP-MS, *J. Anal. At. Spectrom.*, **18** (2003), 431–436.

67. L. Yang, V. Colombini, P. Maxwell, Z. Mester, R. E. Sturgeon, Application of isotope dilution to the determination of methylmercury in fish tissue by solid-phase microextraction gas chromatography–mass spectrometry, *J. Chromatogr. A*, **1011** (2003), 135–142.

68. P. Grinberg, R. C. Campos, Z. Mester, R. E. Sturgeon, Solid phase microextraction capillary gas chromatography combined with furnace atomization plasma emission spectrometry for speciation of mercury in fish tissues, *Spectrochim. Acta*, **58B** (2003), 427–441.

69. A. Taylor, S. Branch, D. Halls, M. Patriarca, M. White, Atomic spectrometry update. Clinical and biological materails, food and beverages, *J. Anal. Atom. Spectrom.*, **19** (2004), 505–556.

70. E. C. O. Santos, V. M. Camara, I. M. Jesus, E. S. Brabo, E. C. B. Loureiro, A. F. S. Mascarenhas, K. F. Fayal, G. C. Sa, F. E. S. Sagica, M. O. Lima, H. Higuchi, I. M. Silveira, A contribution to the establishment of reference values for total mercury levels in hair and fish in *Amazonia, Environ. Res.*, **90** (2002), 6–11.

71. R. Wagemann, E. Trebacz, G. Boila, W. L. Lockhart, Methylmercury and total mercury in tissues of artic marine mammals, *Sci. Total Environ.*, **218** (1998), 19–31.

72. N. Demuth, Massenspektrometrische Isotopenverduennungsanalyse (MSIVA) als Validierungsmethode fuer Bestimmungen von Methylquecksilber mittels GC/ICP-MS (dissertation), University of Mainz, 2001.

73. International Atomic Energy Agency (IAEA), *Austria*, 2004; http://www-naweb.iaea.org/nahu/external/e4/nmrm/met_bio/Hg.htm (accessed 4 April 2004).

74. U.S. Food and Drug Administration, Center for Food Safety and Applied Nutrition, Office of Seafood, *Mercury Levels in Seafood Species*, May 2001; http://www.cfsan.fda.gov/ ~ frf/sea-mehg.htlm (Accessed 29 July, 2003).

75. The Green Lane™, Environment Canada's World Wide Web site, 2004; http://www.ec.gc.ca/MERCURY/EH/EN/eh-b.cfm (accessed 4 April 2004).

76. R. Ritsema, Environmental applications of hyphenated techniques for the speciation of tin, arsenic and mercury (dissertation). University of Pau, 1997.

AUTHOR INDEX

The Determination of Chemical Elements in Food: Applications for Atomic and Mass Spectrometry.
Edited by Sergio Caroli

SUBJECT INDEX

The Determination of Chemical Elements in Food: Applications for Atomic and Mass Spectrometry.
Edited by Sergio Caroli